These safety symbols are used in laboratory and field investigations in this book to indicate possible hazards. Learn the meaning of each symbol and refer to this page often. *Remember to wash your hands thoroughly after completing lab procedures.*

PROTECTIVE EQUIPMENT   Do not begin any lab without the proper protection equipment.

 GOGGLES Proper eye protection must be worn when performing or observing science activities which involve items or conditions as listed below.

 APRON Wear an approved apron when using substances that could stain, wet, or destroy cloth.

 SOAP Wash hands with soap and water before removing goggles and after all lab activities.

 GLOVES Wear gloves when working with biological materials, chemicals, animals, or materials that can stain or irritate hands.

LABORATORY HAZARDS

Symbols	Potential Hazards	Precaution	Response
DISPOSAL	contamination of classroom or environment due to improper disposal of materials such as chemicals and live specimens	• DO NOT dispose of hazardous materials in the sink or trash can. • Dispose of wastes as directed by your teacher.	• If hazardous materials are disposed of improperly, notify your teacher immediately.
EXTREME TEMPERATURE	skin burns due to extremely hot or cold materials such as hot glass, liquids, or metals; liquid nitrogen; dry ice	• Use proper protective equipment, such as hot mitts and/or tongs, when handling objects with extreme temperatures.	• If injury occurs, notify your teacher immediately.
SHARP OBJECTS	punctures or cuts from sharp objects such as razor blades, pins, scalpels, and broken glass	• Handle glassware carefully to avoid breakage. • Walk with sharp objects pointed downward, away from you and others.	• If broken glass or injury occurs, notify your teacher immediately.
ELECTRICAL	electric shock or skin burn due to improper grounding, short circuits, liquid spills, or exposed wires	• Check condition of wires and apparatus for fraying or uninsulated wires, and broken or cracked equipment. • Use only GFCI-protected outlets	• DO NOT attempt to fix electrical problems. Notify your teacher immediately.
CHEMICAL	skin irritation or burns, breathing difficulty, and/or poisoning due to touching, swallowing, or inhalation of chemicals such as acids, bases, bleach, metal compounds, iodine, poinsettias, pollen, ammonia, acetone, nail polish remover, heated chemicals, mothballs, and any other chemicals labeled or known to be dangerous	• Wear proper protective equipment such as goggles, apron, and gloves when using chemicals. • Ensure proper room ventilation or use a fume hood when using materials that produce fumes. • NEVER smell fumes directly. • NEVER taste or eat any material in the laboratory.	• If contact occurs, immediately flush affected area with water and notify your teacher. • If a spill occurs, leave the area immediately and notify your teacher.
FLAMMABLE	unexpected fire due to liquids or gases that ignite easily such as rubbing alcohol	• Avoid open flames, sparks, or heat when flammable liquids are present.	• If a fire occurs, leave the area immediately and notify your teacher.
OPEN FLAME	burns or fire due to open flame from matches, Bunsen burners, or burning materials	• Tie back loose hair and clothing. • Keep flame away from all materials. • Follow teacher instructions when lighting and extinguishing flames. • Use proper protection, such as hot mitts or tongs, when handling hot objects.	• If a fire occurs, leave the area immediately and notify your teacher.
ANIMAL SAFETY	injury to or from laboratory animals	• Wear proper protective equipment such as gloves, apron, and goggles when working with animals. • Wash hands after handling animals.	• If injury occurs, notify your teacher immediately.
BIOLOGICAL	infection or adverse reaction due to contact with organisms such as bacteria, fungi, and biological materials such as blood, animal or plant materials	• Wear proper protective equipment such as gloves, goggles, and apron when working with biological materials. • Avoid skin contact with an organism or any part of the organism. • Wash hands after handling organisms.	• If contact occurs, wash the affected area and notify your teacher immediately.
FUME	breathing difficulties from inhalation of fumes from substances such as ammonia, acetone, nail polish remover, heated chemicals, and mothballs	• Wear goggles, apron, and gloves. • Ensure proper room ventilation or use a fume hood when using substances that produce fumes. • NEVER smell fumes directly.	• If a spill occurs, leave area and notify your teacher immediately.
IRRITANT	irritation of skin, mucous membranes, or respiratory tract due to materials such as acids, bases, bleach, pollen, mothballs, steel wool, and potassium permanganate	• Wear goggles, apron, and gloves. • Wear a dust mask to protect against fine particles.	• If skin contact occurs, immediately flush the affected area with water and notify your teacher.
RADIOACTIVE	excessive exposure from alpha, beta, and gamma particles	• Remove gloves and wash hands with soap and water before removing remainder of protective equipment.	• If cracks or holes are found in the container, notify your teacher immediately.

Your online portal to everything you need

connectED.mcgraw-hill.com

Look for these icons to access
exciting digital resources

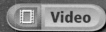 Video

■))) Audio

Review

? Inquiry

WebQuest

✓ Assessment

Concepts in Motion

INTEGRATED

iSCIENCE

Glencoe

Leopard, *Panthera pardus*

Once common across southern Asia and most of Africa, most leopards exist today in sub-Saharan Africa in rain forests and deserts. They are the smallest of the big cats—tigers, lions, jaguars, and leopards. Leopards are known for their ability to climb trees while carrying prey.

The McGraw-Hill Companies

 Education

Send all inquiries to:
McGraw-Hill Education
8787 Orion Place
Columbus, OH 43240-4027

ISBN: 978-0-07-888006-3
MHID: 0-07-888006-8

Printed in the United States of America.

9 10 11 12 13 QVS 15 14

Contents in Brief

Authors and Contributors

Authors

American Museum of Natural History
New York, NY

Michelle Anderson, MS
Lecturer
The Ohio State University
Columbus, OH

Juli Berwald, PhD
Science Writer
Austin, TX

John F. Bolzan, PhD
Science Writer
Columbus, OH

Rachel Clark, MS
Science Writer
Moscow, ID

Patricia Craig, MS
Science Writer
Bozeman, MT

Randall Frost, PhD
Science Writer
Pleasanton, CA

Lisa S. Gardiner, PhD
Science Writer
Denver, CO

Jennifer Gonya, PhD
The Ohio State University
Columbus, OH

Mary Ann Grobbel, MD
Science Writer
Grand Rapids, MI

Whitney Crispen Hagins, MA, MAT
Biology Teacher
Lexington High School
Lexington, MA

Carole Holmberg, BS
Planetarium Director
Calusa Nature Center and
Planetarium, Inc.
Fort Myers, FL

Tina C. Hopper
Science Writer
Rockwall, TX

Jonathan D. W. Kahl, PhD
Professor of Atmospheric Science
University of Wisconsin-
Milwaukee
Milwaukee, WI

Nanette Kalis
Science Writer
Athens, OH

S. Page Keeley, MEd
Maine Mathematics and Science
Alliance
Augusta, ME

Cindy Klevickis, PhD
Professor of Integrated Science
and Technology
James Madison University
Harrisonburg, VA

Kimberly Fekany Lee, PhD
Science Writer
La Grange, IL

Michael Manga, PhD
Professor
University of California, Berkeley
Berkeley, CA

Devi Ried Mathieu
Science Writer
Sebastopol, CA

Elizabeth A. Nagy-Shadman, PhD
Geology Professor
Pasadena City College
Pasadena, CA

William D. Rogers, DA
Professor of Biology
Ball State University
Muncie, IN

Donna L. Ross, PhD
Associate Professor
San Diego State University
San Diego, CA

Marion B. Sewer, PhD
Assistant Professor
School of Biology
Georgia Institute of Technology
Atlanta, GA

Julia Meyer Sheets, PhD
Lecturer
School of Earth Sciences
The Ohio State University
Columbus, OH

Michael J. Singer, PhD
Professor of Soil Science
Department of Land, Air and
Water Resources
University of California
Davis, CA

Karen S. Sottosanti, MA
Science Writer
Pickerington, Ohio

Paul K. Strode, PhD
I.B. Biology Teacher
Fairview High School
Boulder, CO

Jan M. Vermilye, PhD
Research Geologist
Seismo-Tectonic Reservoir
Monitoring (STRM)
Boulder, CO

Judith A. Yero, MA
Director
Teacher's Mind Resources
Hamilton, MT

Dinah Zike, MEd
Author, Consultant, Inventor
of Foldables
Dinah Zike Academy; Dinah-
Might Adventures, LP
San Antonio, TX

Margaret Zorn, MS
Science Writer
Yorktown, VA

Consulting Authors

Alton L. Biggs
Biggs Educational Consulting
Commerce, TX

Ralph M. Feather, Jr., PhD
Assistant Professor
Department of Educational
Studies and Secondary Education
Bloomsburg University
Bloomsburg, PA

Douglas Fisher, PhD
Professor of Teacher Education
San Diego State University
San Diego, CA

Edward P. Ortleb
Science/Safety Consultant
St. Louis, MO

Series Consultants

Science

Solomon Bililign, PhD
Professor
Department of Physics
North Carolina Agricultural and
Technical State University
Greensboro, NC

John Choinski
Professor
Department of Biology
University of Central Arkansas
Conway, AR

Anastasia Chopelas, PhD
Research Professor
Department of Earth and Space
Sciences
UCLA
Los Angeles, CA

David T. Crowther, PhD
Professor of Science Education
University of Nevada, Reno
Reno, NV

A. John Gatz
Professor of Zoology
Ohio Wesleyan University
Delaware, OH

Sarah Gille, PhD
Professor
University of California San
Diego
La Jolla, CA

David G. Haase, PhD
Professor of Physics
North Carolina State University
Raleigh, NC

Janet S. Herman, PhD
Professor
Department of Environmental
Sciences
University of Virginia
Charlottesville, VA

David T. Ho, PhD
Associate Professor
Department of Oceanography
University of Hawaii
Honolulu, HI

Ruth Howes, PhD
Professor of Physics
Marquette University
Milwaukee, WI

**Jose Miguel Hurtado, Jr.,
PhD**
Associate Professor
Department of Geological
Sciences
University of Texas at El Paso
El Paso, TX

Monika Kress, PhD
Assistant Professor
San Jose State University
San Jose, CA

Mark E. Lee, PhD
Associate Chair & Assistant
Professor
Department of Biology
Spelman College
Atlanta, GA

Linda Lundgren
Science writer
Lakewood, CO

Keith O. Mann, PhD
Ohio Wesleyan University
Delaware, OH

Charles W. McLaughlin, PhD
Adjunct Professor of Chemistry
Montana State University
Bozeman, MT

Katharina Pahnke, PhD
Research Professor
Department of Geology and
Geophysics
University of Hawaii
Honolulu, HI

Jesús Pando, PhD
Associate Professor
DePaul University
Chicago, IL

Hay-Oak Park, PhD
Associate Professor
Department of Molecular
Genetics
Ohio State University
Columbus, OH

David A. Rubin, PhD
Associate Professor of Physiology
School of Biological Sciences
Illinois State University
Normal, IL

Toni D. Sauncy
Assistant Professor of Physics
Department of Physics
Angelo State University
San Angelo, TX

Teacher Advisory Board

The Teacher Advisory Board gave the authors, editorial staff, and design team feedback on the content and design of the Student Edition. They provided valuable input in the development of *Glencoe Integrated iScience*.

Frances J. Baldridge
Department Chair
Ferguson Middle School
Beavercreek, OH

Jane E. M. Buckingham
Teacher
Crispus Attucks Medical Magnet
High School
Indianapolis, IN

Elizabeth Falls
Teacher
Blalack Middle School
Carrollton, TX

Nelson Farrier
Teacher
Hamlin Middle School
Springfield, OR

Michelle R. Foster
Department Chair
Wayland Union Middle School
Wayland, MI

Rebecca Goodell
Teacher
Reedy Creek Middle School
Cary, NC

Mary Gromko
Science Supervisor K–12
Colorado Springs District 11
Colorado Springs, CO

Randy Mousley
Department Chair
Dean Ray Stucky Middle School
Wichita, KS

David Rodriguez
Teacher
Swift Creek Middle School
Tallahassee, FL

Derek Shook
Teacher
Floyd Middle Magnet School
Montgomery, AL

Karen Stratton
Science Coordinator
Lexington School District One
Lexington, SC

Stephanie Wood
Science Curriculum Specialist,
K–12
Granite School District
Salt Lake City, UT

Online Guide

Get ConnectED

connectED.mcgraw-hill.com

ConnectED
▷ **Your Digital Science Portal**

 **Video**

See the science in real life through these exciting

Audio

Click the link and you can listen to the text while you

 **Review**

Try these interactive tools to help you review

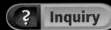 **Inquiry**

Explore concepts through hands–on and virtual labs.

 **WebQuest**

These web-based challenges relate the concepts you're learning

The icons in your online student edition link you to interactive learning opportunities. Browse your online student book to find more.

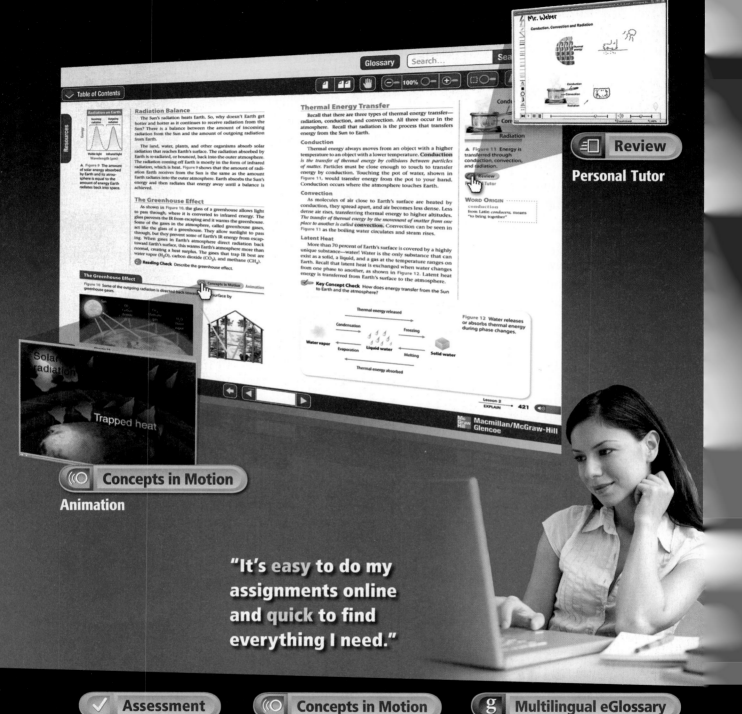

"It's easy to do my assignments online and quick to find everything I need."

✓ **Assessment**

Check how well you

(((◎ **Concepts in Motion**

The textbook comes alive with

g **Multilingual eGlossary**

Read key vocabulary in 13 languages

Treasure Hunt

Your science book has many features that will aid you in your learning. Some of these features are listed below. You can use the activity at the right to help you find these and other special features in the book.

- **BIG IDEA** can be found at the start of each chapter.

- The Reading Guide at the start of each lesson lists 🔑 **Key Concepts,** vocabulary terms, and online supplements to the content.

- **Connect ED** icons direct you to online resources such as animations, personal tutors, math practices, and quizzes.

- **(Inquiry)** Labs and Skill Practices are in each chapter.

- Your **FOLDABLES®** help organize your notes.

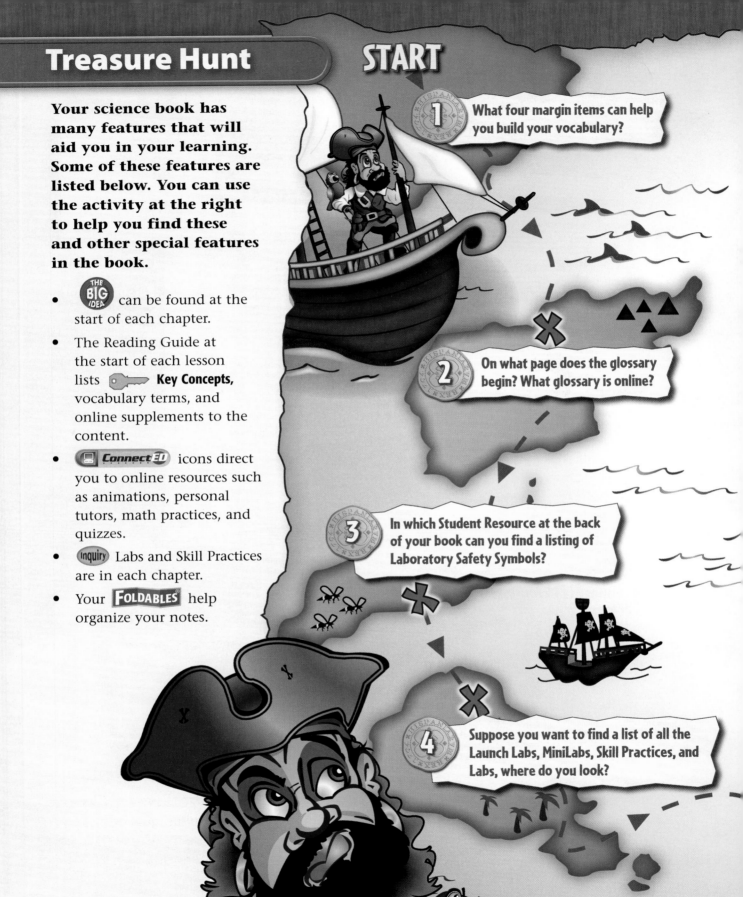

1 What four margin items can help you build your vocabulary?

2 On what page does the glossary begin? What glossary is online?

3 In which Student Resource at the back of your book can you find a listing of Laboratory Safety Symbols?

4 Suppose you want to find a list of all the Launch Labs, MiniLabs, Skill Practices, and Labs, where do you look?

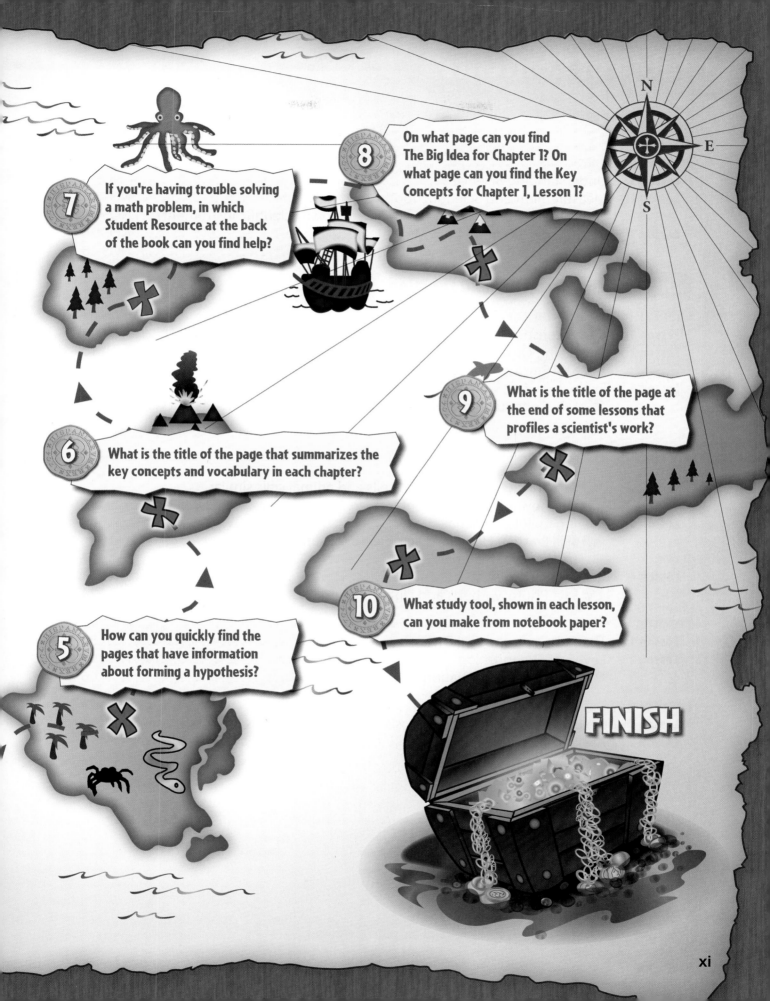

On what page can you find The Big Idea for Chapter 1? On what page can you find the Key Concepts for Chapter 1, Lesson 1?

8

7 If you're having trouble solving a math problem, in which Student Resource at the back of the book can you find help?

9 What is the title of the page at the end of some lessons that profiles a scientist's work?

6 What is the title of the page that summarizes the key concepts and vocabulary in each chapter?

10 What study tool, shown in each lesson, can you make from notebook paper?

5 How can you quickly find the pages that have information about forming a hypothesis?

FINISH

Table of Contents

Table of Contents

Table of Contents

Table of Contents

Table of Contents

Student Resources

Inquiry Launch Labs

Inquiry

Inquiry MiniLabs

TABLE OF CONTENTS

Inquiry Skill Practice

Inquiry

 **Labs**

TABLE OF CONTENTS

Scientific Explanations

THE BIG IDEA How can science provide answers to your questions about the world around you?

Inquiry Vacuuming Corals?

No, these two divers are collecting data about corals in waters near Sulawesi, Indonesia. They are marine biologists, scientists who study living things in oceans and other saltwater environments.

- What information about corals are these scientists collecting?
- What questions do they hope to answer?
- How can science provide answers to their questions and your questions?

Methods of SCIENCE

This chapter begins your study of the nature of science, but there is even more information about the nature of science in this book. Each unit begins by exploring an important topic that is fundamental to scientific study. As you read these topics, you will learn even more about the nature of science.

Reading Guide
Key Concepts 🔑
ESSENTIAL QUESTIONS

- What is scientific inquiry?
- What are the results of scientific investigations?
- How can a scientist minimize bias in a scientific investigation?

Vocabulary

science p. NOS 4
observation p. NOS 6
inference p. NOS 6
hypothesis p. NOS 6
prediction p. NOS 7
technology p. NOS 8
scientific theory p. NOS 9
scientific law p. NOS 9
critical thinking p. NOS 10

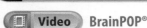 **Multilingual eGlossary**

🖵 **Video** **BrainPOP®**

Understanding Science

What is science?

The last time that you watched squirrels play in a park or in your yard, did you realize that you were practicing science? Every time you observe the natural world, you are practicing science. **Science** *is the investigation and exploration of natural events and of the new information that results from those investigations.*

When you observe the natural world, you might form questions about what you see. While you are exploring those questions, you probably use reasoning, creativity, and skepticism to help you find answers to your questions. People use these behaviors in their daily lives to solve problems, such as how to keep a squirrel from eating bird seed, as shown in **Figure 1.** Similarly, scientists use these behaviors in their work.

Scientists use a reliable set of skills and methods in different ways to find answers to questions. After reading this chapter, you will have a better understanding of how science works, the limitations of science, and scientific ways of thinking. In addition, you will recognize that when you practice science at home or in the classroom, you use scientific methods to answer questions just as scientists do.

Figure 1 Someone used reasoning and creativity to design each of these squirrel-proof bird feeders. However, some solutions don't work. Scientists use similar methods to try to solve problems.

Branches of Science

No one person can study all the natural world. Therefore, people tend to focus their efforts on one of the three fields or branches of science—life science, Earth science, or physical science, as described below. Then people or scientists can seek answers to specific problems within one field of science.

WORD ORIGIN

biology
from Greek *bios*, means "life"; and *logia*, means "study of"

Life Science

Biology, or life science, is the study of all living things. This forest ecologist, a life scientist who studies interactions in forest ecosystems, is studying lichens growing on Douglas firs. Biologists ask questions such as

- How do plants produce their own food?
- Why do some animals give birth to live young and others lay eggs?
- How are reptiles and birds related?

Earth Science

The study of Earth, including its landforms, rocks, soil, and forces that shape Earth's surface, is Earth science. These Earth scientists are collecting soil samples in Africa. Earth scientists ask questions such as

- How do rocks form?
- What causes earthquakes?
- What substances are in soil?

Physical Science

The study of chemistry and physics is physical science. Physical scientists study the interactions of matter and energy. This chemist is preparing antibiotic solutions. Physical scientists ask questions such as

- How do substances react and form new substances?
- Why does a liquid change to a solid?
- How are force and motion related?

Scientific Inquiry

As scientists study the natural world, they ask questions about what they observe. To find the answers to these questions, they usually use certain skills, or methods. The chart in **Figure 2** shows a sequence of the skills that a scientist might use in an investigation. However, it is important to know that, sometimes, not all of these skills are performed in an investigation, or performed in this order. Scientists practice scientific inquiry—a process that uses a variety of skills and tools to answer questions or to test ideas about the natural world.

Ask Questions

Like a scientist, you use scientific inquiry in your life, too. Suppose you decide to plant a vegetable garden. As you plant the vegetable seeds, you water some seeds more than others. Then, you weed part of the garden and mix fertilizer into some of the soil. After a few weeks, you observe that some vegetable plants are growing better than others. An **observation** *is using one or more of your senses to gather information and take note of what occurs.*

Observations often are the beginning of the process of inquiry and can lead to questions such as "Why are some plants growing better than others?" As you are making observations and asking questions, you recall from science class that plants need plenty of water and sunlight to grow. Therefore you infer that perhaps some vegetables are receiving more water or sunlight than others and, therefore, are growing better. An **inference** *is a logical explanation of an observation that is drawn from prior knowledge or experience.*

Hypothesize

After making observations and inferences, you are ready to develop a hypothesis and investigate why some vegetables are growing better than others. *A possible explanation about an observation that can be tested by scientific investigations is a* **hypothesis.** Your hypothesis might be: Some plants are growing taller and more quickly than others because they are receiving more water and sunlight. Or, your hypothesis might be: The plants that are growing quickly have received fertilizer because fertilizer helps plants grow.

Figure 2 This flow chart shows steps you or a scientist might use during a scientific investigation.

Visual Check What happens if a hypothesis is not supported?

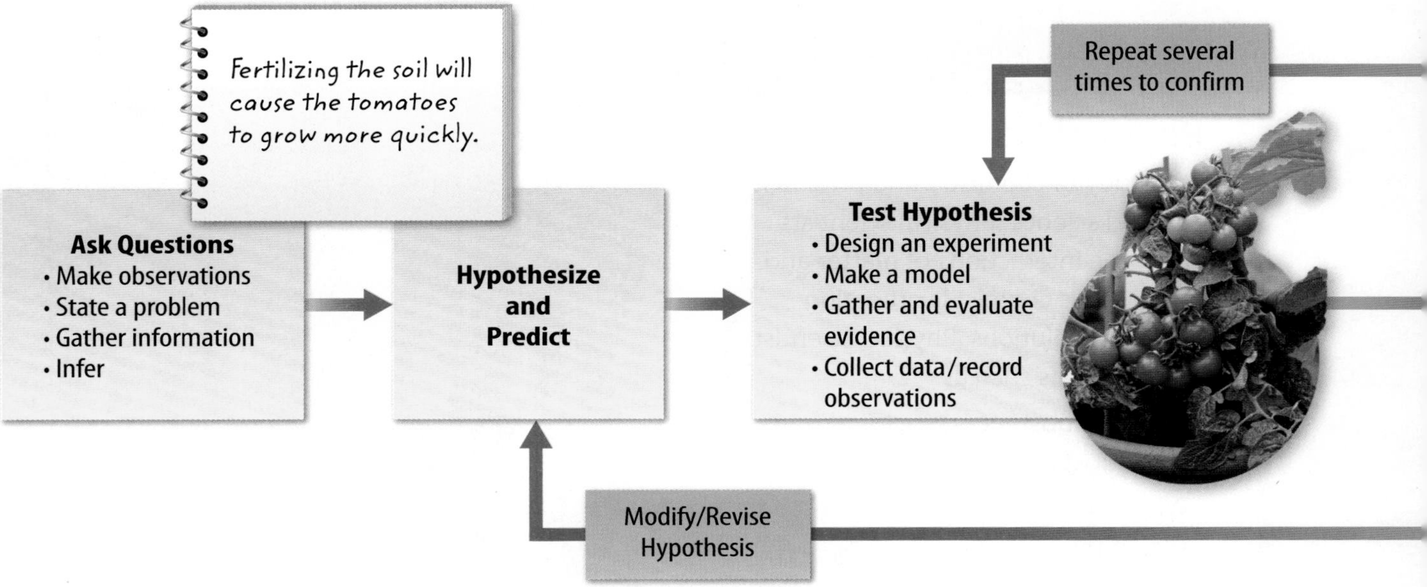

Fertilizing the soil will cause the tomatoes to grow more quickly.

Ask Questions
• Make observations
• State a problem
• Gather information
• Infer

Hypothesize and Predict

Test Hypothesis
• Design an experiment
• Make a model
• Gather and evaluate evidence
• Collect data/record observations

Repeat several times to confirm

Modify/Revise Hypothesis

Predict

After you state a hypothesis, you might make a prediction to help you test your hypothesis. *A* **prediction** *is a statement of what will happen next in a sequence of events.* For instance, based on your hypotheses, you might predict that if some plants receive more water, sunlight, or fertilizer, then they will grow taller and more quickly.

Test your Hypothesis

When you test a hypothesis, you often are testing your predictions. For example, you might design an experiment to test your hypothesis on the fertilizer. You set up an experiment in which you plant seeds and add fertilizer to only some of them. Your prediction is that the plants that get the fertilizer will grow more quickly. If your prediction is confirmed, it supports your hypothesis. If your prediction is not confirmed, your hypothesis might need revision.

Analyze Results

As you are testing your hypothesis, you are probably collecting data about the plants' rates of growth and how much fertilizer each plant receives. Initially, it might be difficult to recognize patterns and relationships in data. Your next step might be to organize and analyze your data.

You can create graphs, classify information, or make models and calculations. Once data are organized, you more easily can study the data and draw conclusions. Other methods of testing a hypothesis and analyzing results are shown in **Figure 2.**

Draw Conclusions

Now you must decide whether your data do or do not support your hypothesis and then draw conclusions. A conclusion is a summary of the information gained from testing a hypothesis. You might make more inferences when drawing conclusions. If your hypothesis is supported, you can repeat your experiment several times to confirm your results. If your hypothesis is not supported, you can modify it and repeat the scientific inquiry process.

Communicate Results

An important step in scientific inquiry is communicating results to others. Professional scientists write scientific articles, speak at conferences, or exchange information on the Internet. This part of scientific inquiry is important because scientists use new information in their research or perform other scientists' investigations to verify results.

 Key Concept Check What is scientific inquiry?

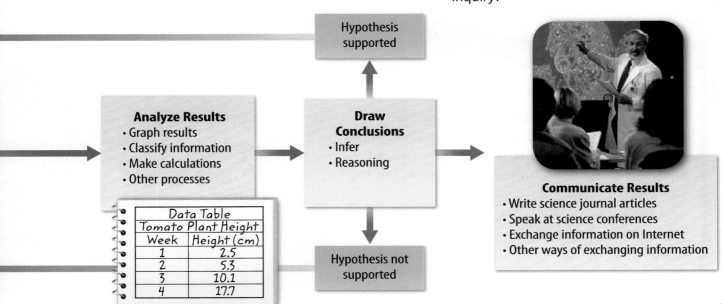

Analyze Results
- Graph results
- Classify information
- Make calculations
- Other processes

Data Table	
Tomato Plant Height	
Week	Height (cm)
1	2.5
2	5.3
3	10.1
4	17.7

Draw Conclusions
- Infer
- Reasoning

Hypothesis supported

Hypothesis not supported

Communicate Results
- Write science journal articles
- Speak at science conferences
- Exchange information on Internet
- Other ways of exchanging information

Results of Scientific Inquiry

Both you and scientists perform scientific inquiry to find answers to questions. There are many outcomes of scientific inquiry, such as technology, materials, and explanations, as shown below.

 Key Concept Check What are the results of scientific investigations?

Technology

The practical use of scientific knowledge, especially for industrial or commercial use is technology. Televisions, MP3 players, and computers are examples of technology. The C-Leg, shown to the right, is one of the latest designs of computer-aided limbs. The prosthetic leg has sensors that anticipate the user's next move, which prevents him or her from stumbling or tripping. In addition, this new technology has several modes that can enable the user to walk, stand for long periods of time, and even ride a bike.

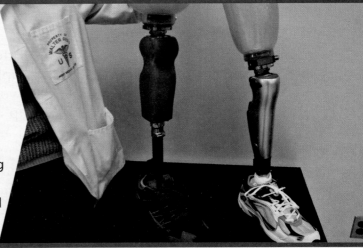

New Materials

Another possible outcome of an investigation is a new material. For example, scientists have developed a bone bioceramic. A bioceramic is a natural calcium-phosphate mineral complex that is part of bones and teeth. This synthetic bone mimics natural bone's structure. Its porous structure allows a type of cell to grow and develop into new bone tissue. The bioceramic can be shaped into implants that are treated with certain cells from the patient's bone marrow. It then can be implanted into the patient's body to replace missing bone.

Possible Explanations

Many times, scientific investigations answer the questions: *who, what, when, where,* or *how.* For example, who left fingerprints at a crime scene? When should fertilizer be applied to plants? What organisms live in rain forests?

In 2007, while exploring in Colombia's tropical rain forests, scientists discovered a new species of poisonous tree frog. The golden frog of Supatá is only 2 cm long.

Scientific Theory and Scientific Laws

Scientists often repeat scientific investigations to verify that the results for a hypothesis or a group of hypotheses are correct. This can lead to a scientific theory.

Scientific Theory The everyday meaning of the word *theory* is an untested idea or an opinion. However, a **scientific theory** *is an explanation of observations or events based on knowledge gained from many observations and investigations.* For example, about 300 years ago, scientists began looking at samples of trees, water, and blood through the first microscopes. They noticed that all of these organisms were made of tinier units, or cells, as shown in **Figure 3.** As more scientists observed cells in other organisms, their observations became known as the cell theory. This theory explains that all living things are made of cells. A scientific theory is assumed to be the best explanation of observations unless it is disproved. The cell theory will continue to explain the makeup of all organisms until an organism is discovered that is not made of cells.

Scientific Laws Scientific laws are different from societal laws, which are agreements on a set of behaviors. A **scientific law** *describes a pattern or an event in nature that is always true.* A scientific theory might explain how and why an event occurs. But a scientific law states only that an event in nature will occur under specific conditions. For example, the law of conservation of mass states that the mass of materials will be the same before and after a chemical reaction. This scientific law does not explain why this occurs—only that it will occur. **Table 1** compares a scientific theory and a scientific law.

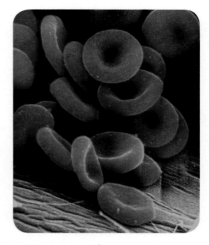

Figure 3 When you view blood using a microscope, you will see that it contains red blood cells.

Make a vertical two-column chart book. Label it as shown. Use it to organize your notes on scientific investigations.

Results of Scientific Investigations | Examples

Table 1 Comparing Scientific Theory and Scientific Law	
Scientific Theory	**Scientific Law**
A scientific theory is based on repeated observations and scientific investigations.	Scientific laws are observations of similar events that have been observed repeatedly.
If new information does not support a scientific theory, the theory will be modified or rejected.	If many new observations do not follow the law, the law is rejected.
A scientific theory attempts to explain why something happens.	A scientific law states that something will happen.
A scientific theory usually is more complex than a scientific law and might contain many well-supported hypotheses.	A scientific law usually is based on one well-supported hypothesis that states that something will happen.

Skepticism in Media

When you see scientific issues in the media, such as newspapers, radio, television, and magazines, it is important to be skeptical. When you are skeptical, you question information that you read or hear, or events you observe. Is the information truthful? Is it accurate? It also is important that you question statements made by people outside their area of expertise, and claims that are based on vague statements.

Evaluating Scientific Evidence

An important skill in scientific inquiry is critical thinking. **Critical thinking** *is comparing what you already know with the information you are given in order to decide whether you agree with it.* Identifying and minimizing bias also is important when conducting scientific inquiry. To minimize bias in an investigation, sampling, repetition, and blind studies can be helpful, as shown below.

 Key Concept Check How can a scientist minimize bias in a scientific investigation?

❶ Sampling

A method of data collection that involves studying small amounts of something in order to learn about the larger whole is sampling. A sample should be a random representation of the whole.

❷ Bias

It is important to reduce bias during scientific investigations. Bias is intentional or unintentional prejudice toward a specific outcome. Sources of bias in an investigation can include equipment choices, hypothesis formation, and prior knowledge.

Suppose you were a part of a taste test for a new cereal. If you knew the price of each cereal, you might think that the most expensive one tastes the best. This is a bias.

❸ Blind Study

A procedure that can reduce bias is a blind study. The investigator, subject, or both do not know which item they are testing. Personal bias cannot affect an investigation if participants do not know what they are testing.

❹ Repetition

If you get different results when you repeat an investigation, then the original investigation probably was flawed. Repetition of experiments helps reduce bias.

Science cannot answer all questions.

You might think that any question can be answered through a scientific investigation. But there are some questions that science cannot answer, such as the one posed in **Figure 4.** Questions about personal opinions, values, beliefs, and feelings cannot be answered scientifically. However, some people use scientific evidence to try to strengthen their claims about these topics.

Safety in Science

Scientists follow safety procedures when they conduct investigations. You too should follow safety procedures when you do any experiments. You should wear appropriate safety equipment and listen to your teacher's instructions. Also, you should learn to recognize potential hazards and to know the meaning of safety symbols. Read more about science laboratory safety in the Science Skill Handbook at the back of this book.

Ethics are especially important when using living things during investigations. Animals should be treated properly. Scientists also should tell research participants about the potential risks and benefits of the research. Anyone can refuse to participate in scientific research.

Figure 4 Science cannot answer questions based on opinions or feelings, such as which paint color is the prettiest.

ACADEMIC VOCABULARY
ethics
(noun) rules of conduct or moral principles

Lesson 1 Review

Assessment Online Quiz
Inquiry Virtual Lab

Use Vocabulary

1 **Explain** the relationship between observations and hypotheses.

2 **Use the terms** *technology, scientific law,* and *scientific theory* in complete sentences.

3 **Contrast** inference and prediction.

4 **Compare and contrast** critical thinking and inference.

Understand Key Concepts

5 Which should NOT be part of scientific inquiry?
  A. bias   C. hypothesis
  B. analysis   D. testing

6 **Describe** four real-life examples of the results of scientific investigations.

7 **Discuss** four ways a scientist can reduce bias in scientific investigations.

Interpret Graphics

8 **Draw** a graphic organizer like the one below. In each oval, list an example of how to test a hypothesis using scientific inquiry.

Test Hypothesis

Critical Thinking

9 **Suggest** Why do you think people believe some theories even if they are not supported by credible evidence?

10 **Evaluate** In a magazine, you read that two scientific investigations attempted to answer the same question. However, the two teams of scientists came to opposite conclusions. How do you decide which investigation was valid?

Measurement and Scientific Tools

Reading Guide

Key Concepts 🔑
ESSENTIAL QUESTIONS

- What is the difference between accuracy and precision?

- Why should you use significant digits?

- What are some tools used by life scientists?

Vocabulary
description p. NOS 12

explanation p. NOS 12

International System of Units (SI) p. NOS 12

accuracy p. NOS 14

precision p. NOS 14

significant digits p. NOS 15

g **Multilingual eGlossary**

Description and Explanation

How would you describe the squirrel's activity in **Figure 5?** A **description** *is a spoken or written summary of observations.* Your description might include information such as: the squirrel buried five acorns near a large tree. A qualitative description uses your senses (sight, sound, smell, touch, taste) to describe an observation. *A large tree* is a qualitative description. However, a quantitative description uses numbers to describe the observation. *Five acorns* is a quantitative description. You can use measuring tools, such as a ruler, a balance, or a thermometer, to make quantitative descriptions.

How would you explain the squirrel's activity? An **explanation** *is an interpretation of observations.* You might explain that the squirrel is storing acorns for food at a later time. When you describe something, you report what you observe. But when you explain something, you try to interpret your observations. This can lead to a hypothesis.

Figure 5 A description and an explanation of a squirrel's activity contain different information.

The International System of Units

Suppose you observed a squirrel searching for buried food and recorded that it traveled about 200 ft from its nest. Someone who measures distances in meters might not understand how far the squirrel traveled. The scientific community solved this problem in 1960. It adopted *an internationally accepted system for measurement called the* **International System of Units (SI).**

SI Base Units and Prefixes

Like scientists and many others around the world, you probably use the SI system in your classroom. All SI units are derived from seven base units, as listed in **Table 2**. For example, the base unit for length, or the unit most commonly used to measure length, is the meter. However, you have probably made measurements in kilometers or millimeters before. Where do these units come from?

A prefix can be added to a base unit's name to indicate either a fraction or a multiple of that base unit. The prefixes are based on powers of ten, such as 0.01 and 100, as shown in **Table 3**. For example, one centimeter (1 cm) is one one-hundredth of a meter and a kilometer (1 km) is 1,000 meters.

Concepts in Motion Interactive Table

Table 2 SI Base Units

Quantity Measured	Unit (symbol)
Length	meter (m)
Mass	kilogram (kg)
Time	second (s)
Electric current	ampere (A)
Temperature	Kelvin (K)
Substance amount	mole (mol)
Light intensity	candela (cd)

Table 3 Prefixes

Prefix	Meaning
Mega– (M)	1,000,000 (10^6)
Kilo– (k)	1,000 (10^3)
Hecto– (h)	100 (10^2)
Deka– (da)	10 (10^1)
Deci– (d)	0.1 (10^{-1})
Centi– (c)	0.01 (10^{-2})
Milli– (m)	0.001 (10^{-3})
Micro– (μ)	0.000 001 (10^{-6})

Conversion

It is easy to convert from one SI unit to another. You either multiply or divide by a power of ten. You also can use proportion calculations to make conversions. For example, a biologist measures an Emperor goose in the field. Her triple-beam balance shows the goose has a mass of 2.8 kg. She could perform the calculation below to find its mass in grams, X.

$$\frac{X}{2.8\ kg} = \frac{1,000\ g}{1\ kg}$$

$$(1\ kg)X = (1,000\ g)(2.8\ kg)$$

$$X = \frac{(1,000\ g)(2.8\ \cancel{kg})}{1\ \cancel{kg}}$$

$$X = 2,800\ g$$

Notice that the answer has the correct units.

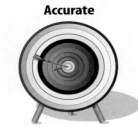

Accurate

An arrow in the center indicates high accuracy.

Precise but not accurate

Arrows far from the center indicate low accuracy. Arrows close together indicate high precision.

Accurate and precise

Arrows in the center indicate high accuracy. Arrows close together indicate high precision.

Not accurate or precise

Arrows far from the center indicate low accuracy. Arrows far apart indicate low precision.

Figure 6 The archery target illustrates accuracy and precision. An accurate shot is in the bull's-eye.

FOLDABLES

Make a horizontal two-tab book with a top-tab. Label it as shown. Use it to compare precision and accuracy.

| Similarities | Differences |

Precision and Accuracy

Precision and Accuracy

Suppose your friend Simon tells you that he will call you in one minute, but he calls you a minute and a half later. Sarah tells you that she will call you in one minute, and she calls exactly 60 seconds later. What is the difference? Sarah is accurate and Simon is not. **Accuracy** *is a description of how close a measurement is to an accepted or true value.* However, if Simon always calls about 30 seconds later than he says he will, then Simon is precise. **Precision** *is a description of how similar or close measurements are to each other,* as shown in **Figure 6**.

Table 4 illustrates the difference between precise and accurate measurements. Students were asked to find the melting point of sucrose, or table sugar. Each student took three temperature readings and calculated the mean, or average, of his or her data. As the recorded data in the table shows, student A had more accurate data. The melting point mean, 184.7°C, is closer to the scientifically accepted melting point, 185°C. Although not accurate, Student C's measurements are the most precise because they are similar in value.

Key Concept Check How do accuracy and precision differ?

Table 4 The data taken by student A are more accurate because each value is close to the accepted value. The data taken by student C are more precise because the data are similar.

Table 4 **Student Melting Point Data**			
	Student A	**Student B**	**Student C**
Trial 1	183.5°C	190.0°C	181.2°C
Trial 2	185.9°C	183.3°C	182.0°C
Trial 3	184.6°C	187.1°C	181.7°C
Mean	184.7°C	186.8°C	181.6°C
Sucrose Melting Point (accepted value) 185°C			

Measurement and Accuracy

The tools used to take measurements can limit the accuracy of the measurements. Suppose you are measuring the temperature at which sugar melts, and the thermometer's measurements are divided into whole numbers. If your sugar sample melts between 183°C and 184°C, you can estimate the temperature between these two numbers. But, if the thermometer's measurements are divided into tenths, and your sample melts between 183.2°C and 183.3°C, your estimate between these numbers would be more accurate.

Significant Digits

In the second example above, you know that the temperature is between 183.2°C and 183.3°C. You could estimate that the temperature is 183.25°C. When you take any measurement, some digits you know for certain and some digits you estimate. **Significant digits** *are the number of digits in a measurement that are known with a certain degree of reliability.* The significant digits in a measurement include all digits you know for certain plus one estimated digit. Therefore, your measurement of 183.25°C would contain five significant digits, as explained in **Table 5.** Using significant digits lets others know how certain your measurements are. **Figure 7** shows an example of rounding to 3 significant digits?

 Key Concept Check Why should you use significant digits?

Figure 7 Since the ruler is divided into tenths, you know the rod is between 5.2 cm and 5.3 cm. You can estimate that the rod is 5.25 cm.

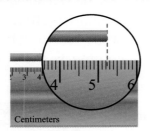

Centimeters

Math Skills

Significant Digits
The number 5,281 has 4 significant digits. Rule 1 in **Table 5** below states that all nonzero numbers are significant.

Practice
Use the rules in **Table 5** to determine the number of significant digits in each of the following numbers: 2.02; 0.0057; 1,500; and 0.500.

 Review
- Math Practice
- Personal Tutor

SCIENCE USE v. COMMON USE

digital
Science Use of, pertaining to, or using numbers (numerical digits)

Common Use of or pertaining to a finger

Table 5 Significant Digits

Rules
1. All nonzero numbers are significant.
2. Zeros between nonzero digits are significant.
3. Final zeros used after the decimal point are significant.
4. Zeros used solely for spacing the decimal point are not significant. The zeros indicate only the position of the decimal point.

* The blue numbers in the examples are the significant digits.

Example	Significant Digits	Applied Rules
1.234	4	1
1.2	2	1
0.023	2	1, 4
0.200	3	1, 3
1,002	4	1, 2
3.07	3	1, 2
0.001	1	1, 4
0.012	2	1, 4
50,600	3	1, 2, 4

Scientific Tools

Scientific inquiry often requires the use of tools. Scientists, including life scientists, might use the tools listed on this page and the next page. You might use one or more of them during a scientific inquiry, too. For more information about the proper use of these tools, see the Science Skill Handbook at the back of this book.

Science Journal ▶

In a science journal, you can record descriptions, explanations, plans, and steps used in a scientific inquiry. A science journal can be a spiral-bound notebook or a loose-leaf binder. It is important to keep your science journal organized so that you can find information when you need it. Make sure you keep thorough and accurate records.

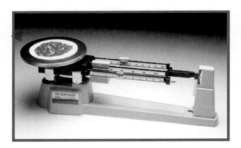

◀ Balances

You can use a triple-beam balance or an electric balance to measure mass. Mass usually is measured in kilograms (kg) or grams (g). When using a balance, do not let objects drop heavily onto the balance. Gently remove an object after you record its mass.

Thermometer ▶

A thermometer measures the temperature of substances. Although the Kelvin (K) is the SI unit for temperature, in the science classroom, you measure temperature in degrees Celsius (°C). Use care when you place a thermometer into a hot substance so that you do not burn yourself. Handle glass thermometers gently so that they do not break. If a thermometer does break, tell your teacher immediately. Do not touch the broken glass or the thermometer's liquid. Never use a thermometer to stir anything.

◀ Glassware

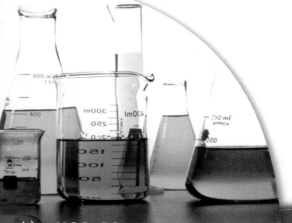

Laboratory glassware is used to hold, pour, heat, and measure liquids. Most labs have many types of glassware. For example, flasks, beakers, petri dishes, test tubes, and specimen jars are used as containers. To measure the volume of a liquid, you use a graduated cylinder. The unit of measure for liquid volume is the liter (L) or milliliter (mL).

◀ Compound Microscope

Microscopes enable you to observe small objects that you cannot observe with just your eyes. Usually, two types of microscopes are in science classrooms—dissecting microscopes and compound light microscopes, such as the one shown to the left. The girl is looking into two eyepieces to observe a magnified image of a small object or organism. However, some microscopes have only one eyepiece.

Microscopes can be damaged easily. It is important to follow your teacher's instructions when carrying and using a microscope. For more information about how to use a microscope, see the Science Skill Handbook at the back of this book.

((O Concepts in Motion   Animation

Computers—Hardware and Software ▶

Computers process information. In science, you can use computers to compile, retrieve, and analyze data for reports. You also can use them to create reports and other documents, to send information to others, and to research information.

The physical components of computers, such as monitors and keyboards, are called hardware. The programs that you run on computers are called software. These programs include word processing, spreadsheets, and presentation programs. When scientists write reports, they use word processing programs. They use spreadsheet programs for organizing and analyzing data. Presentation programs can be used to explain information to others.

Tools Used by Life Scientists

Magnifying Lens

A magnifying lens is a hand-held lens that magnifies, or enlarges, an image of an object. It is not as powerful as a microscope and is useful when great magnification is not needed. Magnifying lenses also can be used outside the lab where microscopes might not be available.

Slide

To observe items using a compound light microscope, you must place it on a thin, rectangular piece of glass called a slide. You must handle slides gently to avoid breaking them.

Dissecting Tools

Scientists use dissecting tools, such scalpels and scissors, to examine tissues, organs, or prepared organisms. Dissecting tools are sharp, so always use extreme caution when handling them.

Pipette

A pipette is similar to an eyedropper. It is a small glass or plastic tube used to draw up and transfer liquids.

 **Key Concept Check** What are some tools used by life scientists?

Lesson 2 Review

Use Vocabulary

1 **Define** *description* and *explanation* in your own words.

2 **Use the term** *International System of Units (SI)* in a sentence.

Understand Key Concepts 🔑

3 Which tool would a scientist use to view a tiny organism?
  A. computer   **C.** test tube
  B. compound   **D.** triple-beam
      microscope   balance

4 **Describe** the difference between accuracy and precision.

5 **Explain** why scientists use significant digits.

Interpret Graphics

6 **Draw** a graphic organizer like the one below. Write the name of an SI base unit in each circle. Add additional circles to the graphic organizer as needed.

Critical Thinking

7 **Recommend** ways that computers can assist life scientists in their work.

Math Skills ✕ ÷   Review
—— Math Practice ——

8 **Suppose** you measure the mass of a book and it is 420.0890 g. How many significant digits are in this measurement?

Materials

500-mL
Erlenmeyer
flask

rubber tubing,
15 cm

2-hole stopper

500-mL beaker

Also needed:
short piece of
plastic tubing,
water, 100-mL
graduated
cylinder,
plastic wrap
(10 cm × 30
cm), bendable
straws, food
coloring
(optional)

Safety

How can you build your own scientific instrument?

All organisms take in and release gases. Your cells take in oxygen and release carbon dioxide just like the cells of other animals, plants, fungi, protists, and some bacteria. However, many plant cells, some protists, and some bacteria also can take in carbon dioxide and release oxygen. In this lab, you will follow a procedure and build your own scientific instrument that measures the change in the volume of a gas.

Learn It

Scientists often **follow procedures** developed by other scientists to collect data. A procedure is a step-by-step explanation of how to accomplish a task. The steps in a procedure tell you what materials to use, how to them, and in what order to perform specific tasks. Some procedures are simple, while others are more complicated and require a lot of practice and skill.

Try It

1. Read and complete a lab safety form.

2. Into each, an Erlenmeyer flask and a beaker, pour 350 mL of water. Pour 100 mL of water into a graduated cylinder.

3. Seal the graduated cylinder with plastic wrap. Place your hand over the plastic wrap and turn the cylinder upside down. Carefully place the sealed end of the graduated cylinder into the beaker of water. Pull off the plastic wrap without losing any water from the graduated cylinder. Have a team member hold the it so that it doesn't tip over.

4. Place one end of a straw in one hole of a 2-hole stopper. Insert the plastic tubing into the other hole. Place one end of the rubber tubing over the plastic tubing.

5. Without lifting the cylinder above the water's surface, insert the free end of the rubber tubing inside the cylinder. Have a team continue to hold the cylinder.

6. Put the stopper in the flask. Record the initial reading of the water in the graduated cylinder in your Science Journal.

7. Gently blow into the straw and watch the change in volume of the water. Continue blowing into the straw until the graduated cylinder contains 50 mL of gas (air).

Apply It

8. **Draw a diagram** of your set up, also known as a eudiometer. Label all the parts, and describe their functions.

9. 🔑 **Key Concept** Describe a scenario in which a life scientist would use this instrument to measure gases.

Reading Guide

Key Concepts 🔑
ESSENTIAL QUESTIONS

- How do independent and dependent variables differ?

- How is scientific inquiry used in a real-life scientific investigation?

Vocabulary

variable p. NOS 20

dependent variable p. NOS 20

independent variable p. NOS 20

constants p. NOS 20

 Multilingual eGlossary

Figure 8 Microalgae are plantlike organisms that can make oils.

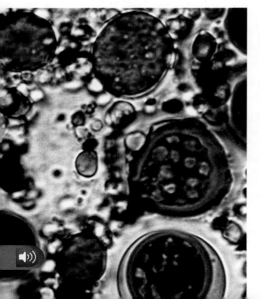

Case Study

Biodiesel from Microalgae

For the last few centuries, fossil fuels have been the main sources of energy for industry and transportation. But, scientists have shown that burning fossil fuels negatively affects the environment. Also, some people are concerned about eventually using up the world's reserves of fossil fuels.

During the past few decades, scientists have explored using protists to produce biodiesel. Biodiesel is a fuel made primarily from living organisms. Protists, shown in **Figure 8,** are a group of microscopic organisms that usually live in water or moist environments. Some of these protists are plantlike because they make their own food using a process called photosynthesis. Microalgae are plantlike protists.

Designing a Controlled Experiment

The scientists in this case study used scientific inquiry to investigate the use of protists to make biodiesel. They designed controlled experiments to test their hypotheses. In the margins of this lesson are examples of how scientists in the study practiced inquiry and the skills you read about in Lesson 1. The notebook pages contain information that a scientist might have written in a science journal.

A controlled experiment is a scientific investigation that tests how one variable affects another. A **variable** *is any factor in an experiment that can have more than one value.* In controlled experiments, there are two types of variables. The **dependent variable** *is the factor measured or observed during an experiment.* The **independent variable** *is the factor that you want to test. It is changed by the investigator to observe how it affects a dependent variable.* **Constants** *are the factors in an experiment that remain the same.*

🔑 **Key Concept Check** How do independent and dependent variables differ?

A controlled experiment has two groups—an experimental group and a control group. The experimental group is used to study how a change in the independent variable changes the dependent variable. The control group contains the same factors as the experimental group, but the independent variable is not changed. Without a control, it is difficult to know whether your experimental observations result from the variable you are testing or from another factor.

Biodiesel

The idea of engines running on fuel made from plant or plantlike sources is not entirely new. Rudolph Diesel, shown in **Figure 9,** invented the diesel engine. He used peanut oil to demonstrate how his engine worked. However, when petroleum was introduced as a diesel fuel source, it was preferred over peanut oil because it was cheaper.

 Reading Check What did Rudolph Diesel use as fuel?

Oil-rich food crops, such as soybeans, can be used as a source of biodiesel. However, some people are concerned that crops grown for fuel sources will replace crops grown for food. If farmers grow more crops for fuel, then the amount of food available worldwide will be reduced. Because of food shortages in many parts of the world, replacing food crops with fuel crops is not a good solution.

Aquatic Species Program

In the late 1970s, the U.S. Department of Energy began funding its Aquatic Species Program (ASP) to investigate ways to remove air pollutants. Coal-fueled power plants produce carbon dioxide (CO_2), a pollutant, as a by-product. In the beginning, the study examined all aquatic organisms that use CO_2 during photosynthesis—their food-making process. These included large plants, commonly know as seaweeds, plants that grow partially underwater, and microalgae. It was hoped these organisms might remove excess CO_2 from the atmosphere. During the studies, however, the project leaders noticed that some microalgae produced large amounts of oil. The program's focus soon shifted to using microalgae to produce oils that could be processed into biodiesel.

Figure 9 Rudolph Diesel invented the first diesel engine in the early 1900s.

Scientific investigations often begin when someone observes an event in nature and wonders why or how it occurs.

A hypothesis is a tentative explanation that can be tested by scientific investigations. A prediction is a statement of what someone expects to happen next in a sequence of events.

Observation A:
While testing microalgae to discover if they would absorb carbon pollutants, ASP project leaders noticed that some species of microalgae had high oil content.

Hypothesis A:
Some microalgae species can be used as a source of biodiesel fuel because the microalgae produce a large amount of oil

Prediction A:
If the correct species is found and the growing conditions are isolated, then large oil amounts will be collected.

Design an Experiment and Collect Data:
The ASP scientists developed a rapid screening test to discover which micro-algae species produced the most oil.
Independent Variable: amount of nitrogen available
Dependent Variable: amount of oil produced
Constants: the growing conditions of algae (temperature, water quality, exposure to the Sun, etc.)

Observation B:
Based on previous microalgae studies, starving microalgae of nutrients could result in more oil production.
Hypothesis B:
Microalgae grown with inadequate amounts of nitrogen alter their growth processes and produce more oil.
Prediction B:
If microalgae receive inadequate amounts of nitrogen then they will produce more oil.

Figure 10 Green microalgae and diatoms showed the most promise during testing for biodiesel production.

Which Microalgae?

Microalgae are microscopic organisms that live in marine (salty) or freshwater environments. Like many plants and other plantlike organisms, they use photosynthesis and make sugar. The process requires light energy. Microalgae make more sugar than they can use as food. They convert excess sugar to oil. Scientists focused on these microalgae because their oil then could be processed into biodiesel.

The scientists began their research by collecting and identifying promising microalgae species. The search focused on microalgae in shallow, inland, saltwater ponds. Scientists predicted that these microalgae were more resistant to changes in temperature and salt content in the water.

By 1985, a test was in place for identifying microalgae with high oil content. Two years later, 3,000 microalgae species had been collected. Scientists checked these samples for tolerance to acidity, salt levels, and temperature and selected 300 species. Of these 300 species, green microalgae and diatoms, as shown in **Figure 10,** showed the most promise. However, it was obvious that no one species was going to be perfect for all climates and water types.

Oil Production in Microalgae

Scientists also began researching how microalgae produce oil. Some studies suggested that starving microalgae of nutrients, such as nitrogen, could increase the amount of oil they produced. However, starving microalgae also caused them to be smaller, resulting in no overall increase in oil production.

Outdoor Testing v. Bioreactors

By the 1980s, the ASP scientists were growing micro-algae in outdoor ponds in New Mexico. However, outdoor conditions were very different from those in the laboratory. Cooler temperatures in the outdoor ponds resulted in smaller microalgae. Native algae species also invaded the ponds, forcing out the high-oil-producing laboratory microalgae species.

The scientists continued to focus on growing microalgae in open ponds, such as the one shown in **Figure 11.** Many scientists still believe that these open ponds are better for producing large quantities of biodiesel from microalgae. But, some researchers are now growing microalgae in closed glass containers called bioreactors, also shown in **Figure 11.** Inside these bioreactors, organisms live and grow under controlled conditions. This method avoids many of the problems associated with open ponds. However, bioreactors are more expensive than open ponds.

A biofuel company in the western United States has been experimenting with a low-cost bioreactor. A scientist at the company explained that they examined the ASP program and hypothesized that they could use long plastic bags, shown in **Figure 11,** instead of closed glass containers.

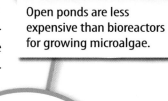

Open ponds are less expensive than bioreactors for growing microalgae.

Figure 11
These three methods of growing microalgae are examples of three different hypotheses that are being tested in controlled experiments.

Microalgae grown in plastic bags are very expensive to harvest.

Microalgae grow under controlled conditions in glass bioreactors.

Why So Many Hypotheses?

According to Dr. Richard Sayre, a biofuel researcher, all the ASP research was based on forming hypotheses. Dr. Sayre says, "It was hypothesis-driven. You just don't go in and say 'Well, I have a feeling this is the right way to do it.' You propose a hypothesis. Then you test it."

Dr. Sayre added, "Biologists have been trained over and over again to develop research strategies based on hypotheses. It's sort of ingrained into our culture. You don't get research support by saying, 'I'm going to put together a system, and it's going to be wonderful.' You have to come up with a question. You propose some strategies for answering the question. What are your objectives? What outcomes do you expect for each objective?"

Reading Check Why is it important for a scientific researcher to develop a good hypothesis?

Increasing Oil Yield

Scientists from a biofuel company in Washington State thought of another way to increase oil production. Researchers knew microalgae use light energy, water, and carbon dioxide and make sugar. The microalgae eventually convert sugar into oil. The scientists wondered if they could increase microalgae oil production by distributing light to all microalgae. The experimental lab setup to test this idea is shown in **Figure 12.**

Observation C:
Microalgae use light energy, water, and carbon dioxide to make sugar, which is converted to oil.
Hypothesis C:
Microalgae will produce more oil if light is distributed evenly throughout because they need light energy to grow and produce more oil.
Prediction C:
If light is distributed more evenly then more microalgae will grow, and more oil will be produced.

Figure 12 Acrylic rods distribute light to microalgae below the water's surface. If microalgae receive light, they can photosynthesize and eventually produce oils. Without light, microalgae are not productive.

Bringing Light to Microalgae

Normally microalgae grow near the surface of a pond. Any microalgae about 5 cm below the pond's surface will grow less. Why is this? First, water blocks light from reaching deep into a pond. Second, microalgae at the top of a pond block light from reaching microalgae below them. Only the top part of a pond is productive.

Experimental Group

Researchers decided to assemble a team of engineers to design a light-distribution system. Light rods distribute artificial light to microalgae in a bioreactor. The bioreactor controls the environmental conditions that affect how the microalgae grow. These conditions include temperature, nutrient levels, carbon dioxide level, airflow, and light.

 Reading Check In the experimental group, what variables are controlled in the bioreactor?

Data from their experiments showed scientists how microalgae in well-lit environments grow compared to how microalgae grow in dimmer environments. Using solar data for various parts of the country, the scientists concluded that the light rod would significantly increase microalgae growth and oil production in outdoor ponds. These scientists next plan to use the light-rod growing method in outdoor ponds.

Field Testing

Scientists plan to take light to microalgae instead of moving microalgae to light. Dr. Jay Burns is chief microalgae scientist at a biofuel company. He said, "What we are proposing to do is to take the light from the surface of a pond and distribute it throughout the depth of the pond. Instead of only the top 5 cm being productive, the whole pond becomes productive."

 Reading Check What is the benefit of the light-distribution system?

Scientists tested their hypothesis, collected data, analyzed the data, and drew conclusions.

Analyze Results:
The experimental results showed that microalgae would produce more oil using a light-rod system than by using just sunlight.
Draw a Conclusion:
The researchers concluded that the light-rod system greatly increased microalgae oil production.

Research scientists and scientists in the field rely on scientific methods and scientific inquiry to solve real-life problems. When a scientific investigation lasts for several years and involves many scientists, such as this study, many hypotheses can be tested. Some hypotheses are supported, and other hypotheses are not. However, information is gathered and lessons are learned. Hypotheses are refined and tested many times. This process of scientific inquiry results in a better understanding of the problem and possible solutions.

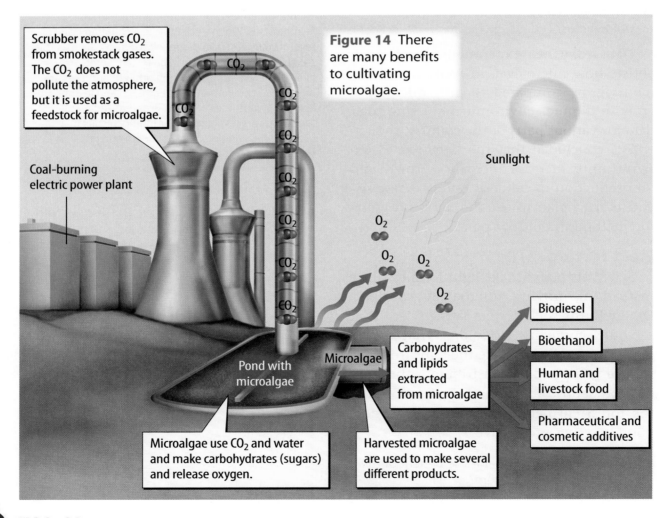

Another Way to Bring Light to Microalgae

Light rods are not the only way to bring light to microalgae. Paddlewheels, as shown in **Figure 13,** can be used to keep the microalgae's locations changing. Paddlewheels continuously rotate microalgae to the surface. This exposes the organisms to more light.

 Key Concept Check Describe three ways in which scientific inquiry was used in this case study.

Why Grow Microalgae?

While the focus of this case study is microalgae growth for biodiesel production, there are other benefits of growing microalgae, as shown in **Figure 14.** Power plants that burn fossil fuels release carbon dioxide into the atmosphere. Evidence indicates that this contributes to global warming. During photosynthesis, microalgae use carbon dioxide and water, release oxygen, and produce sugar, which they convert to oil. Not only do microalgae produce a valuable fuel, they also remove pollutants from and add oxygen to the atmosphere.

Figure 13 During cultivation, paddlewheels bring microalgae to the surface and expose them to light.

Scrubber removes CO_2 from smokestack gases. The CO_2 does not pollute the atmosphere, but it is used as a feedstock for microalgae.

Figure 14 There are many benefits to cultivating microalgae.

Coal-burning electric power plant

Sunlight

O_2

Biodiesel

Bioethanol

Human and livestock food

Pharmaceutical and cosmetic additives

Microalgae

Carbohydrates and lipids extracted from microalgae

Pond with microalgae

Microalgae use CO_2 and water and make carbohydrates (sugars) and release oxygen.

Harvested microalgae are used to make several different products.

Are microalgae the future?

Scientists face many challenges in their quest to produce biodiesel from microalgae. For now, the costs of growing microalgae and extracting their oils are too high to compete with petroleum-based diesel. However, the combined efforts of government-funded programs and commercial biofuel companies might one day make microalgae-based biodiesel an affordable reality in the United States. In fact, a company in Israel has a successful test plant in operation, as shown in **Figure 15.** Plans are underway to build a large-scale industrial facility to convert carbon dioxide gases released from an Israeli coal-powered electrical plants into useful microalgae products. If this technology performs as expected, microalgae cultivation might occur near coal-fueled power plants in other parts of the world, too.

Currently, scientists have no final conclusions about using microalgae as a fuel source. As long as petroleum remains relatively inexpensive and available, it probably will remain the preferred source of diesel fuel. However, if petroleum prices increase or availability decreases, new sources of fuel will be needed. Biodiesel made from microalgae oils might be one of the alternative fuel sources used.

Figure 15 This microalgae test facility in Israel is reducing the amount of carbon dioxide pollution in the atmosphere.

Lesson 3 Review

Assessment   Online Quiz

Use Vocabulary

1 **Define** *variable* in your own words.

2 **Contrast** the terms *dependent variable, independent variable,* and *constants.*

Understand Key Concepts

3 Which factor does the investigator change during an investigation?
   A. constant
   B. dependent variable
   C. independent variable
   D. variable

4 **Give an example** of a scientific inquiry used in a real-life scientific investigation that is not mentioned in this chapter.

Interpret Graphics

5 **Organize Information** Copy and fill in a graphic organizer like the one below with information about the three types of oil production discussed in the study.

Critical Thinking

6 **Hypothesize** other methods to either increase the oil content of microalgae or to grow greater amounts of microalgae for biodiesel production.

7 **Evaluate** scientists' efforts to increase the oil content of microalgae and to grow microalgae more quickly. What would you do differently?

How can you design a bioreactor?

Materials

500-mL Erlenmeyer flask

one-hole stopper with a short pieces of plastic tubing in the hole

500-mL beaker

Also needed: rubber tubing (15 cm), water, 100-mL graduated cylinder, plastic wrap (10 cm × 30 cm), scissors, bendable straw, yeast, sugar, triple-beam balance, stopwatch, ice, thermometer

Safety

You are part of scientific team studying how yeast grows in a bioreactor. In a bioreactor, yeast uses sugar as an energy source and releases carbon dioxide gas as a waste product. One way you can tell how fast yeast grows is to measure the volume of gas the yeast produces.

Ask a Question
How do water temperature and sugar concentration affect yeast growth?

Make Observations

1. Read and complete a lab safety form.
2. Copy the data table shown on the next page into your Science Journal.
3. Place weighing paper or waxed paper on the triple-beam balance, and then zero the balance. Do not place solids directly on the balance. Measure 3 g of yeast. Use the paper to transport the yeast back to your lab station.
4. Repeat step 3 to measure 4 g of sugar.
5. Measure and pour 350 mL of water into both the Erlenmeyer flask and the beaker. Measure 100 mL of water in the graduated cylinder.
6. Seal the graduated cylinder with plastic wrap. Place you hand over the plastic wrap, and turn the graduated cylinder upside down. Carefully place the sealed end of the graduated cylinder into a beaker of water. Pull off the plastic wrap without losing any water from the cylinder. Have a team member hold the graduated cylinder so that it doesn't tip over.
7. Place one end of a 15-cm piece of rubber tubing over the short plastic or glass tubing in the stopper. Without lifting the cylinder above the water's surface, insert the free end of the long piece of tubing inside the graduated cylinder. Have a team continue to hold it. Record the initial reading of the water level in the graduated cylinder in your Science Journal.
8. Add the sugar and then the yeast to the Erlenmeyer flask. Place the stopper in the flask and swirl it to mix the contents. This flask is your bioreactor.
9. Record the volume of gas produced every 10 min for half an hour. To calculate the volume of gas produced for each 10 min time interval, subtract the initial volume from the final volume.

Form a Hypothesis

10 As a class, form a hypothesis that explains how a change in the amount of sugar in your bioreactor affects carbon dioxide production. Form a second hypothesis that explains how a change in temperature of the water affects carbon dioxide production.

Test Your Hypotheses

11 As a class, develop procedures to test your hypotheses. Use a range of temperatures and different amounts of sugar in your tests.

12 With your teammates, set up several bioreactors with the conditions you outlined in your procedures. Record the results from each bioreactor in a separate data table.

13 Using the class data, create two line graphs—one graph for each hypothesis.

Analyze and Conclude

14 **Analyze** What conditions resulted in the fastest growth of yeast?

15 **Compare** Which of the two variables had a greater influence on the growth of yeast? How did you draw that conclusion?

16 **The Big Idea** Which scientific processes did you use in your investigation of bioreactors?

Communicate Your Results

Present your team's results to your class. Include visual aids and at least one graph.

Inquiry Extension

As part of your presentation, propose future research that your team will conduct on bioreactors. Describe other variables or other organisms your team will investigate. Explain the goal of your future research. Will you develop a product that can be marketed? Will you provide an explanation to solve a scientific problem? Will you develop a new technology?

Gas Produced

Temperature of water _____

Amount of sugar _____

Time (min)	Eudiometer Reading (mL)
0	
10	
20	
30	

Lab Tips

☑ Make sure the graduated cylinder is not tilted when you take readings.

☑ If you use a recycled water bottle as your bioreactor, do not squeeze the bottle once you place the stopper in it or you can force air into the eudiometer.

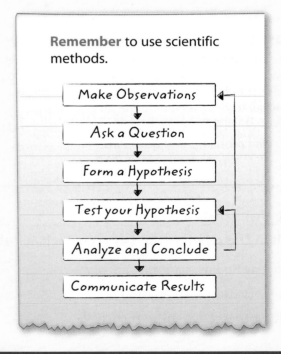

Remember to use scientific methods.

Make Observations

↓

Ask a Question

↓

Form a Hypothesis

↓

Test your Hypothesis

↓

Analyze and Conclude

↓

Communicate Results

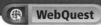

THE BIG IDEA The process of scientific inquiry and performing scientific investigations can provide answers to questions about your world.

Key Concepts Summary

Lesson 1: Understanding Science

- Scientific inquiry, also known as scientific methods, is a collection of skills that scientists use in different combinations to perform scientific investigations.
- Scientific investigations often result in new **technology**, new materials, newly discovered objects or events, or answers to questions.
- A scientist can help minimize bias in a scientific investigation by taking random samples, doing blind studies, repeating an experiment several times, and keeping accurate and honest records.

Lesson 2: Measurement and Scientific Tools

- **Precision** is a description of how similar or close measurements are to each other. **Accuracy** is a description of how close a measurement is to an accepted value.
- **Significant digits** communicate the precision of the tool used to make measurements.
- Life scientists use many tools, such as science journals, microscopes, computers, magnifying lenses, slides, and dissecting tools.

Lesson 3: Case Study: Biodiesel from Microalgae

- The **independent variable** is a factor in an experiment that is manipulated or changed by the investigator to observe how it affects a dependent variable. The **dependent variable** is the factor measured or observed during an experiment.
- Scientific inquiry is used to gain information and find solutions to real-life problems and questions.

Vocabulary

science p. NOS 4
observation p. NOS 6
inference p. NOS 6
hypothesis p. NOS 6
prediction p. NOS 7
technology p. NOS 8
scientific theory p. NOS 9
scientific law p. NOS 9
critical thinking p. NOS 10

description p. NOS 12
explanation p. NOS 12
International System of Units (SI) p. NOS 12
precision p. NOS 14
accuracy p. NOS 14
significant digits p. NOS 15

variable p. NOS 20
dependent variable p. NOS 20
independent variable p. NOS 20
constants p. NOS 20

Use Vocabulary

Explain the relationship between each set of terms.

1. scientific law, scientific theory
2. observation, explanation
3. hypothesis, scientific theory
4. description, explanation
5. International System of Units (SI), significant digits
6. variable, constant

Understand Key Concepts 🔑

7 Which is a quantitative observation?

A. 15 m long

B. red color

C. rough texture

D. strong odor

8 Which is one way scientists indicate how precise and accurate their experimental measurements are?

A. They keep accurate, honest records.

B. They make sure their experiments can be repeated.

C. They use significant figures in their measurements.

D. They record small samples of data.

9 Which is NOT a source of bias?

A. accurate records

B. equipment choice

C. funding source

D. hypothesis formation

Critical Thinking

10 **Explain** What would be the next step in the scientific inquiry process below?

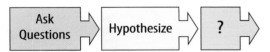

11 **Select** a science career that uses technology. Explain how that career would be different if the technology had not been invented.

12 **Identify** the experimental group, the control group, and controls in the following example. Explain your decision.

A scientist tests a new cough medicine by giving it to a group who have colds. The scientist gives another group with colds a liquid and tells them it is cough medicine. The people in both groups are women between the ages of 20 and 30 who normally are in good health.

Writing in Science

13 **Write** a five-sentence paragraph that includes examples of how bias can be intentional or unintentional and how scientists can reduce bias. Be sure to include topic and concluding sentences in your paragraph.

REVIEW THE **BIG** IDEA

14 What process do scientists use to perform scientific investigations? List a possible sequence of steps in a scientific inquiry and explain your reasoning.

15 What next step of scientific methods might these marine biologists perform?

Math Skills ✕ ÷ **Review**

── Math Practice ──

Significant Digits

16 How many significant figures are in 0.00840, 15.7, and 13.040?

Unit 1

LIFE: Structure & Function

1665
Robert Hooke discovers cells while examining thin slices of cork under a microscope.

1674
Anton van Leeuwenhoek observes living cells under a microscope and names the moving organisms *animalcules*.

1831
The nucleus is given its name by Robert Brown.

1839
Theodor Schwann publishes a book suggesting that the cell is the basic unit of life.

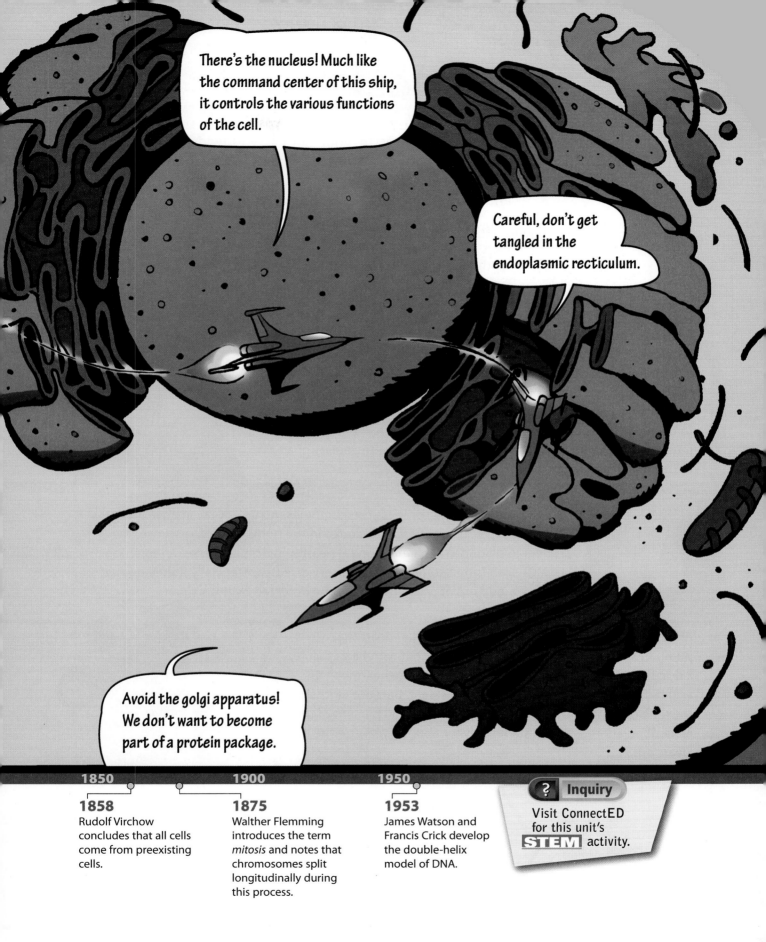

Models

What would you do without your heart—one of the most important muscles in your body? Worldwide, people are on donor lists, patiently waiting for heart transplants because their hearts are not working properly. Today, doctors can diagnose and treat heart problems with the help of models.

A **model** is a representation of an object, a process, an event, or a system that is similar to the physical object or idea being studied. Models can be used to study things that are too big or too small, happen too quickly or too slowly, or are too dangerous or too expensive to study directly. However, some models can replace organs or bones in the body that are not functioning properly.

A magnetic resonance image (MRI) is a type of model created by using a strong magnetic field and radio waves. MRI machines produce high-resolution images of the body from a series of images of different layers of the heart. For example, an MRI model of the heart allows cardiologists to diagnose heart disease or damage. To obtain a clear MRI, the patient must be still. Even the beating of the heart can limit the ability of an MRI to capture clear images.

A computer tomography (CT) scan combines multiple X-ray images into a detailed 3-D visual model of structures in the body. Cardiologists use this model to diagnose a malfunctioning heart or blocked arteries. A limitation of a CT scan is that some coronary artery diseases, especially if they do not involve a buildup of calcium, may not be detected by the scan.

An artificial heart is a physical model of a human heart that can pump blood throughout the body. For a patient with heart failure, a doctor might suggest temporarily replacing the heart with an artificial model while they wait for a transplant. Because of its size, the replacement heart is suitable for about only 50 percent of the male population. And, it is stable only for about 2 years before it wears out.

A cardiologist might use a physical model of a heart to explain a diagnosis to a patient. The parts of the heart can be touched and manipulated to explain how a heart works and the location of any complications. However, this physical model does not function like a real heart, and it cannot be used to diagnose disease.

Maps as Models

One way to think of a computer model, such as an MRI or a CT scan, is as a map. A map is a model that shows how locations are arranged in space. A map can be a model of a small area, such as your street. Or, maps can be models of very large areas, such as a state, a country, or the world.

Biologists study maps to understand where different animal species live, how they interact, and how they migrate. Most animals travel in search of food, water, specific weather, or a place to mate. By placing small electronic tracking devices on migrating animals biologists can create maps of their movements, such as the map of elephant movement in **Figure 1.** These maps are models that help determine how animals survive, repeat the patterns of their life cycle, and respond to environmental changes.

Limitations of Models

It is impossible to include all the details about an object or an idea in one model. A map of elephant migration does not tell you whether the elephant is eating, sleeping, or playing with other elephants. Scientists must consider the limitations of the models they use when drawing conclusions about animal behavior.

All models have limitations. When making decisions about a patient's diagnosis and treatment, a cardiologist must be aware of the information each type of model does and does not provide. CT scans and MRIs each provide different diagnostic information. A doctor needs to know what information is needed before choosing which model to use. Scientists and doctors consider the purpose and limitations of the models they use to ensure that they draw the most accurate conclusions possible.

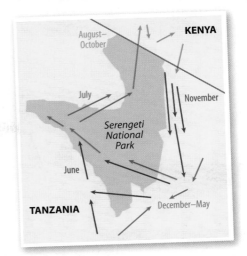

Figure 1 This map is a model of elephants' movements. The colored lines show the paths of three elephants that were equipped with tracking devices for a year.

Inquiry MiniLab

40 minutes

How can you model an elephant enclosure?

You are part of a zoo design firm hired to design a model of a new elephant enclosure that mimics a natural habitat.

1 Read and complete a lab safety form.

2 Research elephants and study the map above to understand the needs of elephants.

3 Create a detailed map of your enclosure using **colored pencils** and a **ruler.** Be sure to include the scale, labels, and a legend.

4 Trade maps with a classmate.

5 Using **salt dough** and **craft supplies,** build a physical 3-D model of the elephant enclosure.

Analyze and Conclude

1. **Describe** How did you decide on the scale for your map?

2. **Compare** What are some similarities between your map and your physical model?

3. **Key Concept** What are the benefits and the limitations of your physical model?

Chapter 1

Classifying and Exploring Life

THE BIG IDEA

What are living things, and how can they be classified?

Inquiry **Dropped Dinner Rolls?**

At first glance, you might think someone dropped dinner rolls on a pile of rocks. These objects might look like dinner rolls, but they're not.

- What do you think the objects are? Do you think they are alive?
- Why do you think they look like this?
- What are living things, and how can they be classified?

Get Ready to Read

What do you think?

Before you read, decide if you agree or disagree with each of these statements. As you read this chapter, see if you change your mind about any of the statements.

1 All living things move.

2 The Sun provides energy for almost all organisms on Earth.

3 A dichotomous key can be used to identify an unknown organism.

4 Physical similarities are the only traits used to classify organisms.

5 Most cells are too small to be seen with the unaided eye.

6 Only scientists use microscopes.

 **ConnectED** Your one-stop online resource

connectED.mcgraw-hill.com

Video

WebQuest

Audio

Assessment

Review

Concepts in Motion

Inquiry

Multilingual eGlossary

Reading Guide

Key Concepts

ESSENTIAL QUESTIONS

- What characteristics do all living things share?

Vocabulary

organism p. 9

cell p. 10

unicellular p. 10

multicellular p. 10

homeostasis p. 13

g **Multilingual eGlossary**

Characteristics of Life

(Inquiry) What's missing?

This toy looks like a dog and can move, but it is a robot. What characteristics are missing to make it alive? Let's find out.

Is it alive?

Living organisms have specific characteristics. Is a rock a living organism? Is a dog? What characteristics describe something that is living?

1. Read and complete a lab safety form.

2. Place three pieces of **pasta** in the bottom of a **clear plastic cup.**

3. Add **carbonated water** to the cup until it is 2/3 full.

4. Observe the contents of the cup for 5 minutes. Record your observations in your Science Journal.

Think About This

1. Think about living things. How do you know they are alive?

2. Which characteristics of life do you think you are observing in the cup?

3. 🔑 **Key Concept** Is the pasta alive? How do you know?

Characteristics of Life

Look around your classroom and then at **Figure 1.** You might see many nonliving things, such as lights and books. Look again, and you might see many living things, such as your teacher, your classmates, and plants. What makes people and plants different from lights and books?

People and plants, like all living things, have all the characteristics of life. All living things are organized, grow and develop, reproduce, respond, maintain certain internal conditions, and use energy. Nonliving things might have some of these characteristics, but they do not have all of them. Books might be organized into chapters, and lights use energy. However, only those things that have all the characteristics of life are living. *Things that have all the characteristics of life are called* **organisms.**

✓ **Reading Check** How do living things differ from nonliving things?

Figure 1 A classroom might contain living and nonliving things.

Organization

Your home is probably organized in some way. For example, the kitchen is for cooking, and the bedrooms are for sleeping. Living things are also organized. Whether an organism is made of one **cell**—*the smallest unit of life*—or many cells, all living things have structures that have specific functions.

Living things that are made of only one cell are called **unicellular** *organisms*. Within a unicellular organism are structures with specialized functions just like a house has rooms for different activities. Some structures take in nutrients or control cell activities. Other structures enable the organism to move.

Living things that are made of two or more cells are called **multicellular** *organisms*. Some multicellular organisms only have a few cells, but others have trillions of cells. The different cells of a multicellular organism usually do not perform the same function. Instead, the cells are organized into groups that have specialized functions, such as digestion or movement.

Growth and Development

The tadpole in **Figure 2** is not a frog, but it will soon lose its tail, grow legs, and become an adult frog. This happens because the tadpole, like all organisms, will grow and develop. When organisms grow, they increase in size. A unicellular organism grows as the cell increases in size. Multicellular organisms grow as the number of their cells increases.

Figure 2 A tadpole grows in size while developing into an adult frog.

Growth and Development   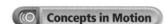 **Concepts in Motion** Animation

☑ **Visual Check** What characteristics of life can you identify in this figure?

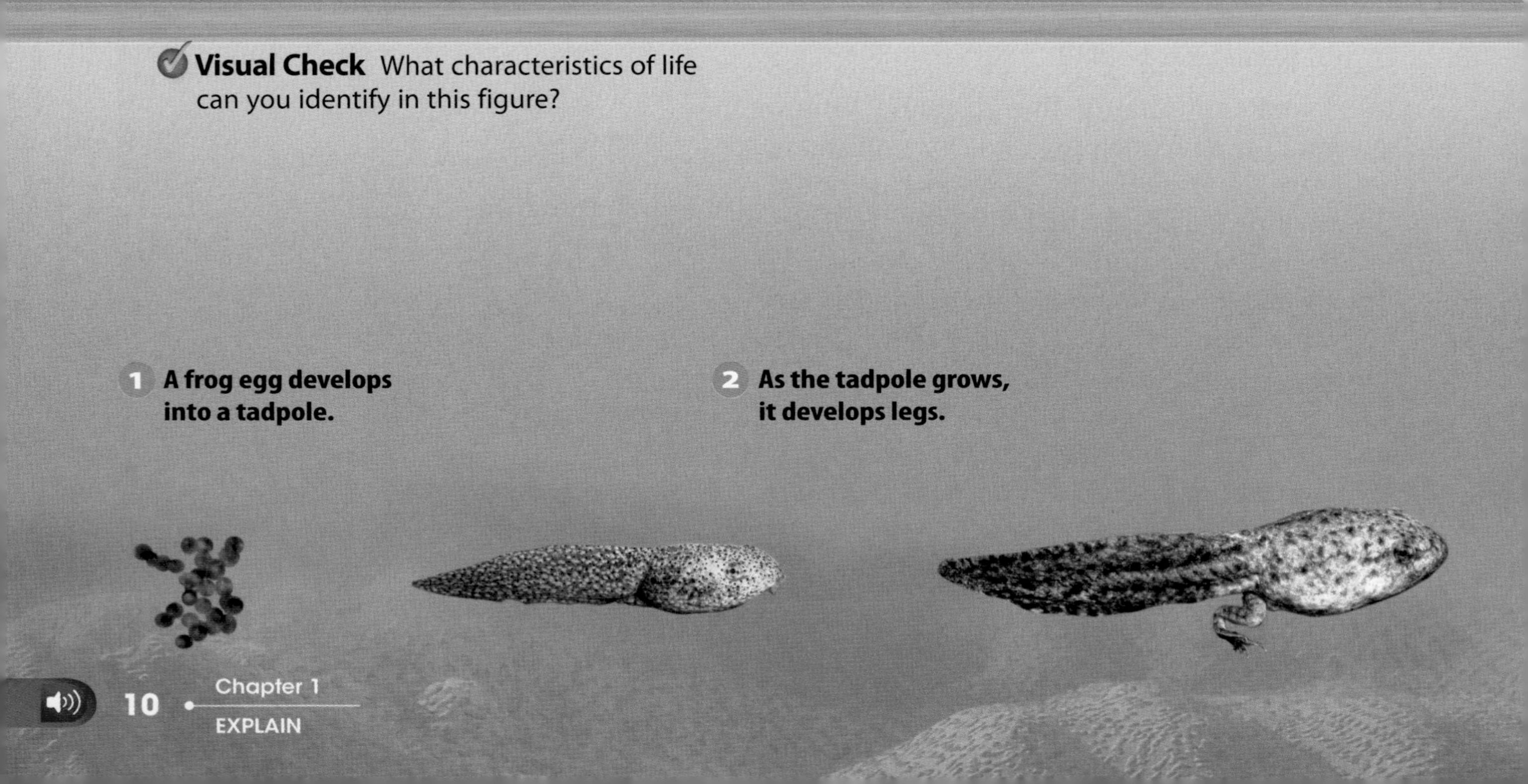

1 A frog egg develops into a tadpole.

2 As the tadpole grows, it develops legs.

Changes that occur in an organism during its lifetime are called development. In multicellular organisms, development happens as cells become specialized into different cell types, such as skin cells or muscle cells. Some organisms undergo dramatic developmental changes over their lifetime, such as a tadpole developing into a frog.

 Reading Check What happens in development?

Reproduction

As organisms grow and develop, they usually are able to reproduce. Reproduction is the process by which one organism makes one or more new organisms. In order for living things to continue to exist, organisms must reproduce. Some organisms within a population might not reproduce, but others must reproduce if the species is to survive.

Organisms do not all reproduce in the same way. Some organisms, like the ones in **Figure 3,** can reproduce by dividing and become two new organisms. Other organisms have specialized cells for reproduction. Some organisms must have a mate to reproduce, but others can reproduce without a mate. The number of offspring produced varies. Humans usually produce only one or two offspring at a time. Other organisms, such as the frog in **Figure 2,** can produce hundreds of offspring at one time.

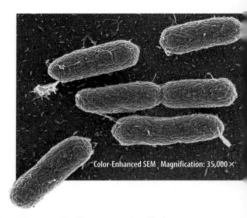
Color-Enhanced SEM Magnification: 35,000×

Figure 3 Some unicellular organisms, like the bacteria shown here, reproduce by dividing. The two new organisms are identical to the original organism.

3 The tadpole continues to grow as it develops into an adult frog.

4 An adult female frog can produce hundreds of eggs.

Responses to Stimuli

If someone throws a ball toward you, you might react by trying to catch it. This is because you, like all living things, respond to changes in the environment. These changes can be internal or external and are called stimuli (STIHM yuh li).

Internal Stimuli

You respond to internal stimuli (singular, stimulus) every day. If you feel hungry and then look for food, you are responding to an internal stimulus—the feeling of hunger. The feeling of thirst that causes you to find and drink water is another example of an internal stimulus.

External Stimuli

Changes in an organism's environment that affect the organism are external stimuli. Some examples of external stimuli are light and temperature.

Many plants, like the one in **Figure 4,** will grow toward light. You respond to light, too. Your skin's response to sunlight might be to darken, turn red, or freckle.

Some animals respond to changes in temperature. The response can be more or less blood flowing to the skin. For example, if the temperature increases, the diameter of an animal's blood vessels increases. This allows more blood to flow to the skin, cooling an animal.

Figure 4 The leaves and stems of plants like this one will grow toward a light source.

Inquiry MiniLab

20 minutes

Did you blink?

Like all living organisms, you respond to changes, or stimuli, in your environment. When you react to a stimulus without thinking, the response is known as a reflex. Let's see what a reflex is like.

1. Read and complete a lab safety form.
2. Sit on a chair with your hands in your lap.
3. Have your partner gently toss a **soft, foam ball** at your face five times. Your partner will warn you when he or she is going to toss the ball. Record your responses in your Science Journal.
4. Have your partner gently toss the ball at your face five times without warning you. Record your responses.
5. Switch places with your partner, and repeat steps 3 and 4.

Analyze and Conclude

1. **Compare** your responses when you were warned and when you were not warned.
2. **Decide** if any of your reactions were reflex responses, and explain your answer.
3. 🔑 **Key Concept** Infer why organisms have reflex responses to some stimuli.

Homeostasis

Have you ever noticed that if you drink more water than usual, you have to go to the bathroom more often? That is because your body is working to keep your internal environment under normal conditions. *An organism's ability to maintain steady internal conditions when outside conditions change is called* **homeostasis** (hoh mee oh STAY sus).

The Importance of Homeostasis

Are there certain conditions you need to do your homework? Maybe you need a quiet room with a lot of light. Cells also need certain conditions to function properly. Maintaining certain conditions—homeostasis—ensures that cells can function. If cells cannot function normally, then an organism might become sick or even die.

Methods of Regulation

A person might not survive if his or her body temperature changes more than a few degrees from 37°C. When your outside environment becomes too hot or too cold, your body responds. It sweats, shivers, or changes the flow of blood to maintain a body temperature of 37°C.

Unicellular organisms, such as the paramecium in **Figure 5**, also have ways of regulating homeostasis. A structure called a contractile vacuole (kun TRAK tul • VA kyuh wohl) collects and pumps excess water out of the cell.

WORD ORIGIN · · · · · · · · · · ·

homeostasis
from Greek *homoios*, means "like, similar"; and *stasis*, means "standing still"

Figure 5 This paramecium lives in freshwater. Water continuously enters its cell and collects in contractile vacuoles. The vacuoles contract and expel excess water from the cell. This maintains normal water levels in the cell.

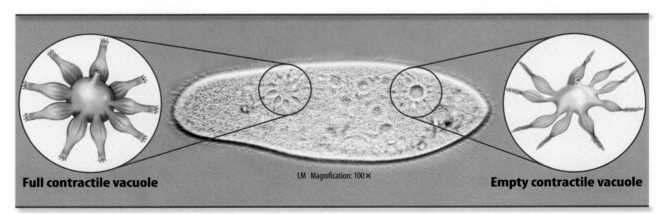

Full contractile vacuole   LM Magnification: 100×   Empty contractile vacuole

There is a limit to the amount of change that can occur within an organism. For example, you are able to survive only a few hours in water that is below 10°C. No matter what your body does, it cannot maintain steady internal conditions, or homeostasis, under these circumstances. As a result, your cells lose their ability to function.

 Reading Check Why is maintaining homeostasis important to organisms?

Energy

Everything you do requires energy. Digesting your food, sleeping, thinking, reading and all of the characteristics of life shown in **Table 1** on the next page require energy. Cells continuously use energy to transport substances, make new cells, and perform chemical reactions. Where does this energy come from?

For most organisms, this energy originally came to Earth from the Sun, as shown in **Figure 6.** For example, energy in the cactus came from the Sun. The squirrel gets energy by eating the cactus, and the coyote gets energy by eating the squirrel.

Key Concept Check What characteristics do all living things share?

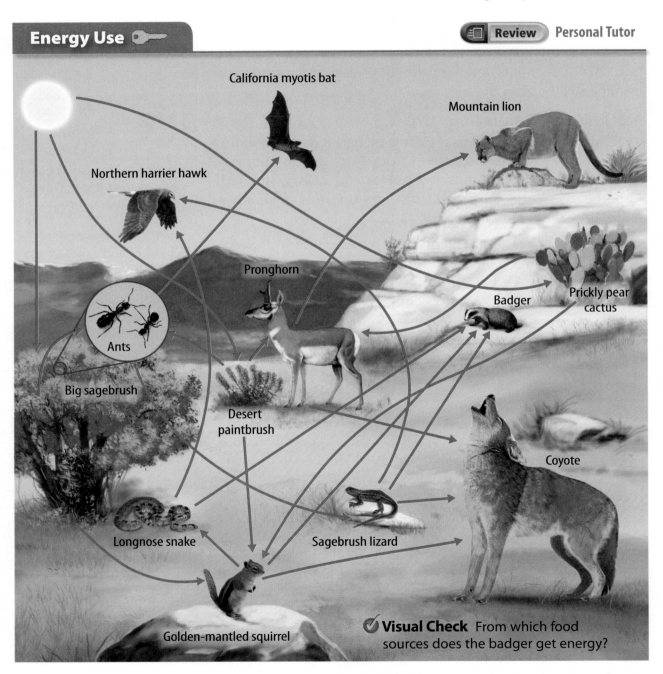

Energy Use

Review   Personal Tutor

- California myotis bat
- Mountain lion
- Northern harrier hawk
- Pronghorn
- Ants
- Badger
- Prickly pear cactus
- Big sagebrush
- Desert paintbrush
- Coyote
- Longnose snake
- Sagebrush lizard
- Golden-mantled squirrel

Visual Check From which food sources does the badger get energy?

Figure 6 All organisms require energy to survive. In this food web, energy passes from one organism to another and to the environment.

Table 1 **Characteristics of Life**   Concepts in Motion Interactive Table

Characteristic	Definition	Example
Organization	Living things have specialized structures with specialized functions. Living things with more than one cell have a greater level of organization because groups of cells function together.	
Growth and development	Living things grow by increasing cell size and/or increasing cell number. Multicellular organisms develop as cells develop specialized functions.	
Reproduction	Living things make more living things through the process of reproduction.	
Response to stimuli	Living things adjust and respond to changes in their internal and external environments.	
Homeostasis	Living things maintain stable internal conditions.	
Use of energy	Living things use energy for all the processes they perform. Living things get energy by making their own food, eating food, or absorbing food.	

Lesson 1 Review

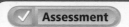
Visual Summary

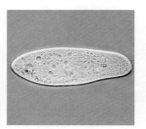

An organism has all the characteristics of life.

Unicellular organisms have specialized structures, much like a house has rooms for different activities.

Homeostasis enables living things to maintain a steady internal environment.

FOLDABLES®

Use your lesson Foldable to review the lesson. Save your Foldable for the project at the end of the chapter.

What do you think NOW?

You first read the statements below at the beginning of the chapter.

1. All living things move.

2. The Sun provides energy for almost all organisms on Earth.

Did you change your mind about whether you agree or disagree with the statements? Rewrite any false statements to make them true.

Use Vocabulary

1 A(n) _____ is the smallest unit of life.

2 **Distinguish** between unicellular and multicellular.

3 **Define** the term *homeostasis* in your own words.

Understand Key Concepts

4 Which is NOT a characteristic of all living things?
   A. breathing   C. reproducing
   B. growing   D. using energy

5 **Compare** the processes of reproduction and growth.

6 **Choose** the characteristic of living things that you think is most important. Explain why you chose that characteristic.

7 **Critique** the following statement: A candle flame is a living thing.

Interpret Graphics

8 **Summarize** Copy and fill in the graphic organizer below to summarize the characteristics of living things.

9 **Describe** all the characteristics of life that are represented in the figure below.

Critical Thinking

10 **Suggest** how organisms would be different if they were not organized.

11 **Hypothesize** what would happen if living things could not reproduce.

AMERICAN
MUSEUM of
NATURAL
HISTORY

CAREERS
in SCIENCE

The Amazing Adaptation of an Air-Breathing Catfish

Discover how some species of armored catfish breathe air.

Have you ever thought about why animals need oxygen? All animals, including you, get their energy from food. When you breathe, the oxygen you take in is used in your cells. Chemical reactions in your cells use oxygen and change the energy in food molecules into energy that your cells can use. Mammals and many other animals get oxygen from air. Most fish get oxygen from water. Either way, after an animal takes in oxygen, red blood cells carry oxygen to cells throughout its body.

Adriana Aquino is an ichthyologist (IHK thee AH luh jihst) at the American Museum of Natural History in New York City. She discovers and classifies species of fish, such as the armored catfish in the family Loricariidae from South America. It lives in freshwater rivers and pools in the Amazon. Its name comes from the bony plates covering its body. Some armored catfish can take in oxygen from water and from air!

Some armored catfish live in fast-flowing rivers. The constant movement of the water evenly distributes oxygen throughout it. The catfish can easily remove oxygen from this oxygen-rich water.

But other armored catfish live in pools of still water, where most oxygen is only at the water's surface. This makes the pools low in oxygen. To maintain a steady level of oxygen in their cells, these fish have adaptations that enable them to take in oxygen directly from air. These catfish can switch from removing oxygen from water through their gills to removing oxygen from air through the walls of their stomachs. They can only do this when they do not have much food in their stomachs. Some species can survive up to 30 hours out of water!

Meet an Ichthyologist

Aquino examines hundreds of catfish specimens. Some she collects in the field, and others come from museum collections. She compares the color, the size, and the shape of the various species. She also examines their internal and external features, such as muscles, gills, and bony plates.

Some armored catfish remove oxygen from air.

It's Your Turn

BRAINSTORM Work with a group. Choose an animal and list five physical characteristics. Brainstorm how these adaptations help the animal be successful in its habitat. Present your findings to the class.

Lesson 2

Reading Guide

Key Concepts 🔑
ESSENTIAL QUESTIONS

- What methods are used to classify living things into groups?
- Why does every species have a scientific name?

Vocabulary

binomial nomenclature p. 21
species p. 21
genus p. 21
dichotomous key p. 22
cladogram p. 23

 Multilingual eGlossary

 Video **BrainPOP®**

Classifying Organisms

Inquiry Alike or Not?

In a band, instruments are organized into groups, such as brass and woodwinds. The instruments in a group are alike in many ways. In a similar way, living things are classified into groups. Why are living things classified?

🔊 **18** •
Chapter 1
ENGAGE

How do you identify similar items?

Do you separate your candies by color before you eat them? When your family does laundry, do you sort the clothes by color first? Identifying characteristics of items can enable you to place them into groups.

1. Read and complete a lab safety form.

2. Examine twelve **leaves.** Choose a characteristic that you could use to separate the leaves into two groups. Record the characteristic in your Science Journal.

3. Place the leaves into two groups, *A* and *B,* using the characteristic you chose in step 2.

4. Choose another characteristic that you could use to further divide group A. Record the characteristic, and divide the leaves.

5. Repeat step 4 with group B.

Think About This

1. What types of characteristics did other groups in class choose to separate the leaves?

2. 🔑 **Key Concept** Why would scientists need rules for separating and identifying items?

Classifying Living Things

How would you find your favorite fresh fruit or vegetable in the grocery store? You might look in the produce section, such as the one shown in **Figure 7.** Different kinds of peppers are displayed in one area. Citrus fruits such as oranges, lemons, and grapefruits are stocked in another area. There are many different ways to organize produce in a grocery store. In a similar way, there have been many different ideas about how to organize, or classify, living things.

A Greek philosopher named Aristotle (384 B.C.–322 B.C.) was one of the first people to classify organisms. Aristotle placed all organisms into two large groups, plants and animals. He classified animals based on the presence of "red blood," the animal's environment, and the shape and size of the animal. He classified plants according to the structure and size of the plant and whether the plant was a tree, a shrub, or an herb.

Figure 7 The produce in this store is classified into groups.

✔️ **Visual Check** What other ways can you think of to classify and organize produce?

Determining Kingdoms

In the 1700s, Carolus Linnaeus, a Swedish physician and botanist, classified organisms based on similar structures. Linnaeus placed all organisms into two main groups, called **kingdoms.** Over the next 200 years, people learned more about organisms and discovered new organisms. In 1969 American biologist Robert H. Whittaker proposed a five-kingdom system for classifying organisms. His system included kingdoms Monera, Protista, Plantae, Fungi, and Animalia.

Determining Domains

The classification system of living things is still changing. The current classification method is called systematics. Systematics uses all the evidence that is known about organisms to classify them. This evidence includes an organism's cell type, its habitat, the way an organism obtains food and energy, structure and function of its features, and the common ancestry of organisms. Systematics also includes molecular analysis—the study of molecules such as DNA within organisms.

Using systematics, scientists identified two distinct groups in Kingdom Monera—Bacteria and Archaea (ar KEE uh). This led to the development of another level of classification called domains. All organisms are now classified into one of three domains—Bacteria, Archaea, or Eukarya (yew KER ee uh) —and then into one of six kingdoms, as shown in **Table 2.**

Key Concept Check What evidence is used to classify living things into groups?

Table 2 Domains and Kingdoms

Domain	Bacteria	Archaea	Eukarya			
Kingdom	Bacteria	Archaea	Protista	Fungi	Plantae	Animalia
Example						
Characteristics	Bacteria are simple unicellular organisms.	Archaea are simple unicellular organisms that often live in extreme environments.	Protists are unicellular and are more complex than bacteria or archaea.	Fungi are unicellular or multicellular and absorb food.	Plants are multicellular and make their own food.	Animals are multicellular and take in their food.

Scientific Names

Suppose you did not have a name. What would people call you? All organisms, just like people, have names. When Linnaeus grouped organisms into kingdoms, he also developed a system for naming organisms. This naming system, called binomial nomenclature (bi NOH mee ul · NOH mun klay chur), is the system we still use today.

Binomial Nomenclature

Linneaus's naming system, **binomial nomenclature,** *gives each organism a two-word scientific name,* such as *Ursus arctos* for a brown bear. This two-word scientific name is the name of an organism's species (SPEE sheez). *A* **species** *is a group of organisms that have similar traits and are able to produce fertile offspring.* In binomial nomenclature, the first word is the organism's genus (JEE nus) name, such as *Ursus*. A **genus** *is a group of similar species.* The second word might describe the organism's appearance or its behavior.

How do species and genus relate to kingdoms and domains? Similar species are grouped into one genus (plural, genera). Similar genera are grouped into families, then orders, classes, phyla, kingdoms, and finally domains, as shown for the grizzly bear in **Table 3.**

WORD ORIGIN · · · · · · · · · ·

genus
from Greek *genos*, means "race, kind"

Table 3 The classification of the brown bear or grizzly bear shows that it belongs to the order Carnivora.

Table 3 Classification of the Brown Bear		
Taxonomic Group	**Number of Species**	**Examples**
Domain Eukarya	About 4–10 million	
Kingdom Animalia	About 2 million	
Phylum Chordata	About 50,000	
Class Mammalia	About 5,000	
Order Carnivora	About 270	
Family Ursidae	8	
Genus *Ursus*	4	
Species *Ursus arctos*	1	

✅ Visual Check What domain does the brown bear belong to?

Uses of Scientific Names

When you talk about organisms, you might use names such as bird, tree, or mushroom. However, these are common names for a number of different species. Sometimes there are several common names for one organism. The animal in **Table 3** on the previous page might be called a brown bear or a grizzly bear, but it has only one scientific name, *Ursus arctos*.

Other times, a common name might refer to several different types of organisms. For example, you might call both of the trees in **Figure 8** pine trees. But these trees are two different species. How can you tell? Scientific names are important for many reasons. Each species has its own scientific name. Scientific names are the same worldwide. This makes communication about organisms more effective because everyone uses the same name for the same species.

▲ **Figure 8** These trees are two different species. *Pinus alba* has long needles, and *Tsuga canadensis* has short needles.

 Key Concept Check Why does every species have a scientific name?

Classification Tools

Suppose you go fishing and catch a fish you don't recognize. How could you figure out what type of fish you have caught? There are several tools you can use to identify organisms.

Dichotomous Keys

A **dichotomous key** *is a series of descriptions arranged in pairs that leads the user to the identification of an unknown organism.* The chosen description leads to either another pair of statements or the identification of the organism. Choices continue until the organism is identified. The dichotomous key shown in **Figure 9** identifies several species of fish.

Dichotomous Key 🔑

1. a. This fish has a mouth that extends past its eye. It is an arrow goby.	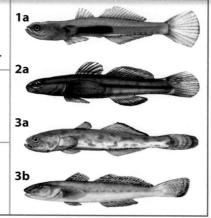
b. This fish does not have a mouth that extends past its eye. Go to step 2.	
2. a. This fish has a dark body with stripes. It is a chameleon goby.	
b. This fish has a light body with no stripes. Go to step 3.	
3. a. This fish has a black-tipped dorsal fin. It is a bay goby.	
b. This fish has a speckled dorsal fin. It is a yellowfin goby.	

▲ **Figure 9** Dichotomous keys include a series of questions to identify organisms.

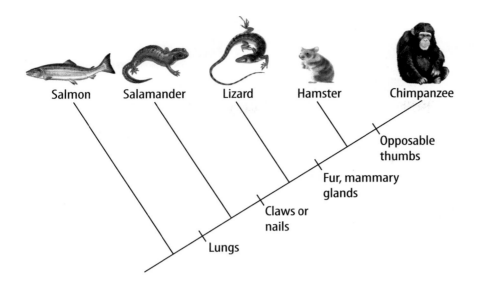

Salmon   Salamander   Lizard   Hamster   Chimpanzee

Opposable thumbs

Fur, mammary glands

Claws or nails

Lungs

Figure 10 A cladogram shows relationships among species. In this cladogram, salamanders are more closely related to lizards than they are to hamsters.

Concepts in Motion
Animation

Cladograms

A family tree shows the relationships among family members, including common ancestors. Biologists use a similar diagram, called a cladogram. *A* **cladogram** *is a branched diagram that shows the relationships among organisms, including common ancestors.* A cladogram, as shown in **Figure 10,** has a series of branches. Notice that each branch follows a new characteristic. Each characteristic is observed in all the species to its right. For example, the salamander, lizard, hamster, and chimpanzee have lungs, but the salmon does not. Therefore, they are more closely related to each other than they are to the salmon.

Inquiry MiniLab   **20 minutes**

How would you name an unknown organism?
Assign scientific names to four unknown alien organisms from a newly discovered planet.

1. Use the table to assign scientific names to identify each alien.
2. Compare your names with those of your classmates.

Analyze and Conclude

1. **Explain** why you chose the two-word names for each organism.

2. **Compare** your names to those of a classmate. Explain any differences.

3. **Key Concept** Discuss how two-word scientific names help scientists identify and organize living things.

Prefix	Meaning	Suffix	Meaning
mon—	one	—antennius	antenna
di—	two	—ocularus	eye
rectanguli—	square	—formus	shape
trianguli—	triangle	—uris	tail

Lesson 2 Review

Visual Summary

All organisms are classified into one of three domains: Bacteria, Archaea, or Eukarya.

Every organism has a unique species name.

A dichotomous key helps to identify an unknown organism through a series of paired descriptions.

FOLDABLES

Use your lesson Foldable to review the lesson. Save your Foldable for the project at the end of the chapter.

What do you think NOW?

You first read the statements below at the beginning of the chapter.

3. A dichotomous key can be used to identify an unknown organism.

4. Physical similarities are the only traits used to classify organisms.

Did you change your mind about whether you agree or disagree with the statements? Rewrite any false statements to make them true.

Use Vocabulary

1 A naming system that gives every organism a two-word name is _____ _____.

2 **Use the term** *dichotomous key* in a sentence.

3 **Organisms** of the same _____ are able to produce fertile offspring.

Understand Key Concepts

4 **Describe** how you write a scientific name.

5 **Compare** the data available today on how to classify things with the data available during Aristotle's time.

6 Which is NOT used to classify organisms?
    A. ancestry
    B. habitat
    C. age of the organism
    D. molecular evidence

Interpret Graphics

7 **Organize Information** Copy and fill in the graphic organizer below to show how organisms are classified.

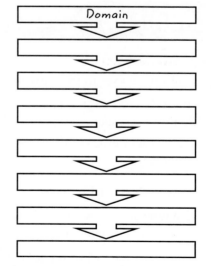

Domain

Critical Thinking

8 **Suggest** a reason scientists might consider changing the current classification system.

9 **Evaluate** the importance of scientific names.

How can you identify a beetle?

A dichotomous key is one of the tools scientists use to identify an unknown organism and **classify** it into a group. To use a dichotomous key, a scientist examines specific characteristics of the unknown organism and compares them to characteristics of known organisms.

Learn It

Sorting objects or events into groups based on common features is called classifying. When classifying, select one feature that is shared by some members of the group, but not by all. Place those members that share the feature in a subgroup. You can **classify** objects or events into smaller and smaller subgroups based on characteristics.

Try It

1 Use the dichotomous key to identify beetle A. Choose between the first pair of descriptions. Follow the instructions for the next choice. Notice that each description either ends in the name of the beetle or instructs you to go on to another set of choices.

2 In your Science Journal, record the identity of the beetle using both its common name and scientific name.

3 Repeat steps 1 and 2 for beetles B, C, and D.

Apply It

4 Think about the choices in each step of the dichotomous key. What conclusion can be made if you arrive at a step and neither choice seems correct?

5 Predict whether a dichotomous key will work if you start at a location other than the first description. Support your reasoning.

6 🔑 **Key Concept** How did the dichotomous key help you classify the unknown beetles?

Dichotomous Key	
1A.	The beetle has long, thin antennae. Go to 5.
1B.	The beetle does not have long, thin antennae. Go to 2.
2A.	The beetle has short antennae that branch. Go to 3.
2B.	The beetle does not have short antennae that branch. It is a stag beetle, *Lucanus cervus*.
3A.	The beetle has a triangular structure between wing covers and upper body. It is a Japanese beetle, *Popillia japonica*.
3B.	The beetle does not have a triangular structure. Go to 4.
4A.	The beetle has a wide, rounded body. It is a June bug, *Cotinis nitida*.
4B.	The beetle does not have a wide, rounded body. It is a death watch beetle, *Xestobium rufovillosum*.
5A.	The beetle has a distinct separation between body parts. Go to 6.
5B.	The beetle has no distinct separation between body parts. It is a firefly, *Photinus pyralis*.
6A.	The beetle has a black, gray, and white body with two black eyespots. It is an eyed click beetle, *Alaus oculatis*.
6B.	The beetle has a dull brown body with light stripes. It is a click beetle, *Chalcolepidius limbatus*.

Lesson 3

Reading Guide

Key Concepts 🔑
ESSENTIAL QUESTIONS

- How did microscopes change our ideas about living things?

- What are the types of microscopes, and how do they compare?

Vocabulary

light microscope p. 28

compound microscope p. 28

electron microscope p. 29

 Multilingual eGlossary

Exploring Life

Inquiry Giant Insect?

Although this might look like a giant insect, it is a photo of a small tick taken with a high-powered microscope. This type of microscope can enlarge an image of an object up to 200,000 times. How can seeing an enlarged image of a living thing help you understand life?

Can a water drop make objects appear bigger or smaller?

For centuries, people have been looking for ways to see objects in greater detail. How can something as simple as a drop of water make this possible?

1 Read and complete a lab safety form.

2 Lay a sheet of **newspaper** on your desk. Examine a line of text, noting the size and shape of each letter. Record your observations in your Science Journal.

3 Add a large drop of **water** to the center of a piece of **clear plastic.** Hold the plastic about 2 cm above the same line of text.

4 Look through the water at the line of text you viewed in step 2. Record your observations.

Think About This

1. Describe how the newsprint appeared through the drop of water.

2. 🔑 **Key Concept** How might microscopes change your ideas about living things?

The Development of Microscopes

Have you ever used a magnifying lens to see details of an object? If so, then you have used a tool similar to the first microscope. The invention of microscopes enabled people to see details of living things that they could not see with the unaided eye. The microscope also enabled people to make many discoveries about living things.

In the mid 1600s English scientist Robert Hooke made one of the most significant discoveries using a microscope. He observed and named cells. Before microscopes, people did not know that living things are made of cells. In the late 1600s the Dutch merchant Anton van Leeuwenhoek (LAY vun hook) made improvements to the first microscopes. His microscope, similar to the one shown in **Figure 11,** had one lens and could magnify an image about 270 times its original size. This made it easier to view organisms.

🔑 **Key Concept Check** How did microscopes change our ideas about living things?

Figure 11 Anton van Leeuwenhoek observed pond water and insects using a microscope like the one shown above.

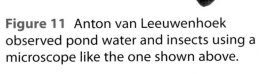

Types of Microscopes

One characteristic of all microscopes is that they magnify objects. Magnification makes an object appear larger than it really is. Another characteristic of microscopes is resolution—how clearly the magnified object can be seen. The two main types of microscopes—light microscopes and electron microscopes—differ in magnification and resolution.

Light Microscopes

If you have used a microscope in school, then you have probably used a light microscope. **Light microscopes** *use light and lenses to enlarge an image of an object.* A simple light microscope has only one lens. *A light microscope that uses more than one lens to magnify an object is called a* **compound microscope.** A compound microscope magnifies an image first by one lens, called the objective lens. The image is then further magnified by another lens, called the ocular lens. The total magnification of the image is equal to the magnifications of the ocular lens and the objective lens multiplied together.

Light microscopes can enlarge images up to 1,500 times their original size. The resolution of a light microscope is about 0.2 micrometers (μm), or two-millionths of a meter. A resolution of 0.2 μm means you can clearly see points on an object that are at least 0.2 μm apart.

Light microscopes can be used to view living or nonliving objects. In some light microscopes, an object is placed directly under the microscope. For other light microscopes, an object must be mounted on a slide. In some cases, the object, such as the white blood cells in **Figure 12,** must be stained with a dye in order to see any details.

Reading Check What are some ways an object can be examined under a light microscope?

Math Skills

Use Multiplication

The magnifying power of a lens is expressed by a number and a multiplication symbol (×). For example, a lens that makes an object look ten times larger has a power of 10×. To determine a microscope's magnification, multiply the power of the ocular lens by the power of the objective lens. A microscope with a 10× ocular lens and a 10× objective lens magnifies an object 10 × 10, or 100 times.

Practice

What is the magnification of a compound microscope with a 10× ocular lens and a 4× objective lens?

 Review

- **Math Practice**
- **Personal Tutor**

Compound Light Microscope

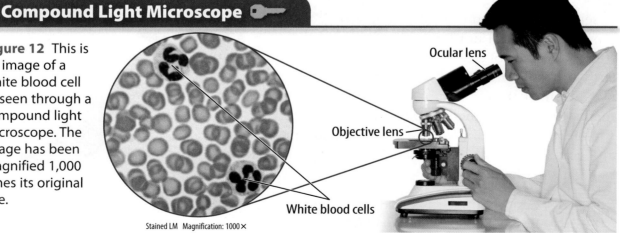

Figure 12 This is an image of a white blood cell as seen through a compound light microscope. The image has been magnified 1,000 times its original size.

Ocular lens

Objective lens

White blood cells

Stained LM Magnification: 1000×

Electron Microscopes

You might know that electrons are tiny particles inside **atoms. Electron microscopes** *use a magnetic field to focus a beam of electrons through an object or onto an object's surface.* An electron microscope can magnify an image up to 100,000 times or more. The resolution of an electron microscope can be as small as 0.2 nanometers (nm), or two-billionths of a meter. This resolution is up to 1,000 times greater than a light microscope. The two main types of electron microscopes are transmission electron microscopes (TEMs) and scanning electron microscopes (SEMs).

TEMs are usually used to study extremely small things such as cell structures. Because objects must be mounted in plastic and then very thinly sliced, only dead organisms can be viewed with a TEM. In a TEM, electrons pass through the object and a computer produces an image of the object. A TEM image of a white blood cell is shown in **Figure 13.**

SEMs are usually used to study an object's surface. In an SEM, electrons bounce off the object and a computer produces a three-dimensional image of the object. An image of a white blood cell from an SEM is shown in **Figure 13.** Note the difference in detail in this image compared to the image in **Figure 12** of a white blood cell from a light microscope.

 Key Concept Check What are the types of microscopes, and how do they compare?

REVIEW VOCABULARY · · · · ·

atom
the building block of matter that is composed of protons, neutrons, and electrons

FOLDABLES®

Make a two-column folded chart. Label the front *Types of Microscopes,* and label the inside as shown. Use it to organize your notes about microscopes.

Figure 13 A TEM greatly magnifies thin slices of an object. An SEM is used to view a three-dimensional image of an object.

Electron Microscopes 🔑

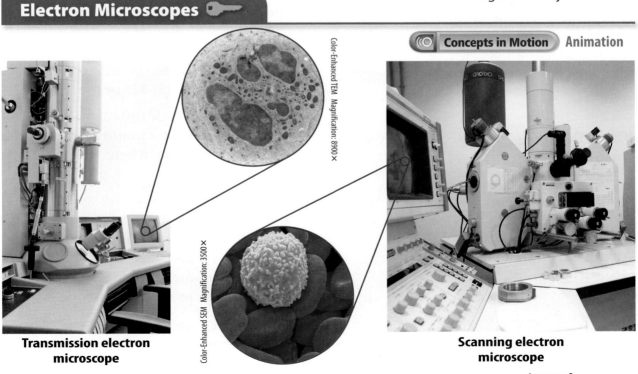

Color-Enhanced TEM Magnification: 8900×

Color-Enhanced SEM Magnification: 3500×

(((O Concepts in Motion))) Animation

Transmission electron microscope

Scanning electron microscope

Inquiry MiniLab 20 minutes

How do microscopes help us compare living things?

A microscope enables scientists to study objects in greater detail than is possible with the unaided eye. Compare what objects look like with the unaided eye to those same objects observed using a microscope.

1 Read and complete a lab safety form.

2 Examine a **sea sponge,** a **leaf,** and **salt crystals.** Draw each object in your Science Journal.

3 Observe **microscope slides of each object** using a **microscope** on low power.

4 Draw each object as it appears under low power.

Analyze and Conclude

1. **Compare** your sketches of the objects observed with your unaided eye and observed with a microscope.

2. 🔑 **Key Concept** Explain how studying an object under a microscope might help you understand it better.

WORD ORIGIN

microscope
from Latin *microscopium,* means "an instrument for viewing what is small"

ACADEMIC VOCABULARY

identify
(verb) to determine the characteristics of a person or a thing

Using Microscopes

The **microscopes** used today are more advanced than the microscopes used by Leeuwenhoek and Hooke. The quality of today's light microscopes and the invention of electron microscopes have made the microscope a useful tool in many fields.

Health Care

People in health-care fields, such as doctors and laboratory technicians, often use microscopes. Microscopes are used in surgeries, such as cataract surgery and brain surgery. They enable doctors to view the surgical area in greater detail. The area being viewed under the microscope can also be displayed on a TV monitor so that other people can watch the procedure. Laboratory technicians use microscopes to analyze body fluids, such as blood and urine. They also use microscopes to determine whether tissue samples are healthy or diseased.

Other Uses

Health care is not the only field that uses microscopes. Have you ever wondered how police determine how and where a crime happened? Forensic scientists use microscopes to study evidence from crime scenes. The presence of different insects can help identify when and where a homicide happened. Microscopes might be used to **identify** the type and age of the insects.

People who study fossils might use microscopes. They might examine a fossil and other materials from where the fossil was found.

Some industries also use microscopes. The steel industry uses microscopes to examine steel for impurities. Microscopes are used to study jewels and identify stones. Stones have some markings and impurities that can be seen only by using a microscope.

✔ **Reading Check** List some uses of microscopes.

Lesson 3 Review

Visual Summary

Living organisms can be viewed with light microscopes.

A compound microscope is a type of light microscope that has more than one lens.

Living organisms cannot be viewed with a transmission electron microscope.

FOLDABLES

Use your lesson Foldable to review the lesson. Save your Foldable for the project at the end of the chapter.

What do you think NOW?

You first read the statements below at the beginning of the chapter.

5. Most cells are too small to be seen with the unaided eye.

6. Only scientists use microscopes.

Did you change your mind about whether you agree or disagree with the statements? Rewrite any false statements to make them true.

Use Vocabulary

1 **Define** the term *light microscope* in your own words.

2 A(n) _____ focuses a beam of electrons through an object or onto an object's surface.

Understand Key Concepts

3 **Explain** how the discovery of microscopes has changed what we know about living things.

4 Which microscope would you use if you wanted to study the surface of an object?
   A. compound microscope
   B. light microscope
   C. scanning electron microscope
   D. transmission electron microscope

Interpret Graphics

5 **Identify** Copy and fill in the graphic organizer below to identify four uses of microscopes.

Microscope Uses

6 **Compare** the images of the white blood cells below. How do they differ?

Critical Thinking

7 **Develop** a list of guidelines for choosing a microscope to use.

Math Skills   **Review** — Math Practice —

8 A student observes a blood sample with a compound microscope that has a 10× ocular lens and a 40× objective lens. How much larger do the blood cells appear under the microscope?

Constructing a Dichotomous Key

Materials

a collection of objects

A dichotomous key is a series of descriptions arranged in pairs. Each description leads you to the name of the object or to another set of choices until you have identified the organism. In this lab, you will create a dichotomous key to classify objects.

Question

How can you create a dichotomous key to identify objects?

Procedure

1. Read and complete a lab safety form.

2. Obtain a container of objects from your teacher.

3. Examine the objects, and then brainstorm a list of possible characteristics. You might look at each object's size, shape, color, odor, texture, or function.

4. Choose a characteristic that would separate the objects into two groups. Separate the objects based on whether or not they have this characteristic. This characteristic will be used to begin a dichotomous key, like the example below.

Dichotomous Key to Identify Office Supplies

The object is made of wood. Go to 1.

The object is not made of wood. Go to 2.

1. The object is longer than 20 cm. Go to 5.

3. The object is not longer than 20 cm. Go to 9.

2. The object is made of metal. Go to 6.

4. The object is not made of metal. Go to 10.

5. Write a sentence to describe the characteristic in step 4, and then write "Go to 1." Write another sentence that has the word "not" in front of the characteristic. Then write "Go to 2."

6. Repeat steps 4 and 5 for the two new groups. Give sentences for new groups formed from the first group consecutive odd numbers. Give sentences for groups formed from the second group consecutive even numbers. Remember to add the appropriate "Go to" directions.

7. Repeat steps 4–6 until there is one object in each group. Give each object an appropriate two-word name.

8. Give your collection of objects and your dichotomous key to another group. Have them identify each object using your dichotomous key. Have them record their answers.

Analyze and Conclude

9. **Evaluate** Was the other team able to correctly identify the collection of objects using your dichotomous key? Why or why not?

10. **The Big Idea** Summarize how dichotomous keys are useful in identifying unknown objects.

Communicate Your Results

Create a poster using drawings or photos of each object you identified. Include your two-word names for the objects.

Inquiry Extension

Teach a peer how to use a dichotomous key. Let the peer use your collection to have a first-hand experience with how a key works.

Lab Tips

☑ Base the questions in your key on observable, measurable, or countable characteristics. Avoid questions that refer to how something is used or how you think or feel about an item.

☑ Remember to start with general questions and then get more and more specific.

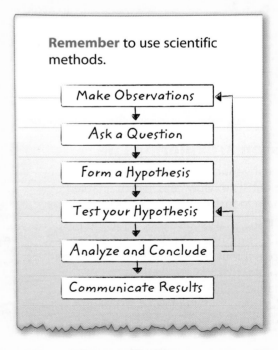

Remember to use scientific methods.

- Make Observations
- Ask a Question
- Form a Hypothesis
- Test your Hypothesis
- Analyze and Conclude
- Communicate Results

Chapter 1 Study Guide

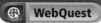

THE BIG IDEA All living things have certain characteristics in common and can be classified using several methods. The invention of the microscope has enabled us to explore life further, which has led to changes in classification.

Key Concepts Summary

| | Vocabulary |

Lesson 1: Characteristics of Life

- An **organism** is classified as a living thing because it has all the characteristics of life.
- All living things are organized, grow and develop, reproduce, respond to stimuli, maintain **homeostasis,** and use energy.

organism p. 9
cell p. 10
unicellular p. 10
multicellular p. 10
homeostasis p. 13

Lesson 2: Classifying Organisms

- Living things are classified into different groups based on physical or molecular similarities.
- Some **species** are known by many different common names. To avoid confusion, every species has a scientific name based on a system called **binomial nomenclature.**

binomial nomenclature p. 21
species p. 21
genus p. 21
dichotomous key p. 22
cladogram p. 23

Lesson 3: Exploring Life

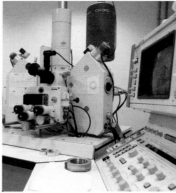

- The invention of microscopes allowed scientists to view cells, which enabled them to further explore and classify life.
- A **light microscope** uses light and has one or more lenses to enlarge an image up to about 1,500 times its original size. An **electron microscope** uses a magnetic field to direct beams of electrons, and it enlarges an image 100,000 times or more.

light microscope p. 28
compound microscope p. 28
electron microscope p. 29

FOLDABLES® Chapter Project

Assemble your lesson Foldables as shown to make a Chapter Project. Use the project to review what you have learned in this chapter.

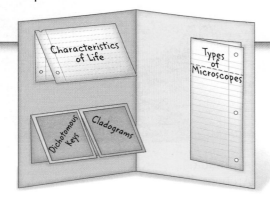

Use Vocabulary

1 A(n) _____ organism is made of only one cell.

2 Something with all the characteristics of life is a(n) _____.

3 A(n) _____ shows the relationships among species.

4 A group of similar species is a(n) _____.

5 A(n) _____ has a resolution up to 1,000 times greater than a light microscope.

6 A(n) _____ is a light microscope that uses more than one lens to magnify an image.

Link Vocabulary and Key Concepts

Concepts in Motion Interactive Concept Map

Copy this concept map, and then use vocabulary terms from the previous page to complete the concept map.

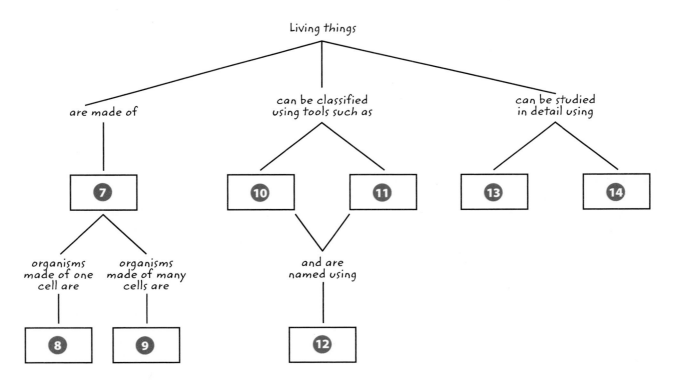

Chapter 1 Review

Understand Key Concepts

1 Which is an internal stimulus?
  A. an increase in moisture
  B. feelings of hunger
  C. number of hours of daylight
  D. the temperature at night

2 Which is an example of growth and development?
  A. a caterpillar becoming a butterfly
  B. a chicken laying eggs
  C. a dog panting
  D. a rabbit eating carrots

3 Based on the food web below, what is an energy source for the mouse?

  A. fox
  B. grass
  C. owl
  D. snake

4 Which shows the correct order for the classification of species?
  A. domain, kingdom, class, order, phylum, family, genus, species
  B. domain, kingdom, phylum, class, order, family, genus, species
  C. domain, kingdom, phylum, class, order, family, species, genus
  D. domain, kingdom, phylum, order, class, family, genus, species

5 The organism shown below belongs in which kingdom?

  A. Animalia
  B. Archaea
  C. Bacteria
  D. Plantae

6 Which was discovered using a microscope?
  A. blood
  B. bones
  C. cells
  D. hair

7 What type of microscope would most likely be used to obtain an image of a live roundworm?
  A. compound light microscope
  B. scanning electron microscope
  C. simple light microscope
  D. transmission electron microscope

8 Which best describes a compound microscope?
  A. uses electrons to magnify the image of an object
  B. uses multiple lenses to magnify the image of an object
  C. uses one lens to magnify the image of an object
  D. uses sound waves to magnify the image of an object

Critical Thinking

9 **Distinguish** between a unicellular organism and a multicellular organism.

10 **Critique** the following statement: An organism that is made of only one cell does not need organization.

11 **Infer** In the figure below, which plant is responding to a lack of water in its environment? Explain your answer.

12 **Explain** how using a dichotomous key can help you identify an organism.

13 **Describe** how the branches on a cladogram show the relationships among organisms.

14 **Assess** the effect of molecular evidence on the classification of organisms.

15 **Compare** light microscopes and electron microscopes.

16 **State** how microscopes have changed the way living things are classified.

17 **Compare** magnification and resolution.

18 **Evaluate** the impact microscopes have on our daily lives.

Writing in Science

19 **Write** a five-sentence paragraph explaining the importance of scientific names. Be sure to include a topic sentence and a concluding sentence in your paragraph.

REVIEW THE BIG IDEA

20 Define the characteristics that all living things share.

21 The photo below shows living and nonliving things. How would you classify the living things by domain and kingdom?

Math Skills ✕∙÷＋

Review — Math Practice —

Use Multiplication

22 A microscope has an ocular lens with a power of 5✕ and an objective lens with a power of 50✕. What is the total magnification of the microscope?

23 A student observes a unicellular organism with a microscope that has a 10✕ ocular lens and a 100✕ objective lens. How much larger does the organism look through this microscope?

24 The ocular lens on a microscope has a power of 10✕. The microscope makes objects appear 500 times larger. What is the power of the objective lens?

Record your answers on the answer sheet provided by your teacher or on a sheet of paper.

Multiple Choice

1 What feature of living things do the terms *unicellular* and *multicellular* describe?

  A how they are organized

  B how they reproduce

  C how they maintain temperature

  D how they produce macromolecules

Use the diagram below to answer question 2.

2 Which characteristic of life does the diagram show?

  A homeostasis

  B organization

  C growth and development

  D response to stimuli

3 A newly discovered organism is 1 m tall, multicellular, green, and it grows on land and performs photosynthesis. To which kingdom does it most likely belong?

  A Animalia

  B Fungi

  C Plantae

  D Protista

4 Unicellular organisms are members of which kingdoms?

  A Animalia, Archaea, Plantae

  B Archaea, Bacteria, Protista

  C Bacteria, Fungi, Plantae

  D Fungi, Plantae, Protista

5 Which microscope would best magnify the outer surface of a cell?

  A compound light

  B scanning electron

  C simple dissecting

  D transmission electron

Use the diagram below to answer question 6.

6 Which discovery was NOT made with the instrument above?

  A Bacterial cells have thick walls.

  B Blood is a mixture of components.

  C Insects have small body parts.

  D Tiny organisms live in pond water.

7 Which statement is false?

  A Binomial names are given to all known organisms.

  B Binomial names are less precise than common names.

  C Binomial names differ from common names.

  D Binomial names enable scientists to communicate accurately.

Use the diagram below to answer question 8.

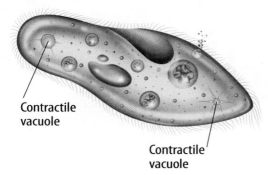

Contractile vacuole

Contractile vacuole

8 Which is the function of the structures in this paramecium?

A growth

B homeostasis

C locomotion

D reproduction

9 Which sequence is from the smallest group of organisms to the largest group of organisms?

A genus → family → species

B genus → species → family

C species → family → genus

D species → genus → family

10 Which information about organisms is excluded in the study of systematics?

A calendar age

B molecular analysis

C energy source

D normal habitat

Constructed Response

11 Copy and complete the table below about the six characteristics of life.

Characteristic	Explanation

12 Choose one characteristic of living things and explain how it affects everyday human life. From your own knowledge, give a specific example.

Use the diagram below to answer question 13.

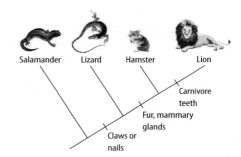

Salamander   Lizard   Hamster   Lion

Carnivore teeth

Fur, mammary glands

Claws or nails

13 Explain why the lion is more closely related to the hamster than the hamster is related to the salamander.

NEED EXTRA HELP?													
If You Missed Question...	1	2	3	4	5	6	7	8	9	10	11	12	13
Go to Lesson...	1	1	2	2	3	3	2	1	2	2	1	1	2

Chapter 2

Cell Structure and Function

THE BIG IDEA

How do the structures and processes of a cell enable it to survive?

Inquiry Alien Life?

You might think this unicellular organism looks like something out of a science-fiction movie. Although it looks scary, the hairlike structures in its mouth enable the organism to survive.

- What do you think the hairlike structures do?

- How might the shape of the hairlike structures relate to their function?

- How do you think the structures and processes of a cell enable it to survive?

Get Ready to Read

What do you think?

Before you read, decide if you agree or disagree with each of these statements. As you read this chapter, see if you change your mind about any of the statements.

1 Nonliving things have cells.

2 Cells are made mostly of water.

3 Different organisms have cells with different structures.

4 All cells store genetic information in their nuclei.

5 Diffusion and osmosis are the same process.

6 Cells with large surface areas can transport more than cells with smaller surface areas.

7 ATP is the only form of energy found in cells.

8 Cellular respiration occurs only in lung cells.

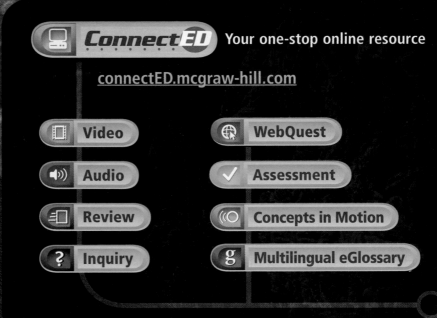

ConnectED Your one-stop online resource

connectED.mcgraw-hill.com

Video

Audio

Review

Inquiry

WebQuest

Assessment

Concepts in Motion

Multilingual eGlossary

Lesson 1

Reading Guide

Key Concepts 🔑
ESSENTIAL QUESTIONS

- How did scientists' understanding of cells develop?

- What basic substances make up a cell?

Vocabulary

cell theory p. 44

macromolecule p. 45

nucleic acid p. 46

protein p. 47

lipid p. 47

carbohydrate p. 47

g **Multilingual eGlossary**

Cells and Life

Inquiry Two of a Kind?

At first glance, the plant and animal in the photo might seem like they have nothing in common. The plant is rooted in the ground, and the iguana can move quickly. Are they more alike than they appear? How can you find out?

What's in a cell?

Most plants grow from seeds. A seed began as one cell, but a mature plant can be made up of millions of cells. How does a seed change and grow into a mature plant?

1. Read and complete a lab safety form.

2. Use a **toothpick** to gently remove the thin outer covering of a **bean seed** that has soaked overnight.

3. Open the seed with a **plastic knife,** and observe its inside with a **magnifying lens.** Draw the inside of the seed in your Science Journal.

4. Gently remove the small, plantlike embryo, and weigh it on a **balance.** Record its mass in your Science Journal.

5. Gently pull a **bean seedling** from the soil. Rinse the soil from the roots. Weigh the seedling, and record the mass.

Think About This

1. How did the mass of the embryo and the bean seedling differ?

2. 🔑 **Key Concept** If a plant begins as one cell, where do all the cells come from?

Understanding Cells

Have you ever looked up at the night sky and tried to find other planets in our solar system? It is hard to see them without using a telescope. This is because the other planets are millions of kilometers away. Just like we can use telescopes to see other planets, we can use microscopes to see the basic units of all living things—cells. But people didn't always know about cells. Because cells are so small, early scientists had no tools to study them. It took hundreds of years for scientists to learn about cells.

More than 300 years ago, an English scientist named Robert Hooke built a microscope. He used the microscope to look at cork, which is part of a cork oak tree's bark. What he saw looked like the openings in a honeycomb, as shown in **Figure 1.** The openings reminded him of the small rooms, called cells, where monks lived. He called the structures cells, from the Latin word *cellula* (SEL yuh luh), which means "small rooms."

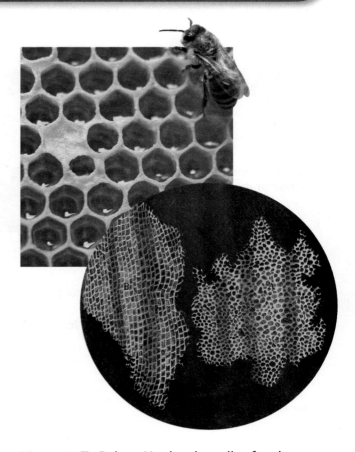

Figure 1 To Robert Hooke, the cells of cork looked like the openings in a honeycomb.

The Cell Theory

After Hooke's discovery, other scientists began making better microscopes and looking for cells in many other places, such as pond water and blood. The newer microscopes enabled scientists to see different structures inside cells. Matthias Schleiden (SHLI dun), a German scientist, used one of the new microscopes to look at plant cells. Around the same time, another German scientist, Theodor Schwann, used a microscope to study animal cells. Schleiden and Schwann realized that plant and animal cells have similar features. You'll read about many of these features in Lesson 2.

Almost two decades later, Rudolf Virchow (VUR koh), a German doctor, proposed that all cells come from preexisting cells, or cells that already exist. The observations made by Schleiden, Schwann, and Virchow were combined into one **theory.** As illustrated in **Table 1,** *the* **cell theory** *states that all living things are made of one or more cells, the cell is the smallest unit of life, and all new cells come from preexisting cells.* After the development of the cell theory, scientists raised more questions about cells. If all living things are made of cells, what are cells made of?

Key Concept Check How did scientists' understanding of cells develop?

REVIEW VOCABULARY

theory
explanation of things or events based on scientific knowledge resulting from many observations and experiments

Table 1 Scientists developed the cell theory after studying cells with microscopes.

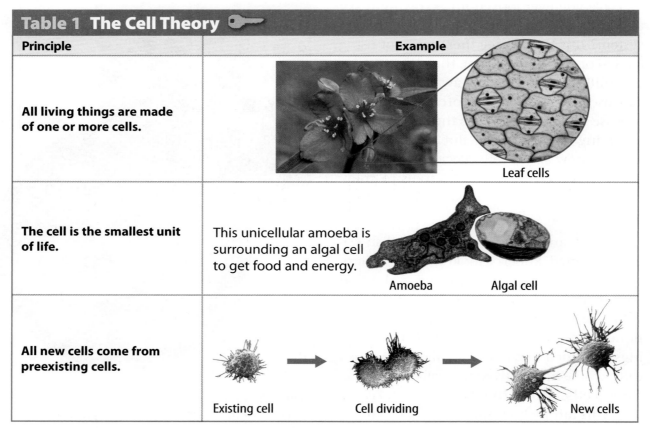

Table 1 The Cell Theory	
Principle	**Example**
All living things are made of one or more cells.	Leaf cells
The cell is the smallest unit of life.	This unicellular amoeba is surrounding an algal cell to get food and energy. Amoeba Algal cell
All new cells come from preexisting cells.	Existing cell Cell dividing New cells

Basic Cell Substances

Have you ever watched a train travel down a railroad track? The locomotive pulls train cars that are hooked together. Like a train, many of the substances in cells are made of smaller parts that are joined together. *These substances, called* **macromolecules,** *form by joining many small molecules together.* As you will read later in this lesson, macromolecules have many important roles in cells. But macromolecules cannot function without one of the most important substances in cells—water.

The Main Ingredient—Water

The main ingredient in any cell is water. It makes up more than 70 percent of a cell's volume and is essential for life. Why is water such an important molecule? In addition to making up a large part of the inside of cells, water also surrounds cells. The water surrounding your cells helps to insulate your body, which maintains homeostasis, or a stable internal environment.

The structure of a water molecule makes it ideal for dissolving many other substances. Substances must be in a liquid to move into and out of cells. A water molecule has two areas:

- An area that is more negative (−), called the negative end; this end can attract the positive part of another substance.

- An area that is more positive (+), called the positive end; this end can attract the negative part of another substance.

Examine **Figure 2** to see how the positive and negative ends of water molecules dissolve salt crystals.

WORD ORIGIN · · · · · · · · · ·

macromolecule
from Greek *makro–*, means "long"; and Latin *molecula,* means "mass"

Figure 2 The positive and negative ends of a water molecule attract the positive and negative parts of another substance, similar to the way magnets are attracted to each other.

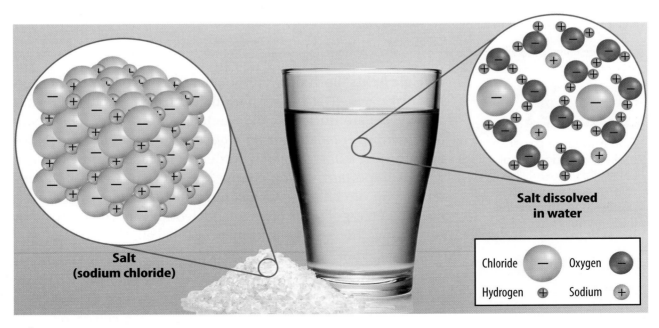

Salt (sodium chloride)

Salt dissolved in water

| Chloride | ⬤ | Oxygen | ⬤ |
| Hydrogen | ⊕ | Sodium | ⊕ |

Visual Check Which part of the salt crystal is attracted to the oxygen in the water molecule?

Macromolecules

Although water is essential for life, all cells contain other substances that enable them to function. Recall that macromolecules are large molecules that form when smaller molecules join together. As shown in **Figure 3,** there are four types of macromolecules in cells: nucleic acids, proteins, lipids, and carbohydrates. Each type of macromolecule has unique functions in a cell. These functions range from growth and communication to movement and storage.

Cell Macromolecules

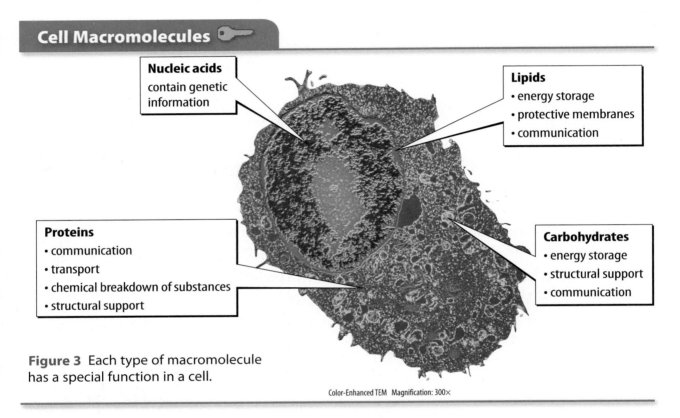

Nucleic acids
contain genetic information

Lipids
• energy storage
• protective membranes
• communication

Proteins
• communication
• transport
• chemical breakdown of substances
• structural support

Carbohydrates
• energy storage
• structural support
• communication

Figure 3 Each type of macromolecule has a special function in a cell.

Color-Enhanced TEM Magnification: 300×

FOLDABLES

Fold a sheet of paper to make a four-door book. Label it as shown. Use it to organize your notes on the macromolecules and their uses in a cell.

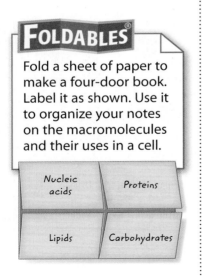

Nucleic Acids Both deoxyribonucleic (dee AHK sih ri boh noo klee ihk) acid (DNA) and ribonucleic (ri boh noo KLEE ihk) acid (RNA) are nucleic acids. **Nucleic acids** *are macromolecules that form when long chains of molecules called nucleotides* (NEW klee uh tidz) *join together.* The order of nucleotides in DNA and RNA is important. If you change the order of words in a sentence, you can change the meaning of the sentence. In a similar way, changing the order of nucleotides in DNA and RNA can change the genetic information in a cell.

Nucleic acids are important in cells because they contain genetic information. This information can pass from parents to offspring. DNA includes instructions for cell growth, cell reproduction, and cell processes that enable a cell to respond to its environment. DNA is used to make RNA. RNA is used to make proteins.

Proteins The macromolecules necessary for nearly everything cells do are proteins. **Proteins** *are long chains of amino acid molecules.* You just read that RNA is used to make proteins. RNA contains instructions for joining amino acids together.

Cells contain hundreds of proteins. Each protein has a unique function. Some proteins help cells communicate with each other. Other proteins transport substances around inside cells. Some proteins, such as amylase (AM uh lays) in saliva, help break down nutrients in food. Other proteins, such as keratin (KER uh tun)—a protein found in hair, horns, and feathers—provide structural support.

Lipids Another group of macromolecules found in cells is lipids. *A* **lipid** *is a large macromolecule that does not dissolve in water.* Because lipids do not mix with water, they play an important role as protective barriers in cells. They are also the major part of cell membranes. Lipids play roles in energy storage and in cell communication. Examples of lipids are cholesterol (kuh LES tuh rawl), phospholipids (fahs foh LIH pids), and vitamin A.

 Reading Check Why are lipids important to cells?

Carbohydrates *One sugar molecule, two sugar molecules, or a long chain of sugar molecules make up* **carbohydrates** (kar boh HI drayts). Carbohydrates store energy, provide structural support, and are needed for communication between cells. Sugars and starches are carbohydrates that store energy. Fruits contain sugars. Breads and pastas are mostly starch. The energy in sugars and starches can be released quickly through chemical reactions in cells. Cellulose is a carbohydrate in the cell walls in plants that provides structural support.

 Key Concept Check What basic substances make up a cell?

 **MiniLab** — 25 minutes

How can you observe DNA?

Nucleic acids are macromolecules that are important in cells because they contain an organism's genetic information. In this lab, you will observe one type of nucleic acid, DNA, in onion root-tip cells using a compound light microscope.

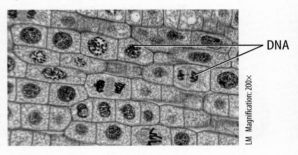

LM Magnification: 200×

1. Read and complete a lab safety form.
2. Obtain a **microscope** and a **slide** from your teacher. Use care and properly handle your microscope.
3. Observe the **onion root-tip cells** at the magnifications assigned by your teacher.
4. Determine the approximate number of cells in your field of view and the number of cells with visible DNA. Record these numbers in your Science Journal.

Analyze and Conclude

1. **Calculate** Using your data, find the percentage of cells with visible DNA that you saw in your microscope's field of view.
2. **Compare** your results with the results of other students. Are all the results the same? Explain.
3. **Create** a data table for the entire class that lists individual results.
4. **Calculate** the total percentage of cells with visible DNA at each magnification.
5. **Key Concept** Did looking at the cells at different magnifications change the percentage of cells with visible DNA? Explain.

Lesson 1 Review

Visual Summary

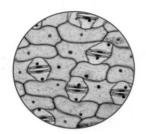

The cell theory summarizes the main principles for understanding that the cell is the basic unit of life.

Water is the main ingredient in every cell.

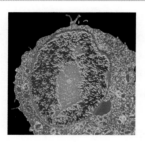

A nucleic acid, such as DNA, contains the genetic information for a cell.

FOLDABLES®

Use your lesson Foldable to review the lesson. Save your Foldable for the project at the end of the chapter.

What do you think NOW?

You first read the statements below at the beginning of the chapter.

1. Nonliving things have cells.

2. Cells are made mostly of water.

Did you change your mind about whether you agree or disagree with the statements? Rewrite any false statements to make them true.

Use Vocabulary

1 The _____ _____ states that the cell is the basic unit of all living things.

2 **Distinguish** between a carbohydrate and a lipid.

3 **Use the term** *nucleic acid* in a sentence.

Understand Key Concepts

4 Which macromolecule is made from amino acids?
   A. lipid   C. carbohydrate
   B. protein   D. nucleic acid

5 **Describe** how the invention of the microscope helped scientists understand cells.

6 **Compare** the functions of DNA and proteins in a cell.

Interpret Graphics

7 **Summarize** Copy and fill in the graphic organizer below to summarize the main principles of the cell theory.

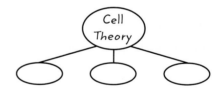

8 **Analyze** How does the structure of the water molecule shown below enable it to interact with other water molecules?

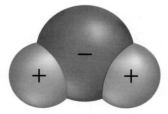

Critical Thinking

9 **Summarize** the functions of lipids in cells.

10 **Hypothesize** why carbohydrates are found in plant cell walls.

A Very Powerful Microscope

Using technology to look inside cells

If Robert Hooke had used an atomic force microscope (AFM), he would have observed more than just cells. He would have seen the macromolecules inside them! An AFM can scan objects that are only nanometers in size. A nanometer is one one-billionth of a meter. That's 100,000 times smaller than the width of a human hair. AFM technology has enabled scientists to better understand how cells function. It also has given them a three-dimensional look at the macromolecules that make life possible. This is how it works.

Photodiode

2 The cantilever can bend up and down, similar to the way a diving board can bend, in response to pushing and pulling forces between the atoms in the tip and the atoms in the sample.

3 A laser beam senses the cantilever's up and down movements. A computer converts these movements into an image of the sample's surface.

1 A probe moves across a sample's surface to identify the sample's features. The probe consists of a cantilever with a tiny, sharp tip. The tip is about 20 nm in diameter at its base.

It's Your Turn

RESEARCH NASA's Phoenix Mars Lander included an atomic force microscope. Find out what scientists discovered on Mars with this instrument.

The Cell

Key Concepts
ESSENTIAL QUESTIONS

- How are prokaryotic cells and eukaryotic cells similar, and how are they different?

- What do the structures in a cell do?

Vocabulary
cell membrane p. 52

cell wall p. 52

cytoplasm p. 53

cytoskeleton p. 53

organelle p. 54

nucleus p. 55

chloroplast p. 57

 Multilingual eGlossary

 Video BrainPOP®

Inquiry Hooked Together?

What do you think happens when one of the hooks in the photo above goes through one of the loops? The two sides fasten together. The shapes of the hooks and loops in the hook-and-loop tape are suited to their function—to hold the two pieces together.

Launch Lab

10 minutes

Why do eggs have shells?

Bird eggs have different structures, such as a shell, a membrane, and a yolk. Each structure has a different function that helps keep the egg safe and assists in development of the baby bird inside of it.

1 Read and complete a lab safety form.

2 Place an **uncooked egg** in a bowl.

3 Feel the shell, and record your observations in your Science Journal.

4 Crack open the egg. Pour the contents into the bowl.

5 Observe the inside of the shell and the contents of the bowl. Record your observations in your Science Journal.

Think About This

1. What do you think is the role of the eggshell?

2. Are there any structures in the bowl that have the same function as the eggshell? Explain.

3. 🔑 **Key Concept** What does the structure of the eggshell tell you about its function?

Cell Shape and Movement

You might recall from Lesson 1 that all living things are made up of one or more cells. As illustrated in **Figure 4,** cells come in many shapes and sizes. The size and shape of a cell relates to its job or function. For example, a human red blood cell cannot be seen without a microscope. Its small size and disk shape enable it to pass easily through the smallest blood vessels. The shape of a nerve cell enables it to send signals over long distances. Some plant cells are hollow and make up tubelike structures that carry materials throughout a plant.

The structures that make up a cell also have unique functions. Think about how the players on a football team perform different tasks to move the ball down the field. In a similar way, a cell is made of different structures that perform different functions that keep a cell alive. You will read about some of these structures in this lesson.

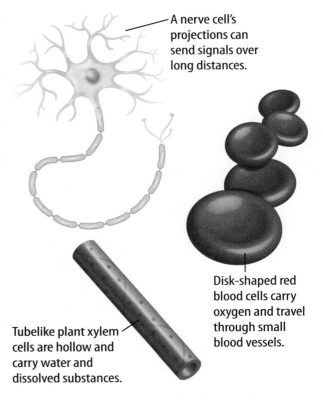

A nerve cell's projections can send signals over long distances.

Disk-shaped red blood cells carry oxygen and travel through small blood vessels.

Tubelike plant xylem cells are hollow and carry water and dissolved substances.

Figure 4 The shape of a cell relates to the function it performs.

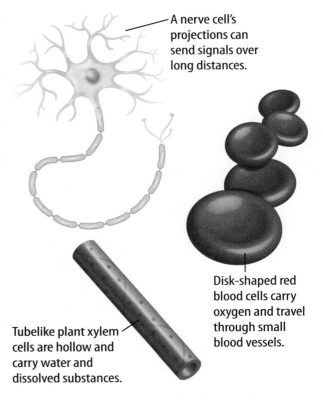

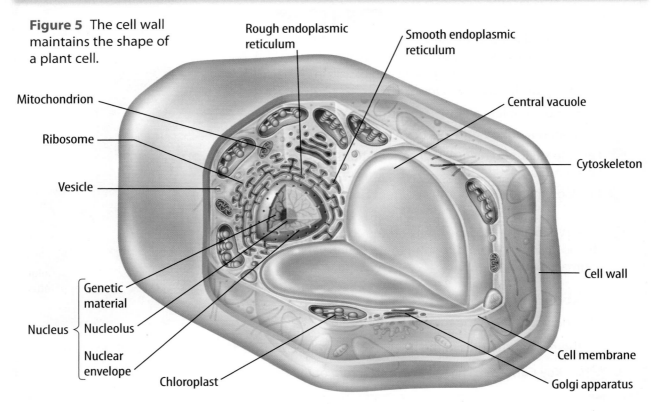

Figure 5 The cell wall maintains the shape of a plant cell.

- Rough endoplasmic reticulum
- Smooth endoplasmic reticulum
- Mitochondrion
- Ribosome
- Vesicle
- Central vacuole
- Cytoskeleton
- Cell wall
- Nucleus
  - Genetic material
  - Nucleolus
  - Nuclear envelope
- Chloroplast
- Cell membrane
- Golgi apparatus

ACADEMIC VOCABULARY

function
(*noun*) the purpose for which something is used

Cell Membrane

Although different types of cells perform different **functions,** all cells have some structures in common. As shown in **Figure 5** and **Figure 6,** every cell is surrounded by a protective covering called a membrane. *The* **cell membrane** *is a flexible covering that protects the inside of a cell from the environment outside a cell.* Cell membranes are mostly made of two different macromolecules—proteins and a type of lipid called phospholipids. Think again about a football team. The defensive line tries to stop the other team from moving forward with the football. In a similar way, a cell membrane protects the cell from the outside environment.

✓ **Reading Check** What are cell membranes made of?

Cell Wall

Every cell has a cell membrane, but some cells are also surrounded by a structure called the cell wall. Plant cells such as the one in **Figure 5,** fungal cells, bacteria, and some types of protists have cell walls. *A* **cell wall** *is a stiff structure outside the cell membrane.* A cell wall protects a cell from attack by viruses and other harmful organisms. In some plant cells and fungal cells, a cell wall helps maintain the cell's shape and gives structural support.

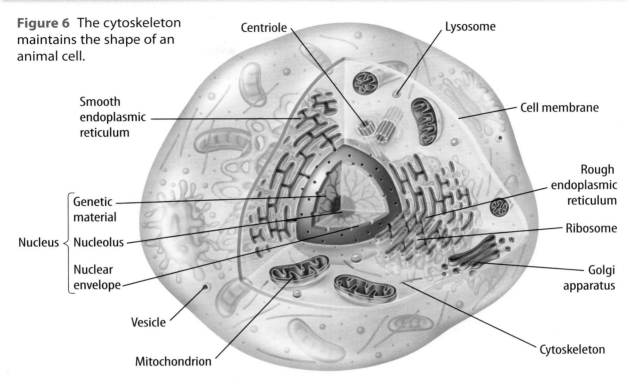

Figure 6 The cytoskeleton maintains the shape of an animal cell.

Centriole

Lysosome

Smooth endoplasmic reticulum

Cell membrane

Rough endoplasmic reticulum

Genetic material

Nucleus
- Nucleolus
- Nuclear envelope

Ribosome

Golgi apparatus

Vesicle

Cytoskeleton

Mitochondrion

✔ **Visual Check** Compare this animal cell to the plant cell in **Figure 5.**

Cell Appendages

Arms, legs, claws, and antennae are all types of appendages. Cells can have appendages too. Cell appendages are often used for movement. Flagella (fluh JEH luh; singular, flagellum) are long, tail-like appendages that whip back and forth and move a cell. A cell can also have cilia (SIH lee uh; singular, cilium) like the ones shown in **Figure 7.** Cilia are short, hairlike structures. They can move a cell or move molecules away from a cell. A microscopic organism called a paramecium (pa ruh MEE shee um) moves around its watery environment using its cilia. The cilia in your windpipe move harmful substances away from your lungs.

Cytoplasm and the Cytoskeleton

In Lesson 1, you read that water is the main ingredient in a cell. Most of this water is in the **cytoplasm,** *a fluid inside a cell that contains salts and other molecules.* The cytoplasm also contains a cell's cytoskeleton. *The* **cytoskeleton** *is a network of threadlike proteins that are joined together.* The proteins form a framework inside a cell. This framework gives a cell its shape and helps it move. Cilia and flagella are made from the same proteins that make up the cytoskeleton.

Color-Enhanced SEM Magnification: Unavailable

Figure 7 Lung cells have cilia that help move fluids and foreign materials.

WORD ORIGIN · · · · · · · · ·

cytoplasm
from Greek *kytos,* means "hollow vessel"; and *plasma,* means "something molded"

Inquiry MiniLab
25 minutes

How do eukaryotic and prokaryotic cells compare?

With the use of better microscopes, scientists discovered that cells can be classified as one of two types—prokaryotic or eukaryotic.

1. Read and complete a lab safety form.

2. Using different **craft items,** make a two-dimensional model of a eukaryotic cell.

3. In your cell model, include the number of cell structures assigned by your teacher.

4. Make each cell structure the correct shape, as shown in this lesson.

5. Make a label for each cell structure of your model.

Analyze and Conclude

1. **Describe** the nucleus of your cell.

2. **Classify** your cell as either a plant cell or an animal cell, and support your classification with evidence.

3. 🔑 **Key Concept** Compare and contrast a prokaryotic cell, as shown in **Figure 8,** with your eukaryotic cell model.

Cell Types

Recall that the use of microscopes enabled scientists to discover cells. With more advanced microscopes, scientists discovered that all cells can be grouped into two types—prokaryotic (proh ka ree AH tihk) cells and eukaryotic (yew ker ee AH tihk) cells.

Prokaryotic Cells

The genetic material in a prokaryotic cell is not surrounded by a membrane, as shown in **Figure 8.** This is the most important feature of a prokaryotic cell. Prokaryotic cells also do not have many of the other cell parts that you will read about later in this lesson. Most prokaryotic cells are unicellular organisms and are called prokaryotes.

Figure 8 In prokaryotic cells, the genetic material floats freely in the cytoplasm.

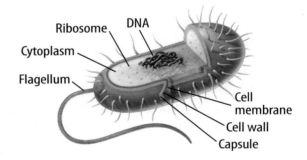

Ribosome · DNA · Cytoplasm · Flagellum · Cell membrane · Cell wall · Capsule

Eukaryotic Cells

Plants, animals, fungi, and protists are all made of eukaryotic cells, such as the ones shown in **Figure 5** and **Figure 6,** and are called eukaryotes. With few exceptions, each eukaryotic cell has genetic material that is surrounded by a membrane. Every eukaryotic cell also has *other structures, called* **organelles,** *which have specialized functions. Most organelles are surrounded by membranes.* Eukaryotic cells are usually larger than prokaryotic cells. About ten prokaryotic cells would fit inside one eukaryotic cell.

🔑 **Key Concept Check** How are prokaryotic cells and eukaryotic cells similar, and how are they different?

Cell Organelles

As you have just read, organelles are eukaryotic cell structures with specific functions. Organelles enable cells to carry out different functions at the same time. For example, cells can obtain energy from food, store information, make macromolecules, and get rid of waste materials all at the same time because different organelles perform the different tasks.

The Nucleus

The largest organelle inside most eukaryotic cells is the nucleus, shown in **Figure 9.** *The **nucleus** is the part of a eukaryotic cell that directs cell activities and contains genetic information stored in DNA.* DNA is organized into structures called chromosomes. The number of chromosomes in a nucleus is different for different species of organisms. For example, kangaroo cells contain six pairs of chromosomes. Most human cells contain 23 pairs of chromosomes.

FOLDABLES®

Fold a sheet of paper into a vertical half book. Use it to record information about cell organelles and their functions.

Cell Organelles and Their Functions

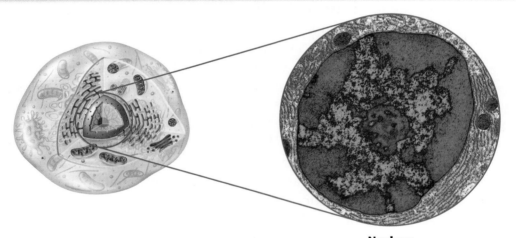

Nucleus
Color-Enhanced TEM   Magnification: 15,500×

Figure 9 The nucleus directs cell activity and is surrounded by a membrane.

In addition to chromosomes, the nucleus contains proteins and an organelle called the nucleolus (new KLEE uh lus). The nucleolus is often seen as a large dark spot in the nucleus of a cell. The nucleolus makes ribosomes, organelles that are involved in the production of proteins. You will read about ribosomes later in this lesson.

Surrounding the nucleus are two membranes that form a structure called the nuclear **envelope.** The nuclear envelope contains many pores. Certain molecules, such as ribosomes and RNA, move into and out of the nucleus through these pores.

✓ **Reading Check** What is the nuclear envelope?

SCIENCE USE V. COMMON USE

envelope
Science Use an outer covering

Common Use a flat paper container for a letter

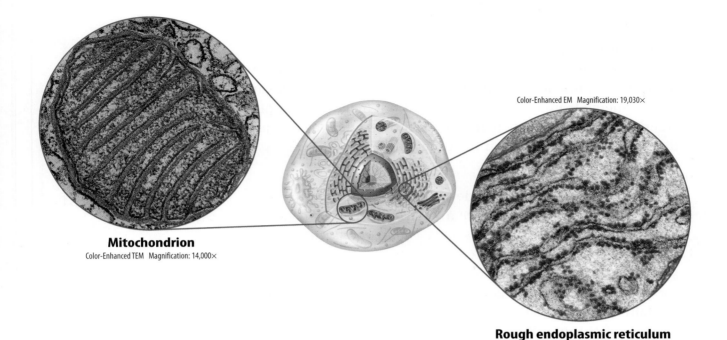

Mitochondrion
Color-Enhanced TEM Magnification: 14,000×

Color-Enhanced EM Magnification: 19,030×

Rough endoplasmic reticulum

Figure 10 The endoplasmic reticulum is made of many folded membranes. Mitochondria provide a cell with usable energy.

Manufacturing Molecules

You might recall from Lesson 1 that proteins are important molecules in cells. Proteins are made on small structures called ribosomes. Unlike other cell organelles, a ribosome is not surrounded by a membrane. Ribosomes are in a cell's cytoplasm. They also can be attached to a weblike organelle called the endoplasmic reticulum (en duh PLAZ mihk • rih TIHK yuh lum), or ER. As shown in **Figure 10,** the ER spreads from the nucleus throughout most of the cytoplasm. ER with ribosomes on its surface is called rough ER. Rough ER is the site of protein production. ER without ribosomes is called smooth ER. It makes lipids such as cholesterol. Smooth ER is important because it helps remove harmful substances from a cell.

 Reading Check Contrast smooth ER and rough ER.

Processing Energy

All living things require energy in order to survive. Cells process some energy in specialized organelles. Most eukaryotic cells contain hundreds of organelles called mitochondria (mi tuh KAHN dree uh; singular, mitochondrion), shown in **Figure 10.** Some cells in a human heart can contain a thousand mitochondria.

Like the nucleus, a mitochondrion is surrounded by two membranes. Energy is released during chemical reactions that occur in the mitochondria. This energy is stored in high-energy molecules called ATP—adenosine triphosphate (uh DEH nuh seen • tri FAHS fayt). ATP is the fuel for cellular processes such as growth, cell division, and material transport.

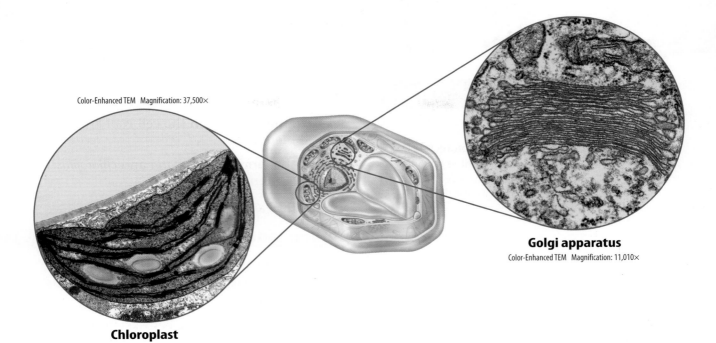

Color-Enhanced TEM Magnification: 37,500×

Golgi apparatus
Color-Enhanced TEM Magnification: 11,010×

Chloroplast

Plant cells and some protists, such as algae, also contain organelles called chloroplasts (KLOR uh plasts), shown in **Figure 11.** **Chloroplasts** *are membrane-bound organelles that use light energy and make food—a sugar called glucose—from water and carbon dioxide in a process known as photosynthesis* (foh toh SIHN thuh sus). The sugar contains stored chemical energy that can be released when a cell needs it. You will read more about photosynthesis in Lesson 4.

 Reading Check Which types of cells contain chloroplasts?

Processing, Transporting, and Storing Molecules

Near the ER is an organelle that looks like a stack of pancakes. This is the Golgi (GAWL jee) apparatus, shown in **Figure 11.** It prepares proteins for their specific jobs or functions. Then it packages the proteins into tiny, membrane-bound, ball-like structures called vesicles. Vesicles are organelles that transport substances from one area of a cell to another area of a cell. Some vesicles in an animal cell are called lysosomes. Lysosomes contain substances that help break down and recycle cellular components.

Some cells also have saclike structures called vacuoles (VA kyuh wohlz). Vacuoles are organelles that store food, water, and waste material. A typical plant cell usually has one large vacuole that stores water and other substances. Some animal cells have many small vacuoles.

 Key Concept Check What is the function of the Golgi apparatus?

Figure 11 Plant cells have chloroplasts that use light energy and make food. The Golgi apparatus packages materials into vesicles.

Visual Summary

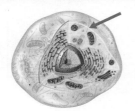

A cell is protected by a flexible covering called the cell membrane.

Cells can be grouped into two types—prokaryotic cells and eukaryotic cells.

In a chloroplast, light energy is used for making sugars in a process called photosynthesis.

FOLDABLES

Use your lesson Foldable to review the lesson. Save your Foldable for the project at the end of the chapter.

What do you think NOW?

You first read the statements below at the beginning of the chapter.

3. Different organisms have cells with different structures.

4. All cells store genetic information in their nuclei.

Did you change your mind about whether you agree or disagree with the statements? Rewrite any false statements to make them true.

Use Vocabulary

1 **Distinguish** between the cell wall and the cell membrane.

2 **Use the terms** *mitochondria* and *chloroplasts* in a sentence.

3 **Define** *organelle* in your own words.

Understand Key Concepts 🔑

4 Which organelle is used to store water?
- **A.** chloroplast
- **B.** lysosome
- **C.** nucleus
- **D.** vacuole

5 **Explain** the role of the cytoskeleton.

6 **Draw** a prokaryotic cell and label its parts.

7 **Compare** the roles of the endoplasmic reticulum and the Golgi apparatus.

Interpret Graphics

8 **Explain** how the structure of the cells below relates to their function.

9 **Compare** Copy the table below and fill it in to compare the structures of a plant cell to the structures of an animal cell.

Structure	Plant Cell	Animal Cell
Cell membrane	yes	yes
Cell wall		
Mitochondrion		
Chloroplast		
Nucleus		
Vacuole		
Lysosome		

Critical Thinking

10 **Analyze** Why are most organelles surrounded by membranes?

11 **Compare** the features of eukaryotic and prokaryotic cells.

How are plant cells and animal cells similar and how are they different?

A light microscope enables you to observe many of the structures in cells. Increasing the magnification means you see a smaller portion of the object, but lets you see more detail. As you see more details, you can **compare and contrast** different cell types. How are they alike? How are they different?

Materials

microscope

microscope slide and coverslip

forceps

dropper

Elodea plant

Prepared slide of human cheek cells

Safety

Learn It

Observations can be analyzed by noting the similarities and differences between two or more objects that you observe. You **compare** objects by noting similarities. You **contrast** objects by looking for differences.

Try It

1 Read and complete a lab safety form.

2 Using forceps, make a wet-mount slide of a young leaf from the tip of an *Elodea* plant.

3 Use a microscope to observe the leaf on low power. Focus on the top layer of cells.

4 Switch to high power and focus on one cell. The large organelle in the center of the cell is the central vacuole. Moving around the central vacuole are green, disklike objects called chloroplasts. Try to find the nucleus. It looks like a clear ball.

5 Draw a diagram of an *Elodea* cell in your Science Journal. Label the cell wall, central vacuole, chloroplasts, cytoplasm, and nucleus. Return to low power and remove the slide. Properly dispose of the slide.

6 Observe the prepared slide of cheek cells under low power.

7 Switch to high power and focus on one cell. Draw a diagram of one cheek cell. Label the cell membrane, cytoplasm, and nucleus. Return to low power and remove the slide.

Apply It

8 Based on your diagrams, how do the shapes of the *Elodea* cell and cheek cell compare?

9 🔑 **Key Concept** Compare and contrast the cell structures in your two diagrams. Which structures did you observe in both cells? Which structures did you observe in only one of the cells?

Reading Guide

Key Concepts 🗝️
ESSENTIAL QUESTIONS

- How do materials enter and leave cells?
- How does cell size affect the transport of materials?

Vocabulary

passive transport p. 61

diffusion p. 62

osmosis p. 62

facilitated diffusion p. 63

active transport p. 64

endocytosis p. 64

exocytosis p. 64

g Multilingual eGlossary

Moving Cellular Material

Inquiry Why the Veil?

A beekeeper often wears a helmet with a face-covering veil made of mesh. The openings in the mesh are large enough to let air through, yet small enough to keep bees out. In a similar way, some things must be allowed in or out of a cell, while other things must be kept in or out. How do the right things enter or leave a cell?

What does the cell membrane do?

All cells have a membrane around the outside of the cell. The cell membrane separates the inside of a cell from the environment outside a cell. What else might a cell membrane do?

1. Read and complete a lab safety form.
2. Place a square of **wire mesh** on top of a **beaker.**
3. Pour a small amount of **birdseed** on top of the wire mesh. Record your observations in your Science Journal.

Think About This

1. What part of a cell does the wire mesh represent?

2. What happened when you poured birdseed on the wire mesh?

3. **Key Concept** How do you think the cell membrane affects materials that enter and leave a cell?

Passive Transport

Recall from Lesson 2 that membranes are the boundaries between cells and between organelles. Another important role of membranes is to control the movement of substances into and out of cells. A cell membrane is semipermeable. This means it allows only certain substances to enter or leave a cell. Substances can pass through a cell membrane by one of several different processes. The type of process depends on the physical and chemical properties of the substance passing through the membrane.

Small molecules, such as oxygen and carbon dioxide, pass through membranes by a process called passive transport. **Passive transport** *is the movement of substances through a cell membrane without using the cell's energy.* Passive transport depends on the amount of a substance on each side of a membrane. For example, suppose there are more molecules of oxygen outside a cell than inside it. Oxygen will move into that cell until the amount of oxygen is equal on both sides of the cell's membrane. Since oxygen is a small molecule, it passes through a cell membrane without using the cell's energy. The different types of passive transport are explained on the following pages.

✓ **Reading Check** Describe a semipermeable membrane.

FOLDABLES

Fold a sheet of paper into a two-tab book. Label the tabs as shown. Use it to organize information about the different types of passive and active transport.

Diffusion

WORD ORIGIN ············

diffusion
from Latin *diffusionem*, means "scatter, pour out"
············

What happens when the concentration, or amount per unit of volume, of a substance is unequal on each side of a membrane? The molecules will move from the side with a higher concentration of that substance to the side with a lower concentration. **Diffusion** *is the movement of substances from an area of higher concentration to an area of lower concentration.*

Usually, diffusion continues through a membrane until the concentration of a substance is the same on both sides of the membrane. When this happens, a substance is in equilibrium. Compare the two diagrams in **Figure 12.** What happened to the red dye that was added to the water on one side of the membrane? Water and dye passed through the membrane in both directions until there were equal concentrations of water and dye on both sides of the membrane.

Diffusion 🔑

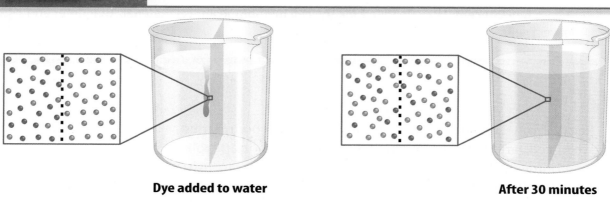

Dye added to water   **After 30 minutes**

✔ **Visual Check** What would the water in the beaker on the right look like if the membrane did not let anything through?

Figure 12 Over time, the concentration of dye on either side of the membrane becomes the same.

Osmosis—The Diffusion of Water

Diffusion refers to the movement of any small molecules from higher to lower concentrations. However, **osmosis** *is the diffusion of water molecules only through a membrane.* Semipermeable cell membranes also allow water to pass through them until equilibrium occurs. For example, the amount of water stored in the vacuoles of plant cells can decrease because of osmosis. That is because the concentration of water in the air surrounding the plant is less than the concentration of water inside the vacuoles of plant cells. Water will continue to diffuse into the air until the concentrations of water inside the plant's cells and in the air are equal. If the plant is not watered to replace the lost water, it will wilt and eventually die.

Facilitated Diffusion

Some molecules are too large or are chemically unable to travel through a membrane by diffusion. *When molecules pass through a cell membrane using special proteins called transport proteins, this is* **facilitated diffusion.** Like diffusion and osmosis, facilitated diffusion does not require a cell to use energy. As shown in **Figure 13,** a cell membrane has transport proteins. The two types of transport proteins are carrier proteins and channel proteins. Carrier proteins carry large molecules, such as the sugar molecule glucose, through the cell membrane. Channel proteins form pores through the membrane. Atomic particles, such as sodium ions and potassium ions, pass through the cell membrane by channel proteins.

Reading Check How do materials move through the cell membrane in facilitated diffusion?

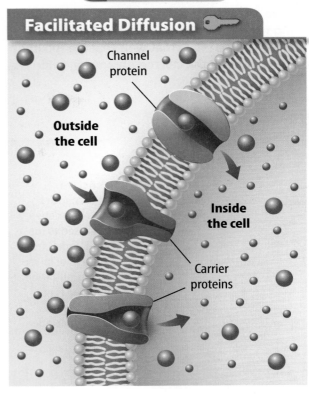

Facilitated Diffusion

Channel protein

Outside the cell

Inside the cell

Carrier proteins

Figure 13 Transport proteins are used to move large molecules into and out of a cell.

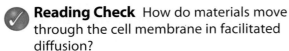

MiniLab

20 minutes

How is a balloon like a cell membrane?

Substances within a cell are constantly in motion. How can a balloon act like a cell membrane?

1 Read and complete a lab safety form.

2 Make a three-column table in your Science Journal to record your data. Label the first column *Balloon Number*, the second column *Substance*, and the third column *Supporting Evidence*.

3 Use your senses to identify what substance is in each of the **numbered balloons.**

4 Record what you think each substance is.

5 Record the evidence supporting your choice.

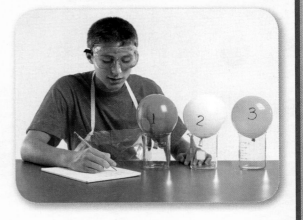

Analyze and Conclude

1. **List** the senses that were most useful in identifying the substances.

2. **Infer** if you could identify the substances if you were blindfolded. If so, how?

3. **Describe** how the substances moved, and explain why they moved this way.

4. **Key Concept** Explain how a balloon is like a cell membrane in terms of the movement of substances.

Figure 14 Active transport is most often used to bring needed nutrients into a cell. Endocytosis and exocytosis move materials that are too large to pass through the cell membrane by other methods.

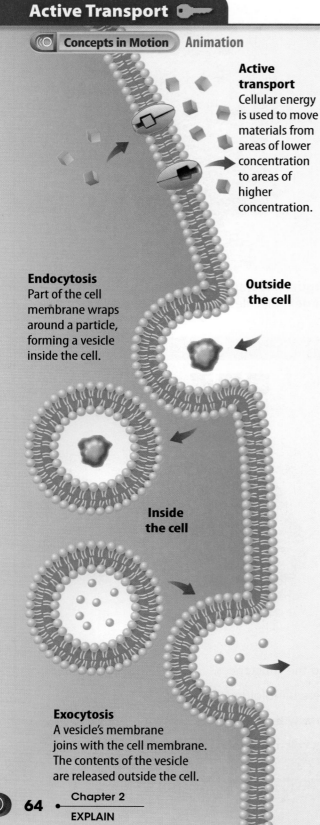

Active Transport 🔑

Concepts in Motion Animation

Active transport Cellular energy is used to move materials from areas of lower concentration to areas of higher concentration.

Outside the cell

Endocytosis Part of the cell membrane wraps around a particle, forming a vesicle inside the cell.

Inside the cell

Exocytosis A vesicle's membrane joins with the cell membrane. The contents of the vesicle are released outside the cell.

Active Transport

Sometimes when cellular materials pass through membranes it requires a cell to use energy. **Active transport** *is the movement of substances through a cell membrane only by using the cell's energy.*

Recall that passive transport is the movement of substances from areas of higher concentration to areas of lower concentration. However, substances moving by active transport move from areas of lower concentration to areas of higher concentration, as shown in **Figure 14.**

Active transport is important for cells and organelles. Cells can take in needed nutrients from the environment through carrier proteins by using active transport. This occurs even when concentrations of these nutrients are lower in the environment than inside the cell. Some other molecules and waste materials also leave cells by active transport.

Endocytosis and Exocytosis

Some substances are too large to enter a cell membrane by diffusion or by using a transport protein. These substances can enter a cell by another process. **Endocytosis** (en duh si TOH sus), shown in **Figure 14,** *is the process during which a cell takes in a substance by surrounding it with the cell membrane.* Many different types of cells use endocytosis. For example, some cells take in bacteria and viruses using endocytosis.

Some substances are too large to leave a cell by diffusion or by using a transport protein. These substances can leave a cell another way. **Exocytosis** (ek soh si TOH sus), shown in **Figure 14,** *is the process during which a cell's vesicles release their contents outside the cell.* Proteins and other substances are removed from a cell through this process.

 Key Concept Check How do materials enter and leave cells?

Cell Size and Transport

Recall that the movement of nutrients, waste material, and other substances into and out of a cell is important for survival. For this movement to happen, the area of the cell membrane must be large compared to its volume. The area of the cell membrane is the cell's surface area. The volume is the amount of space inside the cell. As a cell grows, both its volume and its surface area increase. The volume of a cell increases faster than its surface area. If a cell were to keep growing, it would need large amounts of nutrients and would produce large amounts of waste material. However, the surface area of the cell's membrane would be too small to move enough nutrients and wastes through it for the cell to survive.

 Key Concept Check How does cell size affect the transport of materials?

Review
- **Math Practice**
- **Personal Tutor**

Math Skills ✕÷+ Use Ratios

A ratio is a comparison of two numbers, such as surface area and volume. If a cell were cube-shaped, you would calculate surface area by multiplying its length (ℓ) by its width (w) by the number of sides (6).

Surface area $= \ell \times w \times 6$

You would calculate the volume of the cell by multiplying its length (ℓ) by its width (w) by its height (h).

Volume $= \ell \times w \times h$

To find the surface-area-to-volume ratio of the cell, divide its surface area by its volume.

$$\frac{\text{Surface area}}{\text{Volume}}$$

In the table below, surface-area-to-volume ratios are calculated for cells that are 1 mm, 2 mm, and 4 mm per side. Notice how the ratios change as the cell's size increases.

	1 mm 1 mm 1 mm	2 mm 2 mm 2 mm	4 mm 4 mm 4 mm
Length	1 mm	2 mm	4 mm
Width	1 mm	2 mm	4 mm
Height	1 mm	2 mm	4 mm
Number of sides	6	6	6
Surface area ($\ell \times w \times$ **no. of sides**)	1 mm $\times$ 1 mm $\times$ 6 $= 6$ mm^2	2 mm $\times$ 2 mm $\times$ 6 $= 24$ mm^2	4 mm $\times$ 4 mm $\times$ 6 $= 96$ mm^2
Volume ($\ell \times w \times h$)	1 mm $\times$ 1 mm $\times$ 1 mm $= 1$ mm^3	2 mm $\times$ 2 mm $\times$ 2 mm $= 8$ mm^3	4 mm $\times$ 4 mm $\times$ 4 mm $= 64$ mm^3
Surface-area-to-volume ratio	$\frac{6 \text{ mm}^2}{1 \text{ mm}^3} = \frac{6}{1}$ or 6:1	$\frac{24 \text{ mm}^2}{8 \text{ mm}^3} = \frac{3}{1}$ or 3:1	$\frac{96 \text{ mm}^2}{64 \text{ mm}^3} = \frac{1.5}{1}$ or 1.5:1

Practice

What is the surface-area-to-volume ratio of a cell whose six sides are 3 mm long?

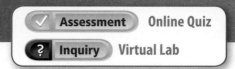

✓ **Assessment** Online Quiz
? **Inquiry** Virtual Lab

Visual Summary

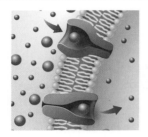

Small molecules can move from an area of higher concentration to an area of lower concentration by diffusion.

In facilitated diffusion, proteins transport larger molecules through a cell membrane.

Some molecules move from areas of lower concentration to areas of higher concentration through active transport.

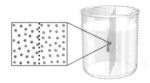

Use your lesson Foldable to review the lesson. Save your Foldable for the project at the end of the chapter.

What do you think NOW?

You first read the statements below at the beginning of the chapter.

5. Diffusion and osmosis are the same process.

6. Cells with large surface areas can transport more than cells with smaller surface areas.

Did you change your mind about whether you agree or disagree with the statements? Rewrite any false statements to make them true.

Use Vocabulary

1 **Use the term** *osmosis* in a sentence.

2 **Distinguish** between active transport and passive transport.

3 The process by which vesicles move substances out of a cell is _____.

Understand Key Concepts

4 **Explain** why energy is needed in active transport.

5 **Summarize** the function of endocytosis.

6 **Contrast** osmosis and diffusion.

7 What is limited by a cell's surface-area-to-volume ratio?
  A. cell shape   C. cell surface area
  B. cell size   D. cell volume

Interpret Graphics

8 **Identify** the process shown below, and explain how it works.

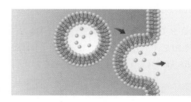

9 **Copy** and fill in the graphic organizer below to describe ways that cells transport substances.

Critical Thinking

10 **Relate** the surface area of a cell to the transport of materials.

Math Skills
Review
Math Practice

11 **Calculate** the surface-area-to-volume ratio of a cube whose sides are 6 cm long.

How does an object's size affect the transport of materials?

Materials

hard-cooked eggs

metric ruler

blue food coloring

250-mL beaker

plastic spoon

plastic knife

paper towels

Safety

Nutrients, oxygen, and other materials enter and leave a cell through the cell membrane. Does the size of a cell affect the transport of these materials throughout the cell? In this lab, you will **analyze and conclude** how the size of a cube of egg white affects material transport.

Learn It

To **analyze** how an object's size affects material transport, you will need to calculate each object's surface-area-to-volume ratio. The following formulas are used to calculate surface area and volume of a cube.

surface area (mm^2) = (length of 1 side)$^2 \times 6$

volume (mm^3) = (length of 1 side)^3

To calculate the ratio of surface area to volume, divide surface area by volume.

Try It

1. Read and complete a lab safety form.

2. Measure and cut one large cube of egg white that is 20 mm on each side. Then, measure and cut one small cube of egg white that is 10 mm on each side.

3. Place 100 mL of water in a plastic cup. Add 10 drops of food coloring. Gently add the egg-white cubes, and soak overnight.

4. Remove the cubes from the cup with a plastic spoon and place them on a paper towel. Cut each cube in half.

5. Examine the inside surface of each cube. Measure and record in millimeters how deep the blue food coloring penetrated into each cube.

Apply It

6. How does the depth of the color compare on the two cubes?

7. Calculate the surface area, the volume, and the surface-area-to-volume ratio of each cube. How do the surface-area-to-volume ratios of the two cubes compare?

8. 🔑 **Key Concept** Would a cell with a small surface-area-to-volume ratio be able to transport nutrients and waste through the cell as efficiently as a cell with a large surface-area-to-volume ratio?

Lesson 4

Reading Guide

Key Concepts 🔑
ESSENTIAL QUESTIONS

- How does a cell obtain energy?
- How do some cells make food molecules?

Vocabulary
cellular respiration p. 69
glycolysis p. 69
fermentation p. 70
photosynthesis p. 71

g Multilingual eGlossary

Cells and Energy

Inquiry Why are there bubbles?

Have you ever seen bubbles on a green plant in an aquarium? Where did the bubbles come from? Green plants use light energy and make sugars and oxygen.

What do you exhale?

Does the air you breathe in differ from the air you breathe out?

1 Read and complete a lab safety form.

2 Unwrap a **straw.** Use the straw to slowly blow into a small **cup** of **bromthymol blue.** Do not splash the liquid out of the cup.

3 In your Science Journal, record any changes in the solution.

Think About This

1. What changes did you observe in the solution?

2. What do you think caused the changes in the solution?

3. 🗝 **Key Concept** Why do you think the air you inhale differs from the air you exhale?

Cellular Respiration

When you are tired, you might eat something to give you energy. All living things, from one-celled organisms to humans, need energy to survive. Recall that cells process energy from food into the energy-storage compound ATP. **Cellular respiration** *is a series of chemical reactions that convert the energy in food molecules into a usable form of energy called ATP.* Cellular respiration is a complex process that occurs in two parts of a cell—the cytoplasm and the mitochondria.

Reactions in the Cytoplasm

The first step of cellular respiration, called glycolysis, occurs in the cytoplasm of all cells. **Glycolysis** *is a process by which glucose, a sugar, is broken down into smaller molecules.* As shown in **Figure 15,** glycolysis produces some ATP molecules. It also uses energy from other ATP molecules. You will read on the following page that more ATP is made during the second step of cellular respiration than during glycolysis.

Reading Check What is produced during glycolysis?

Glycolysis 🗝

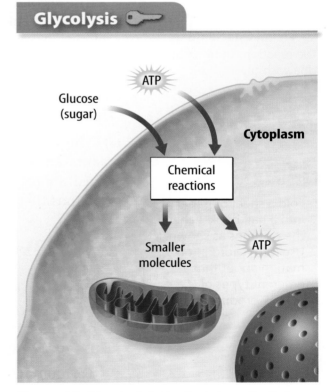

Figure 15 Glycolysis is the first step of cellular respiration.

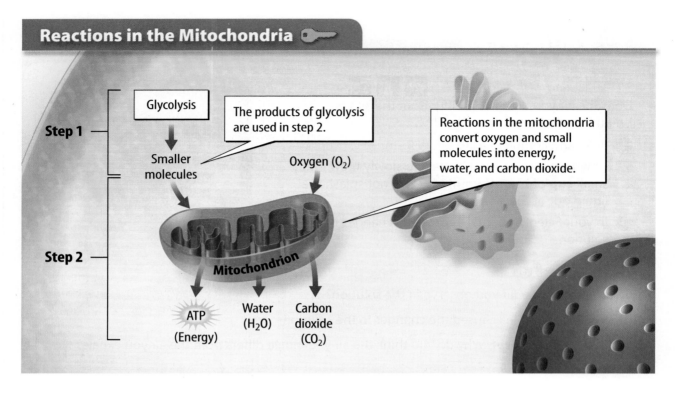

Figure 16 After glycolysis, cellular respiration continues in the mitochondria.

✔️**Visual Check** Compare the reactions in mitochondria with glycolysis.

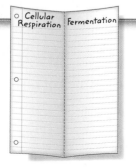
Reactions in the Mitochondria

The second step of cellular respiration occurs in the mitochondria of eukaryotic cells, as shown in **Figure 16.** This step of cellular respiration requires oxygen. The smaller molecules made from glucose during glycolysis are broken down. Large amounts of ATP—usable energy—are produced. Cells use ATP to power all cellular processes. Two waste products—water and carbon dioxide (CO_2)—are given off during this step.

The CO_2 released by cells as a waste product is used by plants and some unicellular organisms in another process called photosynthesis. You will read more about the chemical reactions that take place during photosynthesis in this lesson.

Fermentation

Have you ever felt out of breath after exercising? Sometimes when you exercise, your cells don't have enough oxygen to make ATP through cellular respiration. Then, chemical energy is obtained through a different process called fermentation. This process does not use oxygen.

Fermentation *is a reaction that eukaryotic and prokaryotic cells can use to obtain energy from food when oxygen levels are low.* Because no oxygen is used, fermentation makes less ATP than cellular respiration does. Fermentation occurs in a cell's cytoplasm, not in mitochondria.

✔️ **Key Concept Check** How does a cell obtain energy?

Types of Fermentation

One type of fermentation occurs when glucose is converted into ATP and a waste product called lactic acid, as illustrated in **Figure 17.** Some bacteria and fungi help produce cheese, yogurt, and sour cream using lactic-acid fermentation. Muscle cells in humans and other animals can use lactic-acid fermentation and obtain energy during exercise.

Some types of bacteria and yeast make ATP through a process called alcohol fermentation. However, instead of producing lactic acid, alcohol fermentation produces an alcohol called ethanol and CO_2, also illustrated in **Figure 17.** Some types of breads are made using yeast. The CO_2 produced by yeast during alcohol fermentation makes the dough rise.

✓ **Reading Check** Compare lactic-acid fermentation and alcohol fermentation.

Figure 17 Your muscle cells produce lactic acid as a waste during fermentation. Yeast cells produce carbon dioxide and alcohol as wastes during fermentation.

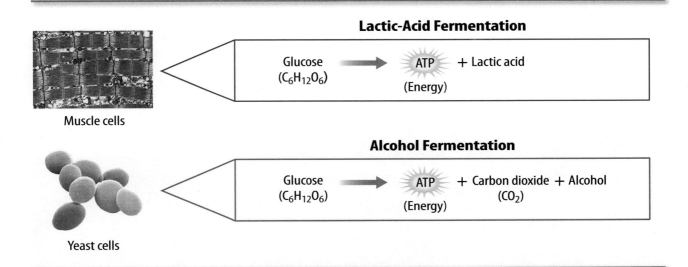

Lactic-Acid Fermentation

Glucose ($C_6H_{12}O_6$) → ATP (Energy) + Lactic acid

Muscle cells

Alcohol Fermentation

Glucose ($C_6H_{12}O_6$) → ATP (Energy) + Carbon dioxide (CO_2) + Alcohol

Yeast cells

Photosynthesis

Humans and other animals convert food energy into ATP through cellular respiration. However, plants and some unicellular organisms obtain energy from light. **Photosynthesis** *is a series of chemical reactions that convert light energy, water, and CO_2 into the food-energy molecule glucose and give off oxygen.*

Lights and Pigments

Photosynthesis requires light energy. In plants, pigments such as chlorophyll absorb light energy. When chlorophyll absorbs light, it absorbs all colors except green. Green light is reflected as the green color seen in leaves. However, plants contain many pigments that reflect other colors, such as yellow and red.

WORD ORIGIN ·············

photosynthesis
from Greek *photo*, means "light"; and *synthesis*, means "composition"

Reactions in Chloroplasts

The light energy absorbed by chlorophyll and other pigments powers the chemical reactions of photosynthesis. These reactions occur in chloroplasts, the organelles in plant cells that convert light energy to chemical energy in food. During photosynthesis, light energy, water, and carbon dioxide combine and make sugars. Photosynthesis also produces oxygen that is released into the atmosphere, as shown in **Figure 18.**

 Key Concept Check How do some cells make food molecules?

Importance of Photosynthesis

Recall that photosynthesis uses light energy and CO_2 and makes food energy and releases oxygen. This food energy is stored in the form of glucose. When an organism, such as the bird in **Figure 18,** eats plant material, such as fruit, it takes in food energy. An organism's cells use the oxygen released during photosynthesis and convert the food energy into usable energy through cellular respiration. **Figure 18** illustrates the important relationship between cellular respiration and photosynthesis.

Figure 18 The relationship between cellular respiration and photosynthesis is important for life.

Cellular Respiration and Photosynthesis

Review   Personal Tutor

Light energy

Chloroplast

Carbon dioxide (CO_2)
Water (H_2O)

Glucose ($C_6H_{12}O_6$)
Oxygen (O_2)

Mitochondrion

ATP

$C_6H_{12}O_6 + 6O_2 \longrightarrow 6CO_2 + 6H_2O +$ ATP (Energy)

Cellular respiration

$$6CO_2 + 6H_2O \xrightarrow[\text{Chlorophyll}]{\text{Light energy}} C_6H_{12}O_6 + 6O_2$$

Photosynthesis

Lesson 4 Review

Visual Summary

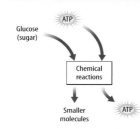

Glycolysis is the first step in cellular respiration.

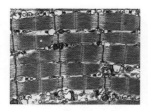

Fermentation provides cells, such as muscle cells, with energy when oxygen levels are low.

Light energy powers the chemical reactions of photosynthesis.

FOLDABLES

Use your lesson Foldable to review the lesson. Save your Foldable for the project at the end of the chapter.

What do you think NOW?

You first read the statements below at the beginning of the chapter.

7. ATP is the only form of energy found in cells.

8. Cellular respiration occurs only in lung cells.

Did you change your mind about whether you agree or disagree with the statements? Rewrite any false statements to make them true.

Use Vocabulary

1 **Define** *glycolysis* using your own words.

2 **Distinguish** between cellular respiration and fermentation.

3 A process used by plants to convert light energy into food energy is _____.

Understand Key Concepts

4 Which contains pigments that absorb light energy?
   A. chloroplast   **C.** nucleus
   B. mitochondrion   **D.** vacuole

5 **Relate** mitochondria to cellular respiration.

6 **Describe** the role of chlorophyll in photosynthesis.

7 **Give an example** of how fermentation is used in the food industry.

Interpret Graphics

8 **Draw** a graphic organizer like the one below. Fill in the boxes with the substances used and produced during photosynthesis.

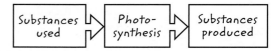

9 **Summarize** the steps of cellular respiration using the figure below.

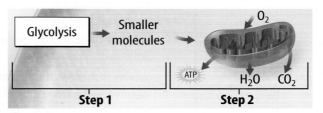

Critical Thinking

10 **Design** a concept map to show the relationship between cellular respiration in animals and photosynthesis in plants.

11 **Summarize** the roles of glucose and ATP in energy processing.

Materials

test tube

Elodea

scissors

beaker

lamp

watch or clock

thermometer

Safety

Photosynthesis and Light

You might think of photosynthesis as a process of give and take. Plant cells take in water and carbon dioxide, and, powered by light energy, make their own food. Plants give off oxygen as a waste product during photosynthesis. Can you determine how the intensity of light affects the rate of photosynthesis?

Ask a Question

How does the intensity of light affect photosynthesis?

Make Observations

1 Read and complete a lab safety form.

2 Cut the bottom end of an *Elodea* stem at an angle, and lightly crush the cut end. Place the *Elodea* in a test tube with the cut side at the top. Fill the test tube with water. Stand the test tube and a thermometer in a beaker filled with water. (The water in the beaker keeps the water in the test tube from getting too warm under the lamp.)

3 Place the beaker containing your test tube on a sheet of paper under a lamp. Measure the temperature of the water in the beaker. Record the temperature in your Science Journal.

4 When bubbles of oxygen begin to rise from the plant, start counting the number of bubbles per minute. Continue to record this data for 10 minutes.

5 Record the temperature of the water in the beaker at the end of the test.

6 Calculate the average number of bubbles produced per minute by your plant.

7 Compare your data with your classmates' data.

Form a Hypothesis

8 Use your data to form a hypothesis relating the amount of light to the rate of photosynthesis.

Test Your Hypothesis

9 Repeat the experiment, changing the light variable so that you are observing your plant's reaction to getting either more or less light. An increase or decrease in water temperature will indicate a change in the amount of light. Keep all other conditions the same.

10 Record your data in a table similar to the one shown at right, and calculate the average number of bubbles per minute.

Analyze and Conclude

11 **Use Variables** How does the amount of light affect photosynthesis? What is your evidence?

12 **The Big Idea** How do plant cells make food? What do they take in and what do they give off? What source of energy do they use?

Communicate Your Results

Compile all the class data on one graph to show the effects of varying amounts of light on the rate of photosynthesis.

Inquiry **Extension**

What other variables might affect the rate of photosynthesis? For example, how does different-colored light or a change in temperature affect the rate of photosynthesis? To investigate your question, design a controlled experiment.

Lab Tips

☑ To calculate the average number of bubbles per minute, add the total number of bubbles observed in 10 minutes, and then divide by 10.

Number of Bubbles per Minute		
Time	Control	Less Light
1		
2		
3		
4		
5		
6		
7		
8		
9		
10		

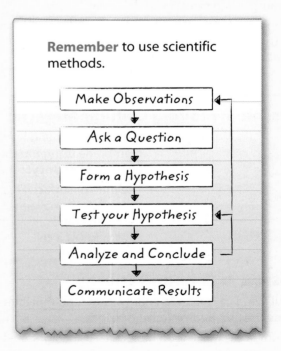

Remember to use scientific methods.

Make Observations
↓
Ask a Question
↓
Form a Hypothesis
↓
Test your Hypothesis
↓
Analyze and Conclude
↓
Communicate Results

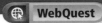
 THE BIG IDEA

A cell is made up of structures that provide support and movement; process energy; and transport materials into, within, and out of a cell.

Key Concepts Summary

Vocabulary

Lesson 1: Cells and Life

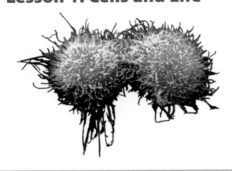

- The invention of the microscope led to discoveries about cells. In time, scientists used these discoveries to develop the **cell theory,** which explains how cells and living things are related.
- Cells are composed mainly of water, **proteins, nucleic acids, lipids,** and **carbohydrates.**

cell theory p. 44
macromolecule p. 45
nucleic acid p. 46
protein p. 47
lipid p. 47
carbohydrate p. 47

Lesson 2: The Cell

- Cell structures have specific functions, such as supporting a cell, moving a cell, controlling cell activities, processing energy, and transporting molecules.
- A prokaryotic cell lacks a nucleus and other **organelles,** while a eukaryotic cell has a nucleus and other organelles.

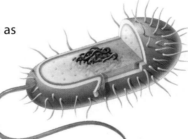

cell membrane p. 52
cell wall p. 52
cytoplasm p. 53
cytoskeleton p. 53
organelle p. 54
nucleus p. 55
chloroplast p. 57

Lesson 3: Moving Cellular Material

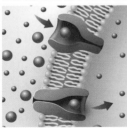

- Materials enter and leave a cell through the cell membrane using **passive transport** or **active transport, endocytosis,** and **exocytosis.**
- The ratio of surface area to volume limits the size of a cell. In a smaller cell, the high surface-area-to-volume ratio allows materials to move easily to all parts of a cell.

passive transport p. 61
diffusion p. 62
osmosis p. 62
facilitated diffusion p. 63
active transport p. 64
endocytosis p. 64
exocytosis p. 64

Lesson 4: Cells and Energy

- All living cells release energy from food molecules through **cellular respiration** and/or **fermentation.**
- Some cells make food molecules using light energy through the process of **photosynthesis.**

$$C_6H_{12}O_6 + 6O_2 \longrightarrow 6CO_2 + 6H_2O + \text{ATP (Energy)}$$

Cellular respiration

$$6CO_2 + 6H_2O \xrightarrow[\text{Chlorophyll}]{\text{Light energy}} C_6H_{12}O_6 + 6O_2$$

Photosynthesis

cellular respiration p. 69
glycolysis p. 69
fermentation p. 70
photosynthesis p. 71

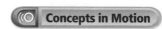
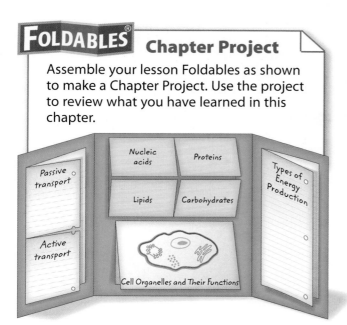

FOLDABLES® Chapter Project

Assemble your lesson Foldables as shown to make a Chapter Project. Use the project to review what you have learned in this chapter.

Passive transport

Active transport

Nucleic acids

Proteins

Lipids

Carbohydrates

Types of Energy Production

Cell Organelles and Their Functions

Use Vocabulary

1 Substances formed by joining smaller molecules together are called _____.

2 The _____ consists of proteins joined together to create fiberlike structures inside cells.

3 The movement of substances from an area of high concentration to an area of low concentration is called _____.

4 A process that uses oxygen to convert energy from food into ATP is _____ _____.

Link Vocabulary and Key Concepts

Concepts in Motion Interactive Concept Map

Copy this concept map, and then use vocabulary terms from the previous page to complete the concept map.

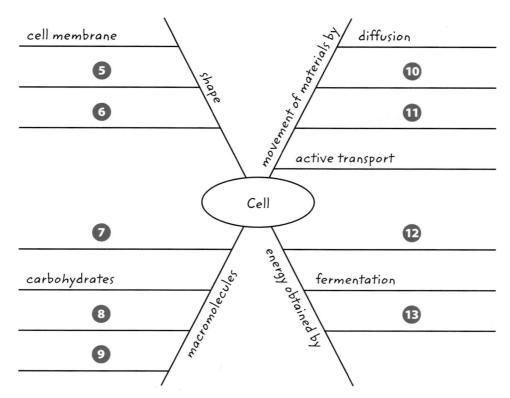

cell membrane

5

6

7

carbohydrates

8

9

shape

macromolecules

Cell

movement of materials by

diffusion

10

11

active transport

12

energy obtained by

fermentation

13

Understand Key Concepts

1 Cholesterol is which type of macromolecule?

A. carbohydrate
B. lipid
C. nucleic acid
D. protein

2 Genetic information is stored in which macromolecule?

A. DNA
B. glucose
C. lipid
D. starch

3 The arrow below is pointing to which cell part?

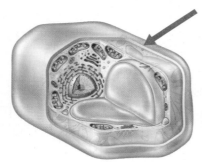

A. chloroplast
B. mitochondrion
C. cell membrane
D. cell wall

4 Which best describes vacuoles?

A. lipids
B. proteins
C. contained in mitochondria
D. storage compartments

5 Which is true of fermentation?

A. does not generate energy
B. does not require oxygen
C. occurs in mitochondria
D. produces lots of ATP

6 Which process eliminates substances from cells in vesicles?

A. endocytosis
B. exocytosis
C. osmosis
D. photosynthesis

7 Which cell shown below can send signals over long distances?

A.

B.

C.

D.

8 The figure below shows a cell. What is the arrow pointing to?

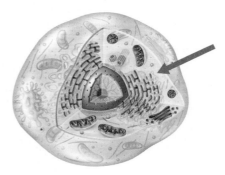

A. chloroplast
B. cytoplasm
C. mitochondrion
D. nucleus

Critical Thinking

9 **Evaluate** the importance of the microscope to biology.

10 **Summarize** the role of water in cells.

11 **Hypothesize** how new cells form from existing cells.

12 **Distinguish** between channel proteins and carrier proteins.

13 **Explain** osmosis.

14 **Infer** Why do cells need carrier proteins that transport glucose?

15 **Compare** the amounts of ATP generated in cellular respiration and fermentation.

16 **Assess** the role of fermentation in baking bread.

17 **Hypothesize** how air pollution like smog affects photosynthesis.

18 **Compare** prokaryotes and eukaryotes by copying and filling in the table below.

Structure	Prokaryote (yes or no)	Eukaryote (yes or no)
Cell membrane		
DNA		
Nucleus		
Endoplasmic reticulum		
Golgi apparatus		
Cell wall		

Writing in Science

19 **Write** a five-sentence paragraph relating the cytoskeleton to the walls of a building. Be sure to include a topic sentence and a concluding sentence in your paragraph.

REVIEW THE BIG IDEA

20 How do the structures and processes of a cell enable it to survive? As an example, explain how chloroplasts help plant cells.

21 The photo below shows a protozoan. What structures enable it to get food into its mouth?

Math Skills

Review
Math Practice

Use Ratios

22 A rectangular solid measures 4 cm long by 2 cm wide by 2 cm high. What is the surface-area-to-volume ratio of the solid?

23 At different times during its growth, a cell has the following surface areas and volumes:

Time	Surface area (μm)	Volume (μm)
1	6	1
2	24	8
3	54	27

What happens to the surface-area-to-volume ratio as the cell grows?

Record your answers on the answer sheet provided by your teacher or on a sheet of paper.

Multiple Choice

1 Which process do plant cells use to capture and store energy from sunlight?

    A endocytosis

    B fermentation

    C glycolysis

    D photosynthesis

Use the diagram below to answer question 2.

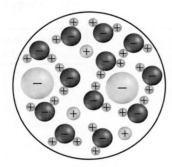

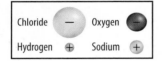

2 The diagram shows salt dissolved in water. What does it show about water molecules and chloride ions?

    A A water molecule consists of oxygen and chloride ions.

    B A water molecule is surrounded by several chloride ions.

    C A water molecule moves away from a chloride ion.

    D A water molecule points its positive end toward a chloride ion.

3 Which transport process requires the use of a cell's energy?

    A diffusion

    B osmosis

    C active transport

    D facilitated diffusion

4 Diffusion differs from active cell transport processes because it

    A forces large molecules from a cell.

    B keeps a cell's boundary intact.

    C moves substances into a cell.

    D needs none of a cell's energy.

Use the diagram below to answer questions 5 and 6.

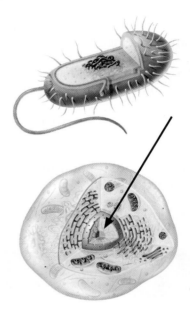

5 Which structure does the arrow point to in the eukaryotic cell?

    A cytoplasm

    B lysosome

    C nucleus

    D ribosome

6 Which feature does a typical prokaryotic cell have that is missing from some eukaryotic cells, like the one above?

    A cytoplasm

    B DNA

    C cell membrane

    D cell wall

7 Which explains why the ratio of cell surface area to volume affects the cell size? Cells with a high surface-to-volume ratio

   A consume energy efficiently.

   B produce waste products slowly.

   C suffer from diseases frequently.

   D transport substances effectively.

Use the diagram below to answer question 8.

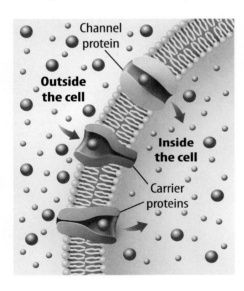

8 Which statement is NOT true of carrier proteins and channel proteins?

   A Carrier proteins change shape as they function but channel proteins do not.

   B Carrier proteins and channel proteins extend through the cell membrane.

   C Channel proteins move items inside a cell but carrier proteins do not.

   D Channel proteins and carrier proteins perform facilitated diffusion.

Constructed Response

9 Copy the table below and complete it using these terms: *cell membrane, cell wall, chloroplast, cytoplasm, cytoskeleton, nucleus.*

Cell Structure	Function
	Maintains the shape of an animal cell
	Controls the activities of a cell
	Traps energy from the Sun
	Controls the materials going in and out of a cell
	Holds the structures of a cell in a watery mix
	Maintains the shape of some plant cells

10 Name the kinds of organisms that have cells with cell walls. Name the kinds of organisms that have cells without cell walls. Briefly describe the benefits of cell walls for organisms.

11 Draw simple diagrams of an animal cell and a plant cell. Label the nucleus, the cytoplasm, the mitochondria, the cell membrane, the chloroplasts, the cell wall, and the central vacuole in the appropriate cells. Briefly describe the main differences between the two cells.

NEED EXTRA HELP?											
If You Missed Question...	1	2	3	4	5	6	7	8	9	10	11
Go to Lesson...	4	1	3	3	2	2	3	3	2	2	2

Chapter 3

From a Cell to an Organism

THE BIG IDEA

How can one cell become a multicellular organism?

Inquiry **What's happening inside?**

From the outside, a chicken egg looks like a simple oval object. But big changes are taking place inside the egg. Over several weeks, the one cell in the egg will grow and divide and become a chick.

- How did the original cell change over time?

- What might have happened to the chick's cells as the chick grew?

- How can one cell become a multicellular chick?

Get Ready to Read

What do you think?

Before you read, decide if you agree or disagree with each of these statements. As you read this chapter, see if you change your mind about any of the statements.

1 Cell division produces two identical cells.

2 Cell division is important for growth.

3 At the end of the cell cycle, the original cell no longer exists.

4 Unicellular organisms do not have all the characteristics of life.

5 All the cells in a multicellular organism are the same.

6 Some organs work together as part of an organ system.

ConnectED Your one-stop online resource

connectED.mcgraw-hill.com

🎞 Video

🔊 Audio

⊟ Review

? Inquiry

⊕ WebQuest

✓ Assessment

《《》 Concepts in Motion

g Multilingual eGlossary

Lesson 1

The Cell Cycle and Cell Division

Reading Guide

Key Concepts 🔑
ESSENTIAL QUESTIONS

- What are the phases of the cell cycle?

- Why is the result of the cell cycle important?

Vocabulary

cell cycle p. 85

interphase p. 86

sister chromatid p. 88

centromere p. 88

mitosis p. 89

cytokinesis p. 89

daughter cell p. 89

 Multilingual eGlossary

 Video BrainPOP®

Inquiry Time to Split?

Unicellular organisms such as these reproduce when one cell divides into two new cells. The two cells are identical to each other. What do you think happened to the contents of the original cell before it divided?

Inquiry Launch Lab

Why isn't your cell like mine? ✂️ 🧴

All living things are made of cells. Some are made of only one cell, while others are made of trillions of cells. Where do all those cells come from?

1. Read and complete a lab safety form.

2. Ask your team members to face away from you. Draw an animal cell on a sheet of **paper.** Include as many organelles as you can.

3. Use **scissors** to cut the cell drawing into equal halves. Fold each sheet of paper in half so the drawing cannot be seen.

4. Ask your team members to face you. Give each team member half of the cell drawing.

5. Have team members sit facing away from each other. Each person should use a **glue stick** to attach the cell half to one side of a sheet of paper. Then, each person should draw the missing cell half.

6. Compare the two new cells to your original cell.

Think About This

1. How did the new cells compare to the original cell?

2. 🔑 **Key Concept** What are some things that might be done in the early steps to produce two new cells that are more like the original cell?

The Cell Cycle

No matter where you live, you have probably noticed that the weather changes in a regular pattern each year. Some areas experience four seasons—winter, spring, summer, and fall. In other parts of the world, there are only two seasons—rainy and dry. As seasons change, temperature, precipitation, and the number of hours of sunlight vary in a regular cycle.

These changes can affect the life cycles of organisms such as trees. Notice how the tree in **Figure 1** changes with the seasons. Like changing seasons or the growth of trees, cells go through cycles. *Most cells in an organism go through a cycle of growth, development, and division called the* **cell cycle.** Through the cell cycle, organisms grow, develop, replace old or damaged cells, and produce new cells.

Figure 1 This maple tree changes in response to a seasonal cycle.

✓ **Visual Check** List the seasonal changes of this maple tree.

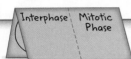

Interphase | Mitotic Phase

Phases of the Cell Cycle

There are two main phases in the cell cycle—interphase and the mitotic (mi TAH tihk) phase. **Interphase** *is the period during the cell cycle of a cell's growth and development.* A cell spends most of its life in interphase, as shown in **Figure 2.** During interphase, most cells go through three stages:

- rapid growth and replication, or copying, of the membrane-bound structures called organelles;

- copying of DNA, the genetic information in a cell; and

- preparation for cell division.

Interphase is followed by a shorter period of the cell cycle known as the mitotic phase. A cell reproduces during this phase. The mitotic phase has two stages, as illustrated in **Figure 2.** The nucleus divides in the first stage, and the cell's fluid, called the cytoplasm, divides in the second stage. The mitotic phase creates two new identical cells. At the end of this phase, the original cell no longer exists.

Key Concept Check What are the two main phases of the cell cycle?

The Cell Cycle

Figure 2 A cell spends most of its life growing and developing during interphase.

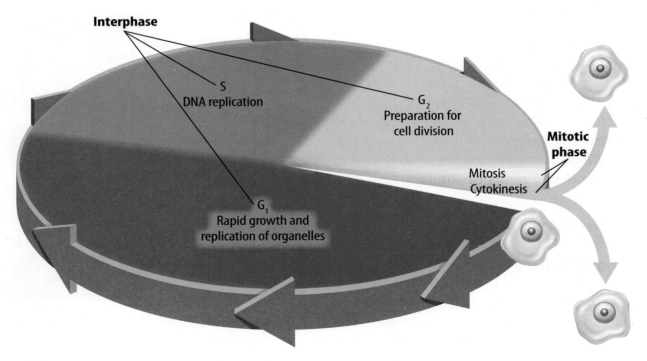

Interphase

S
DNA replication

G₂
Preparation for cell division

G₁
Rapid growth and replication of organelles

Mitotic phase

Mitosis
Cytokinesis

Visual Check Which stage of interphase is the longest?

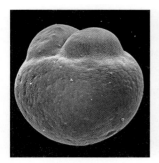

2-cell stage
SEM Magnification: 160×

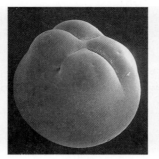

4-cell stage
SEM Magnification: 155×

32-cell stage
SEM Magnification: 150×

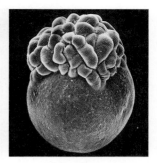

256-cell stage
SEM Magnification: 130×

Length of a Cell Cycle

The time it takes a cell to complete the cell cycle depends on the type of cell that is dividing. Recall that a **eukaryotic** cell has membrane-bound organelles, including a nucleus. For some eukaryotic cells, the cell cycle might last only eight minutes. For other cells, the cycle might take as long as one year. Most dividing human cells normally complete the cell cycle in about 24 hours. As illustrated in **Figure 3,** the cells of some organisms divide very quickly.

Interphase

As you have read, interphase makes up most of the cell cycle. Newly produced cells begin interphase with a period of rapid growth—the cell gets bigger. This is followed by cellular activities such as making proteins. Next, actively dividing cells make copies of their DNA and prepare for cell division. During interphase, the DNA is called chromatin (KROH muh tun). Chromatin is long, thin strands of DNA, as shown in **Figure 4.** When scientists dye a cell in interphase, the nucleus looks like a plate of spaghetti. This is because the nucleus contains many strands of chromatin tangled together.

▲ **Figure 3** The fertilized egg of a zebra fish divides into 256 cells in 2.5 hours.

REVIEW VOCABULARY · · · · ·

eukaryotic
a cell with membrane-bound structures

Figure 4 During interphase, the nuclei of an animal cell and a plant cell contain long, thin strands of DNA called chromatin. ▼

Interphase

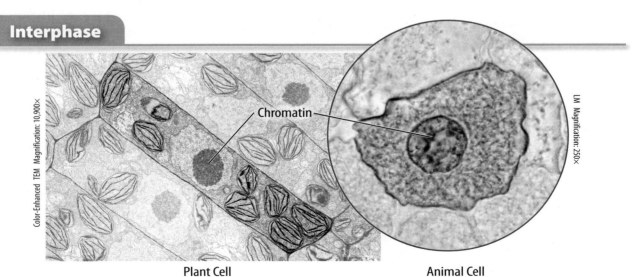

Color-Enhanced TEM Magnification: 10,900×

Chromatin

LM Magnification: 250×

Plant Cell

Animal Cell

Table 1 Phases of the Cell Cycle 🔑　　((O Concepts in Motion) Interactive Table

Phase	Stage	Description
Interphase	G_1	growth and cellular functions; organelle replication
	S	growth and chromosome replication; organelle replication
	G_2	growth and cellular functions; organelle replication
Mitotic phase	mitosis	division of nucleus
	cytokinesis	division of cytoplasm

▲ **Table 1** The two phases of the cell cycle can each be divided into different stages.

Figure 5 The coiled DNA forms a duplicated chromosome made of two sister chromatids connected at the centromere. ▼

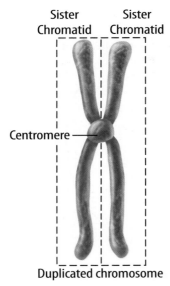

Sister Chromatid　Sister Chromatid

Centromere

Duplicated chromosome

Phases of Interphase

Scientists divide interphase into three stages, as shown in **Table 1.** Interphase begins with a period of rapid growth—the G_1 stage. This stage lasts longer than other stages of the cell cycle. During G_1, a cell grows and carries out its normal cell functions. For example, during G_1, cells that line your stomach make enzymes that help digest your food. Although most cells continue the cell cycle, some cells stop the cell cycle at this point. For example, mature nerve cells in your brain remain in G_1 and do not divide again.

During the second stage of interphase—the S stage—a cell continues to grow and copies its DNA. There are now identical strands of DNA. These identical strands of DNA ensure that each new cell gets a copy of the original cell's genetic information. Each strand of DNA coils up and forms a chromosome. Identical chromosomes join together. The cell's DNA is now arranged as pairs of identical chromosomes. Each pair is called a duplicated chromosome. *Two identical chromosomes, called* sister chromatids, *make up a duplicated chromosome,* as shown in **Figure 5.** Notice that the *sister chromatids are held together by a structure called the* **centromere.**

The final stage of interphase—the G_2 stage—is another period of growth and the final preparation for the mitotic phase. A cell uses energy copying DNA during the S stage. During G_2, the cell stores energy that will be used during the mitotic phase of the cell cycle.

✓ **Reading Check** Describe what happens in the G_2 phase.

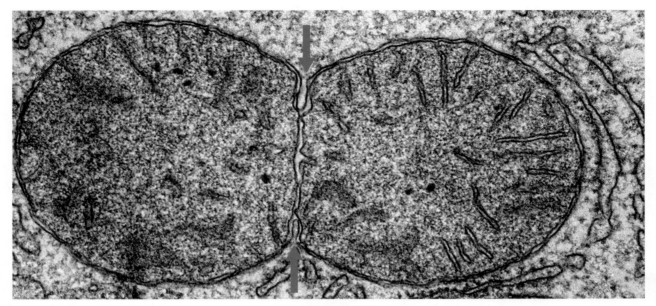

TEM Magnification: Unavailable

Organelle Replication

During cell division, the organelles in a cell are distributed between the two new cells. Before a cell divides, it makes a copy of each organelle. This enables the two new cells to function properly. Some organelles, such as the energy-processing mitochondria and chloroplasts, have their own DNA. These organelles can make copies of themselves on their own, as shown in **Figure 6.** A cell produces other organelles from materials such as proteins and lipids. A cell makes these materials using the information contained in the DNA inside the nucleus. Organelles are copied during all stages of interphase.

The Mitotic Phase

The mitotic phase of the cell cycle follows interphase. It consists of two stages: mitosis (mi TOH sus) and cytokinesis (si toh kuh NEE sus). *In* **mitosis,** *the nucleus and its contents divide. In* **cytokinesis,** *the cytoplasm and its contents divide.* **Daughter cells** *are the two new cells that result from mitosis and cytokinesis.*

During mitosis, the contents of the nucleus divide, forming two identical nuclei. The sister chromatids of the duplicated chromosomes separate from each other. This gives each daughter cell the same genetic information. For example, a cell that has ten duplicated chromosomes actually has 20 chromatids. When the cell divides, each daughter cell will have ten different chromatids. Chromatids are now called chromosomes.

In cytokinesis, the cytoplasm divides and forms the two new daughter cells. Organelles that were made during interphase are divided between the daughter cells.

Figure 6 This mitochondrion is in the final stage of dividing.

WORD ORIGIN

mitosis
from Greek *mitos*, means "warp thread"; and Latin *–osis*, means "process"

Phases of Mitosis

Like interphase, mitosis is a continuous process that scientists divide into different phases, as shown in **Figure 7.**

Prophase During the first phase of mitosis, called prophase, the copied chromatin coils together tightly. The coils form visible duplicated chromosomes. The nucleolus disappears, and the nuclear membrane breaks down. Structures called spindle fibers form in the cytoplasm.

Metaphase During metaphase, the spindle fibers pull and push the duplicated chromosomes to the middle of the cell. Notice in **Figure 7** that the chromosomes line up along the middle of the cell. This arrangement ensures that each new cell will receive one copy of each chromosome. Metaphase is the shortest phase in mitosis, but it must be completed successfully for the new cells to be identical.

Phases of Mitosis

((○ Concepts in Motion · Animation

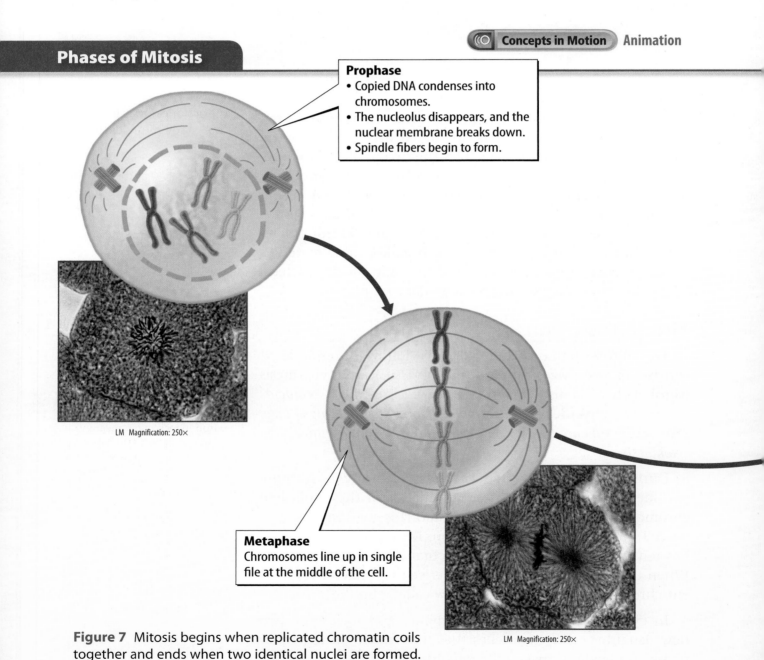

Prophase
• Copied DNA condenses into chromosomes.
• The nucleolus disappears, and the nuclear membrane breaks down.
• Spindle fibers begin to form.

LM Magnification: 250×

Metaphase
Chromosomes line up in single file at the middle of the cell.

LM Magnification: 250×

Figure 7 Mitosis begins when replicated chromatin coils together and ends when two identical nuclei are formed.

Anaphase In anaphase, the third stage of mitosis, the two sister chromatids in each chromosome separate from each other. The spindle fibers pull them in opposite directions. Once separated, the chromatids are now two identical single-stranded chromosomes. As they move to opposite sides of a cell, the cell begins to get longer. Anaphase is complete when the two identical sets of chromosomes are at opposite ends of a cell.

Telophase During telophase, the spindle fibers begin to disappear. Also, the chromosomes begin to uncoil. A nuclear membrane forms around each set of chromosomes at either end of the cell. This forms two new identical nuclei. Telophase is the final stage of mitosis. It is often described as the reverse of prophase because many of the processes that occur during prophase are reversed during telophase.

 Reading Check What are the phases of mitosis?

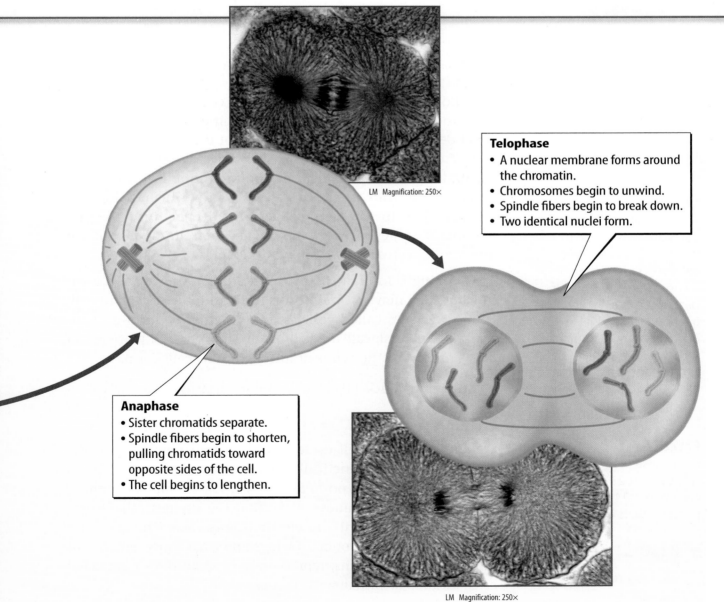

LM Magnification: 250×

Telophase
- A nuclear membrane forms around the chromatin.
- Chromosomes begin to unwind.
- Spindle fibers begin to break down.
- Two identical nuclei form.

Anaphase
- Sister chromatids separate.
- Spindle fibers begin to shorten, pulling chromatids toward opposite sides of the cell.
- The cell begins to lengthen.

LM Magnification: 250×

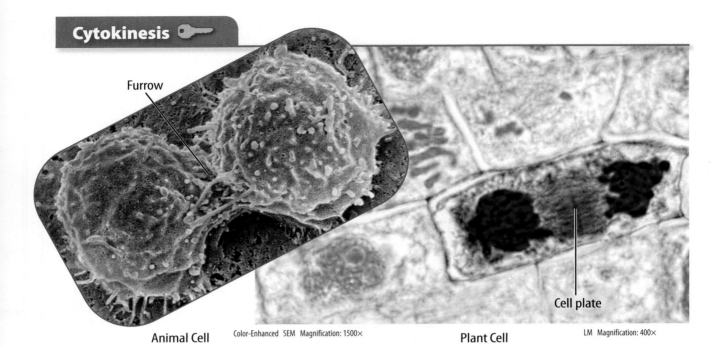

Furrow

Cell plate

Animal Cell

Color-Enhanced SEM Magnification: 1500×

Plant Cell

LM Magnification: 400×

Figure 8 Cytokinesis differs in animal cells and plant cells.

Math Skills

Use Percentages

A percentage is a ratio that compares a number to 100. If the length of the entire cell cycle is 24 hours, 24 hours equals 100%. If part of the cycle takes 6.0 hours, it can be expressed as 6.0 hours/ 24 hours. To calculate percentage, divide and multiply by 100. Add a percent sign.

$$\frac{6.0}{24} = 0.25 \times 100 = 25\%$$

Practice

Interphase in human cells takes about 23 hours. If the cell cycle is 24 hours, what percentage is interphase?

 Review

- **Math Practice**
- **Personal Tutor**

Dividing the Cell's Components

Following the last phase of mitosis, a cell's cytoplasm divides in a process called cytokinesis. The specific steps of cytokinesis differ depending on the type of cell that is dividing. In animal cells, the cell membrane contracts, or squeezes together, around the middle of the cell. Fibers around the center of the cell pull together. This forms a crease, called a furrow, in the middle of the cell. The furrow gets deeper and deeper until the cell membrane comes together and divides the cell. An animal cell undergoing cytokinesis is shown in **Figure 8.**

Cytokinesis in plants happens in a different way. As shown in **Figure 8,** a new cell wall forms in the middle of a plant cell. First, organelles called vesicles join together to form a membrane-bound disk called a cell plate. Then the cell plate grows outward toward the cell wall until two new cells form.

✓ **Reading Check** Compare cytokinesis in plant and animal cells.

Results of Cell Division

Recall that the cell cycle results in two new cells. These daughter cells are genetically identical to each other and to the original cell that no longer exists. For example, a human cell has 46 chromosomes. When that cell divides, it will produce two new cells with 46 chromosomes each. The cell cycle is important for reproduction in some organisms, growth in multicellular organisms, replacement of worn out or damaged cells, and repair of damaged tissues.

Reproduction

In some unicellular organisms, cell division is a form of reproduction. For example, an organism called a paramecium often reproduces by dividing into two new daughter cells or two new paramecia. Cell division is also important in other methods of reproduction in which the offspring are identical to the parent organism.

Growth

Cell division allows multicellular organisms, such as humans, to grow and develop from one cell (a fertilized egg). In humans, cell division begins about 24 hours after fertilization and continues rapidly during the first few years of life. It is likely that during the next few years you will go through another period of rapid growth and development. This happens because cells divide and increase in number as you grow and develop.

Replacement

Even after an organism is fully grown, cell division continues. It replaces cells that wear out or are damaged. The outermost layer of your skin is always rubbing or flaking off. A layer of cells below the skin's surface is constantly dividing. This produces millions of new cells daily to replace the ones that are rubbed off.

Repair

Cell division is also critical for repairing damage. When a bone breaks, cell division produces new bone cells that patch the broken pieces back together.

Not all damage can be repaired, however, because not all cells continue to divide. Recall that mature nerve cells stop the cell cycle in interphase. For this reason, injuries to nerve cells often cause permanent damage.

 Key Concept Check Why is the result of the cell cycle important?

Inquiry MiniLab **20 minutes**

How does mitosis work?

The dolix is a mythical animal whose cells contain just two chromosomes. What happens to a dolix cell nucleus during mitosis?

1. Read and complete a lab safety form.

2. Form four 60-cm lengths of **yarn** into large circles on four separate sheets of **paper.** Each piece of paper represents one phase of mitosis, and the yarn represents the cell membrane.

3. On each sheet of paper, model one phase of mitosis using different colors of yarn to represent the nuclear membrane, the spindles, and the chromosomes. Use **twist ties** to represent centromeres. **Tape** the yarn in place.

4. Label your models, or develop a key to indicate which color is used for which part.

Analyze and Conclude

1. **Identify** If you were to model a dolix cell's nucleus before mitosis began, what would your model look like? Would you be able to see the individual chromosomes?

2. **Integrate** What would a model of your cell look like during the stage immediately following mitosis? What is this stage?

3. 🔑 **Key Concept** During mitosis, a cell forms two new, identical nuclei. Use your models to explain why, in order to do this, mitosis must occur after events in interphase.

Lesson 1 Review

Visual Summary

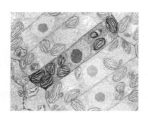

During interphase, most cells go through periods of rapid growth and replication of organelles, copying DNA, and preparation for cell division.

The nucleus and its contents divide during mitosis.

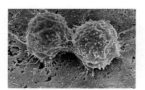

The cytoplasm and its contents divide during cytokinesis.

FOLDABLES

Use your lesson Foldable to review the lesson. Save your Foldable for the project at the end of the chapter.

What do you think **NOW?**

You first read the statements below at the beginning of the chapter.

1. Cell division produces two identical cells.

2. Cell division is important for growth.

3. At the end of the cell cycle, the original cell no longer exists.

Did you change your mind about whether you agree or disagree with the statements? Rewrite any false statements to make them true.

Use Vocabulary

1 **Distinguish** between mitosis and cytokinesis.

2 A duplicated chromosome is made of two _____.

3 **Use the term** *interphase* in a sentence.

Understand Key Concepts

4 Which is NOT part of mitosis?
   A. anaphase   **C.** prophase
   B. interphase   **D.** telophase

5 **Construct** a table to show the different phases of mitosis and what happens during each.

6 **Give three examples** of why the result of the cell cycle is important.

Interpret Graphics

7 **Identify** The animal cell on the right is in what phase of mitosis? Explain your answer.

8 **Organize** Copy and fill in the graphic organizer below to show the results of cell division.

Results of cell division

Critical Thinking

9 **Predict** what might happen to a cell if it were unable to divide by mitosis.

Math Skills

Review
— Math Practice —

10 The mitotic phase of the human cell cycle takes approximately 1 hour. What percentage of the 24-hour cell cycle is the mitotic phase?

DNA
Fingerprinting

Solving Crimes One Strand at a Time

▼ DNA

Every cell in your body has the same DNA in its nucleus. Unless you are an identical twin, your DNA is entirely unique. Identical twins have identical DNA because they begin as one cell that divides and separates. When your cells begin mitosis, they copy their DNA. Every new cell has the same DNA as the original cells. That is why DNA can be used to identify people. Just as no two people have the same fingerprints, your DNA belongs to you alone.

Using scientific methods to solve crimes is called forensics. DNA fingerprinting is now a basic tool in forensics. Samples collected from a crime scene can be compared to millions of samples previously collected and indexed in a computer.

Every day, everywhere you go, you leave a trail of DNA. It might be in skin cells. It might be in hair or in the saliva you used to lick an envelope. If you commit a crime, you will most likely leave DNA behind. An expert crime scene investigator will know how to collect that DNA.

DNA evidence can prove innocence as well. Investigators have reexamined DNA found at old crime scenes. Imprisoned persons have been proven not guilty through DNA fingerprinting methods that were not yet available when a crime was committed.

DNA fingerprinting can also be used to identify bodies that had previously been known only as a John or Jane Doe.

▼ **The Federal Bureau of Investigation (FBI) has a nationwide index of DNA samples called CODIS (Combined DNA Index System).**

It's Your Turn

DISCOVER Your cells contain organelles called mitochondria. They have their own DNA, called mitochondrial DNA. Your mitochondrial DNA is identical to your mother's mitochondrial DNA. Find out how this information is used.

Reading Guide

Key Concepts 🔑
ESSENTIAL QUESTIONS

- How do unicellular and multicellular organisms differ?

- How does cell differentiation lead to the organization within a multicellular organism?

Vocabulary

cell differentiation p. 99

stem cell p. 100

tissue p. 101

organ p. 102

organ system p. 103

g Multilingual eGlossary

▢ Video BrainPOP®

Levels of Organization

Inquiry Scales on Wings?

This butterfly has a distinctive pattern of colors on its wings. The pattern is formed by clusters of tiny scales. In a similar way, multicellular organisms are made of many small parts working together.

How is a system organized?

The places people live are organized in a system. Do you live in or near a city? Cities contain things such as schools and stores that enable them to function on their own. Many cities together make up another level of organization.

1. Read and complete a lab safety form.

2. Using a **metric ruler** and **scissors,** measure and cut squares of **construction paper** that are 4 cm, 8 cm, 12 cm, 16 cm, and 20 cm on each side. Use a different color for each square.

3. Stack the squares from largest to smallest, and glue them together.

4. Cut apart the *City, Continent, Country, County*, and *State* labels your teacher gives you.

5. Use a **glue stick** to attach the *City* label to the smallest square. Sort the remaining labels from smallest to largest, and glue to the corresponding square.

Think About This

1. What is the largest level of organization a city belongs to?

2. Can any part of the system function without the others? Explain.

3. 🔑 **Key Concept** How do you think the system used to organize where people live is similar to how your body is organized?

Life's Organization

You might recall that all matter is made of atoms and that atoms combine and form molecules. Molecules make up cells. A large animal, such as a Komodo dragon, is not made of one cell. Instead, it is composed of trillions of cells working together. Its skin, shown in **Figure 9,** is made of many cells that are specialized for protection. The Komodo dragon has other types of cells, such as blood cells and nerve cells, that perform other functions. Cells work together in the Komodo dragon and enable it to function. In the same way, cells work together in you and in other multicellular organisms.

Recall that some organisms are made of only one cell. These unicellular organisms carry out all the activities necessary to survive, such as absorbing nutrients and getting rid of wastes. But no matter their sizes, all organisms are made of cells.

Color-Enhanced SEM Magnification: 12×

Figure 9 Skin cells are only one of the many kinds of cells that make up a Komodo dragon.

Figure 10 Unicellular organisms carry out life processes within one cell.

Contractile vacuole

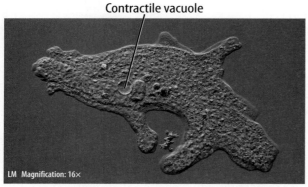

LM Magnification: 16×

This unicellular amoeba captures a desmid for food.

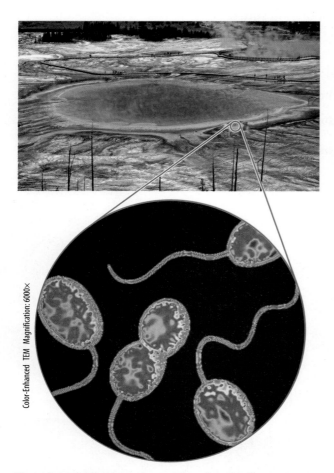

Color-Enhanced TEM Magnification: 6000×

These heat-loving bacteria are often found in hot springs as shown here. They get their energy to produce food from sulfur instead of from light like plants.

Unicellular Organisms

As you read on the previous page, some organisms have only one cell. Unicellular organisms do all the things needed for their survival within that one cell. For example, the amoeba in **Figure 10** is ingesting another unicellular organism, a type of green algae called a desmid, for food. Unicellular organisms also respond to their environment, get rid of waste, grow, and even reproduce on their own. Unicellular organisms include both prokaryotes and some eukaryotes.

Prokaryotes

Recall that a cell without a membrane-bound nucleus is a prokaryotic cell. In general, prokaryotic cells are smaller than eukaryotic cells and have fewer cell structures. A unicellular organism made of one prokaryotic cell is called a prokaryote. Some prokaryotes live in groups called colonies. Some can also live in extreme environments, as shown in **Figure 10**.

Eukaryotes

You might recall that a eukaryotic cell has a nucleus surrounded by a membrane and many other specialized organelles. For example, the amoeba shown in **Figure 10** has an organelle called a contractile vacuole. It functions like a bucket that is used to bail water out of a boat. A contractile vacuole collects excess water from the amoeba's cytoplasm. Then it pumps the water out of the amoeba. This prevents the amoeba from swelling and bursting.

A unicellular organism that is made of one eukaryotic cell is called a eukaryote. There are thousands of different unicellular eukaryotes, such as algae that grow on the inside of an aquarium and the fungus that causes athlete's foot.

✓ **Reading Check** Give an example of a unicellular eukaryotic organism.

Multicellular Organisms

Multicellular organisms are made of many eukaryotic cells working together, like the crew on an airplane. Each member of the crew, from the pilot to the mechanic, has a specific job that is important for the plane's operation. Similarly, each type of cell in a multicellular organism has a specific job that is important to the survival of the organism.

 Key Concept Check How do unicellular and multicellular organisms differ?

Cell Differentiation

As you read in the last lesson, all cells in a multicellular organism come from one cell—a fertilized egg. Cell division starts quickly after fertilization. The first cells made can become any type of cell, such as a muscle cell, a nerve cell, or a blood cell. *The process by which cells become different types of cells is called* **cell differentiation** (dihf uh ren shee AY shun).

You might recall that a cell's instructions are contained in its chromosomes. Also, nearly all the cells of an organism have identical sets of chromosomes. If an organism's cells have identical sets of instructions, how can cells be different? Different cell types use different parts of the instructions on the chromosomes. A few of the many different types of cells that can result from human cell differentiation are shown in **Figure 11.**

FOLDABLES

Make a layered book from three sheets of notebook paper. Label it as shown. Use your book to describe the levels of organization that make up organisms.

Levels of Organization
Cell
Tissue
Organ
Organ System
Organism

Figure 11 A fertilized egg produces cells that can differentiate into a variety of cell types.

Review Personal Tutor

Cell Differentiation in Eukaryotes

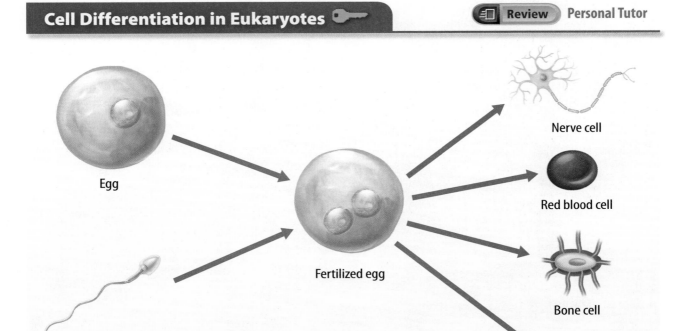

Egg

Sperm

Fertilized egg

Nerve cell

Red blood cell

Bone cell

Muscle cell

Animal Stem Cells Not all cells in a developing animal differentiate. **Stem cells** *are unspecialized cells that are able to develop into many different cell types.* There are many stem cells in embryos but fewer in adult organisms. Adult stem cells are important for the cell repair and replacement you read about in Lesson 1. For example, stem cells in your bone marrow can produce more than a dozen different types of blood cells. These replace ones that are damaged or worn out. Stem cells have also been discovered in skeletal muscles. These stem cells can produce new muscle cells when the **fibers** that make up the muscle are torn.

Plant Cells Plants also have unspecialized cells similar to animal stem cells. These cells are grouped in areas of a plant called meristems (MER uh stemz). Meristems are in different areas of a plant, including the tips of roots and stems, as shown in **Figure 12.** Cell division in meristems produces different types of plant cells with specialized structures and functions, such as transporting materials, making food, storing food, or protecting the plant. These cells might become parts of stems, leaves, flowers, or roots.

SCIENCE USE V. COMMON USE · ·

fiber
Science Use a long muscle cell

Common Use a thread

Figure 12 Plant meristems produce cells that can become part of stems, leaves, flowers, or roots.

Stem meristem

Root meristem

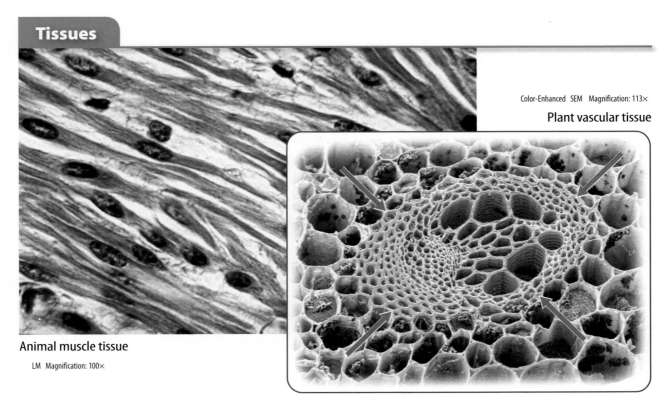

Color-Enhanced SEM Magnification: 113×

Plant vascular tissue

Animal muscle tissue

LM Magnification: 100×

Figure 13 Similar cells work together and form tissues such as this animal muscle tissue that contracts the stomach to help digestion. Plant vascular tissue, indicated by red arrows, moves water and nutrients throughout a plant.

Tissues

In multicellular organisms, similar types of cells are organized into groups. **Tissues** *are groups of similar types of cells that work together to carry out specific tasks.* Humans, like most other animals, have four main types of tissue—muscle, connective, nervous, and epithelial (eh puh THEE lee ul). For example, the animal tissue shown in **Figure 13** is smooth muscle tissue that is part of the stomach. Muscle tissue causes movement. Connective tissue provides structure and support and often connects other types of tissue together. Nervous tissue carries messages to and from the brain. Epithelial tissue forms the protective outer layer of the skin and the lining of major organs and internal body cavities.

Plants also have different types of tissues. The three main types of plant tissue are dermal, vascular (VAS kyuh lur), and ground tissue. Dermal tissue provides protection and helps reduce water loss. Vascular tissue, shown in **Figure 13,** transports water and nutrients from one part of a plant to another. Ground tissue provides storage and support and is where photosynthesis takes place.

 Reading Check Compare animal and plant tissues.

WORD ORIGIN · · · · · · · · · ·

tissue
from Latin *texere*, means "weave"

complex
(adjective) made of two or
more parts

Organs

Complex jobs in organisms require more than one type of tissue. **Organs** *are groups of different tissues working together to perform a particular job.* For example, your stomach is an organ specialized for breaking down food. It is made of all four types of tissue: muscle, epithelial, nervous, and connective. Each type of tissue performs a specific function necessary for the stomach to work properly. Layers of muscle tissue contract and break up pieces of food, epithelial tissue lines the stomach, nervous tissue sends signals to indicate the stomach is full, and connective tissue supports the stomach wall.

Plants also have organs. The leaves shown in **Figure 14** are organs specialized for photosynthesis. Each leaf is made of dermal, ground, and vascular tissues. Dermal tissue covers the outer surface of a leaf. The leaf is a vital organ because it contains ground tissue that produces food for the rest of the plant. Ground tissue is where photosynthesis takes place. The ground tissue is tightly packed on the top half of a leaf. The vascular tissue moves both the food produced by photosynthesis and water throughout the leaf and the rest of the plant.

Figure 14 A plant leaf is an organ made of several different tissues.

 Visual Check Which plant tissue makes up the thinnest layer?

 Reading Check List the tissues in a leaf organ.

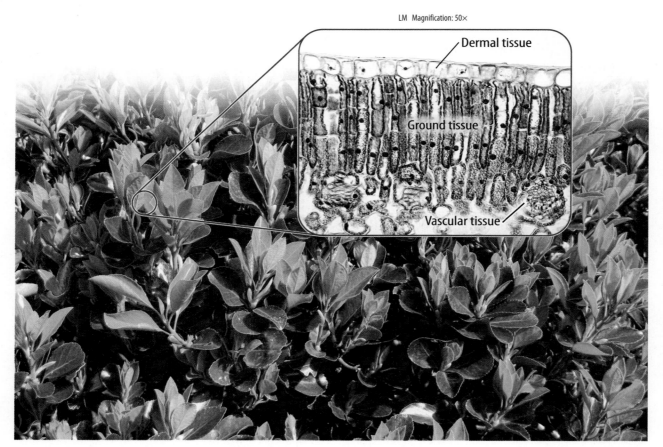

LM Magnification: 50×

Dermal tissue

Ground tissue

Vascular tissue

Organ Systems

Usually organs do not function alone. Instead, **organ systems** *are groups of different organs that work together to complete a series of tasks.* Human organ systems can be made of many different organs working together. For example, the human digestive system is made of many organs, including the stomach, the small intestine, the liver, and the large intestine. These organs and others all work together to break down food and take it into the body. Blood absorbs and transports nutrients from broken down food to cells throughout the body.

Plants have two major organ systems—the shoot system and the root system. The shoot system includes leaves, stems, and flowers. Food and water are transported throughout the plant by the shoot system. The root system anchors the plant and takes in water and nutrients.

 Reading Check What are the major organ systems in plants?

Inquiry **MiniLab**　　　　　　　　　　　　　**25 minutes**

How do cells work together to make an organism?

In a multicellular organism, similar cells work together and make a tissue. A tissue can perform functions that individual cells cannot. Tissues are organized into organs, then organ systems, then organisms. How can you model the levels of organization in an organism?

1. Read and complete a lab safety form.

2. Your teacher will give you a **cardboard shape, macaroni,** and a **permanent marker.**

3. The macaroni represent cells. Use the marker to draw a small circle on each piece of macaroni. This represents the nucleus.

4. Arrange and **glue** enough macaroni on the blank side of the cardboard shape to cover it. Your group of similar cells represents a tissue.

5. One of the squares on the back of your shape is labeled *A, B, C,* or *D*. Find the group with a matching letter. Line up these squares, and use **tape** to connect the two tissues. This represents an organ.

6. Repeat step 4 with the squares labeled *E* or *F.* This represents an organ system.

7. Connect the organ systems by aligning the squares labeled *G* to represent an organism.

Analyze and Conclude

1. Each group had to work with other groups to make a model of an organism. Do cells, tissues, and organs need to work together in organisms? Explain.

2. 🔑 **Key Concept** How does your model show the levels of organization in living things?

Organisms

Multicellular organisms usually have many organ systems. These systems work together to carry out all the jobs needed for the survival of the organisms. For example, the cells in the leaves and the stems of a plant need water to live. They cannot absorb water directly. Water diffuses into the roots and is transported through the stem to the leaves by the transport system.

In the human body, there are many major organ systems. Each organ system depends on the others and cannot work alone. For example, the cells in the muscle tissue of the stomach cannot survive without oxygen. The stomach cannot get oxygen without working together with the respiratory and circulatory systems. **Figure 15** will help you review how organisms are organized.

Key Concept Check How does cell differentiation lead to the organization within a multicellular organism?

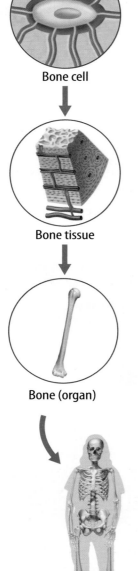

Bone cell

Bone tissue

Bone (organ)

Skeletal system

Figure 15 An organism is made of organ systems, organs, tissues, and cells that all function together and enable the organism's survival.

 Concepts in Motion Animation

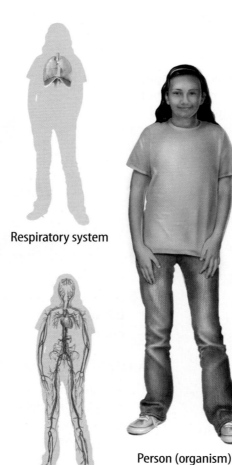

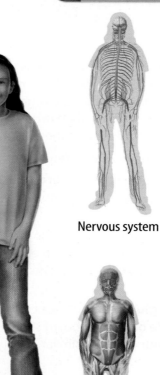

Respiratory system

Nervous system

Digestive system

Circulatory system

Person (organism)

Muscular system

Lesson 2 Review

Visual Summary

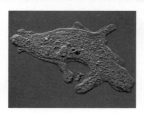

A unicellular organism carries out all the activities necessary for survival within one cell.

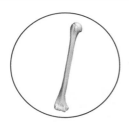

Cells become specialized in structure and function during cell differentiation.

Organs are groups of different tissues that work together to perform a job.

FOLDABLES

Use your lesson Foldable to review the lesson. Save your Foldable for the project at the end of the chapter.

What do you think NOW?

You first read the statements below at the beginning of the chapter.

4. Unicellular organisms do not have all the characteristics of life.

5. All the cells in a multicellular organism are the same.

6. Some organs work together as part of an organ system.

Did you change your mind about whether you agree or disagree with the statements? Rewrite any false statements to make them true.

Use Vocabulary

1 **Define** *cell differentiation* in your own words.

2 **Distinguish** between an organ and an organ system.

Understand Key Concepts

3 **Explain** the difference between a unicellular organism and a multicellular organism.

4 **Describe** how cell differentiation produces different types of cells in animals.

5 Which is the correct sequence of the levels of organization?
A. cell, organ, tissue, organ system, organism
B. organism, organ, organ system, tissue, cell
C. cell, tissue, organ, organ system, organism
D. tissue, organ, organism, organ system, cell

Interpret Graphics

6 **Organize** Copy and fill in the table below to summarize the characteristics of unicellular and multicellular organisms.

Organism Characteristics	
Unicellular	Multicellular

Critical Thinking

7 **Predict** A mistake occurs during mitosis of a muscle stem cell. How might this affect muscle tissue?

8 **Compare** the functions of a cell to the functions of an organism, such as getting rid of wastes.

Cell Differentiation

Materials

cooked eggs

boiled chicken leg

forceps

dissecting scissors

plastic knife

paper towels

Safety

It's pretty amazing that a whole chicken with wings, feet, beak, feathers, and internal organs can come from one cell, a fertilized egg. Shortly after fertilization, the cell begins to divide. The new cells in the developing embryo become specialized both in structure and function. The process by which cells become specialized is called cellular differentiation.

Question

How does a single cell become a multicellular organism?

Procedure

1 Read and complete a lab safety form.

2 Carefully examine the outside of your egg. Remove the shell.

3 Dissect the egg on a paper towel, cutting it in half from tip to rounded end. Examine the inside.

4 Record your observations in your Science Journal. Include a labeled drawing. Infer the function of each part.

5 Discard all your trash in the container provided.

6 Examine the outside of the chicken leg. Describe the skin and its functions.

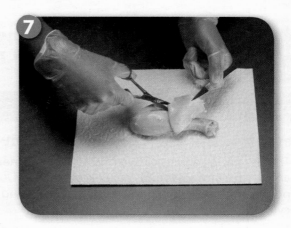

7 Carefully remove the skin using forceps and dissecting scissors. Put the skin in your discard container. Now you should see evidence of fat and muscles. You may also be able to see some blood vessels and tendons, but these are not always visible after cooking. Describe each part that you see and explain its function.

8 Peel back the muscles to reveal the bones. Tendons, ligaments, and cartilage holding the bones in place may also be evident.

9 Put all your trash in the discard container. Your teacher will give you instructions about cleaning up.

Analyze and Conclude

10 **The Big Idea** A single cell can become a multicellular organism through the process of cell differentiation. How do the organization of the egg and the chicken leg compare?

11 **Summarize** How many different types of cell differentiation did you observe in the chicken leg?

Communicate Your Results

Make a poster about how an egg transforms into a chicken through the process of cell differentiation.

 **Extension**

Examine a whole raw chicken or a raw chicken leg that is still attached to a thigh. You might be able to move the muscles in the legs or wings and see parts that were not visible in this lab. Be sure to wear gloves and to wash well with soap and water after touching the raw chicken.

Lab Tips

☑ Work slowly and carefully on your dissections so as not to destroy any structures. Report any accidents to your teacher immediately. Cleaning up is important!

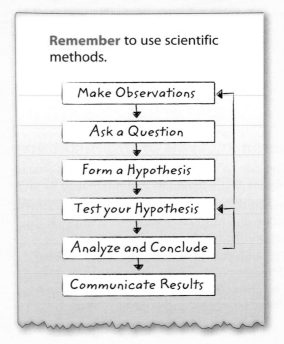

Remember to use scientific methods.

Make Observations
↓
Ask a Question
↓
Form a Hypothesis
↓
Test your Hypothesis
↓
Analyze and Conclude
↓
Communicate Results

Chapter 3 Study Guide

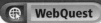 WebQuest

 THE BIG IDEA **Through cell division, one cell can produce new cells to grow and develop into a multicellular organism.**

Key Concepts Summary 🔑

Lesson 1: The Cell Cycle and Cell Division

- The **cell cycle** consists of two phases. During **interphase,** a cell grows and its chromosomes and organelles replicate. During the mitotic phase of the cell cycle, the nucleus divides during **mitosis,** and the cytoplasm divides during **cytokinesis.**
- The cell cycle results in two genetically identical **daughter cells.** The original parent cell no longer exists.
- The cell cycle is important for growth in multicellular organisms, reproduction in some organisms, replacement of worn-out cells, and repair of damaged cells.

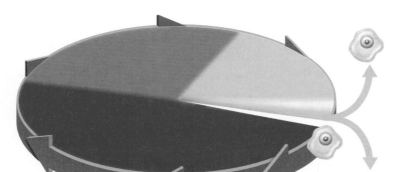

Lesson 2: Levels of Organization

- The one cell of a unicellular organism is able to obtain all the materials that it needs to survive.
- In a multicellular organism, cells cannot survive alone and must work together to provide the organism's needs.
- Through **cell differentiation,** cells become different types of cells with specific functions. Cell differentiation leads to the formation of **tissues, organs,** and **organ systems.**

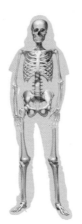

Vocabulary

cell cycle p. 85
interphase p. 86
sister chromatid p. 88
centromere p. 88
mitosis p. 89
cytokinesis p. 89
daughter cell p. 89

cell differentiation p. 99
stem cell p. 100
tissue p. 101
organ p. 102
organ system p. 103

FOLDABLES® Chapter Project

Assemble your lesson Foldables as shown to make a Chapter Project. Use the project to review what you have learned in this chapter.

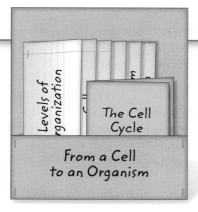

Levels of organization

The Cell Cycle

From a Cell to an Organism

Use Vocabulary

1. Use the term *sister chromatids* in a sentence.
2. Define the term *centromere* in your own words.
3. The new cells formed by mitosis are called _____.
4. Use the term *cell differentiation* in a sentence.
5. Define the term *stem cell* in your own words.
6. Organs are groups of _____ working together to perform a specific task.

Link Vocabulary and Key Concepts

Concepts in Motion Interactive Concept Map

Copy this concept map, and then use vocabulary terms from the previous page and from the chapter to complete the concept map.

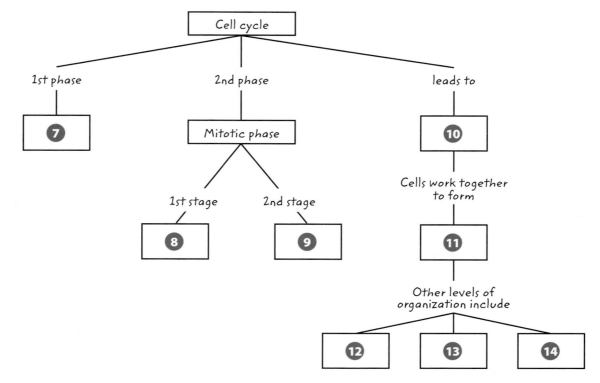

Understand Key Concepts 🔑

1 Chromosomes line up in the center of the cell during which phase?
A. anaphase
B. metaphase
C. prophase
D. telophase

2 Which stage of the cell cycle precedes cytokinesis?
A. G_1
B. G_2
C. interphase
D. mitosis

Use the figure below to answer questions 3 and 4.

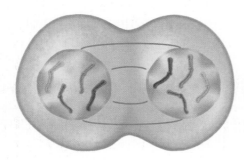

3 The figure represents which stage of mitosis?
A. anaphase
B. metaphase
C. prophase
D. telophase

4 What forms during this phase?
A. centromere
B. furrow
C. sister chromatid
D. two nuclei

5 What is the longest part of the cell cycle?
A. anaphase
B. cytokinesis
C. interphase
D. mitosis

6 A plant's root system is which level of organization?
A. cell
B. organ
C. organ system
D. tissue

7 Where is a meristem often found?
A. liver cells
B. muscle tissue
C. tip of plant root
D. unicellular organism

8 Which is NOT a type of human tissue?
A. connective
B. meristem
C. muscle
D. nervous

9 Which are unspecialized cells?
A. blood cells
B. muscle cells
C. nerve cells
D. stem cells

10 Which level of organization is shown in the figure below?
A. cell
B. organ
C. organ system
D. tissue

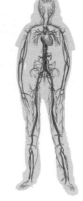

11 Which level of organization completes a series of tasks?
A. cell
B. organ
C. organ system
D. tissue

Critical Thinking

12 Sequence the events that occur during the phases of mitosis.

13 Infer why the chromatin condenses into chromosomes before mitosis begins.

14 Create Use the figure below to create a cartoon that shows a duplicated chromosome separating into two sister chromatids.

15 Classify a leaf as a tissue or an organ. Explain your choice.

16 Distinguish between a tissue and an organ.

17 Construct a table that lists and defines the different levels of organization.

18 Summarize the differences between unicellular organisms and multicellular organisms.

Writing in Science

19 Write a five-sentence paragraph describing a human organ system. Include a main idea, supporting details, and a concluding statement.

REVIEW THE BIG IDEA

20 Why is cell division important for multicellular organisms?

21 The photo below shows a chick growing inside an egg. An egg begins as one cell. How can one cell become a chick?

Math Skills

Review
— Math Practice —

Use Percentages

22 During an interphase lasting 23 hours, the S stage takes an average of 8.0 hours. What percentage of interphase is taken up by the S stage?

Use the following information to answer questions 23 through 25.

During a 23-hour interphase, the G_1 stage takes 11 hours and the S stage takes 8.0 hours.

23 What percentage of interphase is taken up by the G_1 and S stages?

24 What percentage of interphase is taken up by the G_2 phase?

25 How many hours does the G_2 phase last?

Standardized Test Practice

Record your answers on the answer sheet provided by your teacher or on a sheet of paper.

Multiple Choice

1 Which tissue carries messages to and from the brain?

   A connective

   B epithelial

   C muscle

   D nervous

Use the diagram below to answer question 2.

2 What is indicated by the arrow?

   A centromere

   B chromatid

   C chromosome

   D nucleus

3 In which stage of mitosis do spindle fibers form?

   A anaphase

   B metaphase

   C prophase

   D telophase

4 What structures separate during anaphase?

   A centromeres

   B chromatids

   C nuclei

   D organelles

Use the diagram below to answer question 5.

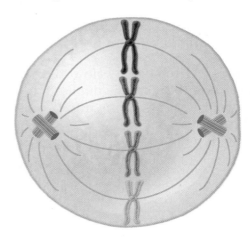

5 What stage of mitosis does the image above represent?

   A anaphase

   B metaphase

   C prophase

   D telophase

6 A plant's dermal tissue

   A produces food for the rest of the plant.

   B provides protection and helps reduce water loss.

   C takes in water and nutrients for use throughout the plant.

   D transports water and nutrients throughout the plant.

7 Which is the most accurate description of a leaf or your stomach?

   A a cell

   B an organ

   C an organ system

   D a tissue

Use the figure below to answer question 8.

8 Which does this figure illustrate?

A an organ

B an organism

C an organ system

D a tissue

9 If a cell has 30 chromosomes at the start of mitosis, how many chromosomes will be in each new daughter cell?

A 10

B 15

C 30

D 60

10 What areas of plants have unspecialized cells?

A flowers

B fruits

C leaves

D meristems

Constructed Response

Use the figure below to answer questions 11 and 12.

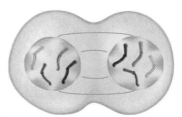

Figure A

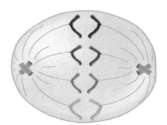

Figure B

11 The figures illustrate two phases of mitosis. Which occurs first: A or B? Explain your reasoning.

12 What stage of the mitotic phase follows those illustrated above? Explain how this stage differs between plant and animal cells.

13 What are some similarities and differences between the G_1 and S stages of interphase?

14 Are all human cells capable of mitosis and cell division? How does this affect the body's ability to repair itself? Support your answer with specific examples.

NEED EXTRA HELP?														
If You Missed Question...	1	2	3	4	5	6	7	8	9	10	11	12	13	14
Go to Lesson...	2	1	1	1	1	2	2	2	1	2	1	1	1	1

Reproduction of Organisms

Why do living things reproduce?

Inquiry **Time to bond?**

Have you ever seen a family of animals, such as the one of penguins shown here? Notice the baby penguin beside its parents. Like all living things, penguins reproduce.

- Do you think all living things have two parents?

- What might happen if the penguins did not reproduce?

- Why do living things reproduce?

Get Ready to Read

What do you think?

Before you read, decide if you agree or disagree with each of these statements. As you read this chapter, see if you change your mind about any of the statements.

1 Humans produce two types of cells: body cells and sex cells.

2 Environmental factors can cause variation among individuals.

3 Two parents always produce the best offspring.

4 Cloning produces identical individuals from one cell.

5 All organisms have two parents.

6 Asexual reproduction occurs only in microorganisms.

ConnectED Your one-stop online resource

connectED.mcgraw-hill.com

Video	WebQuest
Audio	Assessment
Review	Concepts in Motion
Inquiry	Multilingual eGlossary

Lesson 1

Sexual Reproduction and Meiosis

Reading Guide

Key Concepts 🔑
ESSENTIAL QUESTIONS

- What is sexual reproduction, and why is it beneficial?
- What is the order of the phases of meiosis, and what happens during each phase?
- Why is meiosis important?

Vocabulary

sexual reproduction p. 117

egg p. 117

sperm p. 117

fertilization p. 117

zygote p. 117

diploid p. 118

homologous chromosomes p. 118

haploid p. 119

meiosis p. 119

g Multilingual eGlossary

▣ Video BrainPOP®

Inquiry Modern Art?

This photo looks like a piece of modern art. It is actually an image of plant cells. The cells are dividing by a process that occurs during the production of sex cells.

Why do offspring look different?

Unless you're an identical twin, you probably don't look exactly like any siblings you might have. You might have differences in physical characteristics such as eye color, hair color, ear shape, or height. Why are there differences in the offspring from the same parents?

1 Read and complete a lab safety form.

2 Open the **paper bag** labeled *Male Parent,* and, without looking, remove three **beads.** Record the bead colors in your Science Journal, and replace the beads.

3 Open the **paper bag** labeled *Female Parent,* and remove three **beads.** Record the bead colors, and replace the beads.

4 Repeat steps 2 and 3 for each member of the group.

5 After each member has recorded his or her bead colors, study the results. Each combination of male and female beads represents an offspring.

Think About This

1. Compare your group's offspring to another group's offspring. What similarities or differences do you observe?

2. What caused any differences you observed? Explain.

3. 🔑 **Key Concept** Why might this type of reproduction be beneficial to an organism?

What is sexual reproduction?

Have you ever seen a litter of kittens? One kitten might have orange fur like its mother. A second kitten might have gray fur like its father. Still another kitten might look like a combination of both parents. How is this possible?

The kittens look different because of sexual reproduction. **Sexual reproduction** *is a type of reproduction in which the genetic materials from two different cells combine, producing an offspring.* The cells that combine are called sex cells. Sex cells form in reproductive organs. *The female sex cell, an* **egg,** *forms in an ovary. The male sex cell, a* **sperm,** *forms in a testis. During a process called* **fertilization** (fur tuh luh ZAY shun), *an egg cell and a sperm cell join together.* This produces a new cell. *The new cell that forms from fertilization is called a* **zygote.** As shown in **Figure 1,** the zygote develops into a new organism.

Review Personal Tutor

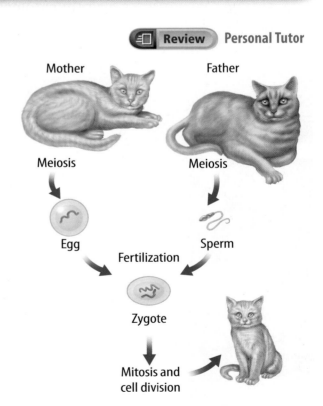

Figure 1 The zygote that forms during fertilization can become a multicellular organism.

Diploid Cells

Following fertilization, a zygote goes through mitosis and cell division. These processes produce nearly all the cells in a multicellular organism. Organisms that reproduce sexually form two kinds of cells—body cells and sex cells. In body cells of most organisms, similar chromosomes occur in pairs. **Diploid** *cells are cells that have pairs of chromosomes.*

Chromosomes

Pairs of chromosomes that have genes for the same traits arranged in the same order are called **homologous** (huh MAH luh gus) **chromosomes.** Because one chromosome is inherited from each parent, the chromosomes are not identical. For example, the kittens mentioned earlier in this lesson inherited a gene for orange fur color from their mother. They also inherited a gene for gray fur color from their father. So, some kittens might be orange, and some might be gray. Both genes for fur color are at the same place on homologous chromosomes, but they code for different colors.

Different organisms have different numbers of chromosomes. Recall that diploid cells have pairs of chromosomes. Notice in **Table 1** that human diploid cells have 23 pairs of chromosomes for a total of 46 chromosomes. A fruit fly diploid cell has 4 pairs of chromosomes, and a rice diploid cell has 12 pairs of chromosomes.

Table 1 An organism's chromosomes can be matched as pairs of chromosomes that have genes for the same traits.

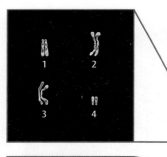

Concepts in Motion Interactive Table

Table 1 Chromosomes of Selected Organisms

Organism	Number of Chromosomes	Number of Homologous Pairs
Fruit fly	8	4
Rice	24	12
Yeast	32	16
Cat	38	19
Human	46	23
Dog	78	39
Fern	1,260	630

Having the correct number of chromosomes is very important. If a zygote has too many or too few chromosomes, it will not develop properly. For example, a genetic condition called Down syndrome occurs when a person has an extra copy of chromosome 21. A person with Down syndrome can have short stature, heart defects, or mental disabilities.

Haploid Cells

Organisms that reproduce sexually also form egg and sperm cells, or sex cells. Sex cells have only one chromosome from each pair of chromosomes. **Haploid** *cells are cells that have only one chromosome from each pair.* Organisms produce sex cells using a special type of cell division called meiosis. *In* **meiosis,** *one diploid cell divides and makes four haploid sex cells.* Meiosis occurs only during the formation of sex cells.

 Reading Check How do diploid cells differ from haploid cells?

The Phases of Meiosis

Next, you will read about the phases of meiosis. Many of the phases might seem familiar to you because they also occur during mitosis. Recall that mitosis and cytokinesis involve one division of the nucleus and the cytoplasm. Meiosis involves two divisions of the nucleus and the cytoplasm. These divisions are called meiosis I and meiosis II. They result in four haploid cells—cells with half the number of chromosomes as the original cell. As you read about meiosis, think about how it produces sex cells with a reduced number of chromosomes.

FOLDABLES

Make a shutter-fold book and label it as shown. Use it to describe and illustrate the phases of meiosis.

Inquiry MiniLab

20 minutes

How does one cell produce four cells?

When a diploid cell goes through meiosis, it produces four haploid cells. How does this happen?

1. Read and complete a lab safety form.

2. Make a copy of the diagram by tracing circles around a **jar lid** on your **paper.** Label as shown.

3. Use **chenille craft wires** to make red and blue duplicated chromosomes 2.5 cm long and green and yellow duplicated chromosomes 1.5 cm long. Recall that a duplicated chromosome has two sister chromatids connected at the centromere.

4. Place the chromosomes in the diploid cell.

5. Move one long chromosome and one short chromosome into each of the middle cells.

6. Separate the two strands of the chromosomes, and place one strand into each of the haploid cells.

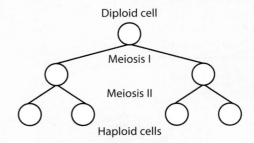

Analyze and Conclude

1. **Describe** What happened to the chromosomes during meiosis I? During meiosis II?

2. **Think Critically** Why are two haploid cells (sperm and egg) needed to form a zygote?

3. **Key Concept** How does one cell form four cells during meiosis?

Phases of Meiosis I

A reproductive cell goes through interphase before beginning meiosis I, which is shown in **Figure 2**. During interphase, the reproductive cell grows and copies, or duplicates, its chromosomes. Each duplicated chromosome consists of two sister chromatids joined together by a centromere.

1 **Prophase I** In the first phase of meiosis I, duplicated chromosomes condense and thicken. Homologous chromosomes come together and form pairs. The membrane surrounding the nucleus breaks apart, and the nucleolus disappears.

2 **Metaphase I** Homologous chromosome pairs line up along the middle of the cell. A spindle fiber attaches to each chromosome.

3 **Anaphase I** Chromosome pairs separate and are pulled toward the opposite ends of the cell. Notice that the sister chromatids stay together.

4 **Telophase I** A nuclear membrane forms around each group of duplicated chromosomes. The cytoplasm divides through cytokinesis and two daughter cells form. Sister chromatids remain together.

Meiosis 🔑   ((◎)) **Concepts in Motion**   **Animation**

Meiosis I

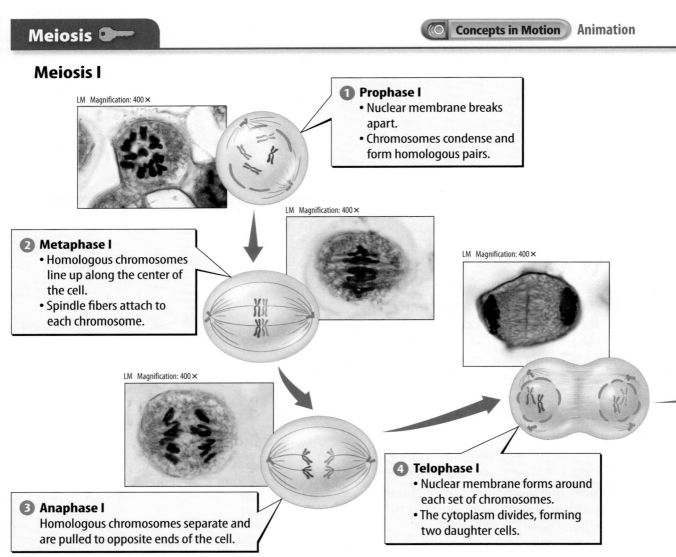

LM Magnification: 400×

1 **Prophase I**
• Nuclear membrane breaks apart.
• Chromosomes condense and form homologous pairs.

LM Magnification: 400×

2 **Metaphase I**
• Homologous chromosomes line up along the center of the cell.
• Spindle fibers attach to each chromosome.

LM Magnification: 400×

LM Magnification: 400×

3 **Anaphase I**
Homologous chromosomes separate and are pulled to opposite ends of the cell.

4 **Telophase I**
• Nuclear membrane forms around each set of chromosomes.
• The cytoplasm divides, forming two daughter cells.

Figure 2 Unlike mitosis, meiosis involves two divisions of the nucleus and the cytoplasm.

Phases of Meiosis II

After meiosis I, the two cells formed during this stage go through a second division of the nucleus and the cytoplasm. This process, shown in **Figure 2,** is called meiosis II.

5 **Prophase II** Chromosomes are not copied again before prophase II. They remain as condensed, thickened sister chromatids. The nuclear membrane breaks apart, and the nucleolus disappears in each cell.

6 **Metaphase II** The pairs of sister chromatids line up along the middle of the cell in single file.

7 **Anaphase II** The sister chromatids of each duplicated chromosome are pulled away from each other and move toward opposite ends of the cells.

8 **Telophase II** During the final phase of meiosis—telophase II—a nuclear membrane forms around each set of chromatids, which are again called chromosomes. The cytoplasm divides through cytokinesis, and four haploid cells form.

 Key Concept Check List the phases of meiosis in order.

Meiosis II

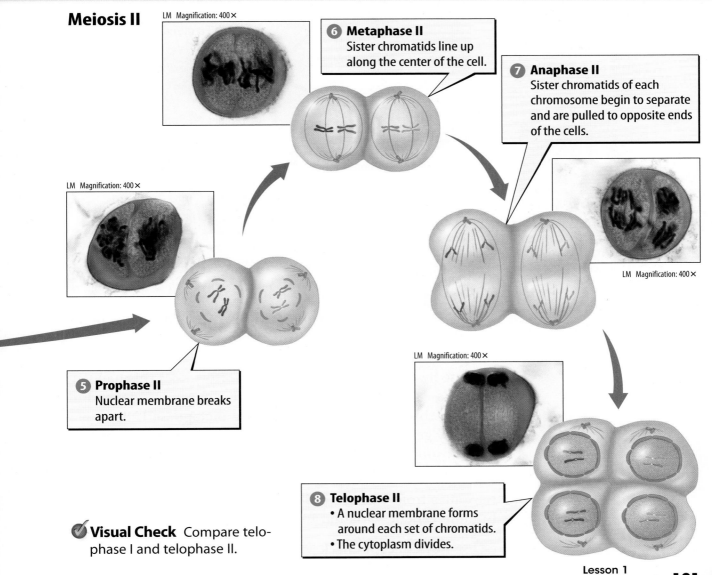

LM Magnification: 400×

6 **Metaphase II**
Sister chromatids line up along the center of the cell.

7 **Anaphase II**
Sister chromatids of each chromosome begin to separate and are pulled to opposite ends of the cells.

LM Magnification: 400×

LM Magnification: 400×

5 **Prophase II**
Nuclear membrane breaks apart.

LM Magnification: 400×

8 **Telophase II**
• A nuclear membrane forms around each set of chromatids.
• The cytoplasm divides.

Visual Check Compare telophase I and telophase II.

Why is meiosis important?

Meiosis forms sex cells with the correct haploid number of chromosomes. This maintains the correct diploid number of chromosomes in organisms when sex cells join. Meiosis also creates genetic variation by producing haploid cells.

Maintaining Diploid Cells

Recall that diploid cells have pairs of chromosomes. Meiosis helps to maintain diploid cells in offspring by making haploid sex cells. When haploid sex cells join together during fertilization, they make a diploid zygote, or fertilized egg. The zygote then divides by mitosis and cell division and creates a diploid organism. **Figure 3** illustrates how the diploid number is maintained in ducks.

Figure 3 Meiosis ensures that the chromosome number of a species stays the same from generation to generation.

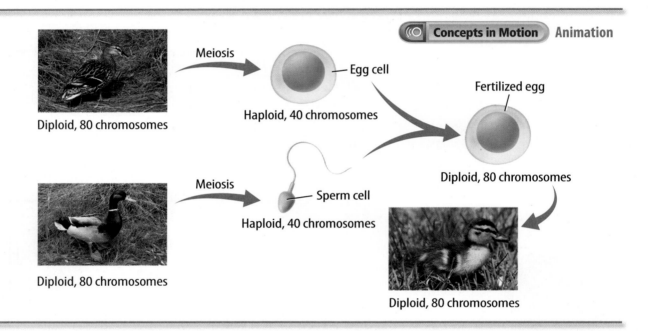

Concepts in Motion **Animation**

Meiosis

Egg cell

Diploid, 80 chromosomes

Haploid, 40 chromosomes

Fertilized egg

Meiosis

Sperm cell

Haploid, 40 chromosomes

Diploid, 80 chromosomes

Diploid, 80 chromosomes

Diploid, 80 chromosomes

Creating Haploid Cells

The result of meiosis is haploid sex cells. This helps maintain the correct number of chromosomes in each generation of offspring. The formation of haploid cells also is important because it allows for genetic variation. How does this happen? Sex cells can have different sets of chromosomes, depending on how chromosomes line up during metaphase I. Because a cell only gets one chromosome from each pair of homologous chromosomes, the resulting sex cells can be different.

The genetic makeup of offspring is a combination of chromosomes from two sex cells. Variation in the sex cells results in more genetic variation in the next generation.

Key Concept Check Why is meiosis important?

How do mitosis and meiosis differ?

Sometimes, it's hard to remember the differences between mitosis and meiosis. Use **Table 2** to review these processes.

During mitosis and cell division, a body cell and its nucleus divide once and produce two identical cells. These processes are important for growth and repair or replacement of damaged tissue. Some organisms reproduce by these processes. The two daughter cells produced by mitosis and cell division have the same genetic information.

During meiosis, a reproductive cell and its nucleus divide twice and produce four cells—two pairs of identical haploid cells. Each cell has half the number of chromosomes as the original cell. Meiosis happens in the reproductive organs of multicellular organisms. Meiosis forms sex cells used for sexual reproduction.

 Reading Check How many cells are produced during mitosis? During meiosis?

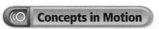 **Concepts in Motion** Interactive Table

Table 2 Comparison of Types of Cell Division

Characteristic	Meiosis	Mitosis and Cell Division
Number of chromosomes in parent cell	diploid	diploid
Type of parent cell	reproductive	body
Number of divisions of nucleus	2	1
Number of daughter cells produced	4	2
Chromosome number in daughter cells	haploid	diploid
Function	forms sperm and egg cells	growth, cell repair, some types of reproduction

Math Skills

Use Proportions

An equation that shows that two ratios are equivalent is a proportion. The ratios $\frac{1}{2}$ and $\frac{3}{6}$ are equivalent, so they can be written as $\frac{1}{2} = \frac{3}{6}$.

You can use proportions to figure out how many daughter cells will be produced during mitosis. If you know that one cell produces two daughter cells at the end of mitosis, you can use proportions to calculate how many daughter cells will be produced by eight cells undergoing mitosis.

Set up an equation of the two ratios. $\quad \frac{1}{2} = \frac{8}{y}$

Cross-multiply. $\quad 1 \times y = 8 \times 2$

$$1y = 16$$

Divide each side by 1. $\quad y = 16$

Practice
You know that one cell produces four daughter cells at the end of meiosis. How many daughter cells would be produced if eight sex cells undergo meiosis?

 Review
- **Math Practice**
- **Personal Tutor**

Advantages of Sexual Reproduction

Did you ever wonder why a brother and a sister might not look alike? The answer is sexual reproduction. The main advantage of sexual reproduction is that offspring inherit half their **DNA** from each parent. Offspring are not likely to inherit the same DNA from the same parents. Different DNA means that each offspring has a different set of traits. This results in genetic variation among the offspring.

 Key Concept Check Why is sexual reproduction beneficial?

Genetic Variation

As you just read, genetic variation exists among humans. You can look at your friends to see genetic variation. Genetic variation occurs in all organisms that reproduce sexually. Consider the plants shown in **Figure 4.** The plants are members of the same species, but they have different traits, such as the ability to resist disease.

Due to genetic variation, individuals within a population have slight differences. These differences might be an advantage if the environment changes. Some individuals might have traits that enable them to survive unusually harsh conditions such as a drought or severe cold. Other individuals might have traits that make them resistant to disease.

Genetic Variation 🔑

Disease-resistant cassava leaves

Cassava leaves with cassava mosaic disease

Figure 4 These plants belong to the same species. However, one is more disease-resistant than the other.

Visual Check How does cassava mosaic disease affect cassava leaves?

Selective Breeding

Did you know that broccoli, kohlrabi, kale, and cabbage all descended from one type of mustard plant? It's true. More than 2,000 years ago farmers noticed that some mustard plants had different traits, such as larger leaves or bigger flower buds. The farmers started to choose which traits they wanted by selecting certain plants to reproduce and grow. For example, some farmers chose only the plants with the biggest flowers and stems and planted their seeds. Over time, the offspring of these plants became what we know today as broccoli, shown in **Figure 5.** This process is called selective breeding. Selective breeding has been used to develop many types of plants and animals with desirable traits. It is another example of the benefits of sexual reproduction.

Figure 5 The wild mustard is the common ancestor to all these plants.

Selective Breeding 🔑

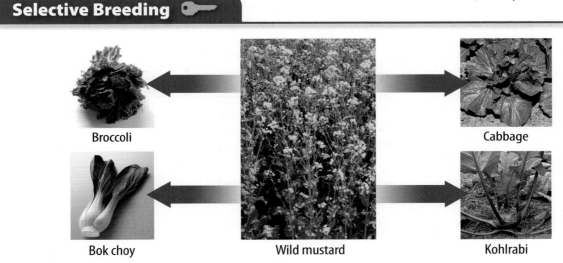

Broccoli

Bok choy

Wild mustard

Cabbage

Kohlrabi

Disadvantages of Sexual Reproduction

Although sexual reproduction produces more genetic variation, it does have some disadvantages. Sexual reproduction takes time and energy. Organisms have to grow and develop until they are mature enough to produce sex cells. Then the organisms have to form sex cells—either eggs or sperm. Before they can reproduce, organisms usually have to find mates. Searching for a mate can take a long time and requires energy. The search for a mate might also expose individuals to predators, diseases, or harsh environmental conditions. In addition, sexual reproduction is limited by certain factors. For example, fertilization cannot take place during pregnancy, which can last as long as two years in some mammals.

✓ **Reading Check** What are the disadvantages of sexual reproduction?

Visual Summary

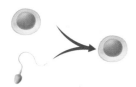

Fertilization occurs when an egg cell and a sperm cell join together.

Organisms produce sex cells through meiosis.

Sexual reproduction results in genetic variation among individuals.

FOLDABLES

Use your lesson Foldable to review the lesson. Save your Foldable for the project at the end of the chapter.

What do you think NOW?

You first read the statements below at the beginning of the chapter.

1. Humans produce two types of cells: body cells and sex cells.

2. Environmental factors can cause variation among individuals.

3. Two parents always produce the best offspring.

Did you change your mind about whether you agree or disagree with the statements? Rewrite any false statements to make them true.

Use Vocabulary

1 **Use the terms** *egg, sperm,* and *zygote* in a sentence.

2 **Distinguish** between haploid and diploid.

3 **Define** *homologous chromosomes* in your own words.

Understand Key Concepts

4 **Define** sexual reproduction.

5 **Draw and label** the phases of meiosis.

6 Homologous chromosomes separate during which phase of meiosis?
  A. anaphase I   C. metaphase I
  B. anaphase II   D. metaphase II

Interpret Graphics

7 **Organize** Copy and fill in the graphic organizer below to sequence the phases of meiosis I and meiosis II.

Meiosis I ☐ → ☐ → ☐ → ☐

Meiosis II ☐ → ☐ → ☐ → ☐

Critical Thinking

8 **Analyze** Why is the result of the stage of meiosis shown below an advantage for organisms that reproduce sexually?

Math Skills ✕ ÷

▭ **Review**
── Math Practice ──

9 If 15 cells undergo meiosis, how many daughter cells would be produced?

10 If each daughter cell from question 9 undergoes meiosis, how many total daughter cells will there be?

The Spider
Mating Dance

Meet Norman Platnick, a scientist studying spiders.

Norman Platnick is fascinated by all spider species—from the dwarf tarantula-like spiders of Panama to the blind spiders of New Zealand. These are just two of the over 1,400 species he's discovered worldwide.

How does Platnick identify new species? One way is the pedipalps. Every spider has two pedipalps, but they vary in shape and size among the over 40,000 species. Pedipalps look like legs but function more like antennae and mouthparts. Male spiders use their pedipalps to aid in reproduction.

Getting Ready When a male spider is ready to mate, he places a drop of sperm onto a sheet of silk he constructs. Then he dips his pedipalps into the drop to draw up the sperm.

Finding a Mate The male finds a female of the same species by touch or by sensing certain chemicals she releases.

Courting and Mating Males of some species court a female with a special dance. For other species, a male might present a female with a gift, such as a fly wrapped in silk. During mating, the male uses his pedipalps to transfer sperm to the female.

What happens to the male after mating? That depends on the species. Some are eaten by the female, while others move on to find new mates.

▲ Spiders reproduce sexually, so each offspring has a unique combination of genes from its parents. Over many generations, this genetic variation has led to the incredible diversity of spiders in the world today.

◄ Norman Platnick is an arachnologist (uh rak NAH luh just) at the American Museum of Natural History. Arachnologists are scientists who study spiders.

It's Your Turn

RESEARCH Select a species of spider and research its mating rituals. What does a male do to court a female? What is the role of the female? What happens to the spiderlings after they hatch? Use images to illustrate a report on your research.

Asexual Reproduction

Reading Guide

Key Concepts 🗝
ESSENTIAL QUESTIONS

- What is asexual reproduction, and why is it beneficial?
- How do the types of asexual reproduction differ?

Vocabulary

asexual reproduction p. 129

fission p. 130

budding p. 131

regeneration p. 132

vegetative reproduction p. 133

cloning p. 134

g Multilingual eGlossary

Inquiry Plants on Plants?

Look closely at the edges of this plant's leaves. Tiny plants are growing there. This type of plant can reproduce without meiosis and fertilization.

How do yeast reproduce?

Some organisms can produce offspring without meiosis or fertilization. You can observe this process when you add sugar and warm water to dried yeast.

1. Read and complete a lab safety form.

2. Pour 125 mL of water into a **beaker.** The water should be at a temperature of 34°C.

3. Add 5 g of **sugar** and 5 g of **yeast** to the water. Stir slightly. Record your observations after 5 minutes in your Science Journal.

4. Using a **dropper,** put a drop of the yeast solution on a **microscope slide.** Place a **coverslip** over the drop.

5. View the yeast solution under a **microscope.** Draw what you see in your Science Journal.

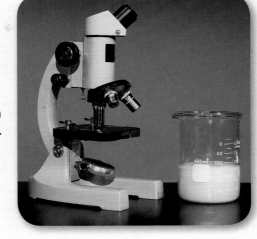

Think About This

1. What evidence did you observe that yeast reproduce?

2. 🔑 **Key Concept** How do you think this process differs from sexual reproduction?

What is asexual reproduction?

Lunch is over and you are in a rush to get to class. You wrap up your half-eaten sandwich and toss it into your locker. A week goes by before you spot the sandwich in the corner of your locker. The surface of the bread is now covered with fuzzy mold—not very appetizing. How did that happen?

The mold on the sandwich is a type of fungus (FUN gus). A fungus releases enzymes that break down organic matter, such as food. It has structures that penetrate and anchor to food, much like roots anchor plants to soil. A fungus can multiply quickly in part because generally a fungus can reproduce either sexually or asexually. Recall that sexual reproduction involves two parent organisms and the processes of meiosis and fertilization. Offspring inherit half their DNA from each parent, resulting in genetic variation among the offspring.

In **asexual reproduction,** *one parent organism produces offspring without meiosis and fertilization.* Because the offspring inherit all their DNA from one parent, they are genetically identical to each other and to their parent.

🔑 **Key Concept Check** Describe asexual reproduction in your own words.

FOLDABLES®

Fold a sheet of paper into a six-celled chart. Label the front "Asexual Reproduction," and label the chart inside as shown. Use it to compare types of asexual reproduction.

Fission	Mitotic cell division	Budding
Animal regeneration	Vegetative reproduction	Cloning

Types of Asexual Reproduction

There are many different types of organisms that reproduce by asexual reproduction. In addition to fungi, bacteria, protists, plants, and animals can reproduce asexually. In this lesson, you will learn how organisms reproduce asexually.

Fission

Recall that prokaryotes have a simpler cell structure than eukaryotes. A prokaryote's DNA is not contained in a nucleus. For this reason, mitosis does not occur and cell division in a prokaryote is a simpler process than in a eukaryote. *Cell division in prokaryotes that forms two genetically identical cells is known as* **fission.**

Fission begins when a prokaryote's DNA molecule is copied. Each copy attaches to the cell membrane. Then the cell begins to grow longer, pulling the two copies of DNA apart. At the same time, the cell membrane begins to pinch inward along the middle of the cell. Finally the cell splits and forms two new identical offspring. The original cell no longer exists.

As shown in **Figure 6,** *E. coli,* a common bacterium, divides through fission. Some bacteria can divide every 20 minutes. At that rate, 512 bacteria can be produced from one original bacterium in about three hours.

Reading Check What advantage might asexual reproduction by fission have over sexual reproduction?

WORD ORIGIN · · · · · · · · · · · ·

fission
from Latin *fissionem*, means "a breaking up, cleaving"

Fission

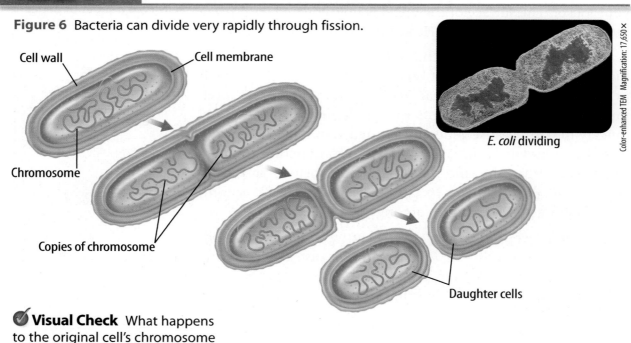

Figure 6 Bacteria can divide very rapidly through fission.

Cell wall

Cell membrane

Chromosome

Copies of chromosome

Daughter cells

E. coli dividing

Color-enhanced TEM Magnification: 17,650×

Visual Check What happens to the original cell's chromosome during fission?

Mitotic Cell Division

Many unicellular eukaryotes reproduce by mitotic cell division. In this type of asexual reproduction, an organism forms two offspring through mitosis and cell division. In **Figure 7,** an amoeba's nucleus has divided by mitosis. Next, the cytoplasm and its contents divide through cytokinesis and two new amoebas form.

Budding

In **budding,** *a new organism grows by mitosis and cell division on the body of its parent.* The bud, or offspring, is genetically identical to its parent. When the bud becomes large enough, it can break from the parent and live on its own. In some cases, an offspring remains attached to its parent and starts to form a colony. **Figure 8** shows a hydra in the process of budding. The hydra is an example of a multicellular organism that can reproduce asexually. Unicellular eukaryotes, such as yeast, can also reproduce through budding, as you saw in the Launch Lab.

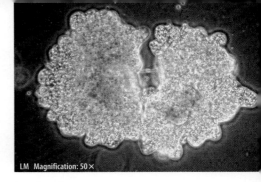
LM Magnification: 50×

▲ **Figure 7** During mitotic cell division, an amoeba divides its chromosomes and cell contents evenly between the daughter cells.

Budding 🔑

Figure 8 The hydra bud has the same genetic makeup as its parent.

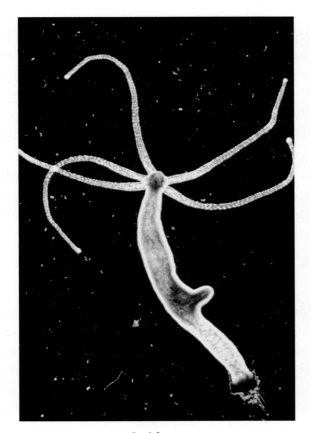

Bud forms.

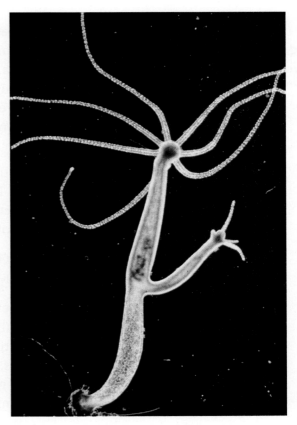

Bud develops a mouth and tentacles.

Figure 9 A planarian can reproduce through regeneration.

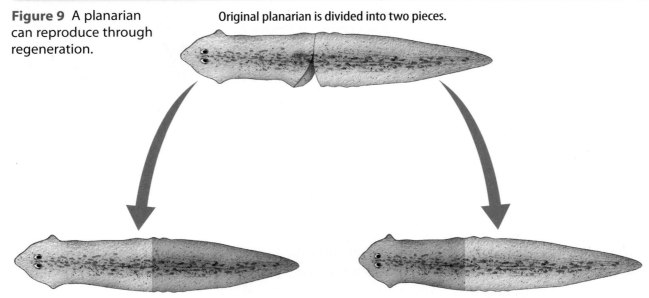

Original planarian is divided into two pieces.

The head end regenerates a new tail.

The tail end regenerates a new head.

Animal Regeneration

Another type of asexual reproduction, **regeneration,** *occurs when an offspring grows from a piece of its parent.* The ability to regenerate a new organism varies greatly among animals.

Producing New Organisms Some sea stars have five arms. If separated from the parent sea star, each arm has the **potential** to grow into a new organism. To regenerate a new sea star, the arm must contain a part of the central disk of the parent. If conditions are right, one five-armed sea star can produce as many as five new organisms.

Sea urchins, sea cucumbers, sponges, and planarians, such as the one shown in **Figure 9,** can also reproduce through regeneration. Notice that each piece of the original planarian becomes a new organism. As with all types of asexual reproduction, the offspring is genetically identical to the parent.

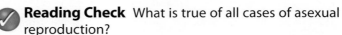 **Reading Check** What is true of all cases of asexual reproduction?

Producing New Parts When you hear the term *regeneration,* you might think about a salamander regrowing a lost tail or leg. Regeneration of damaged or lost body parts is common in many animals. Newts, tadpoles, crabs, hydra, and zebra fish are all able to regenerate body parts. Even humans are able to regenerate some damaged body parts, such as the skin and the liver. This type of regeneration, however, is not considered asexual reproduction. It does not produce a new organism.

ACADEMIC VOCABULARY

potential
(*noun*) possibility

Vegetative Reproduction

Plants can also reproduce asexually in a process similar to regeneration. **Vegetative reproduction** *is a form of asexual reproduction in which offspring grow from a part of a parent plant.* For example, the strawberry plants shown in **Figure 10** send out long horizontal stems called stolons. Wherever a stolon touches the ground, it can produce roots. Once the stolons have grown roots, a new plant can grow—even if the stolons have broken off the parent plant. Each new plant grown from a stolon is genetically identical to the parent plant.

Vegetative reproduction usually involves structures such as the roots, the stems, and the leaves of plants. In addition to strawberries, many other plants can reproduce by this method, including raspberries, potatoes, and geraniums.

Figure 10 The smaller plants were grown from stolons produced by the parent plant.

Visual Check Which plants in the figure are the parent plants?

Inquiry MiniLab

15 minutes

What parts of plants can grow?

You probably know that plants can grow from seeds. But you might be surprised to learn that other parts of plants can grow and produce a new plant.

1. Carefully examine the photos of vegetative reproduction.

2. Create a data chart in your Science Journal to record your observations. Identify which part of the plant (leaf, stem, etc.) would be used to grow a new plant.

Analyze and Conclude

1. **Explain** How is the vegetative reproduction you observed a kind of asexual reproduction?

2. **Infer** how farmers or gardeners might use vegetative reproduction.

3. **Key Concept** Describe a method you might use to produce a new plant using vegetative reproduction.

Cloning

Fission, budding, and regeneration are all types of asexual reproduction that can produce genetically identical offspring in nature. In the past, the term *cloning* described any process that produced genetically identical offspring. Today, however, the word usually refers to a technique developed by scientists and performed in laboratories. **Cloning** *is a type of asexual reproduction performed in a laboratory that produces identical individuals from a cell or from a cluster of cells taken from a multicellular organism.* Farmers and scientists often use cloning to make copies of organisms or cells that have desirable traits, such as large flowers.

Plant Cloning Some plants can be cloned using a method called tissue **culture,** as shown in **Figure 11.** Tissue culture enables plant growers and scientists to make many copies of a plant with desirable traits, such as sweet fruit. Also, a greater number of plants can be produced more quickly than by vegetative reproduction.

Tissue culture also enables plant growers to reproduce plants that might have become infected with a disease. To clone such a plant, a scientist can use cells from a part of a plant where they are rapidly undergoing mitosis and cell division. This part of a plant is called a meristem. Cells in meristems are disease-free. Therefore, if a plant becomes infected with a disease, it can be cloned using meristem cells.

SCIENCE USE V. COMMON USE

culture
Science Use the process of growing living tissue in a laboratory

Common Use the social customs of a group of people

Figure 11 New carrot plants can be produced from cells of a carrot root using tissue culture techniques.

Plant Cloning 🔑

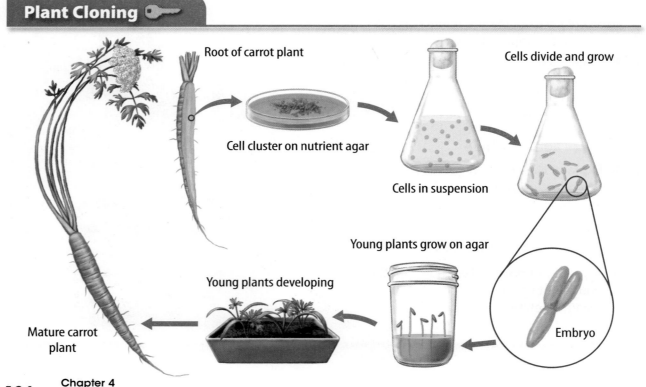

Root of carrot plant

Cells divide and grow

Cell cluster on nutrient agar

Cells in suspension

Young plants grow on agar

Young plants developing

Embryo

Mature carrot plant

Animal Cloning In addition to cloning plants, scientists have been able to clone many animals. Because all of a clone's chromosomes come from one parent (the donor of the nucleus), the clone is a genetic copy of its parent. The first mammal cloned was a sheep named Dolly. **Figure 12** illustrates how this was done.

Scientists are currently working to save some endangered species from extinction by cloning. Although cloning is an exciting advancement in science, some people are concerned about the high cost and the ethics of this technique. Ethical issues include the possibility of human cloning. You might be asked to consider issues like this during your lifetime.

 Key Concept Check Compare and contrast the different types of asexual reproduction.

Figure 12 Scientists used two different sheep to produce the cloned sheep known as Dolly.

Animal Cloning 🔑

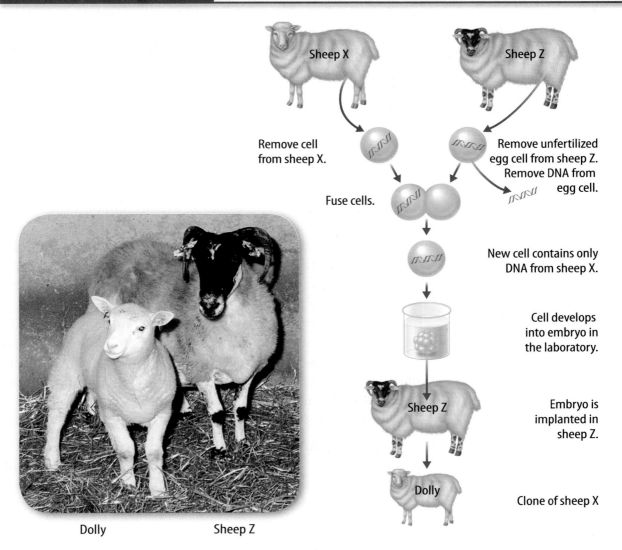

Sheep X

Sheep Z

Remove cell from sheep X.

Remove unfertilized egg cell from sheep Z. Remove DNA from egg cell.

Fuse cells.

New cell contains only DNA from sheep X.

Cell develops into embryo in the laboratory.

Sheep Z

Embryo is implanted in sheep Z.

Dolly

Clone of sheep X

Dolly   Sheep Z

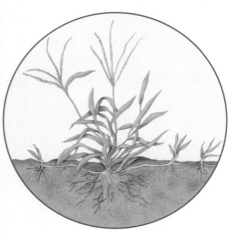

Figure 13 Crabgrass can spread quickly because it reproduces asexually.

Advantages of Asexual Reproduction

What are the advantages to organisms of reproducing asexually? Asexual reproduction enables organisms to reproduce without a mate. Recall that searching for a mate takes time and energy. Asexual reproduction also enables some organisms to rapidly produce a large number of offspring. For example, the crabgrass shown in **Figure 13** reproduces asexually by underground stems called stolons. This enables one plant to spread and colonize an area in a short period of time.

 Key Concept Check How is asexual reproduction beneficial?

Disadvantages of Asexual Reproduction

Although asexual reproduction usually enables organisms to reproduce quickly, it does have some disadvantages. Asexual reproduction produces offspring that are genetically identical to their parent. This results in little genetic variation within a population. Why is genetic variation important? Recall from Lesson 1 that genetic variation can give organisms a better chance of surviving if the environment changes. Think of the crabgrass. Imagine that all the crabgrass plants in a lawn are genetically identical to their parent plant. If a certain weed killer can kill the parent plant, then it can kill all the crabgrass plants in the lawn. This might be good for your lawn, but it is a disadvantage for the crabgrass.

Another disadvantage of asexual reproduction involves genetic changes, called mutations, that can occur. If an organism has a harmful mutation in its cells, the mutation will be passed to asexually reproduced offspring. This could affect the offspring's ability to survive.

Lesson 2 Review

Visual Summary

In asexual reproduction, offspring are produced without meiosis and fertilization.

Cloning is one type of asexual reproduction.

Asexual reproduction enables organisms to reproduce quickly.

FOLDABLES®

Use your lesson Foldable to review the lesson. Save your Foldable for the project at the end of the chapter.

What do you think NOW?

You first read the statements below at the beginning of the chapter.

4. Cloning produces identical individuals from one cell.

5. All organisms have two parents.

6. Asexual reproduction occurs only in microorganisms.

Did you change your mind about whether you agree or disagree with the statements? Rewrite any false statements to make them true.

Use Vocabulary

1 In _____ _____, only one parent organism produces offspring.

2 **Define** the term *cloning* in your own words.

3 **Use the term** *regeneration* in a sentence.

Understand Key Concepts 🔑

4 **State** two reasons why asexual reproduction is beneficial.

5 Which is an example of asexual reproduction by regeneration?
  A. cloning sheep
  B. lizard regrowing a tail
  C. sea star arm producing a new organism
  D. strawberry plant producing stolons

6 **Construct** a chart that includes an example of each type of asexual reproduction.

Interpret Graphics

7 **Examine** the diagram below and write a short paragraph describing the process of tissue culture.

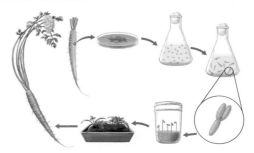

8 **Organize** Copy and fill in the graphic organizer below to list the different types of asexual reproduction that occur in multicellular organisms.

Asexual reproduction

Critical Thinking

9 **Justify** the use of cloning to save endangered animals.

Materials

pool noodles

Safety

Mitosis and Meiosis

During cellular reproduction, many changes occur in the nucleus of cells involving the chromosomes. You could think about these changes as a set of choreographed moves like you would see in a dance. In this lab you will act out the moves that chromosomes make during mitosis and meiosis in order to understand the steps that occur when cells reproduce.

Ask a Question

How do chromosomes change and move during mitosis and meiosis?

Make Observations

1. Read and complete a lab safety form.

2. Form a cell nucleus with four chromosomes represented by students holding four different colors of pool noodles. Other students play the part of the nuclear membrane and form a circle around the chromosomes.

3. The chromosomes duplicate during interphase. Each chromosome is copied, creating a chromosome with two sister chromatids.

4. Perform mitosis.

   a. During prophase, the nuclear membrane breaks apart, and the nucleolus disappears.

   b. In metaphase, duplicated chromosomes align in the middle of the cell.

   c. The sister chromatids separate in anaphase.

   d. In telophase, the nuclear membrane reforms around two daughter cells.

5. Repeat steps 2 and 3. Perform meiosis.

   a. In prophase I, the nuclear membrane breaks apart, the nucleolus disappears, and homologous chromosomes pair up.

   b. In metaphase I, homologous chromosomes line up along the center of the cell.

   c. During anaphase I, the pairs of homologous chromosomes separate.

   d. In telophase I, the nuclear membrane reforms.

   e. Each daughter cell now performs meiosis II independently. In prophase II, the nuclear membrane breaks down, and the nucleolus disappears.

   f. During metaphase II, duplicated chromosomes align in the middle of the cell.

g. Sister chromatids separate in anaphase II.

h. In telophase II, the nuclear membrane reforms.

Form a Hypothesis

6 Use your observations to form a hypothesis about the results of an error in meiosis. For example, you might explain the results of an error during anaphase I.

Test your Hypothesis

7 Perform meiosis, incorporating the error you chose in step 6.

8 Compare the outcome to your hypothesis. Does your data support your hypothesis? If not, revise your hypothesis and repeat steps 6–8.

Analyze and Conclude

9 **Compare and Contrast** How are mitosis and meiosis I similar? How are they different?

10 **The Big Idea** What is the difference between the chromosomes in cells at the beginning and the end of mitosis? At the beginning and end of meiosis?

11 **Critique** How did performing cellular replications using pool noodles help you understand mitosis and meiosis?

Communicate Your Results

Create a chart of the changes and movements of chromosomes in each of the steps in meiosis and mitosis. Include colored drawings of chromosomes and remember to draw the cell membranes.

 **Extension**

Investigate some abnormalities that occur when mistakes are made during mitosis or meiosis. Draw a chart of the steps of reproduction showing how the mistake is made. Write a short description of the problems that result from the mistake.

5

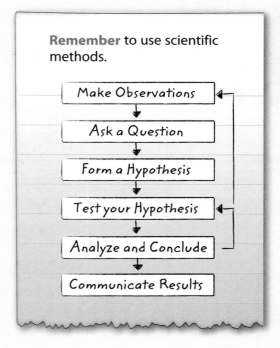

Lab Tips

☑ Figure out where the boundaries of your cell are before you start.

☑ Review the phases of mitosis and meiosis before beginning to act out how the chromosomes move during each process.

Remember to use scientific methods.

> Make Observations
>   ↓
> Ask a Question
>   ↓
> Form a Hypothesis
>   ↓
> Test your Hypothesis
>   ↓
> Analyze and Conclude
>   ↓
> Communicate Results

Chapter 4 Study Guide

Reproduction ensures the survival of species.

Key Concepts Summary 🔑

Lesson 1: Sexual Reproduction and Meiosis

- **Sexual reproduction** is the production of an offspring from the joining of a **sperm** and an **egg.**

- Division of the nucleus and cytokinesis happens twice in **meiosis.** Meiosis I separates homologous chromosomes. Meiosis II separates sister chromatids.

- Meiosis maintains the chromosome number of a species from one generation to the next.

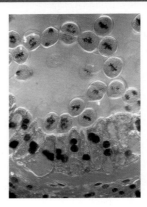

Lesson 2: Asexual Reproduction

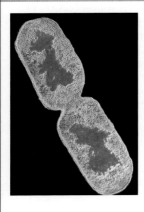

- **Asexual reproduction** is the production of offspring by one parent, which results in offspring that are genetically identical to the parent.

- Types of asexual reproduction include **fission,** mitotic cell division, **budding, regeneration, vegetative reproduction,** and **cloning.**

- Asexual reproduction can produce a large number of offspring in a short amount of time.

Vocabulary

sexual reproduction p. 117

egg p. 117

sperm p. 117

fertilization p. 117

zygote p. 117

diploid p. 118

homologous chromosomes p. 118

haploid p. 119

meiosis p. 119

asexual reproduction p. 129

fission p. 130

budding p. 131

regeneration p. 132

vegetative reproduction p. 133

cloning p. 134

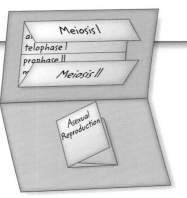
FOLDABLES® Chapter Project

Assemble your lesson Foldables as shown to make a Chapter Project. Use the project to review what you have learned in this chapter.

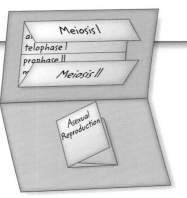

Use Vocabulary

1. Define meiosis in your own words.

2. Distinguish between an egg and a zygote.

3. Use the vocabulary words *haploid* and *diploid* in a sentence.

4. Cell division in prokaryotes is called _____.

5. Define the term *vegetative reproduction* in your own words.

6. Distinguish between regeneration and budding.

7. A type of reproduction in which the genetic materials from two different cells combine, producing an offspring, is called _____ _____.

Link Vocabulary and Key Concepts

Concepts in Motion   Interactive Concept Map

Copy this concept map, and then use vocabulary terms from the previous page to complete the concept map.

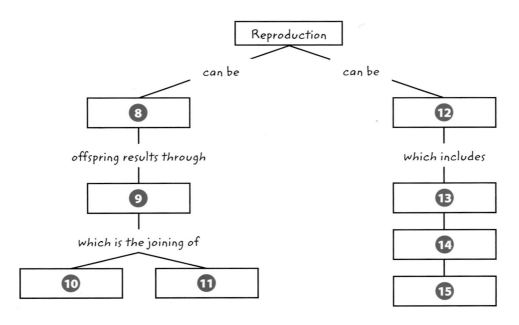

Understand Key Concepts

1 Which is an advantage of sexual reproduction?

A. Offspring are identical to the parents.
B. Offspring with genetic variation are produced.
C. Organisms don't have to search for a mate.
D. Reproduction is rapid.

2 Which describes cells that have only one copy of each chromosome?

A. diploid
B. haploid
C. homologous
D. zygote

Use the figure below to answer questions 3 and 4.

3 Which phase of meiosis I is shown in the diagram?

A. anaphase I
B. metaphase I
C. prophase I
D. telophase I

4 Which phase of meiosis I comes after the phase in the diagram?

A. anaphase I
B. metaphase I
C. prophase I
D. telophase I

5 Tissue culture is an example of which type of reproduction?

A. budding
B. cloning
C. fission
D. regeneration

6 Which type of asexual reproduction is shown in the figure below?

A. budding
B. cloning
C. fission
D. regeneration

7 A bacterium can reproduce by which method?

A. budding
B. cloning
C. fission
D. regeneration

8 Which statement best describes why genetic variation is beneficial to populations of organisms?

A. Individuals look different from one another.
B. Only one parent is needed to produce offspring.
C. Populations of the organism increase more rapidly.
D. Species can better survive environmental changes.

9 In which phase of meiosis II do sister chromatids line up along the center of the cell?

A. anaphase II
B. metaphase II
C. prophase II
D. telophase II

Critical Thinking

10 **Contrast** haploid cells and diploid cells.

11 **Model** Make a model of homologous chromosomes using materials of your choice.

12 **Form a hypothesis** about the effect of a mistake in separating homologous chromosomes during meiosis.

13 **Analyze** Crabgrass reproduces asexually by vegetative reproduction. Use the figure below to explain why this form of reproduction is an advantage for the crabgrass.

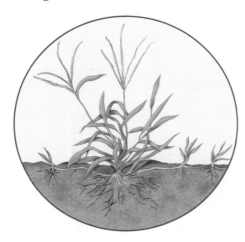

14 **Compare** budding and cloning.

15 **Create** a table showing the advantages and disadvantages of asexual reproduction.

16 **Compare and contrast** sexual reproduction and asexual reproduction.

Writing in Science

17 **Create** a plot for a short story that describes an environmental change and the importance of genetic variation in helping a species survive that change. Include characters, a setting, a climax, and an ending for your plot.

REVIEW THE B**I**G IDEA

18 Think of all the advantages of sexual and asexual reproduction. Use these ideas to summarize why organisms reproduce.

19 The baby penguin below has a mother and a father. Do all living things have two parents? Explain.

Math Skills

 Review
— **Math Practice** —

Use Proportions

20 During mitosis, the original cell produces two daughter cells. How many daughter cells will be produced if 250 mouse cells undergo mitosis?

21 During meiosis, the original reproductive cell produces four daughter cells. How many daughter cells will be produced if 250 mouse reproductive cells undergo meiosis?

22 Two reproductive cells undergo meiosis. Each daughter cell also undergoes meiosis. How many cells are produced when the daughter cells divide?

Record your answers on the answer sheet provided by your teacher or on a sheet of paper.

Multiple Choice

1 How do sea stars reproduce?

    A cloning

    B fission

    C animal regeneration

    D vegetative reproduction

Use the diagram below to answer questions 2 and 3.

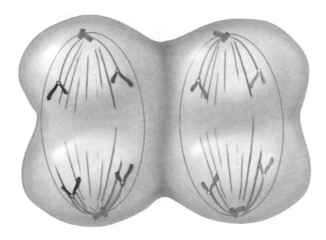

2 What stage of meiosis does the drawing illustrate?

    A anaphase I

    B anaphase II

    C prophase I

    D prophase II

3 Which stage takes place *before* the one in the diagram above?

    A metaphase I

    B metaphase II

    C telophase I

    D telophase II

4 What type of asexual reproduction includes stolons?

    A budding

    B cloning

    C animal regeneration

    D vegetative reproduction

Use the table below to answer question 5.

Comparison of Types of Cell Division		
Characteristic	**Meiosis**	**Mitosis**
Number of divisions of nucleus	2	A
Number of daughter cells produced	B	2

5 Which numbers should be inserted for A and B in the chart?

    A A=1 and B=2

    B A=1 and B=4

    C A=2 and B=2

    D A=2 and B=4

6 Which results in genetic variation?

    A cloning

    B fission

    C sexual reproduction

    D vegetative reproduction

7 Which is NOT true of homologous chromosomes?

    A The are identical.

    B They are in pairs.

    C They have genes for the same traits.

    D They have genes that are in the same order.

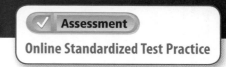
Use the figure below to answer question 8.

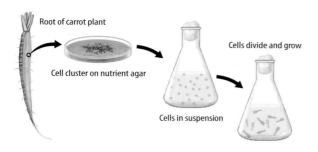

Root of carrot plant

Cell cluster on nutrient agar

Cells in suspension

Cells divide and grow

8 The figure illustrates the first four steps of which reproductive process?

A animal cloning

B regeneration

C tissue culture

D vegetative reproduction

9 If 12 reproductive cells undergo meiosis, how many daughter cells will result?

A 12

B 24

C 48

D 60

10 Which is NOT true of asexual reproduction?

A Many offspring can be produced rapidly.

B Offspring are different from the parents.

C Offspring have no genetic variation.

D Organisms can reproduce without a mate.

Constructed Response

Use the figure below to answer questions 11 and 12.

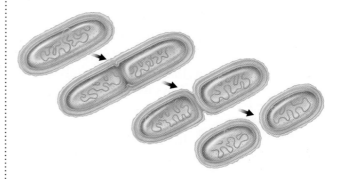

11 Identify the type of asexual reproduction shown in the figure above. How does it differ from sexual reproduction?

12 Compare and contrast budding with the type of asexual reproduction shown in the figure above.

13 What are some differences between the results of selectively breeding plants and cloning them?

14 Use the example of the wild mustard plant to describe the benefits of selective breeding.

15 What are the advantages and disadvantages of cloning animals?

NEED EXTRA HELP?															
If You Missed Question...	1	2	3	4	5	6	7	8	9	10	11	12	13	14	15
Go to Lesson...	2	1	1	2	1	1	1	2	1	2	1,2	2	1,2	1	2

Unit 2

LIFE:
Changes & Interactions

You know, kid, things have changed around here since the days of your great-great-great-great-great-great-great-great-great-great-grandpa Joe. Back then, most of the butterflies in this neighborhood were multicolored.

In those days, there were only a handful of us bark–colored butterflies. Our kind were considered rejects.

Nobody ever gave us a second look.

WHOAH!

1875   1900   1925

1859
Charles Darwin publishes *On the Origin of Species,* in which he explains his theory of natural selection.

1865
Gregor Mendel traces inheritance patterns of certain traits in pea plants and publishes *Experiments on Plant Hybridization.*

1905
Nettie Stevens and Edmund Wilson independently describe the XY sex-determination system, in which males have XY chromosomes and females have XX chromosomes.

1910
Thomas Hunt Morgan determines that genes are carried on chromosomes.

1933
Jean Brachet shows that DNA is found in chromosomes and that RNA is present in the cytoplasm of all cells.

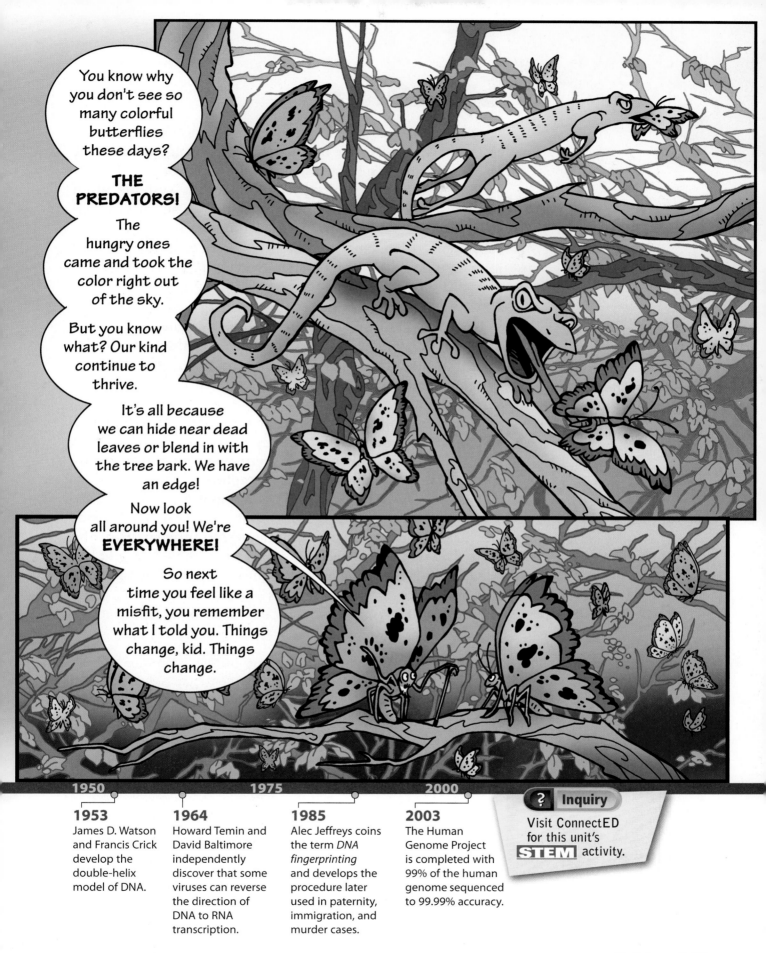

Systems

A **system** is a collection of parts that influence or interact with one another. For example, the human body is a large system made up of many smaller subsystems, such as the ones shown in **Figure 1.**

Like the human body system, complex systems often contain smaller, or less complex, subsystems. The parts of each subsystem interact among themselves, as well as with other subsystems. Each subsystem has a different purpose, but they interact to keep the larger system working properly.

Parts of a System

Systems and subsystems often attempt to achieve a goal. For example, the nervous system, a subsystem of the human body system, regulates your body temperature, as shown in **Figure 2.** Systems and subsystems often are described in terms of their input, processing, and output.

Input is the matter, energy, or information that enters a system. When you exercise, one input to your nervous system is thermal energy. The input, or thermal energy, is detected by special brain and skin cells called receptors.

Processing is the changing of the input to achieve a goal. The hypothalamus processes the input from receptors. It sends electrical signals, carried by the nerves, to other parts of the body. The signals tell the body it is warmer than it should be.

Output is the material, energy, or information that leaves a system. Outputs from the nervous system include sweat, goose bumps, and shivers, all of which can change body temperature.

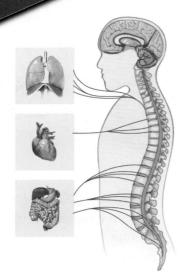

▲ **Figure 1** The nervous system, the respiratory system, the circulatory system, and the digestive system are subsystems of a larger system—the human body.

Figure 2 The nervous system is responsible for regulating body temperature. ▼

Input: Thermal energy released by contracting muscles is detected by receptors.

Processing: Signals from receptors are sent to the brain. The brain then signals glands in the skin to produce sweat.

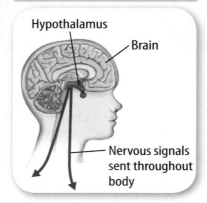

Hypothalamus

Brain

Nervous signals sent throughout body

Output: Sweat forms on the skin. Then, it cools the body as it evaporates.

Figure 3 After a hard race, decreasing levels of carbon dioxide in a rower's blood act as feedback to the nervous system. This feedback signals the rower's brain to restore her breathing to a normal rate.

Feedback in Systems

Many systems use feedback, or information, to monitor and regulate input, process, and output. For example, when you exercise, as shown in **Figure 3,** your muscles produce carbon dioxide as a waste product. Receptors detect this input—high levels of carbon dioxide in your blood. The brain processes this information and signals your nervous system to increase breathing. When you breathe harder and faster, you take in more oxygen and the levels of carbon dioxide in your blood decrease. Once this change is detected by receptors, your brain signals your nervous system to return to normal breathing.

Cooperation, Order, and Change

Body subsystems work together in specific ways to regulate temperature, remove waste from your blood, and respond to other changes in your body. A failure in one subsystem affects other subsystems. For example, if a bad cold causes your lungs to become congested, your respiratory system cannot efficiently exchange oxygen for carbon dioxide. Therefore, muscle cells no longer receive enough oxygen to function normally. As a result, you easily become tired and have trouble catching your breath when you exercise.

Systems Thinking

Thinking in terms of systems might change the way you make choices. For example, some people think that if they reduce the amount they eat, they will lose weight. However, protein is necessary for your muscular system to function properly. Without an input of protein, muscle tissues begin to break down. Someone who does not eat enough protein might not lose weight because they become weak and tired and stop exercising. Thinking about the interactions of the systems and subsystems in your body can lead to decisions that help you achieve long-term goals.

Inquiry MiniLab
25 minutes

Can you analyze a system?

The human body system has many subsystems that respond to environmental changes. How do these subsystems interact?

1. Identify an environmental change that you experience every day.

2. Research one or more subsystems of your body that regulate this change.

3. Design a chart to show the inputs, the processing, the outputs, and the feedback of one of your subsystems.

4. Identify and include another subsystem with which your subsystem interacts.

Analyze and Conclude

1. **Infer** Can a subsystem in the body operate without interacting with one or more other subsystems? Why or why not?

2. **Draw Conclusions** Why do injuries to the brain so often threaten a person's life?

3. **Demonstrate** Use a Venn diagram to show how the subsystem you chose interacts with two other systems in the body.

Chapter 5

Genetics

THE BIG IDEA How are traits passed from parents to offspring?

Inquiry How did this happen?

The color of this fawn is caused by a genetic trait called albinism. Albinism is the absence of body pigment. Notice that the fawn's mother has brown fur, the normal fur color of an adult whitetail deer.

- Why do you think the fawn looks so different from its mother?

- What do you think determines the color of the offspring?

- How do you think traits are passed from generation to generation?

Get Ready to Read

What do you think?

Before you read, decide if you agree or disagree with each of these statements. As you read this chapter, see if you change your mind about any of the statements.

1 Like mixing paints, parents' traits always blend in their offspring.

2 If you look more like your mother than you look like your father, then you received more traits from your mother.

3 All inherited traits follow Mendel's patterns of inheritance.

4 Scientists have tools to predict the form of a trait an offspring might inherit.

5 New DNA is copied from existing DNA.

6 A change in the sequence of an organism's DNA always changes the organism's traits.

ConnectED Your one-stop online resource

connectED.mcgraw-hill.com

- Video
- Audio
- Review
- Inquiry
- WebQuest
- Assessment
- Concepts in Motion
- Multilingual eGlossary

Reading Guide

Key Concepts 🔑

ESSENTIAL QUESTIONS

- Why did Mendel perform cross-pollination experiments?
- What did Mendel conclude about inherited traits?
- How do dominant and recessive factors interact?

Vocabulary

heredity p. 153

genetics p. 153

dominant trait p. 159

recessive trait p. 159

 Multilingual eGlossary

 Video **BrainPOP®**

Mendel and His Peas

Inquiry Same Species?

Have you ever seen a black ladybug? It is less common than the orange variety you might know, but both are the same species of beetle. So why do they look different? Believe it or not, a study of pea plants helped scientists explain these differences.

What makes you unique?

Traits such as eye color have many different types, but some traits have only two types. By a show of hands, determine how many students in your class have each type of trait below.

Student Traits		
Trait	Type 1	Type 2
Earlobes	Unattached	Attached
Thumbs	Curved	Straight
Interlacing fingers	Left thumb over right thumb	Right thumb over left thumb

Think About This

1. Why might some students have types of traits that others do not have?

2. If a person has dimples, do you think his or her offspring will have dimples? Explain.

3. 🔑 **Key Concept** What do you think determines the types of traits you inherit?

Early Ideas About Heredity

Have you ever mixed two paint colors to make a new color? Long ago, people thought an organism's characteristics, or traits, mixed like colors of paint because offspring resembled both parents. This is known as blending inheritance.

Today, scientists know that **heredity** (huh REH duh tee)—*the passing of traits from parents to offspring*—is more complex. For example, you might have blue eyes but both of your parents have brown eyes. How does this happen? More than 150 years ago, Gregor Mendel, an Austrian monk, performed experiments that helped answer these questions and disprove the idea of blending inheritance. Because of his research, Mendel is known as the father of **genetics** (juh NEH tihks)—*the study of how traits are passed from parents to offspring.*

WORD ORIGIN

genetics
from Greek *genesis*, means "origin"

Mendel's Experimental Methods

During the 1850s, Mendel studied genetics by doing controlled breeding experiments with pea plants. Pea plants were ideal for genetic studies because

- they reproduce quickly. This enabled Mendel to grow many plants and collect a lot of data.

- they have easily observed traits, such as flower color and pea shape. This enabled Mendel to observe whether or not a trait was passed from one generation to the next.

- Mendel could control which pairs of plants reproduced. This enabled him to determine which traits came from which plant pairs.

Pollination in Pea Plants

To observe how a trait was inherited, Mendel controlled which plants pollinated other plants. Pollination occurs when pollen lands on the pistil of a flower. **Sperm** cells from the pollen then can fertilize **egg** cells in the pistil. Pollination in pea plants can occur in two ways. Self-pollination occurs when pollen from one plant lands on the pistil of a flower on the same plant, as shown in **Figure 1.** Cross-pollination occurs when pollen from one plant reaches the pistil of a flower on a different plant. Cross-pollination occurs naturally when wind, water, or animals such as bees carry pollen from one flower to another. Mendel allowed one group of flowers to self-pollinate. With another group, he cross-pollinated the plants himself.

Self-Pollination

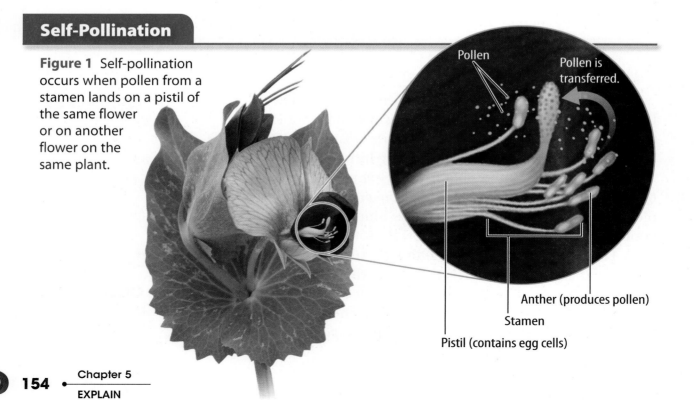

Figure 1 Self-pollination occurs when pollen from a stamen lands on a pistil of the same flower or on another flower on the same plant.

Pollen

Pollen is transferred.

Anther (produces pollen)

Stamen

Pistil (contains egg cells)

True-Breeding Plants

Mendel began his experiments with plants that were true-breeding for the trait he would test. When a true-breeding plant self-pollinates, it always produces offspring with traits that match the parent. For example, when a true-breeding pea plant with wrinkled seeds self-pollinates, it produces only plants with wrinkled seeds. In fact, plants with wrinkled seeds appear generation after generation.

Mendel's Cross-Pollination

By cross-pollinating plants himself, Mendel was able to select which plants pollinated other plants. **Figure 2** shows an example of a manual cross between a plant with white flowers and one with purple flowers.

Figure 2 Mendel removed the stamens of one flower and pollinated that flower with pollen from a flower of a different plant. In this way, he controlled pollination.

Cross-Pollination

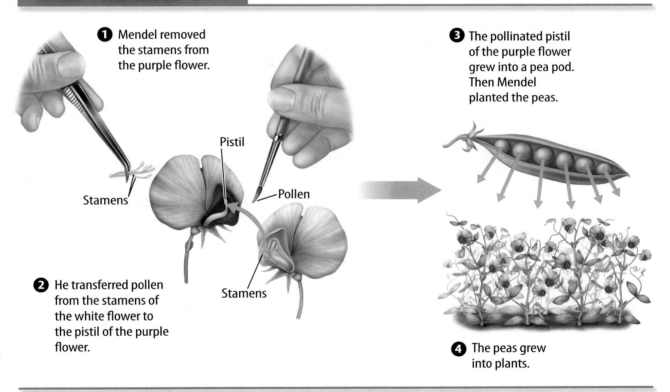

❶ Mendel removed the stamens from the purple flower.

❷ He transferred pollen from the stamens of the white flower to the pistil of the purple flower.

Pistil

Pollen

Stamens

Stamens

❸ The pollinated pistil of the purple flower grew into a pea pod. Then Mendel planted the peas.

❹ The peas grew into plants.

Mendel cross-pollinated hundreds of plants for each set of traits, such as flower color—purple or white; seed color—green or yellow; and seed shape—round or wrinkled. With each cross-pollination, Mendel recorded the traits that appeared in the offspring. By testing such a large number of plants, Mendel was able to predict which crosses would produce which traits.

 Key Concept Check Why did Mendel perform cross-pollination experiments?

Mendel's Results

Once Mendel had enough true-breeding plants for a trait that he wanted to test, he cross-pollinated selected plants. His results are shown in **Figure 3.**

First-Generation Crosses

A cross between true-breeding plants with purple flowers produced plants with only purple flowers. A cross between true-breeding plants with white flowers produced plants with only white flowers. But something unexpected happened when Mendel crossed true-breeding plants with purple flowers and true-breeding plants with white flowers—all the offspring had purple flowers.

New Questions Raised

The results of the crosses between true-breeding plants with purple flowers and true-breeding plants with white flowers led to more questions for Mendel. Why did all the offspring always have purple flowers? Why were there no white flowers? Why didn't the cross produce offspring with pink flowers—a combination of the white and purple flower colors? Mendel carried out more experiments with pea plants to answer these questions.

 Reading Check Predict the offspring of a cross between two true-breeding pea plants with smooth seeds.

First-Generation Crosses

Figure 3 Mendel crossed three combinations of true-breeding plants and recorded the flower colors of the offspring.

Purple × Purple

All purple flowers (true-breeding)

White × White

All white flowers (true-breeding)

Purple (true-breeding) × White (true-breeding)

All purple flowers (hybrids)

Visual Check Suppose you cross hundreds of true-breeding plants with purple flowers with hundreds of true-breeding plants with white flowers. Based on the results of this cross in the figure above, would any offspring produce white flowers? Explain.

Second-Generation (Hybrid) Crosses

The first-generation purple-flowering plants are called **hybrid** plants. This means they came from true-breeding parent plants with different forms of the same trait. Mendel wondered what would happen if he cross-pollinated two purple-flowering hybrid plants.

As shown in **Figure 4,** some of the offspring had white flowers, even though both parents had purple flowers. The results were similar each time Mendel cross-pollinated two hybrid plants. The trait that had disappeared in the first generation always reappeared in the second generation.

The same result happened when Mendel cross-pollinated pea plants for other traits. For example, he found that cross-pollinating a true-breeding yellow-seeded pea plant with a true-breeding green-seeded pea plant always produced yellow-seeded hybrids. A second-generation cross of two yellow-seeded hybrids always yielded plants with yellow seeds and plants with green seeds.

✅ **Reading Check** What is a hybrid plant?

Second-Generation (Hybrid) Crosses

Purple (hybrid) × Purple (hybrid)

Purple and white offspring

Purple (hybrid) × Purple (hybrid)

Purple and white offspring

Figure 4 Mendel cross-pollinated first-generation hybrid offspring to produce second-generation offspring. In each case, the trait that had disappeared from the first generation reappeared in the second generation.

Table 1 When Mendel crossed two hybrids for a given trait, the trait that had disappeared then reappeared in a ratio of about 3:1.

Table 1 Results of Hybrid Crosses

Characteristic	Trait and Number of Offspring		Trait and Number of Offspring		Ratio
Flower color	Purple 705		White 224		3.15:1
Flower position	Axial (Side of stem) 651		Terminal (End of stem) 207		3.14:1
Seed color	Yellow 6,022		Green 2,001		3.01:1
Seed shape	Round 5,474		Wrinkled 1,850		2.96:1
Pod shape	Inflated (Smooth) 882		Constricted (Bumpy) 299		2.95:1
Pod color	Green 428		Yellow 152		2.82:1
Stem length	Long 787		Short 277		2.84:1

Math Skills

Use Ratios

A ratio is a comparison of two numbers or quantities by division. For example, the ratio comparing 6,022 yellow seeds to 2,001 green seeds can be written as follows:

6,022 to 2,001 or

6,022 : 2,001 or

$\frac{6,022}{2,001}$

To simplify the ratio, divide the first number by the second number.

$\frac{6,022}{2,001} = \frac{3}{1}$ or 3:1

Practice

There are 14 girls and 7 boys in a science class. Simplify the ratio.

Review
- Math Practice
- Personal Tutor

More Hybrid Crosses

Mendel counted and recorded the traits of offspring from many experiments in which he cross-pollinated hybrid plants. Data from these experiments are shown in **Table 1.** He analyzed these data and noticed patterns. For example, from the data of crosses between hybrid plants with purple flowers, he found that the ratio of purple flowers to white flowers was about 3:1. This means purple-flowering pea plants grew from this cross three times more often than white-flowering pea plants grew from the cross. He calculated similar ratios for all seven traits he tested.

Mendel's Conclusions

After analyzing the results of his experiments, Mendel concluded that two genetic factors control each inherited trait. He also proposed that when organisms reproduce, each reproductive cell—sperm or egg—contributes one factor for each trait.

 Key Concept Check What did Mendel conclude about inherited traits?

Dominant and Recessive Traits

Recall that when Mendel cross-pollinated a true-breeding plant with purple flowers and a true-breeding plant with white flowers, the hybrid offspring had only purple flowers. Mendel hypothesized that the hybrid offspring had one genetic factor for purple flowers and one genetic factor for white flowers. But why were there no white flowers?

Mendel also hypothesized that the purple factor is the only factor seen or expressed because it blocks the white factor. *A genetic factor that blocks another genetic factor is called a* **dominant** (DAH muh nunt) **trait.** A dominant trait, such as purple pea flowers, is observed when offspring have either one or two dominant factors. *A genetic factor that is blocked by the presence of a dominant factor is called a* **recessive** (rih SE sihv) **trait.** A recessive trait, such as white pea flowers, is observed only when two recessive genetic factors are present in offspring.

From Parents to Second Generation

For the second generation, Mendel cross-pollinated two hybrids with purple flowers. About 75 percent of the second-generation plants had purple flowers. These plants had at least one dominant factor. Twenty-five percent of the second-generation plants had white flowers. These plants had the same two recessive factors.

 Key Concept Check How do dominant and recessive factors interact?

FOLDABLES

Make a vertical two-tab book and label it as shown. Use it to organize your notes on dominant and recessive factors.

Traits
Dominant factors | Recessive factors

Inquiry MiniLab   **20 minutes**

Which is the dominant trait?

Imagine you are Gregor Mendel's lab assistant studying pea plant heredity. Mendel has crossed true-breeding plants with axial flowers and true-breeding plants with terminal flowers. Use the data below to determine which trait is dominant.

Pea Flower Location Results		
Generation	Axial (Number of Offspring)	Terminal (Number of Offspring)
First	794	0
Second	651	207

Analyze and Conclude

1. **Determine** which trait is dominant and which trait is recessive. Support your answer with data.

2. **Key Concept** Analyze the first-generation data. What evidence do you have that one trait is dominant over the other?

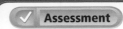
Visual Summary

Genetics is the study of how traits are passed from parents to offspring.

Mendel studied genetics by doing cross-breeding experiments with pea plants.

Purple 705

White 224

Mendel's experiments with pea plants showed that some traits are dominant and others are recessive.

FOLDABLES

Use your lesson Foldable to review the lesson. Save your Foldable for the project at the end of the chapter.

What do you think NOW?

You first read the statements below at the beginning of the chapter.

1. Like mixing paints, parents' traits always blend in their offspring.

2. If you look more like your mother than you look like your father, then you received more traits from your mother.

Did you change your mind about whether you agree or disagree with the statements? Rewrite any false statements to make them true.

Use Vocabulary

1 **Distinguish** between heredity and genetics.

2 **Define** the terms *dominant* and *recessive*.

3 **Use the term** *recessive* in a complete sentence.

Understand Key Concepts

4 A recessive trait is observed when an organism has _____ recessive genetic factor(s).
  A. 0   **C.** 2
  B. 1   **D.** 3

5 **Summarize** Mendel's conclusions about how traits pass from parents to offspring.

6 **Describe** how Mendel cross-pollinated pea plants.

Interpret Graphics

7 **Suppose** the two true-breeding plants shown below were crossed.

 ×

What color would the flowers of the offspring be? Explain.

Critical Thinking

8 **Design an experiment** to test for true-breeding plants.

9 **Examine** how Mendel's conclusions disprove blending inheritance.

Math Skills

Review
— Math Practice —

10 A cross between two pink camellia plants produced the following offspring: 7 plants with red flowers, 7 with white flowers, and 14 with pink flowers. What is the ratio of red to white to pink?

Pioneering
the Science of Genetics

One man's curiosity leads to a branch of science.

Gregor Mendel—monk, scientist, gardener, and beekeeper—was a keen observer of the world around him. Curious about how traits pass from one generation to the next, he grew and tested almost 30,000 pea plants. Today, Mendel is called the father of genetics. After Mendel published his findings, however, his "laws of heredity" were overlooked for several decades.

In 1900, three European scientists, working independently of one another, rediscovered Mendel's work and replicated his results. Then, other biologists quickly began to recognize the importance of Mendel's work.

Gregor Mendel ▶

1902: American physician Walter Sutton demonstrates that Mendel's laws of inheritance can be applied to chromosomes. He concludes that chromosomes contain a cell's hereditary material on genes.

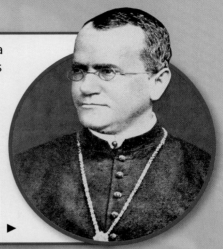

1906: William Bateson, a United Kingdom scientist, coins the term *genetics*. He uses it to describe the study of inheritance and the science of biological inheritance.

1952: American geneticists Martha Chase and Alfred Hershey prove that DNA transmits inherited traits from one generation to the next.

1953: Francis Crick and James Watson determine the structure of the DNA molecule. Their work begins the field of molecular biology and leads to important scientific and medical research in genetics.

2003: The National Human Genome Research Institute (NHGRI) completes mapping and sequencing human DNA. Researchers and scientists are now trying to discover the genetic basis for human health and disease.

It's Your Turn

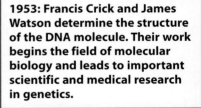

RESEARCH What are some genetic diseases? Report on how genome-based research might help cure these diseases in the future.

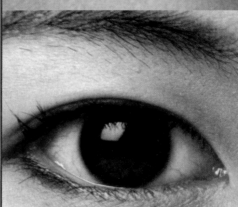

Lesson 2

Understanding Inheritance

Reading Guide

Key Concepts 🔑
ESSENTIAL QUESTIONS

- What determines the expression of traits?
- How can inheritance be modeled?
- How do some patterns of inheritance differ from Mendel's model?

Vocabulary

gene p. 164

allele p. 164

phenotype p. 164

genotype p. 164

homozygous p. 165

heterozygous p. 165

Punnett square p. 166

incomplete dominance p. 168

codominance p. 168

polygenic inheritance p. 169

🅖 **Multilingual eGlossary**

▢ **Video** **BrainPOP®**

Inquiry Make the Connection

Physical traits, such as those shown in these eyes, can vary widely from person to person. Take a closer look at the eyes on this page. What traits can you identify among them? How do they differ?

What is the span of your hand?

Mendel discovered some traits have a simple pattern of inheritance—dominant or recessive. However, some traits, such as eye color, have more variation. Is human hand span a Mendelian trait?

1 Read and complete a lab safety form.

2 Use a **metric ruler** to measure the distance (in cm) between the tips of your thumb and little finger with your hand stretched out.

3 As a class, record everyone's name and hand span in a data table.

Think About This

1. What range of hand span measurements did you observe?

2. 🔑 **Key Concept** Do you think hand span is a simple Mendelian trait like pea plant flower color?

What controls traits?

Mendel concluded that two factors—one from each parent—control each trait. Mendel hypothesized that one factor came from the egg cell and one factor came from the sperm cell. What are these factors? How are they passed from parents to offspring?

Chromosomes

When other scientists studied the parts of a cell and combined Mendel's work with their work, these factors were more clearly understood. Scientists discovered that inside each cell is a nucleus that contains thread-like structures called chromosomes. Over time, scientists learned that chromosomes contain genetic information that controls traits. We now know that Mendel's "factors" are part of chromosomes and that each cell in offspring contains chromosomes from both parents. As shown in **Figure 5,** these chromosomes exist as pairs—one chromosome from each parent.

Figure 5 Humans have 23 pairs of chromosomes. Each pair has one chromosome from the father and one chromosome from the mother.

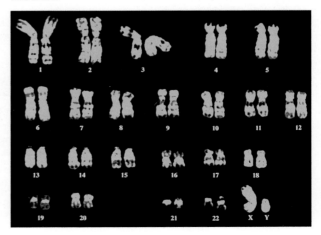

Genes and Alleles

Scientists have discovered that each chromosome can have information about hundreds or even thousands of traits. *A gene (JEEN) is a section on a chromosome that has genetic information for one trait.* For example, a gene of a pea plant might have information about flower color. Recall that an offspring inherits two genes (factors) for each trait—one from each parent. The genes can be the same or different, such as purple or white for pea flower color. *The different forms of a gene are called alleles (uh LEELs).* Pea plants can have two purple alleles, two white alleles, or one of each allele. In **Figure 6,** the chromosome pair has information about three traits—flower position, pod shape, and stem length.

Reading Check How many alleles controlled flower color in Mendel's experiments?

Genotype and Phenotype

Look again at the photo at the beginning of this lesson. What human trait can you observe? You might observe that eye color can be shades of blue or brown. *Geneticists call how a trait appears, or is expressed, the trait's phenotype (FEE nuh tipe).* What other phenotypes can you observe in the photo?

Mendel concluded that two alleles control the expression or phenotype of each trait. *The two alleles that control the phenotype of a trait are called the trait's genotype (JEE nuh tipe).* Although you cannot see an organism's genotype, you can make inferences about a genotype based on its phenotype. For example, you have already learned that a pea plant with white flowers has two recessive alleles for that trait. These two alleles are its genotype. The white flower is its phenotype.

WORD ORIGIN ·············

phenotype
from Greek *phainein,* means "to show"

Figure 6 The alleles for flower position are the same on both chromosomes. However, the chromosome pair has different alleles for pod shape and stem length.

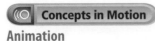

Concepts in Motion

Animation

Chromosome Pair

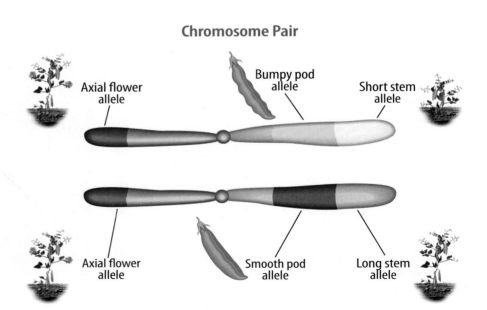

Axial flower allele

Bumpy pod allele

Short stem allele

Axial flower allele

Smooth pod allele

Long stem allele

Symbols for Genotypes Scientists use symbols to represent the alleles in a genotype. In genetics, uppercase letters represent dominant alleles and lowercase letters represent recessive alleles. **Table 2** shows the possible genotypes for both round and wrinkled seed phenotypes. Notice that the dominant allele, if present, is written first.

Table 2 Phenotype and Genotype	
Phenotypes (observed traits)	**Genotypes (alleles of a gene)**
Round	Homozygous dominant (*RR*)
	Heterozygous (*Rr*)
Wrinkled	Homozygous recessive (*rr*)

A round seed can have two genotypes—*RR* and *Rr*. Both genotypes have a round phenotype. Why does *Rr* result in round seeds? This is because the round allele (*R*) is dominant to the wrinkled allele (*r*).

A wrinkled seed has the recessive genotype, *rr*. The wrinkled-seed phenotype is possible only when the same two recessive alleles (*rr*) are present in the genotype.

Homozygous and Heterozygous *When the two alleles of a gene are the same, its genotype is* **homozygous** *(hoh muh ZI gus). Both* RR *and* rr *are homozygous genotypes, as shown in* **Table 2.**

If the two alleles of a gene are different, its genotype is **heterozygous** *(he tuh roh ZI gus).* Rr *is a heterozygous genotype.*

Key Concept Check How do alleles determine the expression of traits?

Punnett Square

Figure 7 A Punnett square can be used to predict the possible genotypes of the offspring. Offspring from a cross between two heterozygous parents can have one of three genotypes.

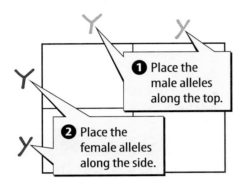

1 Place the male alleles along the top.

2 Place the female alleles along the side.

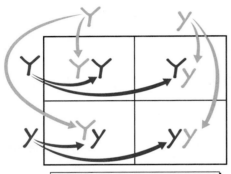

3 Copy female alleles across each row. Copy male alleles down each column. Always list the dominant trait first.

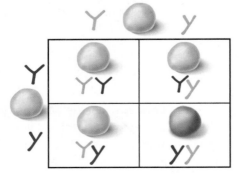

Visual Check What phenotypes are possible for pea offspring of this cross?

Modeling Inheritance

Have you ever flipped a coin and guessed heads or tails? Because a coin has two sides, there are only two possible outcomes—heads or tails. You have a 50 percent chance of getting heads and a 50 percent chance of getting tails. The chance of getting an outcome can be represented by a ratio. The ratio of heads to tails is 50:50 or 1:1.

Reading Check What does a ratio of 2:1 mean?

Plant breeders and animal breeders use a method for predicting how often traits will appear in offspring that does not require performing the crosses thousands of times. Two models—a Punnett square and a pedigree—can be used to predict and identify traits among genetically related individuals.

Punnett Squares

If the genotypes of the parents are known, then the different genotypes and phenotypes of the offspring can be predicted. *A* **Punnett square** *is a model used to predict possible genotypes and phenotypes of offspring.* Follow the steps in **Figure 7** to learn how to make a Punnett square.

Analyzing a Punnett Square

Figure 7 shows an example of a cross between two pea plants that are heterozygous for pea seed color—*Yy* and *Yy.* Yellow is the dominant allele—*Y.* Green is the recessive allele—*y.* The offspring can have one of three genotypes—*YY, Yy,* or *yy.* The ratio of genotypes is written as 1:2:1.

Because *YY* and *Yy* represent the same phenotype—yellow—the offspring can have one of only two phenotypes—yellow or green. The ratio of phenotypes is written 3:1. Therefore, about 75 percent of the offspring of the cross between two heterozygous pea plants will produce yellow seeds. About 25 percent of the plants will produce green seeds.

Using Ratios to Predict

Given a 3:1 ratio, you can expect that an offspring from heterozygous parents has a 3:1 chance of having yellow seeds. But you cannot expect that a group of four seeds will have three yellow seeds and one green seed. This is because one offspring does not affect the phenotype of another offspring. In a similar way, the outcome of one coin toss does not affect the outcome of other coin tosses.

However, if you counted large numbers of offspring from a particular cross, the overall ratio would be close to the ratio predicted by a Punnett square. Mendel did not use Punnett squares. However, by studying nearly 30,000 pea plants, his ratios nearly matched those that would have been predicted by a Punnett square for each cross.

Pedigrees

Another model that can show inherited traits is a pedigree. A pedigree shows phenotypes of genetically related family members. It can also help determine genotypes. In the pedigree in **Figure 8,** three offspring have a trait—attached earlobes—that the parents do not have. If these offspring received one allele for this trait from each parent, but neither parent displays the trait, the offspring must have received two recessive alleles.

 Key Concept Check How can inheritance be modeled?

Pedigree 🔑

Figure 8 In this pedigree, the parents and two offspring have unattached ear lobes—the dominant phenotype. Three offspring have attached ear lobes—the recessive phenotype.

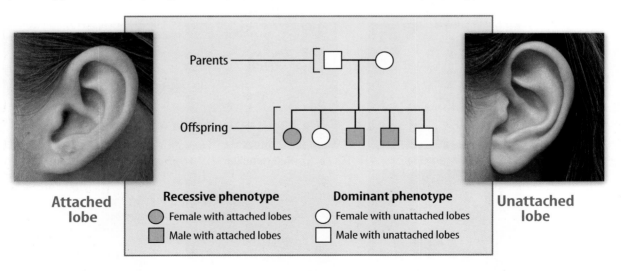

Attached lobe

Recessive phenotype
- 🔴 Female with attached lobes
- 🟦 Male with attached lobes

Dominant phenotype
- ⚪ Female with unattached lobes
- ⬜ Male with unattached lobes

Unattached lobe

✅ **Visual Check** If the genotype of the offspring with attached lobes is *uu,* what is the genotype of the parents? How can you tell?

Complex Patterns of Inheritance

By chance, Mendel studied traits only influenced by one gene with two alleles. However, we know now that some inherited traits have complex patterns of inheritance.

Types of Dominance

Recall that for pea plants, the presence of one dominant allele produces a dominant phenotype. However, not all allele pairs have a dominant-recessive interaction.

Incomplete Dominance Sometimes traits appear to be combinations of alleles. *Alleles show* **incomplete dominance** *when the offspring's phenotype is a combination of the parents' phenotypes.* For example, a pink camellia, as shown in **Figure 9,** results from incomplete dominance. A cross between a camellia plant with white flowers and a camellia plant with red flowers produces only camellia plants with pink flowers.

Codominance The coat color of some cows is an example of another type of interaction between two alleles. *When both alleles can be observed in a phenotype, this type of interaction is called* **codominance.** If a cow inherits the allele for white coat color from one parent and the allele for red coat color from the other parent, the cow will have both red and white hairs.

FOLDABLES

Use two sheets of paper to make a layered book. Label it as shown. Use it to organize your notes on inheritance patterns.

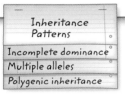

Inheritance
Patterns
Incomplete dominance
Multiple alleles
Polygenic inheritance

Types of Dominance

Figure 9 In incomplete dominance, neither parent's phenotype is visible in the offspring's phenotype. In codominance, both parents' phenotypes are visible separately in the offspring's phenotype.

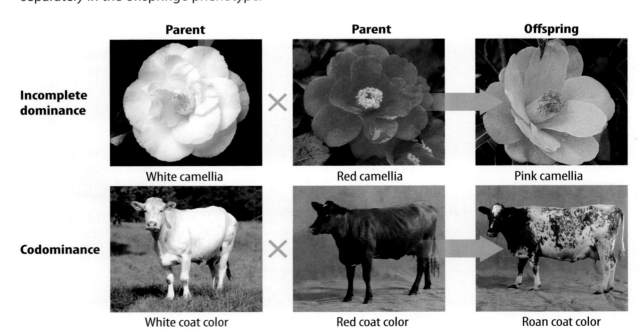

Parent	Parent	Offspring
Incomplete dominance White camellia	Red camellia	Pink camellia
Codominance White coat color	Red coat color	Roan coat color

Table 3 Human ABO Blood Types

Phenotype	Possible Genotypes
Type A	I^AI^A or I^Ai
Type B	I^BI^B or I^Bi
Type O	ii
Type AB	I^AI^B

Multiple Alleles

Unlike the genes in Mendel's pea plants, some genes have more than two alleles, or multiple alleles. Human ABO blood type is an example of a trait that is determined by multiple alleles. There are three different alleles for the ABO blood type—I^A, I^B, and i. The way the alleles combine results in one of four blood types—A, B, AB, or O. The I^A and I^B alleles are codominant to each other, but they both are dominant to the i allele. Even though there are multiple alleles, a person can inherit only two of these alleles—one from each parent, as shown in **Table 3.**

Polygenic Inheritance

Mendel **concluded** that each trait was determined by only one gene. However, we now know that a trait can be affected by more than one gene. **Polygenic inheritance** *occurs when multiple genes determine the phenotype of a trait.* Because several genes determine a trait, many alleles affect the phenotype even though each gene has only two alleles. Therefore, polygenic inheritance has many possible phenotypes.

Look again at the photo at the beginning of this lesson. Eye color in humans is an example of polygenic inheritance. There are also many phenotypes for height in humans, as shown in **Figure 10.** Other human characteristics determined by polygenic inheritance are weight and skin color.

 Key Concept Check How does polygenic inheritance differ from Mendel's model?

ACADEMIC VOCABULARY

conclude
(verb) to reach a logically necessary end by reasoning

Figure 10 The eighth graders in this class have different heights.

Concepts in Motion
Animation

Genes and the Environment

You read earlier in this lesson that an organism's genotype determines its phenotype. Scientists have learned that genes are not the only factors that can affect phenotypes. An organism's environment can also affect its phenotype. For example, the flower color of one type of hydrangea is determined by the soil in which the hydrangea plant grows. **Figure 11** shows that acidic soil produces blue flowers and basic, or alkaline, soil produces pink flowers. Other examples of environmental effects on phenotype are also shown in **Figure 11.**

For humans, healthful choices can also affect phenotype. Many genes affect a person's chances of having heart disease. However, what a person eats and the amount of exercise he or she gets can influence whether heart disease will develop.

Reading Check What environmental factors affect phenotype?

Figure 11 Environmental factors, such as temperature and sunlight, can affect phenotype.

◄ These hydrangea plants are genetically identical. The plant grown in acidic soil produced blue flowers. The plant grown in alkaline soil produced pink flowers.

Siamese cats have alleles that produce a dark pigment only in cooler areas of the body. That's why a Siamese cat's ear tips, nose, paws, and tail are darker than other areas of its body. ►

◄ The wing patterns of the map butterfly, *Araschnia levana,* depend on what time of year the adult develops. Adults that developed in the spring have more orange in their wings than those that developed in the summer.

Lesson 2 Review

Visual Summary

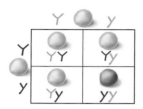

The genes for traits are located on chromosomes.

Geneticists use Punnett squares to predict the possible genotypes and phenotypes of offspring.

In polygenic inheritance, traits are determined by more than one gene and have many possible phenotypes.

FOLDABLES

Use your lesson Foldable to review the lesson. Save your Foldable for the project at the end of the chapter.

What do you think NOW?

You first read the statements below at the beginning of the chapter.

3. All inherited traits follow Mendel's patterns of inheritance.

4. Scientists have tools to predict the form of a trait an offspring might inherit.

Did you change your mind about whether you agree or disagree with the statements? Rewrite any false statements to make them true.

Use Vocabulary

1 **Use** the terms *phenotype* and *genotype* in a complete sentence.

2 **Contrast** homozygous and heterozygous.

3 **Define** *incomplete dominance* in your own words.

Understand Key Concepts

4 How many alleles control a Mendelian trait, such as pea seed color?
  A. one   C. three
  B. two   D. four

5 **Explain** where the alleles for a given trait are inherited from.

6 **Describe** how the genotypes *RR* and *Rr* result in the same phenotype.

7 **Summarize** how polygenic inheritance differs from Mendelian inheritance.

Interpret Graphics

8 **Analyze** this pedigree. If ■ represents a male with the homozygous recessive genotype (*aa*), what is the mother's genotype?

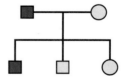

Critical Thinking

9 **Predict** the possible blood genotypes of a child, using the table below, if one parent is type O and the other parent is type A.

Phenotype	Genotype
Blood Type O	*ii*
Blood Type A	$I^A I^A$ or $I^A i$

How can you use Punnett squares to model inheritance?

Geneticists use models to explain how traits are inherited from one generation to the next. A simple model of Mendelian inheritance is a Punnett square. A Punnett square is a model of reproduction between two parents and the possible genotypes and phenotypes of the resulting offspring. It also models the probability that each genotype will occur.

Learn It

In science, a **model** is a representation of how something in the natural world works. A model is used to explain or predict a natural process. Maps, diagrams, three-dimensional representations, and mathematical formulas can all be used to help model nature.

Try It

1 Copy the Punnett square on this page in your Science Journal. Use it to complete a cross between a fruit fly with straight wings *(cc)* and a fruit fly with curly wings *(CC)*.

2 According to your Punnett square, which genotypes are possible in the offspring?

3 Using the information in your Punnett square, calculate the ratio of the dominant phenotype to the recessive phenotype in the offspring.

Apply It

4 Based on the information in your Punnett square, how many offspring will have curly wings? Straight wings?

5 If you switch the locations of the parent genotypes around the Punnett square, does it affect the potential genotypes of their offspring? Explain.

6 🔑 **Key Concept** Design and complete a Punnett square to model a cross between two fruit flies that are heterozygous for the curly wings *(Cc)*. What are the phenotypic ratios of the offspring?

Magnification: 20×

Curly wings *(CC)*

Straight wings *(cc)*

	C	C
c	Cc	
c		

DNA and Genetics

Reading Guide

Key Concepts 🔑
ESSENTIAL QUESTIONS

- What is DNA?
- What is the role of RNA in protein production?
- How do changes in the sequence of DNA affect traits?

Vocabulary

DNA p. 174

nucleotide p. 175

replication p. 176

RNA p. 177

transcription p. 177

translation p. 178

mutation p. 179

g Multilingual eGlossary

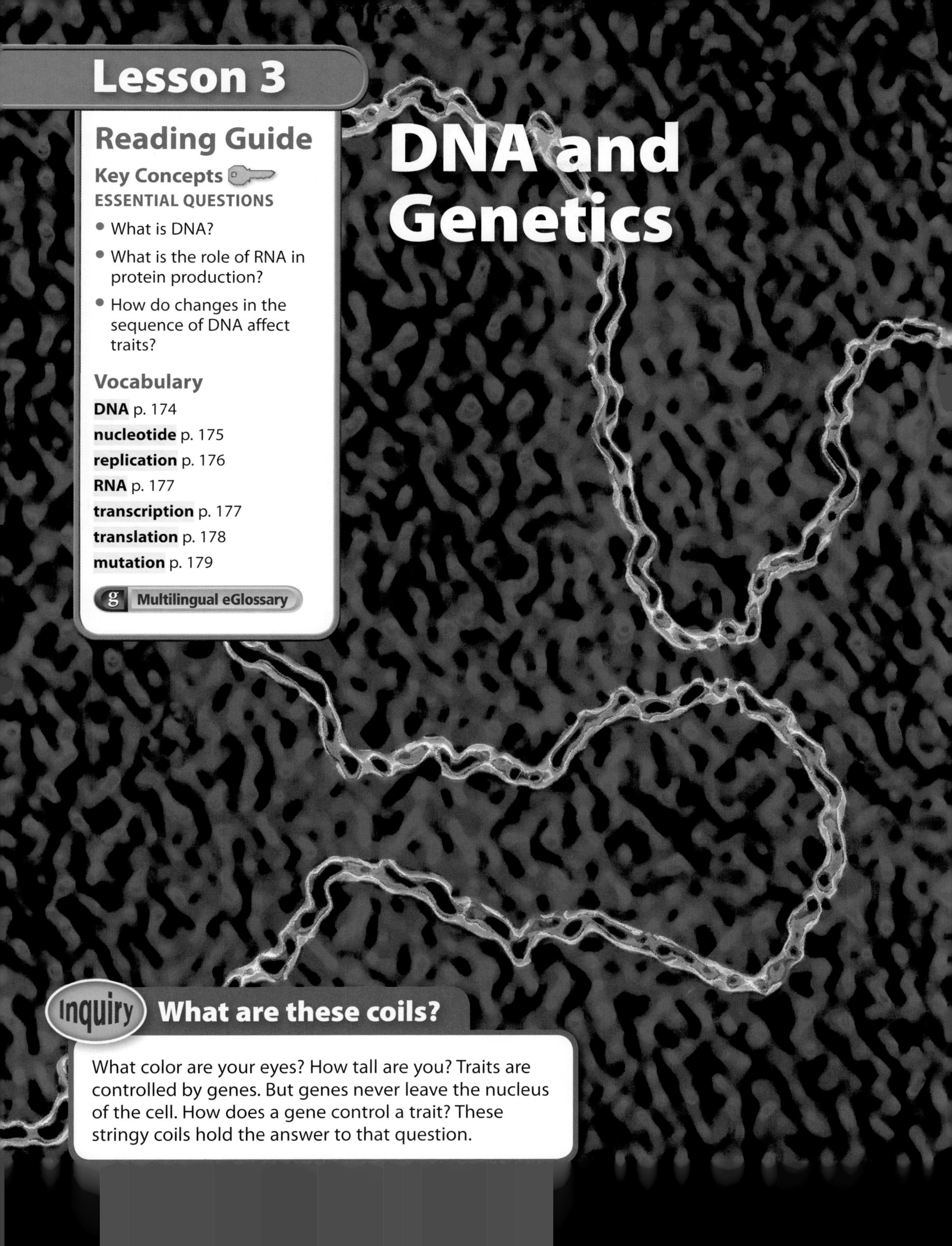

Inquiry What are these coils?

What color are your eyes? How tall are you? Traits are controlled by genes. But genes never leave the nucleus of the cell. How does a gene control a trait? These stringy coils hold the answer to that question.

How are codes used to determine traits?

Interpret this code to learn more about how an organism's body cells use codes to determine genetic traits.

1. Analyze the pattern of the simple code shown to the right. For example, ⟩⟨∟ = DOG

2. In your Science Journal, record the correct letters for the symbols in the code below.

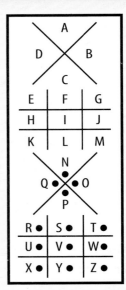

Think About This

1. What do all codes, such as Morse code and Braille, have in common?

2. What do you think might happen if there is a mistake in the code?

3. 🔑 **Key Concept** How do you think an organism's cells might use code to determine its traits?

The Structure of DNA

Have you ever put together a toy or a game for a child? If so, it probably came with directions. Cells put molecules together in much the same way you might assemble a toy. They follow a set of directions.

Genes provide directions for a cell to assemble molecules that express traits such as eye color or seed shape. Recall from Lesson 2 that a gene is a section of a chromosome. Chromosomes are made of proteins and deoxyribonucleic (dee AHK sih ri boh noo klee ihk) acid, or **DNA**—*an organism's genetic material.* A gene is a segment of DNA on a chromosome.

Cells and organisms contain millions of different molecules. Countless numbers of directions are needed to make all those molecules. How do all these directions fit on a few chromosomes? The information, or directions, needed for an organism to grow, maintain itself, and reproduce is contained in DNA. As shown in **Figure 12,** strands of DNA in a chromosome are tightly coiled, like a telephone cord or a coiled spring. This coiling allows more genes to fit in a small space.

🔑 **Key Concept Check** What is DNA?

Figure 12 Strands of DNA are tightly coiled in chromosomes.

A Complex Molecule

What's the best way to fold clothes so they will fit into a drawer or a suitcase? Scientists asked a similar question about DNA. What is the shape of the DNA molecule, and how does it fit into a chromosome? The work of several scientists revealed that DNA is like a twisted zipper. This twisted zipper shape is called a double helix. A model of DNA's double helix structure is shown in **Figure 13.**

How did scientists make this discovery? Rosalind Franklin and Maurice Wilkins were two scientists in London who used X-rays to study DNA. Some of the X-ray data indicated that DNA has a helix shape.

American scientist James Watson visited Franklin and Wilkins and saw one of the DNA X-rays. Watson realized that the X-ray gave valuable clues about DNA's structure. Watson worked with an English scientist, Francis Crick, to build a model of DNA.

Watson and Crick based their work on information from Franklin's and Wilkins's X-rays. They also used chemical information about DNA discovered by another scientist, Erwin Chargaff. After several tries, Watson and Crick built a model that showed how the smaller molecules of DNA bond together and form a double helix.

Four Nucleotides Shape DNA

DNA's twisted-zipper shape is because of molecules called nucleotides. *A* **nucleotide** *is a molecule made of a nitrogen base, a sugar, and a phosphate group.* Sugar-phosphate groups form the sides of the DNA zipper. The nitrogen bases bond and form the teeth of the zipper. As shown in **Figure 13,** there are four nitrogen bases: adenine (A), cytosine (C), thymine (T), and guanine (G). A and T always bond together, and C and G always bond together.

✓ **Reading Check** What is a nucleotide?

Figure 13 A DNA double helix is made of two strands of DNA. Each strand is a chain of nucleotides.

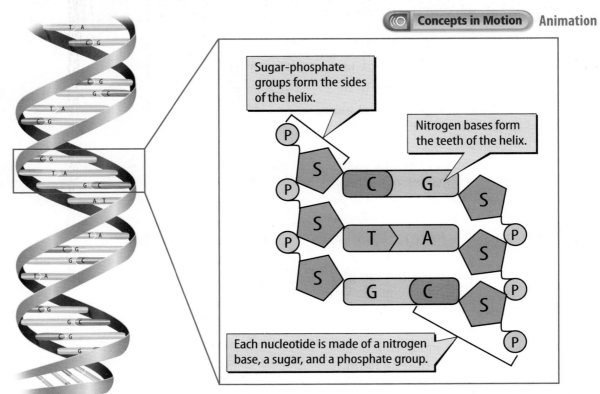

Concepts in Motion Animation

Sugar-phosphate groups form the sides of the helix.

Nitrogen bases form the teeth of the helix.

Each nucleotide is made of a nitrogen base, a sugar, and a phosphate group.

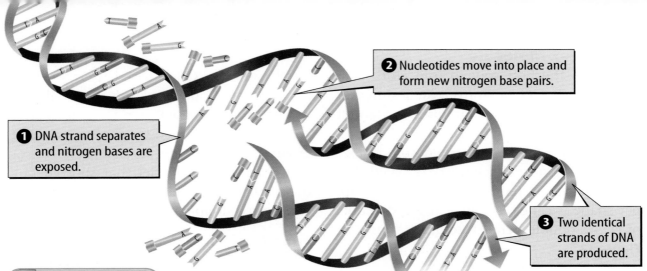

2 Nucleotides move into place and form new nitrogen base pairs.

1 DNA strand separates and nitrogen bases are exposed.

3 Two identical strands of DNA are produced.

Concepts in Motion

Animation

Figure 14 Before a cell divides, its DNA is replicated.

How DNA Replicates

Cells contain DNA in chromosomes. So, every time a cell divides, all chromosomes must be copied for the new cell. The new DNA is identical to existing DNA. *The process of copying a DNA molecule to make another DNA molecule is called* **replication.** You can follow the steps of DNA replication in **Figure 14.** First, the strands separate in many places, exposing individual bases. Then nucleotides are added to each exposed base. This produces two identical strands of DNA.

✓ **Reading Check** What is replication?

Inquiry MiniLab   **25 minutes**

How can you model DNA?

Making a model of DNA can help you understand its structure.

1. Read and complete a lab safety form.

2. Link a **small paper clip** to a **large paper clip.** Repeat four more times, making a chain of 10 paper clips.

3. Choose **four colors of chenille stems.** Each color represents one of the four nitrogen bases. Record the color of each nitrogen base in your Science Journal.

4. Attach a chenille stem to each large paper clip.

5. Repeat step 2 and step 4, but this time attach the corresponding chenille-stem nitrogen bases. Connect the nitrogen bases.

6. Securely insert one end of your double chain into a **block of styrene foam.**

7. Repeat step 6 with the other end of your chain.

8. Gently turn the blocks to form a double helix.

Analyze and Conclude

1. **Explain** which part of a DNA molecule is represented by each material you used.

2. **Predict** what might happen if a mistake were made in creating a nucleotide.

3. 🔑 **Key Concept** How did making a model of DNA help you understand its structure?

Making Proteins

Recall that proteins are important for every cellular process. The DNA of each cell carries a complete set of genes that provides instructions for making all the proteins a cell requires. Most genes contain instructions for making proteins. Some genes contain instructions for when and how quickly proteins are made.

Junk DNA

As you have learned, all genes are segments of DNA on a chromosome. However, you might be surprised to learn that most of your DNA is not part of any gene. For example, about 97 percent of the DNA on human chromosomes does not form genes. Segments of DNA that are not parts of genes are often called junk DNA. It is not yet known whether junk DNA segments have functions that are important to cells.

The Role of RNA in Making Proteins

How does a cell use the instructions in a gene to make proteins? Proteins are made with the help of ribonucleic acid **(RNA)**—*a type of nucleic acid that carries the code for making proteins from the nucleus to the cytoplasm.* RNA also carries amino acids around inside a cell and forms a part of ribosomes.

RNA, like DNA, is made of nucleotides. However, there are key differences between DNA and RNA. DNA is double-stranded, but RNA is single-stranded. RNA has the nitrogen base uracil (U) instead of thymine (T) and the sugar ribose instead of deoxyribose.

The first step in making a protein is to make mRNA from DNA. *The process of making mRNA from DNA is called* **transcription.** **Figure 15** shows how mRNA is transcribed from DNA.

Key Concept Check What is the role of RNA in protein production?

Transcription 🔑

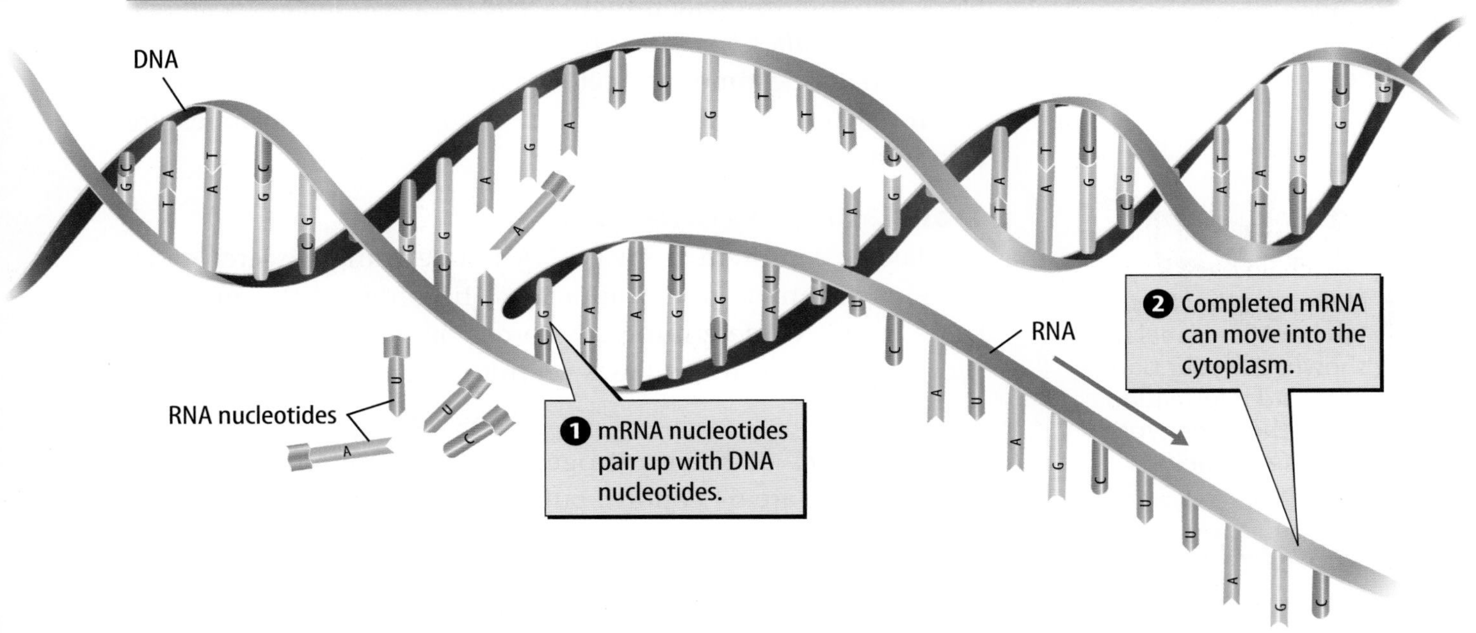

DNA

RNA nucleotides

❶ mRNA nucleotides pair up with DNA nucleotides.

RNA

❷ Completed mRNA can move into the cytoplasm.

Figure 15 Transcription is the first step in making a protein. During transcription, the sequence of nitrogen bases on a gene determines the sequence of bases on mRNA.

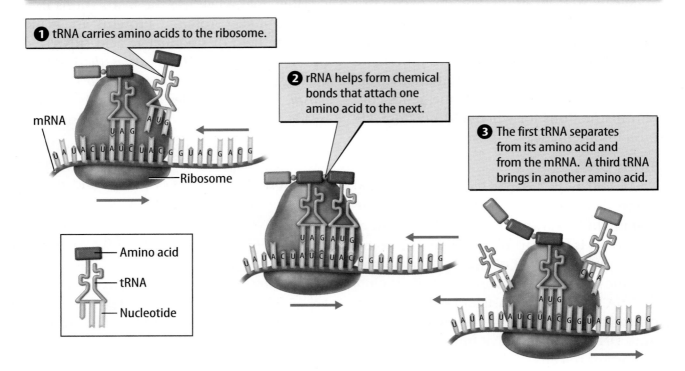

1 tRNA carries amino acids to the ribosome.

2 rRNA helps form chemical bonds that attach one amino acid to the next.

3 The first tRNA separates from its amino acid and from the mRNA. A third tRNA brings in another amino acid.

mRNA

Ribosome

— Amino acid

— tRNA

— Nucleotide

Figure 16 A protein forms as mRNA moves through a ribosome. Different amino acid sequences make different proteins. A complete protein is a folded chain of amino acids.

FOLDABLES

Make a vertical three-tab book and label it as shown. Use your book to record information about the three types of RNA and their functions.

Messenger RNA

Ribosomal RNA

Transfer RNA

Three Types of RNA

On the previous page, you read about messenger RNA (mRNA). There are two other types of RNA, transfer RNA (tRNA) and ribosomal RNA (rRNA). **Figure 16** illustrates how the three work together to make proteins. *The process of making a protein from RNA is called* **translation.** Translation occurs in ribosomes. Recall that ribosomes are cell organelles that are attached to the rough endoplasmic reticulum (rough ER). Ribosomes are also in a cell's cytoplasm.

Translating the RNA Code

Making a protein from mRNA is like using a secret code. Proteins are made of amino acids. The order of the nitrogen bases in mRNA determines the order of the amino acids in a protein. Three nitrogen bases on mRNA form the code for one amino acid.

Each series of three nitrogen bases on mRNA is called a codon. There are 64 codons, but only 20 amino acids. Some of the codons code for the same amino acid. One of the codons codes for an amino acid that is the beginning of a protein. This codon signals that translation should start. Three of the codons do not code for any amino acid. Instead, they code for the end of the protein. They signal that translation should stop.

✓ **Reading Check** What is a codon?

Mutations

You have read that the sequence of nitrogen bases in DNA determines the sequence of nitrogen bases in mRNA, and that the mRNA sequence determines the sequence of amino acids in a protein. You might think these sequences always stay the same, but they can change. *A change in the nucleotide sequence of a gene is called a* **mutation.**

The 46 human chromosomes contain between 20,000 and 25,000 genes that are copied during DNA replication. Sometimes, mistakes can happen during replication. Most mistakes are corrected before replication is completed. A mistake that is not corrected can result in a mutation. Mutations can be triggered by exposure to X-rays, ultraviolet light, radioactive materials, and some kinds of chemicals.

Types of Mutations

There are several types of DNA mutations. Three types are shown in **Figure 17.** In a deletion mutation, one or more nitrogen bases are left out of the DNA sequence. In an insertion mutation, one or more nitrogen bases are added to the DNA. In a substitution mutation, one nitrogen base is replaced by a different nitrogen base.

Each type of mutation changes the sequence of nitrogen base pairs. This can cause a mutated gene to code for a different protein than a normal gene. Some mutated genes do not code for any protein. For example, a cell might lose the ability to make one of the proteins it needs.

WORD ORIGIN

mutation
from Latin *mutare*, means "to change"

Figure 17 Three types of mutations are substitution, insertion, and deletion.

✓ **Visual Check** Which base pairs were omitted during replication in the deletion mutation?

Mutations 🔑

Original DNA sequence

Substitution
The C-G base pair has been replaced with a T-A pair.

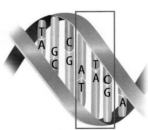

Insertion
Three base pairs have been added.

Deletion
Three base pairs have been removed. Other base pairs will move in to take their place.

Results of a Mutation

The effects of a mutation depend on where in the DNA sequence the mutation happens and the type of mutation. Proteins express traits. Because mutations can change proteins, they can cause traits to change. Some mutations in human DNA cause genetic disorders, such as those described in **Table 4.**

However, not all mutations have negative effects. Some mutations don't cause changes in proteins, so they don't affect traits. Other mutations might cause a trait to change in a way that benefits the organism.

 Key Concept Check How do changes in the sequence of DNA affect traits?

Scientists still have much to learn about genes and how they determine an organism's traits. Scientists are researching and experimenting to identify all genes that cause specific traits. With this knowledge, we might be one step closer to finding cures and treatments for genetic disorders.

Table 4 Genetic Disorders

Defective Gene or Chromosome	Disorder	Description
Chromosome 12, PAH gene	Phenylketonuria (PKU)	People with defective PAH genes cannot break down the amino acid phenylalanine. If phenylalanine builds up in the blood, it poisons nerve cells.
Chromosome 7, CFTR gene	Cystic fibrosis	In people with defective CFTR genes, salt cannot move in and out of cells normally. Mucus builds up outside cells. The mucus can block airways in lungs and affect digestion.
Chromosome 7, elastin gene	Williams syndrome	People with Williams syndrome are missing part of chromosome 7, including the elastin gene. The protein made from the elastin gene makes blood vessels strong and stretchy.
Chromosome 17, BRCA 1; Chromosome 13, BRCA 2	Breast cancer and ovarian cancer	A defect in BRCA1 and/or BRCA2 does not mean the person will have breast cancer or ovarian cancer. People with defective BRCA1 or BRCA2 genes have an increased risk of developing breast cancer and ovarian cancer.

Lesson 3 Review

Visual Summary

DNA is a complex molecule that contains the code for an organism's genetic information.

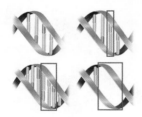

RNA carries the codes for making proteins.

An organism's nucleotide sequence can change through the deletion, insertion, or substitution of nitrogen bases.

Use your lesson Foldable to review the lesson. Save your Foldable for the project at the end of the chapter.

What do you think NOW?

You first read the statements below at the beginning of the chapter.

5. Any condition present at birth is genetic.

6. A change in the sequence of an organism's DNA always changes the organism's traits.

Did you change your mind about whether you agree or disagree with the statements? Rewrite any false statements to make them true.

Use Vocabulary

1 **Distinguish** between transcription and translation.

2 **Use the terms** *DNA* and *nucleotide* in a sentence.

3 A change in the sequence of nitrogen bases in a gene is called a(n) _____.

Understand Key Concepts

4 Where does the process of transcription occur?
  A. cytoplasm   C. cell nucleus
  B. ribosomes   D. outside the cell

5 **Illustrate** Make a drawing that illustrates the process of translation.

6 **Distinguish** between the sides of the DNA double helix and the teeth of the DNA double helix.

Interpret Graphics

7 **Identify** The products of what process are shown in the figure below?

8 **Sequence** Draw a graphic organizer like the one below about important steps in making a protein, beginning with DNA and ending with protein.

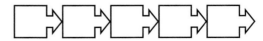

Critical Thinking

9 **Hypothesize** What would happen if a cell were unable to make mRNA?

10 **Assess** What is the importance of DNA replication occurring without any mistakes?

Materials

gummy bears

calculator

paper bag

Safety

Gummy Bear Genetics

Imagine you are on a team of geneticists that is doing "cross-breeding experiments" with gummy bears. Unfortunately, the computer containing your data has crashed. All you have left are six gummy-bear litters that resulted from six sets of parents. But no one can remember which parents produced which litter. You know that gummy-bear traits have either Mendelian inheritance or incomplete dominance. Can you determine which parents produced each set of offspring and how gummy bear traits are inherited?

Ask a Question

What are the genotypes and phenotypes of the parents for each litter?

Make Observations

❶ Obtain a bag of gummy bears. Sort the bears by color (phenotype).

⚠ *Do not eat the gummy bears.*

❷ Count the number (frequency) of bears for each phenotype. Then, calculate the ratio of phenotypes for each litter.

❸ Combine data from your litter with those of your classmates using a data table like the one below.

❹ As a class, select a letter to represent the alleles for color. Record the possible genotypes for your bears in the class data table.

Gummy Bear Cross Data for Lab Group

Cross #	Phenotype Frequencies	Ratio	Possible Genotypes	Mode of Inheritance	Predicted Parental Genotypes
EXAMPLE	15 green/5 pink	3:1	GG or Gg/gg	Mendelian	Gg x Gg
1.					
2.					
3.					
4.					

Form a Hypothesis

5 Use the data to form a hypothesis about the probable genotypes and phenotypes of the parents of your litter and the probable type of inheritance.

Test Your Hypothesis

6 Design and complete a Punnett square using the predicted parental genotypes in your hypothesis.

7 Compare your litter's phenotype ratio with the ratio predicted by the Punnett square. Do your data support your hypothesis? If not, revise your hypothesis and repeat steps 5–7.

Analyze and Conclude

8 **Infer** What were the genotypes of the parents? The phenotypes? How do you know?

9 **THE BIG IDEA** **The Big Idea** Determine the probable modes of inheritance for each phenotype. Explain your reasoning.

10 **Graph** Using the data you collected, draw a bar graph that compares the phenotype frequency for each gummy bear phenotype.

Communicate Your Results

Create a video presentation of the results of your lab. Describe the question you investigated, the steps you took to answer the question, and the data that support your conclusions. Share your video with your classmates.

Inquiry Extension

Think of a question you have about genetics. For example, can you design a pedigree to trace a Mendelian trait in your family? To investigate your question, design a controlled experiment or an observational study.

6

Reminder

Using Ratios
- ☑ A ratio is a comparison of two numbers.
- ☑ A ratio of 15:5 can be reduced to 3:1.

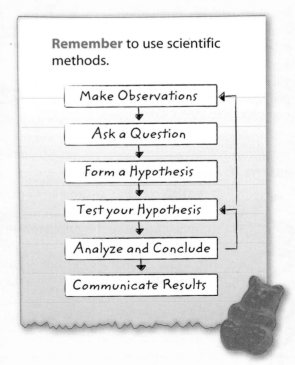

Remember to use scientific methods.

Make Observations → Ask a Question → Form a Hypothesis → Test your Hypothesis → Analyze and Conclude → Communicate Results

Inherited genes are the basis of an organism's traits.

Key Concepts Summary	Vocabulary

Lesson 1: Mendel and His Peas

- Mendel performed cross-pollination experiments to track which traits were produced by specific parental crosses.
- Mendel found that two genetic factors—one from a sperm cell and one from an egg cell—control each trait.
- **Dominant** traits block the expression of **recessive** traits. Recessive traits are expressed only when two recessive factors are present.

heredity p. 153
genetics p. 153
dominant trait p. 159
recessive trait p. 159

Lesson 2: Understanding Inheritance

- **Phenotype** describes how a trait appears.
- **Genotype** describes alleles that control a trait.
- **Punnett squares** and pedigrees are tools to model patterns of inheritance.
- Many patterns of inheritance, such as **codominance** and **polygenic inheritance,** are more complex than Mendel described.

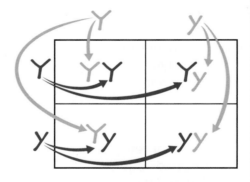

gene p. 164
allele p. 164
phenotype p. 164
genotype p. 164
homozygous p. 165
heterozygous p. 165
Punnett square p. 166
incomplete dominance p. 168
codominance p. 168
polygenic inheritance p. 169

Lesson 3: DNA and Genetics

- **DNA** contains an organism's genetic information.
- **RNA** carries the codes for making proteins from the nucleus to the cytoplasm. RNA also forms part of ribosomes.
- A change in the sequence of DNA, called a **mutation,** can change the traits of an organism.

DNA p. 174
nucleotide p. 175
replication p. 176
RNA p. 177
transcription p. 177
translation p. 178
mutation p. 179

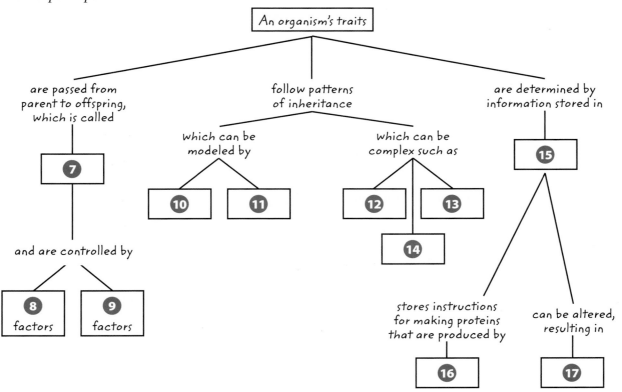
FOLDABLES® Chapter Project

Assemble your lesson Foldables as shown to make a Chapter Project. Use the project to review what you have learned in this chapter.

GENETICS

Traits
- Recessive factors
- Dominant factors

Inheritance Patterns
- Incomplete dominance
- Multiple alleles
- Polygenic inheritance

Messenger RNA
Ribosomal RNA
Transfer RNA

Use Vocabulary

1 The study of how traits are passed from parents to offspring is called _____.

2 The passing of traits from parents to offspring is _____.

3 Human height, weight, and skin color are examples of characteristics determined by _____ _____.

4 A helpful device for predicting the ratios of possible genotypes is a(n) _____.

5 The code for a protein is called a(n) _____.

6 An error made during the copying of DNA is called a(n) _____.

Link Vocabulary and Key Concepts

Concepts in Motion   Interactive Concept Map

Copy this concept map, and then use vocabulary terms from the previous page to complete the concept map.

An organism's traits

- are passed from parent to offspring, which is called
  - **7**
    - and are controlled by
      - **8** factors
      - **9** factors

- follow patterns of inheritance
  - which can be modeled by
    - **10**
    - **11**
  - which can be complex such as
    - **12**
    - **13**
      - **14**

- are determined by information stored in
  - **15**
    - stores instructions for making proteins that are produced by
      - **16**
    - can be altered, resulting in
      - **17**

Understand Key Concepts

1 The process shown below was used by Mendel during his experiments.

What is the process called?

A. cross-pollination
B. segregation
C. asexual reproduction
D. blending inheritance

2 Which statement best describes Mendel's experiments?

A. He began with hybrid plants.
B. He controlled pollination.
C. He observed only one generation.
D. He used plants that reproduce slowly.

3 Before Mendel's discoveries, which statement describes how people believed traits were inherited?

A. Parental traits blend like colors of paint to produce offspring.
B. Parental traits have no effect on their offspring.
C. Traits from only the female parent are inherited by offspring.
D. Traits from only the male parent are inherited by offspring.

4 Which term describes the offspring of a first-generation cross between parents with different forms of a trait?

A. genotype
B. hybrid
C. phenotype
D. true-breeding

5 Which process makes a copy of a DNA molecule?

A. mutation
B. replication
C. transcription
D. translation

6 Which process uses the code on an RNA molecule to make a protein?

A. mutation
B. replication
C. transcription
D. translation

7 The Punnett square below shows a cross between a pea plant with yellow seeds and a pea plant with green seeds.

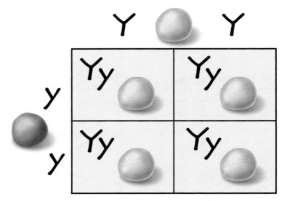

If mating produces 100 offspring, about how many will have yellow seeds?

A. 25
B. 50
C. 75
D. 100

8 Which term describes multiple genes affecting the phenotype of one trait?

A. codominance
B. blending inheritance
C. incomplete dominance
D. polygenic inheritance

Critical Thinking

9 **Compare** heterozygous genotype and homozygous genotype.

10 **Distinguish** between multiple alleles and polygenic inheritance.

11 **Give an example** of how the environment can affect an organism's phenotype.

12 **Predict** In pea plants, the allele for smooth pods is dominant to the allele for bumpy pods. Predict the genotype of a plant with bumpy pods. Can you predict the genotype of a plant with smooth pods? Explain.

13 **Interpret Graphics** In tomato plants, red fruit *(R)* is dominant to yellow fruit *(r)*. Interpret the Punnett square below, which shows a cross between a heterozygous red plant and a yellow plant. Include the possible genotypes and corresponding phenotypes.

	R	r
r	Rr	rr
r	Rr	rr

14 **Compare and contrast** characteristics of replication, transcription, translation, and mutation. Which of these processes takes place only in the nucleus of a cell? Which can take place in both the nucleus and the cytoplasm? How do you know?

Writing in Science

15 **Write** a paragraph contrasting the blending theory of inheritance with the current theory of inheritance. Include a main idea, supporting details, and a concluding sentence.

REVIEW THE BIG IDEA

16 How are traits passed from generation to generation? Explain how dominant and recessive alleles interact to determine the expression of traits.

17 The photo below shows an albino offspring from a non-albino mother. If albinism is a recessive trait, what are the possible genotypes of the mother, the father, and the offspring?

Math Skills

Review

Math Practice

Use Ratios

18 A cross between two heterozygous pea plants with yellow seeds produced 1,719 yellow seeds and 573 green seeds. What is the ratio of yellow to green seeds?

19 A cross between two heterozygous pea plants with smooth green pea pods produced 87 bumpy yellow pea pods, 261 smooth yellow pea pods, 261 bumpy green pea pods, and 783 smooth green pea pods. What is the ratio of bumpy yellow to smooth yellow to bumpy green to smooth green pea pods?

20 A jar contains three red, five green, two blue, and six yellow marbles. What is the ratio of red to green to blue to yellow marbles?

Record your answers on the answer sheet provided by your teacher or on a sheet of paper.

Multiple Choice

Use the diagram below to answer questions 1 and 2.

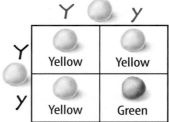

1 Which genotype belongs in the lower right square?

A YY

B Yy

C yY

D yy

2 What percentage of plants from this cross will produce yellow seeds?

A 25 percent

B 50 percent

C 75 percent

D 100 percent

3 When Mendel crossed a true-breeding plant with purple flowers and a true-breeding plant with white flowers, ALL offspring had purple flowers. This is because white flowers are

A dominant.

B heterozygous.

C polygenic.

D recessive.

4 Which process copies an organism's DNA?

A mutation

B replication

C transcription

D translation

Use the chart below to answer question 5.

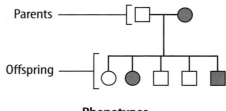

Phenotypes

○ Female, dominant   ● Female, recessive
□ Male, dominant   ■ Male, recessive

5 Based on the pedigree above, how many offspring from this cross had the recessive phenotype?

A 1

B 2

C 3

D 5

6 Which is NOT true of a hybrid?

A It has one recessive allele.

B It has pairs of chromosomes.

C Its genotype is homozygous.

D Its phenotype is dominant.

7 Alleles are different forms of a

A chromosome.

B gene.

C nucleotide.

D protein.

8 Which is true of an offspring with incomplete dominance?

A Both alleles can be observed in its phenotype.

B Every offspring shows the dominant phenotype.

C Multiple genes determine its phenotype.

D Offspring phenotype is a combination of the parents' phenotypes.

Use the diagrams below to answer question 9.

Before Replication

After Replication

9 The diagrams above show a segment of DNA before and after replication. Which occurred during replication?

   A deletion

   B insertion

   C substitution

   D translation

10 Which human characteristic is controlled by polygenic inheritance?

   A blood type

   B earlobe position

   C eye color

   D thumb shape

11 Mendel crossed a true-breeding plant with round seeds and a true-breeding plant with wrinkled seeds. Which was true of every offspring of this cross?

   A They had the recessive phenotype.

   B They showed a combination of traits.

   C They were homozygous.

   D They were hybrid plants.

Constructed Response

Use the diagram below to answer questions 12 and 13.

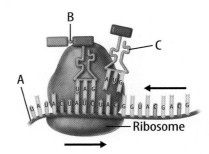

Ribosome

12 Describe what is happening in the phase of translation shown in the diagram.

13 What are the three types of RNA in the diagram? How do these types work together during translation?

14 What is the importance of translation in your body?

15 Mendel began his experiments with true-breeding plants. Why was this important?

16 How did Mendel's experimental methods help him develop his hypotheses on inheritance?

17 What environmental factors affect the phenotypes of organisms other than humans? Provide three examples from nature. What factor, other than genes, affects human phenotype? Give two examples. Why is knowledge of this non-genetic factor helpful?

NEED EXTRA HELP?																	
If You Missed Question...	1	2	3	4	5	6	7	8	9	10	11	12	13	14	15	16	17
Go to Lesson...	2	2	1	3	2	1,2	2	2	3	2	1	3	3	3	1	1	2

The Environment and Change Over Time

THE BIG IDEA How do species adapt to changing environments over time?

Inquiry Swarm of Bees?

A type of orchid plant, called a bee orchid, produces this flower. You might have noticed that the flower looks like a bee.

- What is the advantage to the plant to have flowers that look like bees?

- How did the appearance of the flower develop over time?

- How do species adapt to changing environments over time?

Get Ready to Read

What do you think?

Before you read, decide if you agree or disagree with each of these statements. As you read this chapter, see if you change your mind about any of the statements.

1 Original tissues can be preserved as fossils.

2 Organisms become extinct only in mass extinction events.

3 Environmental change causes variations in populations.

4 Variations can lead to adaptations.

5 Living species contain no evidence that they are related to each other.

6 Plants and animals share similar genes.

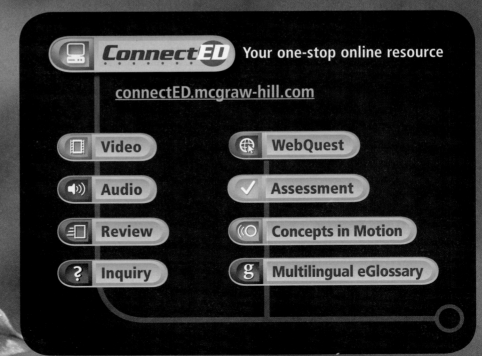

ConnectED Your one-stop online resource

connectED.mcgraw-hill.com

Video

WebQuest

Audio

Assessment

Review

Concepts in Motion

Inquiry

Multilingual eGlossary

Lesson 1

Reading Guide

Key Concepts
ESSENTIAL QUESTIONS

- How do fossils form?
- How do scientists date fossils?
- How are fossils evidence of biological evolution?

Vocabulary

fossil record p. 193

mold p. 195

cast p. 195

trace fossil p. 195

geologic time scale p. 197

extinction p. 198

biological evolution p. 199

 g Multilingual eGlossary

Video BrainPOP®

Fossil Evidence of Evolution

Inquiry What can be learned from fossils?

When scientists find fossils, they use them as evidence to try to answer questions about past life on Earth. When did this organism live? What did this organism eat? How did it move or grow? How did this organism die? To what other organisms is this one related?

How do fossils form?

Evidence from fossils helps scientists understand how organisms have changed over time. Some fossils form when impressions left by organisms in sand or mud are filled in by sediments that harden.

1. Read and complete a lab safety form.

2. Place a **container of moist sand** on top of **newspaper.** Press a **shell** into the moist sand. Carefully remove the shell. Brush any sand on the shell onto the newspaper.

3. Observe the impression, and record your observations in your Science Journal.

4. Pour **plaster of paris** into the impression. Wait for it to harden.
⚠ *The mix gets hot as it sets—do not touch it until it has hardened.*

5. Remove the shell fossil from the sand, and brush it off.

6. Observe the structure of the fossil, and record your observations.

Think About This

1. What effect did the shell have on the sand?

2. 🔑 **Key Concept** What information do you think someone could learn about the shell and the organism that lived inside it by examining the fossil?

The Fossil Record

On your way to school, you might have seen an oak tree or heard a robin. Although these organisms shed leaves or feathers, their characteristics remain the same from day to day. It might seem as if they have been on Earth forever. However, if you were to travel a few million years back in time, you would not see oak trees or robins. You would see different species of trees and birds. That is because species change over time.

You might already know that fossils are the remains or evidence of once-living organisms. *The* **fossil record** *is made up of all the fossils ever discovered on Earth.* It contains millions of fossils that represent many thousands of species. Most of these species are no longer alive on Earth. The fossil record provides evidence that species have changed over time. Fossils help scientists picture what these species looked like. **Figure 1** shows how scientists think the giant bird *Titanus* might have looked when it was alive. The image is based on fossils that have been discovered and are represented in the photo on the previous page.

The fossil record is enormous, but it is still incomplete. Scientists think it represents only a small fraction of all the organisms that have ever lived on Earth.

Figure 1 Based on fossil evidence, scientists can recreate the physical appearance of species that are no longer alive on Earth.

Fossil Formation

If you have ever seen vultures or other animals eating a dead animal, you know they leave little behind. Any soft **tissues** animals do not eat, bacteria break down. Only the dead animal's hard parts, such as bones, shells, and teeth, remain. In most instances, these hard parts also break down over time. However, under rare conditions, some become fossils. The soft tissues of animals and plants, such as skin, muscles, or leaves, can also become fossils, but these are even more rare. Some of the ways that fossils can form are shown in **Table 1.**

Reading Check Why is it rare for soft tissue to become a fossil?

Mineralization

After an organism dies, its body could be buried under mud, sand, or other sediments in a stream or river. If minerals in the water replace the organism's original material and harden into rock, a fossil forms. This process is called mineralization. Minerals in water also can filter into the small spaces of a dead organism's tissues and become rock. Most mineralized fossils are of shell or bone, but wood can also become a mineralized fossil, as shown in **Table 1.**

Carbonization

In carbonization, a fossil forms when a dead organism is compressed over time and pressure drives off the organism's liquids and gases. As shown in **Table 1,** only the carbon outline, or film, of the organism remains.

SCIENCE USE V. COMMON USE

tissue
Science Use similar cells that work together and perform a function

Common Use a piece of soft, absorbent paper

Table 1 Fossils form in several ways.

 Visual Check What types of organisms or tissues are often preserved as carbon films?

Table 1 How Fossils Form		
	Mineralization	**Carbonization**
Description	Rock-forming minerals, such as calcium carbonate ($CaCO_3$), in water filled in the small spaces in the tissue of these pieces of petrified wood. Water also replaced some of the wood's tissue. Mineralization can preserve the internal structures of an organism.	Fossil films made by carbonization are usually black or dark brown. Fish, insects, and plant leaves, such as this fern frond, are often preserved as carbon films.
Example		

Molds and Casts

Sometimes when an organism dies, its shell or bone might make an impression in mud or sand. When the sediment hardens, so does the impression. *The impression of an organism in a rock is called a **mold.*** Sediments can later fill in the mold and harden to form a cast. *A **cast** is a fossil copy of an organism in a rock.* A single organism can form both a mold and a cast, as shown in **Table 1.** Molds and casts show only external features of organisms.

Trace Fossils

Evidence of an organism's movement or behavior—not just its physical structure—also can be preserved in rock. *A **trace fossil** is the preserved evidence of the activity of an organism.* For example, an organism might walk across mud. The tracks, such as the ones shown in **Table 1,** can fossilize if they are filled with sediment that hardens.

Original Material

In rare cases, the original tissues of an organism can be preserved. Examples of original-material fossils include mammoths frozen in ice and saber-toothed cats preserved in tar pits. Fossilized remains of ancient humans have been found in bogs. Most of these fossils are younger than 10,000 years old. However, the insect encased in amber in **Table 1** is millions of years old. Scientists also have found original tissue preserved in the bone of a dinosaur that lived 70 million years ago (mya).

Key Concept Check List the different ways fossils can form.

WORD ORIGIN

fossil
from Latin *fossilis*, means "to obtain by digging"

Molds and Casts	Trace Fossils	Original Material
When sediments hardened around this buried trilobite, a mold formed. Molds are usually of hard parts, such as shells or bone. If a mold is later filled with more sediments that harden, the mold can form a cast.	These footprints were made when a dinosaur walked across mud that later hardened. This trace fossil might provide evidence of the speed and weight of the dinosaur.	If original tissues of organisms are buried in the absence of oxygen for long periods of time, they can fossilize. The insect in this amber became stuck in tree sap that later hardened.

Dating Fossils 🔑

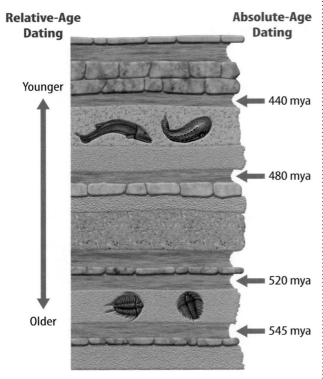

Relative-Age Dating Younger · Older **Absolute-Age Dating**

← 440 mya
← 480 mya
← 520 mya
← 545 mya

Figure 2 If the age of the igneous layers is known, as shown above, it is possible to estimate the age of the sedimentary layers—and the fossils they contain—between them.

✓ **Visual Check** What is the estimated age of the trilobite fossils (bottom layer of fossils)?

REVIEW VOCABULARY

isotopes
atoms of the same element that have different numbers of neutrons

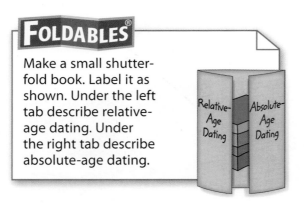

FOLDABLES®

Make a small shutter-fold book. Label it as shown. Under the left tab describe relative-age dating. Under the right tab describe absolute-age dating.

Relative-Age Dating Absolute-Age Dating

Determining a Fossil's Age

Scientists cannot date most fossils directly. Instead, they date the rocks the fossils are embedded inside. Rocks erode or are recycled over time. However, scientists can determine ages for most of Earth's rocks.

Relative-Age Dating

How does your age compare to the ages of those around you? You might be younger than a brother but older than a sister. This is your relative age. Similarly, a rock is either older or younger than rocks nearby. In relative-age dating, scientists determine the relative order in which rock layers were deposited. In an undisturbed rock formation, they know that the bottom layers are oldest and the top layers are youngest, as shown in **Figure 2.** Relative-age dating helps scientists determine the relative order in which species have appeared on Earth over time.

🔑 **Key Concept Check** How does relative-age dating help scientists learn about fossils?

Absolute-Age Dating

Absolute-age dating is more precise than relative-age dating. Scientists take advantage of radioactive decay, a natural clocklike process in rocks, to learn a rock's absolute age, or its age in years. In radioactive decay, unstable **isotopes** in rocks change into stable isotopes over time. Scientists measure the ratio of unstable isotopes to stable isotopes to find the age of a rock. This ratio is best measured in igneous rocks.

Igneous rocks form from volcanic magma. Magma is so hot that it is rare for parts of organisms in it to remain and form fossils. Most fossils form in sediments, which become sedimentary rock. To measure the age of sedimentary rock layers, scientists calculate the ages of igneous layers above and below them. In this way, they can estimate the ages of the fossils embedded within the sedimentary layers, as shown in **Figure 2.**

Fossils over Time

How old do you think Earth's oldest fossils are? You might be surprised to learn that evidence of microscopic, unicellular organisms has been found in rocks 3.4 billion years old. The oldest fossils visible to the unaided eye are about 565 million years old.

The Geologic Time Scale

It is hard to keep track of time that is millions and billions of years long. Scientists organize Earth's history into a time line called the geologic time scale. *The* **geologic time scale** *is a chart that divides Earth's history into different time units.* The longest time units in the geological time scale are eons. As shown in **Figure 3**, Earth's history is divided into four eons. Earth's most recent eon—the Phanerozoic (fa nuh ruh ZOH ihk) eon—is subdivided into three eras, also shown in **Figure 3**.

 Reading Check What is the geologic time scale?

Dividing Time

You might have noticed in **Figure 3** that neither eons nor eras are equal in length. When scientists began developing the geologic time scale in the 1800s, they did not have absolute-age dating methods. To mark time boundaries, they used fossils. Fossils provided an easy way to mark time. Scientists knew that different rock layers contained different types of fossils. Some of the fossils scientists use to mark the time boundaries are shown in **Figure 3**.

Often, a type of fossil found in one rock layer did not appear in layers above it. Even more surprising, entire collections of fossils in one layer were sometimes absent from layers above them. It seemed as if whole communities of organisms had suddenly disappeared.

 Reading Check What do scientists use to mark boundaries in the geologic time scale?

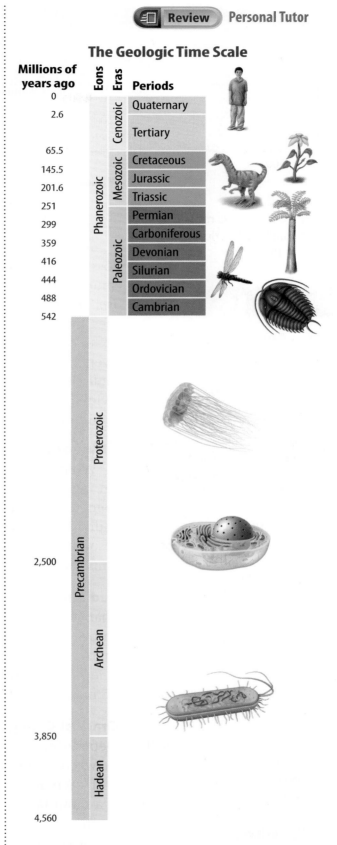

The Geologic Time Scale

Millions of years ago	Eons	Eras	Periods
0		Cenozoic	Quaternary
2.6			Tertiary
65.5	Phanerozoic	Mesozoic	Cretaceous
145.5			Jurassic
201.6			Triassic
251		Paleozoic	Permian
299			Carboniferous
359			Devonian
416			Silurian
444			Ordovician
488			Cambrian
542			
	Proterozoic		
2,500	Precambrian		
	Archean		
3,850			
	Hadean		
4,560			

Figure 3 The Phanerozoic eon began about 540 million years ago and continues to the present day. It contains most of Earth's fossil record.

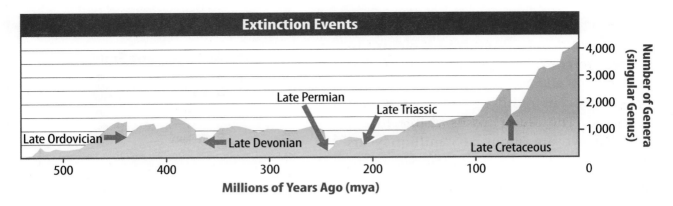

Extinction Events

Late Permian

Late Triassic

Late Ordovician →

← Late Devonian

Late Cretaceous

Number of Genera (singular Genus)

500   400   300   200   100   0

Millions of Years Ago (mya)

Figure 4 Arrows mark the five major extinction events of the Phanerozoic eon.

Math Skills

Use Scientific Notation

Numbers that refer to the ages of Earth's fossils are very large, so scientists use scientific notation to work with them. For example, mammals appeared on Earth about 200 mya or 200,000,000 years ago. Change this number to scientific notation using the following process.

Move the decimal point until only one nonzero digit remains on the left.

200,000,000 = 2.00000000

Count the number of places you moved the decimal point (8) and use that number as a power of ten.

$200,000,000 = 2.0 \times 10^8$ years.

Practice

The first vertebrates appeared on Earth about 490,000,000 years ago. Express this time in scientific notation.

Review

- **Math Practice**
- **Personal Tutor**

Extinctions

Scientists now understand that sudden disappearances of fossils in rock layers are evidence of extinction (ihk STINGK shun) events. **Extinction** *occurs when the last individual organism of a species dies.* A mass extinction occurs when many species become extinct within a few million years or less. The fossil record contains evidence that five mass extinction events have occurred during the Phanerozoic eon, as shown in **Figure 4.** Extinctions also occur at other times, on smaller scales. Evidence from the fossil record suggests extinctions have been common throughout Earth's history.

Environmental Change

What causes extinctions? Populations of organisms depend on resources in their environment for food and shelter. Sometimes environments change. After a change happens, individual organisms of a species might not be able to find the resources they need to survive. When this happens, the organisms die, and the species becomes extinct.

Sudden Changes Extinctions can occur when environments change quickly. A volcanic eruption or a meteorite impact can throw ash and dust into the atmosphere, blocking sunlight for many years. This can affect global climate and food webs. Scientists hypothesize that the impact of a huge meteorite 65 million years ago contributed to the extinction of dinosaurs.

Gradual Changes Not all environmental change is sudden. Depending on the location, Earth's tectonic plates move between 1 and 15 cm each year. As plates move and collide with each other over time, mountains form and oceans develop. If a mountain range or an ocean isolates a species, the species might become extinct if it cannot find the resources it needs. Species also might become extinct if sea level changes.

 Reading Check What is the relationship between extinction and environmental change?

Extinctions and Evolution

The fossil record contains clear evidence of the extinction of species over time. But it also contains evidence of the appearance of many new species. How do new species arise?

Many early scientists thought that each species appeared on Earth independently of every other species. However, as more fossils were discovered, patterns in the fossil record began to emerge. Many fossil species in nearby rock layers had similar body plans and similar structures. It appeared as if they were related. For example, the series of horse fossils in **Figure 5** suggests that the modern horse is related to other extinct species. These species changed over time in what appeared to be a sequence. Change over time is evolution. **Biological evolution** *is the change over time in populations of related organisms.* Charles Darwin developed a theory about how species evolve from other species. You will read about Darwin's theory in the next lesson.

Key Concept Check How are fossils evidence of biological evolution?

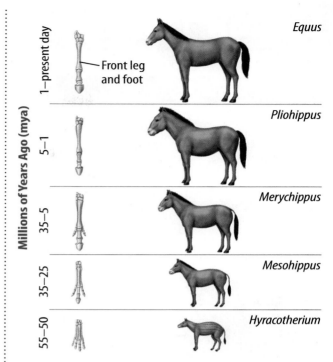

Front leg and foot

Equus

Pliohippus

Merychippus

Mesohippus

Hyracotherium

Figure 5  The fossil record is evidence that horses descended from organisms for which only fossils exist today.

Inquiry MiniLab

20 minutes

How do species change over time?

Over long time periods on Earth, certain individuals within populations of organisms were able to survive better than others.

1. Choose a species from the **Species I.D. Cards.**

2. On **chart paper,** draw six squares in a row and number them 1–6, respectively. Use **colored pencils** and **markers** to make a comic strip showing the ancestral and present-day forms of your species in frames 1 and 6.

3. Use information from the I.D. Card to show what you think would be the progression of changes in the species in frames 2–5.

4. In speech bubbles, explain how each change helped the species to survive.

Analyze and Conclude

1. **Infer** why a scientist would identify a fossil from the species in the first frame of your cartoon as the ancestral form of the present-day species.

2. **Key Concept** How would the fossils of the species at each stage provide evidence of biological change over time?

Lesson 1 Review

Visual Summary

Fossils can consist of the hard parts or soft parts of organisms. Fossils can be an impression of an organism or consist of original tissues.

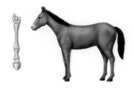

Scientists determine the age of a fossil through relative-age dating or absolute-age dating.

Scientists use fossils as evidence that species have changed over time.

FOLDABLES

Use your lesson Foldable to review the lesson. Save your Foldable for the project at the end of the chapter.

What do you think NOW?

You first read the statements below at the beginning of the chapter.

1. Original tissues can be preserved as fossils.

2. Organisms become extinct only in mass extinction events.

Did you change your mind about whether you agree or disagree with the statements? Rewrite any false statements to make them true.

Use Vocabulary

1 All of the fossils ever found on Earth make up the _____.

2 When the last individual of a species dies, _____ occurs.

3 **Use the term** *biological evolution* in a sentence.

Understand Key Concepts

4 Which is the preserved evidence of the activity of an organism?
   A. cast   **C.** fossil film
   B. mold   **D.** trace fossil

5 **Explain** why the hard parts of organisms fossilize more often than soft parts.

6 **Draw and label** a diagram that shows how scientists date sedimentary rock layers.

Interpret Graphics

7 **Identify** Copy and fill in the table below to provide examples of changes that might lead to an extinction event.

Sudden changes	
Gradual changes	

Critical Thinking

8 **Infer** If the rock layers shown below have not been disturbed, what type of dating method would help you determine which layer is oldest? Explain.

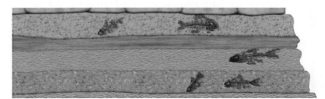

Math Skills

Review
———Math Practice———

9 Dinosaurs disappeared from Earth about 65,000,000 years ago. Express this number in scientific notation.

Can you observe changes through time in collections of everyday objects?

Everyday objects that are invented, designed, and manufactured by humans often exhibit changes over time in both structure and function. How have these changes affected the efficiency and/or safety of some common items?

Materials

picture sets of items that have changed over time

Learn It

When scientists **observe** phenomena, they use their senses, such as sight, hearing, touch, and smell. They examine the entire object or situation first, then look carefully for details. After completing their observations, scientists use words or numbers to describe what they saw.

Try It

1 Working with your group members, choose a set of items that you wish to observe, such as telephones, bicycles, or automobiles.

2 Examine the pictures and observe how the item has changed over time.

3 Record your observations in your Science Journal.

4 Observe details of the structure and function of each of the items. Record your observations.

Apply It

5 **Present** your results in the form of an illustrated time line, a consumer magazine article, a role-play of a person-on-the-street interview, a television advertisement, or an idea of your own approved by your teacher.

6 🔑 **Key Concept** Identify how your product changed over time and in what ways the changes affected the efficiency and/or safety of the product.

Lesson 2

Theory of Evolution by Natural Selection

Reading Guide

Key Concepts

ESSENTIAL QUESTIONS

- Who was Charles Darwin?
- How does Darwin's theory of evolution by natural selection explain how species change over time?
- How are adaptations evidence of natural selection?

Vocabulary

naturalist p. 203

variation p. 205

natural selection p. 206

adaptation p. 207

camouflage p. 208

mimicry p. 208

selective breeding p. 209

 Multilingual eGlossary

Video

What's Science Got to do With It?

inquiry **Are these exactly the same?**

Look closely at these zebras. Are they all exactly the same? How are they different? What accounts for these differences? How do the stripes help these organisms survive in their environment?

Inquiry Launch Lab

20 minutes

Are there variations within your class?

All populations contain variations in some characteristics of their members.

1. Read and complete a lab safety form.
2. Use a **meterstick** to measure the length from your elbow to the tip of your middle finger in centimeters. Record the measurement in your Science Journal.
3. Add your measurement to the class list.
4. Organize all of the measurements from shortest to longest.
5. Break the data into regular increments, such as 31–35 cm, 36–40 cm, and 41–45 cm. Count the number of measurements within each increment.
6. Construct a bar graph using the data. Label each axis and give your graph a title.

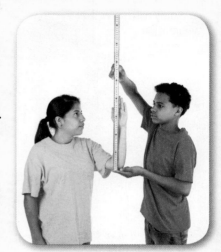

Think About This

1. What are the shortest and longest measurements?
2. How much do the shortest and longest lengths vary from each other?
3. **Key Concept** Describe how your results provide evidence of variations within your classroom population.

Charles Darwin

How many species of birds can you name? You might think of robins, penguins, or even chickens. Scientists estimate that about 10,000 species of birds live on Earth today. Each bird species has similar characteristics. Each has wings, feathers, and a beak. Scientists hypothesize that all birds evolved from an earlier, or ancestral, population of birdlike organisms. As this population evolved into different species, birds became different sizes and colors. They developed different songs and eating habits, but all retained similar bird characteristics.

How do birds and other species evolve? One scientist who worked to answer this question was Charles Darwin. Darwin was an English naturalist who, in the mid-1800s, developed a theory of how evolution works. *A* **naturalist** *is a person who studies plants and animals by observing them.* Darwin spent many years observing plants and animals in their natural habitats before developing his theory. Recall that a theory is an explanation of the natural world that is well supported by evidence. Darwin was not the first to develop a theory of evolution, but his theory is the one best supported by evidence today.

Key Concept Check Who was Charles Darwin?

FOLDABLES

Make a small four-door shutterfold book. Use it to investigate the who, what, when, and where of Charles Darwin, the Galápagos Islands, and the theory of evolution by natural selection.

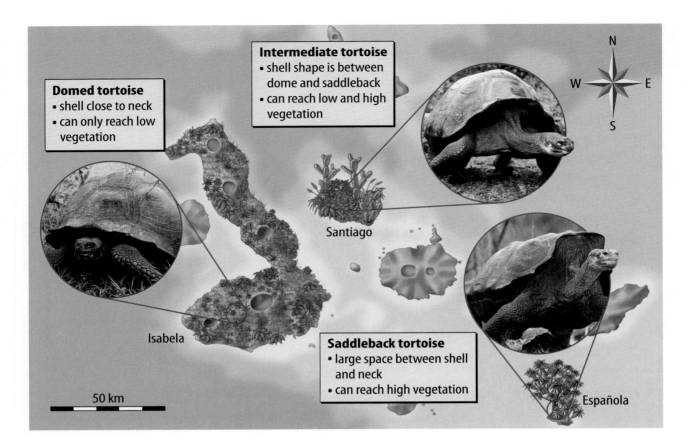

Domed tortoise
- shell close to neck
- can only reach low vegetation

Intermediate tortoise
- shell shape is between dome and saddleback
- can reach low and high vegetation

Saddleback tortoise
- large space between shell and neck
- can reach high vegetation

Santiago

Isabela

Española

50 km

Figure 6 Each island in the Galápagos has a different environment. Tortoises look different depending on which island environment they inhabit.

Visual Check What type of vegetation do domed tortoises eat?

Voyage of the *Beagle*

Darwin served as a naturalist on the HMS *Beagle*, a survey ship of the British navy. During his voyage around the world, Darwin observed and collected many plants and animals.

The Galápagos Islands

Darwin was especially interested in the organisms he saw on the Galápagos (guh LAH puh gus) Islands. The islands, shown in **Figure 6,** are located 1,000 km off the South American coast in the Pacific Ocean. Darwin saw that each island had a slightly different environment. Some were dry. Some were more humid. Others had mixed environments.

Tortoises Giant tortoises lived on many of the islands. When a resident told him that the tortoises on each island looked different, as shown in **Figure 6,** Darwin became curious.

Mockingbirds and Finches Darwin also became curious about the variety of mockingbirds and finches he saw and collected on the islands. Like the tortoises, different types of mockingbirds and finches lived in different island environments. Later, he was surprised to learn that many of these varieties were different enough to be separate species.

Reading Check What made Darwin become curious about the organisms that lived on the Galápagos Islands?

Darwin's Theory

Darwin realized there was a relationship between each species and the food sources of the island it lived on. Look again at **Figure 6.** You can see that tortoises with long necks lived on islands that had tall cacti. Their long necks enabled them to reach high to eat the cacti. The tortoises with short necks lived on islands that had plenty of short grass.

Common Ancestors

Darwin became convinced that all the tortoise species were related. He thought they all shared a common ancestor. He suspected that a storm had carried a small ancestral tortoise population to one of the islands from South America millions of years before. Eventually, the tortoises spread to the other islands. Their neck lengths and shell shapes changed to match their islands' food sources. How did this happen?

Variations

Darwin knew that individual members of a species exhibit slight differences, or variations. *A **variation** is a slight difference in an inherited trait of individual members of a species.* Even though the snail shells in **Figure 7** are not all exactly the same, they are all from snails of the same species. You can also see variations in the zebras in the photo at the beginning of this lesson. Variations arise naturally in populations. They occur in the offspring as a result of sexual reproduction. You might recall that variations are caused by random mutations, or changes, in genes. Mutations can lead to changes in phenotype. Recall that an organism's phenotype is all of the observable traits and characteristics of the organism. Genetic changes to phenotype can be passed on to future generations.

ACADEMIC VOCABULARY

convince
(verb) to overcome by argument

Figure 7 The variations among the shells of a species of tree snail occur naturally within the population.

Visual Check Describe three variations among these snail shells.

Natural Selection

Darwin did not know about genes. But he realized that variations were the key to the puzzle of how populations of tortoises and other organisms evolved. Darwin understood that food is a limiting resource, which means that the food in each island environment could not support every tortoise that was born. Tortoises had to compete with each other for food. As the tortoises spread to the various islands, some were born with random variations in neck length. If a variation benefited a tortoise, allowing it to compete for food better than other tortoises, the tortoise lived longer. Because it lived longer, it reproduced more. It passed on its variations to its offspring.

This is Darwin's theory of evolution by natural selection. **Natural selection** *is the process by which populations of organisms with variations that help them survive in their environments live longer, compete better, and reproduce more than those that do not have the variations.* Natural selection explains how populations change as their environments change. It explains the process by which Galápagos tortoises became matched to their food sources, as illustrated in **Figure 8**. It also explains the diversity of the Galápagos finches and mockingbirds. Birds with beak variations that help them compete for food live longer and reproduce more.

Key Concept Check What role do variations have in the theory of evolution by natural selection?

Natural Selection

Review   Personal Tutor

❶ Reproduction
A population of tortoises produces many offspring that inherit its characteristics.

❷ Variation
A tortoise is born with a variation that makes its neck slightly longer.

❸ Competition
Due to limited resources, not all offspring will survive. An offspring with a longer neck can eat more cacti than other tortoises. It lives longer and produces more offspring.

❹ Selection
Over time, the variation is inherited by more and more offspring. Eventually, all tortoises have longer necks.

Figure 8 A beneficial variation in neck length spreads through a tortoise population by natural selection.

Adaptations

Natural selection explains how all species change over time as their environments change. Through natural selection, a helpful variation in one individual can be passed on to future members of a population. As time passes, more variations arise. The accumulation of many similar variations can lead to an adaptation (a dap TAY shun). *An* **adaptation** *is an inherited trait that increases an organism's chance of surviving and reproducing in its environment.* The long neck of certain species of tortoises is an adaptation to an environment with tall cacti.

WORD ORIGIN · · · · · · · · · ·

adaptation
from Latin *adaptare*, means "to fit"

 Key Concept Check How do variations lead to adaptations?

Types of Adaptations

Every species has many adaptations. Scientists classify adaptations into three categories: structural, behavioral, and functional. Structural adaptations involve color, shape, and other physical characteristics. The shape of a tortoise's neck is a structural adaptation. Behavioral adaptations involve the way an organism behaves or acts. Hunting at night and moving in herds are examples of behavioral adaptations. Functional adaptations involve internal body systems that affect **biochemistry.** A drop in body temperature during hibernation is an example of a functional adaptation. **Figure 9** illustrates examples of all three types of adaptations in the desert jackrabbit.

REVIEW VOCABULARY · · · ·

biochemistry
the study of chemical processes in living organisms

Figure 9 The desert jackrabbit has structural, behavioral, and functional adaptations. These adaptations enable it to survive in its desert environment.

Structural adaptation The jackrabbit's powerful legs help it run fast to escape from predators.

Behavioral adaptation The jackrabbit stays still during the hottest part of the day, helping it conserve energy.

Functional adaptation The blood vessels in the jackrabbit's ears expand to enable the blood to cool before re-entering the body.

Seahorse

Caterpillar

Pelican

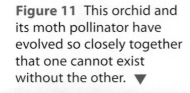
▲ Figure 10
Species evolve adaptations as they interact with their environments, which include other species.

Figure 11 This orchid and its moth pollinator have evolved so closely together that one cannot exist without the other. ▼

Environmental Interactions

Have you ever wanted to be invisible? Many species have evolved adaptations that make them nearly invisible. The seahorse in **Figure 10** is the same color and has a texture similar to the coral it is resting on. This is a structural adaptation called camouflage (KAM uh flahj). **Camouflage** *is an adaptation that enables a species to blend in with its environment.*

Some species have adaptations that draw attention to them. The caterpillar in **Figure 10** resembles a snake. Predators see it and are scared away. *The resemblance of one species to another species is* **mimicry** (MIH mih kree). Camouflage and mimicry are adaptations that help species avoid being eaten. Many other adaptations help species eat. The pelican in **Figure 10** has a beak and mouth uniquely adapted to its food source—fish.

✓ **Reading Check** How do camouflage and mimicry differ?

Environments are complex. Species must adapt to an environment's living parts as well as to an environment's nonliving parts. Nonliving things include temperature, water, nutrients in soil, and climate. Deciduous trees shed their leaves due to changes in climate. Camouflage, mimicry, and mouth shape are adaptations mostly to an environment's living parts. An extreme example of two species adapting to each other is shown in **Figure 11.**

Living and nonliving factors are always changing. Even slight environmental changes affect how species adapt. If a species is unable to adapt, it becomes extinct. The fossil record contains many fossils of species unable to adapt to change.

Artificial Selection

Adaptations provide evidence of how closely Earth's species match their environments. This is exactly what Darwin's theory of evolution by natural selection predicted. Darwin provided many examples of adaptation in *On the Origin of Species,* the book he wrote to explain his theory. Darwin did not write this book until 20 years after he developed his theory. He spent those years collecting more evidence for his theory by studying barnacles, orchids, corals, and earthworms.

Darwin also had a hobby of breeding domestic pigeons. He selectively bred pigeons of different colors and shapes to produce new, fancy varieties. *The breeding of organisms for desired characteristics is called* **selective breeding.** Like many domestic plants and animals produced from selective breeding, pigeons look different from their ancestors, as shown in **Figure 12.** Darwin realized that changes caused by selective breeding were much like changes caused by natural selection. Instead of nature selecting variations, humans selected them. Darwin called this process artificial selection.

Artificial selection explains and supports Darwin's theory. As you will read in Lesson 3, other evidence also supports the idea that species evolve from other species.

Figure 12 The pouter pigeon (bottom left) and the fantail pigeon (bottom right) were derived from the wild rock pigeon (top).

Inquiry MiniLab

20 minutes

Who survives?

Camouflage helps organisms blend in. This can help them avoid predators or sneak up on prey. Camouflage helps organisms survive in their environments.

1. Read and complete a lab safety form.
2. Choose an area of your classroom where your moth will rest with open wings during the day.
3. Use **scissors, paper, markers,** and a **ruler** to design a moth that measures 2–5 cm in width with open wings and will be camouflaged where it is placed. Write the location where the moth is to be placed. Give the location and your completed moth to your teacher.
4. On the following day, you will have 1 minute to spot as many moths in the room as you can.
5. In your Science Journal, record the location of moths spotted by your team.

6. Find the remaining moths that were not spotted. Observe their appearance.

Analyze and Conclude

1. **Compare** the appearances and resting places of the moths that were spotted with those that were not spotted.
2. **Key Concept** Explain how camouflage enables an organism to survive in its environment.

Visual Summary

Charles Darwin developed his theory of evolution partly by observing organisms in their natural environments.

Natural selection occurs when organisms with certain variations live longer, compete better, and reproduce more often than organisms that do not have the variations.

Adaptations occur when a beneficial variation is eventually inherited by all members of a population.

FOLDABLES

Use your lesson Foldable to review the lesson. Save your Foldable for the project at the end of the chapter.

What do you think NOW?

You first read the statements below at the beginning of the chapter.

3. Environmental change causes variations in populations.

4. Variations can lead to adaptations.

Did you change your mind about whether you agree or disagree with the statements? Rewrite any false statements to make them true.

Use Vocabulary

1 A person who studies plants and animals by observing them is a(n) _____.

2 Through _____, populations of organisms adapt to their environments.

3 Some species blend in to their environments through _____.

Understand Key Concepts

4 The observation that the Galápagos tortoises did not all live in the same environment helped Darwin
   A. develop his theory of adaptation.
   B. develop his theory of evolution.
   C. observe mimicry in nature.
   D. practice artificial selection.

5 **Assess** the importance of variations to natural selection.

6 **Compare and contrast** natural selection and artificial selection.

Interpret Graphics

7 **Explain** how the shape of the walking stick at right helps the insect survive in its environment.

8 **Sequence** Copy the graphic organizer below and sequence the steps by which a population of organisms changes by natural selection.

Critical Thinking

9 **Conclude** how Earth's birds developed their diversity through natural selection.

Peter and Rosemary Grant

Observing Natural Selection

Charles Darwin was a naturalist during the mid-1800s. Based on his observations of nature, he developed the theory of evolution by natural selection. Do scientists still work this way—drawing conclusions from observations? Is there information still to be learned about natural selection? The answer to both questions is yes.

Peter and Rosemary Grant are naturalists who have observed finches in the Galápagos Islands for more than 30 years. They have found that variations in the finches' food supply determine which birds will survive and reproduce. They have observed natural selection in action.

The Grants live on Daphne Major, an island in the Galápagos, for part of each year. They observe and take measurements to compare the size and shape of finches' beaks from year to year. They also examine the kinds of seeds and nuts available for the birds to eat. They use this information to relate changes in the birds' food supply to changes in the finch species' beaks.

The island's ecosystem is fragile, so the Grants take great care not to change the environment of Daphne Major as they observe the finches. They carefully plan their diet to avoid introducing new plant species to the island. They bring all the freshwater they need to drink, and they wash in the ocean. For the Grants, it's just part of the job. As naturalists, they try to observe without interfering with the habitat in which they are living.

▲ Peter and Rosemary Grant make observations and collect data in the field.

▲ This large ground finch is one of the kinds of birds studied by the Grants.

It's Your Turn

RESEARCH AND REPORT Find out more about careers in evolution, ecology, or population biology. What kind of work is done in the laboratory? What kind of work is done in the field? Write a report to explain your findings.

Reading Guide

Key Concepts 🔑
ESSENTIAL QUESTIONS

- What evidence from living species supports the theory that species descended from other species over time?

- How are Earth's organisms related?

Vocabulary

comparative anatomy p. 214

homologous structure p. 214

analogous structure p. 215

vestigial structure p. 215

embryology p. 216

g **Multilingual eGlossary**

Biological Evidence of Evolution

Inquiry Does this bird fly?

Some birds, such as the flightless cormorant above, have wings but cannot fly. Their wings are too small to support their bodies in flight. Why do they still have wings? What can scientists learn about the ancestors of present-day birds that have wings but do not fly?

How is the structure of a spoon related to its function?

Would you eat your morning cereal with a spoon that had holes in it? Is using a teaspoon the most efficient way to serve mashed potatoes and gravy to a large group of people? How about using an extra large spoon, or ladle, to eat soup from a small bowl?

1. Read and complete a lab safety form.

2. In a small group, examine your **set of spoons** and discuss your observations.

3. Sketch or describe the structure of each spoon in your Science Journal. Discuss the purpose that each spoon shape might serve.

4. Label the spoons in your Science Journal with their purposes.

Think About This

1. Describe the similarities and differences among the spoons.

2. If spoons were organisms, what do you think the ancestral spoon would look like?

3. 🔑 **Key Concept** Explain how three of the spoons have different structures and functions, even though they are related by their similarities.

Evidence for Evolution

Recall the sequence of horse fossils from Lesson 1. The sequence might have suggested to you that horses evolved in a straight line—that one species replaced another in a series of orderly steps. Evolution does not occur this way. The diagram in **Figure 13** shows a more realistic version of horse evolution, which looks more like a bush than a straight line. Different horse species were sometimes alive at the same time. They are related to each other because each descended from a common ancestor.

Living species that are closely related share a close common ancestor. The degree to which species are related depends on how closely in time they diverged, or split, from their common ancestor. Although the fossil record is incomplete, it contains many examples of fossil sequences showing close ancestral relationships. Living species show evidence of common ancestry, too.

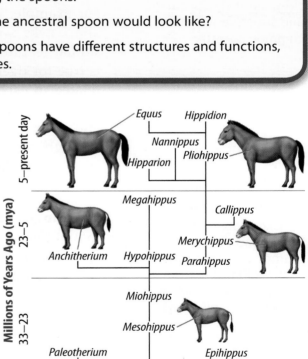

Figure 13 The fossil record indicates that different species of horses often overlapped with each other.

🔍 **Visual Check** Which horse is the common ancestor to all horse species in this graph?

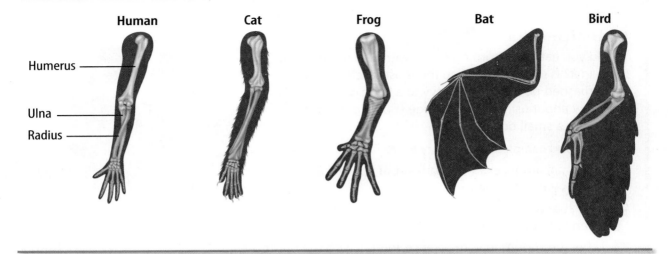

Human **Cat** **Frog** **Bat** **Bird**

Humerus

Ulna

Radius

Figure 14 The forelimbs of these species are different sizes, but their placement and structure suggest common ancestry.

Comparative Anatomy

Common ancestry is not difficult to see in many species. For example, it might seem easy to tell that robins, finches, and hawks evolved from a common ancestor. They all have similar features, such as feathers, wings, and beaks. The same is true for tigers, leopards, and house cats. But how are hawks related to cats? How are both hawks and cats related to frogs and bats? Observations of structural and functional similarities and differences in species that do not look alike are possible through comparative anatomy. **Comparative anatomy** *is the study of similarities and differences among structures of living species.*

Homologous Structures Humans, cats, frogs, bats, and birds look different and move in different ways. Humans use their arms for balance and their hands to grasp objects. Cats use their forelimbs to walk, run, and jump. Frogs use their forelimbs to jump. Bats and birds use their forelimbs as wings for flying. However, the forelimb bones of these species exhibit similar patterns, as shown in **Figure 14**. **Homologous** (huh MAH luh gus) **structures** *are body parts of organisms that are similar in structure and position but different in function.*

Homologous structures, such as the forelimbs of humans, cats, frogs, bats, and birds, suggest that these species are related. The more similar two structures are to each other, the more likely it is that the species have evolved from a recent common ancestor.

 Key Concept Check How do homologous structures provide evidence for evolution?

Analogous Structures Can you think of a body part in two species that serves the same purpose but differs in structure? How about the wings of birds and flies? Both wings in **Figure 15** are used for flight. But bird wings are covered with feathers. Fly wings are covered with tiny hairs. *Body parts that perform a similar function but differ in structure are* **analogous** (uh NAH luh gus) **structures.** Differences in the structure of bird and fly wings indicate that birds and flies are not closely related.

Vestigial Structures

The bird in the photo at the beginning of this lesson has short, stubby wings. Yet it cannot fly. The bird's wings are an example of vestigial structures. **Vestigial** (veh STIH jee ul) **structures** *are body parts that have lost their original function through evolution.* The best explanation for vestigial structures is that the species with a vestigial structure is related to an ancestral species that used the structure for a specific purpose.

The whale shown in **Figure 16** has tiny pelvic bones inside its body. The presence of pelvic bones in whales suggests that whales descended from ancestors that used legs for walking on land. The fossil evidence supports this conclusion. Many fossils of whale ancestors show a gradual loss of legs over millions of years. They also show, at the same time, that whale ancestors became better adapted to their watery environments.

 Key Concept Check How are vestigial structures evidence of descent from ancestral species?

▲ **Figure 15** Though used for the same function—flight—the wings of birds (top) and insects (bottom) are too different in structure to suggest close common ancestry.

Figure 16 Present-day whales have vestigial structures in the form of small pelvic bones. ▼

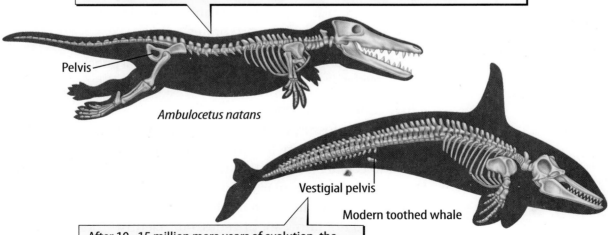

Between 50–40 million years ago, this mammal breathed air and walked clumsily on land. It spent a lot of time in water, but swimming was difficult because of its rear legs. Individuals born with variations that made their rear legs smaller lived longer and reproduced more. This mammal is an ancestor of modern whales.

Pelvis

Ambulocetus natans

Vestigial pelvis

Modern toothed whale

After 10–15 million more years of evolution, the ancestors of modern whales could not walk on land. They were adapted to an aquatic environment. Modern whales have two small vestigial pelvic bones that no longer support legs.

Pharyngeal Pouches

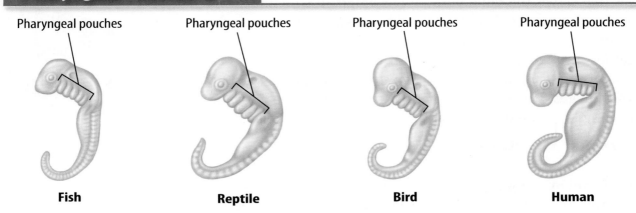

Pharyngeal pouches

Pharyngeal pouches

Pharyngeal pouches

Pharyngeal pouches

Fish

Reptile

Bird

Human

Figure 17 All vertebrate embryos exhibit pharyngeal pouches at a certain stage of their development. These features, which develop into neck and face parts, suggest relatedness.

WORD ORIGIN ············

embryology
from Greek *embryon*, means "to swell" and from Greek *logia*, means "study of"

Developmental Biology

You have just read that studying the internal structures of organisms can help scientists learn more about how organisms are related. Studying the development of embryos can also provide scientists with evidence that certain species are related. *The science of the development of embryos from fertilization to birth is called* **embryology** (em bree AH luh jee).

Pharyngeal Pouches Embryos of different species often resemble each other at different stages of their development. For example, all vertebrate embryos have pharyngeal (fuh rihn JEE ul) pouches at one stage, as shown in **Figure 17.** This feature develops into different body parts in each vertebrate. Yet, in all vertebrates, each part is in the face or neck. For example, in reptiles, birds, and humans, part of the pharyngeal pouch develops into a gland in the neck that regulates calcium. In fish, the same part becomes the gills. One function of gills is to regulate calcium. The similarities in function and location of gills and glands suggest a strong evolutionary relationship between fish and other vertebrates.

 Key Concept Check How do pharyngeal pouches provide evidence of relationships among species?

Molecular Biology

Studies of fossils, comparative anatomy, and embryology provide support for Darwin's theory of evolution by natural selection. Molecular biology is the study of gene structure and function. Discoveries in molecular biology have confirmed and extended much of the data already collected about the theory of evolution. Darwin did not know about genes, but scientists today know that mutations in genes are the source of variations upon which natural selection acts. Genes provide powerful support for evolution.

Reading Check What is molecular biology?

Comparing Sequences All organisms on Earth have genes. All genes are made of DNA, and all genes work in similar ways. This supports the idea that all organisms are related. Scientists can study relatedness of organisms by comparing genes and proteins among living species. For example, nearly all organisms contain a gene that codes for cytochrome *c,* a protein required for cellular respiration. Some species, such as humans and rhesus monkeys, have nearly identical cytochrome *c.* The more closely related two species are, the more similar their genes and proteins are.

 Key Concept Check How is molecular biology used to determine relationships among species?

Divergence Scientists have found that some stretches of shared DNA mutate at regular, predictable rates. Scientists use this "molecular clock" to estimate at what time in the past living species diverged from common ancestors. For example, as shown in **Figure 18,** molecular data indicate that whales and porpoises are more closely related to hippopotamuses than they are to any other living species.

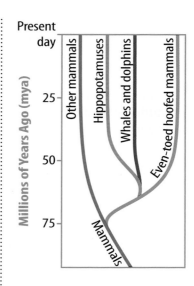

Figure 18 Whales and hippopotamuses share an ancestor that lived 50–60 mya.

Inquiry MiniLab

10 minutes

How related are organisms?

Proteins, such as cytochrome *c,* are made from combinations of just 20 amino acids. The graph below shows the number of amino acid differences in cytochrome *c* between humans and other organisms.

❶ Use the graph at right to answer the questions below.

Analyze and Conclude

1. **Identify** Which organism has the least difference in the number of amino acids in cytochrome *c* compared to humans? Which organism has the most difference?

2. **Infer** Which organisms do you think might be more closely related to each other: a dog and a turtle or a dog and a silkworm? Explain your answer.

3. **Key Concept** Notice the differences in the number of amino acids in cytochrome *c* between each organism and humans. How might these differences explain the relatedness of each organism to humans?

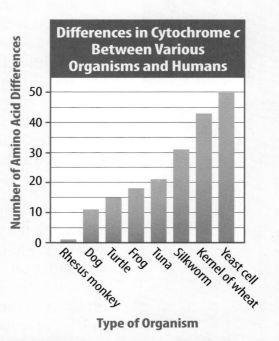

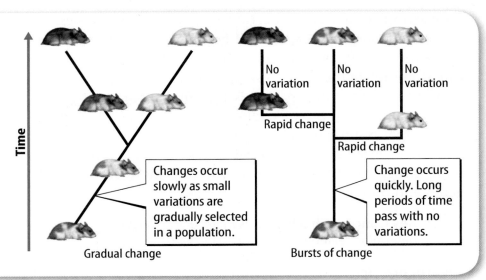

Figure 19 Many scientists think that natural selection produces new species slowly and steadily. Other scientists think species exist stably for long periods, then change occurs in short bursts. ▶

No variation

No variation

No variation

Rapid change

Rapid change

Changes occur slowly as small variations are gradually selected in a population.

Change occurs quickly. Long periods of time pass with no variations.

Gradual change

Bursts of change

Time

Figure 20 *Tiktaalik* lived 385–359 mya. Like amphibians, it had wrists and lungs. Like fish, it had fins, gills, and scales. Scientists think it is an intermediate species linking fish and amphibians. ▼

The Study of Evolution Today

The theory of evolution by natural selection is the cornerstone of modern biology. Since Darwin published his theory, scientists have confirmed, refined, and extended Darwin's work. They have observed natural selection in hundreds of living species. Their studies of fossils, anatomy, embryology, and molecular biology have all provided evidence of relatedness among living and extinct species.

How New Species Form

New evidence supporting the theory of evolution by natural selection is discovered nearly every day. But scientists debate some of the details. **Figure 19** shows that scientists have different ideas about the rate at which natural selection produces new species—slowly and gradually or quickly, in bursts. The origin of a species is difficult to study on human time scales. It is also difficult to study in the incomplete fossil record. Yet, new fossils that have features of species that lived both before them and after them are discovered all the time. For example, the *Tiktaalik* fossil shown in **Figure 20** has both fish and amphibian features. Further fossil discoveries will help scientists study more details about the origin of new species.

Diversity

How evolution has produced Earth's wide diversity of organisms using the same basic building blocks—genes—is an active area of study in evolutionary biology. Scientists are finding that genes can be reorganized in simple ways and give rise to dramatic changes in organisms. Though scientists now study evolution at the molecular level, the basic principles of Darwin's theory of evolution by natural selection have remained unchanged for over 150 years.

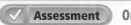
Visual Summary

By comparing the anatomy of organisms and looking for homologous or analogous structures, scientists can determine if organisms had a common ancestor.

Some organisms have vestigial structures, suggesting that they descended from a species that used the structure for a purpose.

Pharyngeal pouches

Human

Scientists use evidence from developmental and molecular biology to help determine if organisms are related.

FOLDABLES

Use your lesson Foldable to review the lesson. Save your Foldable for the project at the end of the chapter.

What do you think NOW?

You first read the statements below at the beginning of the chapter.

5. Living species contain no evidence that they are related to each other.

6. Plants and animals share similar genes.

Did you change your mind about whether you agree or disagree with the statements? Rewrite any false statements to make them true.

Use Vocabulary

1 **Define** *embryology* in your own words.

2 **Distinguish** between a homologous structure and an analogous structure.

3 **Use the term** *vestigial structure* in a complete sentence.

Understand Key Concepts

4 Scientists use molecular biology to determine how two species are related by comparing the genes in one species to genes
   A. in extinct species. **C.** in related species.
   B. in human species. **D.** in related fossils.

5 **Discuss** how pharyngeal pouches provide evidence for biological evolution.

6 **Explain** Some blind cave salamanders have eyes. How might this be evidence that cave salamanders evolved from sighted ancestors?

Interpret Graphics

7 **Interpret** The wings of a flightless cormorant are an example of which type of structure?

8 **Assess** Copy and fill in the graphic organizer below to identify four areas of study that provide evidence for evolution.

Evolution

Critical Thinking

9 **Predict** what a fossil that illustrates the evolution of a bird from a reptile might look like.

Model Adaptations in an Organism

Materials

clay

colored pencils

colored markers

toothpicks

construction paper

Also needed:
creative construction materials, glue, scissors

Safety

Conditions on our planet have changed since Earth formed over 4.5 billion years ago. Changes in the concentrations of gases in the atmosphere, temperature, and the amount of precipitation make Earth different today from when it first formed. Other events, such as volcanic eruptions, meteorite strikes, tsunamis, or wildfires, can drastically and rapidly change the conditions in certain environments. As you have read, Earth's fossil record provides evidence that, over millions of years, many organisms developed adaptations that enabled them to survive as Earth's environmental conditions changed.

Ask a Question

How do adaptations enable an organism to survive changes in the environment?

Make Observations

1 Read and complete a lab safety form.

2 Obtain Version 1.0 of the organism you will model from your teacher.

Volcanic eruption

3 Your teacher will describe Event 1 that has occurred on Earth while your organism is alive. Use markers and a piece of construction paper to design adaptations to your organism that would enable it to survive the changing conditions that result from Event 1. Label the adapted organism *Version 1.1*.

4 For each event that your teacher describes, design and draw the adaptations that would enable your organism to survive the changing conditions. Label each new organism *Version 1.X*, filling in the *X* with the appropriate version number.

5 Use the materials provided to make a model of the final version of your organism, showing all of the adaptations.

Predation

Form a Hypothesis

6 After reviewing and discussing all of the adaptations of your organism, formulate a hypothesis to explain how physical adaptations help an organism survive changes to the environment.

Test Your Hypothesis

7 Research evidence from the fossil record that shows one adaptation that developed and enabled an organism to survive over time under the conditions of one of the environmental events experienced by your model organism.

8 Record the information in your Science Journal.

Analyze and Conclude

9 **Compare** the adaptations that the different groups gave their organisms to survive each event described by your teacher. What kinds of different structures were created to help each organism survive?

10 **The Big Idea** Describe three variations in human populations that would enable some individuals to survive severe environmental changes.

Communicate Your Results

Present your completed organisms to the class and/or judges of "Ultimate Survivor." Explain the adaptations and the reasoning behind them in either an oral presentation or a demonstration, during which classmates and/or judges will review the models.

Inquiry Extension

Compare the organisms made by groups in your class to the organisms created by groups in other sections. Observe the differences in the adaptations of the organisms. In each section, the events were presented in a different order. How might this have affected the final appearance and characteristics of the different organisms?

Meteorite impact

Lab Tips

☑ Make sure you think of all of the implications of an environmental change event before you decide upon an adaptation.

☑ Decide upon your reasoning for the adaptation before putting the adaptation on your model.

Remember to use scientific methods.

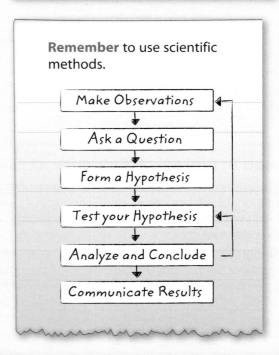

Make Observations
↓
Ask a Question
↓
Form a Hypothesis
↓
Test your Hypothesis
↓
Analyze and Conclude
↓
Communicate Results

Chapter 6 Study Guide

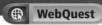
 THE BIG IDEA **Through natural selection, species evolve as they adapt to Earth's changing environments.**

Key Concepts Summary

| Vocabulary |

Lesson 1: Fossil Evidence of Evolution

- Fossils form in many ways, including mineral replacement, carbonization, and impressions in sediment.
- Scientists can learn the ages of fossils by techniques of relative-age dating and absolute-age dating.
- Though incomplete, the **fossil record** contains patterns suggesting the **biological evolution** of related species.

Vocabulary

fossil record p. 193
mold p. 195
cast p. 195
trace fossil p. 195
geologic time scale p. 197
extinction p. 198
biological evolution p. 199

Lesson 2: Theory of Evolution by Natural Selection

- The 19th century **naturalist** Charles Darwin developed a theory of evolution that is still studied today.
- Darwin's theory of evolution by **natural selection** is the process by which populations with **variations** that help them survive in their environments live longer and reproduce more than those without beneficial variations. Over time, beneficial variations spread through populations, and new species that are adapted to their environments evolve.
- **Camouflage, mimicry,** and other **adaptations** are evidence of the close relationships between species and their changing environments.

naturalist p. 203
variation p. 205
natural selection p. 206
adaptation p. 207
camouflage p. 208
mimicry p. 208
selective breeding p. 209

Lesson 3: Biological Evidence of Evolution

- Fossils provide only one source of evidence of evolution. Additional evidence comes from living species, including studies in **comparative anatomy, embryology,** and molecular biology.
- Through evolution by natural selection, all of Earth's organisms are related. The more recently they share a common ancestor, the more closely they are related.

comparative anatomy p. 214
homologous structure p. 214
analogous structure p. 215
vestigial structure p. 215
embryology p. 216

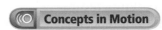
FOLDABLES® Chapter Project

Assemble your lesson Foldables as shown to make a Chapter Project. Use the project to review what you have learned in this chapter.

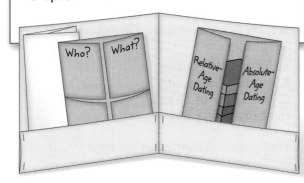

Use Vocabulary

Distinguish between the following terms.

1. *mold* and *cast*
2. *absolute-age dating* and *relative-age dating*
3. *extinction* and *biological evolution*
4. *variations* and *adaptations*
5. *camouflage* and *mimicry*
6. *natural selection* and *selective breeding*
7. *homologous structure* and *analogous structure*
8. *embryology* and *comparative anatomy*
9. *vestigial structure* and *homologous structure*

Link Vocabulary and Key Concepts

Concepts in Motion Interactive Concept Map

Copy this concept map, and then use vocabulary terms from the previous page to complete the concept map.

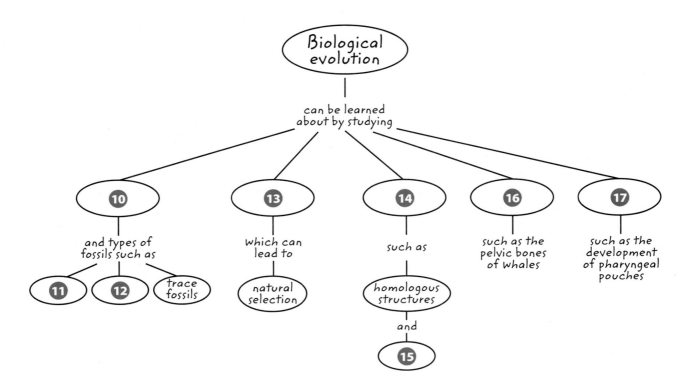

Understand Key Concepts

1 Why do scientists think the fossil record is incomplete?

A. Fossils decompose over time.

B. The formation of fossils is rare.

C. Only organisms with hard parts become fossils.

D. There are no fossils before the Phanerozoic eon.

2 What do the arrows on the graph below represent?

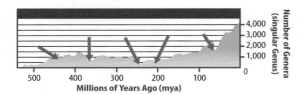

A. extinction events

B. meteorite impacts

C. changes in Earth's temperature

D. the evolution of a new species

3 What can scientists learn about fossils using techniques of absolute-age dating?

A. estimated ages of fossils in rock layers

B. precise ages of fossils in rock layers

C. causes of fossil disappearances in rock layers

D. structural similarities to other fossils in rock layers

4 Which is the sequence by which natural selection works?

A. selection → adaptation → variation

B. selection → variation → adaptation

C. variation → adaptation → selection

D. variation → selection → adaptation

5 Which type of fossil forms through carbonization?

A. cast

B. mold

C. fossil film

D. trace fossil

6 Which is the source of variations in a population of organisms?

A. changes in environment

B. changes in genes

C. the interaction of genes with an environment

D. the interaction of individuals with an environment

7 Which is an example of a functional adaptation?

A. a brightly colored butterfly

B. birds flying south in the fall

C. the spray of a skunk

D. thorns on a rose

8 Which is NOT an example of a vestigial structure?

A. eyes of a blind salamander

B. pelvic bones in a whale

C. thorns on a rose bush

D. wings on a flightless bird

9 Which do the images below represent?

Human

Cat

A. analogous structures

B. embryological structures

C. homologous structures

D. vestigial structures

10 Which is an example of a sudden change that could lead to the extinction of species?

A. a mountain range isolates a species

B. Earth's tectonic plates move

C. a volcano erupts

D. sea level changes

Critical Thinking

11 **Explain** the relationship between fossils and extinction events.

12 **Infer** In 2004, a fossil of an organism that had fins and gills, but also lungs and wrists, was discovered. What might this fossil suggest about evolution?

13 **Summarize** Darwin's theory of natural selection using the Galápagos tortoises or finches as an example.

14 **Assess** how the determination that Earth is 4.6 billion years provided support for the idea that all species evolved from a common ancestor.

15 **Describe** how cytochrome *c* provides evidence of evolution.

16 **Explain** why the discovery of genes was powerful support for Darwin's theory of natural selection.

17 **Interpret Graphics** The diagram below shows two different methods by which evolution by natural selection might proceed. Discuss how these two methods differ.

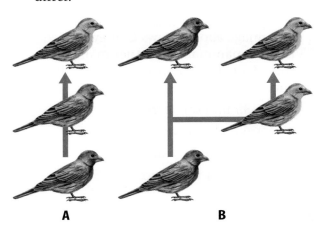

A　　　　　　　**B**

Writing in Science

18 **Write** a paragraph explaining how natural selection and selective breeding are related. Include a main idea, supporting details, and a concluding sentence.

REVIEW THE B**I**G IDEA

19 How do species adapt to changing environments over time? Explain how evidence from the fossil record and from living species suggests that Earth's species are related. List each type of evidence and provide an example of each.

20 The photo below shows an orchid that looks like a bee. How might this adaptation be evidence of evolution by natural selection?

Math Skills ×÷

 Review

Math Practice

Use Scientific Notation

21 The earliest fossils appeared about 3,500,000,000 years ago. Express this number in scientific notation.

22 The oldest fossils visible to the unaided eye are about 565,000,000 years old. What is this time in scientific notation?

23 The oldest human fossils are about 1×10^4 years old. Express this as a whole number.

Standardized Test Practice

Record your answers on the answer sheet provided by your teacher or on a sheet of paper.

Multiple Choice

1 Which may form over time from the impression a bird feather makes in mud?

  A cast

  B mold

  C fossil film

  D trace fossil

2 Which is NOT one of the three main categories of adaptations?

  A behavioral

  B functional

  C pharyngeal

  D structural

Use the figure below to answer question 3.

Bat wing   Insect wing

3 The figure shows the wings of a bat and an insect. Which term describes these structures?

  A analogous

  B developmental

  C homologous

  D vestigial

4 What is an adaptation?

  A a body part that has lost its original function through evolution

  B a characteristic that better equips an organism to survive in its environment

  C a feature that appears briefly during early development

  D a slight difference among the individuals in a species

5 What causes variations to arise in a population?

  A changes in the environment

  B competition for limited resources

  C random mutations in genes

  D rapid population increases

Use the image below to answer question 6.

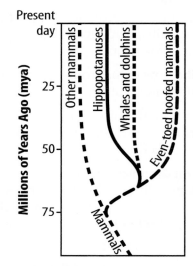

6 The image above shows that even-toed hoofed mammals and other mammals shared a common ancestor. When did this ancestor live?

  A 25–35 million years ago

  B 50–60 million years ago

  C 60–75 million years ago

  D 75 million years ago

7 Which term describes the method Darwin used that resulted in pigeons with desired traits?

  A evolution

  B mimicry

  C natural selection

  D selective breeding

Use the figure below to answer question 8.

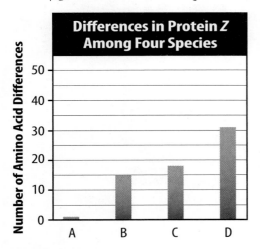

8 The chart shows that species B and C have the fewest amino acid differences for a protein among four species. What does this suggest about their evolutionary relationship?

A They are more closely related to each other than to the other species.

B They evolved at a faster rate when compared to the other species.

C They share a developmental similarity not observed in the other species.

D They do not share a common ancestor with the other species.

9 Which developmental similarity among all vertebrates is evidence that they share a common ancestor?

A analogous structures

B pharyngeal pouches

C variation rates

D vestigial structures

Constructed Response

Use the figure below to answer questions 10 and 11.

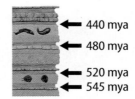

10 What is the approximate age of the fish fossils (top layer of fossils)? Express your answer as a range, and explain how you derived the answer.

11 What type of material or rock most likely forms the layer that contains the fossils? In your response, explain how these fossils formed.

12 Explain how a sudden and drastic environmental change might lead to the extinction of a species.

13 Darwin formulated his theory of evolution by natural selection based on the observation that food is a limiting resource. What did he mean by that? Use the Galápagos tortoises to explain your answer.

14 Explain how the fossil record provides evidence of biological evolution.

NEED EXTRA HELP?														
If You Missed Question...	1	2	3	4	5	6	7	8	9	10	11	12	13	14
Go to Lesson...	1	2	3	2	2	3	2	3	3	1	1	1	2	1

Chapter 7

Human Body Systems

THE BIG IDEA

What are the functions of the human body systems?

This is a photograph of a cross section through a human body. You can see the lower part of a human arm and part of the abdomen. In the abdomen, you might be able to pick out a vertebra, muscles, fat, and part of the intestine.

- What body systems can you identify here?

- What are the functions of the human body systems?

Get Ready to Read

What do you think?

Before you read, decide if you agree or disagree with each of these statements. As you read this chapter, see if you change your mind about any of the statements.

1 A human body has organ systems that carry out specific functions.

2 The body protects itself from disease.

3 All bones in the skeletal system are hollow.

4 The endocrine system makes hormones.

5 The testes produce sperm.

6 Puberty occurs during infancy.

ConnectED Your one-stop online resource

connectED.mcgraw-hill.com

- Video
- Audio
- Review
- Inquiry
- WebQuest
- Assessment
- Concepts in Motion
- Multilingual eGlossary

Reading Guide

Key Concepts
ESSENTIAL QUESTIONS

- How do nutrients enter and leave the body?
- How do nutrients travel through the body?
- How does the body defend itself from harmful invaders?

Vocabulary

organ system p. 231

homeostasis p. 231

nutrient p. 233

Calorie p. 233

lymphocyte p. 239

immunity p. 240

 Multilingual eGlossary

Video

- BrainPOP®
- Science Video
- What's Science Got to do With It?

Transport and Defense

Inquiry Unusual Web?

This branching structure might look like a strange spider web, but it is actually a resin cast of human lungs. The yellowish tubes are large air passages, the white parts are small airways, and the blue parts are blood vessels. Why do the lungs need all these parts?

Which tool can transport water quickly?

You need to transport materials throughout your body. Each cell must receive nutrients and oxygen and get rid of wastes. What kinds of tools do you think would be most effective in moving fluids such as water quickly?

1. Read and complete a lab safety form.
2. Choose one of the **tools** for moving water.
3. Have another student use a **stopwatch** to keep time for 30 s. Use your tool to transport as much water as you can in 30 s from the main **bowl** into a **beaker**.
4. Use a **graduated cylinder** to measure the amount of water you moved from the bowl to the beaker. Record the measurement in your Science Journal.
5. Trade roles with your partner. Repeat steps 2 through 4.
6. Repeat step 5 until you have used all of the tools.

Think About This

1. Which tool was most effective for moving water quickly? Which tool was least effective?

2. **Key Concept** Why do you think moving small items in fluid might be more effective than moving them all individually?

The Body's Organization

Have you ever tried to find a book in a library? Libraries have thousands of books grouped together by subject. Grouping books by subject in a library helps keep them organized and easier to find. Your body's organization helps it function.

All organisms have different parts with special functions. Recall that cells are the basic unit of all living organisms. Organized groups of cells that work together are tissues. Groups of tissues that perform a specific function are organs. *Groups of organs that work together and perform a specific task are* **organ systems.** Organ systems provide movement, transport substances, and perform many other functions that you will read about in this chapter.

Organ systems work together and maintain **homeostasis** (hoh mee oh STAY sus), *or steady internal conditions when external conditions change.* Have you ever jogged, jumped rope, or snowshoed, as shown in **Figure 1,** and started to sweat? When exercising, your body uses stored energy. Your body releases excess energy as thermal energy. Sweat, also called perspiration (pur spuh RAY shun), helps the body release thermal energy and maintain homeostasis.

Figure 1 Sweating helps the body maintain homeostasis by releasing excess thermal energy.

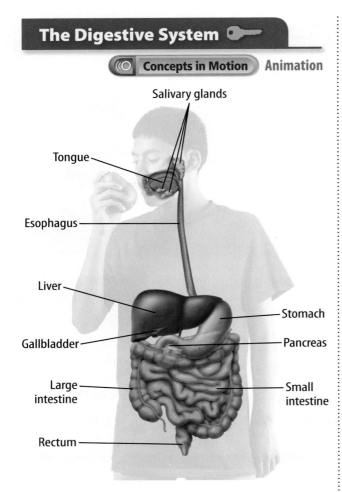

Concepts in Motion **Animation**

Salivary glands

Tongue

Esophagus

Liver

Gallbladder

Large intestine

Rectum

Stomach

Pancreas

Small intestine

Figure 2 Food enters the digestive system through the mouth, and nutrients are absorbed by the small intestine.

FOLDABLES

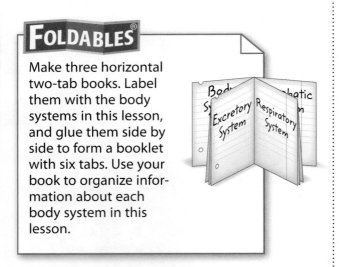

Make three horizontal two-tab books. Label them with the body systems in this lesson, and glue them side by side to form a booklet with six tabs. Use your book to organize information about each body system in this lesson.

Digestion and Excretion

Humans need food, water, and oxygen to survive. Food contains energy that is processed by the body. The process by which food is broken down is called digestion. After digestion, substances that are not used by the body are removed through excretion (ihk SKREE shun).

The Digestive System

As shown in **Figure 2,** the digestive system is made up of several organs. Food and water enter the digestive system through the mouth.

Digestion After food enters the mouth, chewing breaks food into smaller parts. Saliva, which contains enzymes, also helps the mouth break down food. Recall that enzymes are proteins that speed up chemical reactions.

When you swallow, food, water, and other liquids move into a hollow tube called the esophagus (ih SAH fuh gus). The esophagus connects the mouth to the stomach. Digestion continues as food leaves the esophagus and enters the stomach. The stomach is a flexible, baglike organ that contains other enzymes that break down food into smaller parts so that the food can be used by the body.

 Reading Check Identify where food enters the body.

Absorption Next, food moves into the small intestine. By the time food gets to the small intestine, it is a soupy mixture. The small intestine is a tube that has two functions—digestion and absorption. The liver makes a substance called bile. The pancreas makes enzymes. Both bile and enzymes are used in the small intestine to break down food even more. Because the small intestine is very long, it takes food hours to move through it. During that time, particles of food and water are absorbed into the blood.

Excretion The large intestine, or colon (KOH lun), receives digested food that the small intestine did not absorb. The large intestine also absorbs water from the remaining waste material. Most foods are completely digested into smaller parts that can be easily absorbed by the small intestine. However, some foods travel through the entire digestive system without being digested or absorbed. For example, some types of fiber, called insoluble fiber, in vegetables and whole grains are not digested and leave the body through the rectum.

Nutrition

As you have read, one of the functions of the small intestine is absorption. **Nutrients** *are the parts of food used by the body to grow and survive.* There are several types of nutrients. **Proteins,** fats, carbohydrates, vitamins, and minerals are all nutrients. Nutrition labels on food, as shown in **Figure 3,** show the amount of each nutrient in that food. By looking at the labels on packaged foods, you can make sure you get the nutrients you need. Different people need different amounts of nutrients. For example, football players, swimmers, and other athletes need a lot of nutrients for energy. Pregnant women also need lots of nutrients to provide for their developing babies.

Digestion helps release energy from food. *A* **Calorie** *is the amount of energy it takes to raise the temperature of 1 kg of water by 1°C.* The body uses Calories from proteins, fats, and carbohydrates, which each contain a different amount of energy.

 Reading Check Name five types of nutrients.

Figure 3 The information on a nutrition label can help you decide whether a food is healthful to eat.

 **Visual Check** How many servings are in this food container?

Math Skills

Use Proportions

A proportion is an equation of two equal ratios. You can solve a proportion for an unknown value. For example, a 50-g egg provides 70 Calories (C) of energy. How many Calories would you get from 125 g of scrambled eggs?

Write a proportion.
$$\frac{50\text{ g}}{70\text{ C}} = \frac{125\text{ g}}{x}$$

Find the cross products.

$50\text{ g }(x) = 70\text{ C} \times 125\text{ g}$

$50\text{ g }(x) = 8{,}750\text{ C g}$

Divide both sides by 50.
$$\frac{50\text{ g }(x)}{50\text{ g}} = \frac{8{,}750\text{ C g}}{50\text{ g}}$$

Simplify the equation.

$$x = 175\text{ C}$$

Practice

The serving size of a large fast-food hamburger with cheese is 316 g. It contains 790 C of energy. How many Calories would you consume if you ate 100 g of the burger?

Review

- **Math Practice**
- **Personal Tutor**

REVIEW VOCABULARY

protein
a long chain of amino acid molecules

Review **Personal Tutor**

How much water do you lose each day?

Most people lose an average of 2.5 L of water each day. You lose an average of 0.8 L through breathing, 1.5 L through urine, 0.1 L through sweating, and 0.1 L through feces.

1 Read and complete a lab safety form.

2 In your Science Journal, keep track of how much water you drink for 1 day. Include all the liquids you take in, and write down the sources as part of your list. Add and record the total.

3 Use the data above to figure out how much water you lost. First, subtract the amounts for breathing and feces.

4 Next, subtract the amount for sweating. If you did a physical activity that made you sweat, add another 0.5 L to this amount.

5 The remaining amount of water is lost as urine. Record this amount from the data above.

Analyze and Conclude

1. **Calculate** how much of the water you drank was used by your body, and display your data in a table. Did the amount of water lost equal the amount you drank? Why do you think this is so?

2. **Construct** a graph that shows, in percentages, how your body used the water you drank.

3. **Key Concept** Infer how the water you drank must be transported within your body for it to be used in each process.

The Excretory System

The excretory system removes solid, liquid, and gas waste materials from the body. The lungs, skin, liver, kidneys, bladder, and rectum all are parts of the excretory system.

The lungs remove carbon dioxide (CO_2) and excess water as water vapor when you breathe out, or exhale. The skin removes water and salt when you sweat.

The liver removes wastes from the blood. As you have read, the liver also is a part of the digestive system. The digestive and excretory systems work together to break down, absorb, and remove food.

When the liver breaks down proteins, urea forms. Urea is toxic if it stays in the body. The kidneys, shown in **Figure 4,** remove urea from the body by making urine. Urine contains water, urea, and other waste chemicals. Urine leaves each kidney through a tube, called the ureter (YOO ruh tur), and is stored in a flexible sac, called the bladder. Urine is removed from the body through a tube called the urethra (yoo REE thruh).

Like the liver, the rectum is part of the excretory system and the digestive system. Food substances that are not absorbed by the small intestine are mixed with other wastes and form feces. The rectum stores feces until it moves out of the body.

Key Concept Check How does food enter and leave the body?

Figure 4 The kidneys remove waste material from the body.

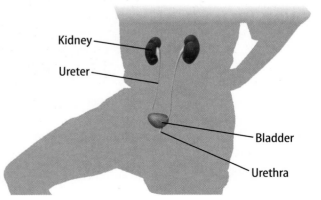

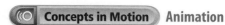

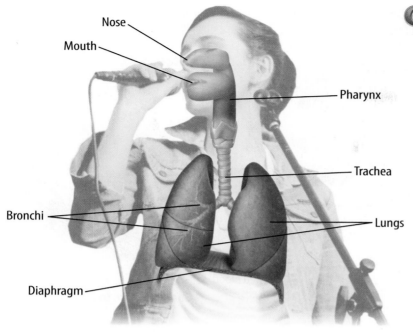

Nose

Mouth

Pharynx

Trachea

Bronchi

Lungs

Diaphragm

Figure 5 Air enters the respiratory system through the nose and the mouth. Oxygen enters the blood in the lungs.

Respiration and Circulation

You have read about how the body converts food into nutrients and how the small intestine absorbs nutrients. But how do the oxygen you breathe in and the nutrients absorbed by the small intestine get to the rest of the body? And how do waste products leave the body?

The Respiratory System

The respiratory system, shown in **Figure 5,** exchanges gases between the body and the environment. As air flows through the respiratory system, it passes through the nose and mouth, pharynx (FER ingks), trachea (TRAY kee uh), bronchi (BRAHN ki; singular, bronchus), and lungs. The parts of the respiratory system work together and supply the body with oxygen. They also rid the body of wastes, such as carbon dioxide.

Pharynx and Trachea Oxygen enters the body when you inhale, or breathe in. Carbon dioxide leaves the body when you exhale. When you inhale, air enters the nostrils and passes through the pharynx. Because the pharynx is part of the throat, it is a part of both the digestive and respiratory systems. Food goes through the pharynx to the esophagus. Air travels through the pharynx to the trachea. The trachea is also called the windpipe because it is a long, tubelike organ that connects the pharynx to the bronchi.

✔ **Reading Check** Which organ is part of both the digestive system and the respiratory system?

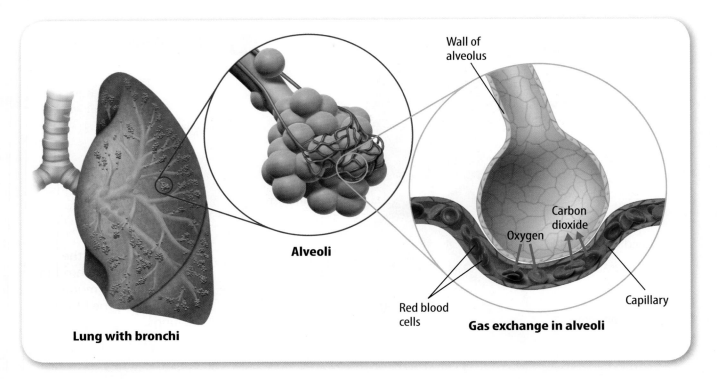

Wall of alveolus

Alveoli

Carbon dioxide

Oxygen

Red blood cells

Gas exchange in alveoli

Capillary

Lung with bronchi

Figure 6 Bronchi divide into smaller tubes that end in clusters of alveoli that are surrounded by capillaries.

Visual Check Which gas leaves the alveoli and enters capillaries?

SCIENCE USE V. COMMON USE

vessel

Science Use a tube in the body that carries fluid such as blood

Common Use a ship

Bronchi and Alveoli There are two bronchi; one enters the left lung, and one enters the right lung. As shown in **Figure 6,** the bronchi divide into smaller tubes that end in tiny groups of cells that look like bunches of grapes. These groups of cells are called alveoli (al VEE uh li). Inside each lung, there are more than 100 million alveoli. The alveoli are surrounded by blood **vessels** called capillaries. Oxygen in the alveoli enters the capillaries. The blood inside capillaries transports oxygen to the rest of the body

Reading Check What are alveoli, and what do they do?

Inhaling and exhaling require the movement of a thin muscle under the lungs called the diaphragm (DI uh fram). As the diaphragm contracts and moves down, air enters the lungs and you inhale. When the diaphragm relaxes and moves up, you exhale.

The Circulatory System

As shown in **Figure 7,** the heart, blood, and blood vessels make up the circulatory system. It transports nutrients, gases, wastes, and other substances through the body. Blood vessels transport blood to all organs in the body. Because your body uses oxygen and nutrients continually, your circulatory system transports blood between the heart, lungs, and other organs more than 1,000 times each day!

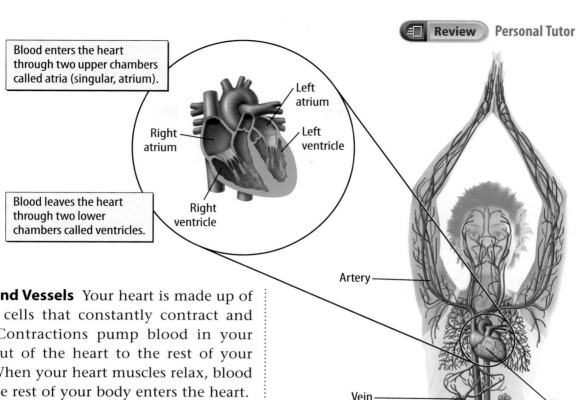

Blood enters the heart through two upper chambers called atria (singular, atrium).

Left atrium

Right atrium

Left ventricle

Right ventricle

Blood leaves the heart through two lower chambers called ventricles.

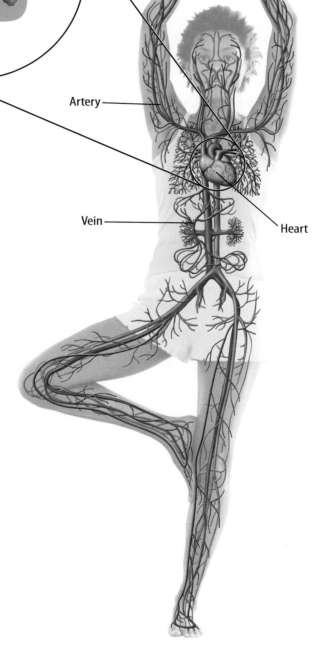

Artery

Vein

Heart

Heart and Vessels Your heart is made up of muscle cells that constantly contract and relax. Contractions pump blood in your heart out of the heart to the rest of your body. When your heart muscles relax, blood from the rest of your body enters the heart.

Blood travels through your body in tiny tubes called vessels. If all the blood vessels in your body were laid end-to-end in a single line, it would be more than 95,000 km long.

The three main types of blood vessels are arteries, veins, and capillaries. Arteries carry blood away from your heart. Usually this blood is oxygen-rich and contains nutrients, except for the blood in the pulmonary arteries that contains CO_2. Arteries are large and surrounded by muscle cells that help blood move through the vessels faster. Veins transport blood that contains CO_2 back to your heart, except for the blood in the pulmonary veins, which is oxygen-rich. Capillaries are very tiny vessels that enable oxygen, CO_2, and nutrients to move between your circulatory system and your entire body.

You just read that capillaries surround the alveoli in your lungs Capillaries also surround the small intestine, where they absorb nutrients and transport them to the rest of the body.

 Key Concept Check How do nutrients travel through the body?

Figure 7 🔑 The circulatory system transports nutrients and oxygen to all parts of the body and removes wastes, such as CO_2.

Table 1 Human Blood Types

Blood Type	Type A	Type B	Type AB	Type O
Antigens on red blood cells	A A A A A A	B B B B B B	B A A B A	(no antigens)
Percentage of U.S. population with this blood type	42	10	4	44
Clumping proteins in plasma	anti-B	anti-A	none	anti-A and anti-B
Blood type(s) that can be RECEIVED in a transfusion	A or O	B or O	A, B, AB, or O	O only
This blood type can DONATE TO these blood types	A or AB	B or AB	AB only	A, B, AB, O

Table 1 The red blood cells of each blood type have different proteins on their surfaces.

✓**Visual Check** To which blood group can type A donate?

Blood The blood that circulates through vessels has several parts. The liquid part of blood is called plasma and contains nutrients, water, and CO_2. Blood also contains red blood cells, platelets, and white blood cells. Red blood cells carry oxygen. Platelets help the body heal when you get a cut. White blood cells help the body defend itself from toxins and diseases. You will read more about white blood cells on the next page.

Everyone has red blood cells. However, different people have different proteins on the surfaces of their red blood cells, as shown in **Table 1.** Scientists classify these different red-blood-cell proteins into groups called blood types.

People with A proteins on their red blood cells have type A blood. People with B proteins on their red blood cells have type B blood. Some people have both A and B proteins on their red blood cells. They have type AB blood. People with type O blood have neither A nor B proteins on the surfaces of their red blood cells.

Medical professionals use blood types to determine which type of blood a person can receive from a blood donor. For example, because people with type O blood have no proteins on the surfaces of their red blood cells, they can receive blood only from a donor who also has type O blood.

The Lymphatic System

Have you ever had a cold and found it painful to swallow? This can happen if your tonsils swell. Tonsils are small organs on both sides of your throat. They are part of the lymphatic (lihm FA tihk) system.

The spleen, the thymus, bone marrow, and lymph nodes also are parts of the lymphatic system. The spleen stores blood for use in an emergency. The thymus, the spleen, and bone marrow make white blood cells.

Your lymphatic system has three main functions: removing excess fluid around organs, producing white blood cells, and absorbing and transporting fats. The lymphatic system helps your body maintain fluid homeostasis. About 65 percent of the human body is water. Most of this water is inside cells. Sometimes, when water, wastes, and nutrients move between capillaries and organs, not all of the fluid is taken up by the organs. When fluid builds up around organs, swelling can occur. To prevent swelling, the lymphatic system removes the fluid.

 Reading Check Identify a function of the lymphatic system.

Lymph vessels are all over your body, as shown in **Figure 8.** Fluid that travels through the lymph vessels flows into organs called lymph nodes. Humans have more than 500 lymph nodes. The lymph nodes work together and protect the body by removing toxins, wastes, and other harmful substances.

The lymyphatic system makes white blood cells. They help the body defend against infection. There are many different types of white blood cells. *A* **lymphocyte** (LIHM fuh sites) *is a type of white blood cell that is made in the thymus, the spleen, or the bone marrow.* Lymphocytes protect the body by traveling through the circulatory system, defending against infection.

WORD ORIGIN ·

lymphocyte
from Latin *lympha*, means "water"; and Greek *kytos*, means "hollow, as a cell or container"

Figure 8 Lymph vessels are throughout your body.

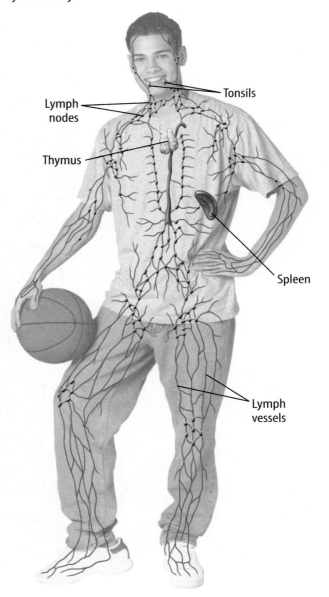

Tonsils

Lymph nodes

Thymus

Spleen

Lymph vessels

Immunity

The lymphatic system protects your body from harmful substances and infection. *The resistance to specific pathogens, or disease-causing agents, is called* **immunity**. The skeletal system produces immune cells, and the circulatory system transports them throughout the body. Immune cells include lymphocytes and other white blood cells. These cells **detect** viruses, bacteria, and other foreign substances that are not normally made in the body. The immune cells attack and destroy them, as shown in **Figure 9.**

If the body is exposed to the same bacteria, virus, or substance later, some immune cells remember and make proteins called antibodies. These antibodies recognize specific proteins on the harmful agent and help the body fight infection faster. Because there are many different types of bacteria and viruses, humans make billions of different types of antibodies. Each type of antibody responds to a different harmful agent.

ACADEMIC VOCABULARY

detect
(verb) to discover the presence of

Figure 9 Lymphocytes surround bacteria and destroy or remove them from the body.

Visual Check How long did it take for the lymphocyte to completely surround the bacterium?

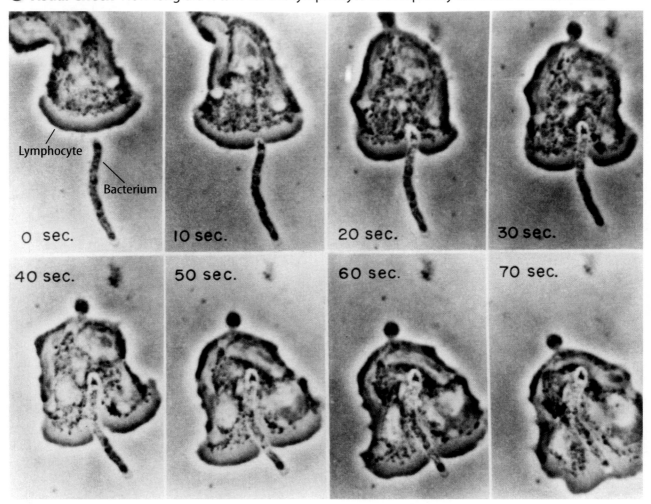

Table 2 Examples of Diseases

Infectious Disease		Noninfectious Disease
Disease	Pathogen	
colds	virus	cancer
AIDS	virus	diabetes
strep throat	bacteria	heart disease
chicken pox	virus	allergy

Types of Diseases

There are two main groups of diseases—infectious and non-infectious—as shown in **Table 2.** Infectious diseases are caused by pathogens, such as bacteria and viruses. Infectious diseases are usually contagious, which means they can be spread from one person to another. The flu is an example of an infectious disease. Viruses that invade organ systems of the body, such as the respiratory system, cause infectious diseases.

A noninfectious disease is caused by the environment or a genetic disorder, not a pathogen. Skin cancer, diabetes, and allergies are examples of noninfectious diseases. Noninfectious diseases are not contagious and cannot be spread from one person to another.

Lines of Defense

The human body has many ways of protecting itself from viruses, bacteria, and harmful substances. Skin and mucus (MYEW kus) are parts of the first line of defense. They prevent toxins and other substances from entering the body. Mucus is a thick, gel-like substance in the nostrils, trachea, and lungs. Mucus traps harmful substances and prevents them from entering your body.

The second line of defense is the immune response. In the immune response, white blood cells attack and destroy harmful substances, as shown in **Figure 9.**

The third line of defense protects your body against substances that have infected the body before. As you have read, immune cells make antibodies that destroy the harmful substances. Vaccines are used to help the body develop antibodies against infectious diseases. For example, many people get an influenza vaccine annually to protect them against the flu.

 Key Concept Check How does the body defend itself from harmful invaders?

Table 2 Diseases are classified into two main groups based on whether they are caused by pathogens.

Lesson 1 Review

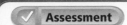
Visual Summary

The kidneys remove liquid wastes from the body.

The circulatory system transports nutrients, gases, wastes, and other substances through the body.

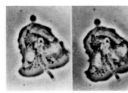

Immune cells detect and destroy viruses, bacteria, and other foreign substances.

FOLDABLES

Use your lesson Foldable to review the lesson. Save your Foldable for the project at the end of the chapter.

What do you think NOW?

You first read the statements below at the beginning of the chapter.

1. A human body has organ systems that carry out specific functions.

2. The body protects itself from disease.

Did you change your mind about whether you agree or disagree with the statements? Rewrite any false statements to make them true.

Use Vocabulary

1 **Use the term** *organ system* in a sentence.

2 **Define** *homeostasis* in your own words.

3 A(n) _____ is a type of white blood cell.

Understand Key Concepts

4 Organs are groups of _____ that work together.
   A. cells   **C.** systems
   B. organisms   **D.** tissues

5 **Differentiate** the role of the liver in the digestive system from its role in the excretory system.

6 **Examine** how the circulatory system and the respiratory system work together and move oxygen through the body.

7 **Contrast** infectious diseases and noninfectious diseases.

Interpret Graphics

8 **Analyze** the nutrition label below. How many Calories, carbohydrates, fats, and proteins are in a serving of this food?

Nutrition Facts	Amount/Serving	%DV*	Amount/Serving	%DV*
Serving Size 20g	**Total Fat** 0g	0%	**Total Carb.** 19g	6%
Servings per Container 2	Sat. Fat 0g	0%	Dietary Fiber 0g	0%
	Trans Fat 0g		Sugars 10g	
Calories 80	**Cholesterol** 0mg	0%	Other Carb. 9g	
Calories from Fat 0	**Sodium** 190mg	8%	**Protein** 0g	
*Percent Daily Values (DV) are	**Potassium** 50mg	2%		
based on a 2,000 calorie diet.	Vit. A 0% • Vit. C 0% • Calcium 0% • Iron 0% • Magnesium 2%			

9 **Summarize** Copy and fill in the graphic organizer below to show how food travels through the digestive system.

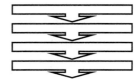

Math Skills

 Review
——— Math Practice ———

10 If 30.5 g of milk contains 18 C, how many Calories will you consume by drinking a glass of milk (244 g)?

How can you model the function of blood cells?

Materials

toy cars and trucks

modeling clay

construction paper

scissors

glue stick

Safety

Your body has different types of cells that perform various functions in the blood. Red blood cells carry oxygen to all the other cells in your body. White blood cells destroy viruses and bacteria that can attack the body and make you sick.

Learn It

In science, a **model** is a representation of how something in the natural world works. A model can be used to demonstrate a process that is difficult to see in action.

Try It

1 Read and complete a lab safety form.

2 Cut out shapes from construction paper to represent the following organs: heart, lungs, stomach, and small intestine. Also cut out a shape to represent a body cell.

3 Draw an outline of a student on a large sheet of paper. Place the organs in the appropriate body position on the outline. Choose a location away from the center of the body, such as an arm or a leg, to place the body cell.

4 Use the modeling clay to create molecules of oxygen, food, and waste materials (carbon dioxide and water). Place the oxygen molecules in the lungs. Place the food molecules in the stomach.

5 Your body gets energy when oxygen helps break down food molecules. Waste products are released during the breakdown of food molecules. Think about how a body cell gets energy. Draw roads to connect the organs and the body cell so that the body cell can get the energy it needs. Select toy vehicles to represent red blood cells and white blood cells.

6 Draw a diagram of your model in your Science Journal.

Apply It

7 How does oxygen reach body cells? Use the appropriate vehicle to model how red blood cells carry oxygen to a body cell. Add the path of the oxygen molecules to your diagram.

8 How do food molecules reach body cells? Use the appropriate vehicle to model how food molecules reach a body cell. Add the path of the food molecules to your diagram.

9 Where are waste materials produced? Use the appropriate vehicle to model how waste materials leave the body. Add the path to your diagram.

10 🔑 **Key Concept** Explain why using police cars and red pickup trucks are appropriate models to represent white blood cells and red blood cells, respectively.

Lesson 2

Reading Guide

Key Concepts
ESSENTIAL QUESTIONS

- How does the body move?
- How does the body respond to changes in its environment?

Vocabulary
compact bone p. 246
spongy bone p. 246
neuron p. 248
reflex p. 249
hormone p. 251

g Multilingual eGlossary

Video BrainPOP®

Structure, Movement, and Control

Inquiry Open wide?

When you have a dental checkup, you are asked to open your mouth. How are you able to open your mouth? What keeps your teeth from falling out when you chew food?

Why is the skeletal system so important?

Your skeletal system protects your body's organs, provides support, helps you move, and stores necessary minerals.

1. Read and complete a lab safety form.

2. Obtain one of the **disassembled human figures** and a **kit of materials.**

3. Use the materials to build a backbone for your figure. Using your backbone, connect the head and the arms to the legs of the figure.

Think About This

1. Which materials did you find helpful in creating a backbone and skeletal structure for your figure? Which were not helpful?

2. What characteristics of the "skeleton" were important as you built it? What problems would be caused by not having a skeleton?

3. 🔑 **Key Concept** Can you make your figure move? How does having a good support structure help it to move?

Structure and Movement

Have you ever had to open your mouth for a dental checkup as shown in the photo on the previous page? The human body can move in many different directions and perform a wide variety of tasks. It is able to do things that require many parts of the body to move, such as shooting a basketball into a hoop or swimming a lap in a pool. The human body also can remain very still, such as when posing for a picture or balancing on one leg.

In this lesson, you will read more about two organ systems—the skeletal system and the muscular system—that give the body structure, help the body move, and protect other organ systems.

The Skeletal System

The skeletal system has four major jobs. It protects internal organs, provides support, helps the body move, and stores minerals. The skeletal system is mostly bones. Adults have 206 bones. Ligaments, tendons, and cartilage are also parts of the skeletal system.

FOLDABLES

Make two horizontal two-tab books. Label them as shown, and glue them side by side to form a booklet with four tabs. Glue this section to the back of the one you made in Lesson 1. Use your book to organize information about the body systems in this lesson.

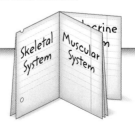

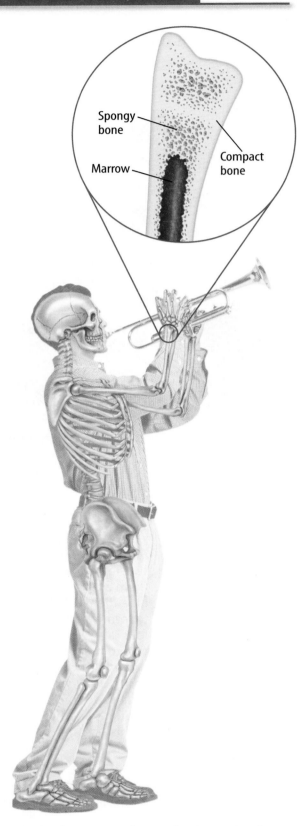

Spongy bone

Marrow

Compact bone

Figure 10 Bone is made up of a dense, hard exterior and a spongy interior.

Storage The skeletal system is also an important storage site for minerals such as calcium. Calcium is essential for life. It has many functions in the body. Muscles require calcium for contractions. The nervous system requires calcium for communication. Most of the calcium in the body is stored in bone. Calcium helps build stronger compact bone. Cheese and milk are good sources of calcium.

Reading Check What mineral is stored by the skeletal system?

Support Without a skeleton, your body would look like a beanbag. Your skeleton gives your body structure and support, as shown in **Figure 10.** Your bones help you stand, sit up, and raise your arms to play an instrument, such as a trumpet.

Protection Many of the bones in the body protect organs that are made of softer tissue. For example, the skull protects the soft tissue of the brain, and the rib cage protects the soft tissue of the lungs and heart.

Movement The skeletal system helps the body move by working with the muscular system. Bones can move because they are attached to muscles. You will read more about the interaction of the skeletal system and the muscular system later in this lesson.

Bone Types Bones are organs that contain two types of tissue. **Compact bone** *is the hard outer layer of bone.* **Spongy bone** *is the interior region of bone that contains many tiny holes.* As shown in **Figure 10,** spongy bone is inside compact bone. Some bones also contain bone marrow. Recall that bone marrow is a part of the lymphatic system and makes white blood cells.

Reading Check How do the two types of bone tissue differ?

The Muscular System

You might already know that there are muscle cells in your arms and legs. But did you know that there are muscle cells in your eyes, heart, and blood vessels? Without muscle cells you would not be able to talk, write, or run.

As shown in **Figure 11,** muscle cells are everywhere in the body. Almost one-half of your body mass is muscle cells. These muscle cells make up the muscular system. By working together, they help the body move.

The muscular system is made of three different types of muscle tissue—skeletal muscle, cardiac muscle, and smooth muscle. Skeletal muscle works with the skeletal system and helps you move. Tendons connect skeletal muscles to bones. Skeletal muscle also gives you the strength to lift heavy objects.

Another type of muscle tissue is cardiac muscle. Cardiac muscle is only in the heart. It continually contracts and relaxes and moves blood throughout your body.

Smooth muscle tissue moves materials through your body. Smooth muscle tissue is in organs such as the stomach and the bladder. Blood vessels also have smooth muscle tissue.

Key Concept Check What systems help the body move?

Figure 11 Cardiac muscle is only in the heart. Organs, such as the stomach, have smooth muscle. Skeletal muscle moves your body.

Visual Check Which type of muscle is in your arms?

The Muscular System

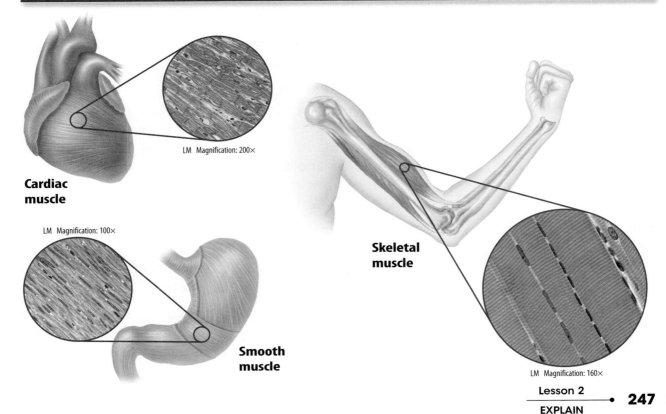

Cardiac muscle

LM Magnification: 200×

LM Magnification: 100×

Smooth muscle

Skeletal muscle

LM Magnification: 160×

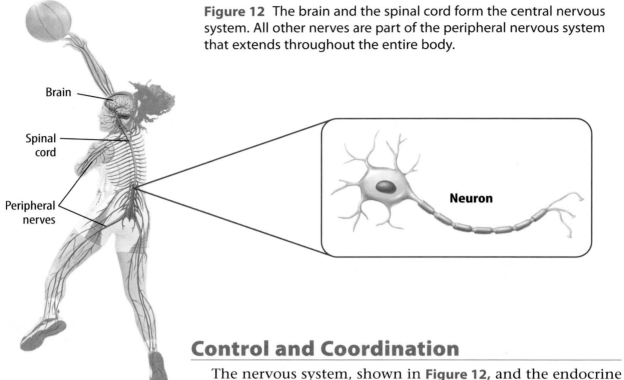

Figure 12 The brain and the spinal cord form the central nervous system. All other nerves are part of the peripheral nervous system that extends throughout the entire body.

Brain

Spinal cord

Peripheral nerves

Neuron

■ Central nervous system (CNS)
■ Peripheral nervous system (PNS)

Control and Coordination

The nervous system, shown in **Figure 12,** and the endocrine system, which you will read about later, receive and process information about your internal and external environments. These two systems control many functions, including movement, communication, and growth, by working with other systems in the body and help maintain homeostasis.

 Key Concept Check How does the body respond to changes in its environment?

The Nervous System

The nervous system is a group of organs and specialized cells that detect, process, and respond to information. The nervous system constantly receives information from your external environment and from inside your body. It can receive information, process it, and produce a response in less than 1 second.

Nerve cells, or **neurons,** *are the basic units of the nervous system.* Neurons can be many different lengths. In adults, some neurons are more than 1 m long. This is about as long as the distance between a toe and the spinal cord.

The nervous system includes the brain, the spinal cord, and nerves. The brain and the spinal cord form the central nervous system. Nerves outside the brain and the spinal cord make up the peripheral nervous system.

WORD ORIGIN · · · · · · · · · ·

neuron
from Greek *neuron*, means "a nerve cell with appendages"
· · · · · · · · · · · · · · · · ·

Processing Information The central nervous system is protected by the skeletal system. Muscles and other organs surround the peripheral nervous system. Information enters the nervous system through neurons in the peripheral nervous system. Most of the information then is sent to the central nervous system for processing. After the central nervous system processes information, it signals the peripheral nervous system to respond.

Voluntary and Involuntary Control The body carries out many functions that depend on the nervous system. Some of these functions such as breathing and digestion are automatic, or involuntary. They do not require you to think about them to make them happen. The nervous system automatically controls these functions and maintains homeostasis.

Most of the other functions of the nervous system are not automatic. They require you to think about them to make them happen. Tasks such as reading, talking, and walking are voluntary. These tasks require input, processing, and a response.

Reflexes Have you ever touched a hot pan with your hand? Touching a hot object sends a rapid signal that your hand is in pain. The signal is so fast that you do not think about moving your hand; it just happens automatically. *Automatic movements in response to a signal are called* **reflexes.** The spinal cord receives and processes reflex signals, as shown in **Figure 13.** Processing the information in the spinal cord instead of the brain helps the body respond more quickly.

Figure 13 🔑 Reflexes happen automatically.

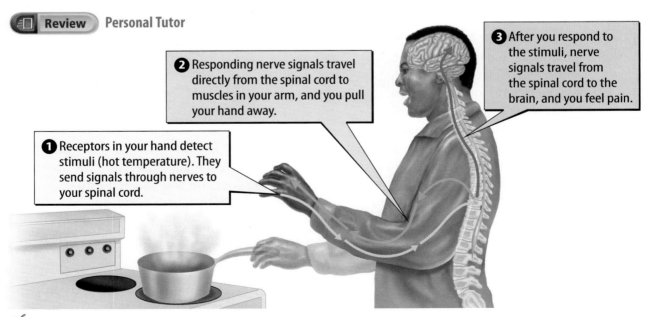

🖥 Review   **Personal Tutor**

② Responding nerve signals travel directly from the spinal cord to muscles in your arm, and you pull your hand away.

③ After you respond to the stimuli, nerve signals travel from the spinal cord to the brain, and you feel pain.

① Receptors in your hand detect stimuli (hot temperature). They send signals through nerves to your spinal cord.

✅ **Visual Check** What detects heat when you touch a hot pan?

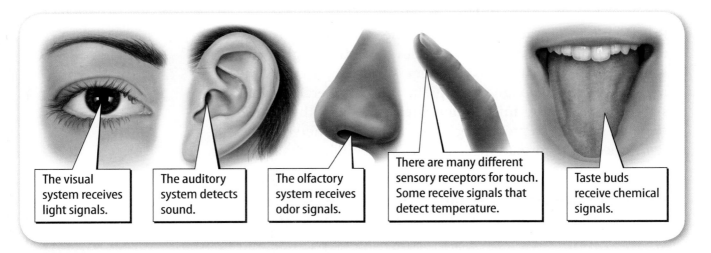

The visual system receives light signals.

The auditory system detects sound.

The olfactory system receives odor signals.

There are many different sensory receptors for touch. Some receive signals that detect temperature.

Taste buds receive chemical signals.

Figure 14 Each sense receives a different type of signal.

The Senses Humans detect their external environment with five senses—vision, hearing, smell, touch, and taste—as shown in **Figure 14.** Each of the five senses has specific neurons that receive signals from the environment. Information detected by the senses is sent to the spinal cord and then to the brain for processing and a response. Responses depend on the specific signal detected. Some responses cause muscles to contract and move such as when you touch a hot surface. The aroma of baking cookies might cause your mouth to produce saliva.

Inquiry MiniLab

15 minutes

Does your sight help you keep your balance?

It can be hard to keep your balance when standing on one leg. Does shutting your eyes make this task easier or more difficult?

1. Read and complete a lab safety form.

2. Stand upright and lift your left leg, balancing yourself on your right leg. Hold your left arm out so it is over your left knee.

3. Move your left leg backward and forward while maintaining your balance. As you move your leg, move your left arm at the same time. Have another student nearby to help you if you lose your balance.

4. Count how many times you are able to move your arm and your leg together before you lose your balance. Record this number in your Science Journal.

5. Repeat steps 2–4 with your eyes closed.

Analyze and Conclude

1. **Compare** How many times were you able to swing your arm and your leg with your eyes open? With your eyes closed?

2. **Analyze** Was it easier to maintain your balance with your eyes open or closed? Explain your answer.

3. **Key Concept** Infer how your vision helps you maintain homeostasis.

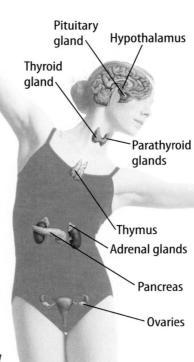

Pituitary gland
Hypothalamus
Thyroid gland
Parathyroid glands
Thymus
Adrenal glands
Pancreas
Ovaries

The Endocrine System

How tall were you in first grade? How tall are you now? From the time you were born until now, your body has changed. These changes are controlled by the endocrine system, shown in **Figure 15.** Like the nervous system, the endocrine system sends signals to the body. *Chemical signals released by the organs of the endocrine system are called* **hormones.** Hormones cause organ systems to carry out specific functions.

Why does your body need two organ systems to process information? The signals sent by the nervous system travel quickly through neurons. Hormones travel in blood through blood vessels in the circulatory system. These messages travel more slowly than nerve messages. A signal sent by the nervous system can travel from your head to your toes in less than 1 s, but a hormone will take about 20 s to make the trip. Although hormones take longer to reach their target organ system, their effects usually last longer.

Many of the hormones made by the endocrine system work with other organ systems and maintain homeostasis. For example, parathyroid hormone works with the skeletal system and controls calcium storage. Insulin is a hormone that is released from the pancreas that signals the digestive system to control nutrient homeostasis. Other hormones, such as growth hormone, work with many organ systems to help you grow. In the next lesson, you will read about another system that the endocrine system works with.

Figure 15 The endocrine system uses hormones to communicate with other organ systems.

✓ **Reading Check** How do hormones help the body maintain homeostasis?

Lesson 2 Review

Visual Summary

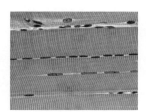

The skeletal system protects organs, provides support, helps the body move, and stores minerals.

Skeletal muscle works with the skeletal system and helps you move.

Reflex signals are received by the spinal cord but are not processed by the brain. This helps the body respond quickly.

FOLDABLES

Use your lesson Foldable to review the lesson. Save your Foldable for the project at the end of the chapter.

What do you think NOW?

You first read the statements below at the beginning of the chapter.

3. All bones in the skeletal system are hollow.

4. The endocrine system makes hormones.

Did you change your mind about whether you agree or disagree with the statements? Rewrite any false statements to make them true.

Use Vocabulary

1 **Distinguish** between compact bone and spongy bone.

2 A chemical signal that is released by the endocrine system is a(n) _____.

3 **Use the term** *neuron* in a sentence.

Understand Key Concepts

4 An automatic movement in response to a signal is called a
   A. hormone.   C. neuron.
   B. muscle.   D. reflex.

5 **Compare** the role of tendons in helping the skeletal system and the muscular system work together to a bridge between two cities.

6 **Infer** How does the skeletal system protect other organ systems in the body?

Interpret Graphics

7 **Summarize** Copy and fill in the graphic organizer below to show the three types of muscle tissue.

8 **Predict** the effect of having less compact bone than normal on the strength of the skeletal system by examining the figure below.

Critical Thinking

9 **Hypothesize** What would be the effect of losing one's sight on the ability to digest food? Explain your answer.

Bone Marrow Transplants

Why might you need new bone marrow?

Healthy blood cells are essential to overall health. Red blood cells carry oxygen throughout the body. Some white blood cells fight infections. Platelets help stop bleeding. A bone marrow transplant is sometimes necessary when a disease interferes with the body's ability to produce healthy blood cells.

Bone marrow is a tissue found inside some of the bones in your body. Healthy bone marrow contains cells that can develop into white blood cells, red blood cells, or platelets. Some diseases, such as leukemia and sickle cell disease, affect bone marrow. Replacing malfunctioning bone marrow with healthy bone marrow can help treat these diseases.

▲ In healthy bone marrow, a stem cell can develop into different types of blood cells.

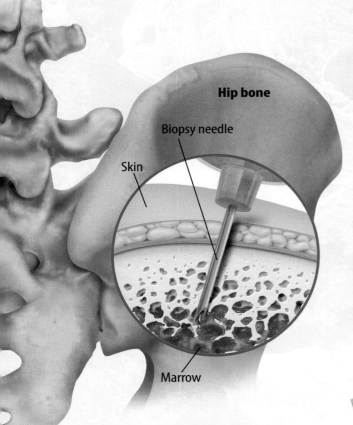

Hip bone

Biopsy needle

Skin

Marrow

▲ Bone marrow is harvested from the pelvic bone. An anesthetic is used to keep the donor from feeling pain during the procedure.

A bone marrow transplant involves several steps. The patient receiving the bone marrow must have treatments to destroy his or her unhealthy bone marrow. Healthy bone marrow must be obtained for the transplant. Sometimes, the patient's own bone marrow can be treated and used for transplant. This transplant has the greatest chance of success. Other transplants involve healthy bone marrow donated by another person. The bone marrow must be tested to ensure that it is a good match for the patient.

The bone marrow donor undergoes a procedure called harvesting. Bone marrow is taken from the donor's pelvic bone. The donor's body replaces the harvested bone marrow, so there are no long-term effects for the donor.

The donated bone marrow is introduced into the patient's bloodstream. If the transplant is successful, the new bone marrow moves into the bone cavities and begins producing healthy blood cells.

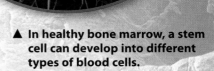

It's Your Turn

NATIONAL
MARROW
DONOR
PROGRAM®

RESEARCH AND REPORT Find out more about bone marrow transplants. What other diseases can be treated using a bone marrow transplant? What is the National Marrow Donor Program? Present your findings to your class.

Lesson 3

Reading Guide

Key Concepts 🔑
ESSENTIAL QUESTIONS

- What do the male and female reproductive systems do?

- How do humans grow and change?

Vocabulary

reproduction p. 255

gamete p. 255

sperm p. 255

ovum p. 255

fertilization p. 255

zygote p. 255

 Multilingual eGlossary

 Video BrainPOP®

Reproduction and Development

Inquiry Strands of Hair?

The things that look like strands of hair are sperm, the male reproductive cells. The red structure is an egg, the female reproductive cell. Why are there so many sperm but only one egg?

How do the sizes of egg and sperm cells compare?

A sperm cell combines with an egg cell to create a zygote that will eventually become a fetus and then a baby. The sperm and egg cells each contribute half the genetic material to the zygote.

1 Read and complete a lab safety form.

2 Select one of the **spheres** to use as a model of an egg cell. With a **ruler,** measure the diameter of the sphere. Record the measurement in your Science Journal.

3 If an average sperm cell is 3–6 microns in diameter, and an average egg cell is 120–150 microns in diameter, determine the diameter of a suitable model for a sperm cell.

4 Find another sphere that is approximately the size needed to create an accurate model to represent a sperm cell. Label both of your models.

Think About This

1. What were the sizes of the spheres you chose to model the sizes of the sperm and egg cells?

2. 🔑 **Key Concept** How do the egg cell and sperm cells interact in reproduction? How do you think size plays a role in this interaction?

Reproduction and Hormones

You have read how the endocrine system works with other organ systems and helps the body grow and maintain homeostasis. The endocrine system has another very important function—to ensure that humans can reproduce. Some of the organs of the endocrine system produce hormones that help humans reproduce. **Reproduction** *is the process by which new organisms are produced.* Reproduction is essential to the continuation of life on Earth.

A male and a female each have special organs for reproduction. Organs in the male reproductive system are different from those in the female reproductive system. *Human reproductive cells,* called **gametes** (GA meets), *are made by the male and female reproductive systems. Male gametes are called* **sperm.** *Female gametes are called* **ova** (OH vah; singular ovum), or eggs.

As shown in the photo at the beginning of the lesson, *a sperm joins with an egg in a reproductive process called* **fertilization.** *The cell that forms when a sperm cell fertilizes an egg cell is called a* **zygote** (ZI goht). A zygote is the first cell of a new human. It contains genetic information from both the sperm and the ovum. The zygote will grow and develop in the female's reproductive system.

✔️ **Reading Check** How do gametes enable humans to reproduce?

FOLDABLES

Make a horizontal two-tab book. Label it as shown, and glue it side by side to the back of the booklet made in Lessons 1 and 2. Use the book to organize information about the male and female reproductive systems.

Male Reproductive System | Female Reproductive System

WORD ORIGIN

zygote
from Greek *zygoun,* means "to join"

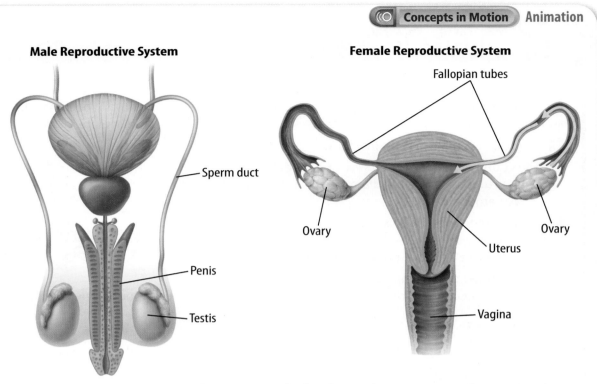

Male Reproductive System

Sperm duct

Penis

Testis

The organs of the male reproductive system produce sperm and deliver it to the female reproductive system.

Female Reproductive System

Fallopian tubes

Ovary

Ovary

Uterus

Vagina

The female reproductive system produces eggs and provides a place for a new human to grow and develop before birth.

Figure 16 🔑 Males and females have specialized organs for reproduction.

The Male Reproductive System

The male reproductive system, shown in **Figure 16,** produces sperm and delivers it to the female reproductive system. Sperm are produced in the testes (TES teez; singular, testis). Sperm develop inside each testis and then are stored in tubes called sperm ducts. Sperm matures in the sperm ducts.

The testes also produce a hormone called testosterone. Testosterone helps sperm change from round cells to long, slender cells that can swim. Once sperm have fully developed, they can travel to the penis. The penis is a tubelike structure that delivers sperm to the female reproductive system. Sperm are transported in a fluid called semen (SEE mun). Semen contains millions of sperm and nutrients that provide the sperm with energy.

🔑 **Key Concept Check** What does the male reproductive system do?

The Female Reproductive System

The female reproductive system contains two ovaries, as shown in **Figure 16.** Eggs grow and mature in the ovaries. Two hormones made by the ovaries, estrogen (ES truh jun) and progesterone (proh JES tuh rohn), help eggs mature. Once mature, eggs are released from the ovaries and enter the fallopian tubes. As shown in **Figure 16,** the fallopian tubes connect the ovaries to the uterus.

If sperm are also present in the fallopian tube, fertilization can occur as the egg enters the fallopian tube. Sperm enter the female reproductive system through the vagina, a tube-shaped organ that leads to the uterus. A fertilized egg, or zygote, can move through the fallopian tube and attach inside the uterus.

If there are no sperm in the fallopian tube, the egg will not be fertilized. However, it will still travel through the fallopian tube and uterus and then break down.

The Menstrual Cycle The endocrine system controls egg maturation and release and thickening of the lining of the uterus in a process called the menstrual (MEN stroo ul) cycle. The menstrual cycle takes about 28 days and has three parts.

During the first part of the cycle, eggs grow and mature and the thickened lining of the uterus leaves the body. In the second part of the cycle, mature eggs are released from the ovaries and the lining of the uterus thickens. In the third part of the cycle, unfertilized eggs and the thickened lining break down. The lining leaves the body in the first part of the next cycle.

 Key Concept Check What does the female reproductive system do?

Human Development

As shown in **Figure 17,** humans develop in many stages. You have read that when a sperm fertilizes an egg, a zygote forms. The zygote develops into an embryo (EM bree oh). An embryo is a ball-shaped structure that attaches inside the uterus and continues to grow.

The embryo develops into a fetus, the last stage before birth. It takes about 38 weeks for a fertilized egg to fully develop. This developmental period is called pregnancy. During this period, the organ systems of the fetus will develop and the fetus will get larger. Pregnancy ends with birth. During birth, the endocrine system releases hormones that help the uterus push the fetus through the vagina and out of the body.

Figure 17 During pregnancy, a unicellular zygote develops into a fetus.

Visual Check When is the heart fully formed?

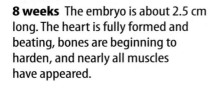

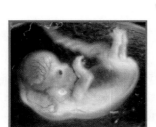

14 weeks Growth and development continue. The fetus is about 6 cm long.

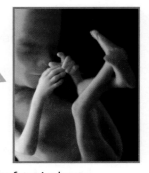

16 weeks The fetus is about 15 cm long and about 140 g. The fetus can make a fist and has a range of facial expressions.

8 weeks The embryo is about 2.5 cm long. The heart is fully formed and beating, bones are beginning to harden, and nearly all muscles have appeared.

5 weeks The embryo is about 7 mm long. The heart and other organs have started to develop. The arms and legs are beginning to bud.

22 weeks The fetus is about 27 cm long and about 430 g. Footprints and fingerprints are forming.

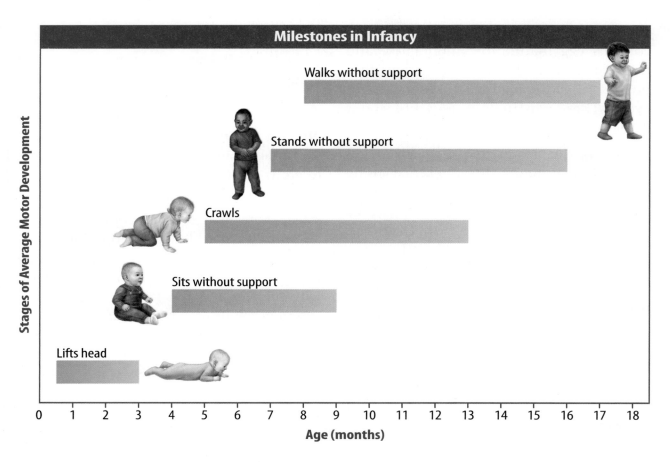

Milestones in Infancy

Stages of Average Motor Development

Walks without support

Stands without support

Crawls

Sits without support

Lifts head

Age (months)
0 1 2 3 4 5 6 7 8 9 10 11 12 13 14 15 16 17 18

▲ **Figure 18** During infancy, a human learns to crawl and walk.

☑ **Visual Check** When does an infant usually crawl?

Figure 19 🔑
Humans continue to change during adolescence and adulthood. ▼

From Birth Through Childhood

The first life stage after birth is infancy, the first 2 years of life. During infancy, the muscular and nervous systems develop and an infant begins walking, as shown in **Figure 18.** Growth and development continue in childhood, which is from about 2 years to about 12 years of age. Bones in the skeletal system grow longer and stronger, and the lymphatic system matures.

Adolescence Through Adulthood

Adolescence follows childhood. During adolescence, growth of the skeletal and muscular systems continues. Organs such as the lungs and kidneys get larger. As the endocrine system develops, the male and female reproductive systems mature. The period of time during which the reproductive system matures is called puberty.

After adolescence is adulthood, as shown in **Figure 19.** During adulthood, humans continue to change. In later adulthood, hair turns gray, wrinkles might form in the skin, and bones become weaker in a process called aging. Aging is a slow process that can last for decades.

🔑 **Key Concept Check** How do humans change during adulthood?

Lesson 3 Review

Visual Summary

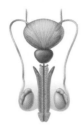

Sperm are produced in the testes and develop inside each testis in the seminiferous tubules.

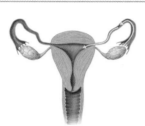

Eggs grow and mature in the ovaries.

During pregnancy, a zygote develops into an embryo and then into a fetus.

FOLDABLES

Use your lesson Foldable to review the lesson. Save your Foldable for the project at the end of the chapter.

What do you think NOW?

You first read the statements below at the beginning of the chapter.

5. The testes produce sperm.

6. Puberty occurs during infancy.

Did you change your mind about whether you agree or disagree with the statements? Rewrite any false statements to make them true.

Use Vocabulary

1 Sperm and ova are types of _____.

2 **Distinguish** between an ovum and a zygote.

3 **Define** *fertilization* in your own words.

Understand Key Concepts 🔑

4 The period between birth and 2 years is called

 A. adolescence. C. childhood.

 B. adulthood. D. infancy.

5 **Compare** the functions of the ovaries and the testes.

6 **Distinguish** between a zygote and a fetus.

Interpret Graphics

7 **Summarize** Copy and fill in the graphic organizer below to show the stages of life.

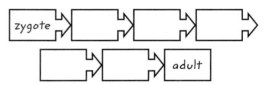

8 **Evaluate** the changes in aging using the photo below.

Critical Thinking

9 **Hypothesize** why development before birth takes a long time, about 38 weeks.

Materials

presentation materials

Model the Body Systems

You have learned about the functions of the different systems of the body. Your task is to find a real-life example of something that can be used as a model for the entire body, including all of its organ systems. You will illustrate this model and use it to describe how the various systems all work together to help the body function as a whole.

Question

What other real-life systems can be used to model the systems of the human body?

Procedure

1. Read and complete a lab safety form.

2. Think about the different systems that make up your body. In your Science Journal, make a table such as the one below. List each body system, and write a description of the role of each system in the body.

3. Discuss a model of the entire body with your teacher and the rest of the class. Note how each system is modeled in the example and how all the systems work together in the model as well.

4. Think of an example of your own. Write out a description of this model in your Science Journal.

5. Complete the last column of your table for the example you chose. Be creative and descriptive.

6. Create a visual display that illustrates your model. Use photos and other pictures to illustrate the different parts of the model system you created. Label each of these pictures to describe which body system they represent. Include a description of the function of each system on the labels.

7. Use the visual of your model to describe how the systems work together and maintain homeostasis. List the events that occur in your model as each system does its job properly.

Body System	Description of Activity	Model System
Circulatory	transport materials	

8 Use the information in your display to predict how your model would be affected if one of the systems did not function properly.

9 Use the model to explain how the entire model would be affected if one of the "body systems" did not work properly. Write out a list of events for this scenario.

Analyze and Conclude

10 Analyze Is there a system that is not included in your model? Explain why some functions are easier to model than others.

11 Assess How do the systems in your model respond to changes in the environment?

12 Evaluate How successful was your model in illustrating the effects of having parts of the system break down?

13 The Big Idea How do all of the systems in your model work together to help the model as a whole?

Communicate Your Results

Share your model with the class. Discuss the parts of your model with other students, and compare the ways you chose to model the same systems.

Inquiry Extension

In real life, human body systems might have problems that cause them to fail. Doctors often fix a body system that no longer functions by replacing failing organs with donated ones. Investigate how doctors use donated organs. Look for a recent news article in which someone's life was saved by one of these procedures. Write a brief summary of what you found, and describe the procedure using your model.

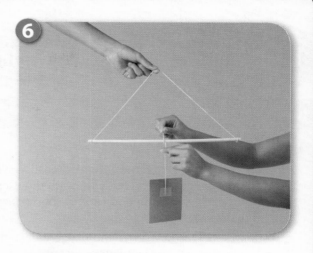

6

Lab Tips

☑ Use the descriptions in your table to help you come up with the different parts of your model. Remember that your model represents general functions.

☑ Use magazines, pictures, or other visuals to illustrate your model. Don't forget to label the parts of the model. Be as descriptive as possible with the labels.

☑ Make sure the parts of your model connect to each other as part of the larger picture.

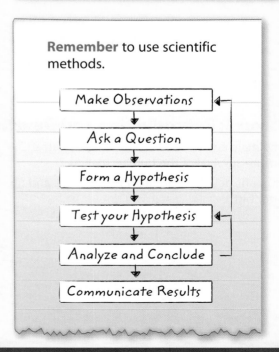

Remember to use scientific methods.

Make Observations
↓
Ask a Question
↓
Form a Hypothesis
↓
Test your Hypothesis
↓
Analyze and Conclude
↓
Communicate Results

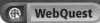
 THE BIG IDEA

Human body systems transport materials; defend against pathogens; provide structure, movement, and control; and enable the species to survive.

Key Concepts Summary 🔑	Vocabulary
Lesson 1: Transport and Defense • Nutrients enter the body through the digestive system. Wastes and water leave the body through the excretory system. Oxygen enters the body and carbon dioxide leaves the body through the respiratory system. • Substances such as **nutrients** and oxygen reach the body's cells through the circulatory system. • The lymphatic system helps the body defend itself against harmful invaders. 	**organ system** p. 231 **homeostasis** p. 231 **nutrient** p. 233 **Calorie** p. 233 **lymphocyte** p. 239 **immunity** p. 240
Lesson 2: Structure, Movement, and Control  • The muscular system and the skeletal system work together and help the body move. The skeletal system provides the body with structure and protects other organ systems. • The nervous system and the endocrine system work together and help the body respond to changes in the environment.	**compact bone** p. 246 **spongy bone** p. 246 **neuron** p. 248 **reflex** p. 249 **hormone** p. 251
Lesson 3: Reproduction and Development • The male and female reproductive systems ensure survival of the human species. • Humans develop and grow both before and after birth. 	**reproduction** p. 255 **gamete** p. 255 **sperm** p. 255 **ovum** p. 255 **fertilization** p. 255 **zygote** p. 255

Review

- Personal Tutor
- Vocabulary eGames
- Vocabulary eFlashcards

FOLDABLES® **Chapter Project**

Assemble your lesson Foldables as shown to make a Chapter Project. Use the project to review what you have learned in this chapter.

Human Body Systems

Use Vocabulary

1. Carbohydrates, fats, and proteins are all _____ and contain _____.

2. The thymus and the spleen produce white blood cells called _____.

3. Bones in the skeletal system are made of a hard exterior called _____.

4. Define the term *hormone* in your own words.

5. Use the terms *reproduction* and *fertilization* in a sentence.

6. Distinguish between ova and sperm.

Link Vocabulary and Key Concepts

Concepts in Motion Interactive Concept Map

Copy this concept map, and then use vocabulary terms from the previous page to complete the concept map.

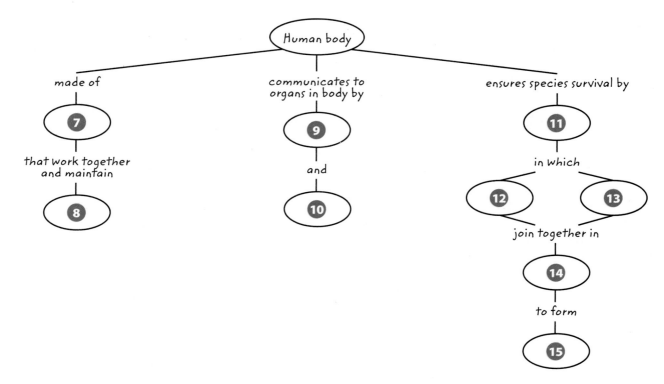

Understand Key Concepts

1. Which body system removes carbon dioxide and waste?
   - A. circulatory
   - B. digestive
   - C. excretory
   - D. lymphatic

2. Which body system makes immune cells?
   - A. circulatory
   - B. digestive
   - C. excretory
   - D. lymphatic

3. Which are bundles of cells in the lungs that take in oxygen?
   - A. alveoli
   - B. bronchi
   - C. nostrils
   - D. trachea

4. Which are proteins that recognize specific proteins on bacteria?
   - A. antibodies
   - B. enzymes
   - C. nutrients
   - D. receptors

5. Which part of the nervous system is shown below?

   - A. brain
   - B. neuron
   - C. peripheral nerve
   - D. spinal cord

6. Which is NOT a type of blood vessel?
   - A. artery
   - B. capillary
   - C. spleen
   - D. vein

7. Which is NOT a type of muscle tissue?
   - A. cardiac
   - B. lymphatic
   - C. skeletal
   - D. smooth

8. Which is a part of the skeletal system?
   - A. ligament
   - B. spleen
   - C. thymus
   - D. trachea

9. Which hormone helps the cells produced in the system below mature?

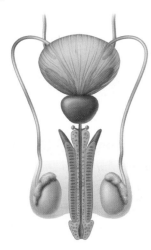

   - A. estrogen
   - B. insulin
   - C. progesterone
   - D. testosterone

10. Which connects the ovaries to the uterus?
    - A. bladder
    - B. cervix
    - C. fallopian tubes
    - D. seminiferous tubules

11. Which system works with the reproductive system?
    - A. endocrine
    - B. excretory
    - C. respiratory
    - D. skeletal

Critical Thinking

12 **Relate** the body's organization to how homeostasis is maintained.

13 **Describe** the roles of the structure shown below in digestion.

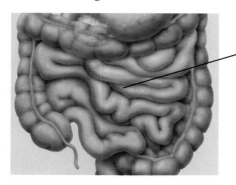

Small intestine

14 **Compare** the functions of lymphatic vessels and blood vessels.

15 **Hypothesize** how an injury to the spinal cord might affect the ability of the nervous system to sense and respond to a change in the environment.

16 **Assess** how the nervous system helps the muscular system control heart rate, digestion, and respiration.

17 **Relate** the organs of the lymphatic system to immunity.

18 **Assess** the role of the skeletal system in the storage of nutrients.

19 **Summarize** the role of puberty in the transition from adolescence to adulthood.

20 **Compare** the functions of the male and female reproductive systems.

Writing in Science

21 **Write** a five-sentence paragraph that distinguishes the two main types of diseases. Be sure to include a topic sentence and a concluding sentence in your paragraph.

REVIEW THE B**I**G IDEA

22 How do organ systems help the body function?

23 The photo below shows parts of the digestive system, the skeletal system, the muscular system, and the nervous system. What are the functions of these systems?

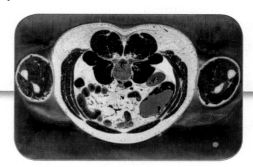

Math Skills ×÷+

📖 **Review**

— **Math Practice** —

Use Proportions

24 Which type of chicken in the table below has the fewest Calories per gram?

Food	Mass (g)	Calories (C)
$\frac{1}{2}$ chicken breast, baked	86	140
1 chicken leg, baked	52	112

25 A small 140-g apple and a 100-g banana each provide 70 C of energy. How many Calories would there be in a 200-g serving of fruit salad that contained equal amounts of apple and banana? [Hint: Add the values for the apple and the banana first.]

Standardized Test Practice

Record your answers on the answer sheet provided by your teacher or on a sheet of paper.

Multiple Choice

1. In which part of the digestive system does absorption of nutrients occur?
   A. esophagus
   B. liver
   C. small intestine
   D. stomach

2. After food enters the mouth, which path does it travel through the digestive system?
   A. esophagus → stomach → small intestine → large intestine → rectum
   B. large intestine → small intestine → stomach → esophagus → rectum
   C. small intestine → large intestine → esophagus → stomach → pancreas
   D. stomach → esophagus → small intestine → large intestine → liver

Use the diagram below to answer question 3.

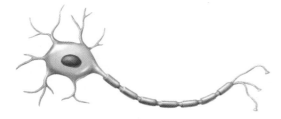

3. What body system is made of the basic unit shown in the diagram?
   A. circulatory system
   B. endocrine system
   C. muscular system
   D. nervous system

4. Which body system provides protection from infection and toxins?
   A. circulatory system
   B. digestive system
   C. excretory system
   D. lymphatic system

5. Which two systems work together to make your body move?
   A. digestive system and skeletal system
   B. lymphatic system and digestive system
   C. nervous system and excretory system
   D. skeletal system and muscular system

Use the image below to answer question 6.

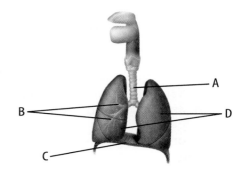

6. Which letter shows the muscle that contracts during inhalation and relaxes during exhalation?
   A. A
   B. B
   C. C
   D. D

7. What role do alveoli play in the body?

   A. They form urea.

   B. They produce lymphocytes.

   C. They connect the right atrium to the right ventricle

   D. They move oxygen into the circulatory system.

8. What is the human development period between 2 years and 12 years called?

   A. adolescence

   B. adulthood

   C. childhood

   D. infancy

Use the diagram below to answer question 9.

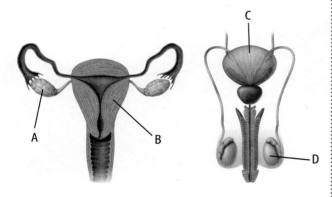

9. Which structure produces male gametes?

   A. A

   B. B

   C. C

   D. D

Constructed Response

Use the figure below to answer questions 10 and 11.

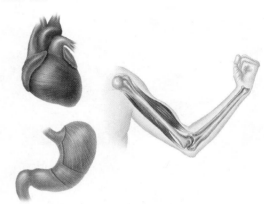

10. Identify the type of muscle shown in each image above.

11. Which muscles are under voluntary control of the nervous system? Which are involuntary? Explain your answer.

12. Which body system would be most likely affected by a diet that includes too little calcium? Explain your answer.

13. Describe how the endocrine system and reproductive system are related.

NEED EXTRA HELP?													
If You Missed Question...	1	2	3	4	5	6	7	8	9	10	11	12	13
Go to Lesson...	1	1	2	1	2	1	1	3	3	2	2	1, 2	2, 3

Chapter 8

Plant Processes and Reproduction

THE BIG IDEA What processes enable plants to survive and reproduce?

Inquiry Holding on for Dear Life?

The tendril of this *Omphalea* (om FAL ee uh) vine grows around a branch in a tropical rain forest.

- How do you think growing around another plant might help the *Omphalea* plant survive?

- Can you think of any other processes that enable plants to survive and reproduce?

Get Ready to Read

What do you think?

Before you read, decide if you agree or disagree with each of these statements. As you read this chapter, see if you change your mind about any of the statements.

1 Plants do not carry on cellular respiration.

2 Plants are the only organisms that carry on photosynthesis.

3 Plants do not produce hormones.

4 Plants can respond to their environments.

5 Seeds contain tiny plant embryos.

6 Flowers are needed for plant reproduction.

ConnectED Your one-stop online resource

connectED.mcgraw-hill.com

Video

WebQuest

Audio

Assessment

Review

Concepts in Motion

Inquiry

Multilingual eGlossary

Reading Guide

Key Concepts 🔑
ESSENTIAL QUESTIONS

- How do materials move through plants?
- How do plants perform photosynthesis?
- What is cellular respiration?
- What is the relationship between photosynthesis and cellular respiration?

Vocabulary
photosynthesis p. 272
cellular respiration p. 274

g **Multilingual eGlossary**

Energy Processing in Plants

Inquiry All Leaf Cells?

You are looking at a magnified cross section of a leaf. As you can see, the cells in the middle of the leaf are different from the cells on the edges. What do you think this might have to do with the cellular processes a leaf carries out that enable a plant's survival?

How can you show the movement of materials inside a plant?

Most parts of plants need water. They also need a system to move water throughout the plant so cells can use it for plant processes.

1 Read and complete a lab safety form.

2 Gently pull two stalks from the base of a bunch of **celery.** Leave one stalk complete. Use a **paring knife** to carefully cut directly across the bottom of the second stalk.

3 Put 100 mL of water into each of two **beakers.** Place 3–4 drops of **blue food coloring** into the water. Place one celery stalk in each beaker.

4 After 20 min, observe the celery near the bottom of each stalk. Observe again after 24 h. Record your observations in your Science Journal.

Think About This

1. What happened in each celery stalk?

2.  **Key Concept** What did the colored water do? Why do you think this occurred?

Materials for Plant Processes

Food, water, and oxygen are three things you need to survive. Some of your organ systems process these materials, and others transport them throughout your body. Like you, plants need food, water, and oxygen to survive. Unlike you, plants do not take in food. Most of them make their own.

Moving Materials Inside Plants

You might recall reading about xylem (ZI lum) and phloem (FLOH em)—the vascular tissue in most plants. These tissues transport materials throughout a plant.

After water enters a plant's roots, it moves into xylem. Water then flows inside xylem to all parts of a plant. Without enough water, plant cells wilt, as shown in **Figure 1.**

Most plants make their own food—a liquid sugar. The liquid sugar moves out of food-making cells, enters phloem, and flows to all plant cells. Cells break down the sugar and release energy. Some plant cells can store food.

Plants require oxygen and carbon dioxide to make food. Like you, plants produce water vapor as a waste product. Carbon dioxide, oxygen, and water vapor pass into and out of a plant through tiny openings in leaves.

 Key Concept Check How do materials move through plants?

Figure 1 This plant wilted due to lack of water in the soil.

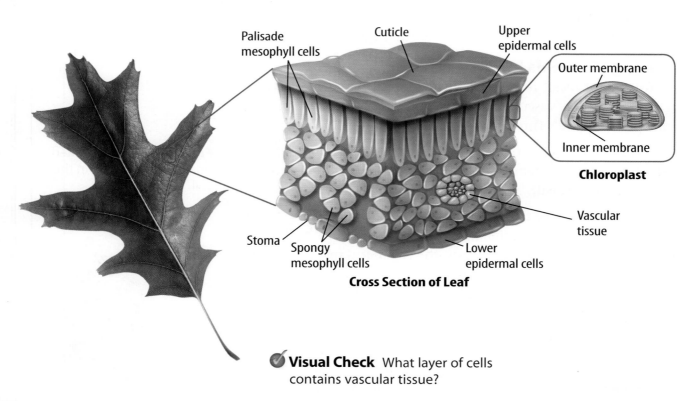

Make a shutterfold book, leaving a 2-cm space between the tabs. Label it as shown. Use the book as a diagram of leaf structure.

Upper Epidermis

Mesophyll Cells

Lower Epidermis

WORD ORIGIN

photosynthesis
from Greek *photo-*, means "light"; and *synthesis*, means "composition"

Figure 2 Photosynthesis occurs inside the chloroplasts of mesophyll cells in most leaves.

Photosynthesis

Plants need food, but they cannot eat as people do. They make their own food, and leaves are the major food-producing organs of plants. This means that leaves are the sites of photosynthesis (foh toh SIHN thuh sus). **Photosynthesis** *is a series of chemical reactions that convert light energy, water, and carbon dioxide into the food-energy molecule glucose and give off oxygen.* The structure of a leaf is well-suited to its role in photosynthesis.

Leaves and Photosynthesis

As shown in **Figure 2,** leaves have many types of cells. The cells that make up the top and bottom layers of a leaf are flat, irregularly shaped cells called epidermal (eh puh DUR mul) cells. On the bottom epidermal layer of most leaves are small openings called stomata (STOH muh tuh). Carbon dioxide, water vapor, and oxygen pass through stomata. Epidermal cells can produce a waxy covering called the cuticle.

Most photosynthesis occurs in two types of mesophyll (ME zuh fil) cells inside a leaf. These cells contain chloroplasts, the organelle where photosynthesis occurs. Near the top surface of the leaf are palisade mesophyll cells. They are packed together. This arrangement exposes the most cells to light. Spongy mesophyll cells have open spaces between them. Gases needed for photosynthesis flow through the spaces between the cells.

Palisade mesophyll cells

Cuticle

Upper epidermal cells

Outer membrane

Inner membrane

Chloroplast

Vascular tissue

Stoma

Spongy mesophyll cells

Lower epidermal cells

Cross Section of Leaf

✓ **Visual Check** What layer of cells contains vascular tissue?

Capturing Light Energy

As you read about the steps of photosynthesis, refer to **Figure 3** to help you understand the process. In the first step of photosynthesis, plants capture the energy in light. This occurs in chloroplasts. Chloroplasts contain plant pigments. Pigments are chemicals that can absorb and reflect light. Chlorophyll, the most common plant pigment, is necessary for photosynthesis. Most plants appear green because chlorophyll reflects green light. Chlorophyll absorbs other colors of light. This light energy is used during photosynthesis.

Once chlorophyll traps and stores light energy, this energy can be transferred to other molecules. During photosynthesis, water molecules are split apart. This releases oxygen into the atmosphere, as shown in **Figure 3**.

 Reading Check How do plants capture light energy?

Making Sugars

Sugars are made in the second step of photosynthesis. This step can occur without light. In chloroplasts, carbon dioxide from the air is converted into sugars by using the energy stored and trapped by chlorophyll. Carbon dioxide combines with hydrogen atoms from the splitting of water molecules and forms sugar molecules. Plants can use this sugar as an energy source or can store it. Potatoes and carrots are examples of plant structures where excess sugar is stored.

 Key Concept Check What are the two steps of photosynthesis?

Why is photosynthesis important?

Try to imagine a world without plants. How would humans or other animals get the oxygen that they need? Plants help maintain the atmosphere you breathe. Photosynthesis produces most of the oxygen in the atmosphere.

Photosynthesis 🔑

Figure 3 Photosynthesis is a series of complex chemical processes. The first step is capturing light energy. In the second step, that energy is used for making glucose, a type of sugar.

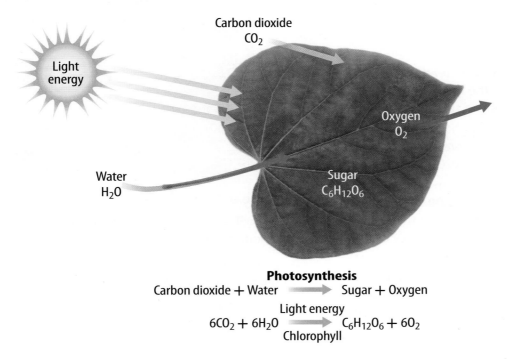

Photosynthesis
Carbon dioxide + Water ⟶ Sugar + Oxygen

Light energy
$$6CO_2 + 6H_2O \xrightarrow[\text{Chlorophyll}]{} C_6H_{12}O_6 + 6O_2$$

Cellular Respiration

ACADEMIC VOCABULARY

energy
(*noun*) usable power

All organisms require **energy** to survive. Energy is in the chemical bonds in food molecules. A process called cellular respiration releases energy. **Cellular respiration** *is a series of chemical reactions that convert the energy in food molecules into a usable form of energy called ATP.*

Releasing Energy from Sugars

Glucose molecules break down during cellular respiration. Much of the energy released during this process is used to make ATP, an energy storage molecule. This process requires oxygen, produces water and carbon dioxide as waste products, and occurs in the cytoplasm and mitochondria of cells.

Why is cellular respiration important?

If your body did not break down the food you eat through cellular respiration, you would not have energy to do anything. Plants produce sugar, but without cellular respiration, plants could not grow, reproduce, or repair tissues.

 Key Concept Check What is cellular respiration?

Inquiry MiniLab
20 minutes

Can you observe plant processes?

Plants perform both photosynthesis and cellular respiration. Can you observe both processes in radish seedlings?

1. Read and complete a lab safety form.
2. Put **potting soil** in the bottom of a **small, self-sealing plastic bag** so that the soil is 3–4 cm deep. Dampen the soil.
3. Drop several **radish seeds** into the bag and close the top, but leave a small opening so air can still get into the bag.
4. Place the bag upright in a place that has a **light source.** Each group should use a different light source. Observe for 4–5 days.
5. Carefully place an open container of **bromthymol blue (0.004%) solution** upright in the bag next to the seedlings. Bromthymol blue turns yellow in the presence of carbon dioxide.
6. Seal the bag. Observe the bag and its contents the next day. Record your observations in your Science Journal.

Analyze and Conclude

1. **Describe** the differences in seedling samples among groups. Why are there differences?
2. **Evaluate** What change in the bromthymol blue solution did you observe? Why?
3. **Key Concept** Explain what processes occurred in the seedlings.

Light energy

Chloroplast

Review
Personal Tutor

Carbon dioxide (CO_2)
Water (H_2O)

Glucose ($C_6H_{12}O_6$)
Oxygen (O_2)

Mitochondrion

ATP

$$6CO_2 + 6H_2O \xrightarrow[\text{Chlorophyll}]{\text{Light energy}} C_6H_{12}O_6 + 6O_2$$

Photosynthesis

$$C_6H_{12}O_6 + 6O_2 \rightarrow 6CO_2 + 6H_2O + \begin{array}{c}\text{ATP}\\\text{(Energy)}\end{array}$$

Cellular respiration

Photosynthesis v. Cellular Respiration

Concepts in Motion
Interactive Table

Process	Photosynthesis	Cellular Respiration
Reactants	light energy, CO_2, H_2O	glucose [sugar], O_2
Products	glucose, O_2	CO_2, H_2O, ATP
Organelle in which it occurs	chloroplasts	mitochondria
Type of organism	photosynthetic organisms including plants and algae	most organisms including plants and animals

Comparing Photosynthesis and Cellular Respiration

Photosynthesis requires light energy and the reactants—substances that react with one another during the process—carbon dioxide and water. Oxygen and the energy-rich molecule glucose are the products, or end substances, of photosynthesis. Most plants, some protists, and some bacteria carry on photosynthesis.

Cellular respiration requires the reactants glucose and oxygen, produces carbon dioxide and water, and releases energy in the form of ATP. Most organisms carry on cellular respiration. Photosynthesis and cellular respiration are interrelated, as shown in **Figure 4.** Life on Earth depends on a balance of these two processes.

Key Concept Check How are photosynthesis and cellular respiration alike, and how are they different?

Figure 4 The relationship between cellular respiration and photosynthesis is important for life.

Visual Check What are the reactants of cellular respiration? What are the products?

Lesson 1 Review

Visual Summary

Materials that a plant requires to survive move through the plant in the vascular tissue, xylem and phloem.

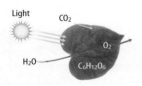

Plants can make their own food by using light energy, water, and carbon dioxide.

The products of photosynthesis are the reactants for cellular respiration.

FOLDABLES®

Use your lesson Foldable to review the lesson. Save your Foldable for the project at the end of the chapter.

What do you think NOW?

You first read the statements below at the beginning of the chapter.

1. Plants do not carry on cellular respiration.

2. Plants are the only organisms that carry on photosynthesis.

Did you change your mind about whether you agree or disagree with the statements? Rewrite any false statements to make them true.

Use Vocabulary

1 A series of chemical reactions that convert the energy in food molecules into a usable form of energy, called ATP, is called _____.

2 Define *photosynthesis* in your own words.

Understand Key Concepts

3 Which structure moves water through plants?
- **A.** chloroplast
- **C.** nucleus
- **B.** mitochondrion
- **D.** xylem

4 Describe how plants use chlorophyll for photosynthesis.

5 Summarize the process of cellular respiration.

Interpret Graphics

6 Explain how the structure shown below is organized for its role in photosynthesis.

7 Compare and Contrast Copy and fill in the table below to compare and contrast photosynthesis and cellular respiration.

Process	Similarities	Differences

Critical Thinking

8 Predict the effect of a plant disease that destroys all of the chloroplasts in a plant.

9 Evaluate why plants perform cellular respiration.

Deforestation and Carbon Dioxide
in the Atmosphere

How does carbon dioxide affect climate?

What do you think when you hear the words *greenhouse gases*? Many people picture pollution from automobiles or factory smokestacks. It might be surprising to learn that cutting down forests affects the amount of one of the greenhouse gases in the atmosphere—carbon dioxide.

Deforestation is the term used to describe the destruction of forests. Deforestation happens because people cut down forests to use the land for other purposes, such as agriculture or building sites, or to use the trees for fuel or building materials.

Trees, like most plants, carry out photosynthesis and make their own food. Carbon dioxide from the atmosphere is one of the raw materials, or reactants, of photosynthesis. When deforestation occurs, trees are unable to remove carbon dioxide from the atmosphere. As a result, the level of carbon dioxide in the atmosphere increases.

Trees affect the amount of atmospheric carbon dioxide in other ways. Large amounts of carbon are stored in the molecules that make up trees. When trees are burned or left to rot, much of this stored carbon is released as carbon dioxide. This increases the amount of carbon dioxide in the atmosphere.

Carbon dioxide in the atmosphere has an impact on climate. Greenhouse gases, such as carbon dioxide, increase the amount of the Sun's energy that is absorbed by the atmosphere. They also reduce the ability of heat to escape back into space. So, when levels of carbon dioxide in the atmosphere increase, more heat is trapped in Earth's atmosphere. This can lead to climate change.

▲ These cattle are grazing on land that was once part of a forest in Brazil.

▲ In a process called slash-and-burn, forest trees are cut down and burned to clear land for agriculture.

It's Your Turn

RESEARCH AND REPORT How can we lower the rate of deforestation? What are some actions you can take that could help slow the rate of deforestation? Research to find out how you can make a difference. Make a poster to share what you learn.

Reading Guide

Key Concepts
ESSENTIAL QUESTIONS

- How do plants respond to environmental stimuli?
- How do plants respond to chemical stimuli?

Vocabulary

stimulus p. 279
tropism p. 280
photoperiodism p. 282
plant hormone p. 283

 Multilingual eGlossary

 Video BrainPOP®

Plant Responses

Inquiry A Meat-Eating Plant?

Venus flytraps have leaves that look like jaws. The leaves close only when a stimulus, such as a fly, brushes against tiny, sensitive hairs on the surface of the leaves. To what other stimuli do you think plants might respond?

Inquiry | Launch Lab

15 minutes

How do plants respond to stimuli?

Plants use light energy and make their own food during photosynthesis. How else do plants respond to light in their environment?

1. Read and complete a lab safety form.
2. Choose a **pot of young radish seedlings.**
3. Place **toothpicks** parallel to a few of the seedlings in the pot in the direction of growth.
4. Place the pot near a **light source**, such as a gooseneck lamp or next to a window. The light source should be to one side of the pot, not directly above the plants.
5. Check the position of the seedlings in relation to the toothpicks after 30 minutes. Record your observations in your Science Journal.
6. Observe the seedlings when you come to class the next day. Record your observations.

Think About This

1. What happened to the position of the seedlings after the first 30 minutes? What is your evidence of change?

2. What happened to the position of the seedlings after a day?

3. 🔑 **Key Concept** Why do you think the position of the seedlings changed?

Stimuli and Plant Responses

Have you ever been in a dark room when someone suddenly turned on the light? You might have reacted by quickly shutting or covering your eyes. **Stimuli** (STIM yuh li; singular, stimulus) *are any changes in an organism's environment that cause a response.*

Often a plant's response to stimuli might be so slow that it is hard to see it happen. The response might occur gradually over a period of hours or days. Light is a stimulus. A plant responds to light by growing toward it, as shown in **Figure 5.** This response occurs over several hours.

In some cases, the response to a stimulus is quick, such as the Venus flytrap's response to touch. When stimulated by an insect's touch, the two sides of the trap snap shut immediately, trapping the insect inside.

> ✓ **Reading Check** Why is it sometimes hard to see a plant's response to a stimulus?

Figure 5 The light is the stimulus, and the seedlings have responded by growing toward the light.

Environmental Stimuli

When it is cold outside, you probably wear a sweatshirt or a coat. Plants cannot put on warm clothes, but they do respond to their environments in a variety of ways. You might have seen trees flower in the spring or drop their leaves in the fall. Both are plant responses to environmental stimuli.

Growth Responses

Plants respond to a number of different environmental stimuli. These include light, touch, and gravity. *A* **tropism** (TROH pih zum) *is a response that results in plant growth toward or away from a stimulus.* When the growth is toward a stimulus, the tropism is called positive. A plant bending toward light is a positive tropism. Growth away from a stimulus is considered negative. A plant's stem growing upward against gravity is a negative tropism.

Light The growth of a plant toward or away from light is a tropism called phototropism. A plant has a light-sensing chemical that helps it detect light. Leaves and stems tend to grow in the direction of light, as shown in **Figure 6.** This response maximizes the amount of light the plant's leaves receive. Roots generally grow away from light. This usually means that the roots grow down into the soil and help anchor the plant.

 Reading Check How is phototropism beneficial to a plant?

Make a horizontal two-tab book and label it as shown. Record what you learn about the two types of stimuli that affect plant growth.

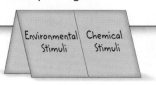

WORD ORIGIN · · · · · · · · · · · ·

tropism
from Greek *tropos*, means "turn" or "turning"

Response to Light 🔑

Figure 6 As a plant's leaves turn toward the light, the amount of light that the leaves can absorb increases.

Touch The response of a plant to touch is called a thigmotropism (thihg MAH truh pih zum). You might have seen vines growing up the side of a building or a fence. This happens because the plant has special structures that respond to touch. These structures, called tendrils, can wrap around or cling to objects, as shown in **Figure 7**. A tendril wrapping around an object is an example of positive thigmotropism. Roots display negative thigmotropism. They grow away from objects in soil, enabling them to follow the easiest path through the soil.

Gravity The response of a plant to gravity is called gravitropism. Stems grow away from gravity, while roots grow toward gravity. The seedlings in **Figure 8** are exhibiting both responses. No matter how a seed lands on soil, when it starts to grow, its roots grow down into the soil. The stem grows up. This happens even when a seed is grown in a dark chamber, indicating that these responses can occur independently of light.

 Key Concept Check What types of environmental stimuli do plants respond to? Give three examples.

Response to Touch

▲ **Figure 7** The tendrils of the vine respond to touch and coil around the blade of grass.

Response to Gravity

Figure 8 Both of these plant stems are growing away from gravity. The upward growth of a plant's stem is negative gravitropism, and the downward growth of its roots is positive gravitropism.

🔍 **Visual Check** How is the plant on the left responding to the pot begin placed on its side?

Flowering Responses

You might think all plants respond to light, but in some plants, flowering is actually a response to darkness! **Photoperiodism** *is a plant's response to the number of hours of darkness in its environment.* Scientists once hypothesized that photoperiodism was a response to light. Therefore, these flowering responses are called long-day, short-day, and day-neutral and relate to the number of hours of daylight in a plant's environment.

Long-Day Plants Plants that flower when exposed to less than 10–12 hours of darkness are called long-day plants. The carnations shown in **Figure 9** are examples of long-day plants. This plant usually produces flowers in summer, when the number of hours of daylight is greater than the number of hours of darkness.

Short-Day Plants Short-day plants require 12 or more hours of darkness for flowering to begin. An example of a short-day plant is the poinsettia, shown in **Figure 9.** Poinsettias tend to flower in late summer or early fall when the number of hours of daylight is decreasing and the number of hours of darkness is increasing.

Day-Neutral Plants The flowering of some plants doesn't seem to be affected by the number of hours of darkness. Day-neutral plants flower when they reach maturity and the environmental conditions are right. Plants such as the roses in **Figure 9** are day-neutral plants.

 Reading Check How is the flowering of day-neutral plants affected by exposure to hours of darkness?

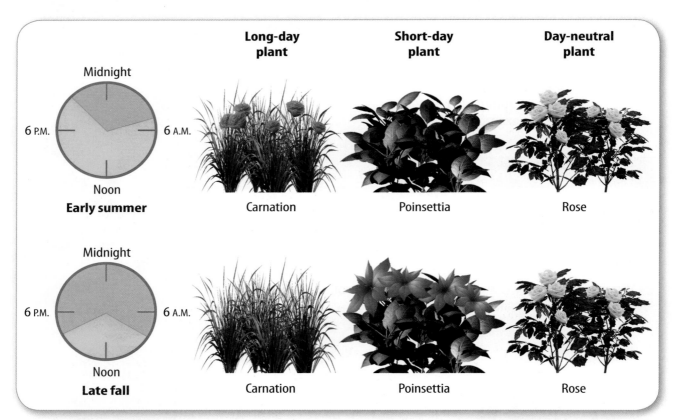

Figure 9 The number of hours of darkness controls flowering in many plants. Long-day plants flower when there are more hours of daylight than darkness, and short-day plants flower when there are more hours of darkness than daylight.

Visual Check What time of year receives more darkness, and what type of plant produces flowers during that season?

Chemical Stimuli

Plants respond to chemical stimuli as well as environmental stimuli. **Plant hormones** *are substances that act as chemical messengers within plants.* These chemicals are produced in tiny amounts. They are called messengers because they usually are produced in one part of a plant and affect another part of that plant.

Auxins

One of the first plant hormones discovered was auxin (AWK sun). There are many different kinds of auxins. Auxins generally cause increased plant growth. They are responsible for phototropism, the growth of a plant toward light. Auxins concentrate on the dark side of a plant's stem, and these cells grow longer. This causes the stem of the plant to grow toward the light, as shown in **Figure 10.**

Ethylene

The plant hormone ethylene helps stimulate the ripening of fruit. Ethylene is a gas that can be produced by fruits, seeds, flowers, and leaves. You might have heard someone say that one rotten apple spoils the whole barrel. This is based on the fact that rotting fruits release ethylene. This can cause other fruits nearby to ripen and possibly rot. Ethylene also can cause plants to drop their leaves.

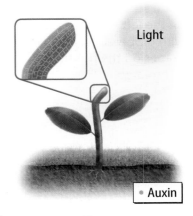

Light

• Auxin

Figure 10 Auxin on the left side of the seedling causes more growth and makes the seedling bend to the right.

 Key Concept Check How do plants respond to the chemical stimuli, or hormones, auxin and ethylene?

Inquiry MiniLab

20 minutes

When will plants flower?

Did you ever think plants could have strategies so that they can germinate, live, grow, reproduce, and continue their species? Photoperiodism is one such strategy.

1 In your Science Journal, copy the table below to classify plants based on their photoperiodisms.

2 Choose 8–10 **pictures of flowers.** Record their names in your table. Use the clues on the back of each photo to determine the correct photoperiodism of each plant.

Analyze and Conclude

1. Interpret Data Based on your table, which plants would flower during the summer?

2. Explain why some plants flower at the same time every year.

3. Infer what might happen if short-day plants were placed under light for an hour or two at night.

4. **Key Concept** Why would photoperiodism be an important strategy for flowering plants?

Plant	Season	Short-Day	Long-Day	Day-Neutral

Figure 11  The grapes on the left were treated with gibberellins, and the grapes on the right were not treated.

Response to Gibberellins

Math Skills

Use Percentages

A percentage is a ratio that compares a number to 100. For example, if a tree grows 2 cm per day with no chemical stimulus and 3 cm per day with a chemical stimulus, what is the percentage increase in growth?

Subtract the original value from the final value.

$$3 \text{ cm} - 2 \text{ cm} = 1 \text{ cm}$$

Set up a ratio between the difference and the original value. Find the decimal equivalent.

$$\frac{1 \text{ cm}}{2 \text{ cm}} = 0.5 \text{ cm}$$

Multiply by 100 and add a percent sign.

$$0.5 \times 100 = 50\%$$

Practice

Without gibberellins, pea seedlings grew to 2 cm in 3 days. With gibberellins, the seedlings grew to 4 cm in 3 days. What was the percentage increase in growth?

 Review

• **Math Practice**
• **Personal Tutor**

Gibberellins and Cytokinins

Rapidly growing areas of a plant, such as roots and stems, produce gibberellins (jih buh REL unz). These hormones increase the rate of cell division and cell elongation. This results in increased growth of stems and leaves. Gibberellins also can be applied to the outside of plants. As shown in **Figure 11**, applying gibberellins to plants can have a dramatic effect.

Root tips produce most of the cytokinins (si tuh KI nunz), another type of hormone. Xylem carries cytokinins to other parts of a plant. Cytokinins increase the rate of cell division, and in some plants, cytokinins slow the aging process of flowers and fruits.

Summary of Plant Hormones

Plants produce many different hormones. The hormones you have just read about are groups of similar compounds. Often, two or more hormones interact and produce a plant response. Scientists are still discovering new information about plant hormones.

Humans and Plant Responses

Humans depend on plants for food, fuel, shelter, and clothing. Humans make plants more productive using plant hormones. Some crops now are easier to grow because humans understand how they respond to hormones. As you study **Figure 12** on the next page, make a list of all the ways humans can benefit from understanding and using plant responses.

✓ **Reading Check** How are humans dependent on plants?

Figure 12 Understanding how plants respond to hormones can benefit people in many ways.

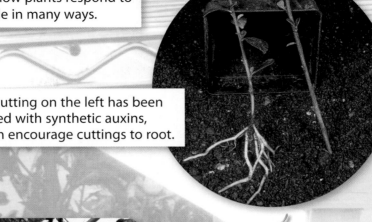

The cutting on the left has been treated with synthetic auxins, which encourage cuttings to root.

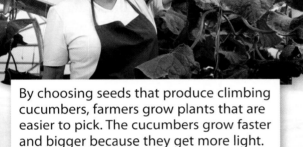

By choosing seeds that produce climbing cucumbers, farmers grow plants that are easier to pick. The cucumbers grow faster and bigger because they get more light.

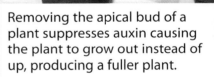

Removing the apical bud of a plant suppresses auxin causing the plant to grow out instead of up, producing a fuller plant.

Bananas can be picked and shipped while still green and then be treated with ethylene to cause them to ripen.

The use of cytokinins helps scientists and horticulturists grow hundreds of identical plants.

Lesson 2 Review

Visual Summary

Plants respond to stimuli in their environments in many ways.

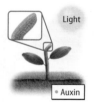

Carnation

Photoperiodism occurs in long-day plants and short-day plants. Day-neutral plants are not affected by the number of hours of darkness.

Light

Auxin

Plant hormones are internal chemical stimuli that produce different responses in plants.

FOLDABLES

Use your lesson Foldable to review the lesson. Save your Foldable for the project at the end of the chapter.

What do you think NOW?

You first read the statements below at the beginning of the chapter.

3. Plants do not produce hormones.

4. Plants can respond to their environments.

Did you change your mind about whether you agree or disagree with the statements? Rewrite any false statements to make them true.

Use Vocabulary

1 **Define** *plant hormone* in your own words.

2 The response of an organism to the number of hours of darkness in its environment is called _____.

3 **Distinguish** between *stimuli* and *tropism*.

Understand Key Concepts

4 **Describe** an example of a plant responding to environmental stimuli.

5 **Distinguish** between a long-day plant and a short-day plant.

6 **Compare** the effect of auxins and gibberellins on plant cells.

7 Which is NOT likely to cause a plant response?
   A. changing the amount of daylight
   B. moving plants away from each other
   C. treating with plant hormones
   D. turning a plant on its side

Interpret Graphics

8 **Identify** Copy the table below and list the plant hormones mentioned in this lesson. Describe the effect of each on plants.

Hormone	Effect on Plants

Critical Thinking

9 **Infer** why the plant shown to the right is growing at an angle.

Math Skills

 Review
Math Practice

10 When sprayed with gibberellins, the diameter of mature grapes increased from 1.0 cm to 1.75 cm. What was the percent increase in size?

What happens to seeds if you change the intensity of light?

Seeds require light, water, gases, and soil to germinate, grow into seedlings, and then grow into mature plants. Different types of seeds require different amounts of each of these factors. What happens if one of these factors is changed?

Materials

plastic tub

potting soil

fast-growing grass seeds

sun shields

light source

metric ruler

mister bottle with water

Safety

Learn It

In any experiment, it is important to keep everything the same except for the item you are testing. The one factor you change, or **manipulate,** is called the independent **variable.** Your experiment should also have a control. The control is an individual instance or experimental subject for which the independent **variable** is not changed.

Try It

1. Read and complete a lab safety form.

2. Fill the plastic tub with potting soil. Water the soil, and then add more soil. Level it to about 1–2 cm from the top. Spread the grass seeds evenly across the soil. Cover the seeds with a thin layer of soil.

3. Obtain the precut shields of vellum, plastic needlepoint grid, and cardboard. These will be used to change the intensity of light coming onto the soil.

4. Cover the soil with the shields by laying them next to each other. Leave one section of soil uncovered.

5. Place the tub on a windowsill or under a growing light.

6. Keep the soil damp, not wet, with a mister. Water gently so the seeds stay in position.

7. Design a table to record observations in your Science Journal. Include columns for day, growth pattern, height, and random sampling counts. Begin observations when seedlings first emerge. Observe seedlings for 3–5 days.

Apply It

8. **Identify** the variables and the controls used in this investigation.

9. **Analyze** the data you collected through your observations. Which light intensity appeared to bring about the fullest, tallest growth?

10. **Draw conclusions** about what would happen if you put one section of seeds in total darkness. Would it germinate? If you changed the light intensity immediately after the seeds germinated, would it survive?

11. 🔑 **Key Concept** Does the amount of light affect the germination and growth of grass seeds? Explain.

Lesson 3

Plant Reproduction

Reading Guide

Key Concepts 🔑
ESSENTIAL QUESTIONS

- What is the alternation of generations in plants?

- How do seedless plants reproduce?

- How do seed plants reproduce?

Vocabulary

alternation of generations p. 290

spore p. 290

pollen grain p. 292

pollination p. 292

ovule p. 292

embryo p. 292

seed p. 292

stamen p. 294

pistil p. 294

ovary p. 294

fruit p. 295

g Multilingual eGlossary

Inquiry A Bee's-Eye View?

Bees can see ultraviolet (UV) light. We see a dandelion as yellow. Because of a bee's ability to see UV light, a bee sees a dandelion like the one above. Why do you think bees see flowers differently than we do? Why do some plants produce flowers while others do not?

Launch Lab

15 minutes

How can you identify fruits?

Flowering plants grow from seeds that they produce. Animals depend on flowering plants for food. The function of the fruit is to disperse the seeds for plant reproduction.

1. Read and complete a lab safety form.

2. Make a two-column table in your Science Journal. Label the columns *Fruits* and *Not Fruits*.

3. Examine a collection of **food items.** Determine whether each item is a fruit. Record your observations in your table.

4. Place each food item on a piece of **plastic wrap.** Use a **plastic or paring knife** to cut the items in half.

5. Examine the inside of each food item. Record your observations.

Think About This

1. What observations did you make about the insides of the food items? Would you reclassify any food item based on your observations? Explain.

2. How can the number of seeds or how they are placed in the fruit help with seed dispersal?

3. **Key Concept** What role do you think a fruit has in a flowering plant's reproduction?

Asexual Reproduction Versus Sexual Reproduction

In early spring, you might see cars or sidewalks covered with a thick, yellow dust. Where did it come from? It probably came from plants that are reproducing. As in all living things, reproduction is part of the life cycles of plants.

Plants can reproduce either asexually, sexually, or both ways. Asexual reproduction occurs when a portion of a plant develops into a separate new plant. This new plant is genetically identical to the original, or parent, plant. Some plants, such as irises and daylilies, can use their underground stems for asexual reproduction. Other plants, such as the houseleeks, or hens and chicks, in **Figure 13,** reproduce asexually using horizontal stems called stolons. One advantage of asexual reproduction is that just one parent organism can produce offspring. However, sexual reproduction in plants usually requires two parent organisms. Sexual reproduction occurs when a plant's sperm combines with a plant's egg. A resulting zygote can grow into a plant. This new plant is a genetic combination of its parents.

Reading Check How are sexual and asexual reproduction different in plants?

Figure 13 Hens and chicks can reproduce without seeds, or asexually. New "chicks" can grow from the stolons on the main "hen" plant.

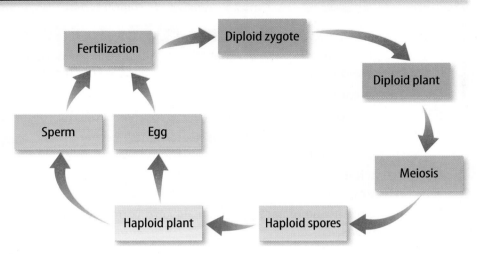

Figure 14 The life cycle of all plants includes an alternation of generations. The diploid generation begins with fertilization. The haploid generation begins with meiosis.

Alternation of Generations

Your body is made of two types of cells—haploid cells and diploid cells. Most of your cells are diploid. The only human haploid cells are sperm and eggs. As a result, you will live your entire life as a diploid organism. To put it another way, your life cycle includes only a diploid stage. That isn't true for all organisms. Some organisms, including plants, have two life stages called **generations**. One generation is almost all diploid cells. The other generation has only haploid cells. **Alternation of generations** *occurs when the life cycle of an organism alternates between diploid and haploid generations,* as shown in **Figure 14.**

 Key Concept Check What is alternation of generations in plants?

The Diploid Generation

When you look at a tree or a flower, you're seeing part of the plant's diploid generation. Meiosis occurs in certain cells in the reproductive structures of a diploid plant. *The daughter cells produced from haploid structures are called* **spores.** Spores grow by mitosis and cell division and form the haploid generation of a plant.

The Haploid Generation

In most plants, the haploid generation is tiny and lives surrounded by special tissues of the diploid plant. In other plants, the haploid generation lives on its own. Certain reproductive cells of the haploid generation produce haploid sperm or eggs by mitosis and cell division. Fertilization takes place when a sperm and an egg fuse and form a diploid zygote. Through mitosis and cell division, the zygote grows into the diploid generation of a plant.

SCIENCE USE V. COMMON USE

generation
Science Use haploid and diploid stages in the life cycle of a plant

Common Use the average span of time between the birth of parents and their offspring

WORD ORIGIN

spore
from Greek *spora*, means "seed, a sowing"

Reproduction in Seedless Plants

Not all plants grow from seeds. The first land plants to inhabit Earth probably were seedless plants—plants that grow from haploid spores, not from seeds. The mosses and ferns in **Figure 15** are examples of seedless plants found on Earth today.

Life Cycle of a Moss

The life cycle of a moss is typical for some seedless plants. The tiny, green moss plants that carpet rocks, bark, and soil in moist areas are haploid plants. These plants grow by **mitosis** and cell division from haploid spores produced by the diploid generation. They have male structures that produce sperm and female structures that produce eggs. Fertilization results in a diploid zygote that grows by mitosis and cell division into the diploid generation of moss, such as the one shown in **Figure 15**. A diploid moss is tiny and not easily seen.

REVIEW VOCABULARY · · · · ·
mitosis
the process during which a nucleus and its contents divide

Life Cycle of a Fern

An alternation of generations is also seen in the life cycle of a fern. The diploid generations are the green leafy plants often seen in forests. These plants produce haploid spores. The spores grow into tiny plants. The haploid plants produce eggs and sperm that can unite and form the diploid generations.

 Key Concept Check How do seedless plants such as mosses and ferns reproduce?

Figure 15 Mosses and ferns usually grow in moist environments. Sperm must swim through a film of water to reach an egg.

Reproduction in Seedless Plants

Fern

Diploid generation of moss

Moss covering log

Concepts in Motion Animation

How do seed plants reproduce?

Most land plants that cover Earth grow from seeds. There are two groups of seed plants—flowerless seed plants and flowering seed plants.

Unlike seedless plants, the haploid generation of a seed plant is within diploid tissue. Separate diploid male and diploid female reproductive structures produce haploid sperm and haploid eggs that join during fertilization.

The Role of Pollen Grains

A **pollen** (PAH lun) **grain** *forms from tissue in a male reproductive structure of a seed plant.* Each pollen grain contains nutrients and has a hard, protective outer covering, as shown in **Figure 16.** Pollen grains produce sperm cells. Wind, animals, gravity, or water currents can carry pollen grains to female reproductive structures.

Plants cannot move and find a mate as most animals can. Do you recall reading about the yellow dust at the beginning of this lesson? That dust is pollen grains. Male reproductive structures produce a vast number of pollen grains. **Pollination** (pah luh NAY shun) *occurs when pollen grains land on a female reproductive structure of a plant that is the same species as the pollen grains.*

The Role of Ovules and Seeds

The female reproductive structure of a seed plant where the haploid egg develops is called the **ovule.** Following pollination, sperm enter the ovule and fertilization occurs. A zygote forms and develops into an **embryo,** *an immature diploid plant that develops from the zygote.* As shown in **Figure 17,** *an embryo, its food supply, and a protective covering make up a* **seed.** A seed's food supply provides the embryo with nourishment for its early growth.

 Key Concept Check How do seed plants reproduce?

Color-enhanced SEM Magnification: 1,100×

▲ **Figure 16** Pollen grains of one type of plant are different from those of any other type of plant.

Visual Check How many different types of pollen are visible in **Figure 16?**

FOLDABLES®

Make a two-tab book and label it as shown. Use it to record information about reproduction in flowerless and flowering plants.

Flowerless Plants | Flowering Plants

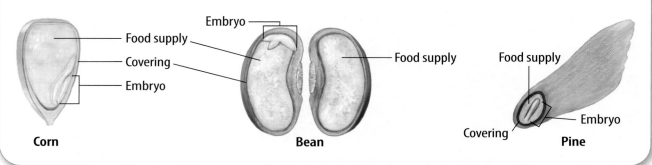

Figure 17 A seed contains a diploid plant embryo and a food supply protected by a hard outer covering.

Corn — Food supply, Covering, Embryo

Bean — Embryo, Food supply

Pine — Food supply, Covering, Embryo

Reproduction in Flowerless Seed Plants

Flowerless seed plants are also known as gymnosperms (JIHM nuh spurmz). The word *gymnosperm* means "naked seed," and gymnosperm seeds are not surrounded by a fruit. The most common gymnosperms are conifers. Conifers, such as pines, firs, cypresses, redwoods, and yews, are trees and shrubs with needlelike or scalelike leaves. Most conifers are evergreens, which means they have leaves all year long. Conifers can live for many years. Bristlecone pines, such as the one shown in **Figure 18,** are among the oldest living trees on Earth.

Life Cycle of a Gymnosperm The life cycle of a gymnosperm, shown in **Figure 19,** includes an alternation of generations. Cones are the male and female reproductive structures of conifers. They contain the haploid generation. Male cones are small, papery structures that produce pollen grains. Female cones can be woody, berrylike, or soft, and they produce eggs. A zygote forms when a sperm from a male cone fertilizes an egg. The zygote is the beginning of the diploid generation. Seeds form as part of the female cone.

 Reading Check Where is the haploid generation of conifers contained?

▲ **Figure 18** Seeds form at the base of each scale on a female cone.

Reproduction in Flowerless Seed Plants 🔑

Concepts in Motion Animation

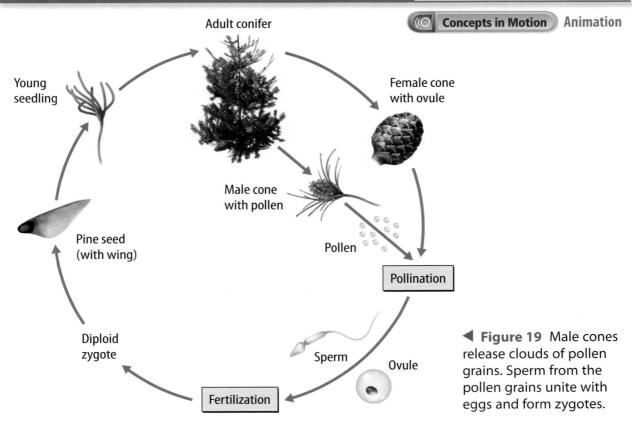

◀ **Figure 19** Male cones release clouds of pollen grains. Sperm from the pollen grains unite with eggs and form zygotes.

Lesson 3
EXPLAIN
293

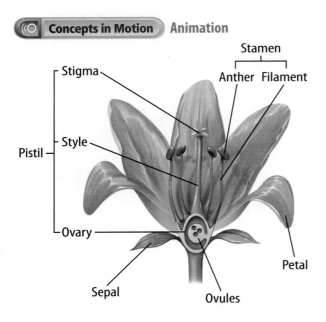

Figure 20 Typical flowers have both male and female structures.

Reproduction in Flowering Seed Plants

Most plants you see around you are angiosperms, or flowering plants. Fruits and vegetables come from angiosperms. Many animals depend on angiosperms for food.

The Flower Reproduction of an angiosperm begins in a flower. Most flowers have male and female reproductive structures, as shown in **Figure 20.**

The male reproductive organ of a flower is the **stamen.** Pollen grains form at the tip of the stamen in the anther. The filament supports the anther and connects it to the base of the flower. *The female reproductive organ of a flower is the* **pistil.** Pollen can land at the tip of the pistil, or stigma. The stigma is at the top of a long tube called the style. *At the base of the style is the* **ovary,** *which contains one or more ovules.* Recall that each ovule eventually will contain a haploid egg and might become a seed if fertilized.

Inquiry MiniLab   **30 minutes**

How can you model a flower?

Imagine that you have just discovered a new species of flowering plant. No one has ever seen this flower before, but it has all the basic flower parts.

1. Read and complete a lab safety form.
2. In your Science Journal, list all the parts your flower has as an angiosperm.
3. Make a large 3-dimensional model of your new flower using **chenille stems, tissue paper, construction paper, tag board, pom poms, plastic beads, scissors,** and **glue.**
4. Check your model to make sure each flower part is in the correct proportion and shows how it interacts with other flower parts.
5. Name your flower. Create a key to identify each part and its function.

Analyze and Conclude

1. **Analyze** Why do flowers have colorful petals and strong scents?
2. **Infer** Why does the end of the stigma feel sticky?
3. **Key Concept** Could your flower be self-pollinating? Explain.

Life Cycle of an Angiosperm A typical life cycle for an angiosperm is shown in **Figure 21.** Pollen grains travel by wind, gravity, water, or animal from the anther to the stigma, where pollination occurs. A pollen tube grows from the pollen grain into the stigma, down the style, to the ovary at the base of the pistil. Sperm develop from a haploid cell in the pollen tube. When the pollen tube enters an ovule, fertilization takes place.

As you read earlier, the zygote that results from fertilization develops into an embryo. Each ovule and its embryo will become a seed. *The ovary, and sometimes other parts of the flower, will develop into a* **fruit** *that contains one or more seeds.* The seeds can grow into new, genetically related plants that produce flowers, and the cycle repeats.

✓ **Reading Check** Do sperm develop before or after pollination?

Reproduction in Flowering Seed Plants 🔑

Concepts in Motion
Animation

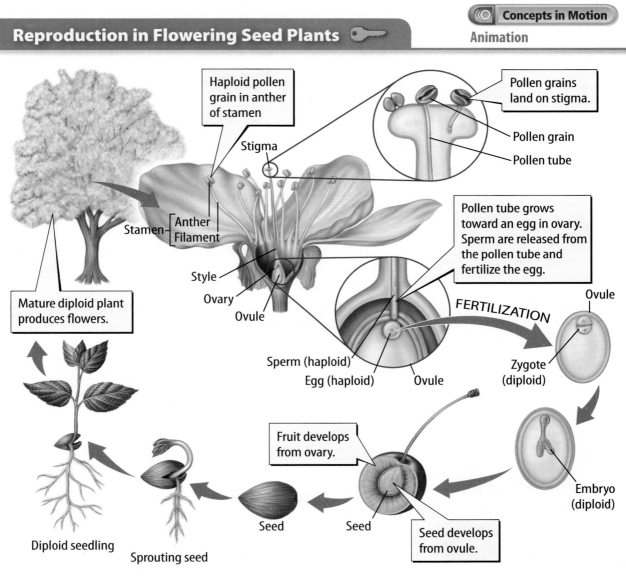

Haploid pollen grain in anther of stamen

Pollen grains land on stigma.

Stigma

Pollen grain

Pollen tube

Stamen — Anther, Filament

Pollen tube grows toward an egg in ovary. Sperm are released from the pollen tube and fertilize the egg.

Style

Ovary

Ovule

Mature diploid plant produces flowers.

FERTILIZATION

Ovule

Sperm (haploid)

Egg (haploid)

Ovule

Zygote (diploid)

Embryo (diploid)

Fruit develops from ovary.

Seed develops from ovule.

Seed

Seed

Diploid seedling

Sprouting seed

Figure 21 During the life cycle of a flowering plant, the haploid generation grows and develops inside the diploid plant.

✓ **Visual Check** How does sperm in a pollen grain reach an egg in the ovule?

Table 1 Flowers, Fruits, and Seeds of Common Plants

Plant	Flower	Fruit	Seed
Pea			
Corn			
Strawberry			
Dandelion			

Table 1 Flowers, fruits, and seeds are important for reproduction in angiosperms.

Visual Check Which of these fruits has seeds on the outside?

Figure 22 The seeds will be excreted by the mouse and might grow into new blackberry bushes.

Fruit and Seed Dispersal Fruits and seeds, such as those in **Table 1,** are important sources of food for people and animals. In most cases, seeds of flowering plants are inside fruits. Pods are the fruits of a pea plant. The peas inside a pod are the seeds. An ear of corn is made up of many fruits, or kernels. The main part of each kernel is the seed. Strawberries have tiny seeds on the outside of the fruit.

Reading Check Where are the seeds of flowering plants usually found?

We usually think of fruits as juicy and edible, such as oranges or watermelons. However, some fruits are hard, dry, and not particularly edible. For example, each parachutelike structure of a dandelion is a dry fruit.

Fruits help protect seeds and disperse them. Some fruits, such as those of a dandelion, are light and float on air currents, which helps to scatter the seeds. Also, when an animal eats a fruit, the fruit's seeds can pass through the animal's digestive system with little or no damage to the seed. Imagine what happens when an animal, such as the mouse shown in **Figure 22,** eats blackberries. The animal digests the juicy fruit and deposits the seeds with its wastes. By this time, the animal might have traveled some distance away from the blackberry bush. This means the animal helped to disperse the seeds away from the blackberry bush.

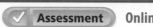

Visual Summary

The life cycle of a plant includes an alternation of generations.

Seedless plants, such as ferns and mosses, grow from haploid spores.

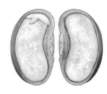

In seed plants, pollination occurs when pollen grains land on the female reproductive structure of a plant of the same species.

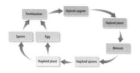

Use your lesson Foldable to review the lesson. Save your Foldable for the project at the end of the chapter.

What do you think NOW?

You first read the statements below at the beginning of the chapter.

5. Seeds contain tiny plant embryos.

6. Flowers are needed for plant reproduction.

Did you change your mind about whether you agree or disagree with the statements? Rewrite any false statements to make them true.

Use Vocabulary

1 The daughter cells produced from haploid structures are called _____.

2 **Distinguish** between an ovule and an ovary.

3 **Define** *pollination* in your own words.

Understand Key Concepts

4 Which is NOT part of the alternation of generations life cycle in plants?
   A. anther   C. haploid
   B. diploid   D. spore

5 **Contrast** the haploid generation of a moss with that of a fern.

6 **Describe** how a pollen tube carries sperm to the ovule in a flower.

7 **Give an example** of a flowerless seed plant.

Interpret Graphics

8 **Examine** the figure below and describe the function of each part of the seed.

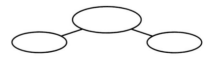

9 **Identify** Copy and fill in the graphic organizer below to identify the female parts of a flower.

Critical Thinking

10 **Create** a picture to show the life cycle of a fern.

11 **Evaluate** the advantages of fruit production in plant reproduction.

Materials

one quad of plants

Also needed
appropriate materials to perform lab

Safety

Design a Stimulating Environment for Plants

Plants usually respond to stimuli in the environment by growing. The response to light is phototropism; plants grow toward the light. The growth response of gravitropism is a little more complicated; stems grow away from the direction of gravity (negative gravitropism), and roots grow in the direction of gravity (positive gravitropism). Thigmotropism is a plant response to touch.

Ask a Question

You have explored tropisms in other labs in this chapter. What questions would you like to answer more thoroughly, or what outcomes would you like to double-check? Do you have another approach in mind to investigate one of the tropisms? Ask a question that you would like to investigate further. Make sure it is testable; think about the variables and equipment you would need.

Make Observations

1. Read and complete a lab safety form.

2. Examine your quad of plants and decide which tropism you want to explore.

3. Make a plan and write it in your Science Journal.

4. Have your teacher approve your plan for your investigation.

5. Choose materials from those provided by your teacher for a simple lab setup.

6. Decide the criteria you will use to show the outcomes you expect.

7. Set up your lab according to your plan.

Form a Hypothesis

8. After observing your plants and lab setup, formulate a hypothesis about the relationship between your selected tropism and a plant's growth. Make a prediction about how the tropism will affect your plants.

Test Your Hypothesis

9 Make any necessary modifications to your setup so your procedure will move toward your expected outcome.

10 Make a data table like the one to the right in your Science Journal.

11 Make your observations as directed by your procedure, and record them in your table.

Analyze and Conclude

12 **Compare** the position of the parts of your plant at the beginning and end of your study. Check to see if the change is easily visible and measurable; try not to jump to conclusions.

13 **Consider** the possible causes of the changes. Determine if it was changing the variable that brought about the effect. Explain.

14 **Relate** how the tropism you modeled could enable plants to meet their needs and survive.

15 **The Big Idea** What might happen if the stimulus you provided for the plant was enlarged, minimized, or eliminated?

Communicate Your Results

Prepare a drama to present your findings. Group members or volunteers from the class can wear pictures or signs to indicate their roles. Begin with students role-playing a healthy plant. Add the role of the stimulus, and be sure to identify the tropism and show results in the plant(s).

Inquiry Extension

Phototropism is one of the plant responses to stimuli that you have been able to explore easily by changing the position of the light source or the position of the plants in relation to the light source. What might happen if you changed the light source itself? What if you put a colored plastic sheet between the light and the plant? Would a red filter cause the same response as a green filter? What if you used different plants? For example, some mustard seeds are fast-germinating. Would these respond the same way as the other plants?

Time Period	[Variable observed] of Plant			
	1	2	3	4
Day 0 prior to tropism				
Day 1				
Day 2				
Day 3				

Lab Tips

☑ Discuss the possible materials you will use with your lab partner. Remember that the materials should help you learn more about the tropism you selected.

☑ Be creative when deciding how to test the tropism you selected.

Remember to use scientific methods.

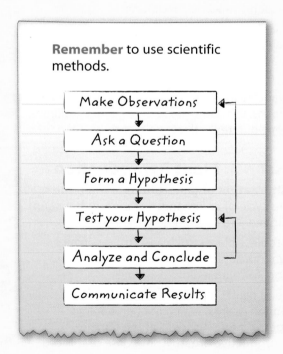

Make Observations → Ask a Question → Form a Hypothesis → Test your Hypothesis → Analyze and Conclude → Communicate Results

Plants transform light energy into chemical energy, respond to stimuli and maintain homeostasis, and reproduce with and without seeds.

Key Concepts Summary 🔑

	Vocabulary
Lesson 1: Energy Processing in Plants • The vascular tissue in most plants, xylem and phloem, move materials throughout plants. • In **photosynthesis,** plants convert light energy, water, and carbon dioxide into the food-energy molecule glucose through a series of chemical reactions. The process gives off oxygen. • **Cellular respiration** is a series of chemical reactions that convert the energy in food molecules into a usable form of energy called ATP. • Photosynthesis and cellular respiration can be considered opposite processes of each other. Sunlight energy · Carbon dioxide CO_2 · Oxygen O_2 · Water H_2O · Sugar $C_6H_{12}O_6$	**photosynthesis** p. 272 **cellular respiration** p. 274
Lesson 2: Plant Responses • Although plants cannot move from one place to another, they do respond to **stimuli,** or changes in their environments. Plants respond to stimuli in different ways. • **Tropisms** are growth responses toward or away from stimuli such as light, touch, and gravity. **Photoperiodism** is a plant's response to the number of hours of darkness in its environment. • Plants respond to chemical stimuli, or **plant hormones,** such as auxins, ethylene, gibberellins, and cytokinins. Different hormones have different effects on plants.	**stimulus** p. 279 **tropism** p. 280 **photoperiodism** p. 282 **plant hormone** p. 283
Lesson 3: Plant Reproduction • **Alternation of generations** is when the life cycle of an organism alternates between diploid and haploid generations. • Seedless plants, such as ferns, reproduce when a haploid sperm fertilizes a haploid egg, forming a diploid zygote. • Seed plants reproduce when **pollen grains,** which contain haploid sperm, land on the tip of the female reproductive organ. At the base of this organ is the **ovary,** which usually contains one or more **ovules.** Each ovule eventually will contain a haploid egg. If the sperm fertilizes the egg, an **embryo** will form within a **seed.** 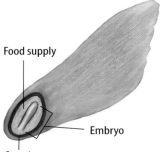 Food supply · Embryo · Covering	**alternation of generations** p. 290 **spore** p. 290 **pollen grain** p. 292 **pollination** p. 292 **ovule** p. 292 **embryo** p. 292 **seed** p. 292 **stamen** p. 294 **pistil** p. 294 **ovary** p. 294 **fruit** p. 295

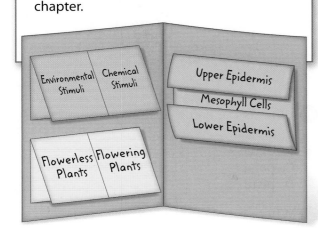

FOLDABLES® Chapter Project

Assemble your lesson Foldables as shown to make a Chapter Project. Use the project to review what you have learned in this chapter.

Environmental Stimuli | Chemical Stimuli

Upper Epidermis
Mesophyll Cells
Lower Epidermis

Flowerless Plants | Flowering Plants

Use Vocabulary

1 Long-day and short-day plants are examples of plants that respond to _____.

2 The process that uses oxygen and produces carbon dioxide is _____.

3 A(n) _____ forms from tissue in a male reproductive structure of a seed plant.

4 A(n) _____ develops from an ovary and surrounding tissue.

5 Sperm travel down the _____ inside the stigma of a flower to reach the ovary.

Link Vocabulary and Key Concepts

 **Concepts in Motion** Interactive Concept Map

Copy this concept map, and then use vocabulary terms from the previous page to complete the concept map.

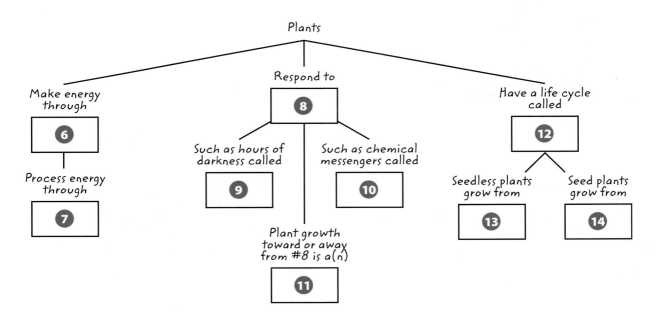

Understand Key Concepts

1 Which material travels from the roots to the leaves through the xylem?
- A. oxygen
- B. sugar
- C. sunlight
- D. water

2 Which organelle is the site of photosynthesis?
- A. chloroplast
- B. mitochondria
- C. nucleus
- D. ribosome

3 Which is a product of cellular respiration?
- A. ATP
- B. light
- C. oxygen
- D. sugar

Use the image below to answer questions 4 and 5.

4 What type of plant-growth response is shown in the photo above?
- A. flowering
- B. gravitropism
- C. photoperiodism
- D. thigmotropism

5 Which stimulus is responsible for this type of growth?
- A. gravity
- B. light
- C. nutrients
- D. touch

Use the image below to answer questions 6–8.

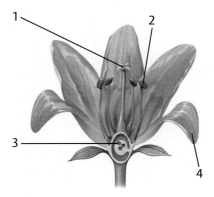

6 What is the name of structure number 3?
- A. anther
- B. ovule
- C. petal
- D. pistil

7 Where is pollen produced?
- A. 1
- B. 2
- C. 3
- D. 4

8 What part of a flower becomes a seed?
- A. 1
- B. 2
- C. 3
- D. 4

9 What plant is shown in the figure below?

- A. diploid fern
- B. diploid moss
- C. haploid fern
- D. haploid moss

Critical Thinking

10 **Infer** which came first—photosynthesis or cellular respiration.

11 **Assess** the importance of material transport in plants.

12 **Construct** a table to compare the reactants and products of photosynthesis and cellular respiration.

13 **Evaluate** the internal structure of a leaf as a location for photosynthesis.

14 **Assess** the need for plants to respond to their environment.

15 **Predict** what would happen if a short-day plant were exposed to more hours of daylight.

16 **Critique** the saying "one rotten apple spoils the whole barrel."

17 **Infer** from the photo below where the light source is in relation to the plant.

18 **Evaluate** the importance of fruit production in flowering plants.

19 **Predict** the effect of cold temperature killing all the flowers on fruit trees.

Writing in Science

20 **Write** a five-sentence paragraph about the importance of plants in your life. Include a main idea, supporting details, and a concluding sentence.

REVIEW THE BIG IDEA

21 Make a list of the plant processes you learned about in this chapter. How do these processes help a plant survive and reproduce?

22 How does the process shown below help a plant survive?

Math Skills

Review

Math Practice

Use Percentages

23 Without treatment with gibberellins, 500 out of 1,000 grass seeds germinated. When sprayed with gibberellins, 875 of the seeds germinated. What was the percentage increase?

24 A bunch of bananas ripens (turns from green to yellow) in 42 hours. When the bananas are placed in a bag with an apple, which releases ethylene, the bananas ripen in 21 hours. What is the percentage change in ripening time?

Standardized Test Practice

Record your answers on the answer sheet provided by your teacher or on a sheet of paper.

Multiple Choice

1 Which structure transports sugars throughout a plant?

  A epidermis

  B phloem

  C stomata

  D xylem

2 What is one similarity between plants and animals?

  A Both plants and animals carry on cellular respiration.

  B Both plants and animals carry on photosynthesis.

  C Both plants and animals have chloroplasts.

  D Both plants and animals use xylem and phloem to transport materials.

Use the diagram below to answer question 3.

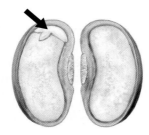

3 Look at the structure that is marked with an arrow in the image above. What will this structure become?

  A a diploid moss

  B a diploid seed plant

  C a haploid fern

  D a haploid flowerless seed plant

4 Which two plant hormones increase the rate of cell division?

  A auxins and cytokinins

  B cytokinins and giberellins

  C ethylene and auxins

  D giberellin and ethylene

5 Which is a product of photosynthesis?

  A carbon dioxide

  B glucose

  C light

  D water

Use the image below to answer question 6.

6 Which cellular process occurs within the organelle shown above?

  A photosynthesis

  B cellular respiration

  C transport of phloem

  D transport of xylem

7 Which plant has a diploid stage that is difficult to see?

  A conifer

  B cherry tree

  C dandelion

  D moss

8 How is cellular respiration related to photosynthesis?

A Animals produce sugars through cellular respiration that are broken down by plants through photosynthesis.

B Animals use cellular respiration while plants use photosynthesis.

C Cellular respiration produces sugars, which are stored through photosynthesis.

D Photosynthesis produces sugars, which are broken down in cellular respiration.

Use the diagram below to answer question 9.

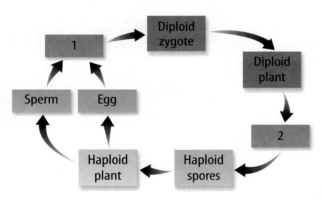

9 Which process occurs at the stage marked *2* on the plant life cycle diagram above?

A asexual reproduction

B fertilization

C meiosis

D mitosis

Constructed Response

Use the figure below to answer questions 10 and 11.

10 Describe what is happening in the image above. In your response, identify the environmental stimulus and the plant growth response.

11 Which plant hormone is involved in the growth response shown in the drawing above? Explain how this hormone causes the growth response.

12 Fruit distributors might use technologies to remove ethylene from the area where fruits are stored. How does this practice affect the fruit? Explain your answer.

13 Plants might reproduce both sexually and asexually. How are these two forms of reproduction similar? How are they different?

NEED EXTRA HELP?													
If You Missed Question...	1	2	3	4	5	6	7	8	9	10	11	12	13
Go to Lesson...	1	1	3	2	1	1	3	1	3	2	2	2	3

Chapter 9

Interactions of Living Things

THE BIG IDEA

How do living things interact with and depend on the other parts of an ecosystem?

Inquiry **Good Neighbors?**

These prairie dogs live together in a network of burrows and tunnels. By cooperating with each other, they can survive more easily than each prairie dog can on its own.

- What types of resources do the prairie dogs need in order to survive?
- How might living together help them survive?
- How do these prairie dogs interact with each other and their environment?

Get Ready to Read

What do you think?

Before you read, decide if you agree or disagree with each of these statements. As you read this chapter, see if you change your mind about any of the statements.

1 An ecosystem contains both living and nonliving things.

2 All changes in an ecosystem occur over a long period of time.

3 Changes that occur in an ecosystem can cause populations to become larger or smaller.

4 Some organisms have relationships with other types of organisms that help them to survive.

5 Most of the energy used by organisms on Earth comes from the Sun.

6 Both nature and humans affect the environment.

ConnectED Your one-stop online resource

connectED.mcgraw-hill.com

Video

Audio

Review

Inquiry

WebQuest

Assessment

Concepts in Motion

Multilingual eGlossary

Lesson 1

Ecosystems and Biomes

Reading Guide

Key Concepts 🔑
ESSENTIAL QUESTIONS

- What are ecosystems?
- What are biomes?
- What happens when environments change?

Vocabulary

ecosystem p. 309
abiotic factor p. 309
biotic factor p. 311
population p. 311
community p. 311
biome p. 312
succession p. 313

 Multilingual eGlossary

 Video BrainPOP®

Inquiry Is anyone home?

This is a photograph of Arctic tundra, a cold region with little rainfall and a short growing season. What kinds of organisms might live here? What might happen if the climate got warmer?

How do environments differ?

Have you ever been enjoying a warm sunny day when it suddenly started to rain? Weather is part of your environment. The different environments on Earth can vary greatly. Some organisms are more comfortable in one environment than another.

1 Read and complete a lab safety form.

2 Your teacher will divide the class into three groups. Students in one group will wear **heavy coats**. Students in another group will hold a **bag of ice**. Students in the third group will have no change to their environment.

3 After five minutes, determine how comfortable you are in your environment.

Think About This

1. What types of environments were represented by the three groups?

2. 🔑 **Key Concept** How did your environment change? What did you do to respond?

What are ecosystems?

How are you similar to a wolf and a pine tree? You, the wolf, and the pine tree are all living things also called organisms. All organisms have some characteristics in common. They all use energy and do certain things to survive. Organisms also interact with their environments. Ecology is the study of how organisms interact with each other and with their environments.

Every organism on Earth lives in an **ecosystem**—*the living and nonliving things in one place*. Different organisms depend on different parts of an ecosystem to survive. For example, the deer in **Figure 1** might eat leaves and drink water from a stream. Although leaves are alive and the water is not, the deer need both to survive. A fish in the stream also needs water to survive, but it interacts differently with water than deer do.

🔑 **Key Concept Check** What is an ecosystem?

Abiotic Factors

Water is just one example of a part of an ecosystem that was never alive. **Abiotic factors** *are the nonliving parts of an ecosystem.* Some important abiotic factors include water, light, temperature, atmosphere, and soil. The types and amounts of these factors available in an ecosystem help determine which organisms can live there.

Figure 1 These deer live in a woodland environment.

✅ **Visual Check** Which abiotic factors can you identify in this photo?

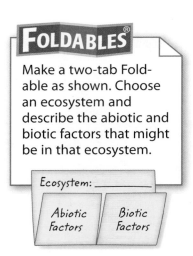
Water

Look at the plants shown in the two environments in **Figure 2**. How do the water requirements of these plants differ? A cactus grows in a desert, where it does not rain often. Ferns and vines live in a rain forest where it is very moist. All organisms need water to live, but some need more water than others do. The type of water in an ecosystem also helps determine which organisms can live there. Some organisms must live in saltwater environments, such as oceans, while others, like humans, must have freshwater to survive.

Light and Temperature

The amount of light available and the temperature of an ecosystem can also determine which organisms can live there. Some organisms, such as plants, require light energy for making food. Temperatures in ecosystems vary, and ecosystems with more sunlight generally have higher temperatures. How are the plants shown in **Figure 2** different? Ferns thrive in a warm rain forest. A cactus survives in a desert environment that can be very hot during the day and very cold at night.

Atmosphere

Very few living things can survive in an ecosystem without oxygen. Earth's **atmosphere** contains oxygen gas as well as other gases, such as water vapor, carbon dioxide, and nitrogen, that organisms need.

SCIENCE USE v. COMMON USE

atmosphere
Science Use the mix of gases surrounding a planet

Common Use a surrounding influence or feeling

Ecosystems 🔑

Figure 2 Deserts and rain forests have different amounts of water. This is one factor that affects which types of plants and animals can live in each environment.

✓ **Visual Check** Which factors determine the types of plants and animals that live in each environment?

Soil

Different ecosystems contain different amounts and types of nutrients, minerals, and rocks in the soil. Soil also can have different textures and hold different amounts of moisture. The depth of soil in an ecosystem can differ as well. All of these factors determine which organisms can live in an ecosystem. How do you think the soils in the environments in **Figure 2** differ?

Biotic Factors

You have read about the nonliving, or abiotic, parts of an ecosystem. These parts are important to the survival of living things. **Biotic factors** *are all of the living or once-living things in an ecosystem.* A parrot is one biotic factor in a rain forest, and so is a fallen tree. Biotic factors can be categorized and studied in several ways.

Populations

Think of the last time you saw a flock of birds. The birds you saw were part of a population. *A* **population** *is made up of all the members of one species that live in an area.* For example, all the gray squirrels living in a neighborhood are a population. Organisms in a population interact and compete for food, shelter, and mates.

Communities

Most ecosystems contain many populations, and these populations form a community. *A* **community** *is all the populations living in an ecosystem at the same time.* For example, populations of trees, worms, insects, and toads are part of a forest community. These populations interact with each other in some way. When trees in a forest ecosystem lose their leaves in the fall, the leaves become food for worms and insects. Toads might use the leaves to hide from predators. Waste from these animals provides nutrients to the trees and insects. You will read more about the types of interactions in communities later in this chapter.

Inquiry) MiniLab   20 minutes

How many living and nonliving things can you find?

Scientists study ecosystems by noting what and how many organisms live in an ecosystem. They also take account of nonliving factors that help make up an ecosystem. You can study a small area of your ecosystem in the same way.

1. Read and complete a lab safety form.

2. In a safe outdoor area, use **string** to section off 1 m² of ground.

3. Inspect your area for 10 minutes with a **magnifying lens.** Create a table like the one below. Record each different living and nonliving thing that you find. Be sure to record if you find the same thing more than once.

Living Things	Nonliving Things	Number of Times Found
	Small rock	5
Beetle		1

Analyze and Conclude

1. **Describe** any trends you see in the numbers and types of organisms that you found.

2. **Predict** how your results might be different if you made your observations during different times of the year.

3. 🔑 **Key Concept** List four factors in the ecosystem you studied that affect the numbers and types of organisms that live there.

WORD ORIGIN · · · · · · · · · · · · · · · · ·

community
from Latin *communitatem,* means "fellowship"

Biomes

The populations and communities that interact in a desert are very different from those that interact in an ocean. Deserts and oceans are different biomes. *A* **biome** *is a geographic area on Earth that contains ecosystems with similar biotic and abiotic features.* Biomes contain ecosystems, populations, and communities, as well as specific biotic and abiotic factors. As a result, biomes can be very different from each other. Some examples of Earth's major biomes are shown in **Figure 3**.

 Key Concept Check What is a biome?

All biomes are part of the biosphere—the part of Earth that supports life. Earth's biomes can be described as either terrestrial or aquatic. *Terrestrial* means related to land, and *aquatic* means related to water. Terrestrial biomes include forests, deserts, tundra, and grasslands. Aquatic biomes include saltwater areas and freshwater areas. Biomes—like communities—can affect each other. For example, a beach ecosystem is part of both a terrestrial and an aquatic biome. Some organisms from each of these biomes interact in the beach ecosystem.

Earth's Major Biomes 🔑

Figure 3 The Earth's biosphere includes all the different ecosystems.

Desert biome

Forest biome

Tundra biome

Biosphere

Freshwater biome

Grassland biome

Saltwater biome

What happens when environments change?

The photographs in **Figure 4** are of Mount St. Helens, an active volcano in the state of Washington. The top two photos were taken only one day apart—before and after a large volcanic explosion. The ecosystem changed dramatically in a very short period of time because of the volcanic eruption. Over time, Earth's ecosystems, including Mount St. Helens, have undergone countless changes ranging from tiny to enormous.

Changes in the environment are caused by both natural processes and human actions. Some of these changes can occur rapidly, like the erupting volcano at Mount St. Helens. Other changes, such as the river flow that slowly carved into the land and created the Grand Canyon in Arizona, can take millions of years.

Response to Change

Sometimes changes have positive effects on ecosystems, such as greater rainfall that results in more plants growing. Other changes can have negative effects. A very dry season could cause plants to die, and animals might starve. Usually, a change in an ecosystem results in both positive and negative effects.

Succession

Over long periods of time, communities can change through succession until they are very different. **Succession** *is the gradual change from one community to another community in an area.* The bottom picture in **Figure 4** shows how Mount St. Helens looked nearly 30 years after its devastating eruption. How did succession change Mount St. Helens and the environment around it?

 Key Concept Check Which biotic and abiotic factors changed after the Mount St. Helens eruption?

 Review Personal Tutor

Figure 4 An unexpected volcanic eruption changed the ecosystem of Mount St. Helens suddenly. Then, over many years, succession occurred and Mount St. Helens changed again.

May 17, 1980

May 18, 1980

Summer 2006

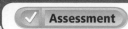
Visual Summary

Biotic factors are the living parts of an ecosystem.

Earth's biosphere contains many different biomes.

Changes in a community can be very slow or very rapid.

FOLDABLES®

Use your lesson Foldable to review the lesson. Save your Foldable for the project at the end of the chapter.

What do you think NOW?

You first read the statements below at the beginning of the chapter.

1. An ecosystem contains both living and nonliving things.

2. All changes in an ecosystem occur over a long period of time.

Did you change your mind about whether you agree or disagree with the statements? Rewrite any false statements to make them true.

Use Vocabulary

1. **Describe** the parts of an ecosystem.
2. **Define** *succession* using your own words.
3. **Distinguish** between biome and biosphere.

Understand Key Concepts

4. **Explain** the difference between biotic and abiotic factors.
5. **Choose** which describes a biotic factor.
    A. community
    B. sunlight
    C. temperature
    D. water
6. **Determine** whether people in your area have changed the local environment. Give some examples.

Interpret Graphics

7. **Compare** two different biomes. Copy the table below and write the name of each biome in the top row. Write the characteristics of each biome in the spaces below its name.

8. **Describe** how the abiotic factors differ between these two communities.

Critical Thinking

9. **Design** your own ecosystem. Tell how the abiotic and biotic factors affect the living things in your ecosystem. Which types of plants and animals live there? How do they interact?

AMERICAN MUSEUM OF NATURAL HISTORY

All for One,
One for All

If you've ever watched ants move single file across a sidewalk, you might have wondered how they know which way to go. This is a question that Dr. Deborah M. Gordon, an ecologist at Stanford University, might ask. She studies the behavior of red harvester ants.

Gordon studies the organization of ant colonies. A colony has one or more reproductive queens and many sterile workers living together. At any given time, ants might be working together on a specific task, such as building new tunnels, protecting the colony, or collecting food. However, no one ant in the colony directs the other ants. Gordon investigates how each ant within a colony takes on different tasks and how they work together as a group.

Ants communicate using chemicals. They release chemicals that other ants smell with their antennae. Each colony has its own unique odor, and only ants from that colony can recognize it. In addition, harvester ants have a chemical "vocabulary." They signal a specific task by communicating with a particular chemical.

To study ant communication, Gordon and her colleagues closely observe red harvester ants in their habitats and conduct experiments. Her team might isolate one of the communication chemicals and then place it in different locations or in the same location at different times of day. Gordon has learned that these ants can change tasks when they meet other ants. When one ant's antennae touch another ant, it can tell by the odor what task the other ant is doing. Gordon is studying how ants use encounters to interact with each other and their environments.

▲ Gordon uses a device called a theodolite to measure locations of ant colonies. She tracks changes in a colony's behavior as it ages and grows.

▲ The red harvester ant is one of only 50 ant species that have been well studied. Scientists have discovered about 10,000 ant species in all.

It's Your Turn

FIELD JOURNAL Find two animals to observe in their natural habitats. Record your observations in your Science Journal. How did they interact with the environment or with other animals? Share your results with classmates.

Reading Guide

Key Concepts 🔑
ESSENTIAL QUESTIONS

- How do individuals and groups of organisms interact?
- What are some examples of symbiotic relationships?

Vocabulary

limiting factor p. 319
biotic potential p. 319
carrying capacity p. 319
habitat p. 320
niche p. 320
symbiotic relationship p. 320

g Multilingual eGlossary

Populations and Communities

Inquiry Community?

These leafcutter ants work together making homes and getting food. The ants in the population interact with each other as well as with other populations in the community. What other populations might the ants interact with?

Inquiry Launch Lab

15 minutes

What is the density of your environment?

Imagine you are in a crowded elevator. Everyone jostles and bumps each other. The temperature increases and ordinary noises seem louder. Like people in an elevator, organisms in an area interact. How does the amount of space available to each organism affect its interaction with other organisms in the same area?

1. Use a **meterstick** to measure the length and width of the classroom.

2. Multiply the length by the width to find the area of the room in square meters.

3. Draw a map of your classroom, including the number and position of desks. Count the number of individuals in your class. Divide the area of the classroom by the number of individuals. In your Science Journal, record how much space each person has.

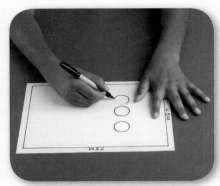

Think About This

1. What would happen if the number of students in your classroom doubled?

2. **Key Concept** How might your interactions with your classmates change if each person had less space?

Populations

Have you ever gone fishing at a lake? If so, recall that each individual fish in the lake is a member of a population. You read in Lesson 1 that a population is all the members of one species that live in an area. The area in which a population lives can be very large, such as the population of all fish in an ocean, as shown at the top of **Figure 5**. The area in which a population lives can also be small, like fish in a lake shown at the bottom of **Figure 5**.

Recall that organisms respond to abiotic and biotic factors in their ecosystem. Sunlight, temperature, and water quality are examples of abiotic factors that affect the fish in the lake and in the ocean. Biotic factors, such as the plants they eat and other organisms that hunt them, also affect the fish. If any of these factors change, the fish population can also change.

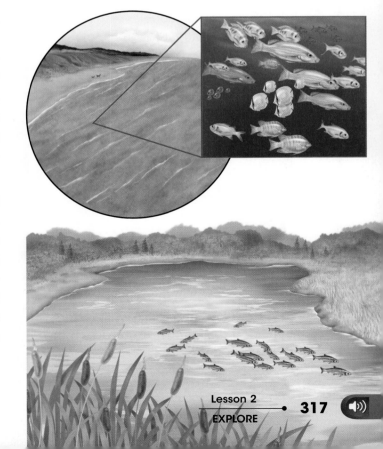

Figure 5 Populations can live in very large areas, such as an ocean, or small areas, such as a lake.

Figure 6 The size of a fish population can change in several different ways.

🔎 **Visual Check** How do you think fish hatching affects the other populations in the community?

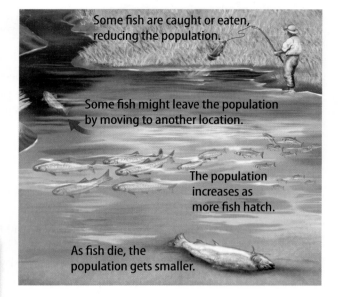

Some fish are caught or eaten, reducing the population.

Some fish might leave the population by moving to another location.

The population increases as more fish hatch.

As fish die, the population gets smaller.

Population Size

Think about the fish in the lake. What might happen to the population if a large number of fish eggs hatched or a hundred people caught fish all at once?

Population size can increase or decrease, as shown in **Figure 6.** Populations can increase when new individuals move into an area or when more individuals are born. Populations can decrease when individuals move away from an area or die. Sometimes the size of a population changes because the ecosystem changes. For example, if there is not much rainfall, a pond might shrink and some fish might die.

Population Density

Are the hallways in your school crowded or is there lots of room? The population density of the hallways reflects how crowded the halls are. Population density describes the number of organisms in the population relative to the amount of space available. A very dense population might have one fish for every few cubic meters of water. If a population is very dense, organisms might have a hard time finding enough resources to survive. They might not grow as large as individuals in less crowded conditions.

Limiting Factors

Populations can increase or decrease in size. What do you think keeps populations from becoming too large? **Limiting factors** *are factors that can limit the growth of a population.* The amounts of available water, space, shelter, and food affect a population's size. With too few resources, some individuals cannot survive. Other factors, like predation, competition, and disease, can also limit how many individuals survive. Some limiting factors in a rabbit population are shown in **Figure 7.**

Biotic Potential Imagine a population of rabbits with an unlimited supply of food, an unlimited amount of land to live on, and no predators. The population would keep growing until it reached its biotic potential. **Biotic potential** *is the potential growth of a population if it could grow in perfect conditions with no limiting factors.* The population's rate of birth is the highest it can be, and its rate of death is the lowest it can be.

Carrying Capacity Almost no population reaches its biotic potential. Instead, it reaches its carrying capacity. **Carrying capacity** *is the largest number of individuals of one species that an ecosystem can support over time.* The limiting factors of an area determine the area's carrying capacity.

Overpopulation Sometimes a population becomes larger than an ecosystem's carrying capacity. Overpopulation is when a population's size grows beyond the ability of the area to support it. This often results in overcrowding, a lack of resources, and an unhealthy environment. For instance, with overcrowding the trout in a lake might not grow very large. Waste from the members of the population might build up faster than it can be broken down, making the population sick.

 Reading Check Why is overpopulation harmful to organisms?

Figure 7 The number of limiting factors in an area limits the growth of a population.

Limiting Factors

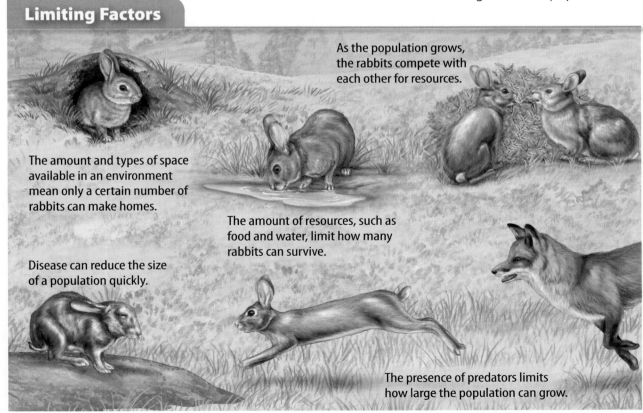

As the population grows, the rabbits compete with each other for resources.

The amount and types of space available in an environment mean only a certain number of rabbits can make homes.

The amount of resources, such as food and water, limit how many rabbits can survive.

Disease can reduce the size of a population quickly.

The presence of predators limits how large the population can grow.

FOLDABLES

Make a three-tab book and label it as shown. Use it to organize your notes on the types of symbiotic relationships that can be in communities.

| Mutualism | Parasitism | Commensalism |

Communities

Recall that populations in the same area interact as a community. Think again about the fish in the lake. Many other populations of organisms also live in and around the lake. These include other species of fish, frogs, algae, bacteria, insects, plants, raccoons, and other organisms.

All the populations in a lake community, such as the one shown in **Figure 8,** interact with each other in different ways. For example, the populations might compete with each other for some of the resources available in the lake. The organisms of each population must have a certain amount of the limited space in the lake in which to live. In some cases, the populations might compete directly as they hunt each other for food or hide from predators.

 Key Concept Check How do the different populations in a lake interact with each other?

Symbiotic Relationships

Populations affect their community by the ways in which they interact with each other. Each population has different ways to stay alive and reproduce. All of the populations in a community share a **habitat,** *the physical place where a population or organism lives.* Each organism also has a niche in the community. A **niche** *is the unique ways an organism survives, obtains food and shelter, and avoids danger in its habitat.*

Some organisms develop relationships with other organisms that help them survive. *A* **symbiotic relationship** *is one in which two different species live together and interact closely over a long period of time.* These relationships can be beneficial to both organisms, beneficial to one and harmful to the other, or beneficial to one and neutral to the other.

WORD ORIGIN · · · · · · · · · ·

habitat
from Latin *habitare,* means "to live, dwell"

Figure 8 Communities are made up of many populations.

Visual Check How might populations of turtles and ducks interact?

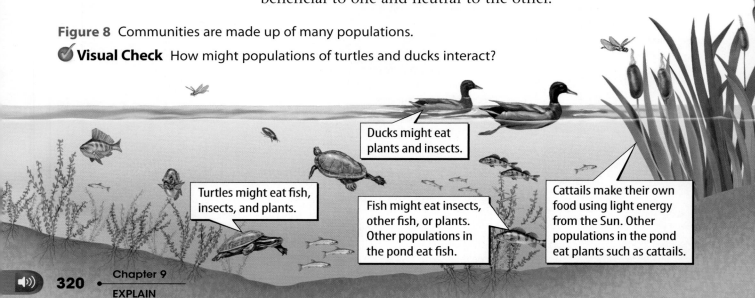

Turtles might eat fish, insects, and plants.

Ducks might eat plants and insects.

Fish might eat insects, other fish, or plants. Other populations in the pond eat fish.

Cattails make their own food using light energy from the Sun. Other populations in the pond eat plants such as cattails.

Mutualism

Recall the leafcutter ants shown at the beginning of this lesson. Leafcutter ants collect plant material and bring it back to their nest. There, they do not eat the leaves, but chew them into small pieces. A fungus grows on the pieces, as shown at the top of **Figure 9.** Some members of the population clean the fungus, removing any molds or other organisms that might harm it. The ants then eat this fungus. The ants provide the fungus with food (the leaves) and in turn use the fungus as a food source. The ants and the fungus have a mutualistic relationship. Mutualism is a symbiotic relationship in which two species in a community benefit from the relationship.

Parasitism

Mistletoe prompts kisses during the holidays—but did you know it is also a parasite? Parasitism is a symbiotic relationship in which one species (the parasite) benefits while another (the host) is harmed. Mistletoe grows in the branches of some trees. It sends its roots into the tissue of its host, the tree. The mistletoe plant takes food and water from the tree. This can weaken the tree. Too many mistletoe plants can eventually kill even strong, old trees like the one shown in the photograph to the right.

Commensalism

Sometimes two organisms can have a symbiotic relationship that affects only one of the organisms. Commensalism is a symbiotic relationship where one species benefits and the other is neither helped nor harmed. Notice the cocklebur attached to the clothing at the bottom of **Figure 9.** Plants produce these cockleburs that will stick to passing animals and humans and spread the plant's seeds around a larger area. The plant benefits from its seeds being spread, while the other organism is not harmed or helped.

 Key Concept Check What is one example of a symbiotic relationship?

Figure 9 Many different types of symbiotic relationships exist among different organisms.

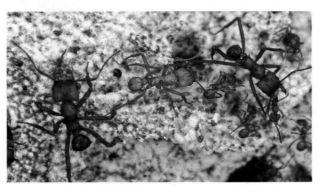

▲ Leafcutter ants and fungi have a mutualistic relationship.

▲ Mistletoe weakens its tree host by taking away vital food and water.

▲ This cocklebur contains the seeds of a plant. By sticking onto an animal or person, it can spread seeds over a larger distance.

Visual Check What makes the cocklebur effective?

Lesson 2 Review

Visual Summary

The factors that limit the size a population of organisms can reach are called limiting factors.

A habitat is the physical environment where a population of organisms lives.

A symbiotic relationship exists when two different species of organisms live together in a close relationship over a long period of time.

FOLDABLES®

Use your lesson Foldable to review the lesson. Save your Foldable for the project at the end of the chapter.

What do you think NOW?

You first read the statements below at the beginning of the chapter.

3. Changes that occur in an ecosystem can cause populations to become larger or smaller.

4. Some organisms form relationships with other types of organisms that help them survive.

Did you change your mind about whether you agree or disagree with the statements? Rewrite any false statements to make them true.

Use Vocabulary

1 **Distinguish** between population and community.

2 **Explain** in your own words how limiting factors affect populations.

3 **Contrast** the terms *parasitism* and *mutualism* in a sentence.

Understand Key Concepts

4 **Summarize** some ways that individuals and groups of organisms interact.

5 **Illustrate** two kinds of symbiotic relationships. Indicate whether a species is helped, harmed, or not affected.

6 Which factor does NOT change the size of a population?
   A. aging   C. death
   B. birth   D. leaving

Interpret Graphics

7 **Summarize** Copy and fill in the graphic organizer below to identify symbiotic relationships.

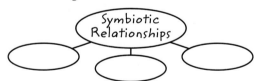

8 **Identify** three populations shown in the community below and explain how they interact with each other.

Critical Thinking

9 **Describe** two problems that occur when a human population grows. Use examples.

10 **Invent** a parasite. You may draw it or write about it. Explain how it uses its host. Tell how it benefits while its host is harmed.

Can you make predictions about a population size?

The table below shows the population size of Soay sheep from the Island of Hirta in Scotland. The data show how the population changed each year between 1986 and 2002. The size of a population can change for many reasons. For example, the population of sheep might increase if the weather is warm, causing more grass to grow for grazing. If there is little rain, grass might not grow as well and fewer sheep will survive. In some populations, cycles occur in which the population grows until resources run out, and the population size begins to decrease. Once resources are available again, the population increases once more.

Learn It

By identifying patterns in the way the sheep population changed, you might be able to guess, or **make a prediction** about, what will happen to the population in the future.

Year	Population Size
1986	700
1987	1,050
1988	1,500
1989	700
1990	900
1991	1,500
1992	950
1993	1,300
1994	1,600
1995	1,200
1996	1,800
1997	1,700
1998	2,000
1999	900
2000	1,500
2001	2,000
2002	900

Try It

1. Using the data in the table, make a graph with population size in the *y*-axis and time on the *x*-axis.

2. Examine the graph you have made. Does the population size increase every year? Does it decrease?

3. On the *x*-axis of your graph, highlight in one color the years that the population decreases. Highlight the years that the population increases in another color.

4. What is the best pattern you could use to describe the data?

Apply It

5. Based on the pattern you identified, would you predict that the population of sheep increased or decreased in 2003? Explain your answer.

6. Were there any years in which the pattern you identified did not occur? Which years were they?

7. 🔑 **Key Concept** Using what you read about limiting factors, what interactions between organisms might cause the pattern you identified in the sheep population?

Lesson 3

Reading Guide

Key Concepts 🔑
ESSENTIAL QUESTIONS

- How does energy move in ecosystems?
- How is the movement of energy in an ecosystem modeled?
- How does matter move in ecosystems?

Vocabulary

producer p. 326

consumer p. 327

food chain p. 328

food web p. 328

energy pyramid p. 329

g Multilingual eGlossary

Energy and Matter

Inquiry Out on a Limb?

This porcupine hopes to avoid the menacing teeth of a mountain lion. A mountain lion must constantly find and catch other animals to eat. How does catching and eating prey help a mountain lion survive? How are the diets of animals different from each other?

Where do you get energy?

You probably eat many different foods to get the energy you need to do things. Have you ever thought about how energy becomes a part of your food? For example, what do cows eat in order to obtain energy for making milk?

① Write down all the foods you ate for dinner last night.

② In your Science Journal, create a table that lists the food you ate in one column. Then write the origin of each food in the second column and the energy source for the food item in the last column. Consider that some animals eat a variety of foods. For instance, cows eat mainly grass, but chickens might eat insects as well as grass.

Food Item	Origin	Energy Source
Steak	Cow	Grass
Broccoli	Plant	the Sun

Think About This

1. Based on your table, do you get most of your energy from plants or animals? Where do these organisms get their energy?

2. Where might humans fit into a food chain?

3. 🔑 **Key Concept** Can you extend the energy source column with another link for any of the foods you ate? What do you think is the ultimate source of energy for foods?

Energy Flow

What did you eat for breakfast this morning? The food you take into your body is your energy **source**—it gives you fuel to walk, play games, read books, sit at a desk, and even to sleep. All life on Earth needs energy for cell processes.

Organisms get energy from food that they make using light or chemical energy or by eating other organisms. When one organism eats another, the energy in the organism that is eaten is transferred to the organism that eats it. In this way, energy travels through organisms, populations, communities, and ecosystems in a flow. When energy moves in a flow, like the one illustrated in **Figure 10**, it does not return to its source, as it does when matter cycles.

Organisms and Energy

Scientists classify organisms by the way they get the energy that they need to survive. Almost all energy on Earth comes from the Sun. Some organisms, such as plants, are able to capture this energy directly and convert it into energy-rich sugars that they use for food. A few organisms are able to capture energy from chemicals in the environment and make food. Other organisms cannot capture energy from sunlight or chemicals. They must obtain their energy by eating food. Organisms that cannot make their own food using the Sun must depend on organisms that can.

ACADEMIC VOCABULARY

source
(noun) a point of origin

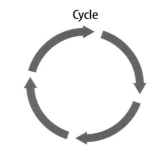

Figure 10 Energy in a flow does not return to its source as it does in a cycle.

Producers

How do some organisms obtain the energy that comes from the Sun? You might recall that energy cannot be created or destroyed, but it can change form. **Producers** *change the energy available in their environment into food energy.* They then use this food energy for living and reproducing. Humans and other organisms can get this energy by eating producers.

 Reading Check Why must producers be present in an environment?

Photosynthesis Energy from the Sun always enters a community through producers. Some producers use a chemical process called photosynthesis and transform light energy from the Sun into food energy. Producers that use light energy include most plants, algae, and some microorganisms. **Figure 11** illustrates how the process of photosynthesis converts light energy into food energy.

WORD ORIGIN ·············

producer
from Latin *producere*, means "to lead"
·············

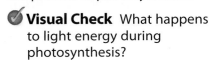

Figure 11 Light energy is changed into food energy by the process of photosynthesis.

 Visual Check What happens to light energy during photosynthesis?

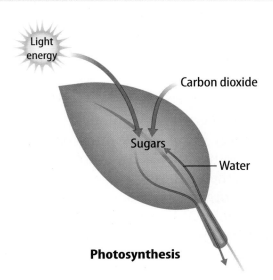

Photosynthesis

Chemosynthesis In some communities, producers get their energy from sources other than light energy. Chemosynthesis is the chemical process some producers use to change chemical energy into food energy.

One example of chemosynthesis occurs in the deep oceans. Bacteria living near volcano vents in the ocean floor use the chemicals as an energy source for making food energy.

Key Concept Check How does energy move from a producer to other organisms?

Consumers

Recall that energy moves through organisms in an ecosystem. Where does light or chemical energy go after producers like shrubs and other plants **transform** it into food energy? Organisms that cannot use light or chemical energy for making food must get their energy by eating other organisms. These organisms are called consumers. **Consumers** *cannot make their own food and get energy by eating other organisms.* Scientists classify organisms as producers or consumers, but they also classify consumers depending on what they eat. Examples of different types of consumers are shown in **Figure 12.** Consumers are classified as either herbivores, omnivores, carnivores, or detritivores based on their diet.

- Herbivores are animals that eat only producers, such as plants.

- Omnivores, such as human beings, are animals that eat both producers and other consumers.

- Carnivores, such as lions, eat only other consumers.

- Detritivores, including some insects, fungi, worms, and some bacteria and protists, eat dead plant or animal material. A type of detritivore called a decomposer breaks down dead material into simple molecules that can be used by other organisms, such as plants.

 Reading Check What are the four types of consumers?

ACADEMIC VOCABULARY

transform
(verb) to change in composition or structure

Figure 12 Consumers are classified according to the type of food they eat.

Consumers 🔑

Herbivore

Omnivore

Carnivore

Detritivore

Lesson 3
EXPLAIN

327

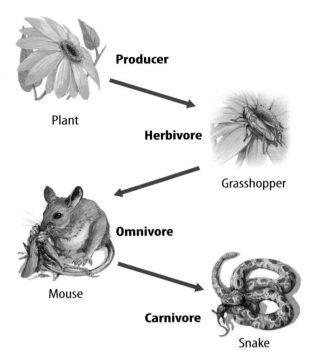

Producer

Plant

Herbivore

Grasshopper

Omnivore

Mouse

Carnivore

Snake

▲ **Figure 13** In this food chain, energy is passed from the plant to the grasshopper, the mouse, and then the snake.

Review · **Personal Tutor**

Food Web 🔑

▲ **Figure 14** A food web conveys more information than a food chain. Notice the arrows that show the many ways energy travels through an ecosystem.

Modeling Energy Flow

Although you cannot see energy, it is always moving through ecosystems. By noticing what different organisms eat, you can study how energy travels through a community. You can use a model to understand how this happens.

A **food chain** *models how energy flows in an ecosystem through feeding relationships.* A food chain is like a connect-the-dots drawing. Food energy moves from a plant to a grasshopper, then to a mouse, and then to a snake in the food chain shown in **Figure 13.**

Each stage of a food chain has less available food energy than the last one because some food energy is converted to thermal energy, commonly called heat, and moves to the environment. For example, it takes a lot of plant material to feed one grasshopper, and then many grasshoppers to feed one mouse. A mouse uses some of that food energy and the rest is transformed into thermal energy. This means less food energy is available to be passed on to the snake.

Food Webs

Food chains are simple models but they do not show all the energy transfers in an ecosystem. *A* **food web** *is a model that shows several connected food chains.*

Food webs show how energy travels through terrestrial ecosystems or aquatic ecosystems. The food web in **Figure 14** models the many ways food energy is transferred throughout a desert ecosystem. In many cases, terrestrial and aquatic food webs interact. For example, terrestrial raccoons might eat crayfish from streams. Food webs also show that food energy can move through several different pathways. For example, the energy in a mouse might be consumed by a fox or a hawk.

🔑 **Key Concept Check** Compare a food chain with a food web.

Modeling Energy Pyramids

As energy travels through different organisms, the amount of available food energy decreases. One way to model the available food energy in an ecosystem is to use an energy pyramid. *An **energy pyramid** shows the amount of energy available at each step of a food chain.*

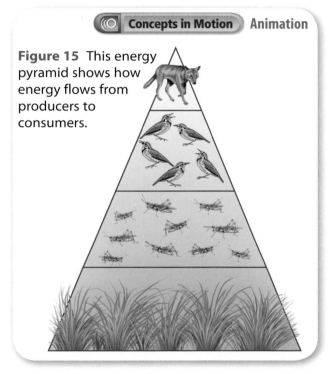

((O) Concepts in Motion   **Animation**

Figure 15 This energy pyramid shows how energy flows from producers to consumers.

As shown in the energy pyramid in **Figure 15,** more food energy is available at the "base" of the pyramid where producers are. The food energy from the producers moves into consumers—herbivores, omnivores, and carnivores—at the next level. Recall that some food energy is transformed into thermal energy by organisms. At each level of the pyramid, the amount of usable food energy decreases.

The top level of an energy pyramid usually shows the carnivores in an ecosystem—those animals in the ecosystem that prey only on other animals. What are some carnivores in the ecosystem around you?

✓ **Reading Check** What happens to the amount of available energy in higher levels of an energy pyramid?

Inquiry **MiniLab**   **20 minutes**

How is energy transferred in a food chain?

Producers get their energy from the environment. Consumers get their energy by eating producers. However, some of that energy is transformed in transfer.

1 Read and complete a lab safety form.
2 Each student is assigned a role. One student is the Sun. Others might be plants, mice, or an owl. The Sun will give one **cup of dried beans** (to represent energy) to each plant.
3 Each plant should keep 20 of its beans and put the rest into a **box.** This represents energy that the organism does not pass on.
4 Each mouse takes 20 beans (representing usable energy) from the plants in its group. Each mouse can use only 8 beans. Each mouse should keep eight beans and put the rest into the box.
5 The owl should take 8 beans from each mouse. The owl uses only one bean and puts the rest of the beans in the box.

Analyze and Conclude

1. **Describe** any patterns you see in how much energy is transferred.

2. 🔑 **Key Concept** Explain why there are fewer organisms in the higher levels of an energy pyramid.

Matter Cycles

You need food for energy, but you also eat and drink to replace vitamins, water, and minerals that leave your body. These substances—food, vitamins, minerals, and water—are examples of matter.

Matter is the physical material that makes up the world around you. Your body—like everything else on Earth—consists of matter. Most of the matter in your body is water. Your body also contains matter in other forms such as carbon and oxygen. Like energy, matter is not created or destroyed. Instead it is transferred through the environment. Unlike energy, though, matter does not flow but moves in a cycle. It is used again and again throughout different parts of an ecosystem. Matter can change forms as it moves through an environment. For example, the liquid form of a type of matter might turn into the gas form. The types of matter found in different environments, as well as the amounts available, determine which organisms can live there.

Water Cycle

Water is important to all life. It moves through every ecosystem on Earth and is in different forms. Its forms include a liquid, a gas (known as water vapor), and a solid (ice). **Figure 16** illustrates how water moves through the environment in the water cycle.

1. Water evaporates from oceans, rivers, and other bodies of water. Plants release water vapor during transpiration and some organisms release water vapor when they breathe out (exhalation).

2. The water vapor then rises into the atmosphere, where it condenses and falls as rain or snow.

3. Water moves across the surface of Earth in lakes, streams and rivers, soaks into the ground, or is taken in by organisms. This water is eventually released again, and the cycle continues.

 Key Concept Check What forms does water take in the water cycle?

Water Cycle (((○ **Concepts in Motion** Animation

Figure 16 Water cycles throughout an environment.

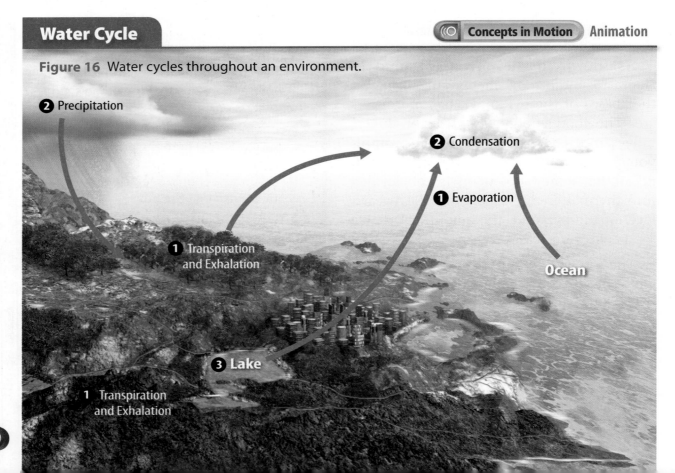

Figure 17 Producers release oxygen gas and consumers take it in.

🔘 **Visual Check** What do the consumers in this image release to the atmosphere?

Oxygen Cycle

Like water, oxygen also cycles through the environment. Oxygen is another example of matter that is important to the survival of many organisms. You take in oxygen when you breathe, and your blood carries the oxygen to all parts of your body. Oxygen is also a part of many molecules that are important to life, such as sugars.

Plants release oxygen as a waste product of photosynthesis. This oxygen enters the atmosphere, and many consumers take it in when they breathe. When these organisms exhale, they release carbon dioxide, which is a by-product of cellular respiration. Carbon dioxide contains oxygen. Some producers take in carbon dioxide, and the cycle continues.

The oxygen cycle in a rain forest is shown in **Figure 17.** There, many plants release large amounts of oxygen. Besides supplying all the rain forest organisms with oxygen, these plants are also an important source of oxygen for other organisms on Earth.

✅ **Reading Check** What organisms are part of the oxygen cycle?

Math Skills ➗

Use Proportions

When two ratios form a proportion, the cross products are equal. Use this method to solve problems such as the following: If there are 48 hours in 2 days, how many hours are there in 5 days?

$$\frac{48 \text{ hr}}{2 \text{ days}} = \frac{n \text{ hr}}{5 \text{ days}}$$

1. Find the cross products.

$$48 \times 5 = n \times 2$$

$$240 = 2n$$

2. Solve for n by dividing both sides by 2.

$$\frac{240}{2} = \frac{2n}{2}$$

$$n = 120 \text{ hours}$$

Practice

If the average human takes 10 breaths per minute, how many breaths will the person take in 1 hour?

📖 **Review**

• **Math Practice**
• **Personal Tutor**

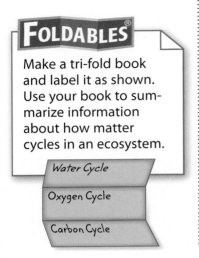
Carbon Cycle

Like bricks used in building, carbon is a fundamental building block for all living things. It is part of molecules such as proteins, carbohydrates, and fats. It is also almost everywhere on Earth in nonliving things. The carbon cycle diagram in **Figure 18** illustrates how carbon moves throughout Earth's environment. When producers use carbon dioxide during photosynthesis, carbon is removed from the atmosphere. Consumers eat these producers and release carbon back into the environment as a waste product. Human activities such as the burning of fossil fuels also add carbon to the atmosphere. Producers again remove the carbon from the atmosphere as they continue making food, and the cycle continues.

Reading Check Where is carbon in living organisms?

Concepts in Motion Animation

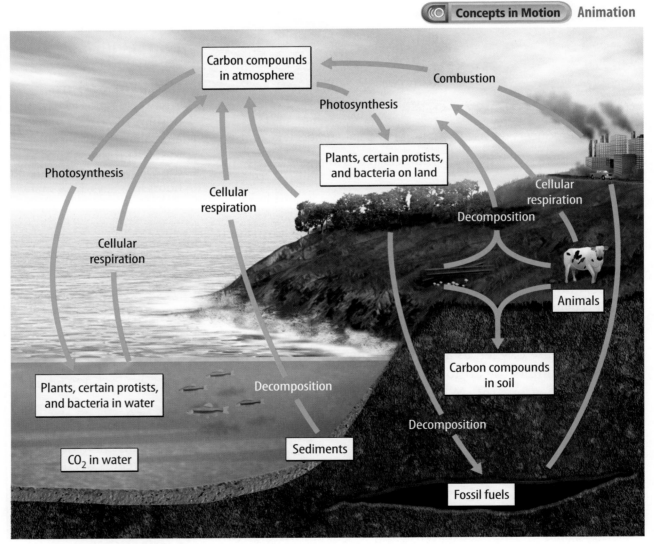

Figure 18 Carbon moves throughout the environment through processes such as photosynthesis.

Visual Check What processes add carbon compounds to the atmosphere?

Lesson 3 Review

Visual Summary

A producer changes the energy available in the environment into food energy.

Consumers must use the energy and nutrients stored in other organisms for living and reproducing.

An energy pyramid shows how much food energy is available to organisms at each level of a community.

FOLDABLES®

Use your lesson Foldable to review the lesson. Save your Foldable for the project at the end of the chapter.

What do you think NOW?

You first read the statements below at the beginning of the chapter.

5. Most of the energy used by organisms on Earth comes from the Sun.

6. Both nature and humans affect the environment.

Did you change your mind about whether you agree or disagree with the statements? Rewrite any false statements to make them true.

Use Vocabulary

1 **Describe** the process of chemosynthesis.

2 **Distinguish** between producers and consumers.

3 **Write** a sentence using the word *decomposer*.

Understand Key Concepts

4 **Explain** how energy from the Sun enters an ecosystem.

5 Which does NOT cycle through an environment?
   A. nitrogen   C. sunlight
   B. oxygen   D. water

6 **State** whether you are an herbivore, an omnivore, or a carnivore. Explain your answer.

Interpret Graphics

7 **Organize Information** Copy and fill in the graphic organizer below for a a food chain of herbivores, carnivores, and producers.

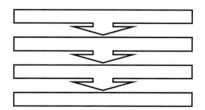

Critical Thinking

8 **Draw** a food web for a local habitat. What are the producers and the consumers? Which are omnivores, carnivores, and herbivores? Draw arrows to connect the organisms and label the different species.

Math Skills ✕ ÷ + −   Review
─── Math Practice ───

9 One liter of air contains about 0.21 L of oxygen (O_2). When filled, the human lungs hold about 6.0 L of air. How much O_2 is in the lungs when they are filled with air?

Materials

bromthymol blue

powdered calcium carbonate or crushed natural chalk

vinegar

250-mL Erlenmeyer flask

20-mL test tube

balloon

filter paper

Safety

Can you observe part of the carbon cycle?

The carbon cycle includes all of the paths that carbon takes as it is transferred through the environment. During one part of the carbon cycle, tiny ocean organisms called phytoplankton (fi toh PLANK tuhn) take in carbon dioxide gas from the air. The carbon dioxide gas is converted to calcium carbonate, which the phytoplankton use to build their skeletons. When phytoplankton die, many of them sink to the bottom of the ocean, where their skeletons become fossilized. Over time, these fossilized skeletons build up and turn into chalk. When this chalk is weathered by rain and waves, it releases carbon dioxide gas into the air. This gas can then move through the many paths of the carbon cycle.

Question

How can you model part of the carbon cycle?

Procedure

1. Read and complete a lab safety form.

2. Measure 15 g of calcium carbonate powder onto filter paper.

3. Pour the calcium carbonate powder into a 250-mL Erlenmeyer flask.

4. Add 50 mL of vinegar to the flask.

5. Quickly stretch the mouth of the balloon over the opening of the flask. Record your observations in your Science Journal.

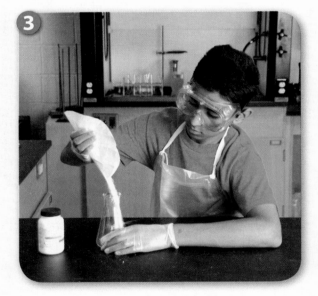

6. Fill the test tube almost to the top with water. Add 15 drops of bromthymol blue, a chemical indicator that turns yellow when exposed to carbon dioxide. Observe the color of the liquid in the test tube.

7. Pinch the neck of your balloon and remove it from the flask. Place the neck of the balloon over the mouth of the test tube and then release the neck, allowing the gas to enter the test tube. Record your observations in an observation table in your Science Journal.

Analyze and Conclude

8. **Interpret Data** What happened to the color of the liquid in the test tube? What did this indicate?

9. **Identify** Where did the gas in the balloon come from?

10. **The Big Idea** Describe the part of the carbon cycle you modeled in this experiment. How do living things affect this part of the carbon cycle? How do nonliving things affect this part of the carbon cycle?

Communicate Your Results

Draw the carbon cycle. Highlight the parts of the carbon cycle you modeled in this experiment.

(inquiry) Extension

When you released carbon from the calcium carbonate, you might have released the same molecules that a dinosaur exhaled millions of years ago. Draw a picture showing the flow of carbon through the environment, including how dinosaur breath could have been captured in calcium carbonate.

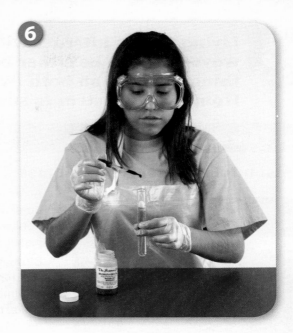

6

Lab Tips

☑ Carefully swirl the calcium carbonate and vinegar in your flask to get the most gas from your reaction.

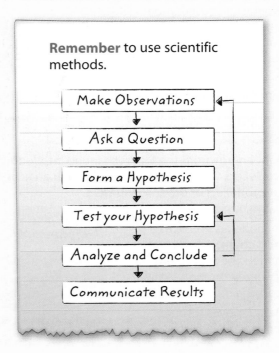

Remember to use scientific methods.

Make Observations
↓
Ask a Question
↓
Form a Hypothesis
↓
Test your Hypothesis
↓
Analyze and Conclude
↓
Communicate Results

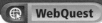
 THE BIG IDEA Living things interact with each other in a variety of ways that can be either beneficial or harmful. Living things depend on both living and nonliving resources from the ecosystem to survive.

Key Concepts Summary 🔑

Vocabulary

Lesson 1: Ecosystems and Biomes

- An **ecosystem** is made up of all the living and nonliving things in a location.
- **Biomes** are large regions that have specific types of climate, physical characteristics, and organisms.
- One environment changes into another in a process called **succession.**

ecosystem p. 309
abiotic factor p. 309
biotic factor p. 311
population p. 311
community p. 311
biome p. 312
succession p. 313

Lesson 2: Populations and Communities

- Organisms must compete with each other to obtain resources, such as food, water, and living space.
- **Symbiotic relationships** include mutualism, parasitism, and commensalism.

limiting factor p. 319
biotic potential p. 319
carrying capacity p. 319
habitat p. 320
niche p. 320
symbiotic relationship p. 320

Lesson 3: Energy and Matter

- Light energy from the Sun is changed into food energy by **producers.** Energy then moves through an ecosystem as organisms eat producers or other **consumers.**

- Energy movement can be modeled simply as a **food chain.** A **food web** models the movement of energy through many food chains in an ecosystem.

Precipitation

Condensation

Transpiration and Exhalation

Evaporation

Ocean

Lake

- Matter moves through ecosystems in cycles. Examples of matter cycles include the carbon, water, and oxygen cycles.

producer p. 326
consumer p. 327
food chain p. 328
food web p. 328
energy pyramid p. 329

 **Chapter Project**

Assemble your lesson Foldables as shown to make a Chapter Project. Use the project to review what you have learned in this chapter.

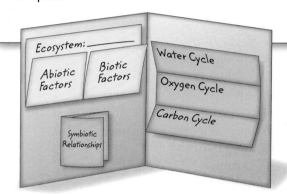

Use Vocabulary

1 A(n) _____ factor is any of the living parts of an ecosystem.

2 The biosphere contains all of Earth's _____.

3 The _____ of a population is the largest number of individuals that can survive in a location over time.

4 Each organism has a(n) _____ that includes the ways it survives, obtains food and shelter, and avoids danger in its habitat.

5 A(n) _____ cannot make food, and must obtain energy from other organisms.

6 A(n) _____ models how much energy is available as it moves through an ecosystem.

Link Vocabulary and Key Concepts

Concepts in Motion Interactive Concept Map

Copy this concept map, and then use vocabulary terms from the previous page to complete the concept map.

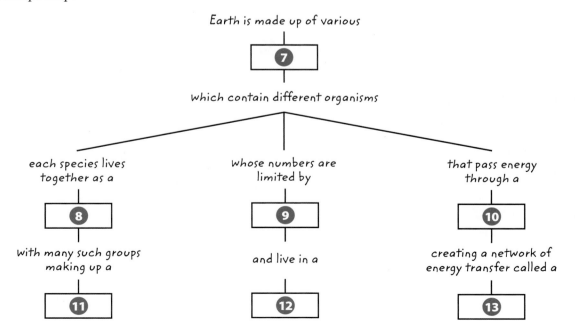

Understand Key Concepts

1 Which is NOT an abiotic factor?
A. atmosphere
B. prey
C. sunlight
D. temperature

2 Which biome is shown below?

A. desert
B. forest
C. grassland
D. tundra

3 Which term describes a slow change in an environment?
A. abiotic
B. biotic
C. regression
D. succession

4 Which is NOT a limiting factor?
A. competition
B. disease
C. amount of resources
D. biotic potential

5 What is a niche?
A. a cycle of matter
B. a source of energy
C. a source of energy
D. where an animal lives

6 Which type of symbiotic relationship is beneficial for both organisms?
A. commensalism
B. competition
C. mutualism
D. parasitism

7 Which includes the process of transpiration?
A. succession
B. condensation
C. oxygen cycle
D. water cycle

8 What is released as a product of photosynthesis?
A. carbon
B. oxygen
C. soil
D. water

9 Which type of consumer gets all the energy it needs from producers?
A. carnivore
B. detritivore
C. herbivore
D. omnivore

10 Which type of process is illustrated by the diagram below?

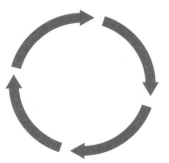

A. a cycle
B. a flow
C. an energy pyramid
D. a food web

Critical Thinking

11 **Contrast** aquatic and terrestrial ecosystems, and explain how they might interact.

12 **List** some of the biotic and abiotic factors you respond to each day. Describe how they impact you.

13 **Consider** the value of the study of ecology. Why is it important or not important?

14 **Differentiate** between a habitat and a niche.

15 **Reflect** on whether human beings have a symbiotic relationship with Earth.

16 **Describe** the symbiotic relationship shown below.

17 **Hypothesize** what might happen to a lake ecosystem if very little rain fell for many years. How would the lake community change? What would happen if the amount of rain greatly increased rather than decreased?

18 **Construct** a food chain that includes you.

Writing in Science

19 **Visualize** succession in your community. What changes have occurred in your community over the last hundred years? What might occur over the next hundred? The next thousand? Write a paragraph describing your predictions.

REVIEW THE BIG IDEA

20 How are the living and nonliving parts of an environment important to an organism? Give an example in which one organism must depend on another organism in order to survive.

21 The photo below shows a population of prairie dogs. How might living together as a population be helpful to the prairie dogs? How might living together be more difficult than living alone?

Math Skills

Review
Math Practice

Use Proportion

22 A large tree releases about 0.31 kg of oxygen per day. A typical adult uses 0.84 kg of oxygen per day. How many trees would provide enough oxygen for the adult?

23 In one year, an acre of trees can provide enough oxygen for about 18 people. How many acres of trees would provide enough oxygen for the population of New York City —about 8,300,000?

Standardized Test Practice

Record your answers on the answer sheet provided by your teacher or on a sheet of paper.

Multiple Choice

1 All of the nonliving and living things in one place are

  A a biome.

  B a population.

  C an ecosystem.

  D an organism.

Use the table below to answer question 2.

Organism	Interaction
Cattails	Make food using sunlight energy
Ducks	Eat plants or insects
Fish	Eat plants, insects, or other fish
Turtles	Eat fish, insects, or plants

2 The organisms in the table above live in or around a lake. Together they form a

  A community.

  B habitat.

  C niche.

  D population.

3 What describes the size of a population in ideal conditions?

  A community

  B biotic potential

  C carrying capacity

  D limiting factor

4 A grassland changes to a desert after many years of little rain. This is an example of what?

  A migration

  B overpopulation

  C predation

  D succession

Use the diagram below to answer question 5.

5 Which limiting factor does the diagram illustrate?

  A carrying capacity

  B overpopulation

  C predation

  D competition for resources

6 Which is true of the movement of matter in ecosystems?

  A Matter is used again and again.

  B Matter retains its original form.

  C Matter increases in amount as it moves.

  D Matter moves in a flow.

7 An owl eats mice and other small rodents in a forest. This owl lives in a hole in the side of a dead tree. What do these two statements describe?

  A a biome

  B a community

  C a niche

  D a population

8 Which illustrates organization from smallest to largest?

  A biome → population → community

  B community → biome → population

  C community → population → biome

  D population → community → biome

Use the diagram below to answer question 9.

9 Which level has the *least* amount of food energy?

  A birds

  B grasses

  C grasshoppers

  D coyote

10 Which is a symbiotic relationship?

  A a hawk eating a mouse

  B a fish laying eggs in a stream

  C a mistletoe plant living on a tree

  D a plant using sunlight and making food

11 Consumers are generally classified into one of four categories based on their

  A appearance.

  B diet.

  C habitat.

  D size.

Constructed Response

Use the chart below to answer question 12.

Abiotic Factor	Importance

12 List the abiotic factors in your ecosystem. Briefly describe their importance.

Use the diagram below to answer questions 13 and 14.

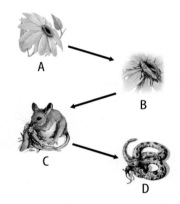

13 Match each letter in the diagram with the word that describes the organism: *carnivore, herbivore, omnivore,* and *producer.* What does the diagram reveal about the movement of energy in an ecosystem?

14 Describe what would happen if (a) mice disappeared from this ecosystem; (b) foxes entered this ecosystem; and (c) drought killed much of the plant life in this ecosystem.

NEED EXTRA HELP?														
If You Missed Question...	1	2	3	4	5	6	7	8	9	10	11	12	13	14
Go to Lesson...	1	1	2	1	2	3	2	1	3	2	3	1	3	3

1895
The first X-ray photograph is taken by Wilhelm Konrad Roentgen of his wife's hand. It is now possible to look inside the human body without surgical intervention.

1898
Chemist Marie Curie and her husband Pierre discover radioactivity. They are later awarded the Nobel Prize in Physics for their discovery.

1917
Ernest Rutherford, the "father of nuclear physics," is the first to split atoms.

1934
Nuclear fission is first achieved experimentally in Rome by Enrico Fermi when his team bombards uranium with neutrons.

1939
The Manhattan Project, a code name for a research program to develop the first atomic bomb, begins. The project is directed by American physicist J. Robert Oppenheimer.

1950   **1975**   **2000**

1945
American-led atomic
bomb attacks on the
Japanese cities of
Hiroshima and
Nagasaki bring
about the end of
World War II.

1954
Obninsk Nuclear Power Plant,
located in the former USSR,
begins operating as the
world's first nuclear power
plant to generate electricity
for a power grid. It produces
around 5 megawatts of
electric power.

2007
Fourteen percent
of the world's
electricity now
comes from
nuclear power.

? Inquiry

**Visit ConnectED for
this unit's
STEM activity.**

Nature of SCIENCE

Patterns

It's a bird! It's a plane! No, it's Venus! Besides the Sun, Venus is brighter than any other star or planet in the sky. It is often seen from Earth without the aid of a telescope, as shown in **Figure 1.** At certain times of the year, Venus can be seen in the early evening. At other times of the year, Venus is best seen in the morning or even during daylight hours.

Astronomers study the patterns of each planet's orbit and rotation. A pattern is a consistent plan or model used as a guide for understanding and predicting things. Studying the orbital patterns of planets allows scientists to predict the future position of each planet. By studying the pattern of Venus's orbit, astronomers can predict when Venus will be most visible from Earth. Astronomers also can predict when Venus will travel between Earth and the Sun, and be visible from Earth, as shown in **Figure 2.** This event is so rare that it has only occurred seven or eight times since the mid-1600s. Using patterns, scientists are able to predict the date when you will be able to see this event in the future.

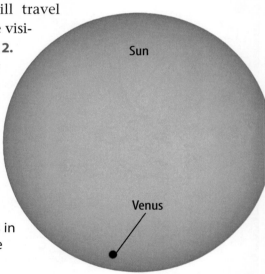

Figure 2 On June 8, 2004, observers around the world watched Venus pass in front of the Sun. This was the first time this event took place since 1882. ▶

▲ **Figure 1** Venus is often so bright in the morning sky that it has been nicknamed the morning star.

Types of Patterns

Physical Patterns

A pattern that you can see and touch is a physical pattern. The crystalline structures of minerals are examples of physical patterns. When atoms form crystals, they produce structural, or physical, patterns. The crystal structure of the Star of India sapphire creates a pattern that reflects light in a stunning star shape.

Cyclic Patterns

An event that repeats many times again in a predictable order has a cyclic pattern. Since Earth's axis is tilted, the angle of the Sun's rays on your location on Earth changes as Earth orbits the Sun. This causes the seasons— winter, spring, summer, and fall— to occur in the same pattern every year.

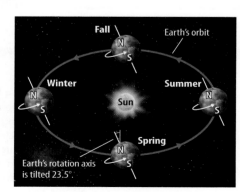

Patterns in Engineering

Engineers study patterns for many reasons, including to understand the physical properties of materials or to optimize the performance of their designs. Have you ever seen bricks with a pattern of holes through them? Clay bricks used in construction are fired, or baked, to make them stronger. Ceramic engineers understand that a regular pattern of holes in a brick assures that the brick is evenly fired and will not easily break.

Maybe you have seen a bridge constructed with a repeating pattern of large, steel triangles. Civil engineers, who design roads and bridges, know that the triangle is one of the strongest shapes in geometry. Engineers often use patterns of triangles in the structure of bridges to make them withstand heavy traffic and high winds.

Patterns in Physical Science

Scientists use patterns to explain past events or predict future events. At one time, only a few chemical elements were known. Chemists arranged the information they knew about these elements in a pattern according to the elements' properties. Scientists predicted the atomic numbers and the properties of elements that had yet to be discovered. These predictions made the discovery of new elements easier because scientists knew what properties to look for.

Look around. There are patterns everywhere—in art and nature, in the motion of the universe, in vehicles traveling on the roads, and in the processes of plant and animal growth. Analyzing patterns helps to understand the universe.

Patterns in Graphs

Scientists often graph their data to help identify patterns. For example, scientists might plot data from experiments on parachute nylon in graphs, such as the one below. Analyzing patterns on graphs then gives engineers information about how to design the strongest parachutes.

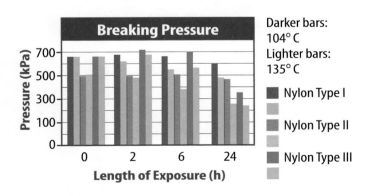

Inquiry MiniLab
15 minutes

How strong is your parachute?
Suppose you need to design a parachute. The graph to the left shows data for three types of parachute nylon. Each was tested to see how it weakens when exposed to different temperatures for different lengths of time. How would you use the patterns in the graph to design your parachute?

1. Write down the different experiments performed and how the variables changed in your Science Journal.

2. Write down all the patterns that you notice in the graph.

Analyze and Conclude

1. **Compare** Which nylon is weakest? What pattern helps you make this comparison?

2. **Identify** Which nylon is most affected by length of exposure to heat? What is its pattern on the graph?

3. **Select** Which nylon would you choose for your parachute? What pattern helped you make your decision?

Chapter 10

Foundations of Chemistry

THE BIG IDEA What is matter, and how does it change?

Why does it glow?

This siphonophore (si FAW nuh fawr) lives in the Arctic Ocean. Its tentacles have a very powerful sting. However, the most obvious characteristic of this organism is the way it glows.

- What might cause the siphonophore to glow?
- How do you think its glow helps the siphonophore survive?
- What changes happen in the matter that makes up the organism?

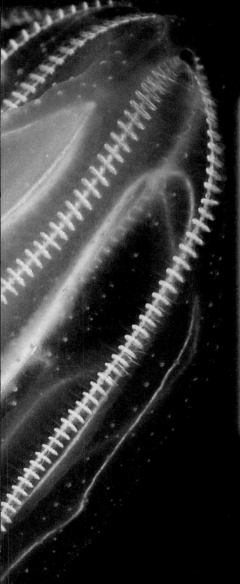

Get Ready to Read

What do you think?

Before you read, decide if you agree or disagree with each of these statements. As you read this chapter, see if you change your mind about any of the statements.

1 The atoms in all objects are the same.

2 You cannot always tell by an object's appearance whether it is made of more than one type of atom.

3 The weight of a material never changes, regardless of where it is.

4 Boiling is one method used to separate parts of a mixture.

5 Heating a material decreases the energy of its particles.

6 When you stir sugar into water, the sugar and water evenly mix.

7 When wood burns, new materials form.

8 Temperature can affect the rate at which chemical changes occur.

ConnectED Your one-stop online resource

connectED.mcgraw-hill.com

- ▣ Video
- ◀)) Audio
- ▤ Review
- ? Inquiry
- ⊕ WebQuest
- ✓ Assessment
- ((O Concepts in Motion
- g Multilingual eGlossary

Lesson 1

Reading Guide

Key Concepts 🔑
ESSENTIAL QUESTIONS

- What is a substance?
- How do atoms of different elements differ?
- How do mixtures differ from substances?
- How can you classify matter?

Vocabulary

matter p. 349
atom p. 349
substance p. 351
element p. 351
compound p. 352
mixture p. 353
heterogeneous mixture p. 353
homogeneous mixture p. 353
dissolve p. 353

g Multilingual eGlossary

▣ Video

- BrainPOP®
- Science Video
- What's Science Got to do With It?

Classifying Matter

Inquiry Making Green?

You probably have mixed paints together. Maybe you wanted green paint and had only yellow paint and blue paint. Perhaps you watched an artist mixing several tints get the color he or she needed. In all these instances, the final color came from mixing colors together and not from changing the color of a paint.

How do you classify matter?

An object made of paper bound together might be classified as a book. Pointed metal objects might be classified as nails or needles. How can you classify an item based on its description?

1 Read and complete a lab safety form.

2 Place the **objects** on a table. Discuss how you might separate the objects into groups with these characteristics:

  a. Every object is the same and has only one part.

  b. Every object is the same but is made of more than one part.

  c. Individual objects are different. Some have one part, and others have more than one part.

3 Identify the objects that meet the requirements for group *a*, and record them in your Science Journal. Repeat with groups *b* and *c*. Any object can be in more than one group.

Think About This

1. Does any object from the bag belong in all three of the groups (*a, b,* and *c*)? Explain.

2. What objects in your classroom would fit into group *b*?

3. 🔑 **Key Concept** What descriptions would you use to classify items around you?

Understanding Matter

Have you ever seen a rock like the one in **Figure 1?** Why are different parts of the rock different in color? Why might some parts of the rock feel harder than other parts? The parts of the rock look and feel different because they are made of different types of matter. **Matter** *is anything that has mass and takes up space.* If you look around, you will see many types of matter. If you are in a classroom, you might see things made of metal, wood, or plastic. If you go to a park, you might see trees, soil, or water in a pond. If you look up at the sky, you might see clouds and the Sun. All of these things are made of matter.

Everything you can see is matter. However, some things you cannot see also are matter. Air, for example, is matter because it has mass and takes up space. Sound and light are not matter. Forces and energy also are not matter. To decide whether something is matter, ask yourself if it has mass and takes up space.

An **atom** *is a small particle that is a building block of matter.* In this lesson, you will explore the parts of an atom and read how atoms can differ. You will also read how different arrangements of atoms make up the many types of matter.

WORD ORIGIN

matter
from Latin *materia*, meaning "material, stuff"

Figure 1 You can see different types of matter in this rock.

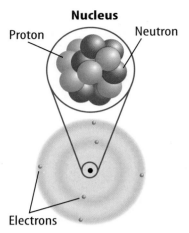

Nucleus

Proton

Neutron

Electrons

Figure 2 An atom has electrons moving in an area outside a nucleus. Protons and neutrons make up the nucleus.

 Review **Personal Tutor**

Atoms

To understand why there are so many types of matter, it helps if you first learn about the parts of an atom. Look at the diagram of an atom in **Figure 2.** At the center of an atom is a nucleus. Protons, which have a positive charge, and neutrons, which have a neutral charge, make up the nucleus. Negatively charged particles, or electrons, move quickly throughout an area around the nucleus called the electron cloud.

✓ **Reading Check** What are the parts of an atom?

Not all atoms have the same number of protons, neutrons, and electrons. Atoms that have different numbers of protons differ in their properties. You will read more about the differences in atoms on the next page.

An atom is almost too small to imagine. Think about how thin a human hair is. The diameter of a human hair is about a million times greater than the diameter of an atom. In addition, an atom is about 10,000 times wider than its nucleus! Even though atoms are so tiny, they determine the properties of the matter they compose.

Inquiry MiniLab  **20 minutes**

How can you model an atom?

How can you model an atom out of its three basic parts?

❶ Read and complete a lab safety form.

❷ Twist the ends of a piece of **florist wire** together to form a ring. Attach two **wires** across the ring to form an *X*.

❸ Use **double-sided tape** to join the **large pom-poms** (protons and neutrons), forming a nucleus. Hang the nucleus from the center of the *X* with **fishing line.**

❹ Use fishing line to suspend each **small pom-pom** (electron) from the ring so they surround the nucleus.

❺ Suspend your model as instructed by your teacher.

Analyze and Conclude

1. **Infer** Based on your model, what can you infer about the relative sizes of protons, neutrons, and electrons?

2. **Model** Why is it difficult to model the location of electrons?

3. 🔑 **Key Concept** Compare your atom with those of other groups. How do they differ?

Substances

You can see that atoms make up most of the matter on Earth. Atoms can combine and arrange in millions of different ways. In fact, these different combinations and arrangements of atoms are what makes up the various types of matter. There are two main classifications of matter—substances and mixtures.

A **substance** *is matter with a composition that is always the same.* This means that a given substance is always make up of the same combination(s) of atoms. Aluminum, oxygen, water, and sugar are examples of substances. Any sample of aluminum is always made up of the same type of atoms, just as samples of oxygen, sugar, and water each are always made of the same combinations of atoms. To gain a better understanding of what makes up substances, let's take a look at the two types of substances—elements and compounds.

Key Concept Check What is a substance?

Elements

Look at the periodic table of elements on the inside back cover of this book. The substances oxygen and aluminum are on the table. They are both elements. *An* **element** *is a substance that consists of just one type of atom.* Because there are about 115 known elements, there are about 115 different types of atoms. Each type of atom contains a different number of protons in its nucleus. For example, each aluminum atom has 13 protons in its nucleus. The number of protons in an atom is the atomic number of the element. Therefore, the atomic number of aluminum is 13, as shown in **Figure 3.**

The atoms of most elements exist as individual atoms. For example, a roll of pure aluminum foil consists of trillions of individual aluminum atoms. However, the atoms of some elements usually exist in groups. For example, the oxygen atoms in air exist in pairs. Whether the atoms of an element exist individually or in groups, each element contains only one type of atom. Therefore, its composition is always the same.

Key Concept Check How do atoms of different elements differ?

Figure 3 Each element on the periodic table consists of just one type of atom.

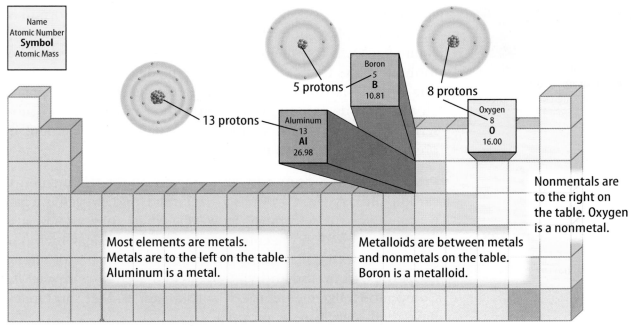

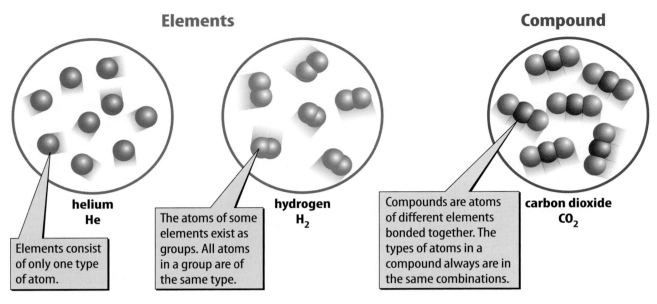

Elements

helium
He

The atoms of some elements exist as groups. All atoms in a group are of the same type.

hydrogen
H₂

Elements consist of only one type of atom.

Compound

Compounds are atoms of different elements bonded together. The types of atoms in a compound always are in the same combinations.

carbon dioxide
CO₂

▲ **Figure 4** 🔑 If a substance contains only one type of atom, it is an element. If it contains more than one type of atom, it is a compound.

Figure 5 Carbon dioxide is a compound composed of carbon and oxygen atoms. ▼

Review Personal Tutor

This subscript means there are two oxygen atoms bonded to one carbon atom.

ACADEMIC VOCABULARY

unique
(adjective) having nothing else like it

Compounds

Water is a substance, but it is not an element. It is a compound. *A **compound** is a type of substance containing atoms of two or more different elements chemically bonded together.* As shown in **Figure 4,** carbon dioxide (CO_2) is also a compound. It consists of atoms of two different elements, carbon (C) and oxygen (O), bonded together. Carbon dioxide is a substance because the C and the O atoms are always combined in the same way.

Chemical Formulas The combination of symbols and numbers that represents a compound is called a chemical formula. Chemical formulas show the different atoms that make up a compound, using their element symbols. Chemical formulas also help explain how the atoms combine. As illustrated in **Figure 5,** CO_2 is the chemical formula for carbon dioxide. The formula shows that carbon dioxide is made of C and O atoms. The small *2* is called a subscript. It means that two oxygen atoms and one carbon atom form carbon dioxide. If no subscript is written after a symbol, one atom of that element is present in the chemical formula.

Properties of Compounds Think again about the elements carbon and oxygen. Carbon is a black solid, and oxygen is a gas that enables fuels to burn. However, when they chemically combine, they form the compound carbon dioxide, which is a gas used to extinguish fires. A compound often has different properties from the individual elements that compose it. Compounds, like elements, are substances, and all substances have their own **unique** properties.

Mixtures

Another classification of matter is mixtures. *A* **mixture** *is matter that can vary in composition.* Mixtures are combinations of two or more substances that are physically blended together. The amounts of the substances can vary in different parts of a mixture and from mixture to mixture. Think about sand mixed with water at the beach. The sand and the water do not bond together. Instead, they form a mixture. The substances in a mixture do not combine chemically. Therefore, they can be separated by physical methods, such as filtering.

Heterogeneous Mixtures

Mixtures can differ depending on how well the substances that make them up are mixed. Sand and water at the beach form a mixture, but the sand is not evenly mixed throughout the water. Therefore, sand and water form a heterogeneous mixture. *A* **heterogeneous mixture** *is a type of mixture in which the individual substances are not evenly mixed.* Because the substances in a heterogeneous mixture are not evenly mixed, two samples of the same mixture can have different amounts of the substances, as shown in **Figure 6.** For example, if you fill two buckets with sand and water at the beach, one bucket might have more sand in it than the other.

Homogeneous Mixtures

Unlike a mixture of water and sand, the substances in mixtures such as apple juice, air, or salt water are evenly mixed. *A* **homogeneous mixture** *is a type of mixture in which the individual substances are evenly mixed.* In a homogeneous mixture, the particles of individual substances are so small and well-mixed that they are not visible, even with most high-powered microscopes.

A homogeneous mixture also is known as a solution. In a solution, the substance present in the largest amount is called the solvent. All other substances in a solution are called solutes. The solutes dissolve in the solvent. *To* **dissolve** *means to form a solution by mixing evenly.* Because the substances in a solution, or homogeneous mixture, are evenly mixed, two samples from a solution will have the same amounts of each substance. For example, imagine pouring two glasses of apple juice from the same container. Each glass will contain the same substances (water, sugar, and other substances) in the same amounts. However, because apple juice is a mixture, the amounts of the substances from one container of apple juice to another might vary.

 Key Concept Check How do mixtures differ from substances?

Figure 6 Types of mixtures differ in how evenly their substances are mixed.

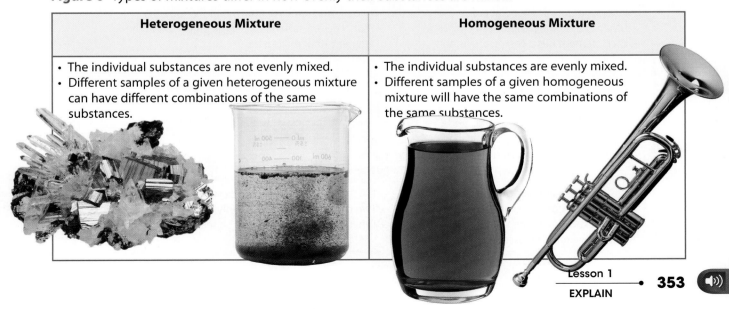

Heterogeneous Mixture	Homogeneous Mixture
• The individual substances are not evenly mixed. • Different samples of a given heterogeneous mixture can have different combinations of the same substances.	• The individual substances are evenly mixed. • Different samples of a given homogeneous mixture will have the same combinations of the same substances.

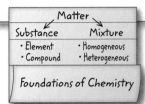

Compounds v. Solutions

If you have a glass of pure water and a glass of salt water, can you tell which is which just by looking at them? You cannot. Both the compound (water) and the solution (salt water) appear identical. How do compounds and solutions differ?

Because water is a compound, its composition does not vary. Pure water is always made up of the same atoms in the same combinations. Therefore, a chemical formula can be used to describe the atoms that make up water (H_2O). Salt water is a homogeneous mixture, or solution. The solute (NaCl) and the solvent (H_2O) are evenly mixed but are not bonded together. Adding more salt or more water only changes the relative amounts of the substances. In other words, the composition varies. Because composition can vary in a mixture, a chemical formula cannot be used to describe mixtures.

Summarizing Matter

You have read in this lesson about classifying matter by the arrangement of its atoms. **Figure 7** is a summary of this classification system.

 Key Concept Check How can you classify matter?

Figure 7 Scientists classify matter according to the arrangement of the atoms that make up the matter.

Classifying Matter 🔑

Matter
- Anything that has mass and takes up space
- Matter on Earth is made up of atoms.
- Two classifications of matter: substances and mixtures

Substances
- Matter with a composition that is always the same
- Two types of substances: elements and compounds

Element
- Consists of just one type of atom
- Organized on the periodic table
- Each element has a chemical symbol.

Compound
- Two or more types of atoms bonded together
- Properties are different from the properties of the elements that make it up
- Each compound has a chemical formula.

Substances physically combine to form mixtures.

Mixtures can be separated into substances by physical methods.

Mixtures
- Matter that can vary in composition
- Substances are not bonded together.
- Two types of mixtures: heterogeneous and homogeneous

Heterogeneous Mixture
- Two or more substances unevenly mixed
- Different substances are visible by an unaided eye or a microscope.

Homogeneous Mixture—Solution
- Two or more substances evenly mixed
- Different substances cannot be seen even by a microscope.

Lesson 1 Review

Visual Summary

A substance has the same composition throughout. A substance is either an element or a compound.

An atom is the smallest part of an element that has its properties. Atoms contain protons, neutrons, and electrons.

The substances in a mixture are not chemically combined. Mixtures can be either heterogeneous or homogeneous.

FOLDABLES

Use your lesson Foldable to review the lesson. Save your Foldable for the project at the end of the chapter.

What do you think NOW?

You first read the statements below at the beginning of the chapter.

1. The atoms in all objects are the same.

2. You cannot always tell by an object's appearance whether it is made of more than one type of atom.

Did you change your mind about whether you agree or disagree with the statements? Rewrite any false statements to make them true.

Use Vocabulary

1 Substances and mixtures are two types of _____.

2 **Use the term** *atom* in a complete sentence.

3 **Define** *dissolve* in your own words.

Understand Key Concepts

4 **Explain** why aluminum is a substance.

5 The number of _____ always differs in atoms of different elements.
   A. electrons   **C.** neutrons
   B. protons   **D.** nuclei

6 **Distinguish** between a heterogeneous mixture and a homogeneous mixture.

7 **Classify** Which term describes matter that is a substance made of different kinds of atoms bonded together?

Interpret Graphics

8 **Describe** what each letter and number means in the chemical formula below.

$$C_6H_{12}O_6$$

9 **Organize Information** Copy and fill in the graphic organizer below to classify matter by the arrangement of its atoms.

Type of Matter	Description

Critical Thinking

10 **Reorder** the elements aluminum, oxygen, fluorine, calcium, and hydrogen from the least to the greatest number of protons. Use the periodic table if needed.

11 **Evaluate** this statement: Substances are made of two or more types of elements.

U.S. Mint

How Coins are Made

In 1793, the U.S. Mint produced more than 11,000 copper pennies and put them into circulation. Soon after, gold and silver coins were introduced as well. Early pennies were made of 95 percent copper and 5 percent zinc. Today's penny contains much more zinc than copper and is much less expensive to produce. Quarters, dimes, and nickels, once made of silver, are now made of copper-nickel alloy.

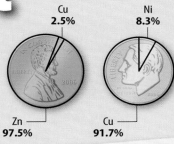

Cu 2.5% Ni 8.3%
Zn 97.5% Cu 91.7%

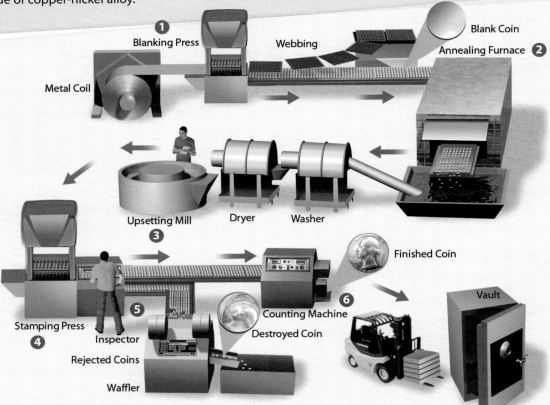

Metal Coil
Blanking Press ①
Webbing
Blank Coin
Annealing Furnace ②
Upsetting Mill ③
Dryer
Washer
Stamping Press ④
Inspector
Rejected Coins
Waffler ⑤
Counting Machine ⑥
Destroyed Coin
Finished Coin
Vault

How Coins Are Made

① **Blanking** For nickels, dimes, quarters, half-dollars, and coin dollars, a strip of 13-inch-wide metal is fed through a blanking press, which punches out round discs called blanks. The leftover webbing strip is saved for recycling. Ready-made blanks are purchased for making the penny.

② **Annealing, Washing, Drying** Blanks are softened in an annealing furnace, which makes the metal less brittle. The blanks are then run through a washer and a dryer.

③ **Upsetting** Usable blanks are put through an upsetting mill, which creates a rim around the edges of each blank.

④ **Striking** The blanks then go to the stamping press, where they are imprinted with designs and inscriptions.

⑤ **Inspection** Once blanks leave the stamping press, inspectors check a few coins from each batch. Coins that are defective go to the waffler in preparation for recycling.

⑥ **Counting and Bagging** A machine counts the finished coins then drops them into large bags that are sealed shut. The coins are then taken to storage before being shipped to Federal Reserve Banks and then to your local bank.

It's Your Turn

COMPARE Collect a variety of coins that includes both older and current coins. Observe and compare their properties. Using the dates of the coins' production, utilize library or Internet sources to research the composition of metals used.

Lesson 2

Reading Guide

Key Concepts 🔑
ESSENTIAL QUESTIONS

- What are some physical properties of matter?
- How are physical properties used to separate mixtures?

Vocabulary
physical property p. 358
mass p. 360
density p. 361
solubility p. 362

g Multilingual eGlossary

▢ **Video** Science Video

Physical Properties

Inquiry Panning by Properties?

The man lowers his pan into the waters of an Alaskan river and scoops up a mixture of water, sediment, and hopefully gold. As he moves the pan in a circle, water sloshes out of it. If he is careful, gold will remain in the pan after the water and sediment are gone. What properties of water, sediment, and gold enable this man to separate this mixture?

Launch Lab

Can you follow the clues?

Clues are bits of information that help you solve a mystery. In this activity, you will use clues to help identify an object in the classroom.

1. Read and complete a lab safety form.

2. Select one **object** in the room. Write a different clue about the object on each of five **index cards.** Clues might include one or two words that describe the object's color, size, texture, shape, or any property you can observe with your senses.

3. Stack your cards face down. Have your partner turn over one card and try to identify the object. Respond either "yes" or "no."

4. Continue turning over cards until your partner identifies your object or runs out of cards. Repeat for your partner's object.

Think About This

1. What kind of clues are the most helpful in identifying an object?

2. How would your clues change if you were describing a substance, such as iron or water, rather than an object?

3. 🔑 **Key Concept** How do you think you use similar clues in your daily life?

REVIEW VOCABULARY

property
a characteristic used to describe something

Physical Properties

As you read in Lesson 1, the arrangement of atoms determines whether matter is a substance or a mixture. The arrangement of atoms also determines the **properties** of different types of matter. Each element and compound has a unique set of properties. When substances mix together and form mixtures, the properties of the substances that make up the mixture are still present.

You can observe some properties of matter, and other properties can be measured. For example, you can see that gold is shiny, and you can find the mass of a sample of iron. Think about how you might describe the different substances and mixtures in the photo on the previous page. Could you describe some of the matter in the photo as a solid or a liquid? Why do the water and the rocks leave the pan before the gold does? Could you describe the mass of the various items in the photo? Each of these questions asks about the physical properties of matter. *A **physical property** is a characteristic of matter that you can observe or measure without changing the identity of the matter.* There are many types of physical properties, and you will read about some of them in this lesson.

States of Matter

How do aluminum, water, and air differ? Recall that aluminum is an element, water is a compound, and air is a mixture. How else do these three types of matter differ? At room temperature, aluminum is a solid, water is a liquid, and air is a gas. Solids, liquids, and gases are called states of matter. The state of matter is a physical property of matter. Substances and mixtures can be solids, liquids, or gases. For example, water in the ocean is a liquid, but water in an iceberg is a solid. In addition, water vapor in the air above the ocean is a gas.

Did you know that the particles, or atoms and groups of atoms, that make up all matter are constantly moving and are attracted to each other? Look at your pencil. It is made up of trillions of moving particles. Every solid, liquid, and gas around you is made up of moving particles that attract one another. What makes some matter a solid and other matter a liquid or a gas? It depends on how close the particles in the matter are to one another and how fast they move, as shown in **Figure 8.**

✓ **Reading Check** How do solids, liquids, and gases differ?

FOLDABLES

Make a vertical two-tab book. Record what you learn about different states of matter under the tabs.

Solid

Liquid

Gas

Figure 8 The three common states of matter on Earth are solid, liquid, and gas.

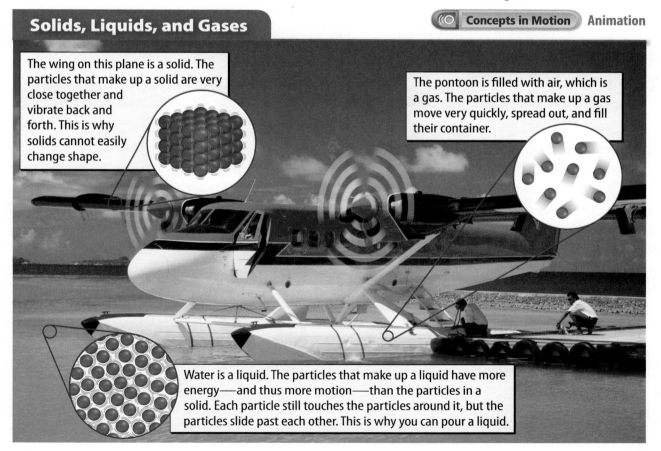

Solids, Liquids, and Gases

《⊙》 Concepts in Motion　Animation

The wing on this plane is a solid. The particles that make up a solid are very close together and vibrate back and forth. This is why solids cannot easily change shape.

The pontoon is filled with air, which is a gas. The particles that make up a gas move very quickly, spread out, and fill their container.

Water is a liquid. The particles that make up a liquid have more energy—and thus more motion—than the particles in a solid. Each particle still touches the particles around it, but the particles slide past each other. This is why you can pour a liquid.

✓ **Visual Check** Which state of matter flows, keeps the same volume, and takes the shape of its container?

Figure 9 The larger dumbbells have greater mass than the smaller dumbbells because they contain more matter.

Size-Dependent Properties

State is only one of many physical properties that you can use to describe matter. Some physical properties, such as mass and volume, depend on the size or amount of matter. Measurements of these properties vary depending on how much matter is in a sample.

Mass Imagine holding a small dumbbell in one hand and a larger one in your other hand. What do you notice? The larger dumbbell seems heavier. The larger dumbbell has more mass than the smaller one. **Mass** *is the amount of matter in an object.* Both small dumbbells shown in **Figure 9** have the same mass because they both contain the same amount of matter. Mass is a size-dependent property of a given substance because its value depends on the size of a sample.

Mass sometimes is confused with weight, but they are not the same. Mass is an amount of matter in something. Weight is the pull of gravity on that matter. Weight changes with location, but mass does not. Suppose one of the dumbbells in the figure was on the Moon. The dumbbell would have the same mass on the Moon that it has on Earth. However, the Moon's gravity is much less than Earth's gravity, so the weight of the dumbbell would be less on the Moon.

Inquiry MiniLab
20 minutes

Can the weight of an object change?

When people go on a diet, both their mass and weight might change. Can the weight of an object change without changing its mass? Let's find out.

1 Read and complete a lab safety form.

2 Use a **balance** to find the mass of five **metal washers.** Record the mass in grams in your Science Journal.

3 Hang the washers from the hook on a **spring scale.** Record the weight in newtons.

4 Lower just the washers into a **500-mL beaker** containing approximately 300 mL water. Record the weight in newtons.

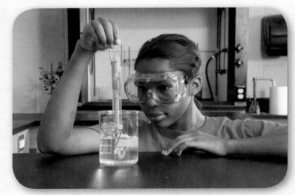

Analyze and Conclude

1. **Draw Conclusions** Did the weight of the washers change during the experiment? How do you know?

2. **Predict** In what other ways might you change the weight of the washers?

3. 🔑 **Key Concept** What factors affect the weight of an object, but not its mass?

Volume Another physical property that depends on the size or the amount of a substance is volume. A unit often used to measure volume is the milliliter (mL). Volume is the amount of space something takes up. Suppose a full bottle of water contains 400 mL of water. If you pour exactly half of the water out, the bottle contains half of the original volume, or 200 mL, of water.

 Reading Check What is a common unit for volume?

Size-Independent Properties

Unlike mass, weight, and volume, some physical properties of a substance do not depend on the amount of matter present. These properties are the same for both small samples and large samples. They are called size-independent properties. Examples of size-independent properties are melting point, boiling point, density, electrical conductivity, and solubility.

Melting Point and Boiling Point The temperature at which a substance changes from a solid to a liquid is its melting point. The temperature at which a substance changes from a liquid to a gas is its boiling point. Different substances have different boiling points and melting points. The boiling point for water is 100°C at sea level. Notice in **Figure 10** that this temperature does not depend on how much water is in the container.

Density Imagine holding a bowling ball in one hand and a foam ball of the same size in the other. The bowling ball seems heavier because the density of the material that makes up the bowling ball is greater than the density of foam. **Density** *is the mass per unit volume of a substance.* Like melting point and boiling point, density is a size-independent property.

Math Skills

Use Ratios

When you compare two numbers by division, you are using a ratio. Density can be written as a ratio of mass and volume. What is the density of a substance if a 5-mL sample has a mass of 25 g?

1. Set up a ratio.

$$\frac{mass}{volume} = \frac{25\ g}{5\ mL}$$

2. Divide the numerator by the denominator to get the mass (in g) of 1 mL.

$$\frac{25\ g}{5\ mL} = \frac{5\ g}{1\ mL}$$

3. The density is 5 g/mL.

Practice

A sample of wood has a mass of 12 g and a volume of 16 mL. What is the density of the wood?

 Review

- **Math Practice**
- **Personal Tutor**

WORD ORIGIN · · · · · · · · · · ·

density
from Latin *densus*, means "compact"; and Greek *dasys*, means "thick"

Figure 10 The boiling point of water is 100°C at sea level. The boiling point does not change for different volumes of water.

 Review **Personal Tutor**

Conductivity Another property that is independent of the sample size is conductivity. Electrical conductivity is the ability of matter to conduct, or carry along, an electric current. Copper often is used for electrical wiring because it has high electrical conductivity. Thermal conductivity is the ability of a material to conduct thermal energy. Metals tend to have high electrical and thermal conductivity. Stainless steel, for example, often is used to make cooking pots because of its high thermal conductivity. However, the handles on the pan probably are made out of wood, plastic, or some other substance that has low thermal conductivity.

 Reading Check What are two types of conductivity?

Solubility Have you ever made lemonade by stirring a powdered drink mix into water? As you stir, the powder mixes evenly in the water. In other words, the powder dissolves in the water.

What do you think would happen if you tried to dissolve sand in water? No matter how much you stir, the sand does not dissolve. **Solubility** *is the ability of one substance to dissolve in another.* The powdered drink mix is soluble in water, but sand is not. **Table 1** explains how physical properties such as conductivity and solubility can be used to identify objects and separate mixtures.

Key Concept Check What are five different physical properties of matter?

Table 1 This table contains the descriptions of several physical properties. It also shows examples of how physical properties can be used to separate mixtures.

Visual Check How might you separate a mixture of iron filings and salt?

Concepts in Motion Interactive Table

Table 1 **Physical Properties of Matter**			
	Property		
	Mass	**Conductivity**	**Volume**
Description of property	The amount of matter in an object	The ability of matter to conduct, or carry along, electricity or heat	The amount of space something occupies
Size-dependent or size independent	Size-dependent	Size-independent	Size-dependent
How the property is used to separate a mixture (example)	Mass typically is not used to separate a mixture.	Conductivity typically is not used to separate a mixture.	Volume could be used to separate mixtures whose parts can be separated by filtration.

Separating Mixtures

In Lesson 1, you read about different types of mixtures. Recall that the substances that make up mixtures are not held together by chemical **bonds.** When substances form a mixture, the properties of the individual substances do not change. One way that a mixture and a compound differ is that the parts of a mixture often can be separated by physical properties. For example, when salt and water form a solution, the salt and the water do not lose any of their individual properties. Therefore, you can separate the salt from the water by using differences in their physical properties. Water has a lower boiling point than salt. If you boil salt water, the water will boil away, and the salt will be left behind. Other physical properties that can be used to separate different mixtures are described in **Table 1.**

Physical properties cannot be used to separate a compound into the elements it contains. The atoms that make up a compound are bonded together and cannot be separated by physical means. For example, you cannot separate the hydrogen atoms from the oxygen atoms in water by boiling water.

 Key Concept Check How are physical properties used to separate mixtures?

SCIENCE USE V. COMMON USE

bond

Science Use a force between atoms or groups of atoms

Common Use a monetary certificate issued by a government or a business that earns interest

Property

Boiling/Melting Points	State of matter	Density	Solubility	Magnetism
The temperature at which a material changes state	Whether something is a solid, a liquid, or a gas	The amount of mass per unit of volume	The ability of one substance to dissolve in another	Attractive force for some metals, especially iron
Size-independent	Size-independent	Size-independent	Size-independent	Size-independent
Each part of a mixture will boil or melt at a different temperature.	A liquid can be poured off a solid.	Objects with greater density sink in objects with less density.	Dissolve a soluble material to separate it from a material with less solubility.	Attract iron from a mixture of materials.

Lesson 2 Review

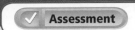

Visual Summary

A physical property is a characteristic of matter that can be observed or measured without changing the identity of the matter.

Examples of physical properties include mass, density, volume, melting point, boiling point, state of matter, and solubility.

Many physical properties can be used to separate the components of a mixture.

FOLDABLES

Use your lesson Foldable to review the lesson. Save your Foldable for the project at the end of the chapter.

What do you think NOW?

You first read the statements below at the beginning of the chapter.

3. The weight of a material never changes, regardless of where it is.

4. Boiling is one method used to separate parts of a mixture.

Did you change your mind about whether you agree or disagree with the statements? Rewrite any false statements to make them true.

Use Vocabulary

1. **Distinguish** between mass and weight.

2. **Use the term** *solubility* in a sentence.

3. An object's _____ is the amount of mass per a certain unit of volume.

Understand Key Concepts

4. **Explain** how to separate a mixture of sand and pebbles.

5. Which physical property is NOT commonly used to separate mixtures?
   - **A.** magnetism
   - **B.** conductivity
   - **C.** density
   - **D.** solubility

6. **Analyze** Name two size-dependent properties and two size-independent properties of an iron nail.

Interpret Graphics

7. **Sequence** Draw a graphic organizer like the one below to show the steps in separating a mixture of sand, iron filings, and salt.

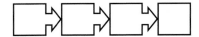

Critical Thinking

8. **Examine** the diagram below.

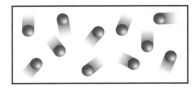

How can you identify the state of matter represented by the diagram?

Math Skills Review
— Math Practice —

9. A piece of copper has a volume of 100.0 cm^3. If the mass of the copper is 890 g, what is the density of copper?

How can following a procedure help you solve a crime?

Materials

Plastic sealable bag

triple-beam balance

50-mL graduated cylinder

paper towels

Also needed:

Crime Scene Objects

Safety

Imagine that you are investigating a crime scene. You find several pieces of metal and broken pieces of plastic that look as if they came from a car's tail light. You also have similar objects collected from the suspect. How can you figure out if they are parts of the same objects?

Learn It

To be sure you do the same tests on each object, it is helpful to **follow a procedure.** A procedure tells you how to use the materials and what steps to take.

Try It

1. Read and complete a lab safety form.

2. Copy the table below into your Science Journal.

3. Use the balance to find the mass of an object from the crime scene. Record the mass in your table.

4. Place about 25 mL of water in a graduated cylinder. Read and record the exact volume. Call this volume V_1.

5. Carefully tilt the cylinder, and allow one of the objects to slide into the water. Read and record the volume. Call this volume V_2.

6. Repeat steps 3–5 for each of the other objects.

Apply It

7. Complete the table by calculating the volume and the density of each object.

8. What conclusions can you draw about the objects collected from the crime scene and those collected from the suspect?

9. 🔑 **Key Concept** How could you use this procedure to help identify and compare various objects?

Object	Mass (M) (g)	V_1 (mL)	V_2 (mL)	Volume of Object (V) ($V_2 - V_1$) (mL)	Density of Object M/V (g/mL)
1					
2					
3					
4					
5					
6					

Lesson 3

Reading Guide

Key Concepts
ESSENTIAL QUESTIONS

- How can a change in energy affect the state of matter?
- What happens when something dissolves?
- What is meant by conservation of mass?

Vocabulary
physical change p. 367

g **Multilingual eGlossary**

Physical Changes

Inquiry Change by Chipping?

This artist is changing a piece of wood into an instrument that will make beautiful music. He planned and chipped, measured and shaped. Chips of wood flew, and rough edges became smooth. Although the wood changed shape, it remained wood. Its identity did not change, just its form.

Where did it go?

When you dissolve sugar in water, where does the sugar go? One way to find out is to measure the mass of the water and the sugar before and after mixing.

1. Read and complete a lab safety form.

2. Add **sugar** to a **small paper cup** until the cup is approximately half full. Bend the cup's opening, and pour the sugar into a **balloon.**

3. With the balloon hanging over the side, stretch the neck of the balloon over a **flask** half full of **water.**

4. Use a **balance** to find the mass of the flask-and-balloon assembly. Record the mass in your Science Journal.

5. Lift the end of the balloon, and empty the sugar into the flask. Swirl until the sugar dissolves. Measure and record the mass of the flask-and-balloon assembly again.

Think About This

1. Is the sugar still present after it dissolves? How do you know?

2. 🔑 **Key Concept** Based on your observations, what do you think happens to the mass of objects when they dissolve? Explain.

Physical Changes

How would you describe water? If you think about water in a stream, you might say that it is a cool liquid. If you think about water as ice, you might describe it as a cold solid. How would you describe the change from ice to water? As ice melts, some of its properties change, such as the state of matter, the shape, and the temperature, but it is still water. In Lesson 2, you read that substances and mixtures can be solids, liquids, or gases. In addition, substances and mixtures can change from one state to another. A **physical change** is *a change in size, shape, form, or state of matter in which the matter's identity stays the same.* During a physical change, the matter does not become something different even though physical properties change.

Change in Shape and Size

Think about changes in the shapes and the sizes of substances and mixtures you experience each day. When you chew food, you are breaking it into smaller pieces. This change in size helps make food easier to digest. When you pour juice from a bottle into a glass, you are changing the shape of the juice. If you fold clothes to fit them into a drawer, you are changing their shapes. Changes in shape and size are physical changes. The identity of the matter has not changed.

WORD ORIGIN

physical
from Greek *physika*, means "natural things"

change
from Latin *cambire*, means "to exchange"

FOLDABLES®

Make a vertical two-tab book. Label the tabs as illustrated. Record specific examples illustrating how adding or releasing thermal energy results in physical change.

Increasing Thermal Energy

Decreasing Thermal Energy

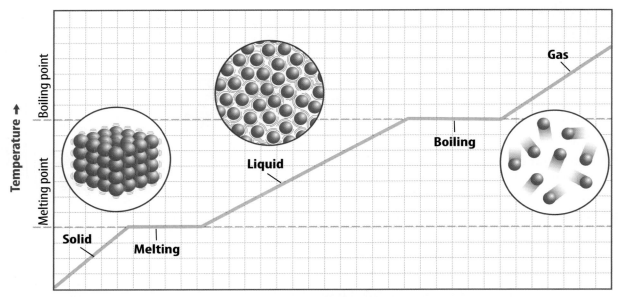

Adding thermal energy →

▲ **Figure 11** 🔑 As thermal energy is added to a material, temperature increases when the state of the material is not changing. Temperature stays the same during a change of state.

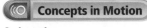

Concepts in Motion

Animation

Figure 12 Solid iodine undergoes sublimation. It changes from a solid to a gas without becoming a liquid. ▼

Change in State of Matter

Why does ice melt in your hand? Or, why does water turn to ice in the freezer? Matter, such as water, can change state. Recall from Lesson 2 how the particles in a solid, a liquid, and a gas behave. To change the state of matter, the movement of the particles has to change. In order to change the movement of particles, thermal energy must be either added or removed.

Adding Thermal Energy When thermal energy is added to a solid, the particles in the solid move faster and faster, and the temperature increases. As the particles move faster, they are more likely to overcome the attractive forces that hold them tightly together. When the particles are moving too fast for attractive forces to hold them tightly together, the solid reaches its melting point. The melting point is the temperature at which a solid changes to a liquid.

After all the solid has melted, adding more thermal energy causes the particles to move even faster. The temperature of the liquid increases. When the particles are moving so fast that attractive forces cannot hold them close together, the liquid is at its boiling point. The boiling point is the temperature at which a liquid changes into a gas and the particles spread out. **Figure 11** shows how temperature and change of state relate to each other when thermal energy is added to a material.

Some solids change directly to a gas without first becoming a liquid. This process is called sublimation. An example of sublimation is shown in **Figure 12.** You saw another example of sublimation in **Figure 5** in Lesson 1.

Removing Thermal Energy When thermal energy is removed from a gas, such as water vapor, particles in the gas move more slowly and the temperature decreases. Condensation occurs when the particles are moving slowly enough for attractive forces to pull the particles close together. Recall that condensation is the process that occurs when a gas becomes a liquid.

After the gas has completely changed to a liquid, removing more thermal energy from the liquid causes particles to move even more slowly. As the motion between the particles slows, the temperature decreases. Freezing occurs when the particles are moving so slowly that attractive forces between the particles hold them tightly together. Now the particles only can vibrate in place. Recall that freezing is the process that occurs when a liquid becomes a solid.

Freezing and melting are reverse processes, and they occur at the same temperature. The same is true of boiling and condensation. Another change of state is deposition. Deposition is the change from a gas directly to a solid, as shown in **Figure 13.** It is the process that is the opposite of sublimation.

 Key Concept Check How can removing thermal energy affect the state of matter?

 **MiniLab**

30 minutes

Can you make ice without a freezer?

What happens when you keep removing energy from a substance?

1. Read and complete a lab safety form.
2. Draw the data table below in your Science Journal. Half fill a **large test tube** with **distilled water.** Use a **thermometer** to measure the temperature of the water, then record it.
3. Place the test tube into a **large foam cup** containing **ice** and **salt.** Use a **stirring rod** to slowly stir the water in the tube.
4. Record the temperature of the water every minute until the water freezes. Continue to record the temperature each minute until it drops to several degrees below 0°C.

Time (min)	0	1	2	3	4	5	6	7	8
Temperature (C)									

Analyze and Conclude

1. **Organize Data** Graph the data in your table. Label time on the *x*-axis and temperature on the *y*-axis.
2. **Interpret Data** According to your data, what is the freezing point of water?
3. **Key Concept** What caused the water to freeze?

Figure 13 When enough thermal energy is removed, one of several processes occurs.

Freezing

Condensation

Deposition

▲ **Figure 14** Salt dissolves when it is added to the water in this aquarium.

Dissolving

Have you ever owned a saltwater aquarium, such as the one shown in **Figure 14?** If you have, you probably had to add certain salts to the water before you added the fish. Can you see the salt in the water? As you added the salt to the water, it gradually disappeared. It was still there, but it dissolved, or mixed evenly, in the water. Because the identities of the substances—water and salt—are not changed, dissolving is a physical change.

Like many physical changes, dissolving is usually easy to reverse. If you boil the salt water, the liquid water will change to water vapor, leaving the salt behind. You once again can see the salt because the particles that make up the substances do not change identity during a physical change.

Key Concept Check What happens when something dissolves?

Conservation of Mass

During a physical change, the physical properties of matter change. The particles in matter that are present before a physical change are the same as those present after the physical change. Because the particles are the same both before and after a physical change, the total mass before and after the change is also the same, as shown in **Figure 15.** This is known as the conservation of mass. You will read in Lesson 4 that mass also is conserved during another type of change—a chemical change.

Figure 15 Mass is conserved during a physical change. ▼

Key Concept Check What is meant by conservation of mass?

Conservation of Mass

Visual Check If a sample of water has a mass of 200 g and the final solution has a mass of 230 g, how much solute dissolved in the water?

Lesson 3 Review

Visual Summary

During a physical change, matter can change form, shape, size, or state, but the identity of the matter does not change.

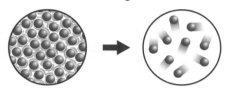

 Matter either changes temperature or changes state when enough thermal energy is added or removed.

 Mass is conserved during physical changes, which means that mass is the same before and after the changes occur.

FOLDABLES

Use your lesson Foldable to review the lesson. Save your Foldable for the project at the end of the chapter.

What do you think NOW?

You first read the statements below at the beginning of the chapter.

5. Heating a material decreases the energy of its particles.

6. When you stir sugar into water, the sugar and water evenly mix.

Did you change your mind about whether you agree or disagree with the statements? Rewrite any false statements to make them true.

Use Vocabulary

1 **Use the term** *physical change* in a sentence.

Understand Key Concepts

2 **Describe** how a change in energy can change ice into liquid water.

3 Which never changes during a physical change?
   A. state of matter   C. total mass
   B. temperature   D. volume

4 **Relate** What happens when something dissolves?

Interpret Graphics

5 **Examine** the graph below of temperature over time as a substance changes from solid to liquid to gas. Explain why the graph has horizontal lines.

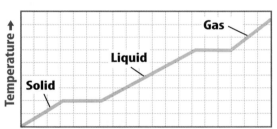

6 **Take Notes** Copy the graphic organizer below. For each heading, summarize the main idea described in the lesson.

Heading	Main Idea
Physical Changes	
Change in State of Matter	
Conservation of Mass	

Critical Thinking

7 **Design** a demonstration that shows that temperature remains unchanged during a change of state.

How can known substances help you identify unknown substances?

While investigating a crime scene, you find several packets of white powder. Are they illegal drugs or just harmless packets of candy? Here's one way to find out.

Materials

plastic spoons

magnifying lens

stirring rod

Also needed: known substances (baking soda, ascorbic acid, sugar, cornstarch) test tubes, test tube rack, watch glass, dropper bottles containing water, iodine, vinegar, and red cabbage indicator

Safety

Learn It

A **control** is something that stays the same. If you determine how a known substance reacts with other substances, you can use it as a control. Unknown substances are **variables.** They might or might not react in the same way.

Try It

1. Read and complete a lab safety form.

2. Copy the data table below into your Science Journal.

3. Use a magnifying lens to observe the appearance of each known substance.

4. Test small samples of each known substance for their reaction with a drop or two of water, vinegar, and iodine solution.

5. Feel the texture of each substance.

6. Mix each substance with water, and add the red cabbage indicator.

7. After you complete your observations, ask your teacher for a mystery powder. Repeat steps 3–6 using the mystery powder. Use the data you collect to identify the powder.

Apply It

8. What test suggests that a substance might be cornstarch?

9. Why should you test the reactions of the substances with many different things?

10. 🔑 **Key Concept** How did you use the properties of the controls to identify your variable?

Substance	Appearance	Texture	Reaction to Water	Reaction to Iodine	Reaction to Vinegar	Red Cabbage Indicator
Baking soda						
Sugar						
Ascorbic acid						
Cornstarch						
Mystery powder						

Lesson 4

Reading Guide

Key Concepts
ESSENTIAL QUESTIONS

- What is a chemical property?
- What are some signs of chemical change?
- Why are chemical equations useful?
- What are some factors that affect the rate of chemical reactions?

Vocabulary

chemical property p. 374

chemical change p. 375

concentration p. 378

 Multilingual eGlossary

▣ Video

- BrainPOP®
- What's Science Got to do With It?

Chemical Properties and Changes

Inquiry A Burning Issue?

As this car burns, some materials change to ashes and gases. The metal might change form or state if the fire is hot enough, but it probably won't burn. Why do fabric, leather, and paint burn? Why do many metals not burn? The properties of matter determine how matter behaves when it undergoes a change.

What can colors tell you?

You mix red and blue paint to get purple paint. Iron changes color when it rusts. Are color changes physical changes?

1. Read and complete a lab safety form.

2. Divide a **paper towel** into thirds. Label one section *RCJ*, the second section *A*, and the third section *B*.

3. Dip one end of three **cotton swabs** into **red cabbage juice** (RCJ). Observe the color, and set the swabs on the paper towel, one in each of the three sections.

4. Add one drop of **substance A** to the swab in the *A* section. Observe any changes, and record observations in your Science Journal.

5. Repeat step 4 with **substance B** and the swab in the *B* section.

6. Observe **substances C** and **D** in their **test tubes.** Then pour C into D. Rock the tube gently to mix. Record your observations.

Think About This

1. What happened to the color of the red cabbage juice when substances A and B were added?

2. 🔑 **Key Concept** Which of the changes you observed do you think was a physical change? Explain your reasoning.

Chemical Properties

Recall that a physical property is a characteristic of matter that you can observe or measure without changing the identity of the matter. However, matter has other properties that can be observed only when the matter changes from one substance to another. *A* **chemical property** *is a characteristic of matter that can be observed as it changes to a different type of matter.* For example, what are some chemical properties of a piece of paper? Can you tell by just looking at it that it will burn easily? The only way to know that paper burns is to bring a flame near the paper and watch it burn. When paper burns, it changes into different types of matter. The ability of a substance to burn is a chemical property. The ability to rust is another chemical property.

Comparing Properties

You now have read about physical properties and chemical properties. All matter can be described using both types of properties. For example, a wood log is solid, rounded, heavy, and rough. These are physical properties that you can observe with your senses. The log also has mass, volume, and density, which are physical properties that can be measured. The ability of wood to burn is a chemical property. This property is obvious only when you burn the wood. It also will rot, another chemical property you can observe when the log decomposes, becoming other substances. When you describe matter, you consider both its physical and its chemical properties.

🔑 **Key Concept Check** What are some chemical properties of matter?

Chemical Changes

Recall that during a physical change, the identity of matter does not change. However, *a* **chemical change** *is a change in matter in which the substances that make up the matter change into other substances with new physical and chemical properties.* For example, when iron undergoes a chemical change with oxygen, rust forms. The substances that undergo a change no longer have the same properties because they no longer have the same identity.

 Reading Check What is the difference between a physical change and a chemical change?

Signs of Chemical Change

How do you know when a chemical change occurs? What signs show you that new types of matter form? As shown in **Figure 16,** signs of chemical changes include the formation of bubbles or a change in odor, color, or energy.

It is important to remember that these signs do not always mean a chemical change occurred. Think about what happens when you heat water on a stove. Bubbles form as the water boils. In this case, bubbles show that the water is changing state, which is a physical change. The evidence of chemical change shown in **Figure 16** means that a chemical change might have occurred. However, the only proof of chemical change is the formation of a new substance.

Key Concept Check What are signs of a chemical change?

FOLDABLES®

Use a sheet of paper to make a chart with four columns. Use the chart throughout this lesson to explain how the identity of matter changes during a chemical change.

Action/ Matter	Signs of Chemical Change	Explain the Chemical Reaction	What affects the reaction rate?

WORD ORIGIN

chemical
from Greek *chemeia,* means "cast together"

Some Signs of Chemical Change 🔑

Figure 16 Sometimes you can observe clues that a chemical change has occurred.

Bubbles   **Energy change**   **Odor change**   **Color change**

 Visual Check What signs show that a chemical change takes place when fireworks explode?

Can you spot the clues for chemical change?

What are some clues that let you know a chemical change might have taken place?

❶ Read and complete a lab safety form.

❷ Add about 25 mL of room-temperature water to a **self-sealing plastic bag.** Add two **dropperfuls** of **red cabbage juice.**

❸ Add one **measuring scoop** of **calcium chloride** to the bag. Seal the bag. Tilt the bag to mix the contents until the solid disappears. Feel the bottom of the bag. Record your observations in your Science Journal.

❹ Open the bag, and add one measuring scoop of **baking soda.** Quickly press the air from the bag and reseal it. Tilt the bag to mix the contents. Observe for several minutes. Record your observations.

Analyze and Conclude

1. **Observe** What changes did you observe?

2. **Infer** Which of the changes suggested that a new substance formed? Explain.

3. **Key Concept** Are changes in energy always a sign of a chemical change? Explain.

Explaining Chemical Reactions

You might wonder why chemical changes produce new substances. Recall that particles in matter are in constant motion. As particles move, they collide with each other. If the particles collide with enough force, the bonded atoms that make up the particles can break apart. These atoms then rearrange and bond with other atoms. When atoms bond together in new combinations, new substances form. This process is called a reaction. Chemical changes often are called chemical reactions.

✓ **Reading Check** What does it mean to say that atoms rearrange during a chemical change?

Using Chemical Formulas

A useful way to understand what happens during a chemical reaction is to write a chemical equation. A chemical equation shows the chemical formula of each substance in the reaction. The formulas to the left of the arrow represent the reactants. Reactants are the substances present before the reaction takes place. The formulas to the right of the arrow represent the products. Products are the new substances present after the reaction. The arrow indicates that a reaction has taken place.

🔑 **Key Concept Check** Why are chemical equations useful?

Figure 17 Chemical formulas and other symbols are parts of a chemical equation.

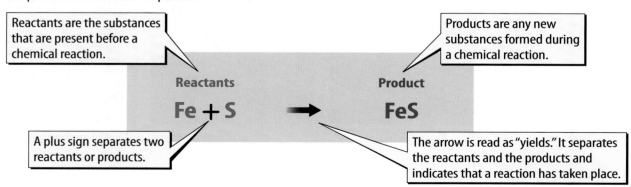

Reactants are the substances that are present before a chemical reaction.

Products are any new substances formed during a chemical reaction.

Reactants

Fe + S ➝ FeS

Product

A plus sign separates two reactants or products.

The arrow is read as "yields." It separates the reactants and the products and indicates that a reaction has taken place.

Balancing Chemical Equations

Look at the equation in **Figure 17**. Notice that there is one iron (Fe) atom on the reactants side and one iron atom on the product side. This is also true for the sulfur (S) atoms. Recall that during both physical and chemical changes, mass is conserved. This means that the total mass before and after a change must be equal. Therefore, in a chemical equation, the number of atoms of each element before a reaction must equal the number of atoms of each element after the reaction. This is called a balanced chemical equation, and it illustrates the conservation of mass. **Figure 18** explains how to write and balance a chemical equation.

When balancing an equation, you cannot change the chemical formula of any reactants or products. Changing a formula changes the identity of the substance. Instead, you can place coefficients, or multipliers, in front of formulas. Coefficients change the amount of the reactants and products present. For example, an H_2O molecule has two H atoms and one O atom. Placing the coefficient *2* before H_2O ($2H_2O$) means that you double the number of H atoms and O atoms present:

$$2 \times 2 \text{ H atoms} = 4 \text{ H atoms}$$
$$2 \times 1 \text{ O atom} = 2 \text{ O atoms}$$

Note that $2H_2O$ is still water. However, it describes two water particles instead of one.

Figure 18 Equations must be balanced because mass is conserved during a chemical reaction.

Balancing Chemical Equations

Review   Personal Tutor

Balancing Chemical Equations Example

When methane (CH_4)—a gas burned in furnaces—reacts with oxygen (O_2) in the air, the reaction produces carbon dioxide (CO_2) and water (H_2O). Write and balance a chemical equation for this reaction.

1 Write the equation, and check to see if it is balanced.

a. Write the chemical formulas with the reactants on the left side of the arrow and the products on the right side.	a. $CH_4 + O_2 \rightarrow CO_2 + H_2O$ not balanced

b. Count the atoms of each element in the reactants and in the products.
■ Note which elements have a balanced number of atoms on each side of the equation.
■ If all elements are balanced, the overall equation is balanced. If not, go to step 2.
	b. reactants → products
	    C=1 C=1 **balanced**
	    H=4 H=2 **not balanced**
	    O=2 O=3 **not balanced**

2 Add coefficients to the chemical formulas to balance the equation.

a. Pick an element in the equation whose atoms are not balanced, such as hydrogen. Write a coefficient in front of a reactant or a product that will balance the atoms of the chosen element in the equation.	a. $CH_4 + O_2 \rightarrow CO_2 + 2H_2O$ not balanced
	b. C=1 C=1 **balanced**
	    H=4 H=4 **balanced**
	    O=2 O=4 **not balanced**

b. Recount the atoms of each element in the reactants and the products, and note which are balanced on each side of the equation.
 	
c. Repeat steps 2a and 2b until all atoms of each element in the reactants equal those in the products.
	c. $CH_4 + 2O_2 \rightarrow CO_2 + 2H_2O$ **balanced**
	    C=1 C=1 **balanced**
	    H=4 H=4 **balanced**
	    O=4 O=4 **balanced**

3 Write the balanced equation that includes the coefficients: $CH_4 + 2O_2 \rightarrow CO_2 + 2H_2O$

Factors that Affect the Rate of Chemical Reactions 🔑

Figure 19 The rate of most chemical reactions increases with an increase in temperature, concentration, or surface area.

1 Temperature

Chemical reactions that occur during cooking happen at a faster rate when temperature increases.

2 Concentration

Acid rain contains a higher concentration of acid than normal rain does. As a result, a statue exposed to acid rain is damaged more quickly than a statue exposed to normal rain.

3 Surface Area

When an antacid tablet is broken into pieces, the pieces have more total surface area than the whole tablet does. The pieces react more rapidly with water because more of the broken tablet is in contact with the water.

The Rate of Chemical Reactions

Recall that the particles that make up matter are constantly moving and colliding with one another. Different factors can make these particles move faster and collide harder and more frequently. These factors increase the rate of a chemical reaction, as shown in **Figure 19.**

1 A higher **temperature** usually increases the rate of reaction. When the temperature is higher, the particles move faster. Therefore, the particles collide with greater force and more frequently.

2 **Concentration** *is the amount of substance in a certain volume.* A reaction occurs faster if the concentration of at least one reactant increases. When concentration increases, there are more particles available to bump into each other and react.

3 **Surface area** also affects reaction rate if at least one reactant is a solid. If you drop a whole effervescent antacid tablet into water, the tablet reacts with the water. However, if you break the tablet into several pieces and then add them to the water, the reaction occurs more quickly. Smaller pieces have more total surface area, so more space is available for reactants to collide.

 Key Concept Check List three factors that affect the rate of a chemical reaction.

Chemistry

To understand chemistry, you need to understand matter. You need to know how the arrangement of atoms results in different types of matter. You also need to be able to distinguish physical properties from chemical properties and describe ways these properties can change. In later chemistry chapters and courses, you will examine each of these topics closely to gain a better understanding of matter.

Lesson 4 Review

✓ **Assessment** Online Quiz

? **Inquiry** Virtual Lab

Visual Summary

A chemical property is observed only as a material undergoes chemical change and changes identity.

Signs of possible chemical change include bubbles, energy change, and change in odor or color.

Chemical equations show the reactants and products of a chemical reaction and that mass is conserved.

Reactants		Product
Fe + S	→	FeS

FOLDABLES

Use your lesson Foldable to review the lesson. Save your Foldable for the project at the end of the chapter.

What do you think NOW?

You first read the statements below at the beginning of the chapter.

7. When wood burns, new materials form.

8. Temperature can affect the rate at which chemical changes occur.

Did you change your mind about whether you agree or disagree with the statements? Rewrite any false statements to make them true.

Use Vocabulary

1 The amount of substance in a certain volume is its _____.

2 **Use the term** *chemical change* in a complete sentence.

Understand Key Concepts

3 **List** some signs of chemical change.

4 Which property of matter changes during a chemical change but does NOT change during a physical change?
- **A.** energy
- **B.** identity
- **C.** mass
- **D.** volume

5 **State** why chemical equations are useful.

6 **Analyze** What affects the rate at which acid rain reacts with a statue?

Interpret Graphics

7 **Examine** Explain how the diagram below shows conservation of mass.

 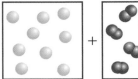

8 **Compare and Contrast** Copy and fill in the graphic organizer to compare and contrast physical and chemical changes.

Physical and Chemical Changes	
Alike	
Different	

Critical Thinking

9 **Compile** a list of three physical changes and three chemical changes you have observed recently.

10 **Recommend** How could you increase the rate at which the chemical reaction between vinegar and baking soda occurs?

Materials

triple-beam
balance

50-mL
graduated
cylinder

magnifying
lens

bar magnet

Also needed:
crime scene
evidence,
unknown
substances,
dropper bottles
containing
water, iodine,
cornstarch, and
red cabbage
indicator,
test tubes,
test tube rack,
stirring rod

Safety

Design an Experiment to Solve a Crime

Recall how you can use properties to identify and compare substances. You now will apply those ideas to solving a crime. You will be given evidence collected from the crime scene and from the suspect's house. As the investigator, decide whether evidence from the crime scene matches evidence from the suspect. What tests will you use? What does the evidence tell you?

Question

Determine which factors about the evidence you would like to investigate further. Consider how you can describe and compare the properties of each piece of evidence. Evaluate the properties you will observe and measure, and decide whether it would be an advantage to classify them as physical properties or chemical properties. Will the changes that the evidence will undergo be helpful to you? Think about controls, variables, and the equipment you have available. Is there any way to match samples exactly?

Procedure

1. Read and complete a lab safety form.

2. In your Science Journal, write the procedures you will use to answer your question. Include the materials and steps you will use to test each piece of evidence. By the appropriate step in the procedure, list any safety procedures you should observe while performing the investigation. Organize your steps by putting them in a graphic organizer, such as the one below. Have your teacher approve your procedures.

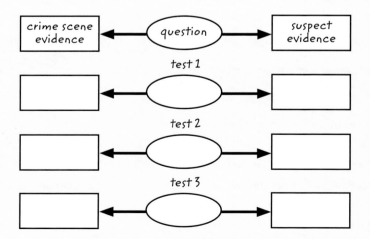

3 Begin by observing and recording your observations on each piece of evidence. What can you learn by comparing physical properties? Are any of the samples made of several parts?

4 Use the available materials to test the evidence. Accurately record all observations and data for each piece of evidence.

5 Add any additional tests you think you need to answer your questions.

Analyze and Conclude

6 Examine the data you have collected. What does the evidence tell you about whether the crime scene and the suspect are related?

7 Write your conclusions in your Science Journal. Be thorough because these are the notes you would use if you had to testify in court about the case.

8 **Analyze** Which data suggest that evidence from the crime scene was or wasn't connected to the suspect?

9 **Draw Conclusions** If you were to testify in court, what conclusions would you be able to state confidently based on your findings?

10 **The Big Idea** How does understanding physical and chemical properties of matter help you to solve problems?

Communicate Your Results

Compare your results with those of other teams. Discuss the kinds of evidence that might be strong enough to convict a suspect.

Research the difference between individual and class evidence used in forensics. Decide which class of evidence your tests provided.

Lab Tips

☑ Don't overlook simple ideas such as matching the edges of pieces.

☑ Can you separate any of the samples into other parts?

☑ Always get your teacher's approval before trying any new test.

Remember to use scientific methods.

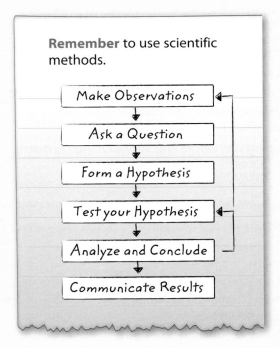

Make Observations

Ask a Question

Form a Hypothesis

Test your Hypothesis

Analyze and Conclude

Communicate Results

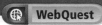
 THE BIG IDEA

Matter is anything that has mass and takes up space. Its physical properties and its chemical properties can change.

Key Concepts Summary

Vocabulary

Lesson 1: Classifying Matter

- A **substance** is a type of **matter** that always is made of atoms in the same combinations.
- **Atoms** of different elements have different numbers of protons.
- The composition of a substance cannot vary. The composition of a **mixture** can vary.
- Matter can be classified as either a substance or a mixture.

8 protons

Oxygen
8
O
16.00

matter p. 349

atom p. 349

substance p. 351

element p. 351

compound p. 352

mixture p. 353

heterogeneous mixture p. 353

homogeneous mixture p. 353

dissolve p. 353

Lesson 2: Physical Properties

- **Physical properties** of matter include size, shape, texture, and state.
- Physical properties such as **density,** melting point, boiling point, and size can be used to separate mixtures.

physical property p. 358

mass p. 360

density p. 361

solubility p. 362

Lesson 3: Physical Changes

- A change in energy can change the state of matter.
- When something dissolves, it mixes evenly in a substance.
- The masses before and after a change in matter are equal.

physical change p. 367

Lesson 4: Chemical Properties and Changes

- **Chemical properties** include ability to burn, acidity, and ability to rust.
- Some signs that might indicate **chemical changes** are the formation of bubbles and a change in odor, color, or energy.

- Chemical equations are useful because they show what happens during a chemical reaction.
- Some factors that affect the rate of chemical reactions are temperature, **concentration,** and surface area.

chemical property p. 374

chemical change p. 375

concentration p. 378

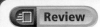

Study Guide

 Review
- **Personal Tutor**
- **Vocabulary eGames**
- **Vocabulary eFlashcards**

 **Chapter Project**

Assemble your lesson Foldables as shown to make a Chapter Project. Use the project to review what you have learned in this chapter. Fasten the Foldable from Lesson 4 on the back of the board.

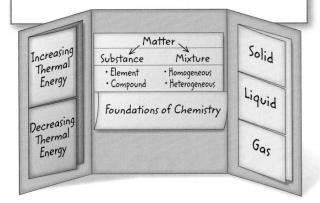

Use Vocabulary

Give two examples of each of the following.

1. element
2. compound
3. homogeneous mixture
4. heterogeneous mixture
5. physical property
6. chemical property
7. physical change
8. chemical change

Link Vocabulary and Key Concepts

Concepts in Motion Interactive Concept Map

Copy this concept map, and then use vocabulary terms from the previous page to complete the concept map.

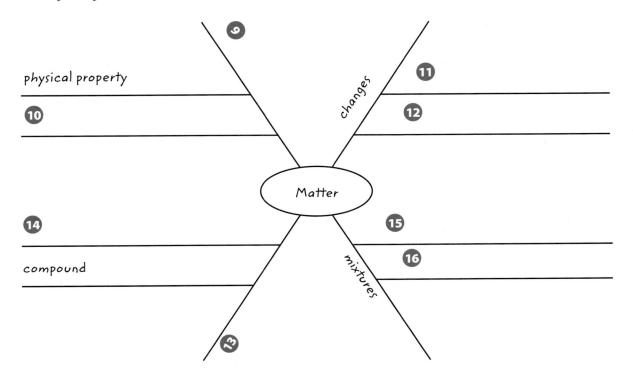

Understand Key Concepts

1 The formula $AgNO_3$ represents a compound made of which atoms?

A. 1 Ag, 1 N, 1 O
B. 1 Ag, 1 N, 3 O
C. 1 Ag, 3 N, 3 O
D. 3 Ag, 3 N, 3 O

2 Which is an example of an element?

A. air
B. water
C. sodium
D. sugar

3 Which property explains why copper often is used in electrical wiring?

A. conductivity
B. density
C. magnetism
D. solubility

4 The table below shows densities for different substances.

Substance	Density (g/cm³)
1	1.58
2	0.32
3	1.52
4	1.62

For which substance would a 4.90-g sample have a volume of 3.10 cm³?

A. substance 1
B. substance 2
C. substance 3
D. substance 4

5 Which would decrease the rate of a chemical reaction?

A. increase in concentration
B. increase in temperature
C. decrease in surface area
D. increase in both surface area and concentration

6 Which physical change is represented by the diagram below?

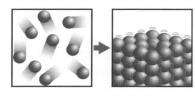

A. condensation
B. deposition
C. evaporation
D. sublimation

7 Which chemical equation is unbalanced?

A. $2KClO_3 \rightarrow 2KCl + 3O_2$
B. $CH_4 + 2O_2 \rightarrow CO_2 + 2H_2O$
C. $Fe_2O_3 + CO \rightarrow 2Fe + 2CO_2$
D. $H_2CO_3 \rightarrow H_2O + CO_2$

8 Which is a size-dependent property?

A. boiling point
B. conductivity
C. density
D. mass

9 Why is the following chemical equation said to be balanced?

$$O_2 + 2PCl_3 \rightarrow 2POCl_3$$

A. There are more reactants than products.
B. There are more products than reactants.
C. The atoms are the same on both sides of the equation.
D. The coefficients are the same on both sides of the equation.

10 The elements sodium (Na) and chlorine (Cl) react and form the compound sodium chloride (NaCl). Which is true about the properties of these substances?

A. Na and Cl have the same properties.
B. NaCl has the properties of Na and Cl.
C. All the substances have the same properties.
D. The properties of NaCl are different from the properties of Na and Cl.

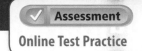
Critical Thinking

11 **Compile** a list of ten materials in your home. Classify each material as an element, a compound, or a mixture.

12 **Evaluate** Would a periodic table based on the number of electrons in an atom be as effective as the one shown in the back of this book? Why or why not?

13 **Develop** a demonstration to show how weight is not the same thing as mass.

14 **Construct** an explanation for how the temperature and energy of a material changes during the physical changes represented by the diagram below.

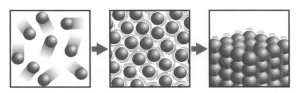

15 **Revise** the definition of physical change given in this chapter so it mentions the type and arrangement of atoms.

16 **Find an example** of a physical change in your home or school. Describe the changes in physical properties that occur during the change. Then explain how you know the change is not a chemical change.

17 **Develop** a list of five chemical reactions you observe each day. For each, describe one way that you could either increase or decrease the rate of the reaction.

Writing in Science

18 **Write** a poem at least five lines long to describe the organization of matter by the arrangement of its atoms. Be sure to include both the names of the different types of matter as well as their meanings.

REVIEW THE BIG IDEA

19 Explain how you are made of matter that undergoes changes. Provide specific examples in your explanation.

20 How does the photo below show an example of a physical change, a chemical change, a physical property, and a chemical property?

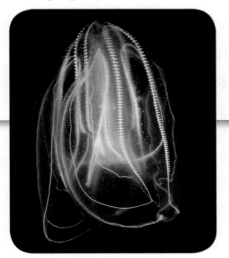

Math Skills

Review
Math Practice

Use Ratios

21 A sample of ice at 0°C has a mass of 23 g and a volume of 25 cm³. Why does ice float on water? (The density of water is 1.00 g/cm³.)

22 The table below shows the masses and the volumes for samples of two different elements.

Element	Mass (g)	Volume (cm³)
Gold	386	20
Lead	22.7	2.0

Which element sample in the table has greater density?

Standardized Test Practice

Record your answers on the answer sheet provided by your teacher or on a sheet of paper.

Multiple Choice

1 Which describes how mixtures differ from substances?

A Mixtures are homogeneous.

B Mixtures are liquids.

C Mixtures can be separated physically.

D Mixtures contain only one kind of atom.

Use the figure below to answer question 2.

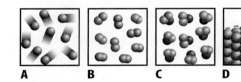

2 Which image in the figure above is a model for a compound?

A A

B B

C C

D D

3 Which is a chemical property?

A the ability to be compressed

B the ability to be stretched into thin wire

C the ability to melt at low temperature

D the ability to react with oxygen

4 You drop a sugar cube into a cup of hot tea. What causes the sugar to disappear in the tea?

A It breaks into elements.

B It evaporates.

C It melts.

D It mixes evenly.

5 Which is an example of a substance?

A air

B lemonade

C soil

D water

Use the figure below to answer question 6.

6 The figure above is a model of atoms in a sample at room temperature. Which physical property does this sample have?

A It can be poured.

B It can expand to fill its container.

C It cannot easily change shape.

D It has a low boiling point.

7 Which observation is a sign of a chemical change?

A bubbles escaping from a carbonated drink

B iron filings sticking to a magnet

C lights flashing from fireworks

D water turning to ice in a freezer

8 Zinc, a solid metal, reacts with a hydro-chloric acid solution. Which will increase the reaction rate?

   A cutting the zinc into smaller pieces

   B decreasing the concentration of the acid

   C lowering the temperature of the zinc

   D pouring the acid into a larger container

Use the figure below to answer question 9.

9 In the figure above, what will be the mass of the final solution if the solid dissolves in the water?

   A 5 g

   B 145 g

   C 150 g

   D 155 g

10 Which is NOT represented in a chemical equation?

   A chemical formula

   B product

   C conservation of mass

   D reaction rate

Constructed Response

Use the graph below to answer questions 11 and 12.

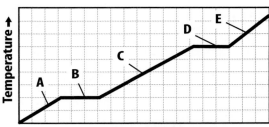

11 Use the graph above to explain why ice will keep water cold on a hot day.

12 Use two sections of the graph to explain what happens when you put a pot of cold water on a stove to boil. Specify which two sections you used.

13 Describe how you would separate a mixture of sugar, sand, and water.

14 The reaction of zinc metal with hydro-chloric acid produces zinc chloride and hydrogen gas. A student writes the following to represent the reaction.

$$Zn + HCl \rightarrow ZnCl_2 + H_2$$

Is the equation correct? Use conservation of mass to support your answer.

NEED EXTRA HELP?														
If You Missed Question...	1	2	3	4	5	6	7	8	9	10	11	12	13	14
Go to Lesson...	1	1	4	3	1	2	4	4	3	4	3	3	2	4

The Periodic Table

THE BIG IDEA How is the periodic table used to classify and provide information about all known elements?

Inquiry What makes this balloon so special?

Things are made out of specific materials for a reason. A weather balloon can rise high in the atmosphere and gather weather information. The plastic that forms this weather balloon and the helium gas that fills it were chosen after scientists researched and studied the properties of these materials.

- What property of helium do you think makes the balloon rise through the air?

- How do you think the periodic table is a useful tool when determining properties of different materials?

Get Ready to Read

What do you think?

Before you read, decide if you agree or disagree with each of these statements. As you read this chapter, see if you change your mind about any of the statements.

1 The elements on the periodic table are arranged in rows in the order they were discovered.

2 The properties of an element are related to the element's location on the periodic table.

3 Fewer than half of the elements are metals.

4 Metals are usually good conductors of electricity.

5 Most of the elements in living things are nonmetals.

6 Even though they look very different, oxygen and sulfur share some similar properties.

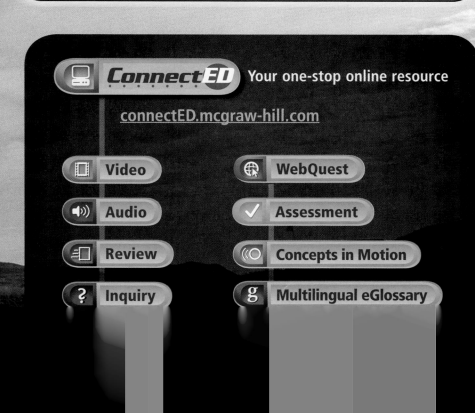

ConnectED Your one-stop online resource

connectED.mcgraw-hill.com

Video   WebQuest

Audio   Assessment

Review   Concepts in Motion

Inquiry   Multilingual eGlossary

Reading Guide

Key Concepts 🔑
ESSENTIAL QUESTIONS

- How are elements arranged on the periodic table?
- What can you learn about elements from the periodic table?

Vocabulary

periodic table p. 391

group p. 396

period p. 396

[g] **Multilingual eGlossary**

[▶] **Video** BrainPOP®

Using the Periodic Table

Inquiry **Same Information?**

You probably have seen a copy of a table that is used to organize the elements. Does it look like this chart? There is no specific shape that a chart of elements must have. However, the relationships among the elements in the chart are important.

Launch Lab

How can objects be organized?

What would it be like to shop at a grocery store where all the products are mixed up on the shelves? Maybe cereal is next to the dish soap and bread is next to the canned tomatoes. It would take a long time to find the groceries that you needed. How does organizing objects help you to find and use what you need?

1 Read and complete a lab safety form.

2 Empty the **interlocking plastic bricks** from the **plastic bag** onto your desk and observe their properties. Think about ways you might group and sequence the bricks so that they are organized.

3 Organize the bricks according to your plan.

4 Compare your pattern of organization with those used by several other students.

Think About This

1. Describe in your Science Journal the way you grouped your bricks. Why did you choose that way of grouping?

2. Describe how you sequenced the bricks.

3. **Key Concept** How does organizing things help you to use them more easily?

What is the periodic table?

The "junk drawer" in **Figure 1** is full of pens, notepads, rubber bands, and other supplies. It would be difficult to find a particular item in this messy drawer. How might you organize it? First, you might dump the contents onto the counter. Then you could sort everything into piles. Pens and pencils might go into one pile. Notepads and paper go into another. Organizing the contents of the drawer makes it easier to find the things you need, also shown in **Figure 1**.

Just as sorting helped to organize the objects in the junk drawer, sorting can help scientists organize information about the elements. Recall that there are more than 100 elements, each with a unique set of physical and chemical properties.

Scientists use a table called the periodic (pihr ee AH dihk) table to organize elements. *The **periodic table** is a chart of the elements arranged into rows and columns according to their physical and chemical properties.* It can be used to determine the relationships among the elements.

In this chapter, you will read about how the periodic table was developed. You will also read about how you can use the periodic table to learn about the elements.

Figure 1 Sorting objects by their similarities makes it easier to find what you need.

Developing a Periodic Table

In 1869 a Russian chemist and teacher named Dimitri Mendeleev (duh MEE tree • men duh LAY uf) was working on a way to classify elements. At that time, more than 60 elements had been discovered. He studied the physical properties such as density, color, melting point, and atomic mass of each element. Mendeleev also noted chemical properties such as how each element reacted with other elements. Mendeleev arranged the elements in a list using their atomic masses. He noticed that the properties of the elements seemed to repeat in a pattern.

When Mendeleev placed his list of elements into a table, he arranged them in rows of increasing atomic mass. Elements with similar properties were grouped the same column. The columns in his table are like the piles of sorted objects in your junk drawer. Both contain groups of things with similar properties.

 Reading Check What physical property did Mendeleev use to place the elements in rows on the periodic table?

Patterns in Properties

The term *periodic* means "repeating pattern." For example, seasons and months are periodic because they follow a repeating pattern every year. The days of the week are periodic since they repeat every seven days.

What were some of the repeating patterns Mendeleev noticed in his table? Melting point is one property that shows a repeating pattern. Recall that melting point is the temperature at which a solid changes to a liquid. The blue line in **Figure 2** represents the melting points of the elements in row 2 of the periodic table. Notice that the melting point of carbon is higher than the melting point of lithium. However, the melting point of fluorine, at the far right of the row, is lower than that of carbon. How do these melting points show a pattern? Look at the red line in **Figure 2.** This line represents the melting points of the elements in row 3 of the periodic table. The melting points follow the same increasing and then decreasing pattern as the blue line, or row 2. Boiling point and reactivity also follow a periodic pattern.

A Periodic Property 🔑

Figure 2 Melting points increase, then decrease, across a period on the periodic table.

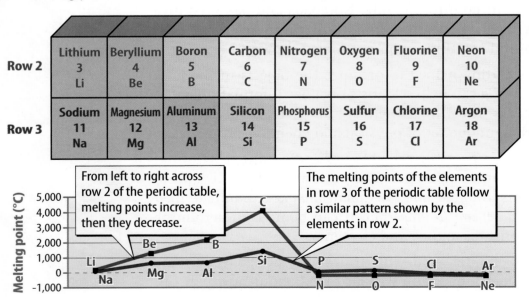

Predicting Properties of Undiscovered Elements

When Mendeleev arranged all known elements by increasing atomic mass, there were large gaps between some elements. He predicted that scientists would discover elements that would fit into these spaces. Mendeleev also predicted that the properties of these elements would be similar to the known elements in the same columns. Both of his predictions turned out to be true.

Changes to Mendeleev's Table

Mendeleev's periodic table enabled scientists to relate the properties of the known elements to their position on the table. However, the table had a problem—some elements seemed out of place. Mendeleev believed that the atomic masses of certain elements must be invalid because the elements appeared in the wrong place on the periodic table. For example, Mendeleev placed tellurium before iodine despite the fact that tellurium has a greater atomic mass than iodine. He did so because iodine's properties more closely resemble those of fluorine and chlorine, just as copper's properties are closer to those of silver and gold, as shown in **Figure 3.**

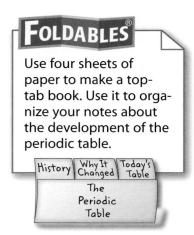
FOLDABLES®

Use four sheets of paper to make a top-tab book. Use it to organize your notes about the development of the periodic table.

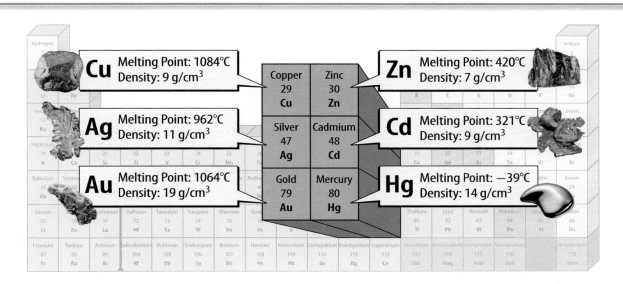

The Importance of Atomic Number

In the early 1900s, the scientist Henry Moseley solved the problem with Mendeleev's table. Moseley found that if elements were listed according to increasing atomic number instead of listing atomic mass, columns would contain elements with similar properties. Recall that the atomic number of an element is the number of protons in the nucleus of each of that element's atoms.

 Key Concept Check What determines where an element is located on the periodic table you use today?

Figure 3 On today's periodic table, copper is in the same column as silver and gold. Zinc is in the same column as cadmium and mercury.

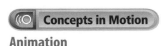 **Concepts in Motion**

Animation

Figure 4 🔑 The periodic table is used to organize elements according to increasing atomic number and properties.

Today's Periodic Table

You can identify many of the properties of an element from its placement on the **periodic** table. The table, as shown in **Figure 4,** is organized into columns, rows, and blocks, which are based on certain patterns of properties. In the next two lessons, you will learn how an element's position on the periodic table can help you interpret the element's physical and chemical properties.

《⊙ **Concepts in Motion** Animation

PERIODIC TABLE OF THE ELEMENTS

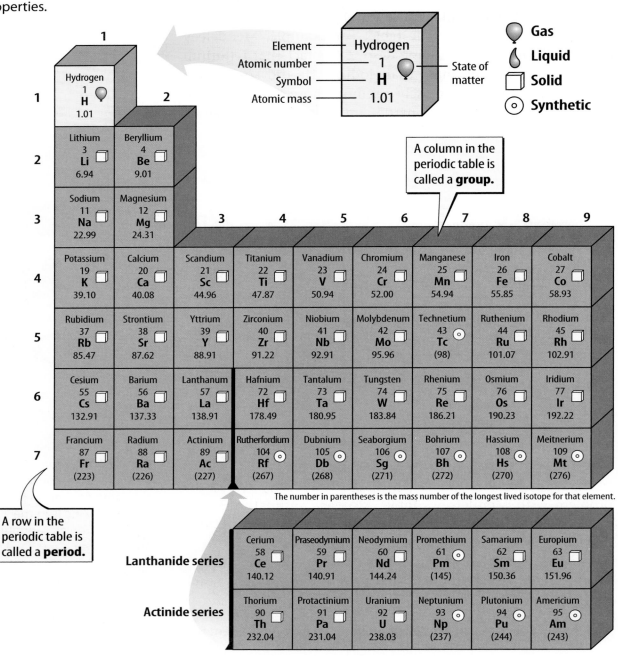

The number in parentheses is the mass number of the longest lived isotope for that element.

A column in the periodic table is called a **group.**

A row in the periodic table is called a **period.**

Gas
Liquid
Solid
Synthetic

What is on an element key?

The element key shows an element's chemical symbol, atomic number, and atomic mass. The key also contains a symbol that shows the state of matter at room temperature. Look at the element key for helium in **Figure 5.** Helium is a gas at room temperature. Some versions of the periodic table give additional information, such as density, conductivity, or melting point.

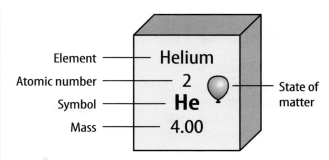

Figure 5 An element key shows important information about each element.

Visual Check What does this key tell you about helium?

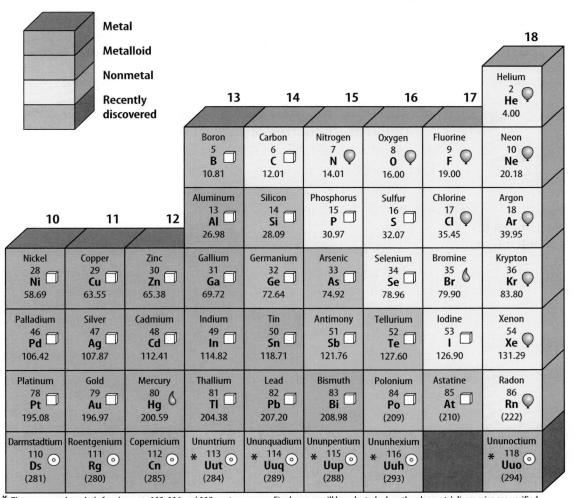

* The names and symbols for elements 113-116 and 118 are temporary. Final names will be selected when the elements' discoveries are verified.

Use Geometry

The distance around a circle is the circumference (*C*). The distance across the circle, through its center, is the diameter (*d*). The radius (*r*) is half of the diameter. The circumference divided by the diameter for any circle is equal to π (pi), or 3.14. The formula for determining the circumference is:

$$C = \pi d \text{ or } C = 2\,\pi r$$

For example, an iron (Fe) atom has a radius of **126 pm** (picometers; 1 picometer = one-trillionth of a meter) The circumference of an iron atom is:

$$C = 2 \times 3.14 \times 126 \text{ pm}$$

$$C = 791 \text{ pm}$$

Practice

The radius of a uranium (U) atom is 156 pm. What is its circumference?

Review

- **Math Practice**
- **Personal Tutor**

Groups

A **group** *is a column on the periodic table.* Elements in the same group have similar chemical properties and react with other elements in similar ways. There are patterns in the physical properties of a group such as density, melting point, and boiling point. The groups are numbered 1–18, as shown in **Figure 4.**

🔑 **Key Concept Check** What can you infer about the properties of two elements in the same group?

Periods

The rows on the periodic table are called **periods.** The atomic number of each element increases by one as you read from left to right across each period. The physical and chemical properties of the elements also change as you move left to right across a period.

Metals, Nonmetals, and Metalloids

Almost three-fourths of the elements on the periodic table are metals. Metals are on the left side and in the middle of the table. Individual metals have some properties that differ, but all metals are shiny and conduct thermal energy and electricity.

With the exception of hydrogen, nonmetals are located on the right side of the periodic table. The properties of nonmetals differ from the properties of metals. Many nonmetals are gases, and they do not conduct thermal energy or electricity.

Between the metals and the nonmetals on the periodic table are the metalloids. Metalloids have properties of both metals and nonmetals. **Figure 6** shows an example of a metal, a metalloid, and a nonmetal.

Figure 6 In period 3, magnesium is a metal, silicon is a metalloid, and sulfur is a nonmetal.

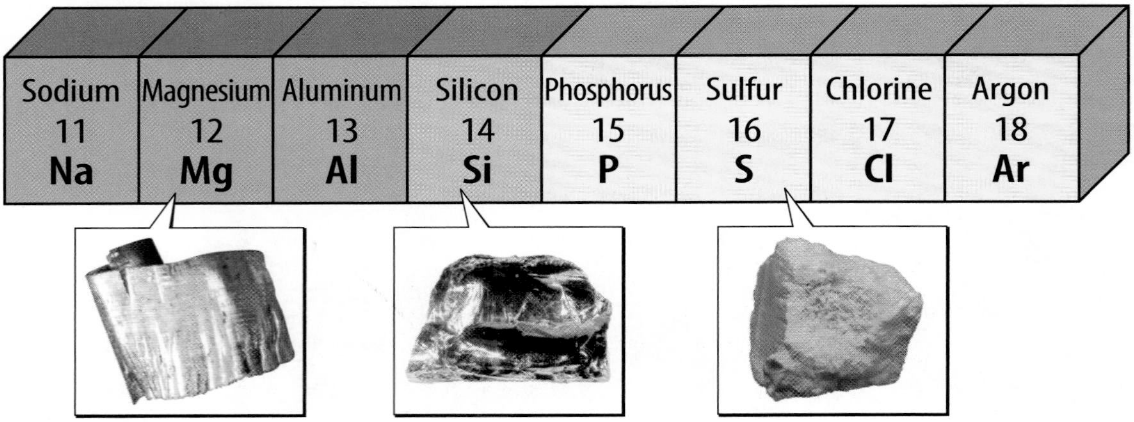

Sodium	Magnesium	Aluminum	Silicon	Phosphorus	Sulfur	Chlorine	Argon
11	12	13	14	15	16	17	18
Na	**Mg**	**Al**	**Si**	**P**	**S**	**Cl**	**Ar**

Glenn T. Seaborg

Niels Bohr

Lise Meitner

| Seaborgium 106 Sg | Bohrium 107 Bh | Hassium 108 Hs | Meitnerium 109 Mt |

Figure 7 Three of these synthetic elements are named to honor important scientists.

How Scientists Use the Periodic Table

Even today, new elements are created in laboratories, named, and added to the present-day periodic table. Four of these elements are shown in **Figure 7.** These elements are all synthetic, or made by people, and do not occur naturally on Earth. Sometimes scientists can create only a few atoms of a new element. Yet scientists can use the periodic table to predict the properties of new elements they create. Look back at the periodic table in **Figure 4.** What group would you predict to contain element 117? You would probably expect element 117 to be in group 17 and to have similar properties to other elements in the group. Scientists hope to one day synthesize element 117.

The periodic table contains more than 100 elements. Each element has unique properties that differ from the properties of other elements. But each element also shares similar properties with nearby elements. The periodic table shows how elements relate to each other and fit together into one organized chart. Scientists use the periodic table to understand and predict elements' properties. You can, too.

 Reading Check How is the periodic table used to predict the properties of an element?

How does atom size change across a period?

One pattern seen on the periodic table is in the radius of different atoms. The figure below shows how atomic radius is measured.

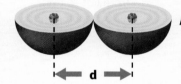
Atomic radius $= \frac{1}{2}d$

1 Read and complete a lab safety form.

2 Using **scissors** and **card stock paper,** cut seven 2-cm × 4-cm rectangles. Using a **marker,** label each rectangle with the atomic symbol of each of the first seven elements in period 2. Obtain the radius for each atom from your teacher.

3 Using a **ruler,** cut **plastic straws** to the same number of millimeters as each atomic radius given in picometers. For example, if the atomic radius is 145 pm, cut a straw 145 mm long.

4 **Tape** each of the labeled rectangles to the top of its appropriate straw.

5 Insert the straws into **modeling clay** according to increasing atomic number.

Analyze and Conclude

1. **Describe** the pattern you see in your model.

2. **Key Concept** Predict the pattern of atomic radii of the elements in period 4.

Lesson 1 Review

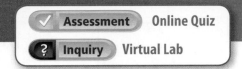
Visual Summary

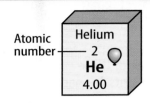

Atomic number — Helium 2 **He** 4.00

On the periodic table, elements are arranged according to increasing atomic number and similar properties.

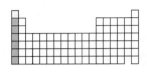

A column of the periodic table is called a group. Elements in the same group have similar properties.

A row of the periodic table is called a period. Properties of elements repeat in the same pattern from left to right across each period.

FOLDABLES®

Use your lesson Foldable to review the lesson. Save your Foldable for the project at the end of the chapter.

What do you think NOW?

You first read the statements below at the beginning of the chapter.

1. The elements on the periodic table are arranged in rows in the order they were discovered.

2. The properties of an element are related to the element's location on the periodic table.

Did you change your mind about whether you agree or disagree with the statements? Rewrite any false statements to make them true.

Use Vocabulary

1 **Identify** the scientific term used for rows on the periodic table.

2 **Name** the scientific term used for columns on the periodic table.

Understand Key Concepts

3 The _____ increases by one for each element as you move left to right across a period.

4 What does the decimal number in an element key represent?
   A. atomic mass   **C.** chemical symbol
   B. atomic number   **D.** state of matter

Interpret Graphics

5 **Classify** each marked element, 1 and 2, as a metal, a nonmetal, or a metalloid.

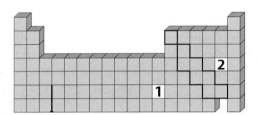

6 **Identify** Copy and fill in the graphic organizer below to identify the color-coded regions of the periodic table.

All Elements — Metals

Critical Thinking

7 **Predict** Look at the perioidic table and predict three elements that have lower melting points than calcium (Ca).

Math Skills

Review
— Math Practice —

8 Carbon (C) and silicon (Si) are in group 4 of the periodic table. The atomic radius of carbon is 77 pm and sulfur is 103 pm. What is the circumference of each atom?

How is the periodic table arranged?

Materials

20 cards

What would happen if schools did not assign students to grades or classes? How would you know where to go on the first day of school? What if your home did not have an address? How could you tell someone where you live? Life becomes easier with organization. The following activity will help you discover how elements are organized on the periodic table.

Learn It

Patterns help you make sense of the world around you. The days of the week follow a pattern, as do the months of the year. **Identifying a pattern** involves organizing things into similar groups and then sequencing the things in the same way in each group.

Try It

1. Obtain cards from your teacher. Turn the cards over so the sides with numbers are facing up.

2. Separate the cards into three or more piles. All of the cards in a pile should have a characteristic in common.

3. Organize each pile into a pattern. Use all of the cards.

4. Lay out the cards into rows and columns based on their characteristics and patterns.

Apply It

5. Describe in your Science Journal the patterns you used to organize your cards. Do other patterns exist in your arrangement?

6. Are there gaps in your arrangement? Can you describe what a card in one of those gaps would look like?

7. 🗝 **Key Concept** What characteristics of elements might you use to organize them in a similar pattern?

Lesson 2

Reading Guide

Key Concepts
ESSENTIAL QUESTIONS

- What elements are metals?
- What are the properties of metals?

Vocabulary

metal p. 401

luster p. 401

ductility p. 402

malleability p. 402

alkali metal p. 403

alkaline earth metal p. 403

transition element p. 404

g **Multilingual eGlossary**

Metals

Inquiry) Where does it strike?

Lightning strikes the top of the Empire State Building approximately 100 times a year. Why does lightning hit the top of this building instead of the city streets or buildings below? Metal lightning rods allow electricity to flow through them more easily than other materials do. Lightning moves through these materials and the building is not harmed.

Launch Lab

20 minutes

What properties make metals useful?

The properties of metals determine their uses. Copper conducts thermal energy, which makes it useful for cookware. Aluminum has low density, so it is used in aircraft bodies. What other properties make metals useful?

1. Read and complete a lab safety form.

2. With your group, observe the **metal objects** in your **container.** For each object, discuss what properties allow the metal to be used in that way.

3. Observe the **photographs of gold and silver jewelry.** What properties make these two metals useful in jewelry?

4. Examine **other objects around the room** that you think are made of metal. Do they share the same properties as the objects in your container? Do they have other properties that make them useful?

Think About This

1. What properties do all the metals share? What properties are different?

2. **Key Concept** In your Science Journal, list at least four properties of metals that determine their uses.

What is a metal?

What do stainless steel knives and forks, copper wire, aluminum foil, and gold jewelry have in common? They are all made from metals.

As you read in Lesson 1, most of the elements on the periodic table are metals. In fact, of all the known elements, more than three-quarters are metals. With the exception of hydrogen, all of the elements in groups 1–12 on the periodic table are metals. In addition, some of the elements in groups 13–15 are metals. To be a metal, an element must have certain properties.

Key Concept Check How does the position of an element on the periodic table allow you to determine if the element is a metal?

Physical Properties of Metals

Recall that physical properties are characteristics used to describe or identify something without changing its makeup. All metals share certain physical properties.

A **metal** *is an element that is generally shiny. It is easily pulled into wires or hammered into thin sheets. A metal is a good conductor of electricity and thermal energy.* Gold exhibits the common properties of metals.

Luster and Conductivity People use gold for jewelry because of its beautiful color and metallic luster. **Luster** *describes the ability of a metal to reflect light.* Gold is also a good conductor of thermal energy and electricity. However, gold is too expensive to use in normal electrical wires or metal cookware. Copper is often used instead.

Figure 8 Gold has many uses based on its properties.

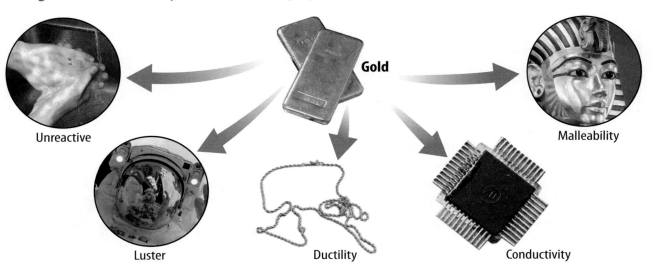

Unreactive

Luster

Gold

Ductility

Malleability

Conductivity

✓ **Visual Check** Analyze why the properties shown in each photo are an advantage to using gold.

WORD ORIGIN · · · · · · · · · · ·

ductility
from Latin *ductilis*, means
"may be led or drawn"

REVIEW VOCABULARY · · · · ·

density
the mass per unit volume of a
substance

Make a two-tab book.
Label it as shown. Use it
to record information
about the properties of
metals.

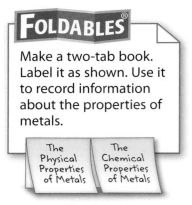

The
Physical
Properties
of Metals

The
Chemical
Properties
of Metals

Ductility and Malleability Gold is the most ductile metal. **Ductility** (duk TIH luh tee) *is the ability to be pulled into thin wires.* A piece of gold with the mass of a paper clip can be pulled into a wire that is more than 3 km long.

Malleability (ma lee uh BIH luh tee) *is the ability of a substance to be hammered or rolled into sheets.* Gold is so malleable that it can be hammered into thin sheets. A pile of a million thin sheets would be only as high as a coffee mug.

Other Physical Properties of Metals In general the **density**, strength, boiling point, and melting point of a metal are greater than those of other elements. Except for mercury, all metals are solid at room temperature. Many uses of a metal are determined by the metal's physical properties, as shown in **Figure 8**.

🔑 **Key Concept Check** What are some physical properties of metals?

Chemical Properties of Metals

Recall that a chemical property is the ability or inability of a substance to change into one or more new substances. The chemical properties of metals can differ greatly. However, metals in the same group usually have similar chemical properties. For example, gold and other elements in group 11 do not easily react with other substances.

Group 1: Alkali Metals

The elements in group 1 are called **alkali** *(AL kuh li)* **metals.** The alkali metals include lithium, sodium, potassium, rubidium, cesium, and francium.

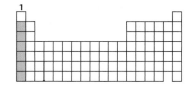

Because they are in the same group, alkali metals have similar chemical properties. Alkali metals react quickly with other elements, such as oxygen. Therefore, in nature, they occur only in compounds. Pure alkali metals must be stored so that they do not come in contact with oxygen and water vapor in the air. **Figure 9** shows potassium and sodium reacting with water.

Alkali metals also have similar physical properties. Pure alkali metals have a silvery appearance. As shown in **Figure 9,** they are soft enough to cut with a knife. The alkali metals also have the lowest densities of all metals. A block of pure sodium metal could float on water because of its very low density.

Figure 9 Alkali metals react violently with water. They are also soft enough to be cut with a knife.

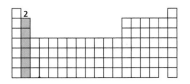

Animation

| **Potassium** | **Sodium** | **Lithium** |

Group 2: Alkaline Earth Metals

The elements in group 2 on the periodic table are called **alkaline** *(AL kuh lun)* **earth metals.** These metals are beryllium, magnesium, calcium, strontium, barium, and radium.

Alkaline earth metals also react quickly with other elements. However, they do not react as quickly as the alkali metals do. Like the alkali metals, pure alkaline earth metals do not occur naturally. Instead, they combine with other elements and form compounds. The physical properties of the alkaline earth metals are also similar to those of the alkali metals. Alkaline earth metals are soft and silvery. They also have low density, but they have greater density than alkali metals.

Reading Check Which element reacts faster with oxygen— barium or potassium?

Figure 10 Transition elements are in blocks at the center of the periodic table. Many colorful materials contain small amounts of transition elements.

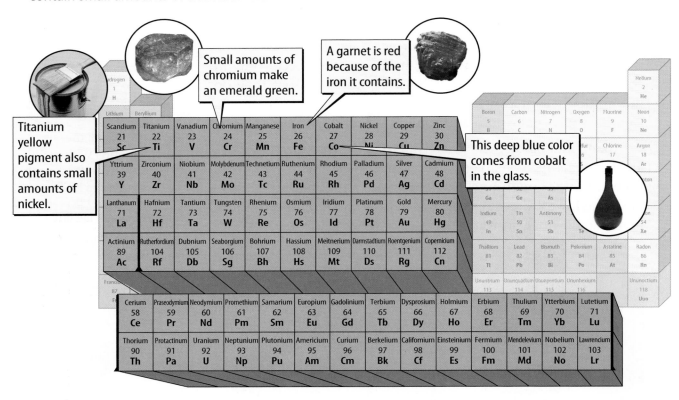

Small amounts of chromium make an emerald green.

A garnet is red because of the iron it contains.

Titanium yellow pigment also contains small amounts of nickel.

This deep blue color comes from cobalt in the glass.

Groups 3–12: Transition Elements

The elements in groups 3–12 are called **transition elements.** The transition elements are in two blocks on the periodic table. The main block is in the center of the periodic table. The other block includes the two rows at the bottom of the periodic table, as shown in **Figure 10.**

Properties of Transition Elements

All transition elements are metals. They have higher melting points, greater strength, and higher densities than the alkali metals and the alkaline earth metals. Transition elements also react less quickly with oxygen. Some transition elements can exist in nature as free elements. An element is a free element when it occurs in pure form, not in a compound.

Uses of Transition Elements

Transition elements in the main block of the periodic table have many important uses. Because of their high densities, strength, and resistance to corrosion, transition elements such as iron make good building materials. Copper, silver, nickel, and gold are used to make coins. These metals are also used for jewelry, electrical wires, and many industrial applications.

Main-block transition elements can react with other elements and form many compounds. Many of these compounds are colorful. Artists use transition-element compounds in paints and pigments. The color of many gems, such as garnets and emeralds, comes from the presence of small amounts of transition elements, as illustrated in **Figure 10.**

Lanthanide and Actinide Series

Two rows of transition elements are at the bottom of the periodic table, as shown in **Figure 10.** These elements were removed from the main part of the table so that periods 6 and 7 were not longer than the other periods. If these elements were included in the main part of the table, the first row, called the lanthanide series, would stretch between lanthanum and halfnium. The second row, called the actinide series, would stretch between actinium and rutherfordium.

Some lanthanide and actinide series elements have valuable properties. For example, lanthanide series elements are used to make strong magnets. Plutonium, one of the actinide series elements, is used as a fuel in some nuclear reactors.

Patterns in Properties of Metals

Recall that the properties of elements follow repeating patterns across the periods of the periodic table. In general, elements increase in metallic properties such as luster, malleability, and electrical conductivity from right to left across a period, as shown in **Figure 11.** The elements on the far right of a period have no metallic properties at all. Potassium (K), the element on the far left in period 4, has the highest luster, is the most malleable, and conducts electricity better than all the elements in this period.

There are also patterns within groups. Metallic properties tend to increase as you move down a group, also shown in **Figure 11.** You could predict that the malleability of gold is greater than the malleability of either silver or copper because it is below these two elements in group 11.

Reading Check Where would you expect to find elements on the periodic table with few or no metallic properties?

How well do materials conduct thermal energy?

How well a material conducts thermal energy can often determine its use.

1. Read and complete a lab safety form.
2. Have your teacher add about 200 mL of very **hot water** to a **250-mL beaker.**
3. Place **rods of metal, plastic, glass, and wood** in the water for 30 seconds.
4. Set four large **ice cubes** on a sheet of **paper towel.** Use **tongs** to quickly remove each rod from the hot water. Place the heated end of the rod on an ice cube.
5. After 30 seconds, remove the rods and examine the ice cubes.

Analyze and Conclude

1. **Conclude** What can you conclude about how well metals conduct thermal energy?

2. **Key Concept** Cookware is often made of metal. What property of metals makes them useful for this purpose?

Figure 11 Metallic properties of elements increase as you move to the left and down on the periodic table.

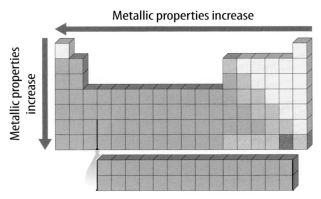

Metallic properties increase

Metallic properties increase

Lesson 2 Review

Visual Summary

Properties of metals include conductivity, luster, malleability, and ductility.

Alkali metals and alkaline earth metals react easily with other elements. These metals make up groups 1 and 2 on the periodic table.

Transition elements make up groups 3–12 and the lanthanide and actinide series on the periodic table.

FOLDABLES

Use your lesson Foldable to review the lesson. Save your Foldable for the project at the end of the chapter.

What do you think NOW?

You first read the statements below at the beginning of the chapter.

3. Fewer than half of the elements are metals.

4. Metals are usually good conductors of electricity.

Did you change your mind about whether you agree or disagree with the statements? Rewrite any false statements to make them true.

Use Vocabulary

1 **Use the term** *luster* in a sentence.

2 **Identify** the property that makes copper metal ideal for wiring.

3 Elements that have the lowest densities of all the metals are called _____.

Understand Key Concepts 🔑

4 **List** the physical properties that most metals have in common.

5 Which is a chemical property of transition elements?
  A. brightly colored
  B. great ductility
  C. denser than alkali metals
  D. reacts little with oxygen

6 **Organize** the following metals from least metallic to most metallic: barium, zinc, iron, and strontium.

Interpret Graphics

7 **Examine** this section of the periodic table. What metal will have properties most similar to those of chromium (Cr)? Why?

Vanadium 23 V	Chromium 24 Cr	Maganese 25 Mn
Niobium 41 Nb	Molybdenum 42 Mo	Technetium 43 Tc

Critical Thinking

8 **Investigate** your classroom and locate five examples of materials made from metal.

9 **Evaluate** the physical properties of potassium, magnesium, and copper. Select the best choice to use for a building project. Explain why this metal is the best building material to use.

Fireworks

Purple: mix of strontium and copper compounds

Metals add a variety of colors to fireworks.

About 1,000 years ago, the Chinese discovered the chemical formula for gunpowder. Using this formula, they invented the first fireworks. One of the primary ingredients in gunpowder is saltpeter, or potassium nitrate. Find potassium on the periodic table. Notice that potassium is a metal. How does the chemical behavior of a metal contribute to a colorful fireworks show?

Blue: copper compounds

Yellow: sodium compounds

Gold: iron burned with carbon

White-hot: barium-oxygen compounds or aluminum or magnesium burn

Orange: calcium compounds

Metal compounds contribute to the variety of colors you see at a fireworks show. Recall that metals have special chemical and physical properties. Compounds that contain metals also have special properties. For example, each metal turns a characteristic color when burned. Lithium, an alkali metal, forms compounds that burn red. Copper compounds burn blue. Aluminum and magnesium burn white.

Green: barium compounds

Red: strontium and lithium compounds

It's Your Turn

FORM AN OPINION Fireworks contain metal compounds. Are they bad for the environment or your health? Research the effects of metals on human health and on the environment. Decide if fireworks are safe to use for holiday celebrations.

Nonmetals and Metalloids

Reading Guide

Key Concepts 🔑
ESSENTIAL QUESTIONS

- Where are nonmetals and metalloids on the periodic table?
- What are the properties of nonmetals and metalloids?

Vocabulary

nonmetal p. 409
halogen p. 411
noble gas p. 412
metalloid p. 413
semiconductor p. 413

 Multilingual eGlossary

Inquiry Why don't they melt?

What do you expect to happen to something when a flame is placed against it? As you can see, the nonmetal material this flower sits on protects the flower from the flame. Some materials conduct thermal energy. Other materials, such as this one, do not.

What are some properties of nonmetals?

You now know what the properties of metals are. What properties do nonmetals have?

1 Read and complete a lab safety form.

2 Examine pieces of **copper, carbon, aluminum,** and **sulfur.** Describe the appearance of these elements in your Science Journal.

3 Use a **conductivity tester** to check how well these elements conduct electricity. Record your observations.

4 Wrap each element sample in a **paper towel.** Carefully hit the sample with a **hammer.** Unwrap the towel and observe the sample. Record your observations.

Think About This

1. Locate these elements on the periodic table. From their locations, which elements are metals? Which elements are nonmetals?

2. **Key Concept** Using your results, compare the properties of metals and nonmetals.

3. **Key Concept** What property of a nonmetal makes it useful to insulate electrical wires?

The Elements of Life

Would it surprise you to learn that more than 96 percent of the mass of your body comes from just four elements? As shown in **Figure 12,** all four of these elements—oxygen, carbon, hydrogen, and nitrogen—are nonmetals. **Nonmetals** *are elements that have no metallic properties.*

Of the remaining elements in your body, the two most common elements also are nonmetals—phosphorus and sulfur. These six elements form the compounds in proteins, fats, nucleic acids, and other large molecules in your body and in all other living things.

Reading Check What are the six most common elements in the human body?

Figure 12 Like other living things, this woman's mass comes mostly from nonmetals.

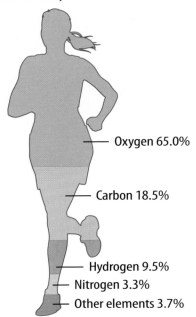

Oxygen 65.0%

Carbon 18.5%

Hydrogen 9.5%
Nitrogen 3.3%
Other elements 3.7%

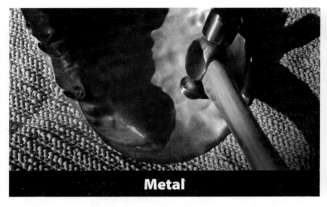

Metal

Nonmetal

▲ **Figure 13** Solid metals, such as copper, are malleable. Solid nonmetals, such as sulfur, are brittle.

Figure 14 Nonmetals have properties that are different from those of metals. Phosphorus and carbon are dull, brittle solids that do not conduct thermal energy or electricity. ▼

How are nonmetals different from metals?

Recall that metals have luster. They are ductile, malleable, and good conductors of electricity and thermal energy. All metals except mercury are solids at room temperature.

The properties of nonmetals are different from those of metals. Many nonmetals are gases at room temperature. Those that are solid at room temperature have a dull surface, which means they have no luster. Because nonmetals are poor conductors of electricity and thermal energy, they are good insulators. For example, nose cones on space shuttles are insulated from the intense thermal energy of reentry by a material made from carbon, a nonmetal. **Figure 13** and **Figure 14** show several properties of nonmetals.

🔑 **Key Concept Check** What properties do nonmetals have?

Properties of Nonmetals 🔑

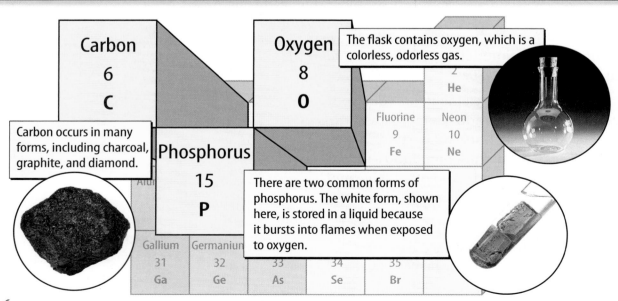

Carbon occurs in many forms, including charcoal, graphite, and diamond.

The flask contains oxygen, which is a colorless, odorless gas.

There are two common forms of phosphorus. The white form, shown here, is stored in a liquid because it bursts into flames when exposed to oxygen.

✓ **Visual Check** Compare the properties of oxygen to those of carbon and phosphorus.

Fluorine   Chlorine   Bromine   Iodine

Figure 15 These glass containers each hold a halogen gas. Although they are different colors in their gaseous state, they react similarly with other elements.

✓ **Visual Check** Compare the colors of these halogens.

Nonmetals in Groups 14–16

Look back at the periodic table in **Figure 4.** Notice that groups 14–16 contain metals, nonmetals, and metalloids. The chemical properties of the elements in each group are similar. However, the physical properties of the elements can be quite different.

Carbon is the only nonmetal in group 14. It is a solid that has different forms. Carbon is in most of the compounds that make up living things. Nitrogen, a gas, and phosphorus, a solid, are the only nonmetals in group 15. These two elements form many different compounds with other elements, such as oxygen. Group 16 contains three nonmetals. Oxygen is a gas that is essential for many organisms. Sulfur and selenium are solids that have the physical properties of other solid nonmetals.

Group 17: The Halogens

An element in group 17 of the periodic table is called a **halogen** (HA luh jun). **Figure 15** shows the halogens fluorine, chlorine, bromine, and iodine. The term *halogen* refers to an element that can react with a metal and form a salt. For example, chlorine gas reacts with solid sodium and forms sodium chloride, or table salt. Calcium chloride is another salt often used on icy roads.

Halogens react readily with other elements and form compounds. They react so readily that halogens only can occur naturally in compounds. They do not exist as free elements. They even form compounds with other nonmetals, such as carbon. In general, the halogens are less reactive as you move down the group.

✓ **Reading Check** Will bromine react with sodium? Explain your answer.

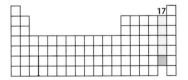

WORD ORIGIN ·············

halogen
from Greek *hals*, means "salt"; and *–gen*, means "to produce"

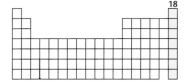

Group 18: The Noble Gases

The elements in group 18 are known as the **noble gases.** The elements helium, neon, argon, krypton, xenon, and radon are the noble gases. Unlike the halogens, the only way elements in this group react with other elements is under special conditions in a laboratory. These elements were not yet discovered when Mendeleev **constructed** his periodic table because they do not form compounds naturally. Once they were discovered, they fit into a group at the far right side of the table.

ACADEMIC VOCABULARY

construct
(verb) to make by combining and arranging parts

Hydrogen

Figure 16 shows the element key for hydrogen. Of all the elements, hydrogen has the smallest atomic mass. It is also the most common element in the universe.

Is hydrogen a metal or a nonmetal? Hydrogen is most often classified as a nonmetal because it has many properties like those of nonmetals. For example, like some nonmetals, hydrogen is a gas at room temperature. However, hydrogen also has some properties similar to those of the group 1 alkali metals. In its liquid form, hydrogen conducts electricity just like a metal does. In some chemical reactions, hydrogen reacts as if it were an alkali metal. However, under conditions on Earth, hydrogen usually behaves like a nonmetal.

 Reading Check Why is hydrogen usually classified as a nonmetal?

Figure 16 More than 90 percent of all the atoms in the universe are hydrogen atoms. Hydrogen is the main fuel for the nuclear reactions that occur in stars.

Hydrogen is a colorless, odorless gas. It is the most common element in the universe.

Hydrogen
1
H

Metalloids

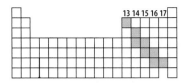

Between the metals and the nonmetals on the periodic table are elements known as metalloids. *A* **metalloid** (MEH tul oyd) *is an element that has physical and chemical properties of both metals and nonmetals.* The elements boron, silicon, germanium, arsenic, antimony, tellurium, polonium, and astatine are metalloids. Silicon is the most abundant metalloid in the universe. Most sand is made of a compound containing silicon. Silicon is also used in many different products, some of which are shown in **Figure 17.**

 Key Concept Check Where are metalloids on the periodic table?

Semiconductors

Recall that metals are good conductors of thermal energy and electricity. Nonmetals are poor conductors of thermal energy and electricity but are good insulators. A property of metalloids is the ability to act as a semiconductor. *A* **semiconductor** *conducts electricity at high temperatures, but not at low temperatures.* At high temperatures, metalloids act like metals and conduct electricity. But at lower temperatures, metalloids act like nonmetals and stop electricity from flowing. This property is useful in electronic devices such as computers, televisions, and solar cells.

WORD ORIGIN · · · · · · · · · · · ·

semiconductor
from Latin *semi-*, means "half"; and *conducere*, means "to bring together"

Figure 17 The properties of silicon make it useful for many different products.

Uses of Silicon

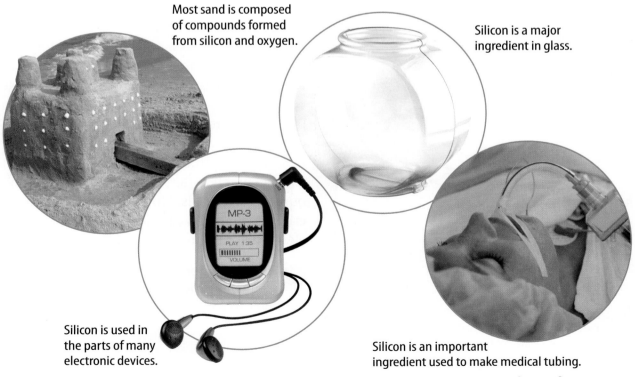

Most sand is composed of compounds formed from silicon and oxygen.

Silicon is a major ingredient in glass.

Silicon is used in the parts of many electronic devices.

MP-3
PLAY 1:35
VOLUME

Silicon is an important ingredient used to make medical tubing.

Figure 18 This microchip conducts electricity at high temperatures using a semiconductor.

📼 Review Personal Tutor

Inquiry MiniLab
15 minutes

Which insulates better?

In this lab, you will compare how well a metal bowl and a nonmetal ball containing a mixture of nonmetals conduct thermal energy.

1. Read and complete a lab safety form.

2. Pour **very warm water** into a **pitcher.**

3. Pour half of the warm water into a **metal bowl.** In your Science Journal, describe how the outside of the bowl feels.

4. Inflate a **beach ball** until it is one-third full. Mold the partially filled beach ball into the shape of a bowl. Pour the remaining warm water into your beach ball bowl. Feel the outside of the bowl. Describe how it feels.

Analyze and Conclude

1. **Explain** the difference in the outside temperatures of the two bowls.

2. **Predict** the results of putting ice in each of the bowls.

3. 🔑 **Key Concept** Make a statement about how well a nonmetal conducts thermal energy.

Properties and Uses of Metalloids

Pure silicon is used in making semiconductor devices for computers and other electronic products. Germanium is also used as a semiconductor. However, metalloids have other uses, as shown in **Figure 18.** Pure silicon and Germanium are used in semiconductors. Boron is used in water softeners and laundry products. Boron also glows bright green in fireworks. Silicon is one of the most abundant elements on Earth. Sand, clay, and many rocks and minerals are made of silicon compounds.

Metals, Nonmetals, and Metalloids

You have read that all metallic elements have common characteristics, such as malleability, conductivity, and ductility. However, each metal has unique properties that make it different from other metals. The same is true for nonmetals and metalloids. How can knowing the properties of an element help you evaluate its uses?

Look again at the periodic table. An element's position on the periodic table tells you a lot about the element. By knowing that sulfur is a nonmetal, for example, you know that it breaks easily and does not conduct electricity. You would not choose sulfur to make a wire. You would not try to use oxygen as a semiconductor or sodium as a building material. You know that transition elements are strong, malleable, and do not react easily with oxygen or water. Because of these characteristics, these metals make good building materials. Understanding the properties of elements can help you decide which element to use in a given situation.

✓ **Reading Check** Why would you not use an element on the right side of the periodic table as a building material?

Visual Summary

A nonmetal is an element that has no metallic properties. Solid nonmetals are dull, brittle, and do not conduct thermal energy or electricity.

Halogens and noble gases are nonmetals. These elements are found in group 17 and group 18 of the periodic table.

Metalloids have some metallic properties and some nonmetallic properties. The most important use of metalloids is as semiconductors.

FOLDABLES

Use your lesson Foldable to review the lesson. Save your Foldable for the project at the end of the chapter.

What do you think NOW?

You first read the statements below at the beginning of the chapter.

5. Most of the elements in living things are nonmetals.

6. Even though they look very different, oxygen and sulfur share some similar properties.

Did you change your mind about whether you agree or disagree with the statements? Rewrite any false statements to make them true.

Use Vocabulary

1. **Distinguish** between a nonmetal and a metalloid.

2. An element in group 17 of the periodic table is called a(n) _____.

3. An element in group 18 of the periodic table is called a(n) _____.

Understand Key Concepts

4. The ability of a halogen to react with a metal to form a salt is an example of a _____ property.
   - **A.** chemical
   - **C.** periodic
   - **B.** noble gas
   - **D.** physical

5. **Classify** each of the following elements as a metal, a nonmetal, or a metalloid: boron, carbon, aluminum, and silicon.

6. **Infer** which group you would expect to contain element 117. Use the periodic table to help you answer this question.

Interpret Graphics

7. **Sequence** nonmetals, metals, and metalloids in order from left to right across the periodic table by copying and completing the graphic organizer below.

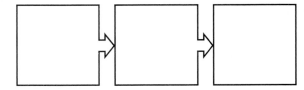

Critical Thinking

8. **Hypothesize** how your classroom would be different if there were no metalloids.

9. **Analyze** why hydrogen is sometimes classified as a metal.

10. **Determine** whether there would be more nonmetals in group 14 or in group 16. Explain your answer.

Alien Insect Periodic Table

The periodic table classifies elements according to their properties. In this lab, you will model the procedure used to develop the periodic table. Your model will include developing patterns using pictures of alien insects. You will then use your patterns to predict what missing alien insects look like.

Question

How can I arrange objects into patterns by using their properties?

Procedure

1. Obtain a set of alien insect pictures. Spread them out so you can see all of them. Observe the pictures with a partner. Look for properties that you might use to organize the pictures.

2. Make a list of properties you might use to group the alien insects. These properties are those that a number of insects have in common.

3. Make a list of properties you might use to sequence the insects. These properties change from one insect to the next in some pattern.

4. With your partner, decide what pattern you will use to arrange the alien insects in an organized rectangular block. All the insects in a vertical column, or group, must be the same in some way. They must also share some feature that changes regularly as you move down the group. All the aliens in a horizontal row, or period, must be the same in some way and must also share some feature that changes regularly as you move across the period.

5. Arrange your insects as you planned. Two insects are missing from your set. Leave empty spaces in your rectangular block for these pictures. When you have finished arranging your insects, have the teacher check your pattern.

6. Write a description of the properties you predict each missing alien insect will have. Then draw a picture of each missing insect.

Analyze and Conclude

7. **Explain** Could you have predicted the properties of the missing insects without placing the others in a pattern? Why or why not?

8. **The Big Idea** How is your arrangement similar to the one developed by Mendeleev for elements? How is it different?

9. **Infer** What properties can you use to predict the identity of one missing insect? What do you not know about that insect?

Communicate Your Results

Create a slide show presentation that demonstrates, step by step, how you grouped and sequenced your insects and predicted the properties of the missing insects. Show your presentation to students in another class.

Inquiry Extension

How could you change the insects so that they better represent the properties of elements, such as atomic mass?

Lab Tips

☑ A property is any observable characteristic that you can use to distinguish between objects.

☑ A pattern is a consistent plan or model used as a guide for understanding or predicting something.

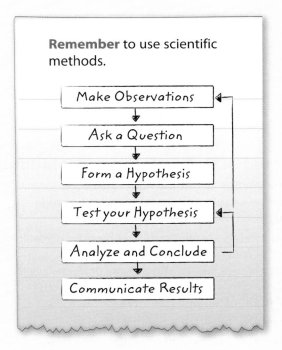

Remember to use scientific methods.

Make Observations
↓
Ask a Question
↓
Form a Hypothesis
↓
Test your Hypothesis
↓
Analyze and Conclude
↓
Communicate Results

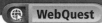
 THE BIG IDEA

Elements are organized on the periodic table according to increasing atomic number and similar properties.

Key Concepts Summary 🔑	Vocabulary
Lesson 1: Using the Periodic Table  • Elements are organized on the **periodic table** by increasing atomic number and similar properties. • Elements in the same **group,** or column, of the periodic table have similar properties. • Elements' properties change across a **period,** which is a row of the periodic table. • Each element key on the periodic table provides the name, symbol, atomic number, and atomic mass for an element.	**periodic table** p. 391 **group** p. 396 **period** p. 396
Lesson 2: Metals  • **Metals** are located on the left and middle side of the periodic table. • Metals are elements that have **ductility, malleability, luster,** and conductivity. • The **alkali metals** are in group 1 of the periodic table, and the **alkaline earth metals** are in group 2. • **Transition elements** are metals in groups 3–12 of the periodic table, as well as the lanthanide and actinide series.	**metal** p. 401 **luster** p. 401 **ductility** p. 402 **malleability** p. 402 **alkali metal** p. 403 **alkaline earth metal** p. 403 **transition element** p. 404
Lesson 3: Nonmetals and Metalloids  • **Nonmetals** are on the right side of the periodic table, and **metalloids** are located between metals and nonmetals. • Nonmetals are elements that have no metallic properties. Solid nonmetals are dull in appearance, brittle, and do not conduct electricity. Metalloids are elements that have properties of both metals and nonmetals. • Some metalloids are **semiconductors.** • Elements in group 17 are called **halogens,** and elements in group 18 are **noble gases.**	**nonmetal** p. 409 **halogen** p. 411 **noble gas** p. 412 **metalloid** p. 413 **semiconductor** p. 413

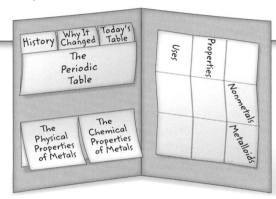
FOLDABLES® Chapter Project

Assemble your lesson Foldables as shown to make a Chapter Project. Use the project to review what you have learned in this chapter.

Use Vocabulary

❶ The element magnesium (Mg) is in _____ 3 of the periodic table.

❷ An element that is shiny, is easily pulled into wires or hammered into thin sheets, and is a good conductor of electricity and heat is a(n) _____.

❸ Copper is used to make wire because it has the property of _____.

❹ An element that is sometimes a good conductor of electricity and sometimes a good insulator is a(n) _____.

❺ An element that is a poor conductor of heat and electricity but is a good insulator is a(n) _____.

Link Vocabulary and Key Concepts

Concepts in Motion Interactive Concept Map

Copy this concept map, and then use vocabulary terms from the previous page to complete the concept map.

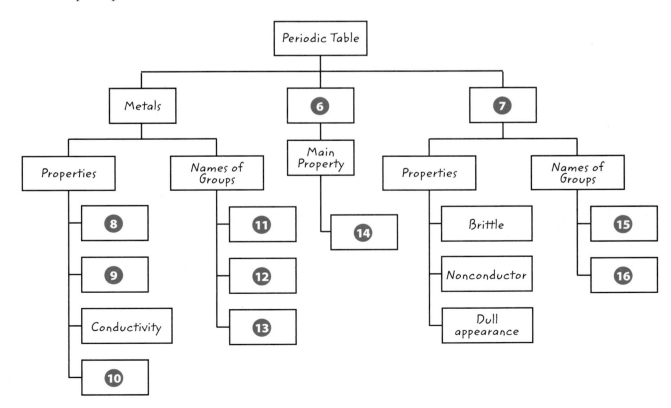

Chapter 11 Review

Understand Key Concepts

1 What determines the order of elements on today's periodic table?

A. increasing atomic mass

B. decreasing atomic mass

C. increasing atomic number

D. decreasing atomic number

2 The element key for nitrogen is shown below.

From this key, determine the atomic mass of nitrogen.

A. 7

B. 7.01

C. 14.01

D. 21.01

3 Look at the periodic table in Lesson 1. Which of the following lists of elements forms a group on the periodic table?

A. Li, Be, B, C, N, O, F, and Ne

B. He, Ne, Ar, Kr, Xe, and Rn

C. B, Si, As, Te, and At

D. Sc, Ti, V, Cr, Mn, Fe, Co, Cu, Ni, and Zn

4 Which is NOT a property of metals?

A. brittleness

B. conductivity

C. ductility

D. luster

5 What are two properties that make a metal a good choice to use as wire in electronics?

A. conductivity, malleability

B. ductility, conductivity

C. luster, malleability

D. malleability, high density

6 Where are most metals on the periodic table?

A. on the left side only

B. on the right side only

C. in the middle only

D. on the left side and in the middle

7 Look at the periodic table in Lesson 1 and determine which element is a metalloid.

A. carbon

B. silicon

C. oxygen

D. aluminum

8 Iodine is a solid nonmetal. What is one property of iodine?

A. conductivity

B. dull appearance

C. malleability

D. ductility

9 The following table lists some information about certain elements in group 17.

Element Symbol	Atomic Number	Melting Point (°C)	Boiling Point (°C)
F	9	−233	−187
Cl	17	−102	−35
Br	35	−7.3	59
I	53	114	183

Which statement describes what happens to these elements as atomic number increases?

A. Both melting point and boiling point decrease.

B. Melting point increases and boiling point decreases.

C. Melting point decreases and boiling point increases.

D. Both melting point and boiling point increase.

Critical Thinking

10 **Recommend** an element to use to fill bottles that contain ancient paper. The element should be a gas at room temperature, should be denser than helium, and should not easily react with other elements.

11 **Apply** Why is mercury the only metal to have been used in thermometers?

12 **Evaluate** the following types of metals as a choice to make a Sun reflector: alkali metals, alkaline earth metals, or transition metals. The metal cannot react with water or oxygen and must be shiny and strong.

13 The figure below shows a pattern of densities.

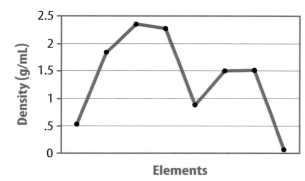

Infer whether you are looking at a graph of elements within a group or across a period. Explain your answer.

14 **Contrast** aluminum and nitrogen. Show why aluminum is a metal and nitrogen is not.

15 **Classify** A student sorted six elements. He placed iron, silver, and sodium in group A. He placed neon, oxygen, and nitrogen in group B. Name one other element that fits in group A and another element that belongs in group B. Explain your answer.

Writing in Science

16 **Write** a plan that shows how a metal, a nonmetal, and a metalloid could be used when constructing a building.

REVIEW THE BIG IDEA

17 Explain how atomic number and properties are used to determine where element 115 is placed on the periodic table.

18 The photo below shows how the properties of materials determine their uses. How can the periodic table be used to help you find elements with properties similar to that of helium?

Math Skills

Use Geometry

19 The table below shows the atomic radii of three elements in group 1 on the periodic table.

Element	Atomic radius
Li	152 pm
Na	186 pm
K	227 pm

a. What is the circumference of each atom?

b. Rubidium (Rb) is the next element in Group 1. What would you predict about the radius and circumference of a rubidium atom?

Standardized Test Practice

Record your answers on the answer sheet provided by your teacher or on a sheet of paper.

Multiple Choice

1 Where are most nonmetals located on the periodic table?

   A in the bottom row

   B on the left side and in the middle

   C on the right side

   D in the top row

Use the figure below to answer question 2.

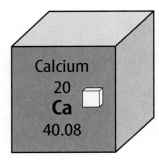

2 What is the atomic mass of calcium?

   A 20

   B 40.08

   C $40.08 \div 20$

   D $40.08 + 20$

3 Which element is most likely to react with potassium?

   A bromine

   B calcium

   C nickel

   D sodium

4 Which group of elements can act as semiconductors?

   A halogens

   B metalloids

   C metals

   D noble gases

Use the table below about group 13 elements to answer question 5.

Element Symbol	Atomic Number	Density (g/cm³)	Atomic Mass
B	5	2.34	10.81
Al	13	2.70	26.98
Ga	31	5.90	69.72
In	49	7.30	114.82

5 How do density and atomic mass change as atomic number increases?

   A Density and atomic mass decrease.

   B Density and atomic mass increase.

   C Density decreases and atomic mass increases.

   D Density increases and atomic mass decreases.

6 Which elements have high densities, strength, and resistance to corrosion?

   A alkali metals

   B alkaline earth metals

   C metalloids

   D transition elements

7 Which is a property of a metal?

   A It is brittle.

   B It is a good insulator.

   C It has a dull appearance.

   D It is malleable.

Use the figure below to answer questions 8 and 9.

17

8 The figure shows a group in the periodic table. What is the name of this group of elements?

 A halogens

 B metalloids

 C metals

 D noble gases

9 Which is a property of these elements?

 A They are conductors.

 B They are semiconductors.

 C They are nonreactive with other elements.

 D They react easily with other elements.

10 What is one similarity among elements in a group?

 A atomic mass

 B atomic weight

 C chemical properties

 D practical uses

Constructed Response

Use the figure below to answer questions 11 and 12.

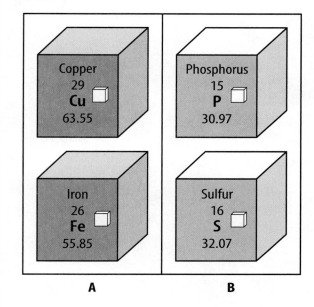

A **B**

11 Groups A and B each contain two elements. Identify each group as metals, nonmetals, or metalloids. Would silicon belong to one of these groups? Why or why not?

12 Which group in the figure above yields the strongest building elements? Why?

13 How does the periodic table of elements help scientists today?

14 What connection does the human body have with the elements on the periodic table?

NEED EXTRA HELP?														
If You Missed Question...	1	2	3	4	5	6	7	8	9	10	11	12	13	14
Go to Lesson...	1	1	3	3	1	2	2	3	3	1	2,3	2	1	3

Chapter 12

Using Energy and Heat

THE BIG IDEA — What are energy transfers and energy transformations?

Inquiry Energy Transformations?

When you look at this photo, do you think of electricity? This power plant transforms the energy stored in coal to the electric energy that people use in their homes and businesses.

- How do you use energy in your home?
- Can you think of other types of energy?
- What are energy transfers and energy transformations?

Get Ready to Read

What do you think?

Before you read, decide if you agree or disagree with each of these statements. As you read this chapter, see if you change your mind about any of the statements.

1. An object sitting on a high shelf has no energy.

2. There are many forms of energy.

3. In most systems, no energy is transferred to the environment.

4. Some forms of energy are replenished naturally.

5. Only particles that make up moving objects are in motion.

6. Thermal energy can be transferred in several ways.

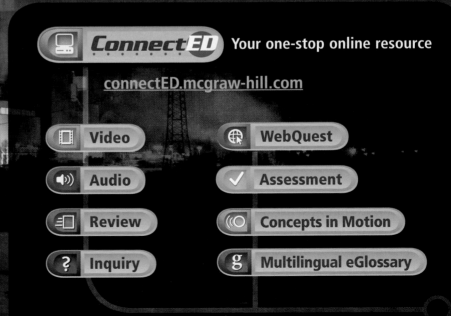

ConnectED Your one-stop online resource

connectED.mcgraw-hill.com

- Video
- Audio
- Review
- Inquiry
- WebQuest
- Assessment
- Concepts in Motion
- Multilingual eGlossary

Lesson 1

Reading Guide

Key Concepts
ESSENTIAL QUESTIONS

- How do potential energy and kinetic energy differ?
- How are mechanical energy and thermal energy similar?
- What two forms of energy are carried by waves?

Vocabulary

energy p. 427

potential energy p. 427

chemical energy p. 428

nuclear energy p. 428

kinetic energy p. 429

electric energy p. 429

mechanical energy p. 430

thermal energy p. 430

wave p. 431

sound energy p. 431

radiant energy p. 432

 Multilingual eGlossary

 Video

- BrainPOP®
- Science Video
- What's Science Got to do With It?

Forms of Energy

Inquiry Got Energy?

This horse and rider need energy to move and to perform their life processes, such as breathing, digesting food, and transporting nutrients and wastes throughout their bodies. The source of energy for both the horse and the rider are the same. It comes from the foods they eat. What form of energy is found in foods? Are there other forms of energy?

Inquiry Launch Lab

15 minutes

Can you transfer energy? 🥽 🧤

You can transfer energy to a ball when you throw it. What are other ways you transfer energy?

1 Read and complete a lab safety form.

2 Hold the handle of a **tuning fork,** and gently strike one arm of the tuning fork with a **mallet.** (Use only the mallet provided for this purpose.)

3 Dip the arms of the tuning fork in a **beaker** of water. Observe what happens, and record your observations in your Science Journal.

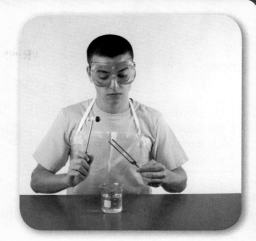

Think About This

1. What is the source of energy that produced sound from the tuning fork? How do you think the tuning fork produced the sound?

2. 🔑 **Key Concept** What happened to the water when the tuning fork was placed in it? Explain what you think happened to the water, the tuning fork, and the energy.

Energy

Some breakfast cereals promise to give you enough energy to get your day off to a great start. News reports often mention the price of oil, which is an energy source that provides fuel for cars and for transporting goods around the world. Meteorologists talk about the approach of a storm system with a lot of energy. News anchors report on earthquakes and tsunamis, which carry so much energy they cause great damage. Politicians talk about the nation's energy policy and the need to conserve energy and to find new energy sources. Energy influences everything in your life, including the climate, the economy, and your body. Scientists define **energy** as *the ability to cause change.*

Potential Energy

Have you ever seen an object perched on the edge of a ledge, such as the one in **Figure 1?** The object's position could easily change, which means it has energy. **Potential energy** *is stored energy due to the interaction between objects or particles.* Objects have potential energy if they have the potential to cause change. Examples of potential energy include objects that could fall due to gravity (interactions between objects—Earth and the egg) and particles that could move because of electric or magnetic forces (interactions between particles—protons and electrons).

Figure 1 🔑 The egg has potential energy because gravitational force can pull it to the ground. A simple nudge could send it falling to the floor.

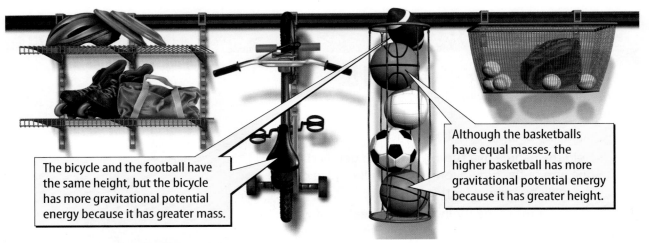

The bicycle and the football have the same height, but the bicycle has more gravitational potential energy because it has greater mass.

Although the basketballs have equal masses, the higher basketball has more gravitational potential energy because it has greater height.

▲ **Figure 2** All of these objects have gravitational potential energy because they have mass and height.

▲ **Figure 3** Your body breaks the chemical bonds in the foods you eat and uses the energy for life processes, including movement.

▲ **Figure 4** In stars, including the Sun, atoms combine or fuse together to form heavier atoms.

Gravitational Potential Energy

Do the items stored on the garage organizer in **Figure 2** have potential energy? The answer is yes, every item— including the shelves and the brackets—has gravitational potential energy. Objects have gravitational potential energy if they have mass and height above Earth's surface.

The gravitational potential of an object energy depends on two factors—the mass of an object and the distance the object is from Earth, as shown in **Figure 2.**

Chemical Energy

Suppose you take the skates from the shelf and play hockey with your friends. Where does your body get the energy it needs? Energy in your body comes from the foods you eat. All objects, including your body and the apple in **Figure 3,** are made of atoms that are joined by chemical bonds. **Chemical energy** *is the energy stored in and released from the bonds between atoms.* Your body breaks chemical bonds in foods and converts the released energy into other forms of energy that your body can use.

Nuclear Energy

The energy stored in and released from the nucleus of an atom is called **nuclear energy.** If you watch a beautiful sunset like the one in **Figure 4,** you experience nuclear energy. The Sun's energy is released through the process of nuclear fusion. During nuclear fusion, the nuclei of atoms join together and release large amounts of energy. Nuclear energy also is released when an atom breaks apart. This breaking apart of an atom is called nuclear fission. Some power plants use nuclear fission to generate, or make, electricity.

mass = 3.6 kg
speed = 0 m/s

mass = 4.5 kg
speed = 8.0 m/s

mass = 5.5 kg
speed = 8.0 m/s

Kinetic Energy

Are you moving your hand as you take notes? Are you squirming in your chair trying to find a comfortable position? If so, you have **kinetic energy**—*energy due to motion.* All objects that have motion have kinetic energy.

Kinetic Energy of Objects

An object's kinetic energy is related to the mass and the **speed** of the object. For example, suppose you are bowling like the people in **Figure 5.** The girl on the left is holding a 3.6-kg bowling ball. Because the ball is not moving, it has no speed and therefore, no kinetic energy. The bowling balls shown in the other two photos have the same speed, 8.0 m/s, but the ball on the right has a greater mass. Therefore, the ball on the right has greater kinetic energy than the ball in the middle.

Electric Energy

Even objects you can't see have kinetic energy. Recall that all materials are made of atoms. In an atom, electrons move around a nucleus. Sometimes electrons even move from one atom to another. Because electrons are moving, they have kinetic energy. When electrons move, they create an electric current. *The energy in an electric current is* **electric energy.** In **Figure 6,** electrons move from one terminal of the battery through the copper wire and bulb to the other terminal of the battery. As the electrons move, their energy is transformed into light. Your brain and the nerves in your body that tell your arm and leg muscles to move also use electric energy.

🔑 **Key Concept Check** How do potential energy and kinetic energy differ?

▲ **Figure 5** The kinetic energy (KE) of an object is related to the object's mass and speed.

✓ **Visual Check** Which of the bowling balls has gravitational potential energy? Explain.

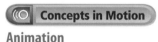 **Concepts in Motion**

Animation

REVIEW VOCABULARY · · · ·
speed
the distance an object moves per unit of time

Figure 6 Electric energy is kinetic energy because the electrons have both mass and motion. ▼

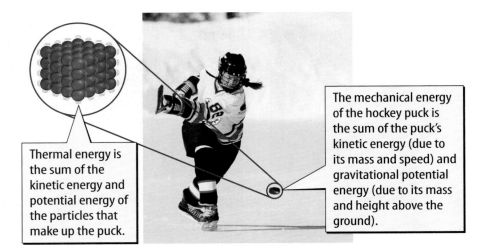

Figure 7 🔑 Mechanical energy is due to large-scale motions and object interactions. Thermal energy is due to atomic-scale motions and particle interactions.

Thermal energy is the sum of the kinetic energy and potential energy of the particles that make up the puck.

The mechanical energy of the hockey puck is the sum of the puck's kinetic energy (due to its mass and speed) and gravitational potential energy (due to its mass and height above the ground).

Combined Kinetic Energy and Potential Energy

Your school is part of an educational system. Earth is part of the solar system. A system is a collection of parts that interact and act together as a whole. In science, everything that is not in a given system is the environment. For example, the hockey player, the hockey stick, the hockey puck, and the ice under the player in **Figure 7** can be considered a system.

Mechanical Energy

Does the hockey puck in **Figure 7** have kinetic energy or potential energy? It has mass and motion, so it has kinetic energy. It also has height above Earth, so it has gravitational potential energy. Scientists often study the energy of systems, such as the one described above. *The sum of the potential energy and the kinetic energy in a system is* **mechanical energy.** You might think of mechanical energy as the ability to move another object. What happens when the hockey puck hits the net? The net moves. The hockey puck has mechanical energy that causes another object to move.

Thermal Energy

Even when the hockey puck is lying on the floor with no obvious motion, the particles that make up the solid puck are in motion—they vibrate back and forth in place. Therefore, the particles have kinetic energy. The particles also have potential energy because of attractive forces between the particles. An object's **thermal energy** *is the sum of the kinetic energy and the potential energy of the particles that make up the object.* Thermal energy of an object increases when the potential energy, the kinetic energy, or both increase.

🔑 **Key Concept Check** Compare mechanical energy and thermal energy.

Figure 8 🔑 When a raindrop falls into water, waves form. The waves carry energy, not matter, across the pond.

Energy Carried by Waves

Have you ever watched a raindrop fall into a still pool of water, as shown in **Figure 8?** The raindrop disturbs the surface of the water and produces waves that move away from the place where the raindrop hit. A **wave** *is a disturbance that transfers energy from one place to another without transferring matter.* Energy, not matter, moves outward from the point where the raindrop hits the water.

✓ **Reading Check** What do waves carry?

Sound Energy

When the raindrop hits the water, you hear a splash. The raindrop not only disturbs the surface of the water, it also disturbs the air. When the raindrop hits the water, it creates a sound wave in the air similar to water waves. Sound waves are waves that move through matter. The wave travels from particle to particle as the particles bump into each other, similar to falling dominoes. **Sound energy** *is energy carried by sound waves.*

As the sound wave travels, it eventually reaches your ear. The sound energy moves tiny hairs inside your ear. This movement is transformed into an electric signal that travels to your brain. Your brain interprets the signal as the sound of a water splash.

Radiant Energy

Have you ever wondered what light is? Light is a form of energy carried by electromagnetic waves, which are electric and magnetic waves moving perpendicularly to one another, as shown in **Figure 9.** *The energy carried by electromagnetic waves is* **radiant energy.** Electromagnetic waves travel through matter and through spaces with little or no matter, such as outer space. Electromagnetic waves often are described by their wavelengths. Wavelength is the distance from one point on a wave to the nearest point just like it.

Visible light is only one form of radiant energy. Gamma rays and X-rays are electromagnetic waves with very short wavelengths. Gamma rays and X-rays often are used in medical procedures. Ultraviolet rays have wavelengths that are a little shorter than those of light. This form of radiant energy is what gives you sunburn. Infrared rays are the energy used by many television remote controls to change channels. They also provide the warmth you feel when the Sun shines on you. Radar, television, and radio waves have long wavelengths compared to visible light.

Key Concept Check What two forms of energy are carried by waves?

Figure 9 Radiant energy is carried by electromagnetic waves (also called rays) with different wavelengths.

Visual Check Determine one type of radiant energy that has a shorter wavelength than visible light and another type that has a longer wavelength.

Electromagnetic Wave

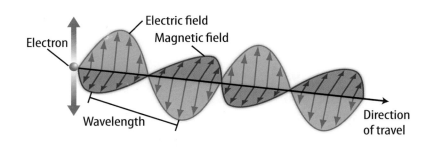

Electron
Electric field
Magnetic field
Wavelength
Direction of travel

Electromagnetic Spectrum

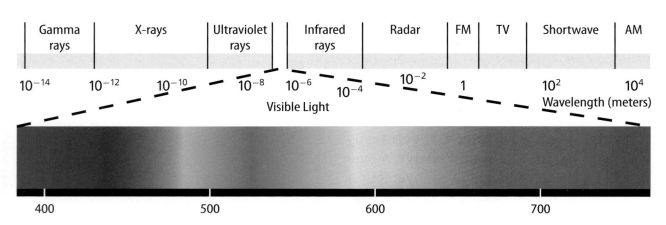

| Gamma rays | X-rays | Ultraviolet rays | Infrared rays | Radar | FM | TV | Shortwave | AM |

10^{-14} 10^{-12} 10^{-10} 10^{-8} 10^{-6} 10^{-4} 10^{-2} 1 10^{2} 10^{4}
Wavelength (meters)

Visible Light

400   500   600   700

Wavelength (nanometers)

Lesson 1 Review

Visual Summary

Objects can have potential energy (stored energy) and kinetic energy (energy due to movement).

Mechanical energy is due to large-scale motions and interactions in a system. Thermal energy is due to atomic-scale motions and interactions in particles.

Two kinds of energy carried by waves are sound energy and radiant energy.

FOLDABLES

Use your lesson Foldable to review the lesson. Save your Foldable for the project at the end of the chapter.

What do you think NOW?

You first read the statements below at the beginning of the chapter.

1. An object sitting on a high shelf has no energy.

2. There are many forms of energy.

Did you change your mind about whether you agree or disagree with the statements? Rewrite any false statements to make them true.

Use Vocabulary

1 The ability to cause change is _____.

2 Energy can be carried by _____, but matter cannot.

Understand Key Concepts

3 **Compare** How are thermal energy and mechanical energy similar?

4 Which form of energy does NOT involve kinetic energy?
 A. chemical   C. mechanical
 B. electric   D. thermal

5 **Explain** why an airplane flying from New York to Los Angeles has both kinetic energy and potential energy.

6 **Compare** how sound waves and electromagnetic waves transfer energy.

Interpret Graphics

7 **Identify** Where are kinetic energy and potential energy the greatest in the loop?

8 **Summarize** Copy and fill in the following graphic organizer to show forms of potential energy.

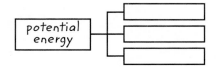

Critical Thinking

9 **List** Which has more kinetic energy: a 5-kg object moving at 5 m/s, 1 m off the ground or a 5-kg object at rest 2 m off the ground?

How can you classify different types of energy?

Materials

magnifying lens

musical greeting card

Safety

When you open a musical greeting card, you probably don't think about how many different forms of energy it uses. What types of energy are used in a musical greeting card?

Learn It

To **classify** means to sort objects into groups based on common features or functions. You can sort different types of energy based on the definitions of each type of energy and your experiences using them. Use this knowledge as you classify the different types of energy used in a musical greeting card.

Try It

1 Read and complete a lab safety form.

2 Carefully observe a musical greeting card before and during its operation. Ask questions such as: "What makes it start? What makes it stop? What is the energy source? What produces the sound?"

3 Write a hypothesis about how you think the greeting card works. What forms of energy are involved in its operation?

4 Carefully dismantle a second musical greeting card. Make a scientific illustration of the electronic parts. Label as many parts as you can. Use a magnifying lens if necessary.

5 Set the dismantled card in operation. Observe any motion. Gently touch parts of the card as it produces sound.

6 Using the card and your illustration, identify and label the different types of energy involved in the card's operation from the time it turns on until it turns off. Record your observations in your Science Journal.

Apply It

7 **Identify Cause and Effect** What causes the card to turn on? What causes it to turn off?

8 **Classify** List all of the types of energy involved in the operation of your card.

9 **Infer** What produces the sound? How does the sound from the card reach your ear?

10 🔑 **Key Concept** What are some examples of potential energy and kinetic energy in the operation of your card? Explain.

Lesson 2

Reading Guide

Key Concepts
ESSENTIAL QUESTIONS

- What is the law of conservation of energy?
- How is energy transformed and transferred?
- What are renewable and nonrenewable energy resources?

Vocabulary

law of conservation of energy p. 436

energy transfer p. 437

energy transformation p. 437

work p. 437

open system p. 439

closed system p. 439

renewable energy resource p. 440

nonrenewable energy resource p. 442

 Multilingual eGlossary

Energy Transfers and Transformations

Inquiry Warm and Cozy?

This penguin chick lives in one of the coldest places on Earth—Antarctica. The chick is standing on its parent's feet to insulate its feet from the ice. This helps prevent thermal energy of the chick's body from transferring to the ice. The chick cuddles with its parent to absorb thermal energy from its parent's body. Without receiving thermal energy from the parent, the baby would quickly die.

How does a flashlight work?

If the lights go out, you might turn on a flashlight. When you push the switch, the light will go on. What happens? How does the flashlight work?

1. Read and complete a lab safety form.

2. Examine a **flashlight.** List the parts that you can see. Predict the types of energy involved in the operation of the flashlight.

3. Use the switch to turn the flashlight on. What do you think happened inside the flashlight to produce the light? Write your ideas in your Science Journal.

4. Take the flashlight apart. Discuss the kinds of energy involved in producing light.

Think About This

1. Was light the only type of energy produced? Why or why not?

2. **Key Concept** Describe the different types of energy involved in a flashlight. Draw a sequence diagram showing how each form of energy changes to the next form.

Figure 10 Several energy changes occur in a flashlight.

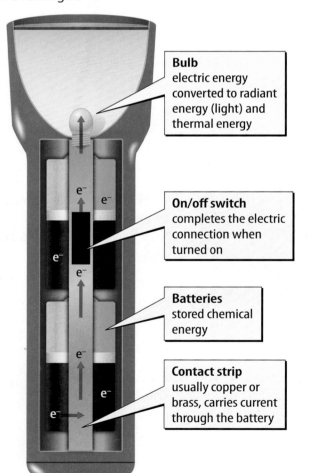

Bulb
electric energy converted to radiant energy (light) and thermal energy

On/off switch
completes the electric connection when turned on

Batteries
stored chemical energy

Contact strip
usually copper or brass, carries current through the battery

Law of Conservation of Energy

Think about turning on the flashlight in the Launch Lab. The **law of conservation of energy** says that *energy can be transformed from one form to another, but it cannot be created or destroyed.* In the flashlight shown in **Figure 10,** chemical energy of the battery is transformed to electric energy (moving electrons) that moves through the contact strip to the bulb. The electric energy is transformed into radiant energy and thermal energy in the lightbulb. The law of conservation of energy indicates that the amount of radiant energy that shines out of the flashlight cannot be greater than the chemical energy stored in the battery.

 Key Concept Check What is the law of conservation of energy?

The amount of radiant energy given off by the flashlight is less than the chemical energy in the battery. Where is the missing energy? As you read this lesson, you will learn that in every energy transformation, some energy transfers to the environment.

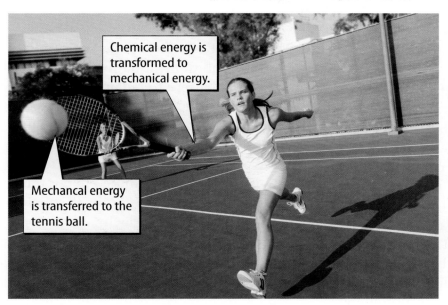

Chemical energy is transformed to mechanical energy.

Mechancal energy is transferred to the tennis ball.

Figure 11 Energy transfers and transformations take place when the tennis player hits the ball.

Visual Check Identify at least one energy transformation that occurs as the ball moves through the air.

Energy Transfer

What happens when the tennis player in **Figure 11** hits the ball with the racket? The mechanical energy of the racket changes the movement of the ball, and the ball's mechanical energy increases. *When energy moves from one object to another without changing form, an* **energy transfer** *occurs.* The tennis racket transfers mechanical energy to the tennis ball.

Energy Transformation

Where does the mechanical energy in the tennis player's racket come from? Chemical energy stored in the player's muscles changes to mechanical energy when she swings her arm. *When one form of energy is converted to another form of energy, an* **energy transformation** *occurs.*

 Key Concept Check Identify an energy transfer and an energy transformation that occurs when someone plays a guitar.

Energy and Work

You might think that reading about energy is a lot of work. But to a scientist, it's not work at all. To a scientist, **work** *is the transfer of energy that occurs when a force makes an object move in the direction of the force. Work is only being done while the force is acting on the object.* As the tennis player swings the racket, the racket applies a force to the ball for about a meter. Although the ball moves 10 m, work is done by the racket only during the time the racket applies a force to the ball. When the ball separates from the racket, the racket no longer does work.

Suppose the tennis player is standing still before she serves the ball. She is using her muscles to hold the ball. Is she doing work on the ball? No; because the ball is not moving, she is not doing work. If a force does not make an object move in the direction of the force, it does no work on the object.

Math Skills

Use a Formula

The amount of work done on an object is calculated using the formula $W = F \times d$, where W = work, F = the force applied to the object, and d = the distance the force moves the object. For example, a student slides a library book across a table. The student pushed the book with a force of **8.5 newtons (N)** a distance of **0.3 m**. The book slides a total distance of 1 m. How much work is done on the book?

$W = F \times d$

$W = 8.5\,N \times 0.30\,m = 2.55\,N{\cdot}m = 2.6\,J$

Recall that the distance used to calculate work is the distance the force was applied to the object.

Note: 1 N·m = 1 J, so 2.6 J of work was done on the book.

Practice

A student lifts a backpack straight up with a force of 53.5 N for a distance of 0.65 m. How much work is done on the backpack?

 Review

- **Math Practice**
- **Personal Tutor**

Inefficiency of Energy Transformations

When a tennis player hits a ball with a racket, most of the mechanical energy of the racket transfers to the ball, but not all of it. You know when a ball hits a racket because you can hear a sound. Some of the mechanical energy of the racket is transformed to sound energy. In addition, some of the mechanical energy of the racket is transformed to thermal energy. The temperature of the racket, the ball, and the air surrounding both objects increases slightly. Anytime there is an energy transformation or energy transfer, some energy is transformed into thermal energy.

 Reading Check Summarize the energy transformations that occur when a tennis racket hits a tennis ball.

Recall the flashlight at the beginning of the lesson. The transformation of chemical energy of the battery to radiant energy from the lightbulb is inefficient, too. As the electric energy moves through the circuit, some electric energy transforms to thermal energy. When electric energy transforms to radiant energy in the lightbulb, more energy transforms to thermal energy. In some flashlights, the bulb is warm to the touch.

Recall that the law of conservation of energy says that energy cannot be created or destroyed. When scientists say that energy transformations are inefficient, they do not mean that energy is destroyed. Energy transformations are inefficient because not all the energy that is transformed to another form of energy is usable.

Inquiry MiniLab   20 minutes

How can you transfer energy?

When you ride your bicycle, pushing the pedals with your feet makes the wheels turn. However, your feet don't touch the wheels. How can you transfer energy to another object without touching it?

1. Read and complete a lab safety form.

2. Place a **cork** at one edge of a **rectangular pan** half-filled with water.

3. Discuss with your teammates how you can make the cork move to the opposite edge of the pan without touching it. Possible tools include a **drinking straw**, a **plastic spoon,** and a **length of string,** but you can't physically touch the cork with these items. You also may use any other methods you can think of without touching the cork.

4. Try each of your ideas. Record your results in your Science Journal.

Analyze and Conclude

1. **Explain** which method that you tried was the most effective and the least effective.

2. **Hypothesize** What would make the cork easier to move? Explain.

3. **Key Concept** What form of energy did you start with to move the cork? In what ways was the energy transferred and transformed during each trial?

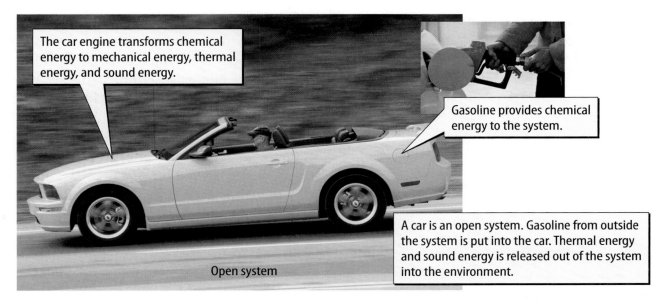

The car engine transforms chemical energy to mechanical energy, thermal energy, and sound energy.

Gasoline provides chemical energy to the system.

A car is an open system. Gasoline from outside the system is put into the car. Thermal energy and sound energy is released out of the system into the environment.

Open system

Open Systems

In Lesson 1, you read that scientists often study the energy of systems. A car, as shown in **Figure 12,** is a system. Chemical energy of the fuel is transformed to mechanical energy of the moving car. Because energy transformations are inefficient, some of the chemical energy transforms to thermal energy and sound energy that are released to the environment. An **open system,** such as a car engine, *is a system that exchanges matter or energy with the environment.*

Closed Systems

Can you think of a system that does not exchange energy with the environment? What about a flashlight? A flashlight releases radiant energy and thermal energy into the environment. What about your body? You eat food, which contains chemical energy and comes from the environment. Your body also releases several types of energy into the environment, including thermal energy, mechanical energy, and sound energy. *A* **closed system** *is a system that does not exchange matter or energy with the environment.* In reality, there are no closed systems. Every physical system transfers some energy to or from its environment. Scientists use the idea of a closed system to study and model the movement of energy.

Energy Transformations and Electric Energy

You probably have heard someone say, "turn off the lights, you're wasting energy." This form of energy is electric energy. Most appliances you use every day require electric energy. Where does this energy come from?

Figure 12 A car is an open system. Gasoline is an input. Thermal energy and sound energy are outputs.

Review
Personal Tutor

Word Origin

system
from Greek *systema,* means "whole made of several parts"

ACADEMIC VOCABULARY··············

resource

(*noun*) a stock or supply of materials, money, or other assets that can be used as needed

·····································

Renewable Energy Resources

If you think about all of the energy used in the United States, you realize that people need a lot of energy to continue living the way they do. This huge demand for energy and the desire to protect the environment has resulted in a search for renewable energy resources. *A* **renewable energy resource** *is an energy resource that is replaced as fast as, or faster than, it is used.* There are several different kinds of renewable energy resources.

Solar Radiant energy from the Sun, or solar energy, is one energy resource that can be converted into electric energy. Some solar energy plants, such as the one in **Figure 13,** transform radiant energy into electric energy with photovoltaic (foh toh vohl TAY ihk), or solar, cells. Photovoltaic cells are made from thin wafers of the element silicon. When radiant energy from the Sun hits the cells, it knocks electrons away from the silicon atoms. This movement of electrons is electric energy. Some homes, businesses, and small appliances, such as calculators, use photovoltaic cells to provide electricity.

▲ **Figure 13** Solar power plants transform radiant energy from the Sun to electric energy.

In some solar energy plants, radiant energy from the Sun is transformed into thermal energy. The thermal energy is used to convert water to steam. The steam turns a generator, which transforms mechanical energy into electric energy.

Wind Have you ever driven along a highway and seen wind turbines such as those in **Figure 14?** Wind turbines are built in places where winds blow almost continuously, such as the vast open spaces of the southwestern United States. Wind moves the blades of the turbine, turning a generator that transforms kinetic energy of the wind to electric energy. One of the drawbacks of wind energy is that wind does not blow steadily at all times. This source of electric energy is not very consistent or predictable.

▲ **Figure 14** Wind turbines transform kinetic energy from the wind to electric energy.

Hydroelectric If you ever have stood underneath a shower, you have felt the energy of falling water. In hydroelectric plants, falling water from rivers and dams is channeled through a turbine. When the turbine spins, mechanical energy is transformed to electric energy. Most of the hydroelectric energy produced in the United States comes from the western part of the country. The hydroelectric plant shown in **Figure 15** is at Shasta Lake in California.

The major drawback of hydroelectric energy is that dams and turbines can interrupt the natural movements of animals in rivers and lakes. Also, there are a limited number of places where rivers are large enough for these energy plants to be built.

Geothermal Earth's temperature increases as you go deeper below the surface. But in a few places, Earth is very hot close to the surface. Geothermal plants are built where thermal energy from Earth is near Earth's surface. In these energy plants, thermal energy is transferred to water creating steam. The steam turns turbines in electric generators. The states with the most geothermal reservoirs are Alaska, Hawaii, and California. The geothermal reservoir shown in **Figure 16** has been producing geothermal power in California since 1921.

Biomass Wood, plants, and even manure and garbage are considered biomass. These sources of stored chemical energy can be transformed to electric energy in energy plants like the one in **Figure 17.** Burning biomass releases carbon dioxide into the atmosphere. Some scientists believe this contributes to climate change and global warming. However, when biomass crops are grown, the plants use carbon dioxide during photosynthesis, reducing the overall amount of carbon dioxide produced in the process.

Reading Check Which energy plant is usually built on a lake or a river?

▲ **Figure 15** Potential energy stored in elevated water is transformed into electric energy in some energy plants.

▲ **Figure 16** In geothermal energy plants such as The Geysers, thermal energy inside Earth is transformed into electric energy.

 **Concepts in Motion** Animation

▲ **Figure 17** Chemical energy stored in biomass, such as wood, plants, manure, and garbage, is used to generate electric energy in some locations.

Electrical Energy Resources

Renewable resources 8.6%

Nonrenewable resources 91.4%

Table 1 Electric Energy Net Generation by Resources as of 2007				
Nonrenewable Resources			**Renewable Resources**	
Resource	**Percentage**		**Resource**	**Percentage**
petroleum	1.6		biomass	about 1.0
natural gas	21.6		hydroelectric	5.8
coal	48.5		geothermal	<1.0
other gases	0.3		wind	<1.0
uranium (nuclear power)	19.4		solar and other	<1.0
Total	**91.4**		**Total**	**About 8.6**

Table 1 Most of the energy used in the United States comes from nonrenewable energy resources.

Visual Check Which resource is used to produce the most electric energy in the United States?

FOLDABLES

Make a vertical two-tab book and label it as shown. Use it to explain renewable and non-renewable resources.

Renewable Resources

Nonrenewable Resources

Nonrenewable Energy Resources

Most of the energy used in homes, schools, stores, and businesses comes from fossil fuels and nuclear energy, as shown in **Table 1.** Fossil fuels and nuclear energy are **nonrenewable energy resources**—*an energy resource that is available in limited amounts or that is used faster than it can be replaced in nature.*

Fossil Fuels Petroleum, natural gas, propane, and coal are fossil fuels. Ancient plants stored radiant energy from the Sun as chemical energy in their molecules. This chemical energy was passed on to the animals that ate the plants. Over millions of years, geological processes converted the remains of these ancient plants and animals into fossil fuels. Fossil fuels are a very concentrated form of chemical energy that easily transforms into other forms of energy. However, when fossil fuels burn, they release harmful wastes such as sulfur dioxide, nitrogen oxide, and carbon dioxide. Sulfur dioxide and nitrogen oxide contribute to acid rain. Carbon dioxide is suspected of contributing to global climate change.

Nuclear Energy In nuclear energy plants, uranium atoms are split apart in a process called nuclear fission. Nuclear fission produces thermal energy, which heats water, producing steam. The steam turns turbines that produce electric energy. While nuclear energy plant emissions are not harmful, the waste from these plants is radioactive. The safe disposal of radioactive waste is a major challenge associated with nuclear energy.

Key Concept Check What are renewable and nonrenewable energy resources?

Lesson 2 Review

Visual Summary

Energy can be transferred and transformed, but it cannot be created or destroyed.

Systems are classified as open systems or closed systems based on their interactions with their environment.

Energy resources are classified as renewable or nonrenewable based on their abundance and availability.

FOLDABLES

Use your lesson Foldable to review the lesson. Save your Foldable for the project at the end of the chapter.

What do you think NOW?

You first read the statements below at the beginning of the chapter.

3. In most systems, no energy is transferred to the environment.

4. Some forms of energy are replenished naturally.

Did you change your mind about whether you agree or disagree with the statements? Rewrite any false statements to make them true.

Use Vocabulary

1 **Define** *energy transformation.*

Understand Key Concepts

2 **Paraphrase** the law of conservation of energy.

3 Which of the following is NOT an example of energy transformations?
   A. A bicyclist pedals to school.
   B. A car's engine moves a car.
   C. A sound wave travels across a hall.
   D. The Sun shines on a tree, and it produces more cells.

4 **Describe** a kind of renewable energy that you could use in your home, and explain why it is a good choice.

Interpret Graphics

5 **Explain** why the object shown here is an open system.

6 **Summarize** Copy and fill in the following graphic organizer to show four possible renewable resources of electric energy.

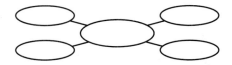

Critical Thinking

7 **Consider** A car company wants to build a wind-powered car that converts 100 percent of the mechanical energy in the wind to the mechanical energy of the moving car. Explain why the company will fail.

Math Skills ✕÷ Review
─── Math Practice ───

8 A child pushes a toy truck with a force of 5.6 N a distance of 3.5 m. How much work is done on the truck?

Biomass

Fresh Ideas About Not-So-Fresh Sources of Fuel

Grass clippings can be broken down to produce methane gas. The device used to convert grass clippings to methane gas is called a digester. The methane gas produced can be used in place of natural gas in appliances. It also can be used to power turbines that produce electricity.

Many people consider grass clippings, dog waste, and used cooking oil garbage. But instead of seeing these materials as garbage, some innovative thinkers see them as sources of biomass energy.

For centuries, humans have used biomass, such as wood, for energy. New technology is expanding the ways biomass can be used. Now biomass can be converted into fuel used to power automobiles, heat homes, and generate electricity.

The use of these fuel sources has several benefits. First, any fuel produced using biomass decreases dependence on nonrenewable resources such as fossil fuels. Also, use of these materials for fuel decreases the amount of waste going to landfills. Technologies to generate fuel from biomass continue to be developed and improved. Who knows what will be used for fuel in the future!

▲ Dog waste can be converted to fuel in a methane digester. Dog waste contains a lot of energy because of the healthy, energy-rich foods fed to most dogs in the United States.

▲ Used cooking oil is an expensive disposal problem for restaurants. Instead, it can be collected and used on its own or combined with diesel fuel to power specially equipped vehicles.

It's Your Turn

RESEARCH Technology has been developed that uses waste from slaughterhouses to make oil. Find out more about this process. Prepare an oral presentation to share what you have learned with other students in your class.

Lesson 3

Particles in Motion

g **Multilingual eGlossary**

Video **BrainPOP®**

Inquiry Catchin' Some Waves?

This Agama lizard regulates its body temperature by absorbing thermal energy from its environment. Some lizards raise their body temperature well above the air temperature by absorbing radiant energy transferred by waves from the Sun.

Where is it the hottest? 🔒 Tie back hair and roll up sleeves.

Would your hands get just as warm if you held them at the sides of a campfire instead of directly over a campfire?

1 Read and complete a lab safety form. Copy the table into your Science Journal. .

2 Use **modeling clay** to hold a **birthday candle** upright. Use a **ring stand and clamp** to mount a **thermometer** horizontally above the candle. The thermometer bulb should be 10 cm above the top of the candle. Record the temperature on the thermometer in your table. Use a **match** to light the candle. Record the temperature every 30 seconds until the temperature reaches 70°C. Add more time columns to the table if needed. Blow out the candle.

⚠ *Do not put thermometer within 10 cm of the flame.*

3 Repeat steps 3–5 with a new candle. This time mount the thermometer 10 cm to the side of the candle flame.

Thermometer Above Flame				
Time (sec)	0	30	60	90
Temp. (°C)				

Thermometer to the Side of Flame				
Time (sec)	0	30	60	90
Temp. (°C)				

Think About This

🔑 **Key Concept** How do you think the energy from the flame traveled to the thermometer in each trial? Explain.

Figure 18 Particles that make up all matter, including carbonated beverages, are in constant motion. On average, solid particles move slowest, liquid particles move faster, and gas particles move the fastest.

⦿ Concepts in Motion Animation

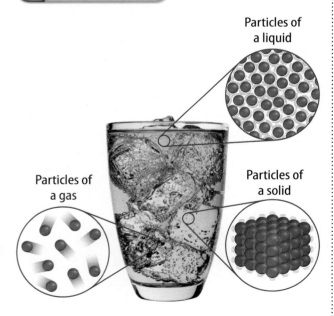

Particles of a liquid

Particles of a gas

Particles of a solid

Kinetic Molecular Theory

You read in Lesson 2 that in every energy transformation some of the energy is transformed into thermal energy. Some of this thermal energy transfers to other materials. The transfer of thermal energy between materials depends on the movement of particles in the materials. The kinetic molecular theory explains how particles move. It has three major points:

- All matter is made of particles.

- Particles are in constant, random motion.

- Particles constantly collide with each other and with the walls of their container.

The kinetic molecular theory explains that the carbonated beverage in **Figure 18** is made of particles. The particles move in different directions and at different speeds. They collide with each other and with the particles that make up the ice and the glass.

🔑 **Key Concept Check** What are the three points of the kinetic molecular theory?

Temperature

When you pick up a glass of ice cold soda, the glass feels cold. Could you estimate its temperature? The temperature of something depends on how much kinetic energy the particles that make up the material have. *The measure of the average kinetic energy of the particles in a material is* **temperature.** If most of the drink particles have little kinetic energy, the drink has a low temperature and the glass feels cold. The SI unit for temperature is the kelvin (K). However, scientists often use the Celsius temperature scale (°C) to measure temperature.

Thermal Expansion

Suppose your teacher told everyone in your classroom to run around. There probably would not be enough space in your classroom for everyone to run as fast as they could. But, if you were in a large gymnasium, then everyone could run very quickly. When the particles that make up a material move slowly, they occupy less volume than they do at a higher temperature. As the temperature of a material increases, particles begin to move faster. They collide with each other more often and push each other farther apart. Thermal expansion is the increase in volume of a material due to a temperature increase, as shown in **Figure 19.** When the temperature of a material decreases, its volume decreases, This is thermal contraction.

Most materials contract as their temperature decreases, but water is an exception. When water is cooled to near its freezing point, interactions between water molecules push the molecules apart. Water expands as it freezes because of these molecular interactions, as shown in **Figure 20.**

Figure 19 The balloon on the top was cooled to −198°C using liquid nitrogen. As the balloon warms to room temperature, the molecules move faster and expand. The balloon undergoes thermal expansion. ▼

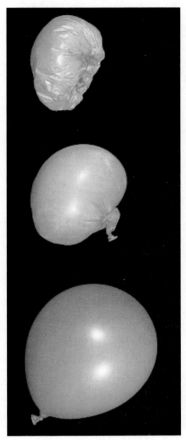

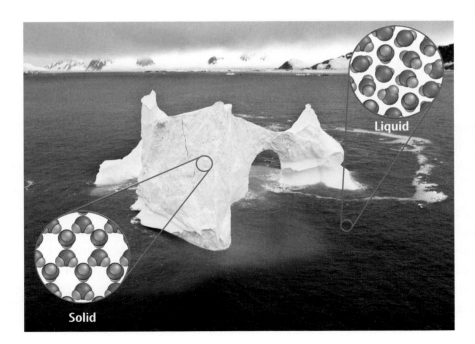

◄ **Figure 20** Because of the structure of a water molecule, as water freezes, the molecules attract in a way that creates empty spaces between them. This makes ice less dense than water. Because ice is less dense than water, ice floats on water.

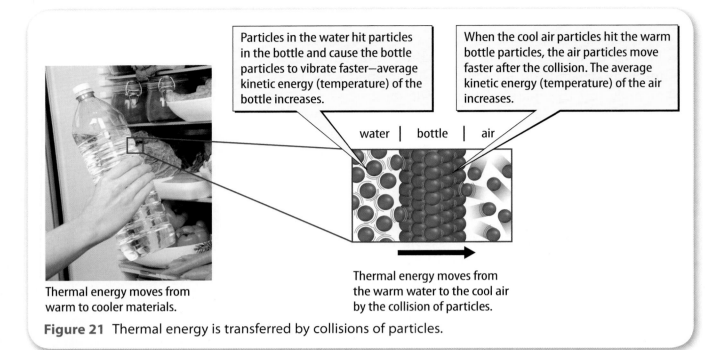

Particles in the water hit particles in the bottle and cause the bottle particles to vibrate faster—average kinetic energy (temperature) of the bottle increases.

When the cool air particles hit the warm bottle particles, the air particles move faster after the collision. The average kinetic energy (temperature) of the air increases.

water | bottle | air

Thermal energy moves from warm to cooler materials.

Thermal energy moves from the warm water to the cool air by the collision of particles.

Figure 21 Thermal energy is transferred by collisions of particles.

Transferring Thermal Energy

Suppose you put a warm bottle of water in the refrigerator. As shown in **Figure 21,** moving water molecules collide with the particles that make up the bottle. These collisions transfer kinetic energy to the particles that make up the bottle, and they vibrate faster. As the particles move faster, their average kinetic energy, or temperature, increases. The particles that make up the bottle then collide with particles that make up the air in the refrigerator.

The average kinetic energy of the particles that make up the air in the refrigerator increases. In other words, the temperature of the air in the refrigerator increases. The average kinetic energy of the particles of water decreases as thermal energy moves from the water to the bottle. Therefore, the temperature of the water decreases.

As the kinetic energy of the particles that make up a material increases, the thermal energy of the particles increases. As the kinetic energy of the particles that make up a material decreases, the thermal energy of the particles decreases. So, when particles transfer kinetic energy, they transfer thermal energy.

Thermal Energy and Heat

Thermal energy moves from warmer materials, such as the warm water in the bottle, to cooler materials, such as the cool air in the refrigerator. *The movement of thermal energy from a region of higher temperature to a region of lower temperature is called* **heat.** Because your hand is warmer than the water bottle, thermal energy moves from your hand to the bottle. When you place the warm bottle in the refrigerator, thermal energy moves from the warm bottle to the cool air in the refrigerator.

Thermal Equilibrium

What happens if you leave the water in the refrigerator for several hours? The temperature of the water, the bottle, and the air in the refrigerator become the same. When the temperatures of materials that are in contact are the same, the materials are said to be in thermal **equilibrium.** After the materials reach thermal equilibrium, the particles that make up the water, the bottle, and the air continue to collide with each other. The particles transfer kinetic energy back and forth, but the average kinetic energy of all the particles remains the same.

Heat Transfer

Suppose you want to heat water to cook pasta, as shown in **Figure 22.** You put a pan of water on the stove and turn the stove on. How is thermal energy transferred to the water?

1 Conduction Fast-moving particles of the gases in the flame collide with the particles that make up the pan. This transfers thermal energy to the pan. Then, the particles that make up the pan collide with particles of water, transferring thermal energy to the water. **Conduction** *is the transfer of thermal energy by collisions between particles in matter.*

2 Radiation If you put your hands near the side of the pan, you feel warmth. The thermal energy you feel is from **radiation**— *the transfer of thermal energy by electromagnetic waves.* All objects emit radiation, but warmer materials, such as hot water, emit more radiation than cooler ones.

3 Convection The flame, or hot gases, heats water at the bottom of the pan. The water at the bottom of the pan undergoes thermal expansion and is now less dense than the water above it. The denser water sinks and forces the less dense, warmer water upward. The water continues this cycle of warming, rising, cooling, and sinking, as thermal energy moves throughout the water. *The transfer of thermal energy by the movement of the particles from one part of a material to another is* **convection.** Convection also occurs in the atmosphere. Warm, less-dense air is forced upward by cooler, more-dense falling air. Thermal energy is transferred as the air rises and sinks.

 Key Concept Check In what three ways is thermal energy transferred?

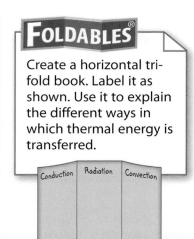
WORD ORIGIN

equilibrium
from Latin *aequus*, means "equal"; and *libra*, means "a balance or scale"

Figure 22 Conduction, radiation, and convection are ways in which thermal energy is transferred.

Review

Personal Tutor

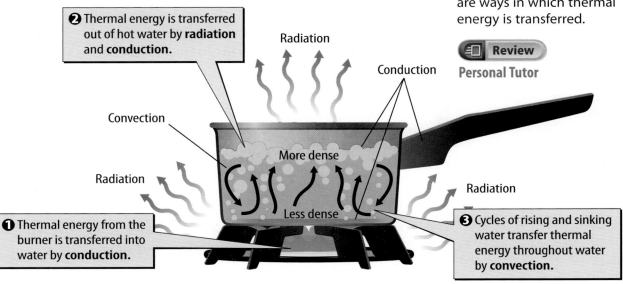

2 Thermal energy is transferred out of hot water by **radiation** and **conduction.**

Radiation

Conduction

Convection

Radiation

More dense

Less dense

Radiation

1 Thermal energy from the burner is transferred into water by **conduction.**

3 Cycles of rising and sinking water transfer thermal energy throughout water by **convection.**

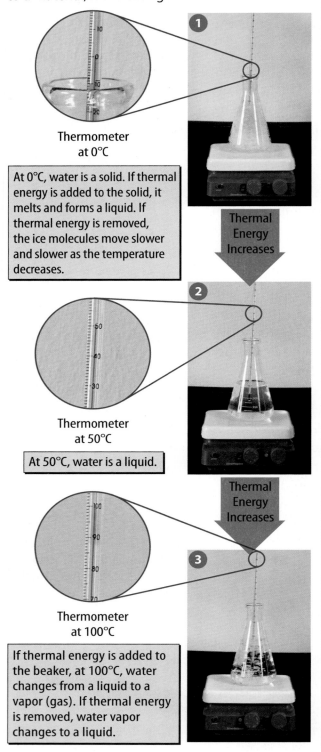

Figure 23 If enough thermal energy is added to a material, it will change state.

Thermometer at 0°C

At 0°C, water is a solid. If thermal energy is added to the solid, it melts and forms a liquid. If thermal energy is removed, the ice molecules move slower and slower as the temperature decreases.

Thermal Energy Increases

Thermometer at 50°C

At 50°C, water is a liquid.

Thermal Energy Increases

Thermometer at 100°C

If thermal energy is added to the beaker, at 100°C, water changes from a liquid to a vapor (gas). If thermal energy is removed, water vapor changes to a liquid.

Heat and Changes of State

When thermal energy is added or removed from a substance, sometimes only the temperature changes. At other times, a more dramatic change occurs—a change of state.

Changes Between Solids and Liquids

What happens if you place a flask of ice on a hot plate, as shown in **Figure 23?** Thermal energy moves from the hot plate to the flask then to the ice. The temperature of the ice increases. When the temperature of the ice reaches the melting point of ice, 0°C, the ice begins to melt. Melting is the change of state from a solid to a liquid. Although ice melts at 0°C, other materials have different melting points. For example, helium melts at −272°C, silver melts at 962°C, and diamonds melt at a temperature over 3,550°C.

As thermal energy transfers to the melting ice, the temperature (average kinetic energy) of the ice does not change. However, the potential energy of the ice increases. As the water molecules move farther apart, the potential energy between the molecules increases.

The reverse process occurs when thermal energy is removed from water. When water is placed in a freezer, thermal energy moves from the water to the colder air in the freezer. The average kinetic energy (temperature) of the water decreases. When the temperature of the water reaches 0°C, the water begins to freeze. Freezing is the change of state from a liquid to a solid. Notice that the freezing point of water is the same as the melting point of ice. Freezing is the opposite of melting.

While water is freezing, the temperature remains at 0°C until all the water is frozen. Once all the water freezes, the temperature of the ice begins to decrease. As the temperature decreases, the water molecules vibrate in place at a slower and slower rate.

✓ **Reading Check** What is a change of state?

Changes Between Liquids and Gases

What happens when ice melts? As thermal energy transfers to the ice, the particles move faster and faster. The average kinetic energy of the water particles that make up ice increases and the ice melts. The temperature of the water continues to increase until it reaches 100°C. At 100°C, water begins to vaporize. **Vaporization** *is the change of state from a liquid to a gas.* While the water is changing state—from a liquid to a gas—the kinetic energy of the particles remains constant.

Liquids vaporize in two ways—boiling and evaporation. Vaporization that occurs within a liquid is called boiling. Vaporization that occurs at the surface of a liquid is called evaporation. Have you heard the term *water vapor?* The gaseous state of a substance that is normally a liquid or a solid at room temperature is called vapor. Because water is liquid at room temperature, its gaseous state is referred to as water vapor.

The reverse process also can occur. Removing thermal energy from a gas changes it to a liquid. The change of state from a gas to a liquid is condensation. The condensation of water vapor that forms on grass overnight is called dew.

Changes Between Solids and Gases

Usually, water transforms from a solid to a liquid and then to a gas as it absorbs thermal energy. However, this is not always the case. On cold winter days, ice often changes directly to water vapor without passing through the liquid state. **Sublimation** is the change of state that occurs when a solid changes to a gas without passing through the liquid state. Dry ice, or solid carbon dioxide, sublimes as shown in **Figure 24.** Dry ice is used to keep foods frozen when they are shipped.

When thermal energy is removed from some materials, they undergo deposition. Deposition is the change of state from a gas directly to a solid without passing through the liquid state. Water vapor undergoes deposition when it freezes and forms frost, as shown in **Figure 24.**

Figure 24 Not all materials go through all three states of matter when they change state. Some materials undergo sublimation (left), and other materials undergo deposition (right).

✅ **Visual Check** How are sublimation and deposition related?

Figure 25 The color variations of this thermogram show the temperature variations in the pan and stove burner. The temperature scale is from white (warmest) through red, yellow, green, cyan, blue, and black (coolest).

✓**Visual Check** Why are the handles black?

Conductors and Insulators

When you put a metal pan on a burner, the pan gets very hot. If the pan has a handle made of wood or plastic, such as the one in **Figure 25,** the handle stays cool. Why doesn't the handle get hot like the pan as a result of thermal conduction?

The metal that makes up the pan is a **thermal conductor,** *a material in which thermal energy moves quickly.* The atoms that make up thermal conductors have electrons that are free to move, transferring thermal energy easily. The material that makes up the pan's handles is a **thermal insulator,** *a material in which thermal energy moves slowly.* The electrons in thermal insulators are held tightly in place and do not transfer thermal energy easily.

 Key Concept Check How do thermal conductors differ from thermal insulators?

Inquiry MiniLab   **25 minutes**

What affects the transfer of thermal energy?

Ice-cold water stays cold longer in a foam cup than in a glass beaker. What other materials keep liquids cold?

1 Read and complete a lab safety form.

2 Place 75 mL of very warm water in each of three **100-mL beakers.**

3 Place a piece of **aluminum foil** over the first beaker and a piece of **cotton batting** over the second beaker. Leave the third beaker open.

4 Place **ice cubes** of equal sizes in three **petri dishes.** Place one dish on top of each beaker. Use a **stopwatch** to measure the time it takes for each ice cube to melt.

5 Make a table in your Science Journal. Label the first column of the rows *Beaker 1, Beaker 2,* and *Beaker 3*. In the second column, record the time it takes each ice cube to melt.

Analyze and Conclude

1. **Identify Cause and Effect** What caused the ice cubes over each beaker to melt? Use the kinetic molecular theory in your explanation.

2. **Identify Relationships** What role did thermal conductors and thermal insulators play in the rate at which the ice cubes melted?

3. **Key Concept** Describe the ways thermal energy transferred from the beakers to the ice.

Lesson 3 Review

Visual Summary

The kinetic molecular theory explains how particles move in matter.

Thermal energy is transferred in various ways by particles and waves.

Materials vary in how well they conduct thermal energy.

FOLDABLES

Use your lesson Foldable to review the lesson. Save your Foldable for the project at the end of the chapter.

What do you think NOW?

You first read the statements below at the beginning of the chapter.

5. Only particles that make up moving objects are in motion.

6. Thermal energy can be transferred in several ways.

Did you change your mind about whether you agree or disagree with the statements? Rewrite any false statements to make them true.

Use Vocabulary

1. **Define** temperature in your own words.

2. **Explain** how heat is related to thermal energy.

Understand Key Concepts

3. **Summarize** the kinetic molecular theory.

4. Which of the following is NOT a way in which thermal energy is transferred?
   - **A.** conduction
   - **C.** radiation
   - **B.** convection
   - **D.** sublimation

5. **Differentiate** between a cloth safety belt and a metal buckle in terms of thermal conductors and insulators.

Interpret Graphics

6. **Explain** how the polar bear below can gain thermal energy if the temperature of the air is below freezing.

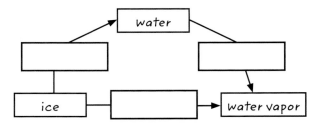

7. **Summarize** Copy and fill in the graphic organizer below showing the state-of-matter changes as thermal energy is added to ice.

```
                    ┌─────────┐
              ┌────▶│  water  │────┐
              │     └─────────┘    │
     ┌────────────┐         ┌────────────┐
     │            │         │            │
     └────────────┘         └────────────┘
          │                        │
     ┌─────────┐  ┌──────────┐  ┌─────────────┐
     │   ice   │──│          │─▶│ water vapor │
     └─────────┘  └──────────┘  └─────────────┘
```

Critical Thinking

8. **Compare** You hold a 65°C cup of cocoa. Your hand is 37°C and the outside air is 6°C. Describe the flow of thermal energy.

Materials

galvanized nails

pennies

LED bulb

Also needed:
potato, alligator clip wires, paper plate, multimeter

Safety

Power a Device with a Potato

In this chapter, you have learned about many types of energy and how energy can be transformed and transferred. Can a common potato transfer energy? Think about the inside of a potato. Is there anything in it that can carry an electric current?

Question

Can potatoes conduct electricity and light a bulb?

Procedure

1. Read and complete a lab safety form.

2. With your teammates, discuss what you know about electric circuits. How can you build an electric circuit using a potato as a battery? Write your ideas in your Science Journal. Draw a diagram of your circuit.

3. Use the materials provided to build the circuit shown in the picture.

4. Place half a potato on a paper plate. Push a galvanized, or zinc-coated, nail and a penny into the potato half.

5. Using two alligator clip wires, attach one end of each wire to the nail and to the penny.

6. Attach the positive probe from the multimeter to the alligator clip wire coming from the penny. Attach the negative probe to the alligator clip wire coming from the nail. Does your battery produce an electric current?

7. Push another galvanized nail and a penny into another potato half. Connect the second potato half to the first, as shown in the diagram, connecting the penny on one potato to the nail on the other. Use the meter to test your battery. Record your data in your Science Journal.

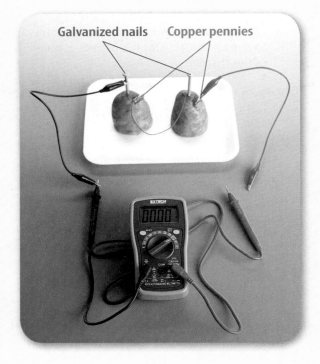

Galvanized nails   Copper pennies

8

Lab Tips

☑ Check the wires in your circuit frequently to make sure they are in tight contact with the nail and the penny.

☑ Set the meter to the lowest range of DC voltage. Some meters require electricity to operate, so the voltage meter might register a lower voltage than is actually in the potato. Use a battery-operated meter, if possible, to avoid this problem.

8 Replace the meter with an LED bulb. Hook one end of the potato battery circuit to each wire coming from the bulb. Does the bulb light?

9 If necessary, redesign your battery circuit until you get the bulb to light. Review all ideas with your teacher before testing your circuit.

Analyze and Conclude

10 **Predict** What sort of devices do you think your potato battery will operate? Explain your answer.

11 **Explain** In this battery, electrons moved from the nails to the pennies. Why did this process light the bulb?

12 🔵 **The Big Idea** Describe, in order, all the energy transfers and transformations in your potato battery.

Communicate Your Results

In small groups, discuss how your battery worked and how you might improve its design. Discuss how changing the distance between the penny and the nail might affect your results.

 **Extension**

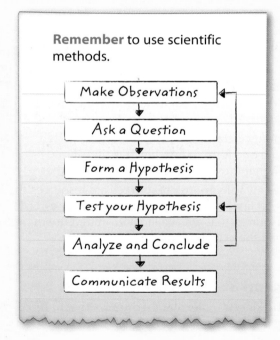

Remember to use scientific methods.

> Make Observations
> Ask a Question
> Form a Hypothesis
> Test your Hypothesis
> Analyze and Conclude
> Communicate Results

Try other types of food, such as a lemon or an apple. Which type of food produces the most electricity? Try other types of nails, such as a steel nail. Replace the penny with a strip of copper or aluminum. What works? What doesn't?

THE BIG IDEA Energy is transferred from object to object when it does not change form; energy is transformed when it changes form.

Key Concepts Summary 🔑	Vocabulary

Lesson 1: Forms of Energy

- **Potential energy** is stored energy, and **kinetic energy** is energy of motion.
- Both **mechanical energy** and **thermal energy** involve kinetic energy and potential energy. Mechanical energy is the sum of the kinetic energy and the potential energy in a system of objects. Thermal energy is the sum of the kinetic energy and the potential energy in a system of particles.
- **Sound energy** and **radiant energy** are carried by waves.

Vocabulary
energy p. 427
potential energy p. 427
chemical energy p. 428
nuclear energy p. 428
kinetic energy p. 429
electric energy p. 429
mechanical energy p. 430
thermal energy p. 430
wave p. 431
sound energy p. 431
radiant energy p. 432

Lesson 2: Energy Transfers and Transformations

- The **law of conservation of energy** says that energy can be transformed from one form to another, but it cannot be created or destroyed.
- Energy is transformed when it is converted from one form to another. It is transferred when it moves from one object to another.
- **Renewable energy resources** are resources that are replaced as fast as, or faster than they are used. **Nonrenewable energy resources** are resources that are available in limited quantities or are used faster than they can be replaced.

law of conservation of energy p. 436
energy transfer p. 437
energy transformation p. 437
work p. 437
open system p. 439
closed system p. 439
renewable energy resource p. 440
nonrenewable energy resource p. 442

Lesson 3: Particles in Motion

- The kinetic molecular theory says that all objects are made of particles; all particles are in constant, random motion; and the particles collide with each other and with the walls of their container.
- Thermal energy is transferred by **conduction, radiation,** and **convection.**
- A **thermal conductor** transfers thermal energy easily and a **thermal insulator** does not transfer thermal energy easily.

temperature p. 447
heat p. 448
conduction p. 449
radiation p. 449
convection p. 449
vaporization p. 451
thermal conductor p. 452
thermal insulator p. 452

FOLDABLES® Chapter Project

Assemble your lesson Foldables as shown to make a Chapter Project. Use the project to review what you have learned in this chapter.

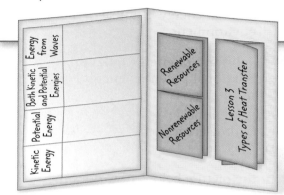

Use Vocabulary

1 Compare and contrast sound energy and radiant energy.

2 Explain why chemical energy and nuclear energy are both considered potential energy.

3 Describe how an open system differs from a closed system.

4 The energy of moving electrons is _____.

5 The energy carried by electromagnetic waves is _____.

6 Define *work* in your own words.

7 Define *conduction* in your own words.

Link Vocabulary and Key Concepts

Concepts in Motion Interactive Concept Map

Copy this concept map, and then use vocabulary terms from the previous page to complete the concept map.

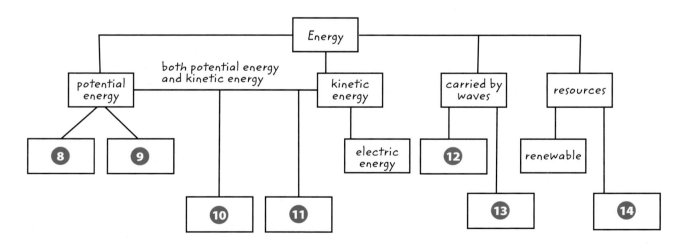

Understand Key Concepts

1 What type of energy does the statue have?

A. electric energy
B. mechanical energy
C. sound energy
D. thermal energy

2 Which is a form of energy that cannot be stored?

A. chemical energy
B. gravitational potential energy
C. nuclear energy
D. sound energy

3 Waves can transfer energy, but they

A. cannot carry sounds.
B. do not move matter.
C. always have the same wavelength.
D. are unable to move through empty space.

4 Which involves ONLY an energy transfer?

A. A boy turns on an electric toaster to warm a piece of bread.
B. A can of juice cools off in a cooler on a hot summer day.
C. A cat jumps down from a tree branch.
D. A truck burns gasoline and moves 20 km.

5 Which could occur in a closed system?

A. Chemical energy is transformed to electric energy in the system.
B. During a transformation of chemical energy to mechanical energy, thermal energy is released to the environment.
C. Electric energy from the environment is transferred to the system.
D. Radiant energy is transferred to the environment.

6 Which form of energy relies on gravitational potential energy?

A. fossil fuels
B. geothermal
C. hydroelectric
D. nuclear

7 In the picture below, how does the thermal energy definitely flow?

A. from the air to the dog
B. from the dog to the snow
C. from the snow to the air
D. from the snow to the dog

8 Which decreases the thermal energy of a can of soup you just took out of your pantry?

A. letting the can sit on the counter for an hour
B. opening the lid of the can of soup with a can opener
C. placing the can of soup under a bright light
D. putting the can of soup in the refrigerator

Critical Thinking

9 **Identify** all the different forms of energy and all the energy transformations that you see in the picture below.

10 **Compare** the energy transformations that take place in the human body to the energy transformations that take place in a gasoline-powered car.

11 **Compare and contrast** each of the following terms: melting and freezing, boiling and evaporation, and sublimation and deposition.

12 **Judge** Determine whether the following statement is correct and explain your reasoning: *The amount of chemical energy in a flashlight's battery is equal to the radiant energy transferred to the environment.*

13 **Evaluate** Using what you read about thermal energy, explain why sidewalks are built as panels with space between them.

14 **Explain** Convection space heaters are small appliances that sit on the floor. Explain how they can heat an entire room.

Writing in Science

15 **Write** a short explanation to a friend explaining the following scenario: Most air-conditioned rooms are set to a temperature of about 22°C. Human body temperature is 37°C. Why don't people in an air-conditioned room come to thermal equilibrium with the room?

REVIEW **THE BIG IDEA**

16 Describe at least four energy transfers or energy transformations that occur in your body.

17 The photo below shows an electrical power plant that uses coal to generate electricity. What type of energy resource is coal? What are the advantages and disadvantages of using coal to generate electricity?

Math Skills

Review — Math Practice —

Use a Formula

18 A child pulls a toy wagon with a force of 25.0 N for a distance of 8.5 m. How much work did the child do on the wagon?

19 A man pushes a box with a force of 75.0 N across a 12.0 m loading dock. How much work did he do on the box?

Standardized Test Practice

Record your answers on the answer sheet provided by your teacher or on a separate sheet of paper.

Multiple Choice

Use the figure to answer questions 1 and 2.

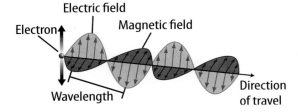

1 Which kind of energy is carried by the wave model shown?

  A chemical energy

  B electric energy

  C radiant energy

  D sound energy

2 Which property does this type of wave have that other kinds of waves do not?

  A It can travel through air.

  B It can travel through water.

  C It can travel through a vacuum.

  D It can travel through metal wires.

3 Which accurately describes potential energy and kinetic energy?

  A A moving soccer ball has potential energy, while a rolling bowling ball has kinetic energy.

  B A rock at the top of a cliff has potential energy, while a moving stream has kinetic energy.

  C The energy stored in chemical bonds is potential energy, while the energy stored in an atom's nucleus is kinetic energy.

  D The energy stored in an atom's nucleus is potential energy, while the energy stored in chemical bonds is kinetic energy.

4 Which describes the sum of potential energy and kinetic energy of objects or systems?

  A nuclear energy and electric energy

  B nuclear energy and mechanical energy

  C thermal energy and electric energy

  D thermal energy and mechanical energy

Use the figure to answer questions 5 and 6.

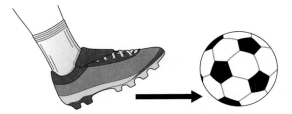

5 The figure above shows someone kicking a soccer ball. Which energy transformation occurs to make the foot move?

  A chemical energy to mechanical energy

  B mechanical energy to chemical energy

  C mechanical energy to mechanical energy

  D thermal energy to mechanical energy

6 Which energy transfer occurs to make the ball move?

  A chemical energy from ball to foot

  B chemical energy from foot to ball

  C mechanical energy from ball to foot

  D mechanical energy from foot to ball

7 Which is a renewable energy resource that is used to produce electricity?

  A biomass

  B coal

  C natural gas

  D petroleum

8 A portable radio transforms chemical energy in batteries into electric energy and then into sound energy. However, not all of the energy from the batteries is converted to sound energy. Which describes how a portable radio still upholds the law of conservation of energy?

A Some energy is destroyed due to inefficiency.

B Some energy goes back into the batteries to be used later.

C Some energy is lost to the surroundings as thermal energy.

D Some energy is lost to the surroundings as chemical energy.

Use the figure to answer question 9.

9 The figure above shows a pan of water being heated on a stove. Which statement is true?

A The pan and the flame are a closed system.

B The natural gas is undergoing an energy transfer.

C Thermal energy is not transferred, and the temperature remains constant.

D This process results in a temperature change and possibly a change of state.

Constructed Response

Use the figure to answer question 10.

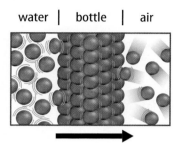

10 Imagine putting a warm bottle of water into a refrigerator. The figure models the particles that make up the water, the bottle, and the air. Explain the energy transfer between the cold refrigerator air outside the bottle and the warm water inside the bottle.

11 How does an open system differ from a closed system?

12 How does the relationship between temperature and volume of water differ from most other materials?

NEED EXTRA HELP?												
If You Missed Question...	1	2	3	4	5	6	7	8	9	10	11	12
Go to Lesson...	1	1	1	1	2	2	2	2	2	3	2	3

Earth: A Dynamic Planet

1450   **1600**   **1700**

1441
Prince Munjong of Korea invents the first rain gauge to gather and measure the amount of liquid precipitation over a period of time.

1450
The first anemometer, a tool to measure wind speed, is developed by Leone Battista Alberti.

1643
Italian physicist Evangelista Torricelli invents the barometer to measure pressure in the air. This tool improves meteorology, which relied on simple sky observations.

1714
German physicist Daniel Fahrenheit develops the mercury thermometer, making it possible to measure temperature.

1752
Swedish astronomer Andres Celsius proposes a centigrade temperature scale where 0° is the freezing point of water and 100° is the boiling point of water.

1806
Francis Beaufort creates a system for naming wind speeds and aptly names it the Beaufort Wind Force Scale. This scale is used mainly to classify sea conditions.

1960
TIROS 1, the world's first weather satellite, is sent into space equipped with a TV camera.

1964
The U.S. National Severe Storms Laboratory begins experimenting with the use of Doppler radar for weather-monitoring purposes.

2006
Meteorologists hold 8,800 jobs in the United States alone. These scientists work in government and private agencies, in research services, on radio and television stations, and in education.

Inquiry
Visit ConnectED for this unit's **STEM** activity.

Nature of SCIENCE

Charts, Tables, and Graphs

Imagine that 3 seconds are left in the semifinal game of your favorite sporting event. The clock runs out, and the buzzer sounds! You cheer as your team advances to the finals! You grab the bracket that you made and record another win.

A bracket organizes and displays the wins and losses of teams in a tournament, as shown in **Figure 1.** Brackets, like maps, tables, and graphs, are a type of chart. A **chart** is a visual display that organizes information. Charts help you organize data. Charts also help you identify patterns, trends, or errors in your data and communicate data to others.

What are tables?

Suppose you volunteer for a cleanup program at a local beach. The organizers need to know the types of debris found at different times of the year. Each month, you collect debris, separate it into categories, and weigh each category of debris. You record your data in a table. A **table** is a type of chart that organizes related data in columns and rows. Titles are usually placed at the top of each column or at the beginning of each row to help organize the data, as shown in **Table 1.**

What are graphs?

A table contains data but it does not clearly show relationships among data. However, displaying data as a graph does clearly show relationships. A **graph** is a type of chart that shows relationships between variables. The organizers of the cleanup program could make different types of graphs from the information in your table to help them better analyze the data.

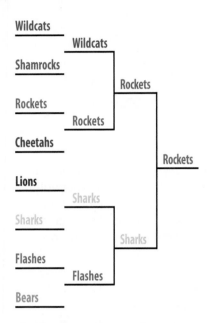

▲ **Figure 1** A sporting bracket is a type of chart that easily enables you to see which team has won the most games in a tournament.

Table 1 This table organizes data on collected debris into rows and columns so measurements can easily be recorded, compared, and used. ▼

Table 1 Types and Amounts of Debris							
Types of Debris	**Jan**	**Mar**	**May**	**July**	**Sept**	**Nov**	**Total for Year**
Plastic	3.0	3.5	3.8	4.0	3.7	3.0	21.0
Polystyrene	0.5	1.3	3.2	4.0	2.5	1.2	12.7
Glass	0.8	1.2	1.5	2.0	1.5	1.0	8.0
Rubber	1.1	1.0	1.3	1.5	1.2	1.3	7.4
Metal	1.0	1.0	1.1	1.4	1.1	1.0	6.6
Paper	1.3	1.1	1.5	1.5	0.8	0.3	6.5
Total for Month	9.4	10.6	13.1	15.1	12.1	9.5	69.8

Circle Graphs

If the cleanup organizers want to know the most common type of debris, they will probably use a circle graph. A circle graph shows the percentage of the total that each category represents. This circle graph shows that plastic makes up the largest percentage of debris. The cleanup organizers could then place plastic recycling barrels on the beach so people can recycle their plastic trash.

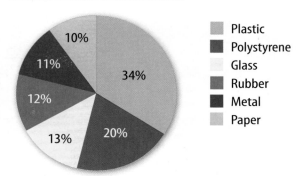

Beach Debris Distribution

- Plastic
- Polystyrene
- Glass
- Rubber
- Metal
- Paper

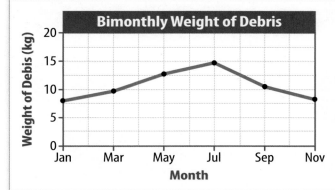

Line Graphs

Suppose the cleanup organizers want to know how the total amount of debris on the beach changes throughout the year. They probably will use a line graph. This line graph shows that volunteers collected more debris in summer than in winter. The cleanup organizers could then create a public service announcement for radio stations that reminds beachgoers to throw trash into trash cans and recycling barrels while visiting the beach.

Bar Graphs

Volunteers collected the most debris in July. The cleanup organizers want to know how much of each type of debris volunteers collected in July. Bar graphs are useful for comparing different categories of measurements. This bar graph shows that 4 kg of both plastic and polystyrene were collected in July. The cleanup organizers could then suggest that beach concession stands use smaller, recyclable food containers.

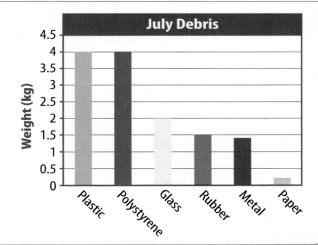

Inquiry MiniLab

25 minutes

How can graphs keep the beach clean?

Suppose you work with the cleanup organizers. What information do you need to make recommendations for keeping the beach cleaner?

1. Based on the type of information in **Table 1,** write a new question about beach debris.

2. Make a graph that allows you to answer your question.

Analyze and Conclude

1. **Distinguish** How did you decide what type of graph to make?

2. **Explain** How did you use your graph to answer your question?

3. **Modify** What recommendations can you make based on your analysis of your graph?

Chapter 13

The Earth System

THE BIG IDEA

How do Earth systems recycle Earth materials?

Inquiry Dinner, Anyone?

Insects, bacteria, and other decomposers help break down a rotting log. Living things such as decomposers make up one of Earth systems. Other Earth systems are made of water, air, and rocks and soil.

- How might the organic matter in the log affect the soil?

- How might wind, air, and rain affect the log?

- How are Earth materials recycled?

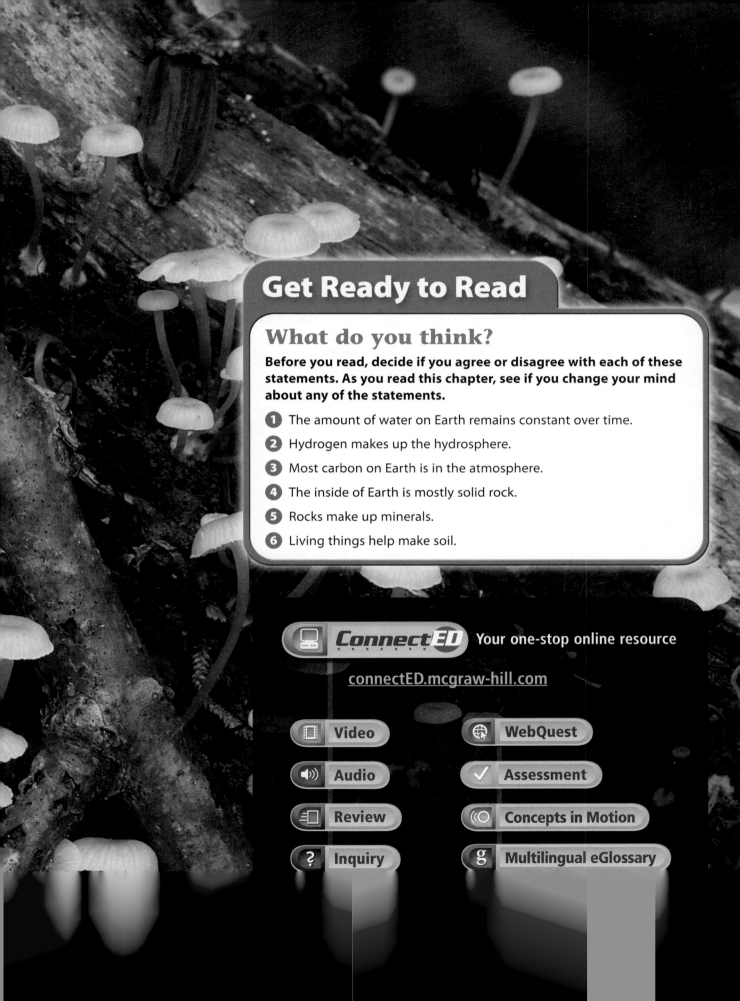

Get Ready to Read

What do you think?

Before you read, decide if you agree or disagree with each of these statements. As you read this chapter, see if you change your mind about any of the statements.

1 The amount of water on Earth remains constant over time.

2 Hydrogen makes up the hydrosphere.

3 Most carbon on Earth is in the atmosphere.

4 The inside of Earth is mostly solid rock.

5 Rocks make up minerals.

6 Living things help make soil.

Reading Guide

Key Concepts 🔑
ESSENTIAL QUESTIONS

- How do Earth systems interact in the carbon cycle?
- How do Earth systems interact in the phosphorus cycle?

Vocabulary

carbon cycle p. 472
greenhouse gas p. 474
phosphorus cycle p. 475

 Multilingual eGlossary

 Video

- BrainPOP®
- Science Video

Earth Systems and Interactions

Inquiry Stormy Weather Ahead?

Water from the ocean evaporates into the atmosphere and condenses into clouds. Rain falls from clouds, often back into the ocean. These are some examples of interactions among Earth's systems. What other interactions can you think of?

inquiry Launch Lab

20 minutes

How do Earth systems interact?

Air, rock and soil, water, and living things make up Earth systems. These systems interact with and affect one another.

1. Read and complete a lab safety form.

2. As a group, discuss how air—represented by a **balloon**—interacts with **rocks,** water, and **plants.**

3. Use **colored pencils** to make a graphic organizer that illustrates the interactions you discussed.

Think About This

1. Describe ways in which air, water, rocks, and plants interact.

2. Do animals and plants interact with water and air in the same ways? Explain.

3. 🔑 **Key Concept** Infer how Earth systems might interact and create a sandy beach.

Earth Systems

You have a respiratory system that moves oxygen into your body and a circulatory system that moves blood throughout your body. Your body contains these two systems and many more that interact. They work together and make one big system—your body.

Earth is a system, too. Like you, Earth has smaller systems that work together. These systems function together and make the larger Earth system. Four of these smaller systems are the atmosphere, the hydrosphere, the geosphere, and the biosphere, as shown in **Figure 1.**

 Reading Check What systems make up the larger Earth system?

The Atmosphere

The outermost Earth system is a mixture of gases and particles of matter called the atmosphere. It forms a layer around the other Earth systems. The atmosphere is mainly nitrogen and oxygen. Gases in the atmosphere move freely, helping transport matter and energy among Earth systems.

Figure 1 Examples of Earth systems are the atmosphere, the hydrosphere, the geosphere, and the biosphere.

Table 1 Interactions Among Earth Systems

	Biosphere	Geosphere	Hydrosphere
Atmosphere	• The ozone layer helps protect organisms from harmful solar radiation. • Plants use oxygen and carbon dioxide during photosynthesis.	• Wind causes weathering and erosion. • Volcanic eruptions eject gas and debris into the air.	• The water cycle influences weather and climate. • Increasing global temperatures lead to melting polar ice caps.
Hydrosphere	• All organisms need water for life functions. • Rising sea levels change habitats.	• Water and ice cause weathering, erosion, and deposition. • Hurricanes and tsunamis change coastal landforms.	
Geosphere	• Materials in the geosphere provide nutrients for life functions. • Organisms contribute to weathering, erosion, and fossil fuel formation.		

Table 1 These are only a few examples of interactions among Earth systems.

Visual Check What other interactions can you name?

 **Interactive Table**

The Hydrosphere

Below the atmosphere is the hydrosphere, the system that contains all of Earth's water. Most of the water is on Earth's surface—in oceans, glaciers, lakes, ice sheets, ponds, and rivers. Smaller amounts of water are deep beneath the surface of the solid Earth, in the atmosphere, and in living things, such as ferns, frogs, cheetahs, and elephants.

Like gases in the atmosphere, water in the hydrosphere continuously moves from place to place. Many substances are easily dissolved in water, such as salts, minerals, and dissolved gases. These dissolved substances move with the water. This results in some of the interactions among Earth systems that are described in **Table 1.** The hydrosphere interacts directly with the geosphere through weathering and erosion. The biosphere and hydrosphere interactions include the change in sea level.

The Geosphere

The largest Earth system is the geosphere, or the solid Earth. The geosphere includes the thin layer of soil and rocks on Earth's surface and all the underlying layers of Earth. Because the geosphere is mainly solid, materials in this system move more slowly than the gases in the atmosphere or the water in the hydrosphere. As the materials move, they slowly transport both energy and matter.

Reading Check Why do materials in the geosphere move slowly?

The Biosphere

All living things on Earth make up the biosphere. Because organisms live in air, water, soil, and rocks, the biosphere is within all other Earth systems. Living things survive using gases from the atmosphere, water from the hydrosphere, and nutrients found in soil and rocks.

Interactions Among Earth Systems

You might interact with your friends by texting or by playing a game. Earth systems also interact. Earth systems interact by exchanging matter and energy. Matter and energy often change in form as they flow between systems.

The Water Cycle

An example of an interaction among Earth systems is the water cycle. The water cycle is the continuous movement of water on, above, and below Earth's surface. As shown in **Figure 2,** water moves within the hydrosphere and into other Earth systems. You might think of water as a liquid, but sometimes it changes state and becomes solid ice or gaseous water vapor. As water flows or changes state, it moves thermal energy within the water cycle and among Earth systems.

The Rock Cycle

Another example of interactions among Earth systems is the rock cycle, shown in **Figure 3.** The rock cycle is the series of processes that change rocks from one form to another. Some processes take place deep within Earth, where few interactions among Earth systems occur. Other processes, such as weathering and erosion, take place on or near Earth's surface. The atmosphere, the hydrosphere, and the biosphere interact with the geosphere through weathering and erosion. For example, rain and plants can weather rocks into sediments. Wind and flowing water can erode rocks and sediment and deposit them in new places.

Reading Check What are some examples of interactions among Earth systems?

Generally, the amount of material cycling through each system remains constant, but it might change state or form. These changes can take place quickly or over millions of years.

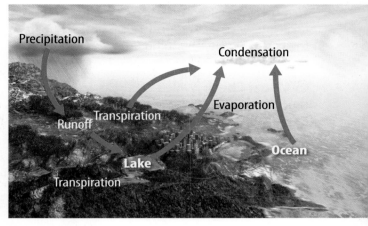

Precipitation
Condensation
Evaporation
Transpiration
Runoff
Ocean
Lake
Transpiration

▲ **Figure 2** Water moves through the hydrosphere and other Earth systems in the water cycle.

Review   Personal Tutor

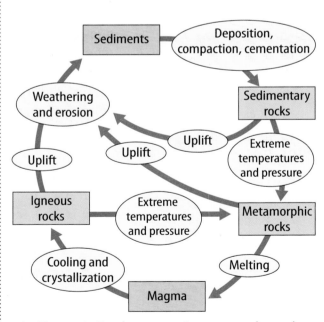

Sediments
Deposition, compaction, cementation
Weathering and erosion
Sedimentary rocks
Uplift
Uplift
Uplift
Extreme temperatures and pressure
Igneous rocks
Extreme temperatures and pressure
Metamorphic rocks
Cooling and crystallization
Melting
Magma

▲ **Figure 3** Earth systems interact and weather and erode rocks in the rock cycle.

Concepts in Motion   Animation

Figure 4 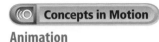 Carbon takes many paths through the carbon cycle, changing form and location.

✓ **Visual Check** Which processes add carbon dioxide to the atmosphere? Which processes remove it?

 **Concepts in Motion**
Animation

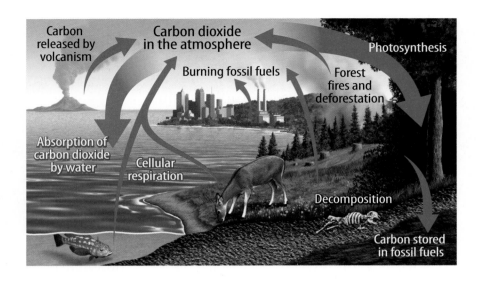

Carbon released by volcanism

Carbon dioxide in the atmosphere

Photosynthesis

Burning fossil fuels

Forest fires and deforestation

Absorption of carbon dioxide by water

Cellular respiration

Decomposition

Carbon stored in fossil fuels

WORD ORIGIN ··············

carbon cycle
from Latin *carbo*, means "glowing coal"; and Greek *kyklos*, means "circle"

FOLDABLES®

Make a horizontal four-column chart book. Label it as shown. Use it to organize your notes on Earth systems and interactions.

Water Cycle	Rock Cycle	Carbon Cycle	Phosphorus Cycle

The Carbon Cycle

Some elements are so important to life that scientists study their individual cycles among Earth systems. One of these elements is carbon. *The **carbon cycle** is the series of processes that continuously move carbon among Earth systems.*

Processes of the Carbon Cycle Trace the path of carbon in **Figure 4** as it is released from the geosphere during a volcanic eruption. Carbon from the geosphere enters the atmosphere as the trace gas carbon dioxide (CO_2). This gas then is removed from the atmosphere in several ways. During photosynthesis, plants use sunlight, CO_2, and water and make simple sugars. As a result, carbon leaves the atmosphere and enters the biosphere. Chemical weathering of rocks moves carbon from the atmosphere and geosphere to the hydrosphere as a dissolved material. Carbon also moves from the atmosphere to the hydrosphere when atmospheric CO_2 dissolves in water.

How does carbon leave the biosphere? Cellular respiration in organisms quickly returns CO_2 to the atmosphere. Even more carbon enters the atmosphere and the soil when organisms die and decay. Sometimes organic matter is buried deep in the geosphere, where it can form fossil fuels.

Carbon leaves the hydrosphere and enters the geosphere when sedimentary rocks form on the ocean floor. Ocean water can warm and release dissolved CO_2 directly into the atmosphere. As carbon moves through Earth systems, the total amount of carbon in the carbon cycle remains the same.

✓ **Key Concept Check** How do Earth systems interact in the carbon cycle?

Table 2 Carbon Reservoirs		
Carbon Reservoirs	Carbon (billions of tons)	Form
Atmosphere	750	CO_2 gas
Biosphere	3,000	organic molecules
Hydrosphere	40,000	dissolved CO_2 gas
Geosphere (crust and upper mantle)	750,000	minerals and rocks
Geosphere (lower mantle)	750,000 +	minerals and rocks

Carbon Reservoirs After water, carbon is the most abundant substance in living organisms. But as you just read, carbon is not limited to the biosphere. Carbon is in reservoirs, or storage places, within all Earth systems, as shown in **Table 2.** On Earth, most carbon is combined with other elements in compounds.

Carbon in the biosphere is stored in organisms. It does not exist as carbon atoms. It is combined with other elements in complex organic molecules, such as sugars and starches. The cells and tissues of all organisms are made of organic compounds.

In both the atmosphere and the hydrosphere, carbon exists as gaseous CO_2. Though the atmosphere is the smallest carbon reservoir, atmospheric CO_2 is important. The amount of CO_2 in the atmosphere affects climate, as you will read later. CO_2 in the hydrosphere is dissolved in water.

Most of Earth's carbon is stored in the geosphere. Carbon is combined with other elements in minerals that form rocks. Limestone contains the mineral calcite, which contains carbon. Carbon also is stored as fossil fuels, such as coal, natural gas, and oil, which form underground.

Table 2 The largest reservoir of carbon is the geosphere.

Visual Check How does the amount of carbon in the atmosphere compare to the amount in the biosphere?

 **Concepts in Motion** Interactive Table

How does the biosphere affect the carbon cycle?

Carbon can leave the atmosphere and enter the biosphere through the process of photosynthesis. During photosynthesis plants use the Sun's energy, carbon dioxide, and water and make sugars. This process releases oxygen into the atmosphere.

1. Read and complete a lab safety form.
2. Use a **knife** to cut off the bottom end of an **Elodea** stem at an angle. Crush the end slightly.
3. Place the Elodea in a **test tube.** Fill the test tube nearly to the top with water.
4. Put the test tube in a **test-tube rack.** Place the rack under a **lamp.**
5. Observe the plant for about 10 min. Record your observations in your Science Journal.

Analyze and Conclude

1. **Describe** what happened to the plant.
2. **Hypothesize** What likely caused the event you observed?
3. **Key Concept** How does photosynthesis affect the carbon cycle?

Humans and the Carbon Cycle Some changes in the amount of CO_2 in the atmosphere occur naturally. For example, a volcanic eruption can release large amounts of CO_2 into the air. However, not all changes in the amount of CO_2 in the atmosphere occur naturally.

Some changes in levels of atmospheric CO_2 are caused by human activities, as shown in **Figure 5.** When people burn **fossil fuels** to generate electricity and to power vehicles, CO_2 is released directly into the atmosphere. Other activities can indirectly increase levels of atmospheric CO_2. For example, large tracts of forests might be cut down to clear land for agriculture or development. More CO_2 remains in the atmosphere because fewer trees are taking in CO_2 during photosynthesis.

CO_2 is a greenhouse gas. **Greenhouse gases** *are gases in the atmosphere that absorb and reradiate thermal energy from the Sun.* These gases keep Earth from becoming too cold to support life. When levels of atmospheric CO_2 increase, more thermal energy is absorbed and reradiated. Earth's average surface temperature increases. This phenomenon, known as global warming, can cause coastal flooding as ice caps melt and sea level rises. These events might cause climates around the world to change, altering habitats and harming living things.

✓ **Reading Check** What are greenhouse gases?

REVIEW VOCABULARY ····

fossil fuels
fuels such as coal, oil, and natural gas that form in the Earth from plant or animal remains

Figure 5 People increase levels of atmospheric CO_2 by burning fossil fuels and cutting down forests.

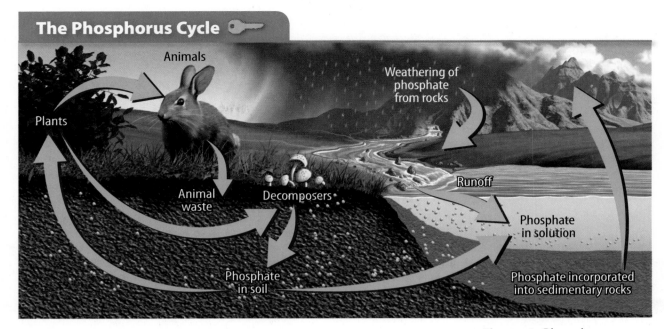

The Phosphorus Cycle

The Phosphorus Cycle

Animals

Plants

Weathering of
phosphate
from rocks

Animal
waste

Decomposers

Runoff

Phosphate
in solution

Phosphate
in soil

Phosphate incorporated
into sedimentary rocks

Figure 6 Phosphorus continuously cycles through Earth systems.

Visual Check How do living things affect the phosphorus cycle?

The Phosphorus Cycle

Some important **elements** do not cycle through all Earth systems. One example is phosphorus (FAHS fuh rus). It does not exist in the atmosphere. *The **phosphorus cycle** is the series of processes that move phosphorus among Earth systems.*

Processes of the Phosphorus Cycle The processes shown in **Figure 6** help move phosphorus through the geosphere, the hydrosphere, and the biosphere. Phosphorus is not found in nature as an element. It forms compounds with oxygen called phosphates (PO_4).

Earth's phosphorus originates in the geosphere. Rocks exposed at Earth's surface release phosphates when they weather. The phosphates either remain in the soil or dissolve and enter the hydrosphere.

Dissolved phosphate molecules move in liquid water through the water cycle. Eventually the phosphates reach lake bottoms or the seafloor and are deposited along with sediment. The phosphorus becomes part of new sedimentary rocks that form from the deposited sediment.

Plants absorb phosphorus from either soil or water. Animals take in phosphorus when they eat plants or when they eat other animals that have eaten plants. These phosphates return to the soil as part of animal waste or as part of decomposing organisms.

Key Concept Check How do Earth systems interact in the phosphorus cycle?

SCIENCE USE V. COMMON USE

element
Science Use a substance that consists of atoms of only one kind

Common Use a part of something

Figure 7 Apatite and turquoise are both phosphate minerals.

Phosphorus Reservoirs Rocks in the geosphere include minerals containing phosphates. The most common of these minerals is apatite (A puh tite). You might have seen jewelry made from turquoise. Turquoise is another phosphate mineral. Examples of the mineral apatite and jewelry made from turquoise are shown in **Figure 7.**

The hydrosphere is another reservoir for phosphorus. As you just read, phosphates dissolve in water. Phosphorus moves through the water cycle as a dissolved substance in liquid water, but it does not evaporate and enter the atmosphere.

The cycling of phosphorus through the geosphere and the hydrosphere occurs over long periods of time. Phosphates incorporated into sedimentary rocks on the ocean floor might not reenter the phosphorus cycle for millions of years.

In contrast, the phosphorus stored in organisms in the biosphere recycles relatively quickly. Like carbon, phosphorus is a necessary element for organisms. It is needed to make cell membranes and transfer energy. It also is an important component of teeth, bones, and shells. Animals store most of their phosphorus in these structures. Animal waste also is a major source of phosphorus.

Reading Check How do living things use the element phosphorus?

Figure 8 Runoff from agricultural fertilizers and other sources can cause rapid algae growth that kills aquatic organisms. ▼

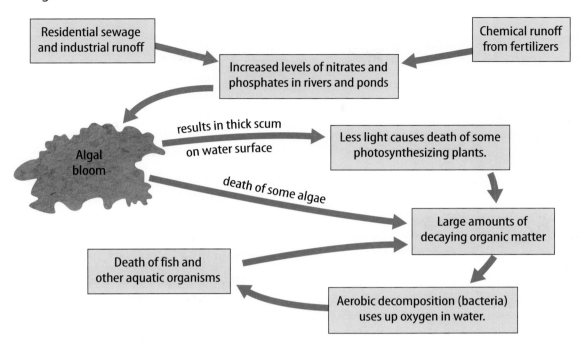

Humans and the Phosphorus Cycle Recall that humans can disturb the carbon cycle. They also can disturb the phosphorus cycle. For example, most of the phosphorus in rain forests is stored in plants. As the plants drop their leaves or die, new plant growth quickly takes up the phosphorus. Clearing the trees in rain forests disturbs the phosphorus cycle.

Clearing forests also exposes soil. Heavy rains wash away the phosphorus released by decaying plants. The lack of this important element makes the soil unproductive for future crops.

Other practices also can impact the phosphorus cycle. As shown in **Figure 8,** runoff from farms, homes, and factories can contain phosphorus. Rain can wash the phosphorus into rivers, streams, and lakes. Algae consume the phosphorus, and the algae population increases. As shown in **Figure 9,** excess algae decompose and use up oxygen in the water. This harms fish and other aquatic organisms.

 Reading Check How can farming affect the phosphorus cycle?

Figure 9 Excessive algal growth can lead to the death of many fish in a short time. ▼

Lesson 1 Review

Visual Summary

Earth systems include the atmosphere, the hydrosphere, the biosphere, and the geosphere.

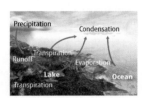

Examples of interactions among Earth systems include the water cycle, the rock cycle, the carbon cycle, and the phosphorus cycle.

Human activities disturb natural cycles.

FOLDABLES

Use your lesson Foldable to review the lesson. Save your Foldable for the project at the end of the chapter.

What do you think NOW?

You first read the statements below at the beginning of the chapter.

1. The amount of water on Earth remains constant over time.

2. Hydrogen makes up the hydrosphere.

3. Most carbon on Earth is in the atmosphere.

Did you change your mind about whether you agree or disagree with the statements? Rewrite any false statements to make them true.

Use Vocabulary

1 Gases that absorb and reradiate thermal energy from the Sun are called _____.

2 **Use the term** *carbon cycle* in a sentence.

Understand Key Concepts

3 What is the form of carbon in the atmosphere?
   A. fossil fuels   **C.** limestone
   B. gas   **D.** simple sugars

4 **Explain** how animals use phosphorus.

5 **Relate** photosynthesis to the amount of CO_2 in the atmosphere.

Interpret Graphics

6 **Interpret** the diagram below. Explain how precipitation affects the phosphorus cycle.

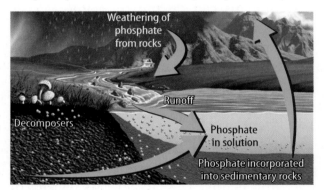

7 **Organize Information** Draw and fill in a graphic organizer like the one below to show the steps of an algal bloom.

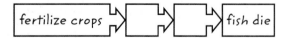

Critical Thinking

8 **Suggest** a way to reduce the amount of CO_2 in the atmosphere.

Math Skills ✕⁒÷   Review

—— Math Practice ——

9 A 50-kg bag of fertilizer contains 3 percent phosphorus. How much phosphorus is in the bag?

GREEN SCIENCE

Tracking Carbon Among Earth Systems

How the Biosphere Affects Levels of Atmospheric Carbon Dioxide

Scientists study Earth systems—atmosphere, hydrosphere, geosphere, and biosphere—to understand how our planet works and changes over time. These studies help scientists predict climate, weather, and natural disasters. They also explore the effects of human actions on the environment.

The National Oceanic and Atmospheric Administration (NOAA) has a research group called the Earth System Research Laboratory (ESRL). This group investigates how weather, air quality, and climate are affected by interactions among Earth systems. Data show that the average temperature in the atmosphere near Earth's surface is increasing. ESRL scientists relate this trend to the increase in carbon dioxide in the atmosphere. How do they know this increase is occurring?

ESRL scientists analyze tens of thousands of air samples collected from eight tall towers across the United States. They analyze samples for levels of carbon dioxide and other greenhouse gases. Using data such as these, scientists have confirmed that carbon dioxide levels are increasing.

But ESRL scientists also discovered good news! Forests and farms are slowing the process. Analyses of air samples collected worldwide show that trees and crops absorb about one-third of the carbon dioxide produced by burning fossil fuels. This shows the interaction between the biosphere and the atmosphere. It also shows that protecting trees and planting new ones can help reduce carbon in the atmosphere.

◀ **The ESRL mounts air-sample collection tubes on existing television, radio, and cell phone towers. The narrow tubes carry the samples from the towers to nearby analysts.**

Trees and the Carbon Cycle

Trees absorb and transfer carbon dioxide from the atmosphere.

Trees take in carbon dioxide through photosynthesis.

Trees release carbon dioxide into the atmosphere through cellular respiration.

The trunk, leaves, and branches store carbon.

Some carbon is carried down to the roots.

Fallen leaves and branches add carbon to the soil.

Carbon is stored in the roots.

When a tree dies, some carbon moves from the roots to the soil.

It's Your Turn

RESEARCH The ocean is a carbon reservoir that soaks up and stores carbon dioxide. Research the role of ocean organisms in reducing atmospheric carbon dioxide. Share your findings with the class.

Lesson 2

Reading Guide

Key Concepts 🔑
ESSENTIAL QUESTIONS

- How do materials in the geosphere differ?
- Why does the geosphere have a layered structure?

Vocabulary

luster p. 482

streak p. 482

cleavage p. 482

fracture p. 482

crust p. 486

mantle p. 487

lithosphere p. 487

asthenosphere p. 487

core p. 488

g Multilingual eGlossary

⬛ Video BrainPOP®

The Geosphere

Inquiry **What weathers rock?**

Some of the rocks that made up this mountain have been broken down by wind, rain, and other agents of weathering. Rocks are one of the materials that make up the geosphere. What other materials are in the geosphere?

Launch Lab

What can you learn from a core sample?

Layers of Earth's crust can be studied easily when they are exposed on the side of a cliff. To learn about layers that are buried, however, scientists often drill into the ground to get a core sample. A core sample is a sample of rock and soil taken from inside Earth's upper crust.

1. Read and complete a lab safety form.
2. Fold a blank sheet of **paper** into four equal squares. Label the squares *A, B, C,* and *D,* respectively.
3. Place thin and thick layers of **different-colored salt dough** in a **bowl.**
4. To get core samples, push a **clear straw** straight down into the clay with a slight twisting motion. Slowly remove the straw.
5. Use **scissors** to snip off the end of the straw that contains the sample. Place the sample on the paper square labeled *A.* Write a description of the sample in your Science Journal.
6. Repeat steps 4 and 5 to take core samples B, C, and D from different parts and different depths of the dough crust.

Think About This

1. What do the layers of salt dough represent?
2. How are the core samples alike? How are they different?
3. Which layers likely represent the oldest and youngest rock layers in your sample? How do you know?
4. 🔑 **Key Concept** Based on your core samples, what can you infer about the materials that make up the upper part of Earth's crust?

Materials in the Geosphere

What materials are in the geosphere? A thin layer of soil covers much of Earth's land. It is on top of a layer of broken rock material. Under the broken rock are layers of mostly solid rock surrounding a hot, metallic center. The basic building blocks for soil, rocks, and metals are minerals. Minerals combine in different ways, forming the other materials in the geosphere.

Minerals

In science, the term *mineral* has a specific definition. A mineral is naturally occurring, inorganic, and solid, and it has a crystal structure and a definite chemical composition. The quartz shown in **Figure 10** is a mineral because it has all five of these characteristics. Quartz formed naturally, was never living, is solid, has a crystal structure, and has a chemical composition of two oxygen atoms for each silicon atom.

✔️ **Reading Check** What are the characteristics of a mineral?

Figure 10 Quartz has the five characteristics of a mineral.

Inquiry MiniLab
20 minutes

Which mineral is which?

Hardness, cleavage, and other properties can be used to distinguish one mineral from another. How can you determine the differences between quartz and calcite?

1. Read and complete a lab safety form.

2. Copy the table below in your Science Journal.

Property	Evidence	Quartz	Calcite
Hardness	scratches glass	yes	no
	scratched by penny		
	scratched by fingernail		
Cleavage	smooth, flat planes visible on surface		
Reacts with HCl	fizzes and produces bubbles		

3. Determine whether your **mineral samples** can be scratched by your fingernail and a **penny.** Record your results in the data table.

4. Examine the samples closely for cleavage planes. Record your observations in the table.

5. Place a few drops of **hydrochloric acid** on each sample. Record the results.

Analyze and Conclude

1. **Summarize** the differences between quartz and calcite.

2. **Key Concept** Why is calcite more likely to break apart than quartz?

Mineral Properties Each mineral has unique physical properties. These properties, such as color and crystal shape, can be used to identify minerals.

Other mineral properties are luster and streak. **Luster** *is the way a mineral's surface reflects light.* Some minerals reflect a lot of light and appear shiny. Other minerals do not reflect a lot of light and appear dull. **Streak** *is the color of a mineral's powder.* It is observed by scratching a mineral across a tile of unglazed porcelain. While the color of a mineral often varies, its streak is always the same.

Hardness is another physical property of minerals. Certain minerals are harder than others. The Mohs scale ranks minerals on a scale of 1 to 10 based on their relative hardness. Talc, the softest mineral, has a hardness of 1. Diamond, the hardest mineral, has a hardness of 10. Hardness is determined by how easily a mineral is scratched by another mineral or by a common object.

Hardness affects how easily a mineral breaks. **Cleavage** *is the tendency of minerals to break along smooth, flat surfaces.* The mineral calcite exhibits cleavage. When calcite breaks, it has defined edges. **Fracture** *is the tendency of minerals to break along irregular surfaces.* Quartz exhibits fracture because it does not break along a flat plane.

Reading Check What properties could you use to identify an unknown mineral?

Mineral Interactions Some minerals tend to break apart and combine with other substances. Other minerals are more stable and durable. Calcite and quartz are two common minerals that display this difference. On the Mohs scale, calcite has a hardness of 3, and quartz has a hardness of 7. Calcite dissolves in water more easily than quartz. This increases its interactions with other materials. Quartz does not dissolve easily and is less likely to break apart.

Rocks

You might know that a rock is a naturally occurring solid composed of minerals and other materials. Rocks are classified according to how they form.

Rock Types The three main types of rocks are igneous (IG nee us), metamorphic, and sedimentary. Igneous rock forms when molten rock material cools and hardens. This can happen deep inside Earth as magma cools, or when molten material called lava flows onto Earth's surface and cools.

Metamorphic rock forms when sedimentary, igneous, or other metamorphic rocks are subjected to high temperatures and extreme pressure. They alter the texture or the chemical composition of the rocks and form new metamorphic rocks.

Rock fragments called sediment make up sedimentary rock. Sedimentary rocks form when sediment is eroded by water, wind, ice, or gravity and deposited in layers. Over time, the weight from upper layers of sediment compresses lower layers. Sediment becomes compacted and cemented together and forms sedimentary rock.

Interactions The formation of sedimentary rocks involves interactions among the geosphere, the atmosphere, the hydrosphere, and the biosphere. Recall that weathering breaks rock into small pieces. Some weathering occurs when rock is physically broken down. Weathering also results from chemical reactions on rock surfaces. Many of these chemical reactions include water. Examples of physical and chemical weathering are shown in **Figure 11.**

 Reading Check Compare physical and chemical weathering.

Figure 11 Some minerals dissolve in water, resulting in chemical weathering. As tree roots grow, they can physically break apart rock.

 Review   Personal Tutor

Soil

Have you ever grown a garden? If so, you have used one of the most important materials in the geosphere—soil. Soil is the loose, weathered material in which plants grow. If you dig into the ground, you would see that soil has a layered **structure,** as shown in **Figure 12.** The layers form as rock is slowly transformed into soil.

How Soil Forms At Earth's surface, interactions among rocks, water, air, and organisms form soil. Soil formation begins when rocks weather into sediment. Water dissolves minerals and other materials from the sediment, and they become part of the developing soil. Animals and plants also affect soil formation. They weather sediment and create open spaces for air and water. In addition, the wastes from organisms add nutrients to soil. Nutrients also enter soil when organisms die and their bodies decay. The organic matter makes soil more fertile and gives it a dark color. It takes hundreds to thousands of years to build thick layers of soil such as those shown in **Figure 12.**

Soil Interactions Soil contains minerals, water, air, and organisms, all in close contact. Therefore, interactions among all Earth systems take place in soil. Recall that plants need phosphorus and carbon to grow. Plants cannot get phosphorus from the air, but they can obtain it from soil or from water in the soil. A major part of the organic matter in soil is carbon that plants obtain from the atmosphere through photosynthesis. Soil plays a major role in the phosphorus and carbon cycles.

 Key Concept Check How do materials in the geosphere differ?

Figure 12 Each layer of soil has different properties.

Visual Check Which soil layer contains the most organic matter?

A-horizon
The A-horizon is the part of the soil that you are the most likely to see when you dig a shallow hole in the soil with your fingers. Organic matter from the decay of roots and the action of soil organisms often makes this horizon excellent for plant growth. Because the A-horizon contains most of the organic matter in the soil, it is usually darker than other horizons.

B-horizon
When water from rain or snow seeps through pores in the A-horizon, it carries clay particles. The clay is then deposited below the upper layer, forming a B-horizon. Other materials also accumulate in B-horizons.

C-horizon
The layer of weathered parent material is called the C-horizon. Parent material can be rock or sediments.

Structure of the Geosphere

What do you think you would see if you could look inside the solid Earth? You would see layers similar to those shown in **Figure 13**. The geosphere has three main layers. Each layer has a different density. Recall that density is a measure of the mass of a material divided by its volume. The densest layer of the geosphere is the center, or core. The least dense layer is the outer crust. The density of the thick mantle varies.

Scientists hypothesize Earth's layers formed early in the planet's history. Ancient Earth was much hotter than it is today. Thermal energy melted some of the rock. Then, gravity pulled denser materials through the melted rock toward Earth's center, forming layers.

In addition to different densities, the layers of the geosphere have different compositions. Most of the geosphere is made of rock, but some of it is made of metal.

You might wonder how scientists know about the density and the makeup of Earth's deep inner layers. After all, humans have never seen them. Scientists gather data by analyzing earthquake waves. When the waves travel through Earth, they change speed and direction as they pass through materials with different densities. Scientists use data about the waves to map Earth's interior.

 Key Concept Why is the geosphere layered?

Layered Structure of Geosphere

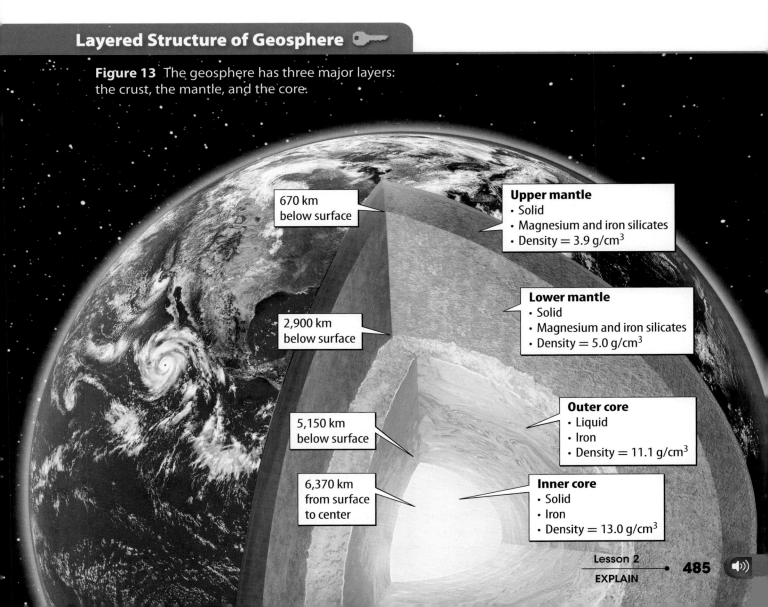

Figure 13 The geosphere has three major layers: the crust, the mantle, and the core.

670 km below surface

2,900 km below surface

5,150 km below surface

6,370 km from surface to center

Upper mantle
- Solid
- Magnesium and iron silicates
- Density = 3.9 g/cm³

Lower mantle
- Solid
- Magnesium and iron silicates
- Density = 5.0 g/cm³

Outer core
- Liquid
- Iron
- Density = 11.1 g/cm³

Inner core
- Solid
- Iron
- Density = 13.0 g/cm³

Figure 14 Oceanic crust is denser than continental crust.

🔵 **Visual Check** How does the thickness of oceanic crust and continental crust compare?

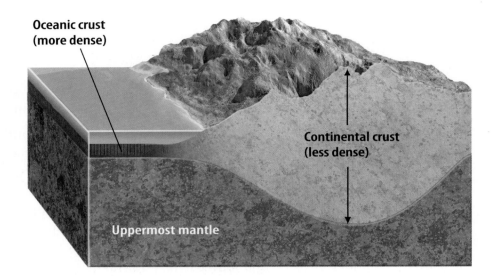
Oceanic crust (more dense)

Continental crust (less dense)

Uppermost mantle

Earth's Crust

Have you ever seen rocky cliffs exposed along the sides of a highway? Those rocks are part of Earth's crust. *The* **crust** *is the thin outer layer of the geosphere.* It is made of brittle rocks. These rocks are made of elements that combine and form minerals. There are approximately 90 naturally occurring elements in Earth's crust. Just eight of these elements make up about 98 percent of the crust. The most common element in Earth's crust is oxygen, followed by silicon, aluminum, iron, calcium, sodium, potassium, and magnesium.

There are two types of crust, as shown in **Figure 14.** One type is under the oceans, and the other type makes up the continents.

Oceanic Crust The crust under the oceans is about 7 km thick. Oceanic crust is made of the dense igneous rocks basalt and gabbro. These rocks are rich in the dense minerals iron and magnesium. This makes oceanic crust denser than continental crust.

Continental Crust The crust that makes up continents is thicker than oceanic crust. Continental crust has an average thickness of about 40 km. Under large mountains, continental crust is as much as 70 km thick. It is not made of the same kinds of rocks as oceanic crust. Continental crust is made of igneous, metamorphic, and sedimentary rocks. Rocks in the continental crust are rich in silicon and oxygen. These elements are less dense than iron and magnesium. This makes continental crust less dense than oceanic crust.

✓ **Reading Check** How do oceanic crust and continental crust differ?

Earth's Mantle

Beneath Earth's crust is the mantle. *The* **mantle** *is the thick, rocky middle layer of the geosphere.* The mantle has the largest volume of any layer of Earth. Much of the mantle is made of the rock peridotite (puh RIH duh tite). Peridotite contains even more iron and magnesium than basalt and, therefore, is denser.

Rocks in the uppermost mantle are more brittle than in the rest of the mantle and similar to rocks in the crust. So the crust and the uppermost mantle are sometimes described as one layer, even though they have different compositions. *The crust and the uppermost mantle form a brittle outer layer called the* **lithosphere.**

Most of the rock below the lithosphere is solid. But high temperatures in the mantle make rock soft enough to flow. This is similar to the way a warm wax candle can bend instead of breaking. The mantle flows slowly, moving about as fast as your fingernails grow.

At a depth of about 100 km is an especially soft layer of the mantle. *This weak, partially melted layer of the mantle is called the* **asthenosphere** (as THEN uh sfihr). Less than 2 percent of the rock in the asthenosphere is melted. This small amount of molten rock makes the asthenosphere weaker than the rest of the mantle. Both the asthenosphere and the lithosphere are shown in **Figure 15.**

WORD ORIGIN ············

lithosphere
from Greek *lithos*, means "stone"; and Greek *sphaira*, means "ball"

 Reading Check Why are the crust and the uppermost mantle sometimes regarded as a single layer?

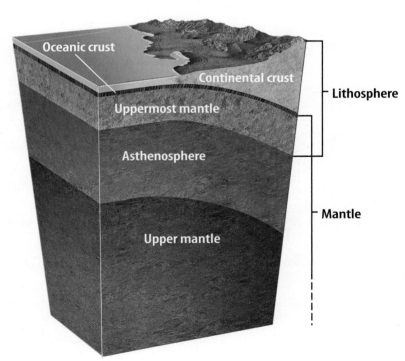

Figure 15 The lithosphere includes the crust and the brittle upper mantle. The asthenosphere is the weak layer of the mantle directly below the lithosphere.

Concepts in Motion

Animation

Earth's Core

The dense, metallic center of Earth is called the **core.** Note that the core is metallic and not rocky like the other layers of the geosphere. Why is the core different? Recall that early in Earth's history, the densest materials in the geosphere sank to the center. Therefore, the core is made mainly of iron with some nickel and traces of other elements. The core is divided into two layers, one liquid and one solid.

Outer Core Due to the high temperatures near the center of Earth, the outer layer of the core is liquid. As Earth spins on its axis, this molten iron flows. Scientists theorize that the movement of liquid iron in the outer core produces Earth's magnetic field. Earth's magnetic field is similar to the magnetic field of a huge bar magnet. The magnetic field, shown in **Figure 16,** protects Earth from charged particles from the Sun. You have interacted with Earth's magnetic field if you have ever used a compass to find a direction. The metal needle in the compass aligns with Earth's magnetic field.

Inner Core Inside the outer core is a sphere of solid metal. Temperatures in this inner core are extremely hot, as high as 4,300°C. Despite the scorching heat, the metal in the inner core is not melted. The high pressure from the masses of all Earth's layers compresses the inner core, making it solid.

 Reading Check What is the structure of the geosphere?

Figure 16 The movement of liquid iron in the outer core produces Earth's magnetic field.

Lesson 2 Review

Visual Summary

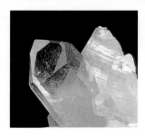

Minerals are the basic building blocks for materials in the geosphere, including soil, rock, and metal.

The three main layers of the geosphere are the crust, the mantle, and the core.

The layers of the geosphere have different densities and chemical compositions.

FOLDABLES

Use your lesson Foldable to review the lesson. Save your Foldable for the project at the end of the chapter.

What do you think NOW?

You first read the statements below at the beginning of the chapter.

4. The inside of Earth is mostly solid rock.

5. Rocks make up minerals.

6. Living things help make soil.

Did you change your mind about whether you agree or disagree with the statements? Rewrite any false statements to make them true.

Use Vocabulary

1 The weak, partially melted layer of the mantle is the _____.

2 **Distinguish** between Earth's crust and its core.

3 **Define** *luster* in your own words.

Understand Key Concepts

4 Which layer is made of metal?
   A. asthenosphere   C. crust
   B. core   D. mantle

5 **Give an example** of an object that has a structure similar to that of the geosphere.

6 **Compare** how minerals can break.

Interpret Graphics

7 **Distinguish** Use the diagram below to distinguish between layers in the geosphere that are classified according to composition and those that are classified according to physical state.

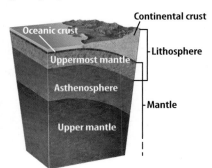

8 **Organize** Copy and fill in the graphic organizer below with details about different types of weathering.

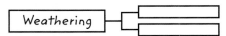

Critical Thinking

9 **Suggest** a way to improve the soil in a garden.

10 **Relate** How do the different densities of Earth materials relate to the layered structure of the geosphere?

Design an Earth-System Game

Materials

poster board

blank index cards

office supplies

construction paper

In this chapter, you learned about interactions among the atmosphere, the hydrosphere, the geosphere, and the biosphere. The four systems interact by exchanging matter and energy. Can you design a game that shows the interactions among Earth systems?

Question

How can a game model interactions among Earth systems?

Procedure

1. Read and complete a lab safety form.

2. The object of the game is to collect or earn points for as many different Earth-system interactions as possible. As a group, decide whether to invent an original game or model it after an existing one.

3. You can design a board game or a card game. A board game should have

- a playing board with interaction squares;
- playing pieces representing air, water, rock, and a living thing;
- written instructions.

A card game should have

- two cards each for air, water, rock, and living things;
- 44 interaction cards (11 for each system).

4. Design and construct your game. Illustrate system interaction squares or cards with drawings, photos, computer graphics, poems, or riddles.

5. Write clear directions for your game. Give your game a title. Have your teacher review your game before playing it.

6. Make a copy of the rules for each group member.

7. Read and follow the instructions carefully. Play trial runs of the game with another group.

8. Evaluate the game, and make any necessary improvements.

9. Copy and complete the *Game Requirements Checklist*. Does your game include all of the required parts? Describe any optional materials that you added to improve the game.

Analyze and Conclude

10. **Evaluate** Did your game accurately represent Earth systems and their interactions? Use examples to support your answer.

11. **Assess** What new insights about interactions among Earth systems did you gain by playing the game?

12. **The Big Idea** Which interactions in the game involved the recycling of Earth materials by Earth systems?

Communicate Your Results

Share your game with another group. Play the other group's game. Then, donate your game to the school library so that other classes can play it.

Inquiry Extension

Select a location outside your school or home. Observe and record Earth-system interactions. Take photos, if possible, or sketch the interactions. Work with other students to create a classroom display of Earth-system interactions.

Game Requirements Checklist
Board Game
1. Illustrated game board ____
2. Playing pieces: Air, Water, Rock, Living Things ____
3. Illustrated system interaction squares ____
4. Optional chance cards ____
5. Other optional materials ____
6. Written instructions ____

Card Game
1. Illustrated Earth system cards (total: 8) ____
2. Illustrated Earth-system interaction cards (total: 44) ____
3. Other optional materials ____
4. Written instructions ____

Lab Tips

☑ Laminate playing cards for durability.

☑ Create playing pieces that clearly represent the systems. For example, use a rock to represent the geosphere.

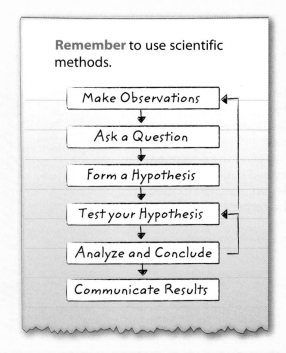

Remember to use scientific methods.

Make Observations

Ask a Question

Form a Hypothesis

Test your Hypothesis

Analyze and Conclude

Communicate Results

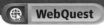

THE BIG IDEA

Interactions among Earth systems that take place near Earth's surface recycle Earth materials through the processes of the rock cycle, the water cycle, the carbon cycle, and the phosphorus cycle.

Key Concepts Summary

Lesson 1: Earth Systems and Interactions

- Earth systems interact in the **carbon cycle** through the weathering of rock and photosynthesis, among other processes.

- Earth systems interact in the **phosphorus cycle** through the water cycle and the decomposition of organisms, among other processes.

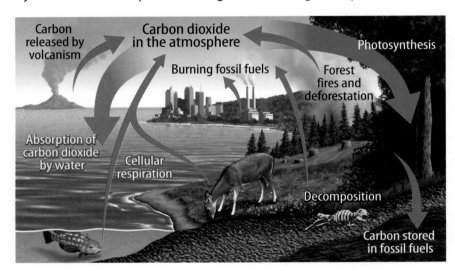

Carbon released by volcanism
Carbon dioxide in the atmosphere
Photosynthesis
Burning fossil fuels
Forest fires and deforestation
Absorption of carbon dioxide by water
Cellular respiration
Decomposition
Carbon stored in fossil fuels

Lesson 2: The Geosphere

- The layers of the geosphere are made of different materials, including soil, rock, and metal.

- The layers of the geosphere have different densities. This gives the geosphere a layered structure, with the denser materials located at the center of the planet and the least dense materials on the surface.

Vocabulary

carbon cycle p. 472
greenhouse gas p. 474
phosphorus cycle p. 475

luster p. 482
streak p. 482
cleavage p. 482
fracture p. 482
crust p. 486
mantle p. 487
lithosphere p. 487
asthenosphere p. 487
core p. 488

FOLDABLES® **Chapter Project**

Assemble your lesson Foldables as shown to make a Chapter Project. Use the project to review what you have learned in this chapter.

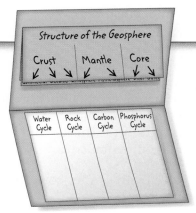

Use Vocabulary

1 Relate the terms *carbon cycle* and *greenhouse gas*.

2 Use the terms *cleavage* and *fracture* in a sentence.

3 What is the color of a mineral's powder called?

4 What word describes how a mineral reflects light?

5 Distinguish between the asthenosphere and the lithosphere.

Link Vocabulary and Key Concepts

Concepts in Motion Interactive Concept Map

Copy this concept map, and then use vocabulary terms from the previous page and other terms from the chapter to complete it.

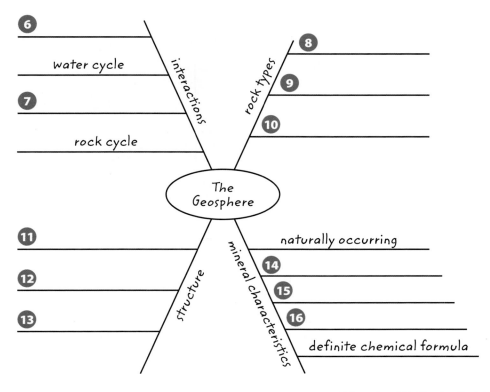

Understand Key Concepts 🗝️

1 Why does carbon move easily through the atmosphere?

   A. Gases flow freely.
   B. Gases are warm.
   C. Water and gas are mixed.
   D. Water flows freely.

2 In the diagram below, which sequence of processes recycles phosphorus from the bottom of the ocean and makes it available to plants?

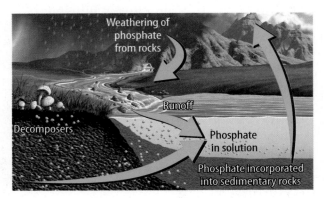

   A. deposition, uplift, weathering
   B. deposition, weathering, uplift
   C. uplift, deposition, weathering
   D. weathering, uplift, deposition

3 Which Earth system is the smallest carbon reservoir?

   A. the atmosphere
   B. the biosphere
   C. the geosphere
   D. the hydrosphere

4 How does too much phosphorus in a body of water kill fish?

   A. It poisons them.
   B. It burns them.
   C. They starve.
   D. They suffocate.

5 What happens when animals on land die?

   A. Phosphorus enters the atmosphere.
   B. Phosphorus enters the soil.
   C. Water enters the geosphere.
   D. Water enters the hydrosphere.

6 Why is the asthenosphere weaker than the rest of the mantle?

   A. It is hotter.
   B. It is partially melted.
   C. It is under less pressure.
   D. It is under greater pressure.

7 What differentiates oceanic crust from continental crust?

   A. Oceanic crust is metal, and continental crust is rock.
   B. Continental crust is solid, and oceanic crust is liquid.
   C. Continental crust is thicker and less dense than oceanic crust.
   D. Oceanic crust is thicker and less dense than continental crust.

8 How does Earth's inner core differ from the outer core?

   A. It is hotter.
   B. It is liquid.
   C. It is cooler.
   D. It is solid.

9 In the diagram below, the arrow is pointing to which layer of the geosphere?

   A. the crust
   B. the inner core
   C. the mantle
   D. the outer core

Critical Thinking

10 **Defend** calling the lithosphere a layer of the geosphere even though it contains two layers that have different compositions.

11 **Suggest** a way to reduce phosphorus runoff into streams and rivers.

12 **Sequence** the steps in soil formation.

13 **Compare** how different types of rocks form.

14 **Construct** Draw and label a diagram of the lithosphere and the asthenosphere.

15 **Predict** what would happen to Earth's magnetic field if the outer core were solid.

16 **Explain** How do humans impact the carbon cycle?

17 **Evaluate** how an activity such as the one shown below might affect soil structure.

Writing in Science

18 **Compare** An analogy compares similarities between two seemingly unlike things or events. For example, an analogy might compare a cell to a factory. Write an analogy that compares Earth's layers to an object of your choice. Include a brief paragraph that explains how these two seemingly different things are similar.

REVIEW THE BIG IDEA

19 How do weathering and erosion recycle Earth materials?

20 How do interactions among Earth's systems help recycle a rotting log?

Math Skills

Review

Math Practice

Use Percentages

21 A 10-kg bag of fertilizer is labeled 20-5-10. How much phosphorus is in the bag?

22 You want to add 0.1 kg of phosphorus to your soil. If a fertilizer contains 4 percent phosphorus, how many kilograms of fertilizer should you use? (Hint: Set up a proportion using percentages and kilograms.)

23 A fertilizer contains 5 percent phosphorus. How many grams of fertilizer would you need to get 10 g of phosphorus? (Hint: Set up a proportion using percentages and grams.)

Record your answers on the answer sheet provided by your teacher or on a sheet of paper.

Multiple Choice

1 How are rocks related to soil?

   A Rocks break up soil horizons.

   B Rocks keep soil from sinking into the mantle of Earth.

   C Soil formation begins with the weathering of rock.

   D Soil turns into rock by erosion.

2 Where is most of Earth's carbon stored?

   A in the atmosphere

   B in the biosphere

   C in the geosphere

   D in the hydrosphere

Use the figure below to answer question 3.

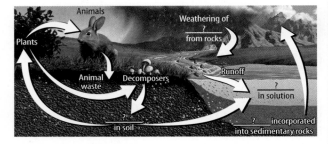

3 Which cycle is shown in the figure above?

   A carbon

   B phosphorus

   C rock

   D water

4 How does carbon in the air enter the biosphere?

   A by cellular respiration

   B by volcanism

   C through decomposition

   D through photosynthesis

5 Which type of rock is formed under extreme pressures and high temperatures?

   A igneous

   B lustrous

   C sedimentary

   D metamorphic

Use the figure below to answer question 6.

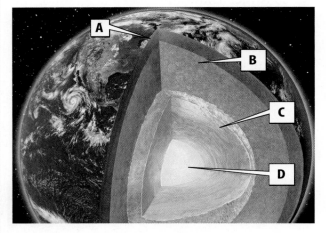

6 Which layer of Earth is liquid?

   A layer A

   B layer B

   C layer C

   D layer D

7 What is one way phosphorus leaves the biosphere?

   A by animals eating plants

   B by forming a solution

   C through animal waste

   D through weathering

8 What is one way the phosphorus cycle differs from the water cycle and the carbon cycle?

A None of the phosphorus cycle occurs in the atmosphere.

B Only the phosphorus cycle has parts that occur in the biosphere.

C Only the phosphorus cycle includes decomposition.

D Water is not part of the phosphorus cycle.

Use the figure below to answer question 9.

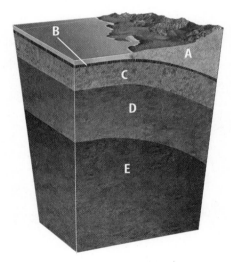

9 Which layers of Earth form the lithosphere?

A layers A and B

B layers A, B, and C

C layers D and E

D layers C, D, and E

Constructed Response

Use the figure below to answer questions 10 and 11.

10 In much of the carbon cycle, carbon moves readily from one form to another. However, in one process carbon is stored in one form for long periods of time. Which letter labels the process that stores carbon? Explain your answer.

11 In what form does carbon exist in the atmosphere? Use the diagram to explain your answer.

12 To identify a mineral, a geologist strikes a sample of the mineral with a small hammer, causing the sample to break apart. What test is the geologist using to help determine the identity of the mineral?

13 Why is the densest layer of Earth found in the center of the planet?

NEED EXTRA HELP?													
If You Missed Question...	1	2	3	4	5	6	7	8	9	10	11	12	13
Go to Lesson...	2	1	1	1	2	2	1	1	2	1	1	2	2

Chapter 14

Earth's Changing Surface

How do natural processes change Earth's surface over time?

Inquiry **What are these plants growing on?**

Mount St. Helens is a volcano in the state of Washington that erupted in 1980. The eruption changed the surface of the land and destroyed many organisms.

- What material buried all of the plants, and where did it come from?

- How can these flowers grow in volcanic material?

- How do natural processes change Earth's surface over time?

Get Ready to Read

What do you think?

Before you read, decide if you agree or disagree with each of these statements. As you read this chapter, see if you change your mind about any of the statements.

1. Continents do not move.

2. Earth's mantle is liquid.

3. Earthquakes occur and volcanoes erupt only near plate boundaries.

4. Volcanoes erupt melted rock.

5. Rocks cannot change.

6. Sediment can be transported by water, wind, and ice.

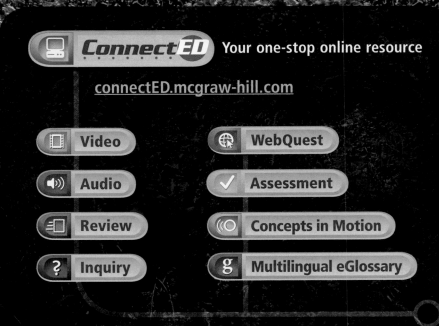

ConnectED Your one-stop online resource

connectED.mcgraw-hill.com

Video

WebQuest

Audio

Assessment

Review

Concepts in Motion

Inquiry

Multilingual eGlossary

Lesson 1

Plate Tectonics

Reading Guide

Key Concepts 🔑
ESSENTIAL QUESTIONS

- What is the theory of plate tectonics?
- What evidence do scientists use to support the theory of plate tectonics?
- How do the forces created by plate motion change Earth's surface?

Vocabulary
plate tectonics p. 501
continental drift p. 502
convergent boundary p. 504
divergent boundary p. 504
transform boundary p. 504
subduction zone p. 504
compression p. 505
tension p. 505
shear p. 505

g Multilingual eGlossary

Video Science Video

Inquiry How do mountains get so big?

Why are some places on Earth very flat and others mountainous? Why are mountains only in certain areas? Processes occur on Earth's surface and below that form mountains such as these in the Himalayas in India.

Can you put the pieces together? ✂

Scientists use different types of evidence to show that Earth's continents once were joined. How have scientists reconstructed Gondwana, a large ancient continent that was made of many continents?

1. Read and complete a lab safety form.

2. Obtain a **fossil evidence handout.** Determine what color or symbol represents each fossil.

3. With **scissors,** carefully cut out each landmass.

4. Use the puzzle pieces of Earth's landmasses to construct a model of Gondwana.

5. Once you are sure of your arrangement, **glue** your model into your Science Journal.

Think About This

1. Which fossils have been found on the landmasses that made up Gondwana? On which present-day continents are they located?

2. How do fossils of *Mesosaurus,* a freshwater reptile, support the existence of Gondwana?

3. 🔑 **Key Concept** How do you think fossil evidence supports the hypothesis that Earth's surface has moved?

Plate Motion

Even though we usually cannot feel it, Earth's surface is always moving. This motion can cause earthquakes and volcanic eruptions. It also can form mountains.

The theory of **plate tectonics** *states that Earth's crust is broken into rigid plates that move slowly over Earth's surface.* The rigid plates are called tectonic plates. You live on the North American Plate, shown in **Figure 1.** Tectonic plates slowly move over Earth's surface. The movement of one plate is described as either moving away from or toward another plate, or sliding past another plate. Plates move at speeds of only a few centimeters per year. At this rate, it takes moving plates millions of years to make new continents, new mountain ranges, or other landforms.

🔑 **Key Concept Check** What is the theory of plate tectonics?

Evidence of Plate Motion

The theory of plate tectonics has enabled geologists to explain many observations about Earth and to predict geologic events. It took scientists studying Earth nearly 100 years to gather the evidence that supports the theory of plate tectonics. It replaced a hypothesis called continental drift.

Figure 1 The continent of North America is part of the North American Plate.

✅ **Visual Check** What plates does the North American Plate interact with to the west?

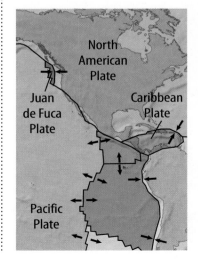

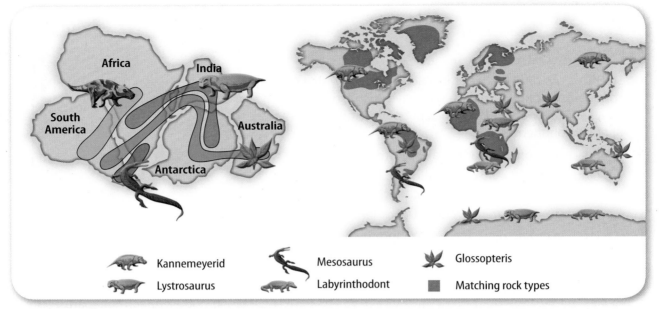

| Kannemeyerid | Mesosaurus | Glossopteris |
| Lystrosaurus | Labyrinthodont | Matching rock types |

▲ **Figure 2** 🔑 Fossil and rock evidence from the Gondwana continents supports the hypothesis of continental drift.

Figure 3 *Mesosaurus* was a freshwater reptile that lived 300 to 270 million years ago. ▼

Review Personal Tutor

Continental Drift Long before geologists proposed the theory of plate tectonics they discovered evidence of continental movement. One piece of evidence is the shape of Earth's continents. Look at the outlines of South America and Africa, shown in **Figure 2.** If you could push these two continents together, they would fit together like two pieces of a puzzle. In 1912, Alfred Wegener developed *the hypothesis that continents move, called* **continental drift.**

Fossil Evidence Different plants and animals live on different continents. For example, lions live in Africa, but not in South America. Many fossils of animals and plants show the same thing—some ancient organisms lived in certain areas, but not in others. However, geologists have discovered the same types of fossils on continents that are now separated by vast oceans.

Fossils of the freshwater reptile *Mesosaurus,* shown in **Figure 3,** have been found in South America and Africa. These two landmasses now are separated by the Atlantic Ocean. So how did a freshwater reptile cross a saltwater ocean? When the two continents were together, as shown in **Figure 2,** *Mesosaurus* likely traveled in freshwater rivers from one area to another.

Geological Evidence Rocks that are made of similar substances and mountains that formed at similar times are present on continents that are now far apart, as shown in **Figure 2.** Scientists can look for similarities in these rocks and mountains, as wells as the locations of ancient glaciers, deserts, and coal swamps, from one continent to the next.

🔑 **Key Concept Check** What evidence do scientists use to support the theory of plate tectonics?

How Plates Move

The hypothesis of continental drift was not accepted for more than 50 years after it was proposed. The main reason was that the hypothesis did not explain how continents could move. Geologists knew that the mantle, the part of Earth underneath continents, was solid. How could a continent push its way through solid rock?

New discoveries during the 1960s led scientists to propose the theory of plate tectonics. Recall that Earth's crust is broken into separate tectonic plates. These plates include the crust beneath the ocean and the continents. Scientists proposed that continents did not just float around the ocean. Instead, they argued, Earth's continents actually are part of tectonic plates. The plates move toward, away from, or past each other, carrying continents with them.

The forces that move plates come from deep within Earth. Earth's mantle is so hot that rocks can deform and move without breaking, much like putty.

Convection affects the mantle underneath tectonic plates. Hotter mantle rises toward Earth's surface and cooler mantle sinks deeper into the mantle, as shown in **Figure 4.** As the mantle moves, it pushes and pulls tectonic plates over Earth's surface.

Reading Check How does Earth's mantle move tectonic plates?

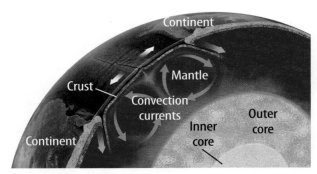

Figure 4 Convection currents cause movement in the mantle.

Inquiry **MiniLab**   **20 minutes**

How do tectonic plates move?

It might be difficult to imagine that continents are able to move slowly over Earth's surface. How can you model plate movement?

1. Read and complete a lab safety form.

2. Obtain a **rectangular, oven-proof pan.** Fill it with water to a depth of 1 cm. Carefully place the pan on a **hot plate,** and wait for the water to stop moving.

3. Use **scissors** to cut 6–8 shapes from a sheet of **construction paper.** Lay them on the top of the water in the pan.

4. Set the temperature of the hot plate to medium. Observe what happens to the shapes as the water warms.

5. Record your observations in your Science Journal. Turn off the hot plate. ⚠ Allow the pan to cool before you move it.

Analyze and Conclude

1. **Model** Which material represents tectonic plates? Which material represents the mantle?

2. **Key Concept** What causes movement in the mantle? How does this movement affect tectonic plates?

Tectonic Plate Boundaries

The edges of tectonic plates are called plate boundaries. *A* **convergent boundary** *is where two plates move toward each other. A* **divergent boundary** *is where two plates move apart from each other. A* **transform boundary** *is where plates slide horizontally past each other.*

Convergent Boundaries

Recall that oceanic crust is denser than continental crust. This difference is important where plates meet. When two plates come together, the denser oceanic plate usually is forced down into the mantle. The less dense continental plate remains on Earth's surface, as shown on the left in **Figure 5.** *The area where one plate slides under another is called a* **subduction zone.** However, when two continents collide at a plate boundary, both continents remain on the surface. As two continents push together, the crust rises up and large mountains form, as shown in the center in **Figure 5.**

 Reading Check What are two ways plates can interact at convergent boundaries?

Divergent Boundaries

When plates move apart at divergent boundaries, a rift forms between the two plates. A rift can form within continents when continental crust moves in opposite directions. A rift also can form at divergent boundaries on the ocean floor, as shown in **Figure 5.** As plates separate, molten rock can erupt from the rift. As the molten rock cools, it forms new crust.

Plate Interactions 🔑   **((◎)) Concepts in Motion** Animation

Figure 5 At convergent boundaries, tectonic plates either subduct or collide. The denser oceanic plate usually is forced into the mantle under the less dense continental plate (left), or two continental plates collide and form mountains (center). At a divergent boundary, molten rock rises up through a rift (right).

✅ **Visual Check** Which type(s) of plate interactions involve magma?

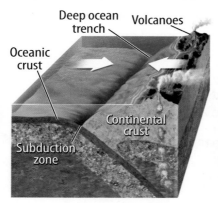

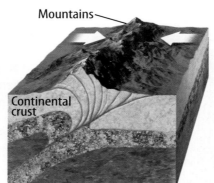

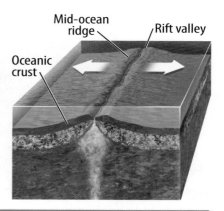

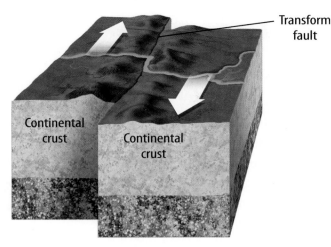
Transform fault

Continental crust

Continental crust

Transform Boundaries

Tectonic plates slide past each other at **transform** boundaries. The two sides of the boundary move in opposite directions. This can deform or break features such as fences, railways, or roads that cross the boundary, as shown in **Figure 6**.

 Reading Check What are the three main types of plate boundaries?

Forces Changing Earth's Surface

Forces within Earth cause plates to move. The three types of plate boundaries experience different types of forces, as illustrated in **Figure 7**. *The squeezing force at a convergent boundary is called* **compression**. *The pulling force at a divergent boundary is called* **tension**. *The side-by-side dragging force at transform boundaries is called* **shear**. These forces result in distinct landforms at plate boundaries.

Even though plates move slowly, the forces at plate boundaries are strong enough to form huge mountains and powerful earthquakes. Tensional forces pull the land apart and form rift valleys and mid-ocean ridges, such as the one shown in **Figure 5**. Compressional forces form mountains such as those shown in the photo at the beginning of this lesson.

Key Concept Check How do the forces created by plate motion change Earth's surface?

Figure 6 Motion of the Pacific Plate relative to the North American Plate has shifted this road along a transform boundary.

Figure 7 The three types of forces—tension, compression, and shear—cause rocks to change shapes in different ways. ▼

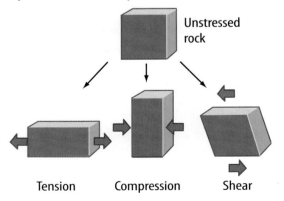

Unstressed rock

Tension   Compression   Shear

Visual Summary

Wegener developed the hypothesis that continents move.

Earth's crust is broken into pieces called tectonic plates. Convection currents in the mantle cause the plates to move.

There are three types of tectonic plate boundaries. Movement occurs at all three boundaries.

FOLDABLES

Use your lesson Foldable to review the lesson. Save your Foldable for the project at the end of the chapter.

What do you think NOW?

You first read the statements below at the beginning of the chapter.

1. Continents do not move.

2. Earth's mantle is liquid.

Did you change your mind about whether you agree or disagree with the statements? Rewrite any false statements to make them true.

Use Vocabulary

1 **Use the terms** *compression* and *convergent boundary* in a sentence.

2 **Explain** the term *continental drift.*

3 **Recall** What theory proposes that Earth's surface is made up of moving, rigid plates?

Understand Key Concepts

4 **Describe** the evidence that supports the hypothesis that continents move with respect to each other.

5 **Contrast** the direction of the plate motion associated with compression and tension.

6 Which is NOT a type of plate boundary?
- **A.** convergent
- **C.** subduction
- **B.** divergent
- **D.** transform

Interpret Graphics

7 **Identify** the types of plate boundaries in shown the graphic below.

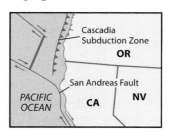

8 **Complete** Copy the graphic organizer below. Fill in the evidence used to support the hypothesis of continental drift.

Critical Thinking

9 **Infer** what happens where two oceanic plates collide.

10 **Explain** how a fossil from a plant or an animal that lived in a warm environment could be found in Antarctica.

Materials

world map

scissors

glue stick

Safety

How are rocks evidence of plate motion?

Several types of evidence have been used to support the idea that Earth's continents were once joined and drifted slowly over time to their present positions. Pangaea (pan JEE uh) was a supercontinent that straddled the equator hundreds of millions of years ago. How can you use rock evidence to reconstruct Pangaea?

Learn It

Scientists learn to **interpret scientific illustrations** by using different types of visual aids. Drawings, diagrams, photographs, and maps can be used to easily show what might take many words to explain. For example, you can reconstruct a map that shows the theoretical Pangaea.

Try It

1 Read and complete a lab safety form.

2 Obtain a world map. Carefully cut out Greenland, North America, South America, Africa, Australia, Eurasia, which is Europe and Asia and Antarctica.

3 Study the maps. Locate mountain ranges on each landmass. Recall that similar rock types and similar geologic structures in certain mountain ranges are evidence that some features of Earth's surface once formed continuous mountain chains.

4 Use the shapes of the landmasses and the locations of the mountain chains to reconstruct Pangaea.

5 When you are sure of your reconstruction, glue the supercontinent into your Science Journal.

6 Use your reconstructed map and what you have learned about plate tectonics to answer the questions below.

Apply It

7 Which mountains likely formed continuous chains long ago?

8 Can you locate any other features that would show how the continents were connected?

9 🗝 **Key Concept** How does geological data support the theory of plate tectonics?

Lesson 2

Reading Guide

Key Concepts 🔑
ESSENTIAL QUESTIONS

- What causes earthquakes?
- What causes volcanoes to form?
- How do earthquakes and volcanoes change Earth's surface?

Vocabulary
earthquake p. 509

fault p. 509

mid-ocean ridge p. 512

hot spot p. 512

lava flow p. 514

volcanic ash p. 514

caldera p. 515

 Multilingual eGlossary

🎞 **Video**

- **Science Video**
- **What's Science Got to do With It?**

Earthquakes and Volcanoes

Inquiry Why is this volcano erupting?

Mount Redoubt in Alaska had been dormant since 1989. In March 2009, it erupted and sent tons of ash and steam into the atmosphere. Mount Redoubt's location and eruption history are related to plate tectonics. Why did Mount Redoubt erupt again after 20 years? Are there other volcanoes nearby?

Launch Lab

15 minutes

Was it built on solid ground? 🥽 🧴 🧤

An earthquake occurs when blocks of crust slide past each other, causing the ground to shake. When the ground shakes, wet soil and sand can behave like a liquid. How can you demonstrate what happens to buildings constructed on ground such as this during an earthquake?

1. Read and complete a lab safety form.
2. Pour **sand** into a **clear container** until it is about two-thirds full.
3. Use a **pitcher** to slowly pour water into the container along one of its edges. Add only enough water so the water level is about 2 or 3 mm below the top of the sand.
4. Place a **brick** in the center of the container to model a tall building. Wait 1 min.
5. In your Science Journal, predict what might happen to a brick building if the land under it vibrates or shakes. Then, use your closed fist to pound the desk near the side of the container several times. Record your observations in your Science Journal.

Think About This

1. **Describe** what happened to the brick when you shook the desk.

2. 🔑 **Key Concept** Describe one way you think an earthquake can change Earth's surface.

Earthquakes

Because tectonic plates move very slowly, most changes to Earth's surface take a long time. But some changes occur very quickly and violently. *An* **earthquake** *is the rupture and sudden movement of rocks along a break or a crack in Earth's crust.* An earthquake can change Earth's surface quickly and dramatically, as shown in **Figure 8.**

Causes of Earthquakes

Slide your textbook across your desk. To make it move, you must push hard enough to overcome the force of friction that keeps your book from sliding. Earth's crust is similar. The forces acting on tectonic plates must be large enough to make blocks of crust move. When these blocks move, earthquakes occur.

The surface along which the crust moves is called a **fault.** Movement along faults occurs when the forces pushing on the rock layers become large enough to cause movement along the fault. Recall that compression and tension cause vertical motion on a fault, and shear forces cause horizontal motion. When pieces of crust slide past each other, energy is released, causing the ground to shake.

🔑 **Key Concept Check** What causes earthquakes?

Figure 8 Movement along the fault in this rice paddy in Japan caused the land to shift during an earthquake.

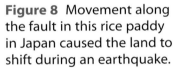

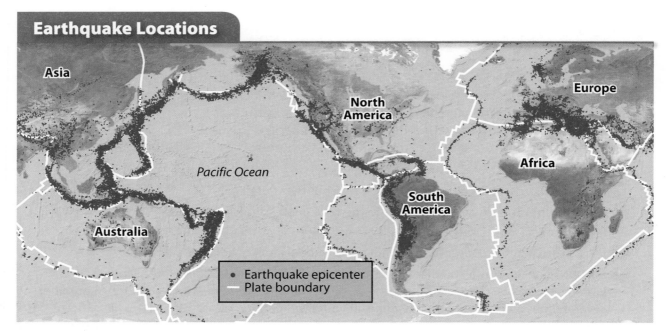

Earthquake Locations

Asia

North America

Europe

Pacific Ocean

Africa

South America

Australia

- Earthquake epicenter
— Plate boundary

▲ **Figure 9** Although most earthquakes occur near plate boundaries, some occasionally occur far from plate boundaries.

Figure 10 This low ridge formed as a result of an earthquake. ▼

Where do earthquakes occur?

Most earthquakes occur near plate boundaries, as shown in **Figure 9.** However, some of the largest earthquakes in the United States occurred very far from plate boundaries. For example, in the winter of 1811–1812, three large earthquakes occurred in Missouri. An explanation of these earthquakes may be their proximity to an old fault system, the New Madrid fault system, which is part of the beginning of a rift valley on the North American Plate.

✓ **Reading Check** Where do most earthquakes occur?

Changing Earth's Surface

You might be familiar with the damage earthquakes can cause. But they also can create landforms. Faults associated with earthquakes can be visible at Earth's surface. Some faults, such as the San Andreas Fault in California, can be more than 1,000 km long. During the massive Sichuan (SI chwan), China earthquake in 2008, blocks of crust moved as much as 9 m along a fault 240 km long and 20 km deep. Earthquakes also form mountains and change Earth's surface in other ways.

Mountains Every time blocks of crust slide past each other along a fault, an earthquake occurs. Blocks might move only 1–2 m, as shown in **Figure 10.** But after hundreds or thousands of earthquakes, blocks of crust will have moved a long distance. Compression and tension forces produce ridges and mountains as crust moves vertically.

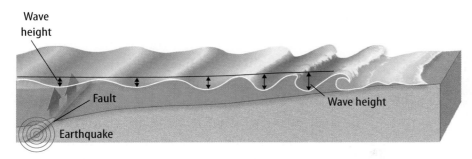

◀ **Figure 11** As the ground pushes the water upward, huge waves form. The waves of a tsunami increase in height as they come closer to shore.

Liquefaction and Landslides Great damage can occur in areas where the ground is made up of loose sediment instead of solid rock. Extreme shaking can cause this material to behave more like a liquid than a solid. This is called liquefaction (li kwuh FAK shun). Since the liquid-like ground is not strong enough to support heavy buildings, part of a building can sink into the ground and can collapse. Liquefaction is responsible for most of the damage to buildings after earthquakes. Shaking caused by earthquakes also can trigger landslides. Landslides bring rocks and soil from the tops of mountains into valleys.

Tsunamis Earthquakes that happen underwater can cause tsunamis (soo NAH meez), as shown in **Figure 11**. Any movement at a fault pushes the water up and creates huge ocean waves. These waves become taller as they reach shallower water near a shore. Tsunamis also can be caused by part of the ocean floor dropping down, or by an underwater volcanic eruption.

🔑 **Key Concept Check** How do earthquakes change Earth's surface?

Inquiry **MiniLab** **20 minutes**

Why is a tsunami so dangerous?

Why are tsunamis so destructive? How can you model a tsunami and explain why these waves are so dangerous?

1. Read and complete a lab safety form.

2. Obtain a **plastic container.** Use **heavy-duty tape** to secure a piece of a **plastic bag** over the hole on the inside of the container.

3. Pour about 500 mL of water into the container.

4. Hold the container steady over a **tub.** Have one member of your group use his or her finger to firmly tap the plastic bag once from below.

5. At the same moment the plastic bag is tapped, have another person drop one drop of **food coloring** directly over the area of water that is pushed upward.

6. Record your observations in your Science Journal.

Analyze and Conclude

1. **Observe** How did the food coloring help model a tsunami?

2. **Draw Conclusions** Why do you think a tsunami is so dangerous?

3. 🔑 **Key Concept** How does an earthquake-generated tsunami change Earth's surface?

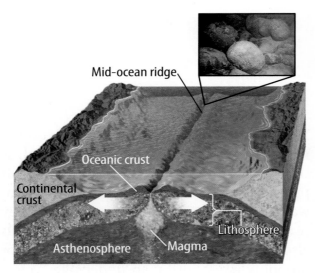

▲ **Figure 12** Mantle that rises below a divergent boundary melts at the low pressures near the surface. The lava erupts into the rift made between the separating plates.

▲ **Figure 13** At a convergent boundary, the mantle above the sinking plate melts and forms magma. The lava erupts onto the plate that remains at Earth's surface.

Volcanoes

Recall that molten rock below Earth's surface is called magma. Because magma is hot, it is less dense than the surrounding rock and rises upward. Volcanoes are landforms that form when magma erupts onto Earth's surface as lava. Volcanoes are common on Earth. Each year about 50–60 different volcanoes erupt somewhere on Earth. There are approximately 1,500 active volcanoes on Earth. Volcanoes can be destructive, but they also make new landforms.

Where Volcanoes Form

Volcanoes can occur at divergent and convergent plate boundaries and at hot spots. At a divergent boundary, lava flows into the rift formed by the separating plates, as shown in **Figure 12.** New crust is made of the rocks that form as this lava cools. *The mountains that form as this lava builds up and cools are called* **mid-ocean ridges.**

At some convergent boundaries, one tectonic plate sinks into the mantle. The sinking plate also carries water into the mantle. This causes the mantle to melt and form magma. The magma rises and erupts onto the plate that does not sink, as shown in **Figure 13.**

 Reading Check How do volcanoes form at convergent boundaries?

Hot Spots Not all volcanoes are near plate boundaries. In a few places, large volcanoes form near the center of a tectonic plate. These volcanoes form at **hot spots,** *locations where volcanoes form far from plate boundaries.* The Hawaiian Islands in the middle of the Pacific Ocean and Yellowstone National Park in Wyoming are hot spots.

The reason hot spots exist is not fully understood. One hypothesis is that hot spots occur above places where the mantle melts. The magma then rises toward the surface and eventually erupts through the crust.

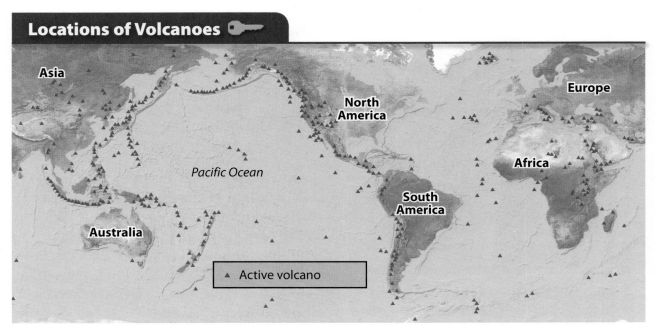

Locations of Volcanoes 🔑

Asia

North America

Europe

Pacific Ocean

Africa

South America

Australia

▲ Active volcano

Figure 14 The locations of Earth's volcanoes form a distinct pattern. Most of Earth's largest volcanoes are located at convergent plate boundaries.

✓**Visual Check** What features on Earth define this distinctive pattern?

Causes of Volcanic Eruptions

In order for magma to form, the crust and the mantle must become hot enough to melt. Rocks melt more easily when pressure is low. Pressure results from the weight of overlying rock, so pressure is lowest at Earth's surface. When hot rocks from deep inside Earth move toward its surface, the decrease in pressure allows these hot rocks to melt.

It also is possible to melt rocks by lowering their melting temperature. The temperature at which rocks melt depends on the makeup of the rock and the presence or absence of water. As shown in **Figure 13,** water enters the mantle at convergent boundaries. This allows the mantle to melt at a lower temperature. This is similar to adding salt to ice. If you put salt on ice, the ice melts at a lower, or colder, temperature.

Because magma is hot, it is also less dense than the rock material around it. It moves upward causing cracks to form in the solid rock. Magma also contains dissolved gases. The rising magma plus the gases cause pressure to build up. Eventually, the magma erupts through cracks in Earth's surface and a volcano is formed. The locations of most of Earth's volcanoes are shown in **Figure 14.**

🔑 **Key Concept Check** What causes volcanoes to form?

FOLDABLES

Make a vertical three-tab Venn book. Label it as shown. Use it to compare the causes and the effects of volcanoes and earthquakes.

Earthquakes

Both

Volcanoes

▲ **Figure 15** These mountains in India formed from large lava flows that erupted about 65 million years ago.

Math Skills

Use Geometry

Geologists estimate the **volume** of a lava flow from a volcano by measuring the average depth and the radius of the hardened lava field. The volume of a cylinder is the area of the base times the height (h). The base of a cylinder is a circle with an area equal to the square of the radius (r^2) times π (3.14). Therefore, $V = \pi \times r^2 \times h$. For example, what is the volume of lava needed to produce a lava field with a radius of 100.0 m and an average depth of 20.0 m?

1. The formula for volume is

$V = \pi \times r^2 \times h$

2. Replace the values in the formula with the given values and calculate.

$V = 3.14 \times (100\text{ m})^2 \times 20.0\text{ m}$

$V = 3.14 \times 10,000\text{ m}^2 \times 20.0\text{ m}$

$V = 628,000\text{ m}^3$

Practice

What is the volume of lava in a field with a radius of 90 m and an average thickness of 10.0 m?

▭ **Review**

- **Math Practice**
- **Personal Tutor**

Changing Earth's Surface

Volcanoes can be as small as a car. They also can be more than 10 km high. The shapes of volcanoes and the way lava erupts depends on where volcanoes form. What comes out of volcanoes, and how do volcanoes change Earth's surface?

Lava Flows Melted mantle material flows easily. When it erupts, it flows over Earth's surface, creating *long streams of molten rock called* **lava flows.** The lava eventually cools and solidifies, forming solid rock. Lava flows can be more than 10 km long. Over time, lava flows build up as flat layers, such as those shown in **Figure 15.**

Explosive Eruptions At convergent plate boundaries, part of the continental crust can become mixed with magma from the mantle. When this mixture of molten materials erupts, it does not flow as easily as lava made only of melted mantle. Instead of forming lava flows, it often solidifies in the atmosphere, where it breaks into *small pieces of lava called* **volcanic ash,** as shown in **Figure 16.** Ash can reach heights greater than 20 km. The ash eventually falls back to Earth's surface. Thick layers of these small pieces of lava can cover large areas that extend more than 100 km from the volcano. Eruptions that eject ash high into the atmosphere are called explosive eruptions. Lava is also produced during these eruptions.

✓ **Reading Check** What are small pieces of solidified magma produced during explosive eruptions called?

Figure 16 Mt. Etna in Sicily is in the middle of a series of eruptions that began in 2001. Explosive eruptions such as this and large lava flows are common. This eruption sent a cloud of ash to a height of 12 km. ▼

▲ **Figure 17** Volcanoes form as lava flows build up. Lava flows piling on top of each other formed Mauna Loa, the highest volcano on Earth. Mount Adams in Washington is made from alternating layers of lava flows and volcanic ash.

Types of Volcanoes

Lava flows can build up and form large volcanoes. Shield volcanoes form after basaltic lava flows have occurred over time. They tend to be large with gentle slopes, such as Mauna Loa in Hawaii, shown in **Figure 17.**

Composite volcanoes also can form as lava flows and ash layers deposited by explosive eruptions build up. These types of volcanoes often have steep sides and are cone-shaped, such as Mount Adams also shown in **Figure 17.** They are most common at convergent boundaries.

Before a volcano erupts, magma builds up in the crust in a reservoir called a magma chamber. What happens when large amounts of magma are removed from this chamber? Sometimes the surface above the chamber collapses. This creates *a large depression in the center of the volcano called a* **caldera** (kal DER uh). Some calderas can be more than 70 km wide.

Effects on the Atmosphere

Volcanoes also change Earth's atmosphere and climate. Volcanic ash and gases from explosive eruptions can blow high into the atmosphere. Some volcanic material remains in the atmosphere for years. This material can block sunlight. This can cause the temperature of the atmosphere near Earth's surface to decrease. Organisms that need sunlight to survive might die.

Key Concept Check How do volcanoes change Earth's surface?

WORD ORIGIN · · · · · · · · · · ·

caldera
from Latin *caldārium,* means "warming"

Lesson 2 Review

Visual Summary

Earthquakes occur mostly along plate boundaries. They can cause great damage.

Volcanos form at two types of plate boundaries. Lava cools and builds up, forming volcanoes and other landforms.

Earthquakes and volcanoes change Earth's surface.

FOLDABLES

Use your lesson Foldable to review the lesson. Save your Foldable for the project at the end of the chapter.

What do you think NOW?

You first read the statements below at the beginning of the chapter.

4. Earthquakes occur and volcanoes erupt only near plate boundaries.

5. Volcanoes erupt melted rock.

Did you change your mind about whether you agree or disagree with the statements? Rewrite any false statements to make them true.

Use Vocabulary

1. Explosive eruptions eject _____ high into the atmosphere.

2. **Distinguish** between a fault and an earthquake.

Understand Key Concepts

3. Which landform is NOT associated with volcanic eruptions?
   - **A.** caldera
   - **B.** fault
   - **C.** lava flow
   - **D.** mid-ocean ridge

4. **State** the type of plate boundary where volcanoes do not normally form.

5. **Contrast** the ways magma forms at mid-ocean ridges and at convergent plate boundaries.

Interpret Graphics

6. **Relate** Copy the graphic organizer below, and use it to relate earthquakes to the changes they produce at Earth's surface.

Critical Thinking

7. **Reconstruct** the stages in the formation of a volcano, beginning with melting the mantle to make magma and ending with the formation of a caldera.

8. **Critique** the claim that all earthquakes and volcanic eruptions occur near plate boundaries.

Math Skills

— Math Practice —

9. During one huge eruption, the Santorini volcano produced a lava field with a radius of 800 m and an average depth of 50 m. What volume of lava did the volcano produce?

Will it erupt again?

Can scientists predict a volcano's future?

About 7,700 years ago, Mount Mazama, a volcano in Oregon's Cascade Mountain Range, erupted so violently it spewed volcanic ash high into the atmosphere. Before the eruption, Mount Mazama was more than 3,600 m tall.

During the eruption, it ejected tremendous amounts of pumice, ash, and rock. The partially emptied magma chamber could no longer support the weight of the volcano, and it collapsed. This created a large depression called a caldera.

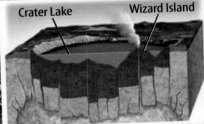

Crater Lake   Wizard Island

Over thousands of years, the caldera filled with rain and snow melt. Today, the caldera contains the beautiful, blue, crystal clear Crater Lake.

At first glance, Crater Lake does not appear to be a volcano, but it is. In the western portion of the lake is a cinder cone called Wizard Island, and two volcanic domes lie submerged beneath the water. These small volcanoes were created by more recent eruptions. One scientist studying the potential for future eruptions is Charlie Mandeville, a geologist at the American Museum of Natural History.

To predict the eruption threat within the Crater Lake caldera, Mandeville must first determine what caused the enormous eruption over 7,000 years ago. He looks for clues in the volcanic rocks from the prehistoric eruption. He searches for tiny pockets of magma solidified inside mineral crystals in the volcanic rocks. Because the crystals grew in the magma before the eruption, they are like microscopic windows into the magma's past. They give scientists valuable information about the magma that caused the volcano to explode so violently. The effects of this explosive eruption were catastrophic, ejecting gas and dust into the atmosphere. This temporarily shielded Earth from solar radiation.

What does the future hold for Crater Lake? Mandeville predicts that the volcano will erupt again. But he also has determined that the magma chamber below the volcano currently does not contain enough magma to cause an explosion as huge as the one that occurred 7,700 years ago.

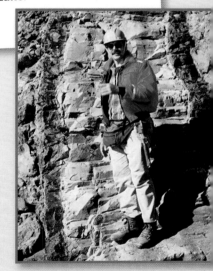

▲ Charlie Mandeville searches rocks for clues to Crater Lake's future.

It's Your Turn

RESEARCH the 1815 explosion of Mount Tambora in Indonesia. What were the effects of this eruption? How does it compare to the eruption of Mount Mazama? Share your findings with your class.

Lesson 3

Reading Guide

Key Concepts 🔑
ESSENTIAL QUESTIONS

- How are weathering and soil formation related?
- How do weathering, erosion, and deposition change Earth's surface?
- How are erosion and deposition related?

Vocabulary

weathering p. 519

erosion p. 519

physical weathering p. 519

chemical weathering p. 520

soil p. 521

sediment p. 522

deposition p. 523

 Multilingual eGlossary

 Video

- BrainPOP®
- What's Science Got to do With It?

Weathering, Erosion, and Deposition

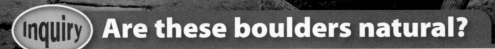

Inquiry Are these boulders natural?

These ball-shaped rocks, called the Moeraki Boulders, can be 3 m in diameter. They formed on the seafloor 60 million years ago. Erosion has exposed them at Earth's surface. How else can erosion and deposition change Earth's surface?

How does temperature affect weathering?

Weathering is any natural process that breaks rocks and minerals into smaller pieces or changes their makeup. How does the same substance weather in water of different temperatures?

1. Read and complete a lab safety form.

2. Carefully pour about 400 mL of hot water into a **500-mL beaker**. Pour about 400 mL of cold water in another beaker.

3. In your Science Journal, predict whether the same substance will dissolve faster in the hot water or the cold water.

4. Use a **spoon** to measure two equal amounts of **salt**. At the same time, add salt to each beaker.

5. Observe the contents of the beakers over the next 5 min. Compare the rates at which the salt dissolves in each beaker.

Think About This

1. How do you think temperature affects the rate of weathering of the salt?

2. 🔑 **Key Concept** How do you think weathering changes Earth's surface?

Weathering

You have read that mountains can form as a result of plate motion and volcanoes. But why don't mountains last forever? **Weathering** *refers to the processes that break down rocks, changing Earth's surface over time.* **Erosion** *is the moving of weathered material, or sediment, from one location to another.* Slowly but surely, weathering and erosion wear down mountains.

Physical Weathering

The process of breaking rock into small pieces without changing the **composition** *of the rock is* **physical weathering.** As you just learned in the Launch Lab, temperature affects physical weathering. Most rocks contain water in cracks and spaces between the particles that make up the rock. During winter or at night, the water in rocks can freeze. When water freezes, it expands. If water in rocks freezes and melts repeatedly, it can break apart rocks. This is called frost wedging.

Plants also can cause physical weathering, as shown in **Figure 18.** The roots of plants grow into cracks in rock, or as in this case, a sidewalk. As the roots grow and take up more space in the cracks, the force they apply to the rock breaks the rock.

✓ **Reading Check** What is the difference between weathering and erosion?

Figure 18 🔑 The roots of plants can break rock in the same way they broke this sidewalk. ▼

▲ **Figure 19** These rocks are red because the minerals in them contain iron that reacted with oxygen in the atmosphere.

Figure 20 This cave in Slovenia was formed when acidic water slowly dissolved the limestone rock around it. ▼

Chemical Weathering

The process of changing the composition of rocks and minerals by exposure to water and the atmosphere is called **chemical weathering.** Some minerals chemically weather more easily than others. For example, calcite, the mineral that makes up limestone, dissolves readily in acidic rainwater. Feldspar, a common mineral in igneous rocks, easily weathers into clay minerals, such as kaolinite. However, other minerals, such as quartz, are resistant to chemical weathering.

Gases in the atmosphere also can cause chemical weathering. Minerals containing iron react with oxygen in the atmosphere and form rust-colored minerals, such as those shown in **Figure 19.** Carbon dioxide in the atmosphere dissolves in water and makes acidic water. Limestone dissolves much faster in acidic water than in nonacidic water, as shown in **Figure 20.**

 **Key Concept Check** How does weathering change Earth's surface?

Temperature also affects the rate of chemical weathering. You might know that chemical reactions happen faster at hot temperatures than at cold temperatures. That is why chemical weathering occurs fastest in hot, wet climates.

Weathering Interactions

Physical weathering exposes more surface area of rocks, allowing more water and atmospheric gases to enter rocks. Recall that water and gases help cause chemical weathering. Chemical weathering weakens rocks by changing the composition of some minerals and dissolving others. For example, clay formed by chemical weathering is weaker than the feldspar from which it formed. This weakening of rocks can increase the rate of physical weathering. In this way, chemical and physical weathering work together.

Reading Check How do chemical and physical weathering work together?

Soil Formation

Soil *consists of weathered rock, mineral material, water, air, and organic matter from the remains of organisms.* Soil forms directly on top of the rock layers from which it is made. The process of soil formation is illustrated in **Figure 21.** Soil formation takes a long time. It is the result of hundreds to thousands of years of weathering. The rock type being weathered, the biological activity, and the climate all affect soil formation.

Biological activity plays an important role in making soil. Worms and other organisms create pathways in soil for water and air. Decaying plants and animals also produce carbon dioxide and other acids that enhance chemical weathering. Eventually, the decayed plants and animals become part of the soil and make it better for plant growth.

Where do you think soil forms the fastest? Warm, wet climates produce soil fastest. Large amounts of rain can speed weathering of rocks. And chemical reactions are faster in warmer temperatures. Weathering also can happen quickly in areas where freezing and thawing break apart rocks.

 Key Concept Check How are weathering and soil formation related?

 **FOLDABLES**

Make a horizontal two-tab book, and label it as shown. Use it to compare the different types of weathering.

Physical Weathering | Chemical Weathering

Soil Formation

Concepts in Motion Animation

Figure 21 Soil formation begins when physical and chemical weathering break down rocks. Organisms add organic matter to the soil. Decaying organic matter makes chemical weathering happen faster.

Visual Check What happens to the solid rock layers during soil formation?

Weathering processes fracture and break down rock. Soil formation can take hundreds or thousands of years.

Plants, bacteria, and burrowing organisms help break down rock.

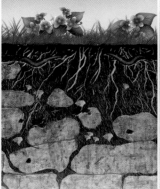

The upper part of the soil contains more organic material than the lower part. The lower part of the soil also can contain weathered rock.

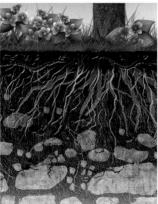

Over time, plants and other organisms in the soil die and decompose. The upper part of the soil contains nutrient-rich organic material

Erosion

Weathering dissolves minerals and produces small particles of rock. *The minerals and small pieces of rock are called* **sediment.** What happens to sediment after it is made? The agents of erosion remove the sediment. Water, ice, and wind can transport sediment from one place to another.

Erosion by Water

Moving water causes erosion. The water picks up rock pieces and sediment. They then scrape along the ground, picking up more material. The faster the water flows, the larger the pieces of sediment the water can carry. Steep mountain streams, such as the one shown to the left, carry away all sediment except large boulders. Water flowing in rivers as well as waves in lakes and oceans all cause erosion.

Erosion by Ice

Glaciers are large masses of ice. As a glacier flows down a mountain, it removes rocks and sediment next to its bottom and sides. This forms a smoothed land surface underneath the ice. Erosion by glaciers makes deep valleys and steep peaks, such as the valley shown to the left. Some glaciers can be large enough to cover continents. The ice covering Antarctica is an example.

Erosion by Wind

Strong winds also can erode sediment. Soil and rock that are not protected by plants can be eroded by wind. In some places, wind has eroded the rocks and made them look so smooth that they seem to have been sculpted by an artist.

Key Concept Check How does erosion change Earth's surface?

Deposition

What happens to eroded sediment? Eventually the moving water, ice, or wind slows down or stops. When this happens, the sediment is deposited. **Deposition** *is the process of laying down eroded material in a new location.*

Deposition by Water

Fast-flowing water carries sediment. If the speed of flowing water decreases, the water can no longer carry the sediment. The sediment will settle at the bottom of the water. Floodplains form when sediment settles out of rivers that flood the areas next to them. The floodplain of the Hatchie River in Tennessee is shown to the right. Sediment also settles out of rivers where they enter lakes and oceans, forming deltas.

Deposition by Ice

When glaciers melt, the water produced by the melting ice does not flow fast enough to carry sediment. The sediment is deposited where the ice melts. Glacial deposits of sediment are called moraines. Some moraines form mounds at the front and sides of glaciers. Other moraines, such as the one shown to the right, can cover the ground that was once under the glacier. The glaciers that once covered much of North America left moraines over most of the areas where they melted.

Deposition by Wind

Wind also can deposit sediment. Sand dunes, such as the ones shown to the right, are landforms made as wind continually moves and deposits sand grains. Wind moves the sand grains up one side of the sand dune and deposits them on the other side. Grain by grain, sand dunes migrate in the direction the wind blows.

 Key Concept Check How does deposition change Earth's surface?

Which will fall first?

When wind lessens, moving water slows, and ice melts, these agents of erosion deposit some of the sediment they carry. How can you observe this deposition? What will you see?

1. Read and complete a lab safety form.

2. Use a **balance** to measure 200 g of each grain-size **sediment**. Tip: Scoop the sediment onto a piece of paper or into a small paper cup on the balance so the sediment will be easy to pour.

3. Use a small **funnel** to pour each type of sediment into a 1-L **plastic bottle.**

4. Slowly add water to the bottle until the water level is just below the neck of the bottle. Screw the cap onto the bottle tightly.

5. In your Science Journal, predict what would happen if you were to mix the sediment and then allow it to settle out of the water. Draw a diagram of what you would expect to see.

6. Gently tip the bottle back and forth 10 times.

7. Stand the bottle on a flat surface. Observe and record what happens.

Analyze and Conclude

1. **Sequence** Explain how the different types of sediment settle out of the water.

2.  **Key Concept** How does deposition change Earth's surface?

Figure 22 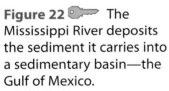 The Mississippi River deposits the sediment it carries into a sedimentary basin—the Gulf of Mexico.

[Review] Personal Tutor

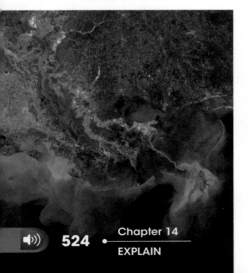

The Erosion-Deposition Cycle

Weathering breaks rock into sediment that can be transported from high mountains to low areas. Sediment builds up on plains, at the bottom of lakes, and at the bottom of the ocean. Over time, thick layers of sediment form. The locations where sediment accumulates are called sedimentary basins. The Gulf of Mexico, shown in **Figure 22,** is a sedimentary basin into which the Mississippi River deposits sediment.

Recall that some minerals dissolve in water. If the water evaporates, the minerals form again. Over time, layers of salt can form in this way as water evaporates in sedimentary basins. The salt surrounding the Great Salt Lake in Utah is an example of minerals reforming as water evaporates.

The cycle of weathering, erosion, and deposition has been repeated many times throughout Earth's history. The cycle continues today. The shapes of continents change. The locations of plate boundaries change. Sediment continues to be deposited in low areas and then forced upward as tectonic activity forms mountains. Earth's surface is continually changing.

 Key Concept Check How are erosion and deposition related?

Lesson 3 Review

Visual Summary

Physical and chemical weathering work together and change Earth's surface. They break down rock and form sediment.

Erosion occurs when sediment is removed and transported from where it formed.

Deposition occurs when sediment is laid down in new locations.

FOLDABLES

Use your lesson Foldable to review the lesson. Save your Foldable for the project at the end of the chapter.

What do you think NOW?

You first read the statements below at the beginning of the chapter.

5. Rocks cannot change composition.

6. Sediment can be transported by water, wind, and ice.

Did you change your mind about whether you agree or disagree with the statements? Rewrite any false statements to make them true.

Use Vocabulary

1 **Define** *sediment* in your own words.

2 Processes that break down rock but do not change its makeup are called _____.

3 Weathered rock, mineral materials, and organic matter make up _____.

Understand Key Concepts

4 **Contrast** physical and chemical weathering.

5 Which process is an example of physical weathering?
- **A.** a nail rusting
- **B.** acidic water dissolving calcite
- **C.** rock weathering to make clay
- **D.** plant roots breaking rock

6 **Identify** three agents of erosion and deposition.

Interpret Graphics

7 **Infer** What type of weathering has the greatest effect in the setting pictured in the figure above? Explain your reasoning.

8 **Sequence** Copy and fill in the graphic organizer below to describe the process of soil formation.

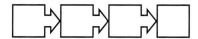

Critical Thinking

9 **Predict** how a marble statue in Alaska will weather compared to one in Florida.

10 **Critique** the claim that soils form within a few years and new soil will quickly replace any soil that is eroded.

Materials

plastic tub

colored pencils

salt dough

Also needed:
any appropriate materials to construct your landform

Safety

Processes That Change Earth's Surface

Have you ever been inside a cave, climbed a mountain, or stood atop a large plateau? How do you think these landforms formed? How long did it take? Many different processes change Earth's surface. How can you design a procedure to show the processes involved in producing such landforms?

Question

Examine the landforms shown in the photographs on these two pages. Choose one landform for further investigation. How do you think your landform formed?

Procedure

1 Closely examine the photograph of your landform. As you do this, think about all the processes that produce landforms. Write a list of these processes in your Science Journal.

2 Decide which processes you think contributed to the formation of your landform. Highlight or place a star next to them in your list from step 1.

3 Use your responses to step 2 to develop a flowchart that explains how you think your landform formed. A flowchart uses boxes to show the steps in a process. Arrows are used to show the correct order of the steps in the process. Include as many steps in the process as you think might have been involved in the formation of your landform.

4 Draw your final flowchart in your Science Journal.

5 Have your teacher approve the steps in your flowchart of your model landform.

6. Use colored pencils to make a detailed sketch of the landform you plan to construct.

7. You will need a large plastic tub to hold your mini-landform. Obtain any other materials you think you will need to make your landform.

8. Construct your landform. Then change your landform by simulating weathering, erosion, or another process you identified.

Analyze and Conclude

9. **Analyze Results** Did the steps in your flowchart produce the landform you expected? Why or why not?

10. **Classify** What Earth processes did you model to produce and change your landform?

11. **Compare and Contrast** How is your mini-landform like the actual landform? In what ways is it different?

12. 🔵 **The Big Idea** How do natural processes change Earth's surface over time?

Communicate Your Results

Present a short travel-show presentation as an Earth science expert, and explain how such a landform forms. Use your flowchart and mini-landform to demonstrate the processes that change Earth's surface. Your presentation should be no more than 2 minutes long.

Inquiry Extension

Research Geographic Information Systems, or GIS, to see what types of geological processes are likely to affect your state. Predict how some local landforms might change over time.

Lab Tips

☑ Frequently refer back to the information in this chapter as you work on this lab.

☑ If one of the steps in your flowchart does not produce the desired results, discuss the problem with others and try to come up with alternative ways to produce the landform.

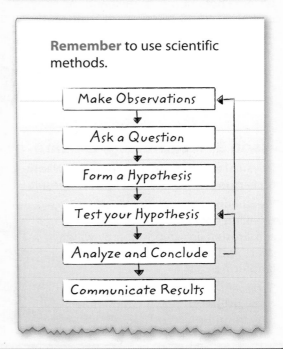

Remember to use scientific methods.

Make Observations

Ask a Question

Form a Hypothesis

Test your Hypothesis

Analyze and Conclude

Communicate Results

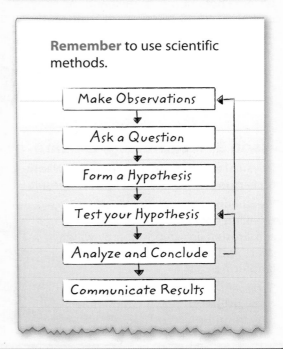

Chapter 14 Study Guide

 THE BIG IDEA

Mountains and valleys form where tectonic plates interact, new crust forms where lava flows from volcanoes, and the land shifts when earthquakes occur.

Key Concepts Summary 🔑

Lesson 1: Plate Tectonics

- The theory of **plate tectonics** states that Earth's surface is broken into rigid pieces, or plates, that move with respect to each other.
- Evidence that continents have moved throughout Earth's history includes the shape of the continents and the presence of similar fossils and geological features on widely separated continents.
- The forces produced by plate motion change Earth's surface by forming mountains and rifts and by causing earthquakes and volcanic eruptions.

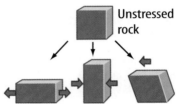

Unstressed rock

Tension Compression Shear

Lesson 2: Earthquakes and Volcanoes

- **Earthquakes** are caused when forces on rocks in the crust become large enough for the rock to move along faults.
- Volcanoes form when rocks in the mantle become hot enough to melt and the molten rock erupts onto Earth's surface.
- Earthquakes and volcanoes change Earth's surface in several ways. They form mountains and new landmasses from **lava flows** and explosive eruptions.

Lesson 3: Weathering, Erosion, and Deposition

- **Physical weathering** and **chemical weathering** break down material on Earth's surface. Soil can form as a result.
- Weathering changes Earth's surface by breaking down material and forming soil. Erosion and deposition by water, wind, and ice create and change landforms.
- Sediment is removed by **erosion,** and **deposition** lays down sediment in a new location.

Vocabulary

plate tectonics p. 501
continental drift p. 502
convergent boundary p. 504
divergent boundary p. 504
transform boundary p. 504
subduction zone p. 504
compression p. 505
tension p. 505
shear p. 505

earthquake p. 509
fault p. 509
mid-ocean ridge p. 512
hot spot p. 512
lava flow p. 514
volcanic ash p. 514
caldera p. 515

weathering p. 519
erosion p. 519
physical weathering p. 519
chemical weathering p. 520
soil p. 521
sediment p. 522
deposition p. 523

FOLDABLES® Chapter Project

Assemble your lesson Foldables as shown to make a Chapter Project. Use the project to review what you have learned in this chapter.

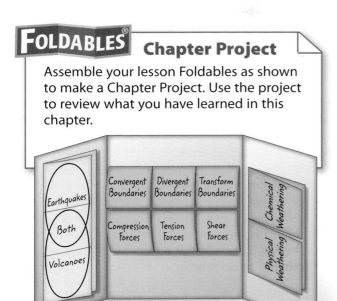

Use Vocabulary

1 What is the name of the structure along which earthquakes occur?

2 New material is added to tectonic plates at _____ boundaries.

3 How are soil and sediment related?

4 What type of force exists at convergent boundaries?

5 Tiny fragments of solidified lava are called _____.

6 What is the name of the process that breaks down rock?

Link Vocabulary and Key Concepts

Concepts in Motion Interactive Concept Map

Copy this concept map, and then use vocabulary terms from the previous page to complete the concept map.

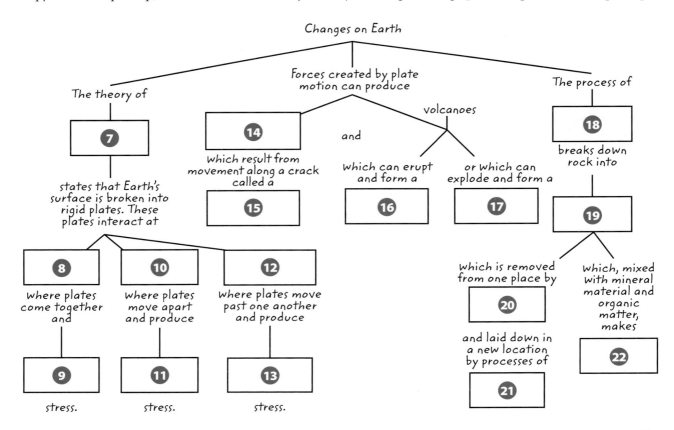

Understand Key Concepts

1 Rifts form at which type of plate boundary?
  A. convergent
  B. divergent
  C. hot spot
  D. transform

2 Sediment is made by
  A. erosion.
  B. deposition.
  C. weathering.
  D. transportation.

3 The force produced when two plates move away from each other is
  A. compression.
  B. shear.
  C. subduction.
  D. tension.

4 What feature is marked by the *X* on the image below?

  A. continental rift
  B. mid-ocean ridge
  C. subduction zone
  D. transform fault

5 Volcanic ash is produced during
  A. explosive eruptions.
  B. lava flows.
  C. liquefaction.
  D. subduction.

6 The sediment deposited by glaciers creates
  A. flood plains.
  B. moraines.
  C. sand dunes.
  D. sedimentary basins.

7 Tectonic plates slide horizontally past each other at
  A. convergent boundaries.
  B. divergent boundaries.
  C. mid-ocean ridges.
  D. transform boundaries.

8 Examine the landform in the figure shown below. What agent of erosion created this feature?

  A. deposition
  B. ice
  C. water
  D. wind

9 Which type of weathering is most common in cold, mountainous areas?
  A. crystallization
  B. dissolving
  C. frost wedging
  D. melting

10 Tectonic plates generally move toward or away from each other at what speed?
  A. centimeters per second
  B. centimeters per day
  C. centimeters per year
  D. centimeters per million years

Critical Thinking

11 **Assess** the evidence used to support the theory of continental drift. Which piece of evidence do you think is most convincing?

12 **Explain** why the theory of plate tectonics resolved the problems with the hypothesis of continental drift.

13 **Rank** the importance of earthquakes and volcanoes in changing Earth's surface. Explain your reasoning.

14 **Hypothesize** Sometimes a combination of forces act on a fault. Examine the figure below. What two forces do you think acted on this fault? Give evidence to support your answer.

15 **Relate** the speed of moving water to the sediment it carries. Why does a river deposit sediment when it enters the ocean or a lake?

Writing in Science

16 **Write** a newspaper story describing an imaginary volcanic eruption or earthquake. Pick a location and a type of event that is consistent with the theory of plate tectonics. First, describe the event and its consequences. Then explain why it happened. Think of a headline that will capture the attention of readers.

REVIEW **THE BIG IDEA**

17 Explain how plate tectonics, volcanic activity, weathering, and erosion interact to make the rock cycle.

18 How do natural processes change Earth's surface over time?

Math Skills

Use Geometry

19 The diameter of a lava field is 1,000 m. The average depth of the lava is 35 m. What is the volume of the lava flow? (Hint: radius = $\frac{\text{diameter}}{2}$)

20 Lava erupts at rates of as little as 0.5 m³/s to 5,000 m³/s. If each eruption took place for 24 h, what is the volume of lava produced in each eruption? (Hint: 24 h = 86,400 seconds)

21 Sometimes, lava flows in only one direction rather than in a circular pattern around the mouth of the volcano. If a lava flow was roughly 100 m wide and 400 m long with a depth of 30 m, what was the volume of the lava? (Hint: The volume of a rectangle is length × width × height (thickness))

Standardized Test Practice

Record your answers on the answer sheet provided by your teacher or on a sheet of paper.

Multiple Choice

1 What causes moraines to form?

  A growing plants

  B melting glaciers

  C running rivers

  D strong winds

2 What happens at divergent and convergent plate boundaries that causes volcanoes to form nearby?

  A Crust and mantle melt into magma.

  B Earthquakes shake magma loose.

  C Magma chambers collapse into calderas.

  D Sediment deposits become volcanic ash.

Use the diagram below to answer question 3.

3 Which can build up and form the type of volcano shown?

  A lava flows and volcanic ash

  B magma trapped in deep chambers

  C sand carried and deposited by wind

  D repeated lava flows

4 The sudden movement of rocks along a fault causes

  A convection.

  B earthquakes.

  C lava flows.

  D volcanic eruptions.

5 Which is part of the theory of plate tectonics?

  A Continents stay in the same place over millions of years.

  B Earthquakes are equally likely at any location on Earth.

  C Earth's crust is broken into rigid plates that move slowly.

  D Earth's plates only can slide past each other because they are rigid.

Use the diagram below to answer question 6.

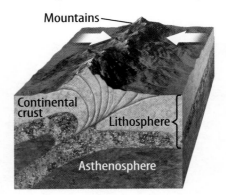

6 What is shown in the diagram above?

  A two continental plates colliding

  B two oceanic plates colliding

  C a rift valley forming as two plates collide

  D an oceanic plate getting subducted beneath a continental plate

7 Which is NOT considered evidence that supports the theory of plate tectonics?

  A identical fossils on distant continents

  B similar rock types on distant continents

  C earthquakes occurring far from plate boundaries

  D the shape of Earth's continents

8 Which two processes form soil?

   A climate and living organisms

   B erosion and deposition

   C glaciers and sediment

   D weathering and biological activity

Use the diagram below to answer question 9.

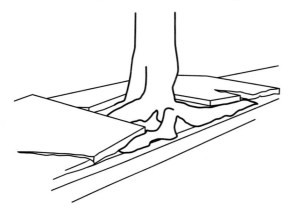

9 In the diagram, the growing tree roots apply enough force to break the sidewalk into pieces. What is the term for this process?

   A chemical weathering

   B erosion

   C liquefaction

   D physical weathering

10 Which can be caused by earthquakes?

   A caldera

   B hot spot

   C tsunami

   D weathering

Constructed Response

Use the diagram below to answer questions 11 and 12.

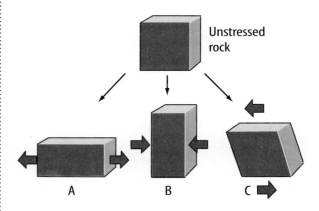

11 Name each force illustrated in the diagram.

12 Identify the type of plate boundary where each force illustrated in the diagram acts. How does each force affect Earth's plates at each boundary? Give an example of a landform created by each force.

13 Describe the main events in the erosion-deposition cycle. What role does weathering play in this cycle?

14 How does the movement of Earth's mantle cause tectonic plate movement?

NEED EXTRA HELP?														
If You Missed Question...	1	2	3	4	5	6	7	8	9	10	11	12	13	14
Go to Lesson...	3	2	2	2	1	1	2	3	3	2	1	1	3	1

Chapter 15

Using Natural Resources

THE BIG IDEA How can people protect Earth's resources?

Inquiry A Typical Day at Work?

These technicians are working on a wind farm off the coast of Denmark. Wind energy meets about 20 percent of Denmark's energy needs.

- How do you think wind energy works?
- What might be some advantages of using wind as an energy source?
- What might be some disadvantages?
- How can people protect Earth's resources?

Get Ready to Read

What do you think?

Before you read, decide if you agree or disagree with each of these statements. As you read this chapter, see if you change your mind about any of the statements.

1 The world's supply of coal will never run out.

2 You should include minerals in your diet.

3 Global warming causes acid rain.

4 Smog can affect human health.

5 Oil left over from frying potatoes can be used as automobile fuel.

6 Hybrid electric vehicles cannot travel far or go fast.

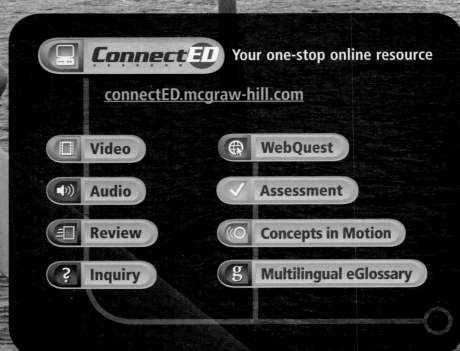

ConnectED Your one-stop online resource

connectED.mcgraw-hill.com

- Video
- WebQuest
- Audio
- Assessment
- Review
- Concepts in Motion
- Inquiry
- Multilingual eGlossary

Reading Guide

Key Concepts 🔑
ESSENTIAL QUESTIONS

- What are natural resources?
- How do the three types of natural resources differ?

Vocabulary

natural resource p. 537
nonrenewable resource p. 538
renewable resource p. 540
inexhaustible resource p. 542
geothermal energy p. 543

g Multilingual eGlossary

▶ Video BrainPOP®

Earth's Resources

Inquiry **A River in the Desert?**

People in dry areas sometimes build structures such as this aqueduct to carry water to their cities. Why do you think water is such an important resource? Do you think Earth's supply of water will ever run out?

Where does it come from?

Almost everything you use comes from natural resources, or materials that come from the environment. Can you identify the natural resources used to make a common object?

1. Read and complete a lab safety form.

2. Choose a common **object** from around your classroom or in your backpack.

3. Create a data table like the one below in your Science Journal. In the first column of your data table, list the object you will investigate.

4. In the second column, determine all the natural resources required to make the object. The example table shown here lists some natural resources used to make a pencil.

Natural Resources in a Pencil	
Object	Natural Resources Required
Pencil	1. wood
	2. graphite
	3.
	4.

Think About This

1. Which natural resource was hardest to identify? How did you figure it out?

2. Compare your data table to a classmate's. Which natural resources were on both lists?

3. 🔑 **Key Concept** What type of natural resource was most common? Why might this be?

Natural Resources

You walk into a room and switch on a light. Where does the electricity to power the light come from? It might come from a power plant that burns coal or natural gas. Or, it might come from rooftop solar panels made with silicon, a mineral found in sand.

The smallest microbe and the largest whale both rely on materials and energy from the environment. The same is true for humans. People depend on the environment for food, clothing, and fuels to heat and light their homes. *Parts of the environment that supply materials useful or necessary for the survival of living things are called* **natural resources.** Natural resources include land, air, minerals, and fuels. The trees and water in **Figure 1** also are natural resources.

🔑 **Key Concept Check** What are natural resources?

Figure 1 🔑 All parts of the environment that are important to living things are natural resources.

Nonrenewable Resources

How often do you travel in a vehicle that runs on gasoline? Do you drink soda from aluminum cans or sip water from plastic bottles? Gasoline, aluminum, and plastic are made from nonrenewable resources.

Nonrenewable resources *are natural resources that are being used up faster than they can be replaced by natural processes.* Nonrenewable resources form slowly, usually over thousands or millions of years. If they are used faster than they form, they will run out. Nonrenewable resources include fossil fuels and minerals.

 Reading Check What characteristic makes a resource nonrenewable?

Fossil Fuels

Fossil fuels include coal, oil, and natural gas. The fossil fuels we use today formed from the decayed remains of organisms that died millions of years ago. Although fossil fuels are forming all the time, we use them much more quickly than nature replaces them. Fossil fuels form underground. As shown in **Figure 2,** coal is mined from the ground. Oil and natural gas are drilled from the ground.

Fossil fuels are used primarily as sources of energy. Many electric power plants burn coal or natural gas to heat water and make steam that powers generators. Natural gas also is used to heat homes and businesses. Gasoline, jet fuel, diesel fuel, kerosene, and other fuels are made from oil. Most plastics also are made from oil.

FOLDABLES

Make a small horizontal tri-fold book. Label it as shown. Use it to identify similarities and differences among the types of resources.

Nonrenewable Resources | Renewable Resources | Inexhaustible Resources

Figure 2 The black rock layer is a seam of coal. It formed from the decayed remains of trees, ferns, and other swamp plants that died 300–400 million years ago.

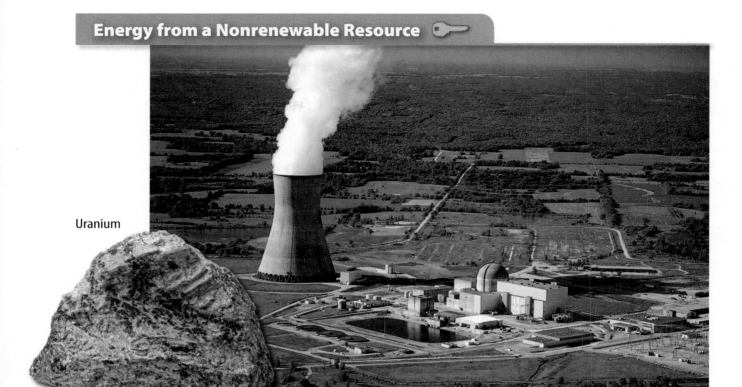

Energy from a Nonrenewable Resource 🔑

Uranium

Concepts in Motion Animation

Figure 3 Uranium is the fuel used to generate electricity in nuclear power plants.

Visual Check What do you think is being emitted by the tower?

Minerals

Have you ever added fertilizer to a plant's soil? Fertilizers contain phosphorus and potassium, two minerals that promote plant growth. The human body also needs minerals for good health, including calcium and magnesium.

Minerals are nonliving substances found in Earth's crust. People use minerals for many purposes. Gypsum is used in wall board and cement. Silicon is important for the manufacture of computers and other electronic devices. Copper is used in electrical wiring.

Uranium is a mineral that can be used as a source of energy. In a nuclear power plant, such as the one shown in **Figure 3**, the nuclei of uranium atoms are split apart in a reaction known as nuclear fission. Some of the energy that held the nuclei together is released as heat, which is then used to boil water and produce steam to generate electricity.

Like fossil fuels, minerals are formed underground by geologic processes that take millions of years. For that reason, most minerals are considered nonrenewable. Some minerals, such as calcium, are plentiful. Others, such as large rubies, are rare.

Reading Check Why are minerals nonrenewable?

Renewable Resources

Supplies of many natural resources are constantly renewed by natural cycles. The water cycle is an example. When liquid water evaporates, it rises into the **atmosphere** as water vapor. Water vapor condenses and falls back to the ground as rain or snow. Water is a renewable resource.

Renewable resources *are natural resources that can be replenished by natural processes at least as quickly as they are used.* These resources do not run out because they are replaced in a relatively short period of time. They include water, air, land, and living things.

Renewable resources are replenished by natural processes. Still, they must be used wisely. If people use any resource faster than it is replaced, it becomes nonrenewable. As shown in **Figure 4,** forests are sometimes nonrenewable resources.

 Reading Check In what way are renewable and nonrenewable resources similar?

Figure 4 Forests can be nonrenewable if trees are cut down faster than they can grow back.

Air

Did you know that plants produce almost all of the oxygen in the air we breathe? Oxygen is a product of photosynthesis. You might recall that photosynthesis is a series of chemical reactions in plants that use energy from light and produce sugars. Without plants, Earth's atmosphere would not contain enough oxygen to support most forms of life.

Air also contains carbon dioxide (CO_2), which plants need for photosynthesis. CO_2 is released into the air when dead plants and animals decay, when fossil fuels or wood are burned, and as a product of cellular respiration in plants and animals. Recall that cellular respiration is a series of chemical reactions that convert energy from food into a form usable by cells. Without CO_2, photosynthesis would not be possible.

Land

Fertile soil is an important resource. Topsoil is the upper layer of soil that contains most of the nutrients plants need. Gardeners know that topsoil can be replenished by the decay of plant material. The carbon, nitrogen, and other elements in the decomposing plants become available for the growth of new plants.

Topsoil can be classified as a renewable resource. However, if it is carried away by water or wind, it can take hundreds of years to rebuild.

Land resources also include wildlife and ecosystems, such as forests, grasslands, deserts, and coral reefs.

 Reading Check How is topsoil replenished by natural processes?

Water

Can you imagine a world without water? All organisms require water to live. People need a reliable supply of freshwater for drinking, washing, and irrigating crops. People also use water to run power plants and factories. Oceans, lakes, and rivers serve as major transportation routes and recreational areas. They are important habitats for many species, including some that people depend on for food.

Most of Earth's surface is covered by water. But only a small amount is freshwater that people can use, and this water must be cleaned before you can drink it. Freshwater is renewed through the water cycle, but the total amount of water on Earth always remains the same.

Has your community ever been asked to conserve water because of a drought? A drought can cause supplies of freshwater to run short. In many large cities, water is transported from hundreds of miles away to meet the needs of residents. In some parts of the world, people travel long distances every day to get the water they need.

Converting a ratio to a percentage often makes it easier to visualize a set of numbers. For example, in 2007, 101.5 quadrillion units (quads) of energy were used in the United States. Of that, 6.813 quads were produced from renewable energy sources. What percentage of U.S. energy was produced from renewable energy sources?

Set up a ratio of the part over the whole.

$$\frac{6.813 \text{ quads}}{101.5 \text{ quads}}$$

Rewrite the fraction as a decimal.

$$\frac{6.813 \text{ quads}}{101.5 \text{ quads}} = 0.0671$$

Multiply by 100 and add %.

$$0.0671 \times 100 = 6.71\%$$

Practice

Of the 101.5 quads of energy used in 2007, 0.341 quads were from wind energy. What percentage of U.S. energy came from wind?

 Review

- **Math Practice**
- **Personal Tutor**

Inexhaustible Resources

An **inexhaustible resource** *is a natural resource that will not run out, no matter how much of it people use.* Energy from the Sun, solar energy, is inexhaustible. So is wind, which is generated by the Sun's uneven heating of Earth's lower atmosphere. Another inexhaustible resource is thermal energy from within Earth.

Key Concept Check How do inexhaustible resources differ from renewable and nonrenewable resources?

Solar Energy

Without heat and light from the Sun, life as it is on Earth would not be possible. If you've studied food chains, you know that energy from the Sun is used by plants and other producers during photosynthesis to make food. Consumers are organisms that get energy by eating producers or other consumers. The energy in food chains always is traced back to the Sun.

Solar energy can be harnessed for many uses. Greenhouses trap heat. They make it possible to grow warm-weather plants in cool climates. Solar cookers concentrate the Sun's heat to cook food. Large solar-power plants provide electricity to many homes. Solar energy also can be used to heat water for individual homes, as shown in **Figure 5.**

Energy from an Inexhaustible Resource

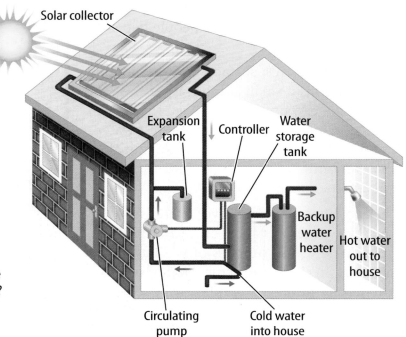

Figure 5 A solar water heater uses energy from the Sun to heat water. The hot water can be stored in a tank until it is needed.

Visual Check In which part of the system is water heated by the Sun?

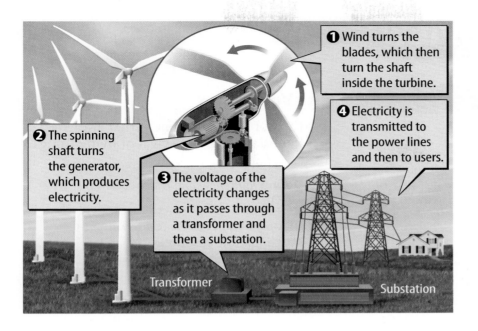

① Wind turns the blades, which then turn the shaft inside the turbine.

② The spinning shaft turns the generator, which produces electricity.

③ The voltage of the electricity changes as it passes through a transformer and then a substation.

④ Electricity is transmitted to the power lines and then to users.

Transformer   Substation

◀ **Figure 6** The wind spins the blades of a wind turbine, which in turn powers a generator that produces electricity.

Visual Check What happens to the electricity as it passes through a transformer and a substation?

Wind Power

What do sailboats, kites, and windmills have in common? All are powered by wind—the movement of air over Earth's surface. Wind is an inexhaustible resource produced by the uneven heating of the atmosphere by the Sun.

If you live in an area with frequent, strong winds, you might have seen giant wind turbines. These turbines, such as the ones shown in **Figure 6,** can be used to produce electricity.

Geothermal Energy

Another inexhaustible resource is geothermal energy. **Geothermal energy** is *thermal energy from within Earth.* Pockets of molten rock, or magma, rise close to the surface of some parts of Earth's crust. The magma heats underground water and rocks. The heated water produces steam used to generate electricity. In California and other regions, geothermal energy produces electricity on a large scale, as shown in **Figure 7.**

WORD ORIGIN · · · · · · · · · · · ·

geothermal
from Greek *geo-*, means "earth"; and Greek *therme*, means "heat"

Figure 7 Geothermal power plants use heat from within Earth to generate electricity. ▼

(((O) **Concepts in Motion** **Animation**

Visual Summary

Living things depend on natural resources such as water, air, and land to meet their needs.

Water is considered a renewable resource.

Wind energy can be transformed into electricity.

FOLDABLES

Use your lesson Foldable to review the lesson. Save your Foldable for the project at the end of the chapter.

What do you think NOW?

You first read the statements below at the beginning of the chapter.

1. The world's supply of coal will never run out.

2. You should include minerals in your diet.

Did you change your mind about whether you agree or disagree with the statements? Rewrite any false statements to make them true.

Use Vocabulary

1. Parts of the environment that are important to the survival of living things are _____.

2. **Define** *nonrenewable resource* in your own words.

3. **Distinguish** between renewable and inexhaustible resources.

Understand Key Concepts

4. Which is a nonrenewable resource?
   - **A.** freshwater
   - **B.** natural gas
   - **C.** sunlight
   - **D.** wood

5. **Explain** why inexhaustible resources also could be considered renewable.

Interpret Graphics

6. **Identify** the natural resource the device below uses, and explain how it works.

7. **Organize Information** Copy the graphic organizer below, and use it to list ways people use sunlight as a natural resource.

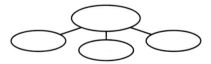

Critical Thinking

8. **Hypothesize** What measures could be taken on a farm to ensure that topsoil remains renewable?

Math Skills ✕➗
—— Math Practice ——

9. Of the 101.5 quads of energy used in 2007, only 0.081 quads were from solar energy. What percentage of U.S. energy came from solar energy?

Clean Energy from Underground

AMERICAN MUSEUM OF NATURAL HISTORY

Using Geothermal Energy to Heat—and Cool

Most of the energy we use comes from burning fossil fuels such as coal and oil. This releases carbon dioxide (CO_2) into the atmosphere, which causes global surface temperatures to rise. To lower CO_2 emissions, some people are switching to clean, renewable energy sources. One source is geothermal energy, thermal energy from inside Earth.

Geothermal Heat Pumps Even as the air cools in winter and warms in summer, temperatures a few meters below Earth's surface stay pretty much the same—around 13°C. Geothermal heat pumps use the temperature difference between air and ground to heat or cool buildings, depending on the season. A geothermal heat pump moves fluid through pipes from a building into the ground and then back again. In winter, the fluid carries thermal energy from the ground to warm the building. In summer, it moves thermal energy from the building to the ground and returns with cooler fluid. Each year, approximately 50,000 geothermal heat pumps are installed in the United States.

Outside air temperature less than 13°C

Warmed air circulates through the house.

Cooler air returns to furnace.

Warmed air circulates through the house.

Cooler air returns to furnace.

Fluid releases thermal energy to the circulating air.

Cold fluid enters ground pipes.

Ground temperature around 13°C

Fluid absorbs thermal energy from the warm ground and flows back into the house.

Geothermal Power Plants Geothermal energy also can produce electricity. At one time, geothermal power plants were located only near geysers and hot springs where geothermal energy is released near Earth's surface. Today, they can be almost anywhere. These plants pump water far underground where temperatures reach up to 200°C. The water boils, and the steam is captured and brought to the surface. The steam turns turbines, which run generators that produce electricity. And it's clean energy, too: geothermal plants release only 1 percent of the carbon dioxide that coal-burning power plants produce.

It's Your Turn

RESEARCH Geothermal heat pumps have been available since the 1940s. Why do you think more homes are not using them? Research this question, and write a short report about your findings.

Pollution

Reading Guide

Key Concepts 🔑
ESSENTIAL QUESTIONS

- How does pollution affect air resources?
- How does pollution affect water resources?
- How does pollution affect land resources?

Vocabulary

pollution p. 547

ozone layer p. 548

photochemical smog p. 548

global warming p. 549

acid precipitation p. 549

g Multilingual eGlossary

🎬 Video

- BrainPOP®
- What's Science Got to do With It?

Inquiry Orange Drink?

Runoff from a mine turned the water in this stream orange. How do you think the runoff affects the organisms that live in the stream? How do you think it affects the organisms that rely on the stream as a source of freshwater?

How do air pollutants move?

Small particles of pollutants can be transported by air movement. Once a pollutant is in the air, how far can it travel?

1 Read and complete a lab safety form.

2 Use a **tape measure** to determine the distance from your desk to the **lab candle.** Record your measurement in your Science Journal.

3 As soon as your teacher blows out the candle, start a **timer.**

4 Stop the timer when you smell the blown-out candle. Record the time in your Science Journal.

Think About This

1. Divide your distance from the lab candle by the time it took you to smell the blown-out candle. How fast did the smell move?

2. Compare your results with students in different parts of the room. Why do you think the speeds varied?

3. 🔑 **Key Concept** How do you think the movement of the smell from the blown-out candle is similar to the movement of a pollutant in the air?

What is pollution?

What happens when smoke gets in the air or toxic chemicals leak into soil? Smoke is a mixture of gases and tiny particles that make breathing difficult, especially for people who have health problems. Toxic chemicals that leak into soil can kill plants and soil organisms. These substances cause pollution. **Pollution** *is the contamination of the environment with substances that are harmful to life.* An example of pollution is shown in **Figure 8.** The oil-covered animals might not survive. Other wildlife also are affected negatively, including fish that people rely on for food.

Most pollution occurs because of human actions, such as burning fossil fuels or spilling toxic materials. However, pollution also can come from natural disasters. Wildfires create smoke. Volcanic eruptions send ash and toxic gases into the atmosphere. Regardless of its source, pollution affects air, water, and land resources.

Figure 8 Oil spills pollute water and harm wildlife.

Air Pollution

Many large cities issue alerts about air quality when air pollution levels are high. On such days, people are asked to avoid activities that contribute to air pollution, such as driving cars, using gasoline-powered lawn mowers, or cooking on charcoal grills. To avoid breathing problems, people also are advised to exercise in the early morning when the air is cleaner. Air pollution that can affect human health and recreational activities can be caused by ozone loss, photochemical smog, global warming, and acid precipitation.

Ozone Loss

Ozone is a molecule composed of three oxygen atoms. In the upper atmosphere, it forms a protective layer around Earth. *The **ozone layer** prevents most harmful ultraviolet (UV) radiation from reaching Earth.* UV radiation from the Sun can cause cancer and cataracts and can damage crops.

In the 1980s, scientists warned that Earth's protective ozone layer was getting thinner. The problem was caused primarily by chlorofluorocarbons (CFCs). CFCs are compounds used in refrigerators, air conditioners, and aerosol sprays. Governments around the world have phased out the use of CFCs and other ozone-depleting gases. As a result, the ozone layer is expected to recover within several decades.

Photochemical Smog

*Sunlight reacts with waste gases from the burning of fossil fuels and forms a type of air pollution called **photochemical smog.*** As shown in **Figure 9**, smog darkens the air and also can smell bad. It is formed of particles and gases that irritate the respiratory system. One of the gases in smog is ozone. In the upper atmosphere, ozone is helpful. But in the lower atmosphere, it is a pollutant that can harm organisms and cause lung damage.

FOLDABLES®

Make a horizontal three-tab book with a tab top. Label it as shown. Use it to explain the effects of pollution.

Effects of Pollution on...

| Air Resources | Water Resources | Land Resources |

WORD ORIGIN ············

photochemical smog
from Greek *photo-*, means "light"; Latin *chemic*, means "alchemy"; and modern English *smog*, blend of "smoke" and "fog"

Figure 9 🔑
Photochemical smog can worsen throughout the day as chemicals continue to react with sunlight.

✓ **Visual Check** What activities contribute to the formation of smog?

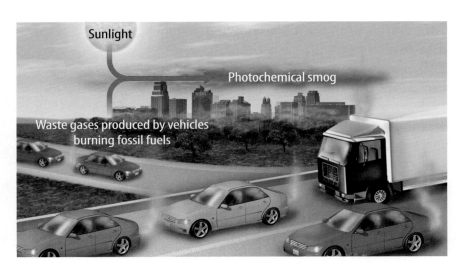

Sunlight

Photochemical smog

Waste gases produced by vehicles burning fossil fuels

Global Warming

You might have heard news reports about the melting of glaciers and sea ice. Earth is getting warmer. **Global warming** *is the scientific observation that Earth's average surface temperature is increasing.* Global warming can lead to climate change—changing weather conditions, changes to ecosystems and food webs, increases in the number and severity of floods and droughts, and increased coastal flooding as sea ice melts and sea level rises.

Data indicate that Earth's average surface temperature and increases in atmospheric carbon dioxide (CO_2) follow the same general trend. CO_2 is a greenhouse gas. This means it traps heat, helping to keep Earth warm. Greenhouse gases occur naturally. Without them, Earth would be too cold to support life. But human activities add greenhouse gases to the atmosphere, especially CO_2 from the burning of fossil fuels. Most scientists, including those on the United Nations Intergovernmental Panel on Climate Change, agree that increases in atmospheric CO_2 are contributing to global warming.

Acid Precipitation

Gases produced by the burning of fossil fuels also create other forms of air pollution, including acid precipitation. **Acid precipitation** *is acidic rain or snow that forms when waste gases from automobiles and power plants combine with moisture in the air.* Coal-burning power plants produce sulfur dioxide gas that combines with moisture to form sulfuric acid. Cars and trucks produce nitrous oxide gases that form nitric acid. Acid precipitation pollutes soil and can kill plants, including trees, as shown in **Figure 10**. It also contributes to water pollution and can damage buildings.

Key Concept Check How does pollution affect air resources?

ACADEMIC VOCABULARY

occur
(verb) to appear or happen

Figure 10 Acid rain can harm soil organisms and plant roots.

Visual Check How did acid rain affect this ecosystem?

Review Personal Tutor

MiniLab
20 minutes

How fast can you turn a sand castle into sediment?

Runoff can move sediment into streams. Sediment blocks stream flow, clogs the feeding structures of animals, and decreases the amount of light for aquatic plants. How does the flow of water affect rates of sedimentation?

1. Read and complete a lab safety form.

2. Use a **foam cup** to build a **sand** castle in a **plastic container.** Measure its height with a **metric ruler.** Record the data in your Science Journal.

3. Fill a **spray bottle** with water. Adjust the tip of the bottle so it sprays a mist.

4. Using a **timer,** spray your sand castle for 30 s. Measure and record the height of your sand castle.

5. Readjust the tip of the spray bottle so it sprays a stream of water. Then, rebuild your castle with fresh sand and repeat step 4.

6. Rebuild your sand castle with fresh sand. Poke three holes in the bottom of the foam cup with a **pencil.** Put your finger over the holes and fill the cup with water. Repeat step 4, letting water run out of the holes onto your castle.

Analyze and Conclude

1. **Evaluate** Which trial caused the largest change in the height of the sand castle?

2. **Model** What natural events could each of your trials represent?

3. **Key Concept** How might these natural events affect the quality of water resources?

Water Pollution

Have you ever seen a stream covered with thick green algae? The stream might have been polluted with fertilizers from nearby lawns or farms. It might contain chemicals from nearby factories. Water pollution can come from chemical runoff and other agricultural, residential, and industrial sources.

Wastewater

You might have been warned not to pour paint or used motor oil into storm drains. In most cities, rainwater that flows into storm drains goes directly into nearby waterways. Materials that go in the drain, including grease and oil washed from the street, can contribute to water pollution.

The wastewater that drains from showers, sinks, and toilets contains harmful viruses and bacteria. To safeguard health, this wastewater usually is purified in a sewage treatment plant before it is released into streams or used to irrigate crops. In some parts of the world, there is little or no sewage treatment. People might have to use polluted water.

Wastewater that comes from industries and mining operations also contains pollutants. It requires treatment before it can be returned to the environment. Even after treatment, some harmful substances might remain and impact water quality.

Runoff and Sediments

When it rains, water can flow over the land. This water, called runoff, flows across lawns and farmland. Along the way, it picks up pesticides, herbicides, and fertilizers. Runoff carries these pollutants into streams, where they can harm insects, fish, and other organisms. Runoff also carries sediment particles into streams. Too much sediment can damage stream habitats, clog waterways, and cause flooding.

Key Concept Check How does pollution affect water resources?

Figure 11 Fertilizers and irrigation water contain salts that can build up in soil, as shown in the photo on the left. The photo on the right shows a mining technique that disturbs ecosystems.

Land Pollution

Have you ever helped clean up litter? Foam containers, plastic bags, bottles, cans, and even furniture and appliances get dumped along roadsides. Litter is more than an eyesore. It can pollute soil and water and disturb wildlife. Sources of land pollution include homes, farms, industry, and mines.

Agriculture

Farmers use pesticides and other agricultural chemicals to help plants grow. But these chemicals become pollutants if they are used in excess or disposed of improperly. Herbicides kill weeds. But if they flow into streams, they can kill algae and plants, and harm fish and amphibians. Some farming practices contaminate soil, as shown in **Figure 11.**

Industry and Mining

Many industrial facilities, including oil refineries and ore processors, produce toxic wastes. For example, coal ash sludge is produced when coal is burned in power plants. The sludge contains mercury, lead, arsenic, and other potentially harmful metals. If toxic wastes such as these are incorrectly stored or disposed of, they contaminate soil and water. The health of people, plants, and wildlife can be affected.

Mining of fossil fuels and minerals can disturb or destroy entire ecosystems, as shown in **Figure 11.** Some coal-mining techniques can release toxic substances that were buried in rock. After the coal has been removed, the area can be restored. But it is difficult or impossible to replace the original ecosystem.

Key Concept Check How does pollution affect land resources?

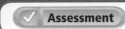
Visual Summary

Pollution, the introduction of harmful substances into the environment, can harm humans and other living things.

Smog, ozone loss, global warming, and acid precipitation are caused by air pollutants.

Land and water can be polluted by littering and chemical runoff from homes, factories, mines, and farms.

FOLDABLES

Use your lesson Foldable to review the lesson. Save your Foldable for the project at the end of the chapter.

What do you think NOW?

You first read the statements below at the beginning of the chapter.

3. Global warming causes acid rain.

4. Smog can affect human health.

Did you change your mind about whether you agree or disagree with the statements? Rewrite any false statements to make them true.

Use Vocabulary

1 **Define** *photochemical smog* in your own words.

2 **Distinguish** between global warming and acid precipitation.

3 The layer of atmosphere that prevents UV radiation from reaching Earth is the _____.

Understand Key Concepts

4 Which is NOT a source of pollution?
    A. burning coal   **C.** photosynthesis
    B. mining minerals   **D.** volcanic eruptions

5 **Explain** the difference between ozone in the lower atmosphere and ozone in the upper atmosphere.

6 **Describe** how pollution affects water resources.

Interpret Graphics

7 **Compare and Contrast** Copy and fill in the table below to compare and contrast air, water, and land pollution.

Type of Pollution	Similarities	Differences
Air		
Water		
Land		

Critical Thinking

8 **Hypothesize** The water in a stream that flows through farmland has always been clear. After a hard rain, the water in the stream became muddy. What caused the change?

9 **Apply** How do trees and other plants help lessen global warming? How might deforestation, or cutting down trees, contribute to global warming?

How can you communicate about pollution?

You have read about different types of pollutants in this chapter. Now it's your turn to communicate what you have learned. A public service announcement (PSA) is like a commercial that explains an important issue.

Materials

stopwatch

computer

Learn It

Communication of ideas is an important part of the work of scientists. A scientific idea that is not reported will not advance scientific knowledge or the public's understanding of science. Scientists often **communicate** their ideas in presentations.

Try It

1. Read and complete a lab safety form.

2. Choose a pollutant you read about in this chapter or a different pollutant in which you have an interest.

3. Research your pollutant. Find out as much as you can about how it is produced, how it enters the environment, what problems it causes, and how its effects can be reduced.

4. Write a 1-min script for a PSA that communicates the information you gathered in step 3.

5. Practice your script until you feel comfortable speaking it before a group. If recording equipment is available, record your PSA.

Apply It

6. Present your PSA to your class.

7. Take questions from the class. Ask your classmates what they learned. Record their comments in your Science Journal.

8. **Critique** your PSA. Did the class understand the message you were trying to communicate? How could you improve your presentation?

9. **Key Concept** How does the pollutant you researched affect natural resources?

Reading Guide

Key Concepts
ESSENTIAL QUESTIONS

- How can people monitor resource use?

- How can people conserve resources?

Vocabulary

sustainability p. 559

recycling p. 560

(g) **Multilingual eGlossary**

Protecting Earth

Inquiry A Better View?

This image taken from a satellite shows an eruption on a volcanic island. What parts of the environment do you see in the image? How can images taken from satellites help scientists study Earth?

Inquiry Launch Lab

How can you turn trash into art?

Reusing materials helps reduce the natural resources needed to make something new. It also reduces the amount of trash discarded in landfills. Some environmentally friendly artists reuse materials to create new art. In this lab, you will create something new from objects you usually might throw away.

1. Read and complete a lab safety form.
2. Carefully consider the materials in the **trash collection.**
3. Using **craft materials,** create a piece of art out of the trash. Try to convey a message with your art. For example, your message might be "Protect Earth."
4. Display your artwork to the class.

Think About This

1. Describe your artwork. What kind of trash did you use? What kind of art did you create?

2. What message did you try to convey?

3. **Key Concept** How do you think reusing materials helps conserve resources?

Monitoring Human Impact on Earth

Earth's human population is expected to grow to 7 billion by 2012. As the population increases, so does humans' impact on the planet. Scientists, governments, and concerned citizens around the world are working to identify environmental problems, educate the public about them, and help find solutions.

Scientists collect data on a variety of environmental conditions by placing detectors on satellites, aircraft, high-altitude balloons, and ground-based monitoring stations. For example, the United States and the European Union have launched satellites into orbit around Earth to gather data on greenhouse gases, ozone, ecosystem changes, melting glaciers and sea ice, climate patterns, and ocean health.

The U.S. Environmental Protection Agency (EPA) is a government organization that monitors the health of the environment and looks for ways to reduce human impacts. The EPA enforces environmental laws and supports research at universities and national laboratories. It also works with citizens and organizations to identify superfund sites—abandoned areas that have been contaminated by toxic wastes—and develops plans for cleaning them up.

Key Concept Check How can people monitor resource use?

FOLDABLES

Make a small shutter-fold book. Label it as shown. Use it to identify technology and methods that protect natural resources.

> Monitoring Natural Resources

> Conserving Natural Resources

Developing Technologies

Many technologies have been developed to protect Earth's resources, and more are on the way. These advances often focus on saving energy and reducing pollution.

Water-Saving Technologies

It takes energy to clean water and to transport it to homes and businesses. So technologies that conserve water also save energy. Low-flow showerheads and toilets help reduce water use. Drip irrigation systems, such as the one shown in **Figure 12,** decrease water waste.

Energy-Saving Technologies

Saving energy can make Earth's supply of fossil fuels last longer. Relying more on renewable energy sources can reduce fossil fuel use. Some of these sources can be expensive, but designs are constantly improving and costs are going down. Researchers estimate that solar electricity soon will cost no more than electricity produced by burning fossil fuels. Burning fewer fossil fuels also creates less pollution.

Other energy-saving advances include compact fluorescent lightbulbs (CFLs). They use about one-fourth the energy of incandescent bulbs and can last ten times longer. In 2007, Americans reduced greenhouse gas emissions by an amount equal to removing 2 million cars from the road just by switching to CFLs.

Figure 12 Drip irrigation slowly delivers water to the roots of plants. Less water is lost to runoff and evaporation.

Inquiry MiniLab

20 minutes each day

What's in the air?

Air pollution made of particles that float in the air is called particulate matter, or PM. The Clean Air Act requires the EPA to monitor PM.

1. Read and complete a lab safety form.
2. To make a PM collector, coat two **plastic container lids** with a layer of **petroleum jelly.**

3. Leave each PM collector in a different location around your school. Record the location, the date, and the time in your Science Journal.

4. On the following day, retrieve the collectors. Record the date and time.
5. Use a **magnifying lens** to observe the PM. Record your observations.

Analyze and Conclude

1. **Describe** What types of PM did you find on your collectors?

2. **Compare** the amount and type of PM found in the different locations. Formulate a reason for any differences you observe.

3. **Key Concept** What conclusions about air quality in your school can you draw from your data?

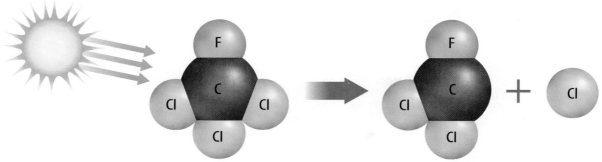

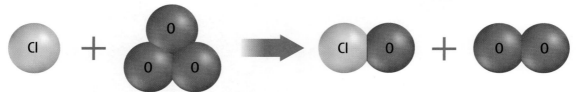

Sunlight reacts with a CFC molecule, causing a chlorine atom to break away.

The chlorine atom reacts with and breaks apart an ozone molecule.

CFC Replacements

You have read that CFCs cause thinning of the ozone layer. How does this happen? The chlorine atoms in CFC molecules react with sunlight to destroy ozone, as shown in **Figure 13.** All CFCs soon will be phased out and replaced with chemicals that do not contain chlorine. Replacements include hydrofluorocarbons (HFCs) and perfluorocarbons (PFCs). Even after CFCs are no longer in use, it will take decades for the ozone layer to recover.

 Reading Check What steps have been taken to reverse the thinning of the ozone layer?

Alternative Fuels

Gasohol and biodiesel are alternative fuels that help reduce humans' use of fossil fuels. They also help reduce air pollution.

Gasohol is a mixture of 90 percent gasoline and 10 percent ethanol. Ethanol is alcohol made from corn, sugar cane, or other plants. Using gasohol helps reduce emissions of carbon monoxide, an air pollutant that contributes to smog. The carbon in ethanol comes from plants rather than fossil fuels. So, using gasohol can help reduce emissions that contribute to global warming.

Biodiesel is made from renewable resources, primarily vegetable oils and animal fats—including oil left over from frying foods in restaurants. Biodiesel can be burned in diesel engines in farm and industrial machinery, trucks, and cars. It produces fewer pollutants than regular diesel fuel, and it reduces CO_2 emissions by 78 percent.

Figure 13 When CFC molecules reach the upper atmosphere, sunlight breaks off chlorine atoms. Each free chlorine atom can destroy an ozone molecule and prevent another from forming.

Visual Check How do CFCs affect ozone molecules?

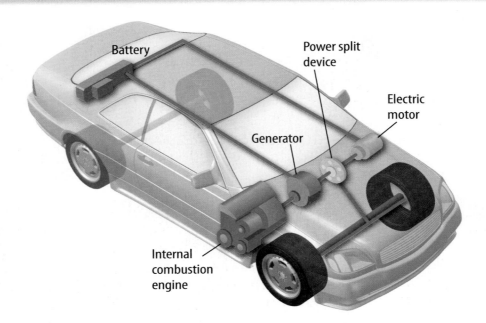

Battery

Power split device

Electric motor

Generator

Internal combustion engine

Figure 14 A hybrid vehicle uses a battery to power an electric motor. A small gasoline engine provides additional power.

✅ **Visual Check** What are the power sources in a hybrid vehicle?

SCIENCE USE v. COMMON USE

hybrid
Science Use an offspring of two animals or plants of different breeds or species

Common Use something that has two different components performing essentially the same task

Automobile Technologies

If you were buying a car, you would want to know how many miles it travels per gallon of fuel—miles per gallon, or mpg. The higher a car's mpg, the less pollution it will add to the environment. A car with a high mpg also will use up fewer fossil-fuel resources.

One of your choices might be a **hybrid** electric vehicle (HEV). HEVs combine a small gasoline engine with an electric motor powered by batteries, as shown in **Figure 14.** HEVs run on battery power as much as possible, with a boost from the engine for longer trips, higher speeds, and steep hills. The engine also charges the batteries. HEVs can get up to twice the mileage of a conventional car—close to 50 mpg in some recent car models.

In the future, another automobile alternative might be a fuel-cell vehicle (FCV). Inside a fuel cell, oxygen from the air chemically combines with hydrogen to produce electricity. The primary waste product is water. Tailpipe emissions from FCVs are nearly pollution-free. However, obtaining hydrogen fuel requires using methane or other fossil fuels. Researchers are looking for alternatives.

✅ **Reading Check** Compare HEVs and FCVs.

Making a Difference

Do you turn off the lights when you leave a room or recycle bottles and cans? If so, you are helping reduce your impact on the environment. You can help protect Earth's resources in other ways as well, such as cleaning up a stream, educating others about environmental issues, analyzing the choices you make as a consumer, and following some of the suggestions you will read about next.

Sustainability

When people talk about environmental issues, they often use the word *sustainability*. **Sustainability** *means meeting human needs in ways that ensure future generations also will be able to meet their needs*. When you turn off the lights as you leave a room, you are saving energy—and you are also helping to ensure a sustainable future. **Figure 15** shows other actions that lead toward a sustainable future.

 Reading Check What is sustainability?

Sustainable Actions 🔑

Figure 15 Planting trees, composting, and picking up litter help sustain the environment.

Restore and Rethink

Restoring damaged habitats and ecosystems to their original state is one way to make a difference. For example, picking up trash can restore water habitats.

You also can rethink the way you perform everyday activities. Instead of riding in a vehicle to nearby places, you could ride your bike or walk.

Reduce and Reuse

You can reduce the amount of waste you create simply by reducing the amount of material you use. For example, avoid products with too much packaging. Or, bring your own bags for carrying purchases, as shown in **Figure 16.**

Reusing items also helps reduce waste. Instead of buying new, reuse something that will work just as well. You also can donate used items to charities or resell them.

Figure 16 🔑 Reusable bags help save energy and reduce waste.

WORD ORIGIN ·············

recycle
from Latin *re-*, means "again"; and Greek *kyklos*, means "circle"

Recycle

If an item cannot be reused, you might be able to recycle it. **Recycling** is *manufacturing new products out of used products.* This process reduces wastes and extends our supply of natural resources. Computers and other electronics contain valuable metals that can be recycled, as well as toxic materials that can contribute to pollution. So recycling also helps makes sure that toxins are properly disposed of.

Compost Leaves, grass clippings, and vegetable scraps can be recycled by composting. In a compost pile, these materials decay into nutrient-rich soil that goes back into the garden.

Buy Recycled Separating recyclables from the rest of the trash is just one step. To keep the cycle going, buy and use recycled products. You can find shoes, clothing, paper, and carpets made from recycled materials.

🔑 **Key Concept Check** How can you conserve resources?

Lesson 3 Review

Visual Summary

Scientists use a variety of techniques to monitor the use of natural resources, including satellites, aircraft, high-altitude balloons, and ground-based monitoring stations.

New technologies such as HEVs and alternative fuels conserve resources and produce less pollution.

People can help protect resources by reducing their use of resources, reusing products, and recycling products.

FOLDABLES

Use your lesson Foldable to review the lesson. Save your Foldable for the project at the end of the chapter.

What do you think NOW?

You first read the statements below at the beginning of the chapter.

5. Oil left over from frying potatoes can be used as automobile fuel.

6. Hybrid electric vehicles cannot travel far or go fast.

Did you change your mind about whether you agree or disagree with the statements? Rewrite any false statements to make them true.

Use Vocabulary

1 **Define** *sustainability* in your own words.

Understand Key Concepts

2 Which produces water as its primary waste product?
- **A.** biodiesel
- **C.** gasohol
- **B.** FCV
- **D.** HEV

3 **Analyze** Compare the tailpipe emissions of an HEV with a car that has only a gasoline engine.

4 **Apply** What water-saving and energy-saving techniques could you use in your kitchen?

Interpret Graphics

5 **Explain** how the process below affects the environment.

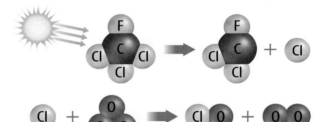

6 **Identify** Copy and fill in the graphic organizer below to identify three ways people can limit waste production.

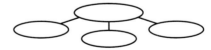

Critical Thinking

7 **Hypothesize** The same delivery truck passes you almost every morning as you walk to school. Each time it goes by, you smell fried potatoes. What could be the reason?

8 **Recommend** The school cafeteria throws out mounds of vegetable peelings and leftover food every day. The school gardener would like to plant vegetables, but the soil is too thin. What sustainable actions could you recommend?

How can you conserve a natural resource?

You have read about natural resources and about how they can be affected by human activities. Even though you are just one person, your actions can help conserve natural resources. Your task is to develop a realistic plan to conserve a natural resource.

Ask Questions

How do your daily activities affect natural resources? What actions could you take to conserve a natural resource?

Make Observations

1. Read and complete a lab safety form.

2. Select the natural resource you want to help conserve. In your Science Journal, describe how the resource is used. Also discuss the problems that threaten the natural resource.

3. Review the list of conservation activities below. Choose activities from the list or come up with your own ideas. Explain how the activities would conserve the resource you have chosen.

4. Write a plan that details how you will enact these activities in your everyday life. Describe the materials you will need and the steps you will follow. Make a time line showing how you will implement the plan.

5. Remember, your plan must be realistic. Conduct research as you do this investigation to help you learn more about natural resources and conservation plans.

Conservation Activities
Change showerheads and faucets to low-flow.
Install rain barrels to use for watering gardens.
Start a compost pile for food scraps or yard waste.
Provide recycling bins at sporting events.
Replace incandescent lights with CFLs or LEDs.
Clean up a local road or pond.
Replace car trips with bicycling or walking whenever possible.
Lower the heating thermostat in winter and raise the air-conditioner thermostat in summer.

Form a Hypothesis

6 State the major goal or goals of your plan in the form of a hypothesis.

Test your Hypothesis

7 Discuss your plan with a classmate. Does he or she think it will work? Is the plan realistic, or will it require large amounts of money, time, and resources? Modify your plan based on your classmate's input.

8 Implement your plan over a scheduled period of time. Follow your time line. Record your observations and any quantifiable data.

Analyze and Conclude

9 **Assess** How did your plan affect the natural resource? Which impacts were you able to quantify? Which impacts were difficult to quantify?

10 **Evaluate** Did you see any limitations to your plan? For example, can it be implemented on a larger scale?

11 **The Big Idea** Why is it important to actively work to conserve Earth's resources?

Communicate Your Results

Create an assessment report of your plan. Make your report engaging and informative. Provide a comprehensive description of your results. Include a proposal for extending the time period of the plan.

Inquiry Extension

Create a 1-min PSA describing your plan. The PSA should encourage others to follow the plan. Include a graph or other visual that shows the positive impact of your plan on the natural resource. Project how this impact would increase if everyone in the class followed your plan. Project the impact of your plan on county, state, and national levels.

Lab Tips

☑ If you are having trouble deciding on a topic, think about your activities as you go about your day. What things do you do that use a lot of natural resources?

☑ Be flexible. Propose more than one way to solve a problem.

☑ Be creative, but keep your plan realistic.

Remember to use scientific methods.

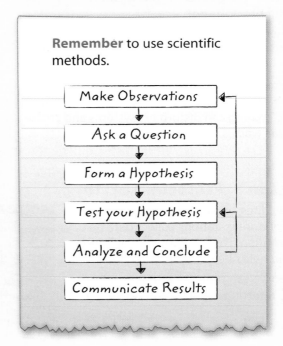

Make Observations

Ask a Question

Form a Hypothesis

Test your Hypothesis

Analyze and Conclude

Communicate Results

Chapter 15 Study Guide

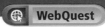

 THE BIG IDEA People can protect Earth's resources by understanding how their use of natural resources affects the environment, knowing which natural resources are in limited supply, and making decisions toward a more sustainable future.

Key Concepts Summary 🔑

Lesson 1: Earth's Resources

- **Natural resources** are raw materials and forms of energy that are important to living things.
- Resources can be **renewable** or **nonrenewable.** Some renewable resources are **inexhaustible.**

Lesson 2: Pollution

- Air pollutants cause **photochemical smog,** ozone loss, **global warming,** and **acid precipitation.**
- Chemical runoff can damage lakes, streams, and water supplies. Sediment runoff from land can disturb aquatic habitats.
- Litter and pollutants can contaminate soil, harm organisms, and reduce land's ability to support life. Mining can disturb ecosystems and create toxic wastes.

Lesson 3: Protecting Earth

- Satellites, aircraft, and ground-based monitoring stations collect data on pollution. The EPA monitors pollution and helps develop clean-up plans.
- People can protect Earth's resources by reducing, reusing, and **recycling.**

Vocabulary

natural resource p. 537
nonrenewable resource p. 538
renewable resource p. 540
inexhaustible resource p. 542
geothermal energy p. 543

pollution p. 547
ozone layer p. 548
photochemical smog p. 548
global warming p. 549
acid precipitation p. 549

sustainability p. 559
recycling p. 560

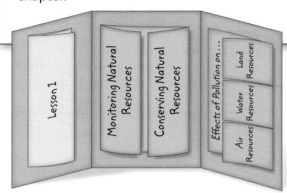

FOLDABLES® **Chapter Project**

Assemble your lesson Foldables as shown to make a Chapter Project. Use the project to review what you have learned in this chapter.

Use Vocabulary

1 Sunshine, oil, coal, uranium, trees, oxygen, and streams are examples of _____.

2 Distinguish between recycling, reusing, and reducing.

3 A tropical forest that takes 1,000 years to recover from being burned down is a(n) _____ resource.

4 Use the term *sustainability* in a sentence.

5 Rainfall that keeps a pond full is a(n) _____ resource.

6 Define *global warming* in your own words.

Link Vocabulary and Key Concepts

⦿ **Concepts in Motion**   **Interactive Concept Map**

Copy this concept map, and then use vocabulary terms from the previous page and other terms from the chapter to complete the concept map.

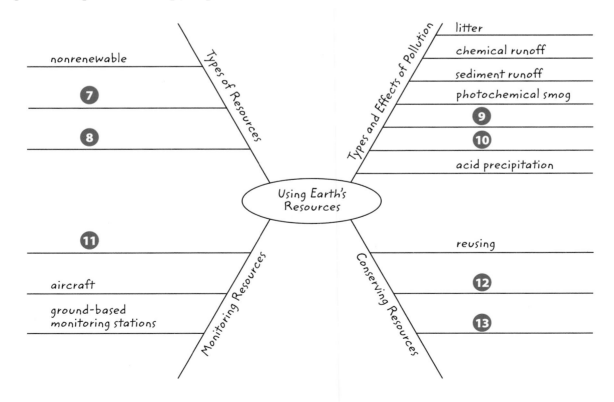

Understand Key Concepts 🔑

1 Which is an inexhaustible resource?
A. air
B. land
C. water
D. wind

2 What are the items below made of?

A. inexhaustible resources
B. nonrecyclable resources
C. nonrenewable resources
D. renewable resources

3 Greenhouse gases contribute to which environmental problem?
A. acid rain
B. global warming
C. ozone depletion
D. photochemical smog

4 Biodiesel can be made from which resource?
A. ethanol
B. gasoline
C. hydrogen gas
D. vegetable oil

5 Which technology produces electricity using sunlight?
A. fuel cells
B. solar cookers
C. solar power plants
D. solar water heaters

6 What kind of pollution results when sunlight reacts with sulfur dioxide produced by the burning of fossil fuels?
A. acid rain
B. global warming
C. ozone depletion
D. photochemical smog

7 Which is a sustainable action?
A. carpooling to a game
B. riding in a car to school
C. running water when brushing teeth
D. throwing out a slightly used shirt

8 What process is illustrated in the diagram below?

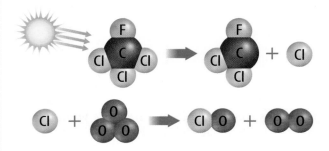

A. acid rain
B. global warming
C. ozone depletion
D. smog formation

9 Uranium is classified as what?
A. a fossil fuel
B. a greenhouse gas
C. a mineral resource
D. a renewable resource

10 Which is a renewable resource?
A. coal
B. sunlight
C. water
D. wind

Critical Thinking

11 **Classify** each of the following as renewable, nonrenewable, or inexhaustible: notebook paper, flashlight batteries, and heat from a volcano.

12 **Explain** why geothermal energy is considered an inexhaustible resource.

13 **Design** an experiment to determine how acid rain affects plants.

14 **Interpret Scientific Illustrations** Describe how the process below affects human health.

15 **Create** a poster explaining why items such as used motor oil or leftover paint should be recycled rather than poured down a storm drain or dumped on soil.

16 **Summarize** how people can help prevent land pollution.

17 **Compare** the emissions of a gasoline-powered automobile, a hybrid electric vehicle, and a fuel-cell vehicle. Which vehicle contributes the least amount of air pollution?

18 **Give** an example of how people can reduce the amount of trash they produce.

Writing in Science

19 **Write** a letter to a younger student about sustainability. Explain what it means, why it is important to his or her generation, and how it involves considering future needs, not just current ones.

REVIEW THE BIG IDEA

20 Give examples of what a government and an individual could do to protect Earth's resources.

21 How can using wind energy help conserve Earth's resources?

Math Skills ×÷+

Review — Math Practice —

Use Percentages

22 Between 2003 and 2006, the amount of U.S. energy produced from renewable sources increased from 6.15 quads to 6.91 quads. What was the percentage increase?

23 Wind energy usage increased from 0.115 quads to 0.341 quads between 2003 and 2007. What was the percentage increase?

24 In 2006, only 6.92 percent of U.S. energy was produced from renewable sources. If the total energy consumption was 99.8 quads, how much energy was produced from renewable sources?

Standardized Test Practice

Record your answers on the answer sheet provided by your teacher or on a sheet of paper.

Multiple Choice

1 Why is coal a nonrenewable resource?

   A Coal is used faster than it forms.

   B Coal cannot be recycled like glass or plastic.

   C Humans do not know how to make coal.

   D Coal formed in the past and no longer forms today.

2 Which is an effect of photochemical smog?

   A global warming

   B lung damage

   C skin cancer

   D acid precipitation

Use the diagram to answer questions 3 and 4.

3 The diagram shows a chlorofluorocarbon (CFCs) reaction. Where does this reaction occur?

   A in water

   B on land

   C in an automobile

   D in the atmosphere

4 How does this reaction affect the environment?

   A It depletes the ozone layer.

   B It pollutes water.

   C It produces acid rain.

   D It produces photochemical smog.

5 Which government agency monitors the health of the environment and works to reduce human impacts?

   A CFC

   B CFL

   C EPA

   D HEV

6 Which material is NOT a fossil fuel?

   A coal

   B copper

   C natural gas

   D oil

Use the diagram to answer question 7.

7 Nika is reading a pamphlet about the proper disposal of different materials. The pamphlet includes the image shown above. What would happen if someone performed the action shown in the image?

   A The oil would cause photochemical smog.

   B The oil would be washed into nearby streams or lakes.

   C The oil would evaporate and cause ozone depletion.

   D The oil would remain in the storm drain and not cause harm.

8 Which action recycles yard waste and kitchen waste?

   A composting

   B planting trees

   C using drip irrigation

   D using compact fluorescent lightbulbs

Use the diagram to answer question 9.

9 What type of energy resource is shown in the diagram?

   A geothermal

   B solar

   C water

   D wind

10 Which is a renewable resource that can become nonrenewable if it is used up too quickly?

   A oil

   B forests

   C coal

   D natural gas

Constructed Response

11 What action could you take to help conserve resources and reduce the amount of waste that enters landfills?

Use the diagram to answer question 12.

12 Which part of the environment is most affected by the activity shown here? Explain your answer.

13 Describe how rethinking your everyday activities contributes to a sustainable future.

14 Identify one renewable resource and one inexhaustible resource. Then, describe similarities and differences between the two resources.

NEED EXTRA HELP?														
If You Missed Question...	1	2	3	4	5	6	7	8	9	10	11	12	13	14
Go to Lesson...	1	2	3	3	3	1	3	3	1,3	1	3	2,3	3	1

Chapter 16

Earth's Atmosphere

How does Earth's atmosphere affect life on Earth?

 What's in the atmosphere?

Earth's atmosphere is made up of gases and small amounts of liquid and solid particles. Earth's atmosphere surrounds and sustains life.

- What type of particles make up clouds in the atmosphere?

- How do conditions in the atmosphere change as height above sea level increases?

- How does Earth's atmosphere affect life on Earth?

Get Ready to Read

What do you think?

Before you read, decide if you agree or disagree with each of these statements. As you read this chapter, see if you change your mind about any of the statements.

1 Air is empty space.

2 Earth's atmosphere is important to living organisms.

3 All the energy from the Sun reaches Earth's surface.

4 Earth emits energy back into the atmosphere.

5 Uneven heating in different parts of the atmosphere creates air circulation patterns.

6 Warm air sinks and cold air rises.

7 If no humans lived on Earth, there would be no air pollution.

8 Pollution levels in the air are not measured or monitored.

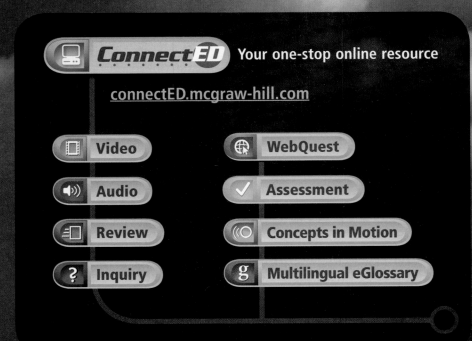

ConnectED Your one-stop online resource

connectED.mcgraw-hill.com

Video

Audio

Review

Inquiry

WebQuest

Assessment

Concepts in Motion

Multilingual eGlossary

Describing Earth's Atmosphere

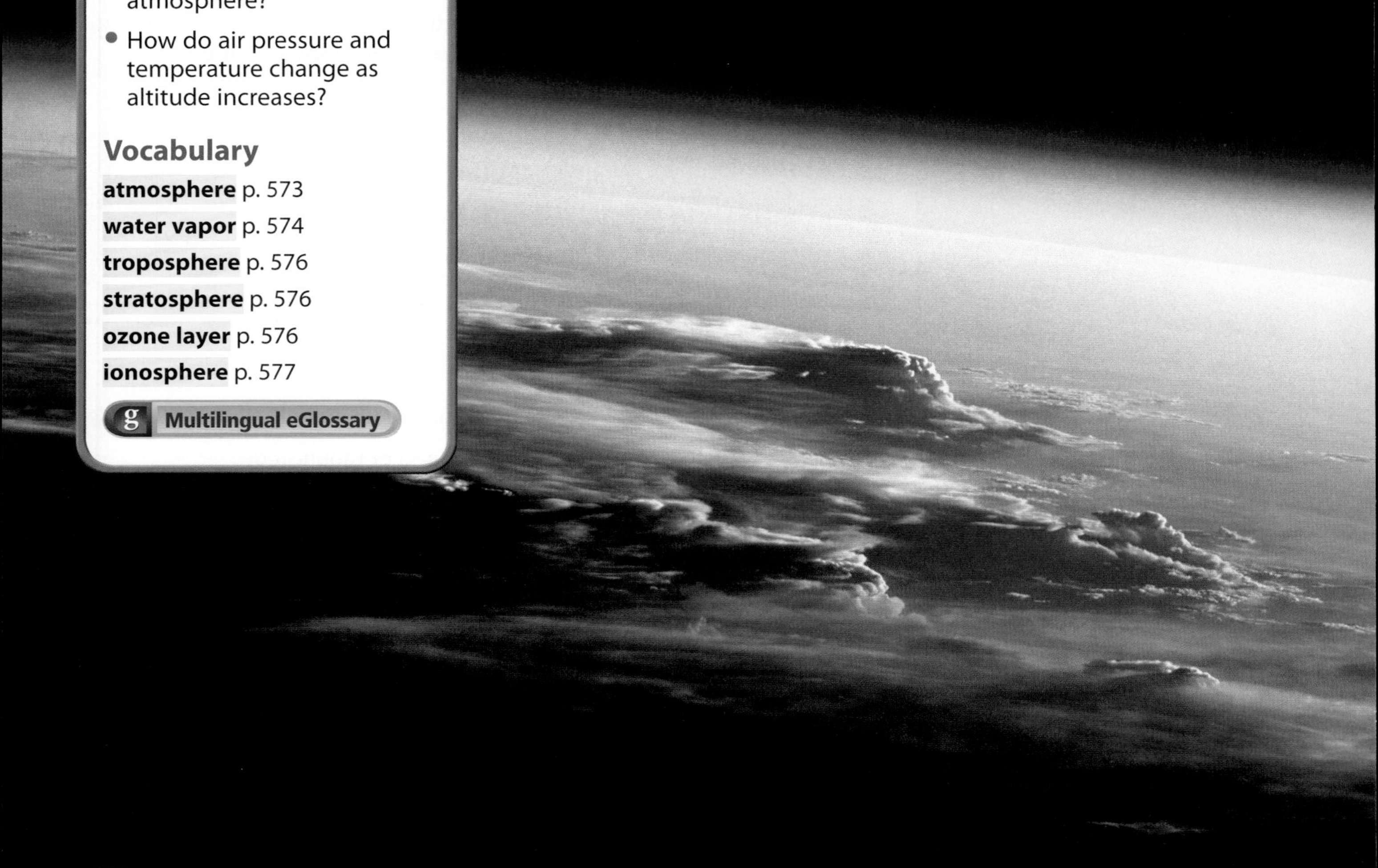

Reading Guide

Key Concepts 🔑
ESSENTIAL QUESTIONS

- How did Earth's atmosphere form?
- What is Earth's atmosphere made of?
- What are the layers of the atmosphere?
- How do air pressure and temperature change as altitude increases?

Vocabulary

atmosphere p. 573

water vapor p. 574

troposphere p. 576

stratosphere p. 576

ozone layer p. 576

ionosphere p. 577

g Multilingual eGlossary

Inquiry **Why is the atmosphere important?**

What would Earth be like without its atmosphere? Earth's surface would be scarred with craters created from the impact of meteorites. Earth would experience extreme daytime-to-nighttime temperature changes. How would changes in the atmosphere affect life? What effect would atmospheric changes have on weather and climate?

Where does air apply pressure?

With the exception of Mercury, most planets in the solar system have some type of atmosphere. However, Earth's atmosphere provides what the atmospheres of other planets cannot: oxygen and water. Oxygen, water vapor, and other gases make up the gaseous mixture in the atmosphere called air. In this activity, you will explore air's effect on objects on Earth's surface.

1. Read and complete a lab safety form.
2. Add **water** to a **cup** until it is two-thirds full.
3. Place a large **index card** over the opening of the cup so that it is completely covered.
4. Hold the cup over a tub or a large bowl.
5. Place one hand on the index card to hold it in place as you quickly turn the cup upside down. Remove your hand.

Think About This

1. What happened when you turned the cup over?

2. How did air play a part in your observation?

3. 🔑 **Key Concept** How do you think these results might differ if you repeated the activity in a vacuum?

Importance of Earth's Atmosphere

The photo on the previous page shows Earth's atmosphere as seen from space. How would you describe the atmosphere? *The* **atmosphere** (AT muh sfihr) *is a thin layer of gases surrounding Earth.* Earth's atmosphere is hundreds of kilometers high. However, when compared to Earth's size, it is about the same relative thickness as an apple's skin to an apple.

The atmosphere contains the oxygen, carbon dioxide, and water necessary for life on Earth. Earth's atmosphere also acts like insulation on a house. It helps keep temperatures on Earth within a range in which living organisms can survive. Without it, daytime temperatures would be extremely high and nighttime temperatures would be extremely low.

The atmosphere helps protect living organisms from some of the Sun's harmful rays. It also helps protect Earth's surface from being struck by meteors. Most meteors that fall toward Earth burn up before reaching Earth's surface. Friction with the atmosphere causes them to burn. Only the very largest meteors strike Earth.

✔️ **Reading Check** Why is Earth's atmosphere important to life on Earth?

WORD ORIGIN ·············

atmosphere
from Greek *atmos*, means "vapor"; and Latin *sphaera*, means "sphere"

Origins of Earth's Atmosphere

Most scientists agree that when Earth formed, it was a ball of molten rock. As Earth slowly cooled, its outer surface hardened. Erupting volcanoes emitted hot gases from Earth's interior. These gases surrounded Earth, forming its atmosphere.

Ancient Earth's atmosphere was thought to be water vapor with a little carbon dioxide (CO_2) and nitrogen. **Water vapor** *is water in its gaseous form.* This ancient atmosphere did not have enough oxygen to support life as we know it. As Earth and its atmosphere cooled, the water vapor condensed into **liquid.** Rain fell and then evaporated from Earth's surface repeatedly for thousands of years. Eventually, water accumulated on Earth's surface, forming oceans. Most of the original CO_2 that dissolved in rain is in rocks on the ocean floor. Today the atmosphere has more nitrogen than CO_2.

Earth's first organisms could undergo photosynthesis, which changed the atmosphere. Recall that photosynthesis uses light energy to produce sugar and oxygen from carbon dioxide and water. The organisms removed CO_2 from the atmosphere and released oxygen into it. Eventually the levels of CO_2 and oxygen supported the development of other organisms.

 Key Concept Check How did Earth's present atmosphere form?

 **MiniLab** 　　　　　　　　　　　　　　　**20 minutes**

Why does the furniture get dusty?

Have you ever noticed that furniture gets dusty? The atmosphere is one source for dirt and dust particles. Where can you find dust in your classroom?

1. Read and complete a lab safety form.

2. Choose a place in your classroom to collect a sample of dust.

3. Using a **duster,** collect dust from about a 50-cm^2 area.

4. Examine the duster with a **magnifying lens.** Observe any dust particles. Some might be so small that they only make the duster look gray.

5. Record your observations in your Science Journal.

6. Compare your findings with those of other members of your class.

Analyze and Conclude

1. **Analyze** how the area surrounding your collection site might have influenced how much dust you observed on the duster.

2. **Infer** the source of the dust.

3. **Key Concept** Other than gases and water droplets, predict what Earth's atmosphere might contain.

Figure 1 Oxygen and nitrogen make up most of the atmosphere, with the other gases making up only 1 percent. ▼

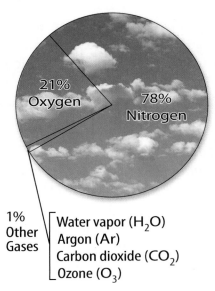

21% Oxygen

78% Nitrogen

1% Other Gases
- Water vapor (H_2O)
- Argon (Ar)
- Carbon dioxide (CO_2)
- Ozone (O_3)

Visual Check What percent of the atmosphere is made up of oxygen and nitrogen?

▲ **Figure 2** One way solid particles enter the atmosphere is from volcanic eruptions.

Composition of the Atmosphere

Today's atmosphere is mostly made up of invisible gases, including nitrogen, oxygen, and carbon dioxide. Some solid and liquid particles, such as ash from volcanic eruptions and water droplets, are also present.

Gases in the Atmosphere

Study **Figure 1.** Which gas is the most abundant in Earth's atmosphere? Nitrogen makes up about 78 percent of Earth's atmosphere. About 21 percent of Earth's atmosphere is oxygen. Other gases, including argon, carbon dioxide, and water vapor, make up the remaining 1 percent of the atmosphere.

The amounts of water vapor, carbon dioxide, and ozone vary. The concentration of water vapor in the atmosphere ranges from 0 to 4 percent. Carbon dioxide is 0.038 percent of the atmosphere. A small amount of ozone is at high altitudes. Ozone also occurs near Earth's surface in urban areas.

Solids and Liquids in the Atmosphere

Tiny solid particles are also in Earth's atmosphere. Many of these, such as pollen, dust, and salt, can enter the atmosphere through natural processes. **Figure 2** shows another natural source of particles in the atmosphere—ash from volcanic eruptions. Some solid particles enter the atmosphere because of human activities, such as driving vehicles that release soot.

The most common liquid particles in the atmosphere are water droplets. Although microscopic in size, water droplets are visible when they form clouds. Other atmospheric liquids include acids that result when volcanoes erupt and fossil fuels are burned. Sulfur dioxide and nitrous oxide combine with water vapor in the air and form the acids.

Key Concept Check What is Earth's atmosphere made of?

Layers of the Atmosphere

The atmosphere has several different layers, as shown in **Figure 3**. Each layer has unique properties, including the composition of gases and how temperature changes with altitude. Notice that the scale between 0–100 km in **Figure 3** is not the same as the scale from 100–700 km. This is so all the layers can be shown in one image.

Troposphere

The atmospheric layer closest to Earth's surface is called the **troposphere** (TRO puh sfihr). Most people spend their entire lives within the troposphere. It extends from Earth's surface to altitudes between 8–15 km. Its name comes from the Greek word *tropos,* which means "change." The temperature in the troposphere decreases as you move away from Earth. The warmest part of the troposphere is near Earth's surface. This is because most sunlight passes through the atmosphere and warms Earth's surface. The warmth is radiated to the troposphere, causing weather.

 Reading Check Describe the troposphere.

Stratosphere

The atmospheric layer directly above the troposphere is the **stratosphere** (STRA tuh sfihr). The stratosphere extends from about 15 km to about 50 km above Earth's surface. The lower half of the stratosphere contains the greatest amount of ozone gas. *The area of the stratosphere with a high concentration of ozone is referred to as the* **ozone layer.** The presence of the ozone layer causes increasing stratospheric temperatures with increasing altitude.

An ozone (O_3) molecule differs from an oxygen (O_2) molecule. Ozone has three oxygen atoms instead of two. This difference is important because ozone absorbs the Sun's ultraviolet rays more effectively than oxygen does. Ozone protects Earth from ultraviolet rays that can kill plants, animals, and other organisms and cause skin cancer in humans.

Layers of Atmosphere

Concepts in Motion Animation

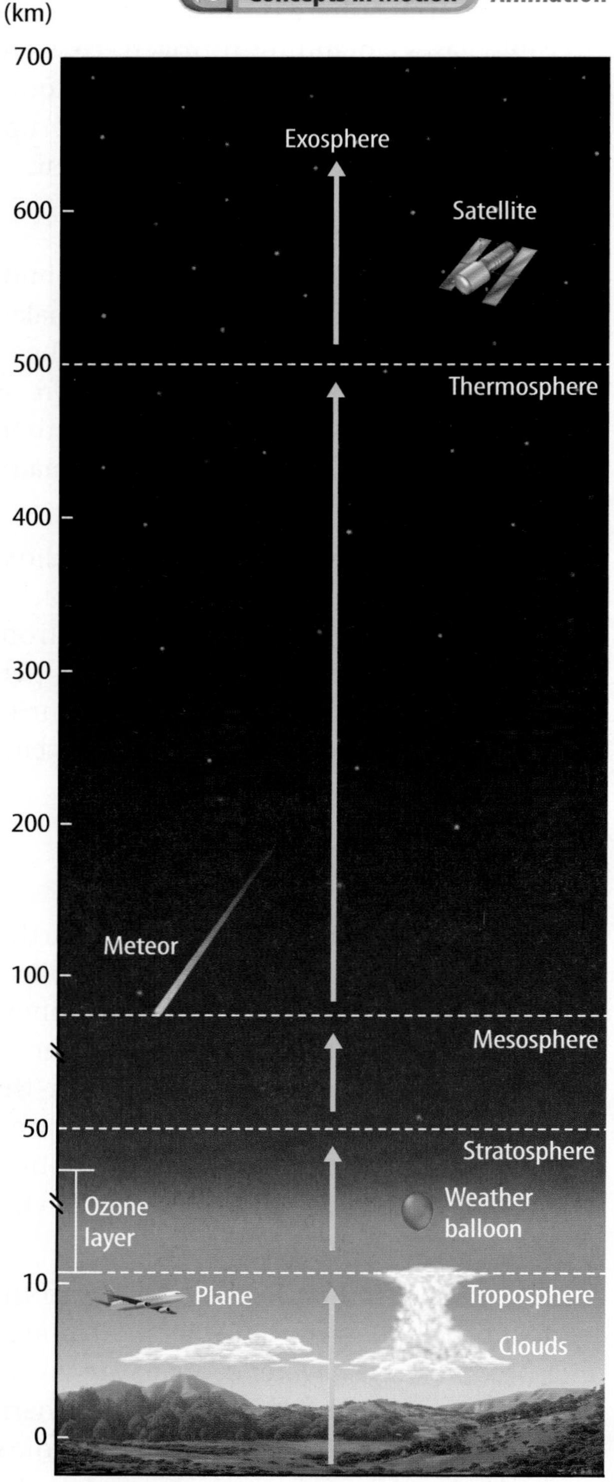

(km)

- 700 — Exosphere
- 600 — Satellite
- 500 — Thermosphere
- 400
- 300
- 200
- 100 — Meteor — Mesosphere
- 50 — Stratosphere
- Ozone layer — Weather balloon
- 10 — Plane — Troposphere — Clouds
- 0

Figure 3 Scientists divide Earth's atmosphere into different layers.

Visual Check In which layer of the atmosphere do planes fly?

Mesosphere and Thermosphere

As shown in **Figure 3,** the mesosphere extends from the stratosphere to about 85 km above Earth. The thermosphere can extend from the mesopshere to more than 500 km above Earth. Combined, these layers are much broader than the troposphere and the stratosphere, yet only 1 percent of the atmosphere's gas molecules are found in the mesosphere and the thermosphere. Most meteors burn up in these layers instead of striking Earth.

Ionosphere *The* **ionosphere** *is a region within the mesosphere and thermosphere that contains ions.* Between 60 km and 500 km above Earth's surface, the ionosphere's ions reflect AM radio waves transmitted at ground level. After sunset when ions recombine, this reflection increases. **Figure 4** shows how AM radio waves can travel long distances, especially at night, by bouncing off Earth and the ionosphere.

Radio Waves and the Ionosphere

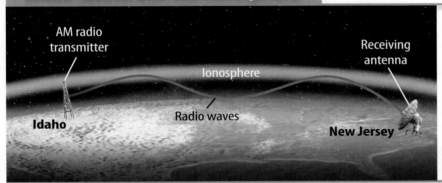

AM radio transmitter

Receiving antenna

Ionosphere

Radio waves

Idaho

New Jersey

Figure 4 Radio waves can travel long distances in the atmosphere.

Auroras The ionosphere is where stunning displays of colored lights called auroras occur, as shown in **Figure 5.** Auroras are most frequent in the spring and fall, but are best seen when the winter skies are dark. Auroras occur when ions from the Sun strike air molecules, causing them to emit vivid colors of light. People who live in the higher latitudes, nearer to the North Pole and the South Pole, are most likely to see auroras.

Exosphere

The exosphere is the atmospheric layer farthest from Earth's surface. Here, pressure and density are so low that individual gas molecules rarely strike one another. The molecules move at incredibly fast speeds after absorbing the Sun's radiation. The atmosphere does not have a definite edge, and molecules that are part of it can escape the pull of gravity and travel into space.

 Key Concept Check What are the layers of the atmosphere?

▲ **Figure 5** Auroras occur in the ionosphere.

Figure 6 Molecules in the air are closer together near Earth's surface than they are at higher altitudes. ▼

Increasing altitude

Figure 7 Temperature differences occur within the layers of the atmosphere. ▼

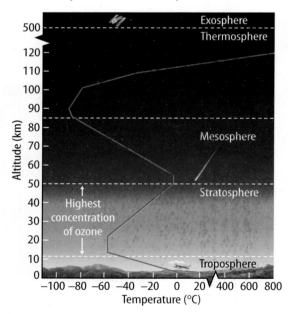

Altitude (km)

Exosphere
Thermosphere

Mesosphere

Highest
concentration
of ozone

Stratosphere

Troposphere

−100 −80 −60 −40 −20 0 20 400 600 800
Temperature (°C)

 Visual Check Which temperature pattern is most like the troposphere's?

Air Pressure and Altitude

Gravity is the force that pulls all objects toward Earth. When you stand on a scale, you can read your weight. This is because gravity is pulling you toward Earth. Gravity also pulls the atmosphere toward Earth. The pressure that a column of air exerts on anything below it is called air pressure. Gravity's pull on air increases its density. At higher altitudes, the air is less dense. **Figure 6** shows that air pressure is greatest near Earth's surface because the air molecules are closer together. This dense air exerts more force than the less dense air near the top of the atmosphere. Mountain climbers sometimes carry oxygen tanks at high altitudes because fewer oxygen molecules are in the air at high altitudes.

Reading Check How does air pressure change as altitude increases?

Temperature and Altitude

Figure 7 shows how temperature changes with altitude in the different layers of the atmosphere. If you have ever been hiking in the mountains, you have experienced the temperature cooling as you hike to higher elevations. In the troposphere, temperature decreases as altitude increases. Notice that the opposite effect occurs in the stratosphere. As altitude increases, temperature increases. This is because of the high concentration of ozone in the stratosphere. Ozone absorbs energy from sunlight, which increases the temperature in the stratosphere.

In the mesosphere, as altitude increases, temperature again decreases. In the thermosphere and exosphere, temperatures increase as altitude increases. These layers receive large amounts of energy from the Sun. This energy is spread across a small number of particles, creating high temperatures.

Key Concept Check How does temperature change as altitude increases?

Lesson 1 Review

Visual Summary

21% Oxygen 78% Nitrogen

Earth's atmosphere consists of gases that make life possible.

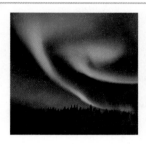

Layers of the atmosphere include the troposphere, the stratosphere, the mesosphere, the thermosphere, and the exosphere.

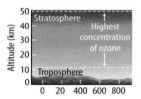

The ozone layer is the area in the stratosphere with a high concentration of ozone.

FOLDABLES

Use your lesson Foldable to review the lesson. Save your Foldable for the project at the end of the chapter.

What do you think NOW?

You first read the statements below at the beginning of the chapter.

1. Air is empty space.

2. Earth's atmosphere is important to living organisms.

Did you change your mind about whether you agree or disagree with the statements? Rewrite any false statements to make them true.

Use Vocabulary

1. The _____ is a thin layer of gases surrounding Earth.

2. The area of the stratosphere that helps protect Earth's surface from harmful ultraviolet rays is the _____.

3. **Define** Using your own words, define *water vapor*.

Understand Key Concepts

4. Which atmospheric layer is closest to Earth's surface?
   - **A.** mesosphere
   - **C.** thermosphere
   - **B.** stratosphere
   - **D.** troposphere

5. **Identify** the two atmospheric layers in which temperature decreases as altitude increases.

Interpret Graphics

6. **Contrast** Copy and fill in the graphic organizer below to contrast the composition of gases in Earth's early atmosphere and its present-day atmosphere.

Atmosphere	Gases
Early	
Present-day	

7. **Determine** the relationship between air pressure and the water in the glass in the photo below.

Critical Thinking

8. **Explain** three ways the atmosphere is important to living things.

A Crack in Earth's Shield

AMERICAN MUSEUM OF NATURAL HISTORY

Scientists discover an enormous hole in the ozone layer that protects Earth.

The ozone layer is like sunscreen, protecting Earth from the Sun's ultraviolet rays. But not all of Earth is covered. Every spring since 1985, scientists have been monitoring a growing hole in the ozone layer above Antarctica.

This surprising discovery was the outcome of years of research from Earth and space. The first measurements of polar ozone levels began in the 1950s, when a team of British scientists began launching weather balloons in Antarctica. In the 1970s, NASA started using satellites to measure the ozone layer from space. Then, in 1985 a close examination of the British team's records indicated a large drop in ozone levels during the Antarctic spring. The levels were so low that the scientists checked and rechecked their instruments before they reported their findings. NASA scientists quickly confirmed the discovery—an enormous hole in the ozone layer over the entire continent of Antarctica. They reported that the hole might have originated as far back as 1976.

Human-made compounds found mostly in chemicals called chlorofluorocarbons, or CFCs, are destroying the ozone layer. During cold winters, molecules released from these compounds are transformed into new compounds by chemical reactions on ice crystals that form in the ozone layer over Antarctica. In the spring, warming by the Sun breaks down the new compounds and releases chlorine and bromine. These chemicals break apart ozone molecules, slowly destroying the ozone layer.

In 1987, CFCs were banned in many countries around the world. Since then, the loss of ozone has slowed and possibly reversed, but a full recovery will take a long time. One reason is that CFCs stay in the atmosphere for more than 40 years. Still, scientists predict the hole in the ozone layer will eventually mend.

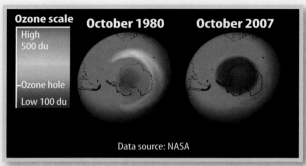

Ozone scale | October 1980 | October 2007
High 500 du
Ozone hole
Low 100 du

Data source: NASA

▲ **A hole in the ozone layer has developed over Antarctica. Even though it has gotten worse over the years, the hole has not grown as fast as scientists initially thought it would.**

Global Warming and the Ozone

Drew Shindell is a NASA scientist investigating the connection between the ozone layer in the stratosphere and the buildup of greenhouse gases throughout the atmosphere. Surprisingly, while these gases warm the troposphere, they are actually causing temperatures in the stratosphere to become cooler. As the stratosphere cools above Antarctica, more clouds with ice crystals form—a key step in the process of ozone destruction. While the buildup of greenhouse gases in the atmosphere may slow the recovery, Shindell still thinks that eventually the ozone layer will heal itself.

It's Your Turn

NEWSCAST Work with a partner to develop three questions about the ozone layer. Research to find the answers. Take the roles of reporter and scientist. Present your findings to the class in a newscast format.

Lesson 2

Reading Guide

Key Concepts 🔑
ESSENTIAL QUESTIONS

- How does energy transfer from the Sun to Earth and the atmosphere?
- How are air circulation patterns within the atmosphere created?

Vocabulary
radiation p. 582
conduction p. 585
convection p. 585
stability p. 586
temperature inversion p. 587

g Multilingual eGlossary

Energy Transfer in the Atmosphere

Inquiry What's really there?

Mirages are created as light passes through layers of air that have different temperatures. How does energy create the reflections? What other effects does energy have on the atmosphere?

What happens to air as it warms?

Light energy from the Sun is converted to thermal energy on Earth. Thermal energy powers the weather systems that impact your everyday life.

1. Read and complete a lab safety form.
2. Turn on a **lamp** with an incandescent lightbulb.
3. Place your hands under the light near the lightbulb. What do you feel?
4. Dust your hands with **powder.**
5. Place your hands below the lightbulb and clap them together once.
6. Observe what happens to the particles.

Think About This

1. How might the energy in step 3 move from the lightbulb to your hand?

2. How did the particles move when you clapped your hands?

3. **Key Concept** How did particle motion show you how the air was moving?

Energy from the Sun

ACADEMIC VOCABULARY

process
(noun) an ordered series of actions

The Sun's energy travels 148 million km to Earth in only 8 minutes. How does the Sun's energy get to Earth? It reaches Earth through the **process** of radiation. **Radiation** *is the transfer of energy by electromagnetic waves.* Ninety-nine percent of the radiant energy from the Sun consists of visible light, ultraviolet light, and infrared radiation.

Visible Light

The majority of sunlight is visible light. Recall that visible light is light that you can see. The atmosphere is like a window to visible light, allowing it to pass through. At Earth's surface it is converted to thermal energy, commonly called heat.

Near-Visible Wavelengths

The wavelengths of ultraviolet (UV) light and infrared radiation (IR) are just beyond the range of visibility to human eyes. UV light has short wavelengths and can break chemical bonds. Excess exposure to UV light will burn human skin and can cause skin cancer. Infrared radiation (IR) has longer wavelengths than visible light. You can sense IR as thermal energy or warmth. Earth absorbs energy from the Sun and then radiates it into the atmosphere as IR.

Reading Check Contrast visible light and ultraviolet light.

Energy on Earth

As the Sun's energy passes through the atmosphere, some of it is absorbed by gases and particles, and some of it is reflected back into space. As a result, not all the energy coming from the Sun reaches Earth's surface.

Absorption

Study **Figure 8.** Gases and particles in the atmosphere absorb about 20 percent of incoming solar radiation. Oxygen, ozone, and water vapor all absorb incoming ultraviolet light. Water and carbon dioxide in the troposphere absorb some infrared radiation from the Sun. Earth's atmosphere does not absorb visible light. Visible light must be converted to infrared radiation before it can be absorbed.

Reflection

Bright surfaces, especially clouds, **reflect** incoming radiation. Study **Figure 8** again. Clouds and other small particles in the air reflect about 25 percent of the Sun's radiation. Some radiation travels to Earth's surface and is then reflected by land and sea surfaces. Snow-covered, icy, or rocky surfaces are especially reflective. As shown in **Figure 8,** this accounts for about 5 percent of incoming radiation. In all, about 30 percent of incoming radiation is reflected into space. This means that, along with the 20 percent of incoming radiation that is absorbed in the atmosphere, Earth's surface only receives and absorbs about 50 percent of incoming solar radiation.

Figure 8 Some of the energy from the Sun is reflected or absorbed as it passes through the atmosphere.

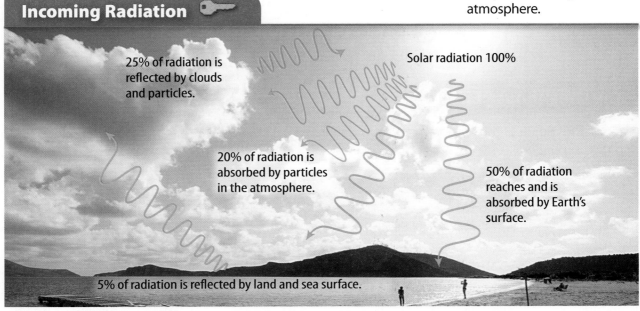

Incoming Radiation

25% of radiation is reflected by clouds and particles.

Solar radiation 100%

20% of radiation is absorbed by particles in the atmosphere.

50% of radiation reaches and is absorbed by Earth's surface.

5% of radiation is reflected by land and sea surface.

Visual Check What percent of incoming radiation is absorbed by gases and particles in the atmosphere?

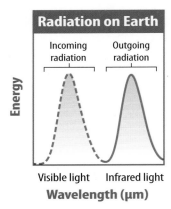

Radiation on Earth

Incoming radiation   Outgoing radiation

Energy

Visible light   Infrared light

Wavelength (μm)

▲ **Figure 9** The amount of solar energy absorbed by Earth and its atmosphere is equal to the amount of energy Earth radiates back into space.

Radiation Balance

The Sun's radiation heats Earth. So, why doesn't Earth get hotter and hotter as it continues to receive radiation from the Sun? There is a balance between the amount of incoming radiation from the Sun and the amount of outgoing radiation from Earth.

The land, water, plants, and other organisms absorb solar radiation that reaches Earth's surface. The radiation absorbed by Earth is then re-radiated, or bounced back, into the atmosphere. Most of the energy radiated from Earth is infrared radiation, which heats the atmosphere. **Figure 9** shows that the amount of radiation Earth receives from the Sun is the same as the amount Earth radiates into the outer atmosphere. Earth absorbs the Sun's energy and then radiates that energy away until a balance is achieved.

The Greenhouse Effect

As shown in **Figure 10,** the glass of a greenhouse allows light to pass through, where it is converted to infrared energy. The glass prevents the IR from escaping and it warms the greenhouse. Some of the gases in the atmosphere, called greenhouse gases, act like the glass of a greenhouse. They allow sunlight to pass through, but they prevent some of Earth's IR energy from escaping. Greenhouse gases in Earth's atmosphere trap IR and direct it back to Earth's surface. This causes an additional buildup of thermal energy at Earth's surface. The gases that trap IR best are water vapor (H_2O), carbon dioxide (CO_2), and methane (CH_4).

 Reading Check Describe the greenhouse effect.

The Greenhouse Effect

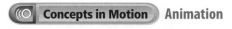 (((o))) **Concepts in Motion** Animation

Figure 10 Some of the outgoing radiation is directed back toward Earth's surface by greenhouse gases.

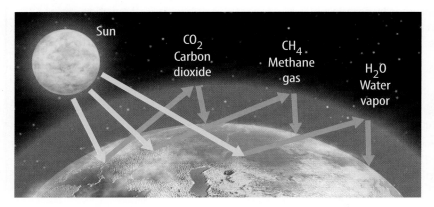

Sun

CO_2 Carbon dioxide

CH_4 Methane gas

H_2O Water vapor

Thermal Energy Transfer

Recall that there are three types of thermal energy transfer—radiation, conduction, and convection. All three occur in the atmosphere. Recall that radiation is the process that transfers energy from the Sun to Earth.

Conduction

Thermal energy always moves from an object with a higher temperature to an object with a lower temperature. **Conduction** *is the transfer of thermal energy by collisions between particles of matter.* Particles must be close enough to touch to transfer energy by conduction. Touching the pot of water, shown in **Figure 11**, would transfer energy from the pot to your hand. Conduction occurs where the atmosphere touches Earth.

Convection

As molecules of air close to Earth's surface are heated by conduction, they spread apart, and air becomes less dense. Less dense air rises, transferring thermal energy to higher altitudes. *The transfer of thermal energy by the movement of particles within matter is called* **convection.** Convection can be seen in **Figure 11** as the boiling water circulates and steam rises.

Latent Heat

More than 70 percent of Earth's surface is covered by a highly unique substance—water! Water is the only substance that can exist as a solid, a liquid, and a gas within Earth's temperature ranges. Recall that latent heat is exchanged when water changes from one phase to another, as shown in **Figure 12**. Latent heat energy is transferred from Earth's surface to the atmosphere.

 Key Concept Check How does energy transfer from the Sun to Earth and the atmosphere?

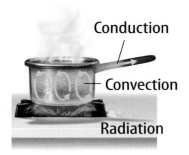

▲ **Figure 11** Energy is transferred through conduction, convection, and radiation.

 Review
Personal Tutor

Word Origin
conduction
from Latin *conducere*, means "to bring together"

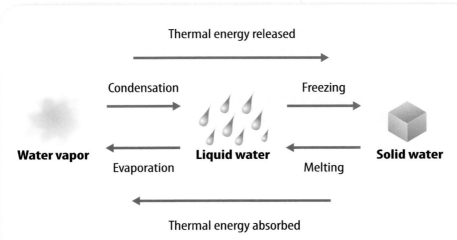

Figure 12 Water releases or absorbs thermal energy during phase changes.

Thermal energy released

Condensation Freezing

Water vapor   Liquid water   Solid water

Evaporation   Melting

Thermal energy absorbed

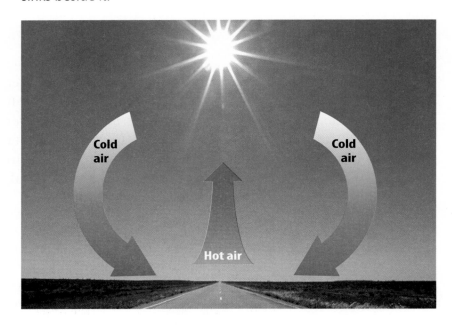

Figure 13 Rising warm air is replaced by cooler, denser air that sinks beside it.

Cold air

Cold air

Hot air

Fold a sheet of paper to make a four-column, four-row table and label as shown. Use it to record information about thermal energy transfer.

Energy Transfer by	Description	Everyday Example	Effect on the Atmosphere
Radiation			
Convection			
Conduction			

Figure 14 Lens-shaped lenticular clouds form when air rises with a mountain wave. ▼

Mountain Wave

Circulating Air

You've read that energy is transferred through the atmosphere by convection. On a hot day, air that is heated becomes less dense. This creates a pressure difference. Cool, denser air pushes the warm air out of the way. The warm air is replaced by the more dense air, as shown in **Figure 13.** The warm air is often pushed upward. Warmer, rising air is always accompanied by cooler, sinking air.

Air is constantly moving. For example, wind flowing into a mountain range rises and flows over it. After reaching the top, the air sinks. This up-and-down motion sets up an atmospheric phenomenon called a mountain wave. The upward moving air within mountain waves creates lenticular (len TIH kyuh lur) clouds, shown in **Figure 14.** Circulating air affects weather and climate around the world.

🔑 **Key Concept Check** How are air circulation patterns within the atmosphere created?

Stability

When you stand in the wind, your body forces some of the air to move above you. The same is true for hills and buildings. Conduction and convection also cause air to move upward. **Stability** *describes whether circulating air motions will be strong or weak.* When air is unstable, circulating motions are strong. During stable conditions, circulating motions are weak.

Cold air
Warm air

Normal conditions

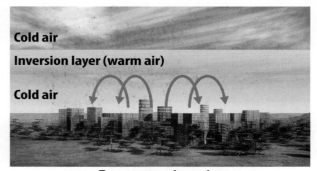

Cold air
Inversion layer (warm air)
Cold air

Temperature inversion

Unstable Air and Thunderstorms Unstable conditions often occur on warm, sunny afternoons. During unstable conditions, ground-level air is much warmer than higher-altitude air. As warm air rises rapidly in the atmosphere, it cools and forms large, tall clouds. Latent heat, released as water vapor changes from a gas to a liquid, adds to the instability, and produces a thunderstorm.

Reading Check Relate unstable air to the formation of thunderstorms.

Stable Air and Temperature Inversions Sometimes ground-level air is nearly the same temperature as higher-altitude air. During these conditions, the air is stable, and circulating motions are weak. A temperature inversion can occur under these conditions. A **temperature inversion** *occurs in the troposphere when temperature increases as altitude increases.* During a temperature inversion, a layer of cooler air is trapped by a layer of warmer air above it, as shown in **Figure 15.** Temperature inversions prevent air from mixing and can trap pollution in the air close to Earth's surface.

Figure 15 A temperature inversion occurs when cooler air is trapped beneath warmer air.

Visual Check How do conditions during a temperature inversion differ from normal conditions?

Inquiry MiniLab
20 minutes

Can you identify a temperature inversion?
You've read that a temperature inversion is a reversal of normal temperature conditions in the troposphere. What do data from a temperature inversion look like on a graph?

Analyze and Conclude

1. **Describe** the information presented in the graph. How do the graph's lines differ?

2. **Analyze** Which graph line represents normal conditions in the troposphere? Which represents a temperature inversion? Explain your answers in your Science Journal.

3. **Key Concept** From the graph, what pattern does a temperature inversion have?

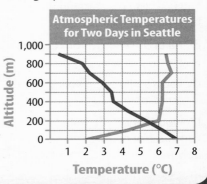

Atmospheric Temperatures for Two Days in Seattle

Altitude (m)
Temperature (°C)

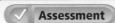
Visual Summary

Not all radiation from the Sun reaches Earth's surface.

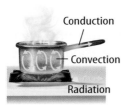

Thermal energy transfer in the atmosphere occurs through radiation, conduction, and convection.

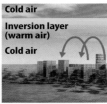

Temperature inversions prevent air from mixing and can trap pollution in the air close to Earth's surface.

FOLDABLES

Use your lesson Foldable to review the lesson. Save your Foldable for the project at the end of the chapter.

What do you think NOW?

You first read the statements below at the beginning of the chapter.

3. All of the energy from the Sun reaches Earth's surface.

4. Earth emits energy back into the atmosphere.

Did you change your mind about whether you agree or disagree with the statements? Rewrite any false statements to make them true.

Use Vocabulary

1 The property of the atmosphere that describes whether circulating air motions will be strong or weak is called _____.

2 **Define** *conduction* in your own words.

3 _____ is the transfer of thermal energy by the movement of particles within matter.

Understand Key Concepts

4 Which statement is true?
  A. The Sun's energy is completely blocked by Earth's atmosphere.
  B. The Sun's energy passes through the atmosphere without warming it significantly.
  C. The Sun's IR energy is absorbed by greenhouse gases.
  D. The Sun's energy is primarily in the UV range.

5 **Distinguish** between conduction and convection.

Interpret Graphics

6 **Explain** how greenhouses gases affect temperatures on Earth.

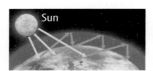

7 **Sequence** Copy and fill in the graphic organizer below to describe how energy from the Sun is absorbed in Earth's atmosphere.

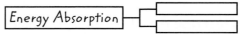

Critical Thinking

8 **Suggest** a way to keep a parked car cool on a sunny day.

9 **Relate** temperature inversions to air stability.

Can you conduct, convect, and radiate?

After solar radiation reaches Earth, the molecules closest to Earth transfer thermal energy from molecule to molecule by conduction. The newly warmed air becomes less dense and moves through the process of convection.

Materials

candle

metal rod

glass rod

wooden dowel

500-mL beaker

ice

bowls (2)

lamp

glass cake pan

food coloring

250-mL beaker

Safety

Learn It

When you **compare and contrast** two or more things, you look for similarities and differences between them. When you **compare** two things, you look for the similarities, or how they are the same. When you **contrast** them, you look for how they are different from each other.

Try It

1. Read and complete a lab safety form.
2. Drip a small amount of melted candle wax onto one end of a metal rod, a glass rod, and a wooden dowel.
3. Place a 500-mL beaker on the lab table. Have your teacher add 350 mL of very hot water. Place the ends of the rods without candle wax in the water. Set aside.
4. Place an ice cube into each of two small bowls labeled A and B.
5. Place bowl A under a lamp with a 60- or 75-watt lightbulb. Place the light source 10 cm above the bowl. Turn on the lamp. Set bowl B aside.
6. Fill a glass cake pan with room-temperature water to a level of 2 cm. Put 2–3 drops of red food coloring into a 250-mL beaker of very hot water. Put 2–3 drops of blue food coloring into a 250-mL beaker of very cold water and ice cubes. Carefully pour the hot water into one end of the pan. Slowly pour the very cold water into the same end of the pan. Observe what happens from the side of the pan. Record your observations in your Science Journal.
7. Observe the candle wax on the rods in the hot water and the ice cubes in the bowls.

Apply It

8. What happened to the candle wax? Identify the type of energy transfer.
9. Which ice cube melted the most in the bowls? Identify the type of energy transfer that melted the ice.
10. Compare and contrast how the hot and cold water behaved in the pan. Identify the type of energy transfer.
11. 🔑 **Key Concept** Explain how each part of the lab models radiation, conduction, or convection.

Lesson 3

Air Currents

Reading Guide

Key Concepts
ESSENTIAL QUESTIONS

- How does uneven heating of Earth's surface result in air movement?
- How are air currents on Earth affected by Earth's spin?
- What are the main wind belts on Earth?

Vocabulary

wind p. 591

trade winds p. 593

westerlies p. 593

polar easterlies p. 593

jet stream p. 593

sea breeze p. 594

land breeze p. 594

 Multilingual eGlossary

Video

- Science Video
- What's Science Got to do With It?

Inquiry How does air push these blades?

If you have ever ridden a bicycle into a strong wind, you know the movement of air can be a powerful force. Some areas of the world have more wind than others. What causes these differences? What makes wind?

Why does air move?

Early sailors relied on wind to move their ships around the world. Today, wind is used as a renewable source of energy. In the following activity, you will explore what causes air to move.

1. Read and complete a lab safety form.
2. Inflate a **balloon.** Do not tie it. Hold the neck of the balloon closed.
3. Describe how the inflated balloon feels.
4. Open the neck of the balloon without letting go of the balloon. Record your observations of what happens in your Science Journal.

Think About This

1. What caused the inflated balloon surface to feel the way it did when the neck was closed?

2. What caused the air to leave the balloon when the neck was opened?

3. **Key Concept** Why didn't outside air move into the balloon when the neck was opened?

Global Winds

There are great wind belts that circle the globe. The energy that causes this massive movement of air originates at the Sun. However, wind patterns can be global or local.

Unequal Heating of Earth's Surface

The Sun's energy warms Earth. However, the same amount of energy does not reach all of Earth's surface. The amount of energy an area gets depends largely on the Sun's angle. For example, energy from the rising or setting Sun is not very intense. But Earth heats up quickly when the Sun is high in the sky.

In latitudes near the equator—an area referred to as the tropics—sunlight strikes Earth's surface at a high angle—nearly 90°— year round. As a result, in the tropics there is more sunlight per unit of surface area. This means that the land, the water, and the air at the equator are always warm.

At latitudes near the North Pole and the South Pole, sunlight strikes Earth's surface at a low angle. Sunlight is now spread over a larger surface area than in the tropics. As a result, the poles receive very little energy per unit of surface area and are cooler.

Recall that differences in density cause warm air to rise. Warm air puts less pressure on Earth than cooler air. Because it's so warm in the tropics, air pressure is usually low. Over colder areas, such as the North Pole and the South Pole, air pressure is usually high. This difference in pressure creates wind. **Wind** *is the movement of air from areas of high pressure to areas of low pressure.* Global wind belts influence both climate and weather on Earth.

Key Concept Check How does uneven heating of Earth's surface result in air movement?

Figure 16 Three cells in each hemisphere move air through the atmosphere.

✔ **Visual Check** Which wind belt do you live in?

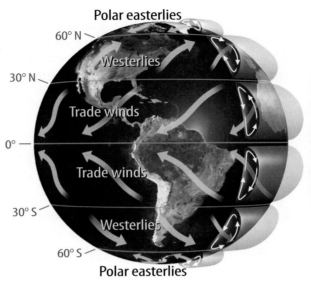

Polar easterlies
60° N
Westerlies
30° N
Trade winds
0°
Trade winds
30° S
Westerlies
60° S
Polar easterlies

FOLDABLES

Make a shutterfold. As illustrated, draw Earth and the three cells found in each hemisphere on the inside of the shutterfold. Describe each cell and explain the circulation of Earth's atmosphere. On the outside, label the global wind belts.

Polar Easterlies
Westerlies
Trade Winds
Trade Winds
Westerlies
Polar Easterlies

Global Wind Belts

Figure 16 shows the three-cell model of circulation in Earth's atmosphere. In the northern hemisphere, hot air in the cell nearest the equator moves to the top of the troposphere. There, the air moves northward until it cools and moves back to Earth's surface near 30° latitude. Most of the air in this convection cell then returns to Earth's surface near the equator.

The cell at the highest northern latitudes is also a convection cell. Air from the North Pole moves toward the equator along Earth's surface. The cooler air pushes up the warmer air near 60° latitude. The warmer air then moves northward and repeats the cycle. The cell between 30° and 60° latitude is not a convection cell. Its motion is driven by the other two cells, in a motion similar to a pencil that you roll between your hands. Three similar cells exist in the southern hemisphere. These cells help generate the global wind belts.

The Coriolis Effect

What happens when you throw a ball to someone across from you on a moving merry-go-round? The ball appears to curve because the person catching the ball has moved. Similarly, Earth's rotation causes moving air and water to appear to move to the right in the northern hemisphere and to the left in the southern hemisphere. This is called the Coriolis effect. The contrast between high and low pressure and the Coriolis effect creates distinct wind patterns, called prevailing winds.

Key Concept Check How are air currents on Earth affected by Earth's spin?

Prevailing Winds

The three global cells in each hemisphere create northerly and southerly winds. When the Coriolis effect acts on the winds, they blow to the east or the west, creating relatively steady, predictable winds. Locate the trade winds in **Figure 16**. *The* **trade winds** *are steady winds that flow from east to west between 30°N latitude and 30°S latitude.*

At about 30°N and 30°S air cools and sinks. This creates areas of high pressure and light, calm winds at the equator called the doldrums. Sailboats without engines can be stranded in the doldrums.

The prevailing **westerlies** *are steady winds that flow from west to east between latitudes 30°N and 60°N, and 30°S and 60°S.* This region is also shown in **Figure 16**. *The* **polar easterlies** *are cold winds that blow from the east to the west near the North Pole and the South Pole.*

 Key Concept Check What are the main wind belts on Earth?

Jet Streams

Near the top of the troposphere is a narrow band of high winds called the **jet stream**. Shown in **Figure 17,** jet streams flow around Earth from west to east, often making large loops to the north or the south. Jet streams influence weather as they move cold air from the poles toward the tropics and warm air from the tropics toward the poles. Jet streams can move at speeds up to 300 km/h and are more unpredictable than prevailing winds.

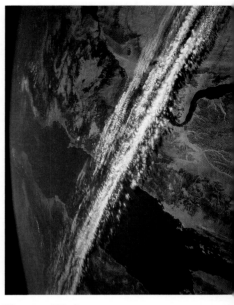

Figure 17 Jet streams are thin bands of high wind speed. The clouds seen here have condensed within a cooler jet stream.

Inquiry MiniLab

20 minutes

Can you model the Coriolis effect?

Earth's rotation causes the Coriolis effect. It affects the movement of water and air on Earth.

1. Read and complete a lab safety form.

2. Draw dot A in the center of a piece of **foamboard.** Draw dot B along the outer edge of the foamboard.

3. Roll a **table-tennis ball** from dot A to dot B. Record your observations in your Science Journal.

4. Center the foamboard on a **turntable**. Have your partner rotate the foamboard at a medium speed. Roll the ball along the same path. Record your observations.

Analyze and Conclude

1. **Contrast** the path of the ball when the foamboard was not moving to when it was spinning.

2. **Key Concept** How might air moving from the North Pole to the equator travel due to Earth's rotation?

Local Winds

You have just read that global winds occur because of pressure differences around the globe. In the same way, local winds occur whenever air pressure is different from one location to another.

Sea and Land Breezes

Anyone who has spent time near a lake or an ocean shore has probably experienced the connection between temperature, air pressure, and wind. *A* **sea breeze** *is wind that blows from the sea to the land due to local temperature and pressure differences.* **Figure 18** shows how sea breezes form. On sunny days, land warms up faster than water does. The air over the land warms by conduction and rises, creating an area of low pressure. The air over the water sinks, creating an area of high pressure because it is cooler. The differences in pressure over the warm land and the cooler water result in a cool wind that blows from the sea onto land.

A **land breeze** *is a wind that blows from the land to the sea due to local temperature and pressure differences.* **Figure 18** shows how land breezes form. At night, the land cools more quickly than the water. Therefore, the air above the land cools more quickly than the air over the water. As a result, an area of lower pressure forms over the warmer water. A land breeze then blows from the land toward the water.

 Reading Check Compare and contrast sea breezes and land breezes.

Figure 18 Sea breezes and land breezes are created as part of a large reversible convection current.

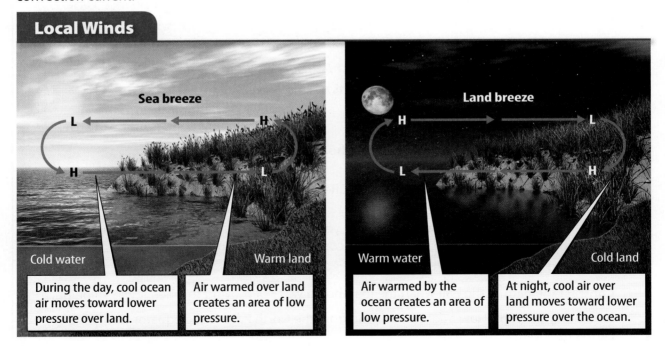

Local Winds

Sea breeze

During the day, cool ocean air moves toward lower pressure over land.

Air warmed over land creates an area of low pressure.

Cold water

Warm land

Land breeze

Air warmed by the ocean creates an area of low pressure.

At night, cool air over land moves toward lower pressure over the ocean.

Warm water

Cold land

Visual Check Sequence the steps involved in the formation of a land breeze.

✓ **Assessment**   Online Quiz

Visual Summary

Wind is created by pressure differences between one location and another.

Prevailing winds in the global wind belts are the trade winds, the westerlies, and the polar easterlies.

Sea breezes and land breezes are examples of local winds.

FOLDABLES

Use your lesson Foldable to review the lesson. Save your Foldable for the project at the end of the chapter.

What do you think NOW?

You first read the statements below at the beginning of the chapter.

5. Uneven heating in different parts of the atmosphere creates air circulation patterns.

6. Warm air sinks and cold air rises.

Did you change your mind about whether you agree or disagree with the statements? Rewrite any false statements to make them true.

Use Vocabulary

1. The movement of air from areas of high pressure to areas of low pressure is _____.

2. A(n) _____ is wind that blows from the sea to the land due to local temperature and pressure differences.

3. **Distinguish** between westerlies and trade winds.

Understand Key Concepts

4. Which does NOT affect global wind belts?
   A. air pressure
   B. land breezes
   C. the Coriolis effect
   D. the Sun

5. **Relate** Earth's spinning motion to the Coriolis effect.

Interpret Graphics

Use the image below to answer question 6.

6. **Explain** a land breeze.

7. **Organize** Copy and fill in the graphic organizer below to summarize Earth's global wind belts.

Wind Belt	Description
Trade winds	
Westerlies	
Polar easterlies	

Critical Thinking

8. **Infer** what would happen without the Coriolis effect.

9. **Explain** why the wind direction is often the same in Hawaii as it is in Greenland.

Can you model global wind patterns?

In each hemisphere, air circulates in specific patterns. Recall that scientists use the three-cell model to describe these circulation cells. General circulation of the atmosphere produces belts of prevailing winds around the world. In this activity, you will make a **model** of the main circulation cells in Earth's atmosphere.

Materials

ribbons

globe

permanent marker

scissors

transparent tape

Safety

Learn It

Making a **model** can help you visualize how a process works. Scientists use models to represent processes that may be difficult to see in real time. Sometimes a model represents something too small to see with the unaided eye, such as a model of an atom. Other models, such as one of the solar system, represent something that is too large to see from one location.

Try It

1. Read and complete a lab safety form.

2. Refer to **Figure 16** to make your model.

3. Choose one color of ribbon for the circulation cells. Make a separate loop of ribbon long enough to cover the latitude boundaries of each cell. Draw arrows on each ribbon to show the direction that the air flows in that cell. Make one loop for each cell in the northern hemisphere and one for each in the southern hemisphere. Tape your "cells" onto the globe.

4. Choose different-colored ribbons to model each of these wind belts: trade winds, westerlies, and polar easterlies, in both hemispheres. Draw arrows on each ribbon to show the direction that the wind blows. Tape the ribbons on the globe.

5. Create a color key to identify each cell and its corresponding wind type.

Apply It

6. Explain how your model represents the three-cell model used by scientists. How does your model differ from actual air movement in the atmosphere?

7. Explain why you cannot accurately model the global winds with this model.

8. 🔑 **Key Concept** Explain how latitude affects global winds.

Lesson 4

Air Quality

Reading Guide

Key Concepts
ESSENTIAL QUESTIONS

- How do humans impact air quality?
- Why do humans monitor air quality standards?

Vocabulary

air pollution p. 598

acid precipitation p. 599

photochemical smog p. 599

particulate matter p. 600

 Multilingual eGlossary

Video BrainPOP®

Inquiry How did this happen?

Air pollution can be trapped near Earth's surface during a temperature inversion. This is especially common in cities located in valleys and surrounded by mountains. What do you think the quality of the air is like on a day like this one? Where does pollution come from?

How does acid rain form?

Vehicles, factories, and power plants release chemicals into the atmosphere. When these chemicals combine with water vapor, they can form acid rain.

1. Read and complete a lab safety form.
2. Half-fill a **plastic cup** with **distilled water.**
3. Dip a strip of **pH paper** into the water. Use a **pH color chart** to determine the pH of the distilled water. Record the pH in your Science Journal.
4. Use a **dropper** to add **lemon juice** to the water until the pH equals that of acid rain. Swirl and test the pH each time you add 5 drops of the lemon juice to the mixture.

Think About This

1. A strong acid has a pH between 0 and 2. How does the pH of lemon juice compare to the pH of other substances? Is acid rain a strong acid?

2. **Key Concept** Why might scientists monitor the pH of rain?

Substances	pH
Hydrochloric acid	0.0
Lemon juice	2.3
Vinegar	2.9
Tomato juice	4.1
Coffee (black)	5.0
Acid rain	5.6
Rainwater	6.5
Milk	6.6
Distilled water	7.0
Blood	7.4
Baking soda solution	8.4
Toothpaste	9.9
Household ammonia	11.9
Sodium hydroxide	14.0

Sources of Air Pollution

The contamination of air by harmful substances including gases and smoke is called **air pollution.** Air pollution is harmful to humans and other living things. Years of exposure to polluted air can weaken a human's immune system. Respiratory diseases such as asthma can be caused by air pollution.

Air pollution comes from many sources. Point-source pollution is pollution that comes from an identifiable source. Examples of point sources include smokestacks of large factories, such as the one shown in **Figure 19,** and electric power plants that burn fossil fuels. They release tons of polluting gases and particles into the air each day. An example of natural point-source pollution is an erupting volcano.

Nonpoint-source pollution is pollution that comes from a widespread area. One example of pollution from a nonpoint-source is air pollution in a large city. This is considered nonpoint-source pollution because it cannot be traced back to one source. Some bacteria found in swamps and marshes are examples of natural sources of nonpoint-source pollution.

Key Concept Check Compare point-source and nonpoint-source pollution.

Figure 19 One example of point-source pollution is a factory smoke stack.

Causes and Effects of Air Pollution

The harmful effects of air pollution are not limited to human health. Some pollutants, including ground-level ozone, can damage plants. Air pollution can also cause serious damage to human-made structures. Sulfur dioxide pollution can discolor stone, corrode metal, and damage paint on cars.

Acid Precipitation

When sulfur dioxide and nitrogen oxides combine with moisture in the atmosphere and form precipitation that has a pH lower than that of normal rainwater, it is called **acid precipitation.** Acid precipitation includes acid rain, snow, and fog. It affects the chemistry of water in lakes and rivers. This can harm the organisms living in the water. Acid precipitation damages buildings and other structures made of stone. Natural sources of sulfur dioxide include volcanoes and marshes. However, the most common sources of sulfur dioxide and nitrogen oxides are automobile exhausts and factory and power plant smoke.

Smog

Photochemical smog *is air pollution that forms from the interaction between chemicals in the air and sunlight.* Smog forms when nitrogen dioxide, released in gasoline engine exhaust, reacts with sunlight. A series of chemical reactions produces ozone and other compounds that form smog. Recall that ozone in the stratosphere helps protect organisms from the Sun's harmful rays. However, ground-level ozone can damage the tissues of plants and animals. Ground-level ozone is the main component of smog. Smog in urban areas reduces visibility and makes air difficult to breathe. **Figure 20** shows New York City on a clear day and on a smoggy day.

 Key Concept Check How do humans impact air quality?

Figure 20 Smog can be observed as haze or a brown tint in the atmosphere.

Smog

Particulate Pollution

Although you can't see them, over 10,000 solid or liquid particles are in every cubic centimeter of air. A cubic centimeter is about the size of a sugar cube. This type of pollution is called particulate matter. **Particulate** (par TIH kyuh lut) **matter** *is a mixture of dust, acids, and other chemicals that can be hazardous to human health.* The smallest particles are the most harmful. These particles can be inhaled and can enter your lungs. They can cause asthma, bronchitis, and lead to heart attacks. Children and older adults are most likely to experience health problems due to particulate matter.

Particulate matter in the atmosphere absorbs and scatters sunlight. This can create haze. Haze particles scatter light, make things blurry, and reduce visibility.

Word Origin · · · · · · · · · · · ·

particulate
from Latin *particula*, means
"small part"
· · · · · · · · · · · · · · · ·

Movement of Air Pollution

Wind can influence the effects of air pollution. Because air carries pollution with it, some wind patterns cause more pollution problems than others. Weak winds or no wind prevents pollution from mixing with the surrounding air. During weak wind conditions, pollution levels can become dangerous.

For example, the conditions in which temperature inversions form are weak winds, clear skies, and longer winter nights. As land cools at night, the air above it also cools. Calm winds, however, prevent cool air from mixing with warm air above it. **Figure 21** shows how cities located in valleys experience a temperature inversion. Cool air, along with the pollution it contains, is trapped in valleys. More cool air sinks down the sides of the mountain, further preventing layers from mixing. The pollution in the photo at the beginning of the lesson was trapped due to a temperature inversion.

Figure 21 At night, cool air sinks down the mountain sides, trapping pollution in the valley below.

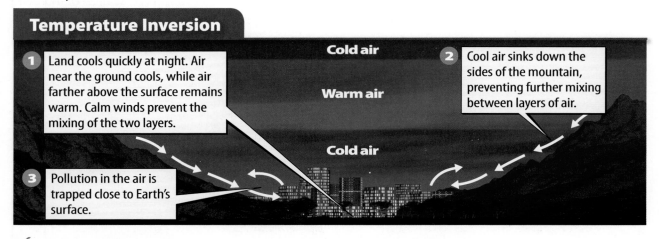

Temperature Inversion

1. Land cools quickly at night. Air near the ground cools, while air farther above the surface remains warm. Calm winds prevent the mixing of the two layers.

2. Cool air sinks down the sides of the mountain, preventing further mixing between layers of air.

3. Pollution in the air is trapped close to Earth's surface.

Cold air

Warm air

Cold air

✓ **Visual Check** How is pollution trapped by a temperature inversion?

Maintaining Healthful Air Quality

Preserving the quality of Earth's atmosphere requires the cooperation of government officials, scientists, and the public. The Clean Air Act is an example of how government can help fight pollution. Since the Clean Air Act became law in 1970, steps have been taken to reduce automobile emissions. Pollutant levels have decreased significantly in the United States. Despite these advances, serious problems still remain. The amount of ground-level ozone is still too high in many large cities. Also, acid precipitation produced by air pollutants continues to harm organisms in lakes, streams, and forests.

Air Quality Standards

The Clean Air Act gives the U.S. government the power to set air quality standards. The standards protect humans, animals, plants, and buildings from the harmful effects of air pollution. All states are required to make sure that pollutants, such as carbon monoxide, nitrogen oxides, particulate matter, ozone, and sulfur dioxide, do not exceed harmful levels.

 Reading Check What is the Clean Air Act?

Monitoring Air Pollution

Pollution levels are continuously monitored by hundreds of instruments in all major U.S. cities. If the levels are too high, authorities may advise people to limit outdoor activities.

Inquiry MiniLab 15 minutes

Can being out in fresh air be harmful to your health?

Are you going to be affected if you play tennis for a couple hours, go biking with your friends, or even just lie on the beach? Even if you have no health problems related to your respiratory system, you still need to be aware of the quality of air in your area of activity for the day.

Analyze and Conclude

1. Which values on the AQI indicate that the air quality is good?

2. At what value is the air quality unhealthful for anyone who may have allergies and respiratory disorders?

3. Which values would be considered as warnings of emergency conditions?

4. **Key Concept** The quality of air in different areas changes throughout the day. Explain how you can use the AQI to help you know when you should limit your outdoor activity.

Air Quality Index (AQI) Values	Levels of Health Concern
0 to 50	Good
51 to 100	Moderate
101 to 150	Unhealthful for Sensitive Groups
151 to 200	Unhealthful
201 to 300	Very Unhealthful
301 to 500	Hazardous

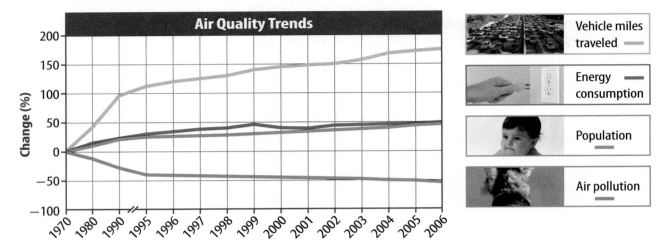

Air Quality Trends

	Vehicle miles traveled
	Energy consumption
	Population
	Air pollution

Figure 22 Pollution emissions have declined, even though the population is increasing.

Math Skills

Use Graphs
The graph above shows the percent change in four different pollution factors from 1970 through 2006. All values are based on the 0 percent amount in 1970. For example, from 1970 to 1990, the number of vehicle miles driven increased by 100 percent, or the vehicle miles doubled. Use the graph to infer which factors might be related.

Practice
1. What was the percent change in population between 1970 and 2006?

2. What other factor changed by about the same amount during that period?

Review
• **Math Practice**
• **Personal Tutor**

Air Quality Trends

Over the last several decades, air quality in U.S. cities has improved, as shown in **Figure 22.** Even though some pollution-producing processes have increased, such as burning fossil fuels and traveling in automobiles, levels of certain air pollutants have decreased. Airborne levels of lead and carbon monoxide have decreased the most. Levels of sulfur dioxide, nitrogen oxide, and particulate matter have also decreased.

However, ground-level ozone has not decreased much. Why do ground-level ozone trends lag behind those of other pollutants? Recall that ozone can be created from chemical reactions involving automobile exhaust. The increase in the amount of ground-level ozone is because of the increase in the number of miles traveled by vehicles.

Key Concept Check Why do humans monitor air quality standards?

Indoor Air Pollution

Not all air pollution is outdoors. The air inside homes and other buildings can be as much as 50 times more polluted than outdoor air! The quality of indoor air can impact human health much more than outdoor air quality.

Indoor air pollution comes from many sources. Tobacco smoke, cleaning products, pesticides, and fireplaces are some common sources. Furniture upholstery, carpets, and foam insulation also add pollutants to the air. Another indoor air pollutant is radon, an odorless gas given off by some soil and rocks. Radon leaks through cracks in a building's foundation and sometimes builds up to harmful levels inside homes. Harmful effects of radon come from breathing its particles.

Lesson 4 Review

Visual Summary

Air pollution comes from point sources, such as factories, and nonpoint sources, such as automobiles.

Photochemical smog contains ozone, which can damage tissues in plants and animals.

FOLDABLES

Use your lesson Foldable to review the lesson. Save your Foldable for the project at the end of the chapter.

What do you think NOW?

You first read the statements below at the beginning of the chapter.

7. If no humans lived on Earth, there would be no air pollution.

8. Pollution levels in the air are not measured or monitored.

Did you change your mind about whether you agree or disagree with the statements? Rewrite any false statements to make them true.

Use Vocabulary

1 **Define** *acid precipitation* in your own words.

2 _____ forms when chemical reactions combine pollution with sunlight.

3 The contamination of air by harmful substances, including gases and smoke, is _____.

Understand Key Concepts

4 Which is NOT true about smog?
   A. It contains nitrogen oxide.
   B. It contains ozone.
   C. It reduces visibility.
   D. It is produced only by cars.

5 **Describe** two ways humans add pollution to the atmosphere.

6 **Assess** whether urban or rural areas are more likely to have high levels of smog.

7 **Identify** and describe the law designed to reduce air pollution.

Interpret Graphics

8 **Compare and Contrast** Copy and fill in the graphic organizer below to compare and contrast details of smog and acid precipitation.

	Similarities	Differences
Smog		
Acid Precipitation		

Critical Thinking

9 **Describe** how conduction and convection are affected by paving over a grass field.

Math Skills ×÷   📖 Review
—— Math Practice ——

10 Based on the graph on the opposite page, what was the total percent change in air pollution between 1970 and 2006?

Radiant Energy Absorption

Materials

thermometer

sand

500-mL beaker

lamp

stopwatch

paper towels

spoon

potting soil

clay

Safety

Ultimately, the Sun is the source of energy for Earth. Energy from the Sun moves through the atmosphere and is absorbed and reflected from different surfaces on Earth. Light surfaces reflect energy, and dark surfaces absorb energy. Both land and sea surfaces absorb energy from the Sun, and air in contact with these surfaces is warmed through conduction.

Ask a Question

Which surfaces on Earth absorb the most energy from the Sun?

Make Observations

1. Read and complete a lab safety form.

2. Make a data table in your Science Journal to record your observations of energy transfer. Include columns for Type of Surface, Temperature Before Heating, and Temperature After Heating.

3. Half-fill a 500-mL beaker with sand. Place a thermometer in the sand and carefully add enough sand to cover the thermometer bulb— about 2 cm deep. Keep the bulb under the sand for 1 minute. Record the temperature in the data table.

4. Place the beaker under the light source. Record the temperature after 10 minutes.

5. Repeat steps 3 and 4 using soil and water.

Form a Hypothesis

6. Use the data in your table to form a hypothesis stating which surfaces on Earth, such as forests, wheat fields, lakes, snowy mountain tops, and deserts, will absorb the most radiant energy.

Test Your Hypothesis

7 Decide what materials could be used to mimic the surfaces on Earth from your hypothesis.

8 Repeat the experiment with materials approved by the teacher to test your hypothesis.

9 Examine your data. Was your hypothesis supported? Why or why not?

Analyze and Conclude

10 **Infer** which types of areas on Earth absorb the most energy from the Sun.

11 **Think Critically** When areas of Earth are changed so they become more likely to reflect or absorb energy from the Sun, how might these changes affect conduction and convection in the atmosphere?

12 **The Big Idea** Explain how thermal energy from the Sun being received by and reflected from Earth's surface is related to the role of the atmosphere in maintaining conditions suitable for life.

Communicate Your Results

Display data from your initial observations to compare your findings with your classmates' findings. Explain your hypothesis, experiment results, and conclusions to the class.

Inquiry Extension

What could you add to this investigation to show how cloud cover changes the amount of radiation that will reach Earth's surfaces? Design a study that could test the effect of cloud cover on radiation passing through Earth's atmosphere. How could you include a way to show that clouds also reflect radiant energy from the Sun?

6

Lab Tips

☑ If possible, use leaves, straw, shaved ice, and other natural materials to test your hypothesis.

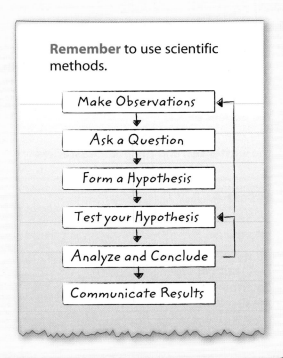

Remember to use scientific methods.

Make Observations
↓
Ask a Question
↓
Form a Hypothesis
↓
Test your Hypothesis
↓
Analyze and Conclude
↓
Communicate Results

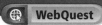

THE BIG IDEA

The gases in Earth's atmosphere, some of which are needed by organisms to survive, affect Earth's temperature and the transfer of thermal energy to the atmosphere.

Key Concepts Summary

Vocabulary

Lesson 1: Describing Earth's Atmosphere

- Earth's **atmosphere** formed as Earth cooled and chemical and biological processes took place.
- Earth's atmosphere consists of nitrogen, oxygen, and a small amount of other gases, such as CO_2 and **water vapor**.
- The atmospheric layers are the **troposphere**, the **stratosphere**, the mesosphere, the thermosphere, and the exosphere.
- Air pressure decreases as altitude increases. Temperature either increases or decreases as altitude increases, depending on the layer of the atmosphere.

21% Oxygen 78% Nitrogen

atmosphere p. 573
water vapor p. 574
troposphere p. 576
stratosphere p. 576
ozone layer p. 576
ionosphere p. 577

Lesson 2: Energy Transfer in the Atmosphere

Conduction
Convection
Radiation

- The Sun's energy is transferred to Earth's surface and the atmosphere through **radiation**, **conduction**, **convection**, and latent heat.
- Air circulation patterns are created by convection currents.

radiation p. 582
conduction p. 585
convection p. 585
stability p. 586
temperature inversion p. 587

Lesson 3: Air Currents

- Uneven heating of Earth's surface creates pressure differences. **Wind** is the movement of air from areas of high pressure to areas of low pressure.
- Air currents curve to the right or to the left due to the Coriolis effect.
- The main wind belts on Earth are the **trade winds**, the **westerlies**, and the **polar easterlies**.

wind p. 591
trade winds p. 593
westerlies p. 593
polar easterlies p. 593
jet stream p. 593
sea breeze p. 594
land breeze p. 594

Lesson 4: Air Quality

- Some human activities release pollution into the air.
- Air quality standards are monitored for the health of organisms and to determine if anti-pollution efforts are successful.

air pollution p. 598
acid precipitation p. 599
photochemical smog p. 599
particulate matter p. 600

FOLDABLES® Chapter Project

Assemble your lesson Foldables® as shown to make a Chapter Project. Use the project to review what you have learned in this chapter.

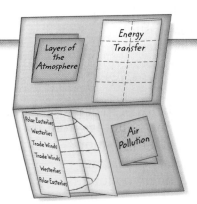

Use Vocabulary

1 Radio waves travel long distances by bouncing off electrically charged particles in the _____.

2 The Sun's thermal energy is transferred to Earth through space by _____.

3 Rising currents of warm air transfer energy from Earth to the atmosphere through _____.

4 A narrow band of winds located near the top of the troposphere is a(n) _____.

5 _____ are steady winds that flow from east to west between 30°N latitude and 30°S latitude.

6 In large urban areas, _____ forms when pollutants in the air interact with sunlight.

7 A mixture of dust, acids, and other chemicals that can be hazardous to human health is called _____.

Link Vocabulary and Key Concepts

Concepts in Motion Interactive Concept Map

Copy this concept map, and then use vocabulary terms from the previous page to complete the concept map.

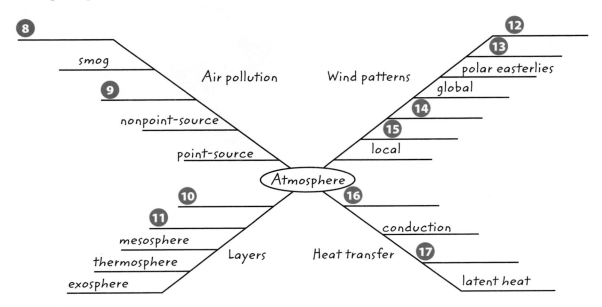

Understand Key Concepts

1 Air pressure is greatest
  A. at a mountain base.
  B. on a mountain top.
  C. in the stratosphere.
  D. in the ionosphere.

2 In which layer of the atmosphere is the ozone layer found?
  A. troposphere
  B. stratosphere
  C. mesosphere
  D. thermosphere

Use the image below to answer question 3.

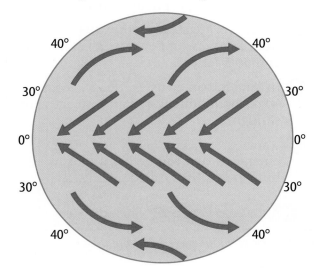

3 This diagram represents the atmosphere's
  A. air masses.
  B. global wind belts.
  C. inversions.
  D. particulate motion.

4 The Sun's energy
  A. is completely absorbed by the atmosphere.
  B. is completely reflected by the atmosphere.
  C. is in the form of latent heat.
  D. is transferred to the atmosphere after warming Earth.

5 Which type of energy is emitted from Earth to the atmosphere?
  A. ultraviolet radiation
  B. visible radiation
  C. infrared radiation
  D. aurora borealis

6 Which is a narrow band of high winds located near the top of the troposphere?
  A. polar easterly
  B. a jet stream
  C. a sea breeze
  D. a trade wind

7 Which helps protect people, animals, plants, and buildings from the harmful effects of air pollution?
  A. primary pollutants
  B. secondary pollutants
  C. ozone layer
  D. air quality standards

Use the photo below to answer question 8.

8 This photo shows a potential source of
  A. ultraviolet radiation.
  B. indoor air pollution.
  C. radon.
  D. smog.

Critical Thinking

9 **Predict** how atmospheric carbon dioxide levels might change if more trees were planted on Earth. Explain your prediction.

10 **Compare** visible and infrared radiation.

11 **Assess** whether your home is heated by conduction or convection.

12 **Sequence** how the unequal heating of Earth's surface leads to the formation of wind.

13 **Evaluate** whether a sea breeze could occur at night.

14 **Interpret Graphics** What are the top three sources of particulate matter in the atmosphere? What could you do to reduce particulate matter from any of the sources shown here?

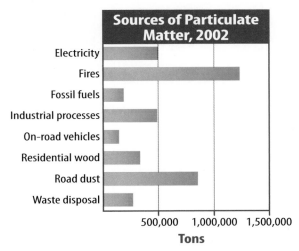

Sources of Particulate Matter, 2002

Electricity
Fires
Fossil fuels
Industrial processes
On-road vehicles
Residential wood
Road dust
Waste disposal

500,000   1,000,000   1,500,000
Tons

15 **Diagram** how acid precipitation forms. Include possible sources of sulfur dioxide and nitrogen oxide and organisms that can be affected by acid precipitation.

Writing in Science

16 **Write** a paragraph explaining whether you think it would be possible to permanently pollute the atmosphere with particulate matter.

REVIEW **THE BIG IDEA**

17 Review the title of each lesson in the chapter. List all of the characteristics and components of the troposphere and the stratosphere that affect life on Earth. Describe how life is impacted by each one.

18 Discuss how energy is transferred from the Sun throughout Earth's atmosphere.

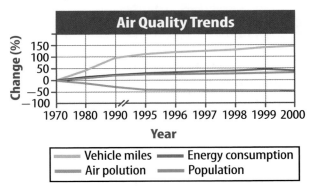

Math Skills   Review

Math Practice

Use Graphs

Air Quality Trends

Change (%)
150
100
50
0
−50
−100

1970 1980 1990 1995 1996 1997 1998 1999 2000
Year

— Vehicle miles — Energy consumption
— Air polution — Population

19 What was the percent change in energy use between 1996 and 1999?

20 What happened to energy use between 1999 and 2000?

21 What was the total percentage change between vehicle miles traveled and air pollution from 1970 to 2000?

Standardized Test Practice

Record your answers on the answer sheet provided by your teacher or on a sheet of paper.

Multiple Choice

1 What causes the phenomenon known as a mountain wave?

A radiation imbalance

B rising and sinking air

C temperature inversion

D the greenhouse effect

Use the diagram below to answer question 2.

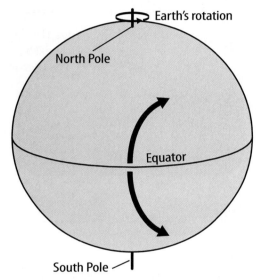

2 What phenomenon does the diagram above illustrate?

A radiation balance

B temperature inversion

C the Coriolis effect

D the greenhouse effect

3 Which do scientists call greenhouse gases?

A carbon dioxide, hydrogen, nitrogen

B carbon dioxide, methane gas, water vapor

C carbon monoxide, oxygen, argon

D carbon monoxide, ozone, radon

4 In which direction does moving air appear to turn in the northern hemisphere?

A down

B up

C right

D left

Use the diagram below to answer question 5.

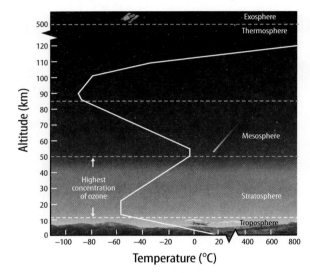

5 Which layer of the atmosphere has the widest range of temperatures?

A mesosphere

B stratosphere

C thermosphere

D troposphere

6 Which was the main component of Earth's original atmosphere?

A carbon dioxide

B nitrogen

C oxygen

D water vapor

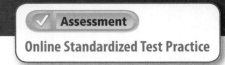
7 Which is the primary cause of the global wind patterns on Earth?

   A ice cap melting

   B uneven heating

   C weather changing

   D waves breaking

Use the diagram below to answer question 8.

Energy Transfer Methods

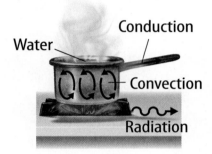

8 In the diagram above, which transfers thermal energy in the same way the Sun's energy is transferred to Earth?

   A the boiling water

   B the burner flame

   C the hot handle

   D the rising steam

9 Which substance in the air of U.S. cities has decreased least since the Clean Air Act began?

   A carbon monoxide

   B ground-level ozone

   C particulate matter

   D sulfur dioxide

Constructed Response

Use the table below to answer questions 10 and 11.

Layer	Significant Fact

10 In the table above, list in order the layers of Earth's atmosphere from lowest to highest. Provide one significant fact about each layer.

11 Explain how the first four atmospheric layers are important to life on Earth.

Use the table below to answer question 12.

Heat Transfer	Explanation
Conduction	
Convection	
Latent heat	
Radiation	

12 Complete the table to explain how heat energy transfers from the Sun to Earth and its atmosphere.

13 What are temperature inversions? How do they form? What is the relationship between temperature inversions and air pollution?

NEED EXTRA HELP?													
If You Missed Question...	1	2	3	4	5	6	7	8	9	10	11	12	13
Go to Lesson...	2	3	2	3	1	1	3	2	4	1	1	2	2, 4

Chapter 17

Weather

THE BIG IDEA

How do scientists describe and predict weather?

Inquiry Is this a record snowfall?

Buffalo, New York, is famous for its snowstorms, averaging 3 m of snow each year. Other areas of the world might only get a few centimeters of snow a year. In some parts of the world, it never snows.

- Why do some areas get less snow than others?

- How do scientists describe and predict weather?

Get Ready to Read

What do you think?

Before you read, decide if you agree or disagree with each of these statements. As you read this chapter, see if you change your mind about any of the statements.

1 Weather is the long-term average of atmospheric patterns of an area.

2 All clouds are at the same altitude within the atmosphere.

3 Precipitation often occurs at the boundaries of large air masses.

4 There are no safety precautions for severe weather, such as tornadoes and hurricanes.

5 Weather variables are measured every day at locations around the world.

6 Modern weather forecasts are done using computers.

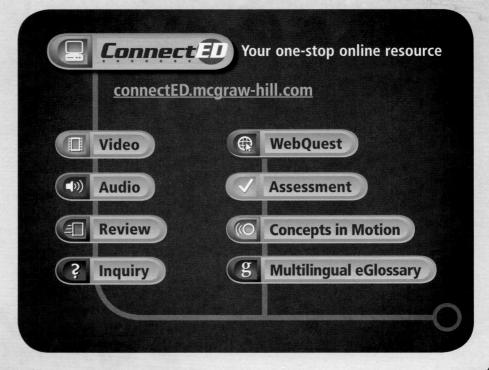

ConnectED Your one-stop online resource

connectED.mcgraw-hill.com

- Video
- Audio
- Review
- Inquiry
- WebQuest
- Assessment
- Concepts in Motion
- Multilingual eGlossary

Lesson 1

Reading Guide

Key Concepts 🔑
ESSENTIAL QUESTIONS

- What is weather?
- What variables are used to describe weather?
- How is weather related to the water cycle?

Vocabulary
weather p. 615

air pressure p. 616

humidity p. 616

relative humidity p. 617

dew point p. 617

precipitation p. 619

water cycle p. 619

  Multilingual eGlossary

🎞 Video

- BrainPOP®
- Science Video

Describing Weather

Inquiry Why are clouds different?

If you look closely at the photo, you'll see that there are different types of clouds in the sky. How do clouds form? If all clouds consist of water droplets and ice crystals, why do they look different? Are clouds weather?

Can you make clouds in a bag?

When water vapor in the atmosphere cools, it condenses. The resulting water droplets make up clouds.

1 Read and complete a lab safety form.

2 Half-fill a **500-mL beaker** with **ice** and **cold water.**

3 Pour 125 mL of **warm water** into a **resealable plastic bag** and seal the bag.

4 Carefully lower the bag into the ice water. Record your observations in your Science Journal.

Think About This

1. What did you observe when the warm water in the bag was put into the beaker?

2. What explanation can you give for what happened?

3. 🔑 **Key Concept** What could you see in the natural world that results from the same process?

What is weather?

Everybody talks about the weather. "Nice day, isn't it?" "How was the weather during your vacation?" Talking about weather is so common that we even use weather terms to describe unrelated topics. "That homework assignment was a breeze." Or "I'll take a rain check."

Weather *is the atmospheric conditions, along with short-term changes, of a certain place at a certain time.* If you have ever been caught in a rainstorm on what began as a sunny day, you know the weather can change quickly. Sometimes it changes in just a few hours. But other times your area might have the same sunny weather for several days in a row.

Weather Variables

Perhaps some of the first things that come to mind when you think about weather are temperature and rainfall. As you dress in the morning, you need to know what the temperature will be throughout the day to help you decide what to wear. If it is raining, you might cancel your picnic.

Temperature and rainfall are just two of the **variables** used to describe weather. Meteorologists, scientists who study and predict weather, use several specific variables that describe a variety of atmospheric conditions. These variables include air temperature, air pressure, wind speed and direction, humidity, cloud coverage, and precipitation.

🔑 **Key Concept Check** What is weather?

REVIEW VOCABULARY · · · · ·
variable
a quantity that can change

Air Temperature

The measure of the average **kinetic energy** of molecules in the air is air temperature. When the temperature is high, molecules have a high kinetic energy. Therefore, molecules in warm air move faster than molecules in cold air. Air temperatures vary with time of day, season, location, and altitude.

Air Pressure

The force that a column of air applies on the air or a surface below it is called **air pressure.** Study **Figure 1.** Is air pressure at Earth's surface more or less than air pressure at the top of the atmosphere? Air pressure decreases as altitude increases. Therefore, air pressure is greater at low altitudes than at high altitudes.

You might have heard the term *barometric pressure* during a weather forecast. Barometric pressure refers to air pressure. Air pressure is measured with an instrument called a barometer, shown in **Figure 2.** Air pressure is typically measured in millibars (mb). Knowing the barometric pressure of different areas helps meteorologists predict the weather.

 Reading Check What instrument measures air pressure?

Wind

As air moves from areas of high pressure to areas of low pressure, it creates wind. Wind direction is the direction from which the wind is blowing. For example, winds that blow from west to east are called westerlies. Meteorologists measure wind speed using an instrument called an anemometer (a nuh MAH muh tur). An anemometer is also shown in **Figure 2.**

Humidity

The amount of water vapor in the air is called **humidity** (hyew MIH duh tee). Humidity can be measured in grams of water per cubic meter of air (g/m^3). When the humidity is high, there is more water vapor in the air. On a day with high humidity, your skin might feel sticky, and sweat might not evaporate from your skin as quickly.

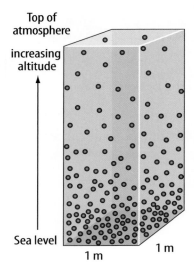

Figure 1 Increasing air pressure comes from having more molecules overhead.

Visual Check What happens to air pressure as altitude decreases?

Figure 2 Barometers, left, and anemometers, right, are used to measure weather variables.

Relative Humidity

Think about how a sponge can absorb water. At some point, it becomes full and cannot absorb any more water. In the same way, air can only contain a certain amount of water vapor. When air is saturated, it contains as much water vapor as possible. Temperature determines the maximum amount of water vapor air can contain. Warm air can contain more water vapor than cold air. *The amount of water vapor present in the air compared to the maximum amount of water vapor the air could contain at that temperature is called* **relative humidity.**

Relative humidity is measured using an instrument called a psychrometer and is given as a percent. For example, air with a relative humidity of 100 percent cannot contain any more moisture and dew or rain will form. Air that contains only half the water vapor it could hold has a relative humidity of 50 percent.

 Reading Check Compare and contrast humidity and relative humidity.

Dew Point

When a sponge becomes saturated with water, the water starts to drip from the sponge. Similarly, when air becomes saturated with water vapor, the water vapor will condense and form water droplets. When air near the ground becomes saturated, the water vapor in air will condense to a liquid. If the temperature is above 0°C, dew forms. If the temperature is below 0°C, ice crystals, or frost, form. Higher in the atmosphere clouds form. The graph in **Figure 3** shows the total amount of water vapor that air can contain at different temperatures.

When the temperature decreases, the air can hold less moisture. As you just read, the air becomes saturated, condensation occurs, and dew forms. *The temperature at which air is saturated and condensation can occur is called the* **dew point.**

When will dew form?

The relative humidity on a summer day is 80 percent. The temperature is 35°C. Will the dew point be reached if the temperature drops to 25°C later in the evening? Use **Figure 3** below to find the amount of water vapor needed for saturation at each temperature.

1. Calculate the amount of water vapor in air that is 35°C and has 80 percent relative humidity. (Hint: multiply the amount of water vapor air can contain at 35°C by the percent of relative humidity.)

2. At 25°C, air can hold 2.2 g/cm^3 of water vapor. If your answer from step 1 is less than 2.2 g/cm^3, the dew point is not reached and dew will not form. If the number is greater, dew will form.

Analyze and Conclude

Key Concept After the Sun rises in the morning the air's temperature increases. How does the relative humidity change after sunrise? What does the line represent?

Figure 3 As air temperature increases, the air can contain more water vapor.

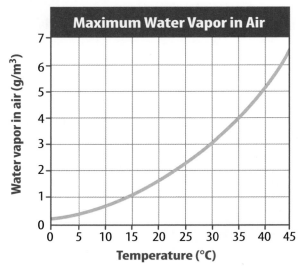

Figure 4 Clouds have different shapes and can be found at different altitudes.

Stratus clouds
- flat, white, and layered
- altitude up to 2,000 m

Cumulus clouds
- fluffy, heaped, or piled up
- 2,000 to 6,000 m altitude

Cirrus clouds
- wispy
- above 6,000 m

Clouds and Fog

When you exhale outside on a cold winter day, you can see the water vapor in your breath condense into a foggy cloud in front of your face. This also happens when warm air containing water vapor cools as it rises in the atmosphere. When the cooling air reaches its dew point, water vapor condenses on small particles in the air and forms droplets. Surrounded by thousands of other droplets, these small droplets block and reflect light. This makes them visible as clouds.

Clouds are water droplets or ice crystals suspended in the atmosphere. Clouds can have different shapes and be present at different altitudes within the atmosphere. Different types of clouds are shown in **Figure 4.** Because we observe that clouds move, we recognize that water and thermal energy are transported from one location to another. Recall that clouds are also important in reflecting some of the Sun's incoming radiation.

A cloud that forms near Earth's surface is called fog. Fog is a suspension of water droplets or ice crystals close to or at Earth's surface. Fog reduces visibility, the distance a person can see into the atmosphere.

Reading Check What is fog?

WORD ORIGIN · · · · · · · · · · ·

precipitation
from Latin *praecipitationem*,
means "act or fact of falling
headlong"

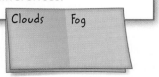

Make a horizontal two-tab book and label the tabs as illustrated. Use it to collect information on clouds and fog. Find similarities and differences.

Clouds   Fog

Precipitation

Recall that droplets in clouds form around small solid particles in the atmosphere. These particles might be dust, salt, or smoke. Precipitation occurs when cloud droplets combine and become large enough to fall back to Earth's surface. **Precipitation** *is water, in liquid or solid form, that falls from the atmosphere.* Examples of precipitation—rain, snow, sleet, and hail—are shown in **Figure 5.**

Rain is precipitation that reaches Earth's surface as droplets of water. Snow is precipitation that reaches Earth's surface as solid, frozen crystals of water. Sleet may originate as snow. The snow melts as it falls through a layer of warm air and refreezes when it passes through a layer of below-freezing air. Other times it is just freezing rain. Hail reaches Earth's surface as large pellets of ice. Hail starts as a small piece of ice that is repeatedly lifted and dropped by an updraft within a cloud. A layer of ice is added with each lifting. When it finally becomes too heavy for the updraft to lift, it falls to Earth.

 Key Concept Check What variables are used to describe weather?

The Water Cycle

Precipitation is an important process in the water cycle. Evaporation and condensation are phase changes that are also important to the water cycle. *The* **water cycle** *is the series of natural processes by which water continually moves among oceans, land, and the atmosphere.* As illustrated in **Figure 6,** most water vapor enters the atmosphere when water at the ocean's surface is heated and evaporates. Water vapor cools as it rises in the atmosphere and condenses back into a liquid. Eventually, droplets of liquid and solid water form clouds. Clouds produce precipitation, which falls to Earth's surface and later evaporates, continuing the cycle.

 Key Concept Check How is weather related to the water cycle?

Types of Precipitation

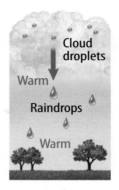

Rain   **Snow**

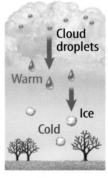
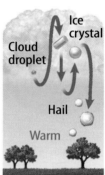

Sleet   **Hail**

▲ **Figure 5** Rain, snow, sleet, and hail are forms of precipitation.

Visual Check What is the difference between snow and sleet?

 Review

The Water Cycle   **Personal Tutor**

Figure 6 The Sun's energy powers the water cycle, which is the continual movement of water between the ocean, the land, and the atmosphere.

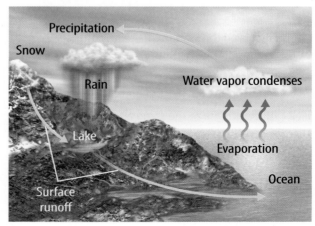

Lesson 1 Review

Visual Summary

Weather is the atmospheric conditions, along with short-term changes, of a certain place at a certain time.

Meteorologists use weather variables to describe atmospheric conditions.

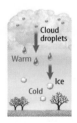

Forms of precipitation include rain, sleet, snow, and hail.

FOLDABLES®

Use your lesson Foldable to review the lesson. Save your Foldable for the project at the end of the chapter.

What do you think NOW?

You first read the statements below at the beginning of the chapter.

1. Weather is the long-term average of atmospheric patterns of an area.

2. All clouds are at the same altitude within the atmosphere.

Did you change your mind about whether you agree or disagree with the statements? Rewrite any false statements to make them true.

Use Vocabulary

1 **Define** *humidity* in your own words.

2 **Use the term** *precipitation* in a sentence.

3 _____ is the pressure that a column of air exerts on the surface below it.

Understand Key Concepts

4 Which is NOT a standard weather variable?
  A. air pressure
  B. moon phase
  C. temperature
  D. wind speed

5 **Identify** and describe the different variables used to describe weather.

6 **Relate** humidity to cloud formation.

7 **Describe** how processes in the water cycle are related to weather.

Interpret Graphics

8 **Identify** Which type of precipitation is shown in the diagram below? How does this precipitation form?

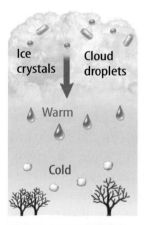

Critical Thinking

9 **Analyze** Why would your ears pop if you climbed a tall mountain?

10 **Differentiate** among cloud formation, fog formation, and dew point.

AMERICAN
MUSEUM OF
NATURAL
HISTORY

Flooding caused widespread devastation in New Orleans, a city that lies below sea level. The storm surge broke through levees that had protected the city.

Is there a link between hurricanes and global warming?

Scientists worry that hurricanes might be getting bigger and happening more often.

On August 29, 2005, Hurricane Katrina roared through New Orleans, Louisiana. The storm destroyed homes and broke through levees, flooding most of the low-lying city. In the wake of the disaster, many wondered whether global warming was responsible. If warm oceans are the fuel for hurricanes, could rising temperatures cause stronger or more frequent hurricanes?

Climate scientists have several ways to investigate this question. They examine past hurricane activity, sea surface temperature, and other climate data. They compare these different types of data and look for patterns. Based on the laws of physics, they put climate and hurricane data into equations. A computer solves these equations and makes computer models. Scientists analyze the models to see whether there is a connection between hurricane activity and different climate variables.

What have scientists learned? So far they have not found a link between warming oceans and the frequency of hurricanes. However, they have found a connection between warming oceans and hurricane strength. Models suggest that rising ocean temperatures might create more destructive hurricanes with stronger winds and more rainfall.

The warm waters of the Gulf of Mexico fueled Hurricane Katrina as it spun toward Louisiana.

But global warming is not the only cause of warming oceans. As the ocean circulates, it goes through cycles of warming and cooling. Data show that the Atlantic Ocean has been in a warming phase for the past few decades.

Whether due to global warming or natural cycles, ocean temperatures are expected to rise even more in coming years. While rising ocean temperatures might not produce more hurricanes, climate research shows they could produce more powerful hurricanes. Perhaps the better question is not what caused Hurricane Katrina, but how we can prepare for equal-strength or more destructive hurricanes in the future.

It's Your Turn

DIAGRAM With a partner, create a storyboard with each frame showing one step in hurricane formation. Label your drawings. Share your storyboard with the class.

Lesson 2

Weather Patterns

Reading Guide

Key Concepts
ESSENTIAL QUESTIONS

- What are two types of pressure systems?
- What drives weather patterns?
- Why is it useful to understand weather patterns?
- What are some examples of severe weather?

Vocabulary

high-pressure system p. 623

low-pressure system p. 623

air mass p. 624

front p. 626

tornado p. 629

hurricane p. 630

blizzard p. 631

 Multilingual eGlossary

Video

What's Science Got to do With It?

Inquiry What caused this flooding?

Surging waves and rain from Hurricane Katrina caused flooding in New Orleans, Louisiana. Why are flooding and other types of severe weather dangerous? How does severe weather form?

Launch Lab

10 minutes

How can temperature affect pressure?

Air molecules that have low energy can be packed closely together. As energy is added to the molecules they begin to move and bump into one another.

1. Read and complete a lab safety form.

2. Close a **resealable plastic bag** except for a small opening. Insert a **straw** through the opening and blow air into the bag until it is as firm as possible. Remove the straw and quickly seal the bag.

3. Submerge the bag in a **container** of **ice water** and hold it there for 2 minutes. Record your observations in your Science Journal.

4. Remove the bag from the ice water and submerge it in **warm water** for 2 minutes. Record your observations.

Think About This

1. What do the results tell you about the movement of air molecules in cold air and in warm air?

2. 🔑 **Key Concept** What property of the air is demonstrated in this activity?

Pressure Systems

Weather is often associated with pressure systems. Recall that air pressure is the weight of the molecules in a large mass of air. When air molecules are cool, they are closer together than when they are warm. Cool air masses have high pressure, or more weight. Warm air masses have low pressure.

A **high-pressure system**, shown in **Figure 7**, *is a large body of circulating air with high pressure at its center and lower pressure outside of the system.* Because air moves from high pressure to low pressure, the air inside the system moves away from the center. Dense air sinks, bringing clear skies and fair weather.

A **low-pressure system**, also shown in **Figure 7**, *is a large body of circulating air with low pressure at its center and higher pressure outside of the system.* This causes air inside the low pressure system to rise. The rising air cools and the water vapor condenses, forming clouds and sometimes precipitation—rain or snow.

🔑 **Key Concept Check** Compare and contrast two types of pressure systems.

Figure 7 Air moving from areas of high pressure to areas of low pressure is called wind.

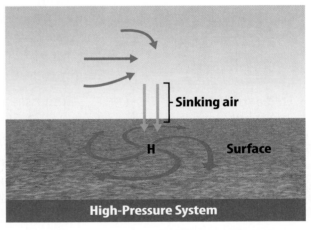

High-Pressure System

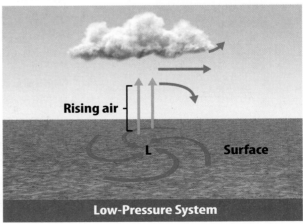

Low-Pressure System

Figure 8 Five main air masses impact climate across North America.

✅ **Visual Check** Where does continental polar air come from?

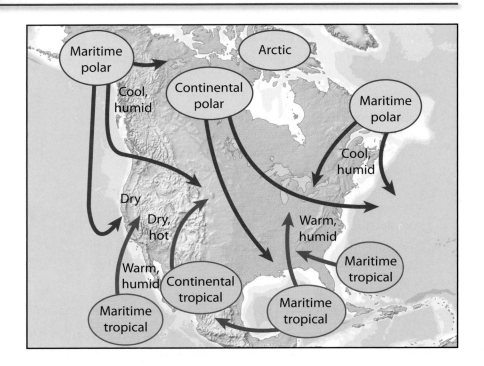

Air Masses

Have you ever noticed that the weather sometimes stays the same for several days in a row? For example, during winter in the northern United States, extremely cold temperatures often last for three or four days in a row. Afterward, several days might follow with warmer temperatures and snow showers.

Air masses are responsible for this pattern. **Air masses** *are large bodies of air that have uniform temperature, humidity, and pressure.* An air mass forms when a large high pressure system lingers over an area for several days. As a high pressure system comes in contact with Earth, the air in the system takes on the temperature and moisture characteristics of the surface below it.

Like high- and low-pressure systems, air masses can extend for a thousand kilometers or more. Sometimes one air mass covers most of the United States. Examples of the main air masses that affect weather in the United States are shown in **Figure 8**.

Air Mass Classification

Air masses are classified by their temperature and moisture characteristics. Air masses that form over land are referred to as continental air masses. Those that form over water are referred to as maritime masses. Warm air masses that form in the equatorial regions are called tropical. Those that form in cold regions are called polar. Air masses near the poles, over the coldest regions of the globe, are called arctic and antarctic air masses.

FOLDABLES

Fold a sheet of paper into thirds along the long axis. Label the outside *Air Masses*. Make another fold about 2 inches from the long edge of the paper to make a three-column chart. Label as shown.

Arctic Air Masses Forming over Siberia and the Arctic are arctic air masses. They contain bitterly cold, dry air. During winter, an arctic air mass can bring temperatures down to −40°C.

Continental Polar Air Masses Because land cannot transfer as much moisture to the air as oceans can, air masses that form over land are drier than air masses that form over oceans. Continental polar air masses are fast-moving and bring cold temperatures in winter and cool weather in summer. Find the continental polar air masses over Canada in **Figure 8.**

Maritime Polar Air Masses Forming over the northern Atlantic and Pacific Oceans, maritime polar air masses are cold and humid. They often bring cloudy, rainy weather.

Continental Tropical Air Masses Because they form in the tropics over dry, desert land, continental tropical air masses are hot and dry. They bring clear skies and high temperatures. Continental tropical air masses usually form during the summer.

Maritime Tropical Air Masses As shown in **Figure 8,** maritime tropical air masses form over the western Atlantic Ocean, the Gulf of Mexico, and the eastern Pacific Ocean. These moist air masses bring hot, humid air to the southeastern United States during summer. In winter, they can bring heavy snowfall.

Air masses can change as they move over the land and ocean. Warm, moist air can move over land and become cool and dry. Cold, dry air can move over water and become moist and warm.

Key Concept Check What drives weather patterns?

Math Skills

Conversions
To convert Fahrenheit (°F) units to Celsius (°C) units, use this equation:

$$°C = \frac{(°F - 32)}{1.8}$$

Convert **76°F** to °C

1. Always perform the operation in parentheses first.

    (**76°F** − 32 = **44°F**)

2. Divide the answer from Step 1 by 1.8.

    $$\frac{44°F}{1.8} = 24°C$$

To convert °C to °F, follow the same steps using the following equation:

$$°F = (°C \times 1.8) + 32$$

Practice
1. Convert 86°F to °C.
2. Convert 37°C to °F.

Review

- **Math Practice**
- **Personal Tutor**

Inquiry MiniLab   **20 minutes**

How can you observe air pressure?
Although air seems very light, air molecules do exert pressure. You can observe air pressure in action in this activity.

1. Read and complete a lab safety form.
2. Tightly cap the empty **plastic bottle.**
3. Place the bottle in a **bucket of ice** for 10 minutes. Record your observations in your Science Journal.

Analyze and Conclude

1. **Interpret** how air pressure affected the bottle.

2. **Key Concept** Discuss how changing air pressure in Earth's atmosphere affects other things on Earth, such as weather.

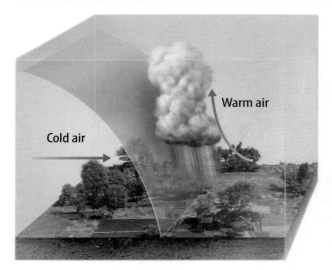

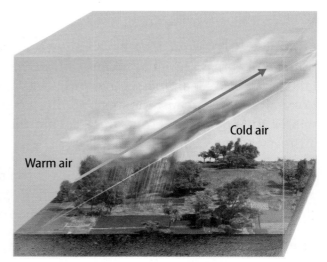

Cold

Warm

Figure 9 Certain types of fronts are associated with specific weather.

✔ **Visual Check** Describe the difference between a cold front and a warm front.

SCIENCE USE V. COMMON USE · ·

front
Science Use a boundary between two air masses

Common Use the foremost part or surface of something

Fronts

In 1918, Norwegian meteorologist Jacob Bjerknes (BYURK nehs) and his coworkers were busy developing a new method for forecasting the weather. Bjerknes noticed that specific types of weather occur at the boundaries between different air masses. Because he was trained in the army, Bjerknes used a military term to describe this boundary—front.

A military front is the boundary between opposing armies in a battle. *A weather* **front***, however, is a boundary between two air masses.* Drastic weather changes often occur at fronts. As wind carries an air mass away from the area where it formed, the air mass will eventually collide with another air mass. Changes in temperature, humidity, cloud types, wind, and precipitation are common at fronts.

Cold Fronts

When a colder air mass moves toward a warmer air mass, a cold front forms, as shown in **Figure 9.** The cold air, which is denser than the warm air, pushes underneath the warm air mass. The warm air rises and cools. Water vapor in the air condenses and clouds form. Showers and thunderstorms often form along cold fronts. It is common for temperatures to decrease as much as 10°C when a cold front passes through. The wind becomes gusty and changes direction. In many cases, cold fronts give rise to severe storms.

✔ **Reading Check** What types of weather are associated with cold fronts?

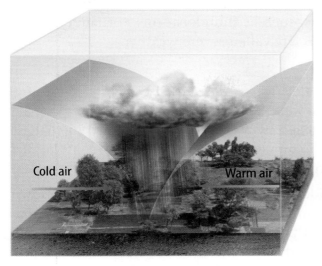

Stationary   Occluded

Warm Fronts

As shown in **Figure 9,** a warm front forms when less dense, warmer air moves toward colder, denser air. The warm air rises as it glides above the cold air mass. When water vapor in the warm air condenses, it creates a wide blanket of clouds. These clouds often bring steady rain or snow for several hours or even days. A warm front not only brings warmer temperatures, but it also causes the wind to shift directions.

Both a cold front and a warm front form at the edge of an approaching air mass. Because air masses are large, the movement of fronts is used to make weather forecasts. When a cold front passes through your area, temperatures will remain low for the next few days. When a warm front arrives, the weather will become warmer and more humid.

Stationary and Occluded Fronts

Sometimes an approaching front will stall for several days with warm air on one side of it and cold air on the other side. When the boundary between two air masses stalls, the front is called a stationary front. Study the stationary front shown in **Figure 9.** Cloudy skies and light rain are found along stationary fronts.

Cold fronts move faster than warm fronts. When a fast-moving cold front catches up with a slow-moving warm front, an occluded or blocked front forms. Occluded fronts, shown in **Figure 9,** usually bring precipitation.

 Key Concept Check Why is it useful to understand weather patterns associated with fronts?

Severe Weather

Some weather events can cause major damage, injuries, and death. These events, such as thunderstorms, tornadoes, hurricanes, and blizzards, are called severe weather.

Thunderstorms

Also known as electrical storms because of their lightning, thunderstorms have warm temperatures, moisture, and rising air, which may be supplied by a low-pressure system. When these conditions occur, a cumulus cloud can grow into a 10-km-tall thundercloud, or cumulonimbus cloud, in as little as 30 minutes.

A typical thunderstorm has a three-stage life cycle, shown in **Figure 10.** The cumulus stage is **dominated** by cloud formation and updrafts. Updrafts are air currents moving vertically away from the ground. After the cumulus cloud has been created, downdrafts begin to appear. Downdrafts are air currents moving vertically toward the ground. In the mature stage, heavy winds, rain, and lightning dominate the area. Within 30 minutes of reaching the mature stage, the thunderstorm begins to fade, or dissipate. In the dissipation stage, updrafts stop, winds die down, lightning ceases, and precipitation weakens.

Strong updrafts and downdrafts within a thunderstorm cause millions of tiny ice crystals to rise and sink, crashing into each other. This creates positively and negatively charged particles in the cloud. The difference in the charges of particles between the cloud and the charges of particles on the ground eventually creates electricity. This is seen as a bolt of lightning. Lightning can move from cloud to cloud, cloud to ground, or ground to cloud. It can heat the nearby air to more than 27,000°C. Air molecules near the bolt rapidly expand and then contract, creating the sound identified as thunder.

ACADEMIC VOCABULARY

dominate
(verb) to exert the guiding influence on

Figure 10 Thunderstorms have distinct stages characterized by the direction in which air is moving.

Thunderstorms 🔑

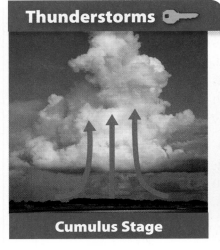

Cumulus Stage

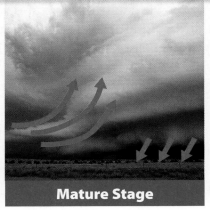

Mature Stage

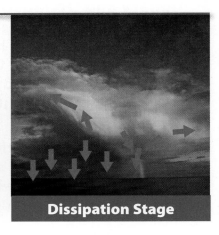

Dissipation Stage

✔ **Visual Check** Describe what happens during each stage of a thunderstorm.

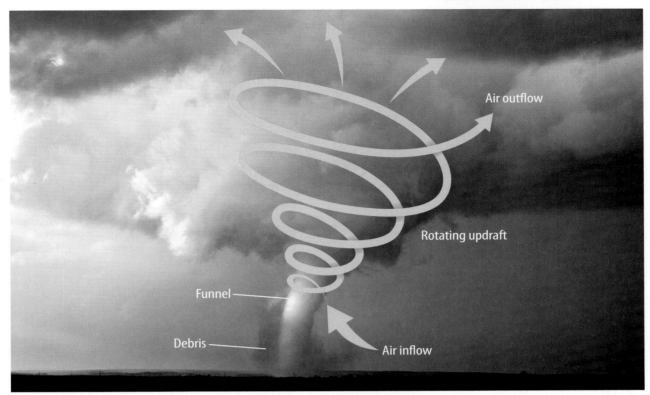

Air outflow

Rotating updraft

Funnel

Debris

Air inflow

Figure 11 A funnel cloud forms when updrafts within a thunderstorm begin rotating.

Tornadoes

Perhaps you have seen photos of the damage from a tornado. *A* **tornado** *is a violent, whirling column of air in contact with the ground.* Most tornadoes have a diameter of several hundred meters. The largest tornadoes exceed 1,500 m in diameter. The intense, swirling winds within tornadoes can reach speeds of more than 400 km/h. These winds are strong enough to send cars, trees, and even entire houses flying through the air. Tornadoes usually last only a few minutes. More destructive tornadoes, however, can last for several hours.

Formation of Tornadoes When thunderstorm updrafts begin to rotate, as shown in **Figure 11,** tornadoes can form. Swirling winds spiral downward from the thunderstorm's base, creating a funnel cloud. When the funnel reaches the ground, it becomes a tornado. Although the swirling air is invisible, you can easily see the debris lifted by the tornado.

✓ **Reading Check** How do tornadoes form?

Tornado Alley More tornadoes occur in the United States than anywhere else on Earth. The central United States, from Nebraska to Texas, experiences the most tornadoes. This area has been nicknamed Tornado Alley. In this area, cold air blowing southward from Canada frequently collides with warm, moist air moving northward from the Gulf of Mexico. These conditions are ideal for severe thunderstorms and tornadoes.

Classifying Tornadoes Dr. Ted Fujita developed a method for classifying tornadoes based on the damage they cause. On the modified Fujita intensity scale, F0 tornadoes cause light damage, breaking tree branches and damaging billboards. F1 though F4 tornadoes cause moderate to devastating damage, including tearing roofs from homes, derailing trains, and throwing vehicles in the air. F5 tornadoes cause incredible damage, such as demolishing concrete and steel buildings and pulling the bark from trees.

Figure 12 Hurricanes consist of alternating bands of heavy precipitation and sinking air.

Hurricane Formation

Low pressure

1 As warm, moist air rises into the atmosphere, it cools, water vapor condenses, and clouds form. As more air rises, it creates an area of low pressure over the ocean.

2 As air continues to rise, a tropical depression forms. Tropical depressions bring thunderstorms with winds between 37–62 km/h.

3 Air continues to rise, rotating counterclockwise. The storm builds to a tropical storm with winds in excess of 63 km/h. It produces strong thunderstorms.

4 When winds exceed 119 km/h, the storm becomes a hurricane. Only one percent of tropical storms become hurricanes.

Inside a Hurricane

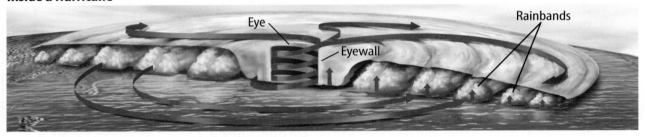

Eye

Eyewall

Rainbands

✓ **Visual Check** How do hurricanes form?

WORD ORIGIN · · · · · · · · · · · ·

hurricane
from Spanish *huracan*, means "tempest"

Hurricanes

An intense tropical storm with winds exceeding 119 km/h is a **hurricane.** Hurricanes are the most destructive storms on Earth. Like tornadoes, hurricanes have a circular shape with intense, swirling winds. However, hurricanes do not form over land. Hurricanes typically form in late summer over warm, tropical ocean water. **Figure 12** sequences the steps in hurricane formation. A typical hurricane is 480 km across, more than 150 thousand times larger than a tornado. At the center of a hurricane is the eye, an area of clear skies and light winds.

Damage from hurricanes occurs as a result of strong winds and flooding. While still out at sea, hurricanes create high waves that can flood coastal areas. As a hurricane crosses the coastline, or makes landfall, strong rains intensify and can flood and devastate entire areas. But once a hurricane moves over land or colder water, it loses its energy and dissipates.

In other parts of the world, these intense tropical storms have other names. In Asia, the same type of storm is called a typhoon. In Australia it is called a tropical cyclone.

Winter Storms

Not all severe weather occurs when temperatures are warm. Winter weather can also be severe. Snow and ice can make driving difficult and dangerous. When temperatures are close to freezing (0°C), rain can freeze when it hits the ground. Ice storms coat the ground, trees, and buildings with a layer of ice, as shown in **Figure 13.**

Freezing Rain

Figure 13 The weight of ice from freezing rain can cause trees, power lines, and other structures to break.

A **blizzard** *is a violent winter storm characterized by freezing temperatures, strong winds, and blowing snow.* During blizzards, blowing snow often reduces visibility to a few meters or even less. If you are outside during a blizzard, strong winds and very cold temperatures can rapidly cool exposed skin. Windchill, the combined cooling effect of cold temperature and wind on exposed skin, can cause frostbite and hypothermia (hi poh THER mee uh).

 Key Concept Check What are examples of severe weather?

Severe Weather Safety

To help keep people safe, the U.S. National Weather Service issues watches and warnings during severe weather events. A watch means that severe weather is possible. A warning means that severe weather is already occurring. Heeding severe weather watches and warnings is important and could save your life.

It is also important to know how to protect yourself during dangerous weather. During thunderstorms, you should stay inside if possible, and stay away from metal objects and electrical cords. If you are outside, stay away from water, high places and isolated trees. Dressing properly is important in all kinds of weather. When windchill temperatures are below −20°C you should dress in layers, keep your head and fingers covered, and limit your time outdoors.

Visual Summary

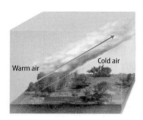

Low-pressure systems, high-pressure systems, and air masses all influence weather.

Weather often changes as a front passes through an area.

The National Weather Service issues warnings about severe weather such as thunderstorms, tornadoes, hurricanes, and blizzards.

FOLDABLES

Use your lesson Foldable to review the lesson. Save your Foldable for the project at the end of the chapter.

What do you think NOW?

You first read the statements below at the beginning of the chapter.

3. Precipitation often occurs at the boundaries of large air masses.

4. There are no safety precautions for severe weather, such as tornadoes and hurricanes.

Did you change your mind about whether you agree or disagree with the statements? Rewrite any false statements to make them true.

Use Vocabulary

1. **Distinguish** between an air mass and a front.

2. **Define** *low-pressure system* using your own words.

3. **Use the term** *high-pressure system* in a sentence.

Understand Key Concepts

4. Which air mass is humid and warm?
   A. continental polar
   B. continental tropical
   C. maritime polar
   D. maritime tropical

5. **Give an example** of cold-front weather.

6. **Compare and contrast** hurricanes and tornadoes.

7. **Explain** how thunderstorms form.

Interpret Graphics

8. **Compare and Contrast** Copy and fill in the graphic organizer below to compare and contrast high-pressure and low-pressure systems.

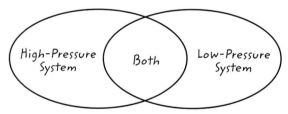

Critical Thinking

9. **Suggest** a reason that low-pressure systems are cloudy and rainy or snowy.

10. **Design** a pamphlet that contains tips on how to stay safe during different types of severe weather.

Math Skills

Review
— Math Practice —

11. Convert 212°F to °C.

12. Convert 20°C to °F.

Why does the weather change?

One day it is sunny, the next day it is pouring rain. If you look at only one location, the patterns that cause the weather to change are difficult to see. However, when you look on the large scale, the patterns become apparent.

Learn It

Recognizing cause and effect is an important part of science and conducting experiments. Scientists look for cause-and-effect relationships between variables. The maps below show the movement of fronts and pressure systems over a two-day period. What effect will these systems have on the weather as they move across the United States?

Try It

1 Examine the weather maps below. The thin black lines on each map represent areas where the barometric pressure is the same. The pressure is indicated by the number on the line. The center of a low- or high-pressure system is indicated by the word LOW or HIGH. Identify the location of low- and high- pressure systems on each map. Use the key below the maps to the identify the location of warm and cold fronts.

2 Find locations A, B, C, and where you live on the map. For each location, describe how the systems change positions over the two days.

3 What is the cause of and effect on precipitation and temperature at each location?

Apply It

4 The low-pressure system produced several tornadoes. Which location did they occur closest to? Explain.

5 The weather patterns generally move from west to east. Predict the weather on the third day for each location.

6 One day it is clear and sunny, but you notice that the pressure is less than it was the day before. What weather might be coming? Why?

7 🔑 **Key Concept** How does understanding weather patterns help make predicting the weather more accurate?

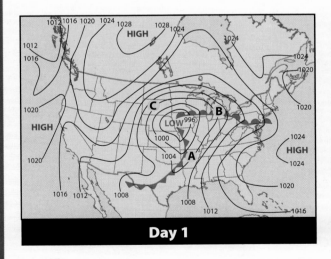

Day 1

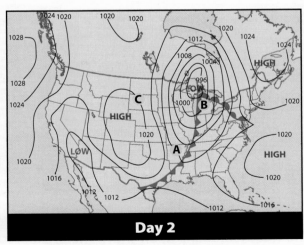

Day 2

▲▲▲▲ Cold front

●●●● Warm front

Lesson 3

Reading Guide

Key Concepts
ESSENTIAL QUESTIONS

- What instruments are used to measure weather variables?
- How are computer models used to predict the weather?

Vocabulary

surface report p. 635

upper-air report p. 635

Doppler radar p. 636

isobar p. 637

computer model p. 638

Weather Forecasts

Inquiry What's inside?

Information about weather variables is collected by the weather radar station shown here. Data, such as the amount of rain falling in a weather system, help meteorologists make accurate predictions about severe weather. What other instruments do meteorologists use to forecast weather? How do they collect and use data?

Launch Lab

Can you understand the weather report?

Weather reports use numbers and certain vocabulary terms to help you understand the weather conditions in a given area for a given time period. Listen to a weather report for your area. Can you record all the information reported?

1. In your Science Journal, make a list of data you would expect to hear in a weather report.

2. Listen carefully to a **recording of a weather report** and jot down numbers and measurements you hear next to those on your list.

3. Listen a second time and make adjustments to your original notes, such as adding more data, if necessary.

4. Listen a third time, then share the weather forecast as you heard it.

Think About This

1. What measurements were difficult for you to apply to understanding the weather report?

2. Why are so many different types of data needed to give a complete weather report?

3. List the instruments that might be used to collect each kind of data.

4. **Key Concept** Where do meteorologists obtain the data they use to make a weather forecast?

Measuring the Weather

Being a meteorologist is like being a doctor. Using specialized instruments and visual observations, the doctor first measures the condition of your body. The doctor later combines these measurements with his or her knowledge of medical science. The result is a forecast of your future health, such as, "You'll feel better in a few days if you rest and drink plenty of fluids."

Similarly, meteorologists, scientists who study weather, use specialized instruments to measure conditions in the atmosphere, as you read in Lesson 1. These instruments include thermometers to measure temperature, barometers to measure air pressure, psychrometers to measure relative humidity, and anemometers to measure wind speed.

Surface and Upper-Air Reports

A **surface report** *describes a set of weather measurements made on Earth's surface.* Weather variables are measured by a weather station—a collection of instruments that report temperature, air pressure, humidity, precipitation, and wind speed and direction. Cloud amounts and visibility are often measured by human observers.

An **upper-air report** *describes wind, temperature, and humidity conditions above Earth's surface.* These atmospheric conditions are measured by a radiosonde (RAY dee oh sahnd), a package of weather instruments carried many kilometers above the ground by a weather balloon. Radiosonde reports are made twice a day simultaneously at hundreds of locations around the world.

Satellite and Radar Images

Images taken from satellites orbiting about 35,000 km above Earth provide information about weather conditions on Earth. A visible light image, such as the one shown in **Figure 14,** shows white clouds over Earth. The infrared image, also shown in **Figure 14,** shows infrared energy in false color. The infrared energy comes from Earth and is stored in the atmosphere as thermal energy. Monitoring infrared energy provides information about cloud height and atmospheric temperature.

Figure 14 Meteorologists use visible light and infrared satellite images to identify fronts and air masses.

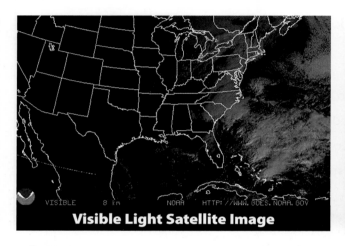

Visible Light Satellite Image

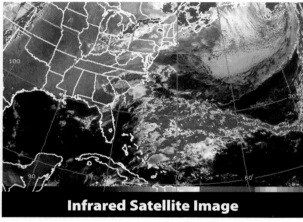

Infrared Satellite Image

✓ **Visual Check** How is an infrared satellite image different from a visible light satellite image?

Radar measures precipitation when radio waves bounce off raindrops and snowflakes. **Doppler radar** *is a specialized type of radar that can detect precipitation as well as the movement of small particles, which can be used to approximate wind speed.* Because the movement of precipitation is caused by wind, Doppler radar can be used to estimate wind speed. This can be especially important during severe weather, such as tornadoes or thunderstorms.

🔑 **Key Concept Check** Identify the weather variables that radiosondes, infrared satellites, and Doppler radar measure.

Weather Maps

Every day, thousands of surface reports, upper-air reports, and satellite and radar observations are made around the world. Meteorologists have developed tools that help them simplify and understand this enormous amount of weather data.

FOLDABLES

Make a horizontal two-tab book and label the tabs as illustrated. Use it to collect information on satellite and radar images. Compare and contrast these information tools.

Weather Satellites | Doppler Radar

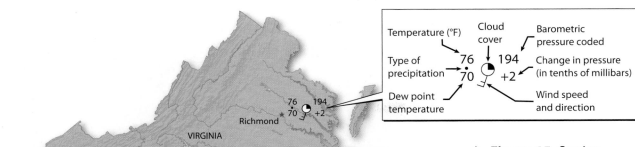

Temperature (°F)
Cloud cover
Barometric pressure coded
Type of precipitation
Change in pressure (in tenths of millibars)
Dew point temperature
Wind speed and direction

76
194
70
+2

▲ **Figure 15** Station models contain information about weather variables.

The Station Model

As shown in **Figure 15,** the station model diagram displays data from many different weather measurements for a particular location. It uses numbers and symbols to display data and observations from surface reports and upper-air reports.

Mapping Temperature and Pressure

In addition to station models, weather maps also have other symbols. For example, **isobars** *are lines that connect all places on a map where pressure has the same value.* Locate an isobar on the map in **Figure 16.** Isobars show the location of high- and low-pressure systems. Isobars also provide information about wind speed. Winds are strong when isobars are close together. Winds are weaker when isobars are farther apart.

In a similar way, isotherms (not shown) are lines that connect places with the same temperature. Isotherms show which areas are warm and which are cold. Fronts are represented as lines with symbols on them, as indicated in **Figure 16.**

✓ **Reading Check** Compare isobars and isotherms.

WORD ORIGIN · · · · · · · · · · ·

isobar
from Greek *isos*, means "equal"; and *baros*, means "heavy"

Weather Map

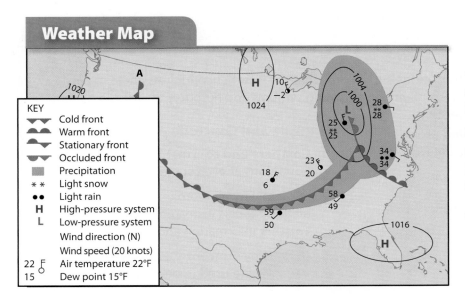

KEY

▽▽	Cold front
△△	Warm front
△▽	Stationary front
▽▽	Occluded front
▨	Precipitation
✳ ✳	Light snow
• •	Light rain
H	High-pressure system
L	Low-pressure system
	Wind direction (N)
	Wind speed (20 knots)
22	Air temperature 22°F
15	Dew point 15°F

◀ **Figure 16** Weather maps contain symbols that provide information about the weather.

✓ **Visual Check** Which symbols represent high-pressure and low-pressure systems?

 **Concepts in Motion**

Animation

Figure 17 Meteorologists analyze data from various sources—such as radar and computer models—in order to prepare weather forecasts.

Predicting the Weather

Modern weather forecasts are made with the help of computer models, such as the ones shown in **Figure 17. Computer models** *are detailed computer programs that solve a set of complex mathematical formulas.* The formulas predict what temperatures and winds might occur, when and where it will rain and snow, and what types of clouds will form.

Government meteorological offices also use computers and the Internet to exchange weather measurements continuously throughout the day. Weather maps are drawn and forecasts are made using computer models. Then, through television, radio, newspapers, and the Internet, the maps and forecasts are made available to the public.

Key Concept Check How are computers used to predict the weather?

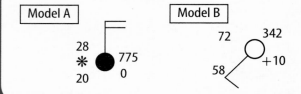

Inquiry MiniLab

20 minutes

How is weather represented on a map?

Meteorologists often use station models to record what the weather conditions are for a particular location. A station model is a diagram containing symbols and numbers that displays many different weather measurements.

Use the **station model legend** provided by your teacher to interpret the data in each station model shown here.

Model A

28
* ● 775
20   0

Model B

72 ◯ 342
        +10
58

Analyze and Conclude

1. **Compare and contrast** the weather conditions at each station model.

2. **Explain** why meteorologists might use station models instead of reporting weather information another way.

3. **Key Concept** Discuss what variables are used to describe weather.

Lesson 3 Review

Visual Summary

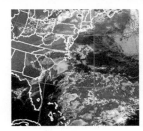

Weather variables are measured by weather stations, radiosondes, satellites, and Doppler radar.

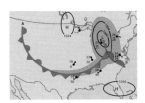

Weather maps contain information in the form of a station model, isobars and isotherms, and symbols for fronts and pressure systems.

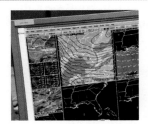

Meteorologists use computer models to help forecast the weather.

FOLDABLES

Use your lesson Foldable to review the lesson. Save your Foldable for the project at the end of the chapter.

What do you think NOW?

You first read the statements below at the beginning of the chapter.

5. Weather variables are measured every day at locations around the world.

6. Modern weather forecasts are done using computers.

Did you change your mind about whether you agree or disagree with the statements? Rewrite any false statements to make them true.

Use Vocabulary

1 **Define** *computer model* in your own words.

2 A line connecting places with the same pressure is called a(n) _____.

3 **Use the term** *surface report* in a sentence.

Understand Key Concepts

4 Which diagram shows surface weather measurements?
  A. an infrared satellite image
  B. an upper air chart
  C. a station model
  D. a visible light satellite image

5 **List** two ways that upper-air weather conditions are measured.

6 **Describe** how computers are used in weather forecasting.

7 **Distinguish** between isobars and isotherms.

Interpret Graphics

8 **Identify** Copy and fill in the graphic organizer below to identify the components of a surface map.

Symbol	Meaning
●▬●▬●	
H	

Critical Thinking

9 **Suggest** ways to forecast the weather without using computers.

10 **Explain** why isobars and isotherms make it easier to understand a weather map.

Materials

graph paper

local weather maps

outdoor thermometer

barometer

Can you predict the weather?

Weather forecasts are important—not just so you are dressed right when you leave the house, but also to help farmers know when to plant and harvest, to help cities know when to call in the snow plows, and to help officials know when and where to evacuate in advance of severe weather.

Ask a Question

Can you predict the weather?

Make Observations

1 Read and complete a lab safety form.

2 Collect weather data daily for a period of one week. Temperature and pressure should be recorded as a number, but precipitation, wind conditions, and cloud cover can be described in words. Make your observations at the same time each day.

3 Graph temperature in degrees and air pressure in millibars on the same sheet of paper, placing the graphs side by side, as shown on the next page. Beneath the graphs, for each day, add notes that describe precipitation, wind conditions, and cloud cover.

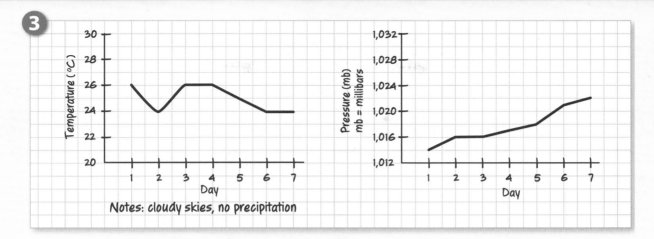

3

Notes: cloudy skies, no precipitation

Form a Hypothesis

4 Examine your data and the weather maps. Look for factors that appear to be related. For example, your data might suggest that when the pressure decreases, clouds follow.

5 Find three sets of data pairs that seem to be related. Form three hypotheses, one for each set of data pairs.

Test Your Hypothesis

6 Look at your last day of data. Using your hypotheses, predict the weather for the next day.

7 Collect weather data the next day and evaluate your predictions.

8 Repeat steps 6 and 7 for at least two more days.

Analyze and Conclude

9 **Analyze** Compare your hypotheses with the results of your predictions. How successful were you? What additional information might have improved your predictions?

10 **The Big Idea** Scientists have more complex and sophisticated tools to help them predict their weather, but with fairly simple tools, you can make an educated guess. Write a one-paragraph summary of the data you collected and how you interpreted it to predict the weather.

Communicate Your Results

For each hypothesis you generated, make a small poster that states the hypothesis, shows a graph that supports it, and shows the results of your predictions. Write a concluding statement about the reliability of your hypothesis. Share your results with the class.

 **Extension**

Investigate other forms of data you might collect and find out how they would help you to make a forecast. Try them out for a week and see if your ability to make predictions improves.

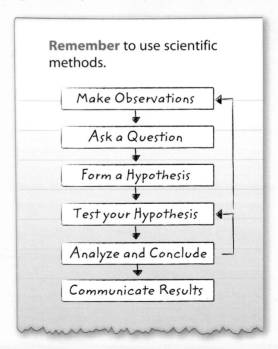

Remember to use scientific methods.

Make Observations → Ask a Question → Form a Hypothesis → Test your Hypothesis → Analyze and Conclude → Communicate Results

WebQuest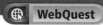

THE BIG IDEA Scientists use weather variables to describe weather and study weather systems. Scientists use computers to predict the weather.

Key Concepts Summary 🔑

Lesson 1: Describing Weather

- **Weather** is the atmospheric conditions, along with short-term changes, of a certain place at a certain time.
- Variables used to describe weather are air temperature, **air pressure**, wind, **humidity**, and **relative humidity**.
- The processes in the water cycle—evaporation, condensation, and **precipitation**—are all involved in the formation of different types of weather.

Vocabulary

weather p. 615
air pressure p. 616
humidity p. 616
relative humidity p. 617
dew point p. 617
precipitation p. 619
water cycle p. 619

Lesson 2: Weather Patterns

- **Low-pressure systems** and **high-pressure systems** are two systems that influence weather.
- Weather patterns are driven by the movement of **air masses**.
- Understanding weather patterns helps make weather forecasts more accurate.
- Severe weather includes thunderstorms, **tornadoes**, **hurricanes**, and **blizzards**.

high-pressure system p. 623
low-pressure system p. 623
air mass p. 624
front p. 626
tornado p. 629
hurricane p. 630
blizzard p. 631

Lesson 3: Weather Forecasts

- Thermometers, barometers, anemometers, radiosondes, satellites, and **Doppler radar** are used to measure weather variables.
- **Computer models** use complex mathematical formulas to predict temperature, wind, cloud formation, and precipitation.

surface report p. 635
upper-air report p. 635
Doppler radar p. 636
isobar p. 637
computer model p. 638

FOLDABLES® Chapter Project

Assemble your lesson Foldables as shown to make a Chapter Project. Use the project to review what you have learned in this chapter.

Use Vocabulary

1 The pressure that a column of air exerts on the area below it is called _____.

2 The amount of water vapor in the air is called _____.

3 The natural process in which water constantly moves among oceans, land, and the atmosphere is called the _____.

4 A(n) _____ is a boundary between two air masses.

5 At the center of a(n) _____, air rises and forms clouds and precipitation.

6 A continental polar _____ brings cold temperatures during winter.

7 When the same _____ passes through two locations on a weather map, both locations have the same pressure.

8 The humidity in the air compared to the amount air can hold is the _____.

Link Vocabulary and Key Concepts

((O) **Concepts in Motion** Interactive Concept Map

Copy this concept map, and then use vocabulary terms from the previous page to complete the concept map.

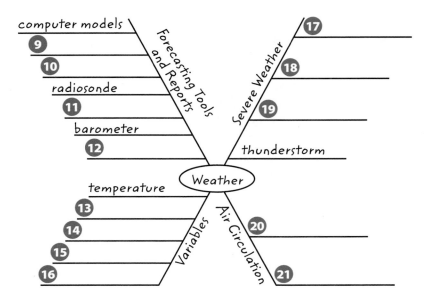

Understand Key Concepts 🔑

1 Clouds form when water changes from
   A. gas to liquid.
   B. liquid to gas.
   C. solid to gas.
   D. solid to liquid.

2 Which type of precipitation reaches Earth's surface as large pellets of ice?
   A. hail
   B. rain
   C. sleet
   D. snow

3 Which of these sinking-air situations usually brings fair weather?
   A. air mass
   B. cold front
   C. high-pressure system
   D. low-pressure system

4 Which air mass contains cold, dry air?
   A. continental polar
   B. continental tropical
   C. maritime tropical
   D. maritime polar

5 Study the front below.

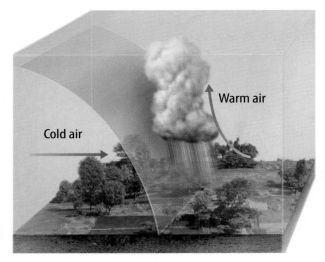

Warm air

Cold air

How does this type of front form?
   A. A cold front overtakes a warm front.
   B. Cold air moves toward warmer air.
   C. The boundary between two fronts stalls.
   D. Warm air moves toward colder air.

6 Which is an intense tropical storm with winds exceeding 119 km/h?
   A. blizzard
   B. hurricane
   C. thunderstorm
   D. tornado

7 Which contains measurements of temperature, air pressure, humidity, precipitation, and wind speed and direction?
   A. a radar image
   B. a satellite image
   C. a surface report
   D. a weather station

8 What does Doppler radar measure?
   A. air pressure
   B. air temperature
   C. the rate at which air pressure changes
   D. the speed at which precipitation travels

9 Study the station model below.

What is the temperature according to the station model?
   A. 3°F
   B. 55°F
   C. 81°F
   D. 138°F

10 Which describes cirrus clouds?
   A. flat, white, and layered
   B. fluffy, at middle altitudes
   C. heaped or piled up
   D. wispy, at high altitudes

11 Which instrument measures wind speed?
   A. anemometer
   B. barometer
   C. psychrometer
   D. thermometer

Critical Thinking

12 Predict Suppose you are on a ship near the equator in the Atlantic Ocean. You notice that the barometric pressure is dropping. Predict what type of weather you might experience.

13 Compare a continental polar air mass with a maritime tropical air mass.

14 Assess why clouds usually form in the center of a low-pressure system.

15 Predict how maritime air masses would change if the oceans froze.

16 Compare two types of severe weather.

17 Interpret Graphics Identify the front on the weather map below. Predict the weather for areas along the front.

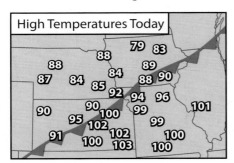

18 Assess the validity of the weather forecast: "Tomorrow's weather will be similar to today's weather."

19 Compare and contrast surface weather reports and upper-air reports. Why is it important for meterologists to monitor weather variables high above Earth's surface?

Writing in Science

20 Write a paragraph about the ways computers have improved weather forecasts. Be sure to include a topic sentence and a concluding sentence.

REVIEW THE **BIG** IDEA

21 Identify the instruments used to measure weather variables.

22 How do scientists use weather variables to describe and predict weather?

23 Describe the factors that influence weather.

24 Use the factors listed in question 23 to describe how a continental polar air mass can change to a maritime polar air mass.

Math Skills

 Review

— **Math Practice** —

Use Conversions

25 Convert from Fahrenheit to Celsius.
   a. Convert 0°F to °C.
   b. Convert 104°F to °C.

26 Convert from Celsius to Fahrenheit.
   a. Convert 0°C to °F.
   b. Convert −40°C to °F.

27 The Kelvin scale of temperature measurement starts at zero and has the same unit size as Celsius degrees. Zero degrees Celsius is equal to 273 kelvin.

Convert 295 K to Fahrenheit.

Standardized Test Practice

Record your answers on the answer sheet provided by your teacher or on a sheet of paper.

Multiple Choice

1 Which measures the average kinetic energy of air molecules?

   A humidity

   B pressure

   C speed

   D temperature

Use the diagram below to answer question 2.

2 Which weather system does the above diagram illustrate?

   A high pressure

   B hurricane

   C low pressure

   D tornado

3 What causes weather to remain the same several days in a row?

   A air front

   B air mass

   C air pollution

   D air resistance

4 Which lists the stages of a thunderstorm in order?

   A cumulus, dissipation, mature

   B cumulus, mature, dissipation

   C dissipation, cumulus, mature

   D dissipation, mature, cumulus

5 What causes air to reach its dew point?

   A decreasing air currents

   B decreasing humidity

   C dropping air pressure

   D dropping temperatures

6 Which measures air pressure?

   A anemometer

   B barometer

   C psychrometer

   D thermometer

Use the diagram below to answer question 7.

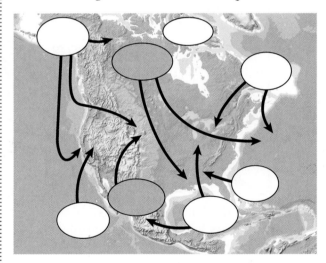

7 Which type of air masses do the shaded ovals in the diagram depict?

   A antarctic

   B arctic

   C continental

   D maritime

8 Which BEST expresses moisture saturation?

   A barometric pressure

   B relative humidity

   C weather front

   D wind direction

Use the diagram below to answer question 9.

Maximum Water Vapor in Air

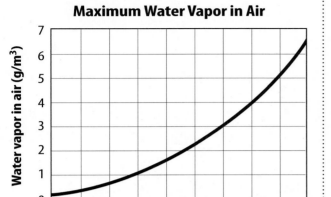

9 What happens to maximum moisture content when air temperatures increase from 15°C to 30°C?

  A increases from 1 to 2 g/m^3

  B increases from 1 to 3 g/m^3

  C increases from 2 to 3 g/m^3

  D increases from 2 to 4 g/m^3

10 When isobars are close together on a weather map,

  A cloud cover is extensive.

  B temperatures are high.

  C warm fronts prevail.

  D winds are strong.

11 Which provides energy for the water cycle?

  A air currents

  B Earth's core

  C ocean currents

  D the Sun

Constructed Response

Use the table below to answer question 12.

Weather Variable	Measurement

12 In the table above, list the variables weather scientists use to describe weather. Then describe the unit of measurement for each variable.

Use the diagram below to answer questions 13 and 14.

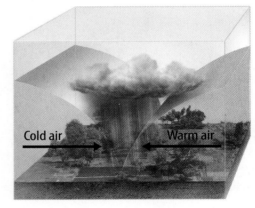

Cold air   Warm air

13 What does the diagram above depict?

14 Describe the weather conditions associated with the diagram.

15 How do weather fronts form?

NEED EXTRA HELP?															
If You Missed Question...	1	2	3	4	5	6	7	8	9	10	11	12	13	14	15
Go to Lesson...	1	2	2	2	1	1, 3	2	1	1	3	1	1	2	2	2

Chapter 18

Climate

 THE BIG IDEA

What is climate and how does it impact life on Earth?

Inquiry **What happened to this tree?**

Climate differs from one area of Earth to another. Some areas have little rain and high temperatures. Other areas have low temperatures and lots of snow. Where this tree grows—on Humphrey Head Point in England—there is constant wind.

- What are the characteristics of different climates?

- What factors affect the climate of a region?

- What is climate and how does it impact life on Earth?

Get Ready to Read

What do you think?

Before you read, decide if you agree or disagree with each of these statements. As you read this chapter, see if you change your mind about any of the statements.

1 Locations at the center of large continents usually have the same climate as locations along the coast.

2 Latitude does not affect climate.

3 Climate on Earth today is the same as it has been in the past.

4 Climate change occurs in short-term cycles.

5 Human activities can impact climate.

6 You can help reduce the amount of greenhouse gases released into the atmosphere.

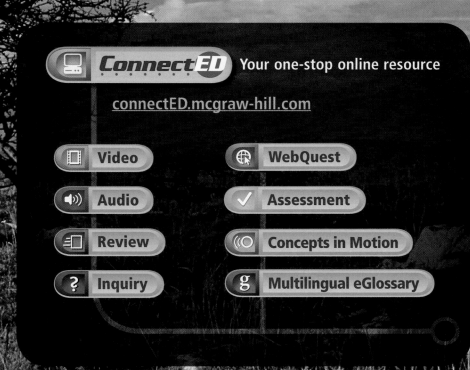

ConnectED Your one-stop online resource

connectED.mcgraw-hill.com

Video

WebQuest

Audio

Assessment

Review

Concepts in Motion

Inquiry

Multilingual eGlossary

Reading Guide

Key Concepts 🔑
ESSENTIAL QUESTIONS

- What is climate?
- Why is one climate different from another?
- How are climates classified?

Vocabulary

climate p. 651
rain shadow p. 653
specific heat p. 653
microclimate p. 655

g Multilingual eGlossary

Climates of Earth

Inquiry **What makes a desert a desert?**

How much precipitation do deserts get? Are deserts always hot? What types of plants grow in the desert? Scientists look at the answers to all these questions to determine if an area is a desert.

How do climates compare?

Climate describes long-term weather patterns for an area. Temperature and precipitation are two factors that help determine climate.

1. Read and complete a lab safety form.
2. Select a location on a **globe.**
3. Research the average monthly temperatures and levels of precipitation for this location.
4. Record your data in a chart like the one shown here in your Science Journal.

Think About This

1. Describe the climate of your selected location in terms of temperature and precipitation.

2. Compare your data to Omsk, Russia. How do the climates differ?

3. **Key Concept** Mountains, oceans, and latitude can affect climates. Do any of these factors account for the differences you observed? Explain.

Omsk, Russia 73.5° E, 55° N		
Month	Average Monthly Temperature	Average Monthly Level of Precipitation
January	−14°C	13 mm
February	−12°C	9 mm
March	−5°C	9 mm
April	8°C	18 mm
May	18°C	31 mm
June	24°C	52 mm
July	25°C	61 mm
August	22°C	50 mm
September	17°C	32 mm
October	7°C	26 mm
November	−4°C	19 mm
December	−12°C	15 mm

What is climate?

You probably already know that the term *weather* describes the atmospheric conditions and short term changes of a certain place at a certain time. The weather changes from day to day in many places on Earth. Other places on Earth have more constant weather. For example, temperatures in Antarctica rarely are above 0°C, even in the summer. Areas in Africa's Sahara, shown in the photo on the previous page, have temperatures above 20°C year-round.

Climate *is the long-term average weather conditions that occur in a particular region.* A region's climate depends on average temperature and precipitation, as well as how these variables change throughout the year.

What affects climate?

Several factors determine a region's climate. The latitude of a location affects climate. For example, areas close to the equator have the warmest climates. Large bodies of water, including lakes and oceans, also influence the climate of a region. Along coastlines, weather is more constant throughout the year. Hot summers and cold winters typically happen in the center of continents. The altitude of an area affects climate. Mountainous areas are often rainy or snowy. Buildings and concrete, which retain solar energy, cause temperatures to be higher in urban areas. This creates a special climate in a small area.

Key Concept Check What is climate?

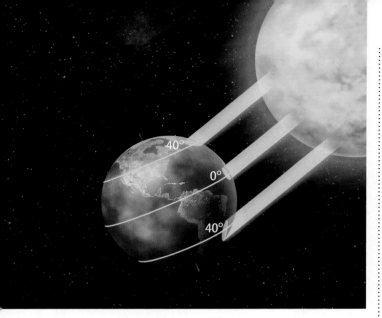

Figure 1 Latitudes near the poles receive less solar energy and have lower average temperatures.

Latitude

Recall that, starting at the equator, latitude increases from 0° to 90° as you move toward the North Pole or the South Pole. The amount of solar energy per unit of Earth's surface area depends on latitude. **Figure 1** shows that locations close to the equator receive more solar energy per unit of surface area annually than locations located farther north or south. This is due mainly to the fact that Earth's curved surface causes the angle of the Sun's rays to spread out over a larger area. Locations near the equator also tend to have warmer climates than locations at higher latitudes. Polar regions are colder because annually they receive less solar energy per unit of surface area. In the middle latitudes, between 30° and 60°, summers are generally hot and winters are usually cold.

Altitude

Climate is also influenced by altitude. Recall that temperature decreases as altitude increases in the troposphere. So, as you climb a tall mountain you might experience the same cold, snowy climate that is near the poles. **Figure 2** shows the difference in average temperatures between two cities in Colorado at different altitudes.

Altitude and Climate 🔑

Figure 2 As altitude increases, temperature decreases.

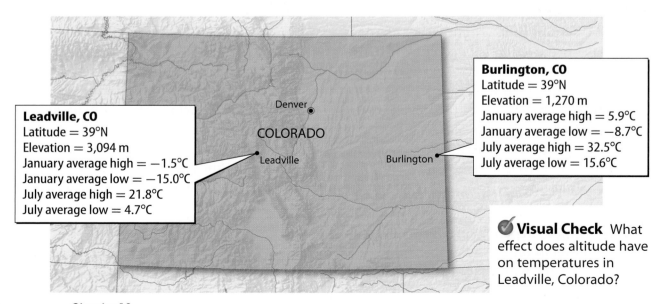

Leadville, CO
Latitude = 39°N
Elevation = 3,094 m
January average high = −1.5°C
January average low = −15.0°C
July average high = 21.8°C
July average low = 4.7°C

Burlington, CO
Latitude = 39°N
Elevation = 1,270 m
January average high = 5.9°C
January average low = −8.7°C
July average high = 32.5°C
July average low = 15.6°C

✅ **Visual Check** What effect does altitude have on temperatures in Leadville, Colorado?

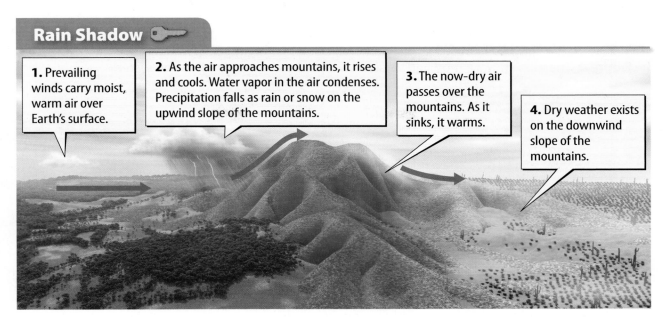

1. Prevailing winds carry moist, warm air over Earth's surface.

2. As the air approaches mountains, it rises and cools. Water vapor in the air condenses. Precipitation falls as rain or snow on the upwind slope of the mountains.

3. The now-dry air passes over the mountains. As it sinks, it warms.

4. Dry weather exists on the downwind slope of the mountains.

Rain Shadows

Mountains influence climate because they are barriers to prevailing winds. This leads to unique **precipitation** patterns called rain shadows. *An area of low rainfall on the downwind slope of a mountain is called a* **rain shadow**, as shown in **Figure 3.** Different amounts of precipitation on either side of a mountain range influence the types of vegetation that grow. Abundant amounts of vegetation grow on the side of the mountain exposed to the precipitation. The amount of vegetation on the downwind slope is sparse due to the dry weather.

Large Bodies of Water

On a sunny day at the beach, why does the sand feel warmer than the water? It is because water has a high specific heat. **Specific heat** *is the amount (joules) of thermal energy needed to raise the temperature of 1 kg of a material by 1°C.* The specific heat of water is about six times higher than the specific heat of sand. This means the ocean water would have to absorb six times as much thermal energy to be the same temperature as the sand.

The high specific heat of water causes the climates along coastlines to remain more constant than those in the middle of a continent. For example, the West Coast of the United States has moderate temperatures year-round.

Ocean currents can also modify climate. The Gulf Stream is a warm current flowing northward along the coast of eastern North America. It brings warmer temperatures to portions of the East Coast of the United States and parts of Europe.

✔️ **Reading Check** How do large bodies of water influence climate?

Figure 3 Rain shadows form on the downwind slope of a mountain.

✅ **Visual Check** Why don't rain shadows form on the upwind slope of mountains?

REVIEW VOCABULARY

precipitation
water, in liquid or solid form, that falls from the atmosphere

Figure 4 The map shows a modified version of Köppen's climate classification system.

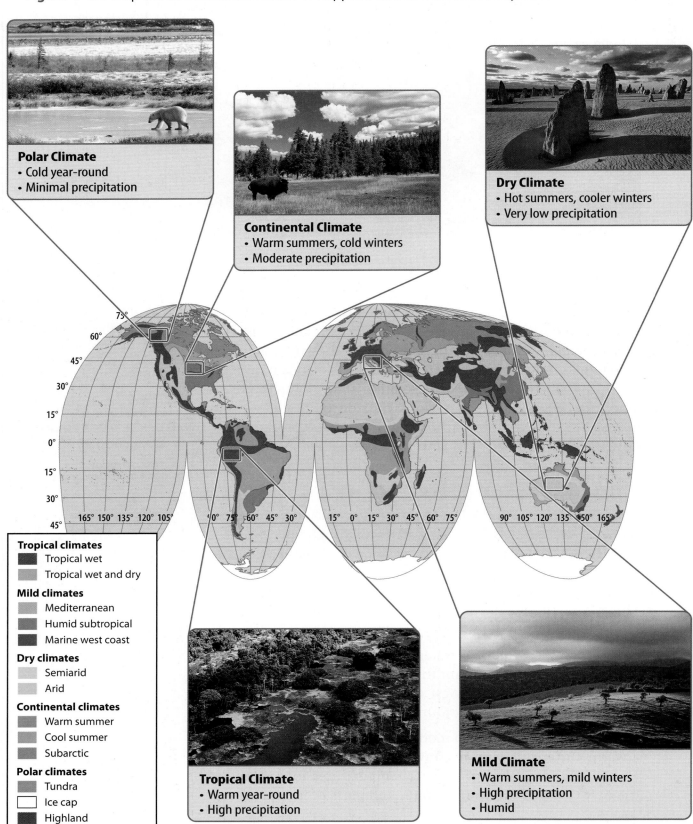

Polar Climate
- Cold year-round
- Minimal precipitation

Continental Climate
- Warm summers, cold winters
- Moderate precipitation

Dry Climate
- Hot summers, cooler winters
- Very low precipitation

Tropical climates
- Tropical wet
- Tropical wet and dry

Mild climates
- Mediterranean
- Humid subtropical
- Marine west coast

Dry climates
- Semiarid
- Arid

Continental climates
- Warm summer
- Cool summer
- Subarctic

Polar climates
- Tundra
- Ice cap
- Highland

Tropical Climate
- Warm year-round
- High precipitation

Mild Climate
- Warm summers, mild winters
- High precipitation
- Humid

Classifying Climates

What is the climate of any particular region on Earth? This can be a difficult question to answer because many factors affect climate. In 1918 German scientist Wladimir Köppen (vlah DEE mihr • KAWP pehn) developed a system for classifying the world's many climates. Köppen classified a region's climate by studying its temperature, precipitation, and native vegetation. Native vegetation is often limited to particular climate conditions. For example, you would not expect to find a warm-desert cactus growing in the cold, snowy arctic. Köppen identified five climate types. A modified version of Köppen's classification system is shown in **Figure 4.**

 Key Concept Check How are climates classified?

Microclimates

Roads and buildings in cities have more concrete than surrounding rural areas. The concrete absorbs solar radiation, causing warmer temperatures than in the surrounding countryside. The result is a common microclimate called the urban heat island, as shown in **Figure 5.** *A* **microclimate** *is a localized climate that is different from the climate of the larger area surrounding it.* Other examples of microclimates include forests, which are often cooler and less windy than the surrounding countryside, and hilltops, which are windier than nearby lower land.

 Key Concept Check Why is one climate different from another?

FOLDABLES®

Use three sheets of notebook paper to make a layered book. Label it as shown. Use it to organize your notes on the factors that determine a region's climate.

Factors that Determine Climate
Latitude
Rain Shadows
Altitude
Water
Local Effects (microclimates)

WORD ORIGIN · · · · · · · · · ·

microclimate
from Greek *mikros*, means "small"; and *klima*, means "region, zone"

· · · · · · · · · · · · ·

Microclimate

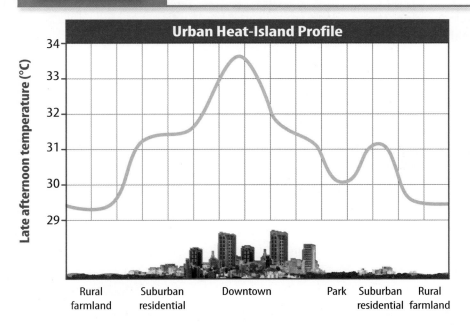

Urban Heat-Island Profile

Late afternoon temperature (°C)

34 — 33 — 32 — 31 — 30 — 29

Rural farmland · Suburban residential · Downtown · Park · Suburban residential · Rural farmland

Figure 5 The temperature is often warmer in urban areas when compared to temperatures in the surrounding countryside.

✓ **Visual Check** What is the temperature difference between downtown and rural farmland?

How Climate Affects Living Organisms

Organisms have adaptations for the climates where they live. For example, polar bears have thick fur and a layer of fat that helps keep them warm in the Arctic. Many animals that live in deserts, such as the camels in **Figure 6,** have adaptations for surviving in hot, dry conditions. Some desert plants have extensive shallow root systems that collect rainwater. Deciduous trees, found in continental climates, lose their leaves during the winter, which reduces water loss when soils are frozen.

Climate also influences humans in many ways. Average temperature and rainfall in a location help determine the type of crops humans grow there. Thousands of orange trees grow in Florida, where the climate is mild. Wisconsin's continental climate is ideal for growing cranberries.

Climate also influences the way humans design buildings. In polar climates, the soil is frozen year-round—a condition called permafrost. Humans build houses and other buildings in these climates on stilts. This is done so that thermal energy from the building does not melt the permafrost.

Figure 6 Camels are adapted to dry climates and can survive up to three weeks without drinking water.

✓ **Reading Check** How are organisms adapted to different climates?

Inquiry) MiniLab 40 minutes

Where are microclimates found?

Microclimates differ from climates in the larger region around them. In this lab, you will identify a microclimate.

1. Read and complete a lab safety form.
2. Select two areas near your school. One area should be in an open location. The other area should be near the school building.
3. Make a data table like the one at the right in your Science Journal.
4. Measure and record data at the first area. Find wind direction using a **wind sock,** temperature using a **thermometer,** and relative humidity using a **psychrometer** and a **relative humidity chart.**
5. Repeat step 4 at the second area.

	Sidewalk	Soccer Fields
Temperature		
Wind direction		
Relative humidity		

Analyze and Conclude

1. **Graph Data** Make a bar graph showing the temperature and relative humidity at both sites.

2. **Use** the data in your table to compare wind direction.

3. **Interpret Data** How did weather conditions at the two sites differ? What might account for these differences?

4. **Key Concept** How might you decide which site is a microclimate? Explain.

Lesson 1 Review

Visual Summary

Climate is influenced by several factors including latitude, altitude, and an area's location relative to a large body of water or mountains.

Rain shadows occur on the downwind slope of mountains.

Microclimates can occur in urban areas, forests, and hilltops.

FOLDABLES®

Use your lesson Foldable to review the lesson. Save your Foldable for the project at the end of the chapter.

What do you think NOW?

You first read the statements below at the beginning of the chapter.

1. Locations at the center of large continents usually have the same climate as locations along the coast.

2. Latitude does not affect climate.

Did you change your mind about whether you agree or disagree with the statements? Rewrite any false statements to make them true.

Use Vocabulary

1 The amount of thermal energy needed to raise the temperature of 1 kg of a material by 1°C is called _____.

2 **Distinguish** between climate and microclimate.

3 **Use the term** *rain shadow* in a sentence.

Understand Key Concepts

4 How are climates classified?
   A. by cold- and warm-water ocean currents
   B. by latitude and longitude
   C. by measurements of temperature and humidity
   D. by temperature, precipitation, and vegetation

5 **Describe** the climate of an island in the tropical Pacific Ocean.

6 **Compare** the climates on either side of a large mountain range.

7 **Distinguish** between weather and climate.

Interpret Graphics

8 **Summarize** Copy and fill in the graphic organizer below to summarize information about the different types of climate worldwide.

Climate Type	Description
Tropical	
Dry	
Mild	
Continental	
Polar	

Critical Thinking

9 **Distinguish** between the climates of a coastal location and a location in the center of a large continent.

10 **Infer** how you might snow ski on the island of Hawaii.

Materials

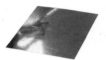

bowl

polyester film

transparent tape

stopwatch

light source

thermometer

Safety

Can reflection of the Sun's rays change the climate?

Albedo is the term used to refer to the percent of solar energy that is reflected back into space. Clouds, for example, reflect about 50 percent of the solar energy they receive, whereas dark surfaces on Earth might reflect as little as 5 percent. Snow has a very high albedo and reflects 75 to 90 percent of the solar energy it receives. The differences in how much solar energy is reflected back into the atmosphere from different regions of Earth can cause differences in climate. Also, changes in albedo can affect the climate of that region.

Learn It

When an observation cannot be made directly, a simulation can be used to draw reasonable conclusions. This strategy is known as **inferring.** Simulating natural occurrences on a small scale can provide indirect observations so realistic outcomes can be inferred.

Try It

1. Read and complete a lab safety form.

2. Make a data table for recording temperatures in your Science Journal.

3. Cover the bottom of a bowl with a sheet of polyester film. Place a thermometer on top of the sheet. Record the temperature in the bottom of the bowl.

4. Put the bowl under the light source and set the timer for 5 minutes. After 5 minutes, record the temperature. Remove the thermometer and allow it to return to its original temperature. Repeat two more times.

5. Repeat the experiment, but this time tape the sheet of polyester film over the top of the bowl and the thermometer.

Apply It

6. **Analyze** the data you collected. What difference did you find when the polyester film covered the bowl?

7. **Conclude** What can you conclude about the Sun's rays reaching the bottom of the bowl when it was covered by the polyester film?

8. **Infer** what happens to the Sun's rays when they reach clouds in the atmosphere. Explain.

9. **Describe** how the high albedo of the ice and snow in the polar regions contribute to the climate there.

10. 🔑 **Key Concept** If a region of Earth were to be covered most of the time by smog or clouds, would the climate of that region change? Explain your answer.

Lesson 2

Reading Guide

Key Concepts
ESSENTIAL QUESTIONS

- How has climate varied over time?
- What causes seasons?
- How does the ocean affect climate?

Vocabulary

ice age p. 660

interglacial p. 660

El Niño/Southern Oscillation p. 664

monsoon p. 665

drought p. 665

g Multilingual eGlossary

Climate Cycles

Inquiry How did this lake form?

A melting glacier formed this lake. How long ago did this happen? What type of climate change occurred to cause a glacier to melt? Will it happen again?

Inquiry Launch Lab

20 minutes

How does Earth's tilted axis affect climate?

Earth's axis is tilted at an angle of 23.5°. This tilt influences climate by affecting the amount of sunlight that reaches Earth's surface.

1. Read and complete a lab safety form.
2. Hold a **penlight** about 25 cm above a sheet of paper at a 90° angle. Use a **protractor** to check the angle.
3. Turn off the overhead lights and turn on the penlight. Your partner should trace the circle of light cast by the penlight onto the paper.
4. Repeat steps 2 and 3, but this time hold the penlight at an angle of 23.5° from perpendicular.

Think About This

1. How did the circles of light change during each trial?
2. Which trial represented the tilt of Earth's axis?
3. 🔑 **Key Concept** How might changes in the tilt of Earth's axis affect climate? Explain.

Figure 7 Scientists study the different layers in an ice core to learn more about climate changes in the past.

Long-Term Cycles

Weather and climate have many cycles. In most areas on Earth, temperatures increase during the day and decrease at night. Each year, the air is warmer during summer and colder during winter. But climate also changes in cycles that take much longer than a lifetime to complete.

Much of our knowledge about past climates comes from natural records of climate. Scientists study ice cores, shown in **Figure 7,** drilled from ice layers in glaciers and ice sheets. Fossilized pollen, ocean sediments, and the growth rings of trees also are used to gain information about climate changes in the past. Scientists use the information to compare present-day climates to those that occurred many thousands of years ago.

✓ **Reading Check** How do scientists find information about past climates on Earth?

Ice Ages and Interglacials

Earth has experienced many major atmospheric and climate changes in its history. **Ice ages** *are cold periods lasting from hundreds to millions of years when glaciers cover much of Earth.* Glaciers and ice sheets advance during cold periods and retreat during **interglacials**—*the warm periods that occur during ice ages or between ice ages.*

Major Ice Ages and Warm Periods

The most recent ice age began about 2 million years ago. The ice sheets reached maximum size about 20,000 years ago. At that time, about half the northern hemisphere was covered by ice. About 10,000 years ago, Earth entered its current inter-glacial period, called the Holocene Epoch.

Temperatures on Earth have fluctuated during the Holocene. For example, the period between 950 and 1100 was one of the warmest in Europe. The Little Ice Age, which lasted from 1250 to about 1850, was a period of bitterly cold temperatures.

 Key Concept Check How has climate varied over time?

Causes of Long-Term Climate Cycles

As the amount of solar energy reaching Earth changes, Earth's climate also changes. One factor that affects how much energy Earth receives is the shape of its orbit. The shape of Earth's orbit appears to vary between elliptical and circular over the course of about 100,000 years. As shown in **Figure 8,** when Earth's orbit is more circular, Earth averages a greater distance from the Sun. This results in below-average tempera-tures on Earth.

Another factor that scientists suspect influences climate change on Earth is changes in the tilt of Earth's axis. The tilt of Earth's axis changes in 41,000-year cycles. Changes in the angle of Earth's tilt affect the range of temperatures throughout the year. For example, a decrease in the angle of Earth's tilt, as shown in **Figure 8,** could result in a decrease in temperature differences between summer and winter. Long-term climate cycles are also influenced by the slow movement of Earth's continents, as well as changes in ocean circulation.

WORD ORIGIN ·············

interglacial
from Latin *inter–*, means "among, between"; and *glacialis*, means "icy, frozen"

Figure 8 This exaggerated image shows how the shape of Earth's orbit varies between elliptical and circular. The angle of the tilt varies from 22° to 24.5° about every 41,000 years. Earth's current tilt is 23.5°.

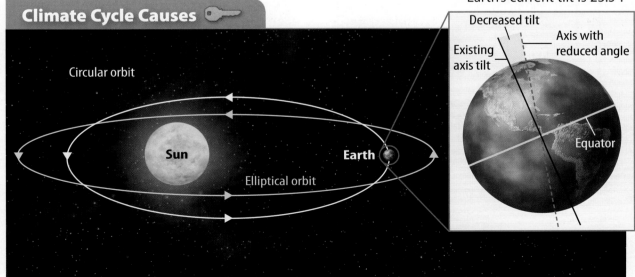

Climate Cycle Causes

Circular orbit

Sun

Earth

Elliptical orbit

Decreased tilt

Axis with reduced angle

Existing axis tilt

Equator

FOLDABLES

Make a horizontal three-tab book and label it as shown. Use your book to organize information about short-term climate cycles. Fold the book into thirds and label the outside *Short Term Climate Cycles.*

Short-term Climate cycles

| Seasons | ENSO | Monsoons |

Short-Term Cycles

In addition to its long-term cycles, climate also changes in short-term cycles. Seasonal changes and changes that result from the interaction between the ocean and the atmosphere are some examples of short-term climate change.

Seasons

Changes in the amount of solar energy received at different latitudes during different times of the year give rise to the seasons. Seasonal changes include regular changes in temperature and the number of hours of day and night.

Recall from Lesson 1 that the amount of solar energy per unit of Earth's surface is related to latitude. Another factor that affects the amount of solar energy received by an area is the tilt of Earth's axis. **Figure 9** shows that when the northern hemisphere is tilted toward the Sun, the angle at which the Sun's rays strike Earth's surface is higher. There are more daylight hours than dark hours. During this time, temperatures are warmer, and the northern hemisphere experiences summer. At the same time, the southern hemisphere is tilted away from the Sun and the angle at which the Sun's rays strike Earth's surface is lower. There are fewer hours of daylight, and the southern hemisphere experiences winter.

Figure 9 shows that the opposite occurs when six months later the northern hemisphere is tilted away from the Sun. The angle at which Sun's rays strike Earth's surface is lower, and temperatures are colder. During this time, the northern hemisphere experiences winter. The southern hemisphere is tilted toward the Sun and the angle between the Sun's rays and Earth's surface is higher. The southern hemisphere experiences summer.

Figure 9 The solar energy rays reaching a given area of Earth's surface is more intense when tilted toward the Sun.

Review   Personal Tutor

🔑 **Key Concept Check** What causes seasons?

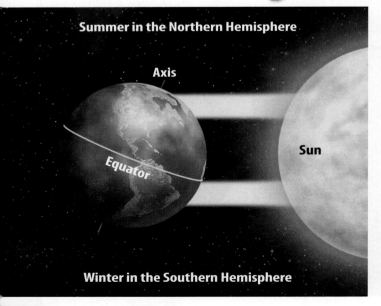

Summer in the Northern Hemisphere

Axis

Equator

Sun

Winter in the Southern Hemisphere

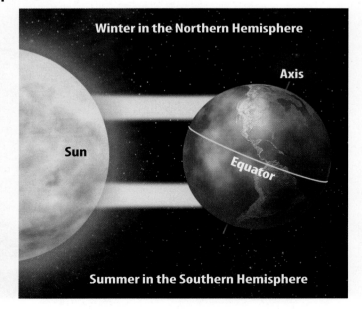

Winter in the Northern Hemisphere

Axis

Sun

Equator

Summer in the Southern Hemisphere

Northern Hemisphere Seasons 🔑

Spring equinox
March 21

Spring

Winter solstice
December 22

Winter

Sun

Summer solstice
June 21

Summer

Fall

Fall equinox
September 23

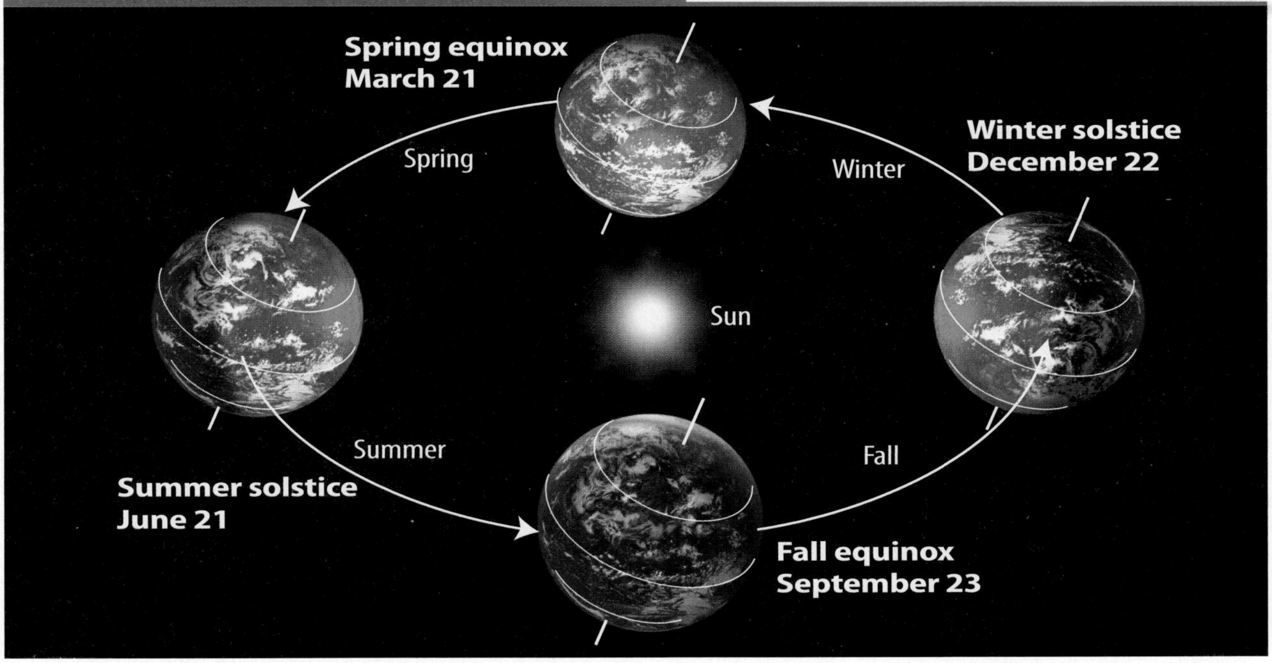

Figure 10 Seasons change as Earth completes its yearly revolution around the Sun.

✅ **Visual Check** How does the amount of sunlight striking the North Pole change from summer to winter?

Solstices and Equinoxes

Earth revolves around the Sun once about every 365 days. During Earth's **revolution,** there are four days that mark the beginning of each of the seasons. These days are a summer solstice, a fall equinox, a winter solstice, and a spring equinox.

As shown in **Figure 10,** the solstices mark the beginnings of summer and winter. In the northern hemisphere, the summer solstice occurs on June 21 or 22. On this day, the northern hemisphere is tilted toward the Sun. In the southern hemisphere, this day marks the beginning of winter. The winter solstice begins on December 21 or 22 in the northern hemisphere. On this day, the northern hemisphere is tilted away from the Sun. In the southern hemisphere, this day marks the beginning of summer.

Equinoxes, also shown in **Figure 10,** are days when Earth is positioned so that neither the northern hemisphere nor the southern hemisphere is tilted toward or away from the Sun. The equinoxes are the beginning of spring and fall. On equinox days, the number of daylight hours almost equals the number of nighttime hours everywhere on Earth. In the northern hemisphere, the spring equinox occurs on March 21 or 22. This is the beginning of fall in the southern hemisphere. On September 22 or 23, fall begins in the northern hemisphere and spring begins in the southern hemisphere.

✅ **Reading Check** Compare and contrast solstices and equinoxes.

SCIENCE USE V. COMMON USE

revolution

Science Use the action by a celestial body of going around in an orbit or an elliptical course

Common Use a sudden, radical, or complete change

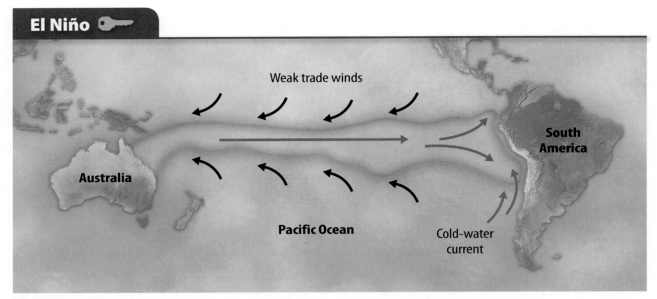

Weak trade winds

South America

Australia

Pacific Ocean

Cold-water current

Figure 11 During El Niño, the trade winds weaken and warm water surges toward South America.

✔ **Visual Check** Where is the warm water during normal conditions?

ACADEMIC VOCABULARY

phenomenon
(noun) an observable fact or event

El Niño and the Southern Oscillation

Close to the equator, the trade winds blow from east to west. These steady winds push warm surface water in the Pacific Ocean away from the western coast of South America. This allows cold water to rush upward from below—a process called upwelling. The air above the cold, upwelling water cools and sinks, creating a high-pressure area. On the other side of the Pacific Ocean, air rises over warm, equatorial waters, creating a low-pressure area. This difference in air pressures across the Pacific Ocean helps keep the trade winds blowing.

As **Figure 11** shows, sometimes the trade winds weaken, reversing the normal pattern of high and low pressures across the Pacific Ocean. Warm water surges back toward South America, preventing cold water from upwelling. This **phenomenon,** called El Niño, shows the connection between the atmosphere and the ocean. During El Niño, the normally dry, cool western coast of South America warms and receives lots of precipitation. Climate changes can be seen around the world. Droughts occur in areas that are normally wet. The number of violent storms in California and the southern United States increases.

✔ **Reading Check** How do conditions in the Pacific Ocean differ from normal during El Niño?

The combined ocean and atmospheric cycle that results in weakened trade winds across the Pacific Ocean is called **El Niño/ Southern Oscillation,** *or ENSO.* A complete ENSO cycle occurs every 3–8 years. The North Atlantic Oscillation (NAO) is another cycle that can change the climate for decades at a time. The NAO affects the strength of storms throughout North America and Europe by changing the position of the jet stream.

Monsoons

Another climate cycle involving both the atmosphere and the ocean is a monsoon. A **monsoon** *is a wind circulation pattern that changes direction with the seasons.* Temperature differences between the ocean and the land cause winds, as shown in **Figure 12**. During summer, warm air over land rises and creates low pressure. Cooler, heavier air sinks over the water, creating high pressure. The winds blow from the water toward the land, bringing heavy rainfall. During winter, the pattern reverses and winds blow from the land toward the water.

The world's largest monsoon is found in Asia. Cherrapunji, India, is one of the world's wettest locations—receiving an average of 10 m of monsoon rainfall each year. Precipitation is even greater during El Niño events. A smaller monsoon occurs in southern Arizona. As a result, weather is dry during spring and early summer with thunderstorms occurring more often from July to September.

 Key Concept Check How does the ocean affect climate?

Droughts, Heat Waves, and Cold Waves

A **drought** *is a period with below-average precipitation.* A drought can cause crop damage and water shortages.

Droughts are often accompanied by heat waves—periods of unusually high temperatures. Droughts and heat waves occur when large hot-air masses remain in one place for weeks or months. Cold waves are long periods of unusually cold temperatures. These events occur when a large continental polar air mass stays over a region for days or weeks. Severe weather of these kinds can be the result of climatic changes on Earth or just extremes in the average weather of a climate.

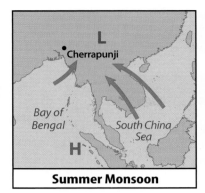
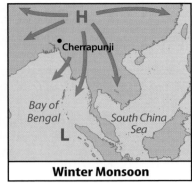

Figure 12 Monsoon winds reverse with the change of seasons.

Inquiry MiniLab   **20 minutes**

How do climates vary?

Unlike El Niño, La Niña is associated with cold ocean temperatures in the Pacific Ocean.

1. As the map shows, average temperatures change during a La Niña winter.
2. The color key shows the range of temperature variation from normal.
3. Find a location on the map. How much did temperatures during La Niña depart from average temperatures?

Temperature Change During La Niña

Temperature change (°C)

Analyze and Conclude

1. **Recognize Cause and Effect** Did La Niña affect the climate in your chosen area?

2. **Key Concept** Describe any patterns you see. How did La Niña affect climate in your chosen area? Use data from the map to support your answer.

Lesson 2 Review

Visual Summary

Scientists learn about past climates by studying natural records of climate, such as ice cores, fossilized pollen, and growth rings of trees.

Long-term climate changes, such as ice ages and interglacials, can be caused by changes in the shape of Earth's orbit and the tilt of its axis.

Short-term climate changes include seasons, El Niño/Southern Oscillation, and monsoons.

FOLDABLES

Use your lesson Foldable to review the lesson. Save your Foldable for the project at the end of the chapter.

What do you think NOW?

You first read the statements below at the beginning of the chapter.

3. Climate on Earth today is the same as it has been in the past.

4. Climate change occurs in short-term cycles.

Did you change your mind about whether you agree or disagree with the statements? Rewrite any false statements to make them true.

Use Vocabulary

1 **Distinguish** an ice age from an interglacial.

2 A(n) _____ is a period of unusually high temperatures.

3 **Define** *drought* in your own words.

Understand Key Concepts

4 What happens during El Niño/Southern Oscillation?
   A. An interglacial climate shift occurs.
   B. The Pacific pressure pattern reverses.
   C. The tilt of Earth's axis changes.
   D. The trade winds stop blowing.

5 **Identify** causes of long-term climate change.

6 **Describe** how upwelling can affect climate.

Interpret Graphics

7 **Sequence** Copy and fill in the graphic organizer below to describe the sequence of events during El Niño/Southern Oscillation.

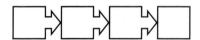

Critical Thinking

8 **Assess** the possibility that Earth will soon enter another ice age.

9 **Evaluate** the relationship between heat waves and drought.

10 **Identify** and explain the climate cycle shown below. Illustrate how conditions change during the summer.

Winter Monsoon

Frozen in Time

Looking for clues to past climates, Lonnie Thompson races against the clock to collect ancient ice from melting glaciers.

Earth's climate is changing. To understand why, scientists investigate how climates have changed throughout Earth's history by looking at ancient ice that contains clues from past climates. Scientists collected these ice samples only from glaciers at the North Pole and the South Pole. Then, in the 1970s, geologist Lonnie Thompson began collecting ice from a new location—the tropics.

◄ Thompson has led expeditions to 15 countries and Antarctica.

Thompson, a geologist from the Ohio State University, and his team scale glaciers atop mountains in tropical regions. On the Quelccaya ice cap in Peru, they collect ice cores—columns of ice layers that built up over hundreds to thousands of years. Each layer is a capsule of a past climate, holding dust, chemicals, and gas that were trapped in the ice and snow during that period.

To collect ice cores, they drill hundreds of feet into the ice. The deeper they drill, the further back in time they go. One core is nearly 12,000 years old!

Collecting ice cores is not easy. The team hauls heavy equipment up rocky slopes in dangerous conditions—icy windstorms, thin air, and avalanche threats. Thompson's greatest challenge is the warming climate. The Quelccaya ice cap is melting. It has shrunk by 30 percent since Thompson's first visit in 1974. It's a race against time to collect ice cores before the ice disappears. When the ice is gone, so are the secrets it holds about climate change.

Thousands of ice core samples are stored in deep freeze at Thompson's lab. One core from Antarctica is over 700,000 years old, which is well before the existence of humans. ▶

Secrets in the Ice

In the lab, Thompson and his team analyze the ice cores to determine

- **Age of ice:** Every year, snow accumulations form a new layer. Layers help scientists date the ice and specific climate events.

- **Precipitation:** Each layer's thickness and composition help scientists determine the amount of snowfall that year.

- **Atmosphere:** As snow turns to ice, it traps air bubbles, providing samples of the Earth's atmosphere. Scientists can measure the trace gases from past climates.

- **Climate events:** The concentration of dust particles helps scientists determine periods of increased wind, volcanic activity, dust storms, and fires.

It's Your Turn

WRITE AN INTRODUCTION Imagine Lonnie Thompson is giving a speech at your school. You have been chosen to introduce him. Write an introduction highlighting his work and achievements.

Lesson 3

Reading Guide

Key Concepts
ESSENTIAL QUESTIONS

- How can human activities affect climate?
- How are predictions for future climate change made?

Vocabulary

global warming p. 670

greenhouse gas p. 670

deforestation p. 671

global climate model p. 673

 g Multilingual eGlossary

▢ Video BrainPOP®

Recent Climate Change

Inquiry Will Tuvalu sink or swim?

This small island sits in the middle of the Pacific Ocean. What might happen to this island if the sea level rose? What type of climate change might cause sea level to rise?

What changes climates?

Natural events such as volcanic eruptions spew dust and gas into the atmosphere. These events can cause climate change.

1. Read and complete a lab safety form.
2. Place a **thermometer** on a sheet of **paper.**
3. Hold a **flashlight** 10 cm above the paper. Shine the light on the thermometer bulb for 5 minutes. Observe the light intensity. Record the temperature in your Science Journal.
4. Use a **rubber band** to secure 3–4 layers of **cheesecloth or gauze** over the bulb end of the flashlight. Repeat step 3.

Think About This

1. Describe the effect of the cheesecloth on the flashlight in terms of brightness and temperature.

2. 🔑 **Key Concept** Would a volcanic eruption cause temperatures to increase or decrease? Explain.

Regional and Global Climate Change

Average temperatures on Earth have been increasing for the past 100 years. As the graph in **Figure 13** shows, the warming has not been steady. Globally, average temperatures were fairly steady from 1880 to 1900. From 1900 to 1945, they increased by about 0.5°C. A cooling period followed, ending in 1975. Since then, average temperatures have steadily increased. The greatest warming has been in the northern hemisphere. However, temperatures have been steady in some areas of the southern hemisphere. Parts of Antarctica have cooled.

✓ **Reading Check** How have temperatures changed over the last 100 years?

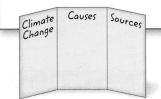

FOLDABLES

Make a tri-fold book from a sheet of paper. Label it as shown. Use it to organize your notes about climate change and the possible causes.

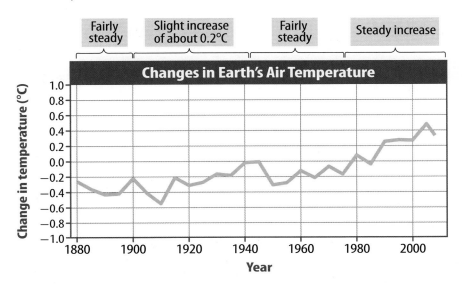

Figure 13 Temperature change has not been constant throughout the past 100 years.

✓ **Visual Check** What 20-year period has seen the most change?

Human Impact on Climate Change

The rise in Earth's average surface temperature during the past 100 years is often referred to as **global warming.** Scientists have been studying this change and the possible causes of it. In 2007, the Intergovernmental Panel on Climate Change (IPCC), an international organization created to study global warming, concluded that most of this temperature increase is due to human activities. These activities include the release of increasing amounts of greenhouse gases into the atmosphere through burning fossil fuels and large-scale cutting and burning of forests. Although many scientists agree with the IPCC, some scientists propose that global warming is due to natural climate cycles.

WORD ORIGIN

deforestation
from Latin *de–*, means "down from, concerning"; and *forestum silvam*, means "the outside woods"

Greenhouse Gases

Gases in the atmosphere that absorb Earth's outgoing infrared radiation are **greenhouse gases.** Greenhouse gases help keep temperatures on Earth warm enough for living things to survive. Recall that this phenomenon is referred to as the greenhouse effect. Without greenhouse gases, the average temperature on Earth would be much colder, about –18°C. Carbon dioxide (CO_2), methane, and water vapor are all greenhouse gases.

Study the graph in **Figure 14.** What has happened to the levels of CO_2 in the atmosphere over the last 120 years? Levels of CO_2 have been increasing. Higher levels of greenhouse gases create a greater greenhouse effect. Most scientists suggest that global warming is due to the greater greenhouse effect. What are some sources of the excess CO_2?

 Reading Check How do greenhouse gases affect temperatures on Earth?

Climate Change

Figure 14 Over the recent past, globally averaged temperatures and carbon dioxide concentration in the atmosphere have both increased.

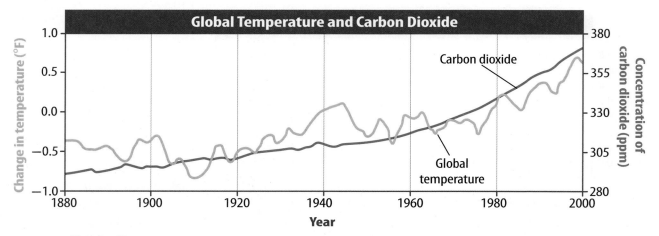

Human-Caused Sources Carbon dioxide enters the atmosphere when fossil fuels, such as coal, oil, and natural gas, burn. Burning fossil fuels releases energy that provides electricity, heats homes and buildings, and powers automobiles.

Deforestation *is the large-scale cutting and/or burning of forests.* Forest land is often cleared for agricultural and development purposes. Deforestation, shown in **Figure 15,** affects global climate by increasing carbon dioxide in the atmosphere in two ways. Living trees remove carbon dioxide from the air during photosynthesis. Cut trees, however, do not. Sometimes cut trees are burned to clear a field, adding carbon dioxide to the atmosphere as the trees burn. According to the Food and Agriculture Organization of the United Nations, deforestation makes up about 25 percent of the carbon dioxide released from human activities.

Natural Sources Carbon dioxide occurs naturally in the atmosphere. Its sources include volcanic eruptions and forest fires. Cellular respiration in organisms contributes additional CO_2.

Aerosols

The burning of fossil fuels releases more than just greenhouse gases into the atmosphere. Aerosols, tiny liquid or solid particles, are also released. Most aerosols reflect sunlight back into space. This prevents some of the Sun's energy from reaching Earth, potentially cooling the climate over time.

Aerosols also cool the climate in another way. When clouds form in areas with large amounts of aerosols, the cloud droplets are smaller. Clouds with small droplets, as shown in **Figure 16,** reflect more sunlight than clouds with larger droplets. By preventing sunlight from reaching Earth's surface, small-droplet clouds help cool the climate.

 Key Concept Check How can human activities affect climate?

▲ **Figure 15** When forests are cut down, trees can no longer use carbon dioxide from the atmosphere. In addition, any wood left rots and releases more carbon dioxide into the atmosphere.

Figure 16 Clouds made up of small droplets reflect more sunlight than clouds made up of larger droplets. ▼

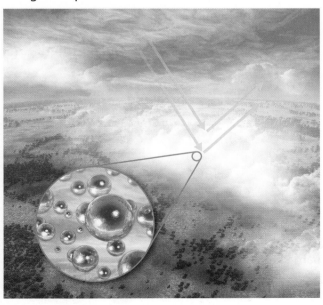

Climate and Society

A changing climate can present serious problems for society. Heat waves and droughts can cause food and water shortages. Excessive rainfall can cause flooding and mudslides. However, climate change can also benefit society. Warmer temperatures can mean longer growing seasons. Farmers can grow crops in areas that were previously too cold. Governments throughout the world are responding to the problems and opportunities created by climate change.

Environmental Impacts of Climate Change

Recall that ENSO cycles can change the amount of precipitation in some areas. Warmer ocean surface temperatures can cause more water to evaporate from the ocean surface. The increased water vapor in the atmosphere can result in heavy rainfall and frequent storms in North and South America. Increased precipitation in these areas can lead to decreased precipitation in other areas, such as parts of southern Africa, the Mediterranean, and southern Asia.

Increasing temperatures can also impact the environment in other ways. Melting glaciers and polar ice sheets can cause the sea level to rise. Ecosystems can be disrupted as coastal areas flood. Coastal flooding is a serious concern for the one billion people living in low-lying areas on Earth.

Extreme weather events are also becoming more common. What effect will heat waves, droughts, and heavy rainfall have on infectious disease, existing plants and animals, and other systems of nature? Will increased CO_2 levels work similarly?

The annual thawing of frozen ground has caused the building shown in **Figure 17** to slowly sink as the ground becomes soft and muddy. Permanently higher temperatures would create similar events worldwide. This and other ecosystem changes can affect migration patterns of insects, birds, fish, and mammals.

Figure 17 Buildings in the Arctic that were built on frozen soil are now being damaged by the constant freezing and thawing of the soil.

Predicting Climate Change

Weather forecasts help people make daily choices about their clothing and activities. In a similar way, climate forecasts help governments decide how to respond to future climate changes.

A **global climate model,** *or GCM, is a set of complex equations used to predict future climates.* GCMs are similar to models used to forecast the weather. GCMs and weather forecast models are different. GCMs make long-term, global predictions, but weather forecasts are short-term and can be only regional predictions. GCMs combine mathematics and physics to predict temperature, amount of precipitation, wind speeds, and other characteristics of climate. Powerful supercomputers solve mathematical equations and the results are displayed as maps. GCMs include the effects of greenhouse gases and oceans in their calculations. In order to test climate models, past records of climate change can and have been used.

✓ **Reading Check** What is a GCM?

One drawback of GCMs is that the forecasts and predictions cannot be immediately compared to real data. A weather forecast model can be analyzed by comparing its predictions with meteorological measurements made the next day. GCMs predict climate conditions for several decades in the future. For this reason, it is difficult to evaluate the accuracy of climate models.

Most GCMs predict further global warming as a result of greenhouse gas emissions. By the year 2100, temperatures are expected to rise by between 1°C and 4°C. The polar regions are expected to warm more than the tropics. Summer arctic sea ice is expected to completely disappear by the end of the twenty-first century. Global warming and sea-level rise are predicted to continue for several centuries.

🔑 **Key Concept Check** How are predictions for future climate change made?

How much CO_2 do vehicles emit?

Much of the carbon dioxide emitted into the atmosphere by households comes from gasoline-powered vehicles. Different vehicles emit different amounts of CO_2.

1. To calculate the amount of CO_2 given off by a vehicle, you must know how many miles per gallon of gasoline the vehicle gets. This information is shown in the chart below.

2. Assume that each vehicle is driven about 15,000 miles annually. Calculate how many gallons each vehicle uses per year. Record your data in your Science Journal in a chart like the one below.

3. One gallon of gasoline emits about 20 lbs of CO_2. Calculate and record how many pounds of CO_2 are emitted by each vehicle annually.

	Estimated MPG	Gallons of Gas Used Annually	Amount of CO_2 Emitted Annually (lbs)
SUV	15		
Hybrid	45		
Compact car	25		

Analyze and Conclude

1. **Compare and contrast** the amount of CO_2 emitted by each vehicle.

2. 🔑 **Key Concept** Write a letter to a person who is planning to buy a vehicle. Explain which vehicle would have the least impact on global warming and why.

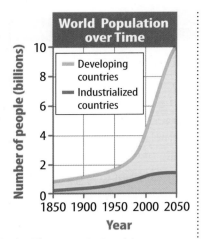

World Population over Time

- Developing countries
- Industrialized countries

Number of people (billions)

1850 1900 1950 2000 2050
Year

▲ **Figure 18** Earth's population is predicted to increase to more than 9 billion people by 2050.

Human Population

In 2000, more than 6 billion people inhabited Earth. As shown in **Figure 18,** Earth's population is expected to increase to 9 billion by the year 2050. What effects will a 50-percent increase in population have on Earth's atmosphere?

It is predicted that by the year 2030, two of every three people on Earth will live in urban areas. Many of these areas will be in developing countries in Africa and Asia. Large areas of forests are already being cleared to make room for expanding cities. Significant amounts of greenhouse gases and other air pollutants will be added to the atmosphere.

 Reading Check How could an increase in human population affect climate change?

Ways to Reduce Greenhouse Gases

People have many options for reducing levels of pollution and greenhouse gases. One way is to develop alternative sources of energy that do not release carbon dioxide into the atmosphere, such as solar energy or wind energy. Automobile emissions can be reduced by as much as 35 percent by using hybrid vehicles. Hybrid vehicles use an electric motor part of the time, which reduces fuel use.

Emissions can be further reduced by green building. Green building is the practice of creating energy-efficient buildings, such as the one shown in **Figure 19.** People can also help remove carbon dioxide from the atmosphere by planting trees in deforested areas.

You can also help control greenhouse gases and pollution by conserving fuel and recycling. Turning off lights and electronic equipment when you are not using them reduces the amount of electricity you use. Recycling metal, paper, plastic, and glass reduces the amount of fuel required to manufacture these materials.

Figure 19 Solar heating, natural lighting, and water recycling are some of the technologies used in green buildings. ▶

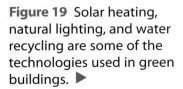

Lesson 3 Review

Visual Summary

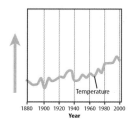

Many scientists suggest that global warming is due to increased levels of greenhouse gases in atmosphere.

Human activities, such as deforestation and burning fossil fuels, can contribute to global warming.

Ways to reduce greenhouse gas emissions include using solar and wind energy, and creating energy-efficient buildings.

FOLDABLES®

Use your lesson Foldable to review the lesson. Save your Foldable for the project at the end of the chapter.

What do you think NOW?

You first read the statements below at the beginning of the chapter.

5. Human activities can impact climate.

6. You can help reduce the amount of greenhouse gases released into the atmosphere.

Did you change your mind about whether you agree or disagree with the statements? Rewrite any false statements to make them true.

Use Vocabulary

1 **Define** *global warming* in your own words.

2 A set of complex equations used to predict future climates is called _____.

3 **Use the term** *deforestation* in a sentence.

Understand Key Concepts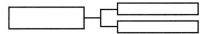

4 Which human activity can have a cooling effect on climate?
- **A.** release of aerosols
- **B.** global climate models
- **C.** greenhouse gas emission
- **D.** large area deforestation

5 **Describe** how human activities can impact climate.

6 **Identify** the advantages and disadvantages of global climate models.

7 **Describe** two ways deforestation contributes to the greenhouse effect.

Interpret Graphics

8 **Determine Cause and Effect** Draw a graphic organizer like the one below to identify two ways burning fossil fuels impacts climate.

Critical Thinking

9 **Suggest** ways you can reduce greenhouse gas emissions.

10 **Assess** the effects of global warming in the area where you live.

Math Skills   Review

— Math Practice —

11 A 32-inch LCD flat-panel TV uses about 125 watts of electricity. If the screen size is increased to 40 inches, the TV uses 200 watts of electricity. What is the percent reduction of electricity if you use a 32-inch TV instead of a 40-inch TV?

Materials

plastic wrap

2 jars with lids

sand

thermometer

desk lamp

stopwatch

rubber band

Safety

The greenhouse effect is a gas!

Human survival on Earth depends on the greenhouse effect. How can you model the greenhouse effect to help understand how it keeps Earth's temperature in balance?

Ask a Question

How will the temperature in a greenhouse compare to that of an open system when exposed to solar energy?

Make Observations

1. Read and complete a lab safety form.

2. Decide which type of container you think will make a good model of a greenhouse. Make two identical models.

3. Place equal amounts of sand in the bottom of each greenhouse.

4. Place a thermometer in each greenhouse in a position where you can read the temperature. Secure it on the wall of the container so you are not measuring the temperature of the sand.

5. Leave one container open, and close the other container.

6. Place the greenhouses under a light source—the Sun or a lamp. Have the light source the same distance from each greenhouse and at the same angle.

7. Read the starting temperature and then every 5–10 minutes for at least three readings. Record the temperatures in your Science Journal and organize them in a table like the one shown on the next page.

Form a Hypothesis

8. Think about some adjustments you could make to your greenhouses to model other components of the greenhouse effect. For example, translucent tops, or white tops, could represent materials that would reflect more light and thermal energy.

9. Based on your observations, form a hypothesis about what materials would most accurately model the greenhouse effect.

Temperature (°C)			
	Reading 1	Reading 2	Reading 3
Greenhouse 1			
Greenhouse 2			

Test Your Hypothesis

10 Set up both greenhouse models in the same way for the hypothesis you are testing. Determine how many trials are sufficient for a valid conclusion. Graph your data to give a visual for your comparison.

Analyze and Conclude

11 Did thermal energy escape from either model? How does this compare to solar energy that reaches Earth and radiates back into the atmosphere?

12 If the greenhouse gases trap thermal energy and keep Earth's temperature warm enough, what would happen if they were not in the atmosphere?

13 If too much of a greenhouse gas, such as CO_2, entered the atmosphere, would the temperature rise?

14 **The Big Idea** If you could add water vapor or CO_2 to your model greenhouses to create an imbalance of greenhouse gases, would this affect the temperature of either system? Apply this to Earth's greenhouse gases.

Communicate Your Results

Discuss your findings with your group and organize your data. Share your graphs, models, and conclusions with the class. Explain why you chose certain materials and how these related directly to your hypothesis.

 **Extension**

Now that you understand the importance of the function of the greenhouse effect, do further investigating into what happens when the balance of greenhouse gases changes. This could result in global warming, which can have a very negative impact on Earth and the atmosphere. Design an experiment that could show how global warming occurs.

Lab Tips

☑ Focus on one concept in designing your lab so you do not get confused with the complexities of materials and data.

☑ Do not add clouds to your greenhouse as part of your model. Clouds are condensed water; water vapor is a gas.

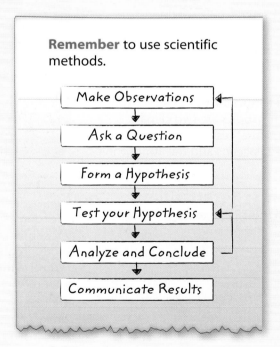

Remember to use scientific methods.

Make Observations → Ask a Question → Form a Hypothesis → Test your Hypothesis → Analyze and Conclude → Communicate Results

Climate is the long-term average weather conditions that occur in an area. Living things have adaptations to the climate in which they live.

Key Concepts Summary 🔑

Vocabulary

Lesson 1: Climates of Earth

- **Climate** is the long-term average weather conditions that occur in a particular region.
- Climate is affected by factors such as latitude, altitude, **rain shadows** on the downwind slope of mountains, vegetation, and the **specific heat** of water.
- Climate is classified based on precipitation, temperature, and native vegetation.

climate p. 651
rain shadow p. 653
specific heat p. 653
microclimate p. 655

Lesson 2: Climate Cycles

- Over the past 4.6 billion years, climate on Earth has varied between **ice ages** and warm periods. **Interglacials** mark warm periods on Earth during or between ice ages.
- Earth's axis is tilted. This causes seasons as Earth revolves around the Sun.
- The **El Niño/Southern Oscillation** and **monsoons** are two climate patterns that result from interactions between oceans and the atmosphere.

ice age p. 660
interglacial p. 660
El Niño/Southern Oscillation p. 664
monsoon p. 665
drought p. 665

Lesson 3: Recent Climate Change

- Releasing carbon dioxide and aerosols into the atmosphere through burning fossil fuels and **deforestation** are two ways humans can affect climate change.
- Predictions about future climate change are made using computers and **global climate models.**

global warming p. 670
greenhouse gas p. 670
deforestation p. 671
global climate model p. 673

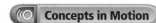
FOLDABLES **Chapter Project**

Assemble your lesson Foldables as shown to make a Chapter Project. Use the project to review what you have learned in this chapter.

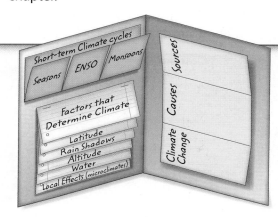

Use Vocabulary

1 A(n) _____ is an area of low rainfall on the downwind slope of a mountain.

2 Forests often have their own _____, with cooler temperatures than the surrounding countryside.

3 The lower _____ of land causes it to warm up faster than water.

4 A wind circulation pattern that changes direction with the seasons is a(n) _____.

5 Upwelling, trade winds, and air pressure patterns across the Pacific Ocean change during a(n) _____.

6 Earth's current _____ is called the Holocene Epoch.

7 A(n) _____ such as carbon dioxide absorbs Earth's infrared radiation and warms the atmosphere.

8 Additional CO_2 is added to the atmosphere when _____ of large land areas occurs.

Link Vocabulary and Key Concepts

Concepts in Motion Interactive Concept Map

Copy this concept map, and then use vocabulary terms from the previous page and other terms in this chapter to complete the concept map.

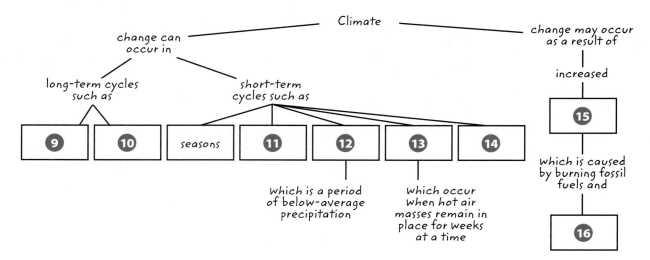

Understand Key Concepts

1 The specific heat of water is _____ than the specific heat of land.

A. higher
B. lower
C. less efficient
D. more efficient

2 The graph below shows average monthly temperature and precipitation of an area over the course of a year.

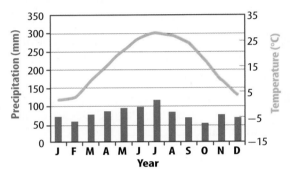

Which is the most likely location of the area?

A. in the middle of a large continent
B. in the middle of the ocean
C. near the North Pole
D. on the coast of a large continent

3 Which are warm periods during or between ice ages?

A. ENSO
B. interglacials
C. monsoons
D. Pacific oscillations

4 Long-term climate cycles are caused by all of the following EXCEPT

A. changes in ocean circulation.
B. Earth's revolution of the Sun.
C. the slow movement of the continents.
D. variations in the shape of Earth's orbit.

5 A rain shadow is created by which factor that affects climate?

A. a large body of water
B. buildings and concrete
C. latitude
D. mountains

6 During which event do trade winds weaken and the usual pattern of pressure across the Pacific Ocean reverses?

A. drought
B. El Niño/Southern Oscillation event
C. North Atlantic Oscillation event
D. volcanic eruption

7 The picture below shows Earth as it revolves around the Sun.

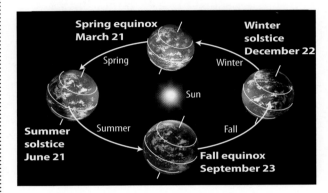

Which season is it in the southern hemisphere in July?

A. fall
B. spring
C. summer
D. winter

8 Which is not a greenhouse gas?

A. carbon dioxide
B. methane
C. oxygen
D. water vapor

9 Which cools the climate by preventing sunlight from reaching Earth's surface?

A. aerosols
B. greenhouse gases
C. lakes
D. water vapor molecules

10 Which action can reduce greenhouse gas emissions?

A. building houses on permafrost
B. burning fossil fuels
C. cutting down forests
D. driving a hybrid vehicle

Critical Thinking

11 Hypothesize how the climate of your town would change if North America and Asia moved together and became one enormous continent.

12 Interpret Graphics Identify the factor that affects climate, as shown in this graph. How does this factor affect climate?

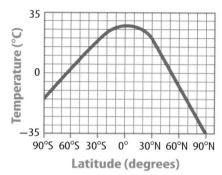

13 Diagram Draw a diagram that explains the changes that occur during an El Niño/ Southern Oscillation event.

14 Evaluate which would cause more problems for your city or town: a drought, a heat wave, or a cold wave. Explain.

15 Recommend a life change you could make if the climate in your city were to change.

16 Formulate your opinion about the cause of global warming. Use facts to support your opinion.

17 Predict the effects of population increase on the climate where you live.

18 Compare how moisture affects the climates on either side of a mountain range.

Writing in Science

19 Write a short paragraph that describes a microclimate near your school or your home. What is the cause of the microclimate?

REVIEW THE BIG IDEA

20 What is climate? Explain what factors affect climate and give three examples of different types of climate.

21 Explain how life on Earth is affected by climate.

Math Skills

Use Percentages

22 Fred switches from a sport-utility vehicle that uses 800 gal of gasoline a year to a compact car that uses 450 gal.

  a. By what percent did Fred reduce the amount of gasoline used?

  b. If each gallon of gasoline released 20 pounds of CO_2, by what percent did Fred reduce the released CO_2?

23 Of the 186 billion tons of CO_2 that enter Earth's atmosphere each year from all sources, 6 billion tons are from human activity. If humans reduced their CO_2 production by half, what percentage decrease would it make in the total CO_2 entering the atmosphere?

Standardized Test Practice

Record your answers on the answer sheet provided by your teacher or on a sheet of paper.

Multiple Choice

1 Which is a drawback of a global climate model?

   A Its accuracy is nearly impossible to evaluate.

   B Its calculations are limited to specific regions.

   C Its predictions are short-term only.

   D Its results are difficult to interpret.

Use the diagram below to answer question 2.

2 What kind of climate would you expect to find at position 4?

   A mild

   B continental

   C tropical

   D dry

3 The difference in air temperature between a city and the surrounding rural area is an example of a(n)

   A inversion.

   B microclimate.

   C seasonal variation.

   D weather system.

4 Which does NOT help explain climate differences?

   A altitude

   B latitude

   C oceans

   D organisms

5 What is the primary cause of seasonal changes on Earth?

   A Earth's distance from the Sun

   B Earth's ocean currents

   C Earth's prevailing winds

   D Earth's tilt on its axis

Use the diagram below to answer question 6.

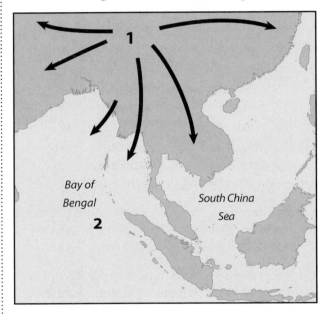

6 In the above diagram of the Asian winter monsoon, what does 1 represent?

   A high pressure

   B increased precipitation

   C low temperatures

   D wind speed

7 Climate is the _____ average weather conditions that occur in a particular region. Which completes the definition of *climate*?

   A global

   B long-term

   C mid-latitude

   D seasonal

Use the diagram below to answer question 8.

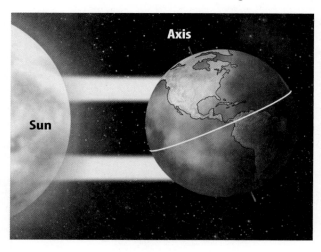

8 In the diagram above, what season is North America experiencing?

A fall

B spring

C summer

D winter

9 Which climate typically has warm summers, cold winters, and moderate precipitation?

A continental

B dry

C polar

D tropical

10 Which characterizes interglacials?

A earthquakes

B monsoons

C precipitation

D warmth

Constructed Response

Use the diagram below to answer question 11.

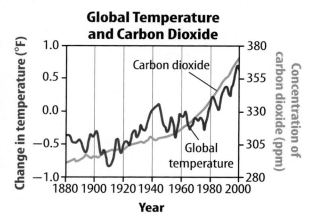

11 Compare the lines in the graph above. What does this graph suggest about the relationship between global temperature and atmospheric carbon dioxide?

Use the table below to answer questions 12 and 13.

Human Sources	Natural Sources

12 List two human and three natural sources of carbon dioxide. How do the listed human activities increase carbon dioxide levels in the atmosphere?

13 Which human activity listed in the table above also produces aerosols? What are two ways aerosols cool Earth?

NEED EXTRA HELP?													
If You Missed Question...	1	2	3	4	5	6	7	8	9	10	11	12	13
Go to Lesson...	3	1	1	1	2	2	1	2	1	2	3	3	3

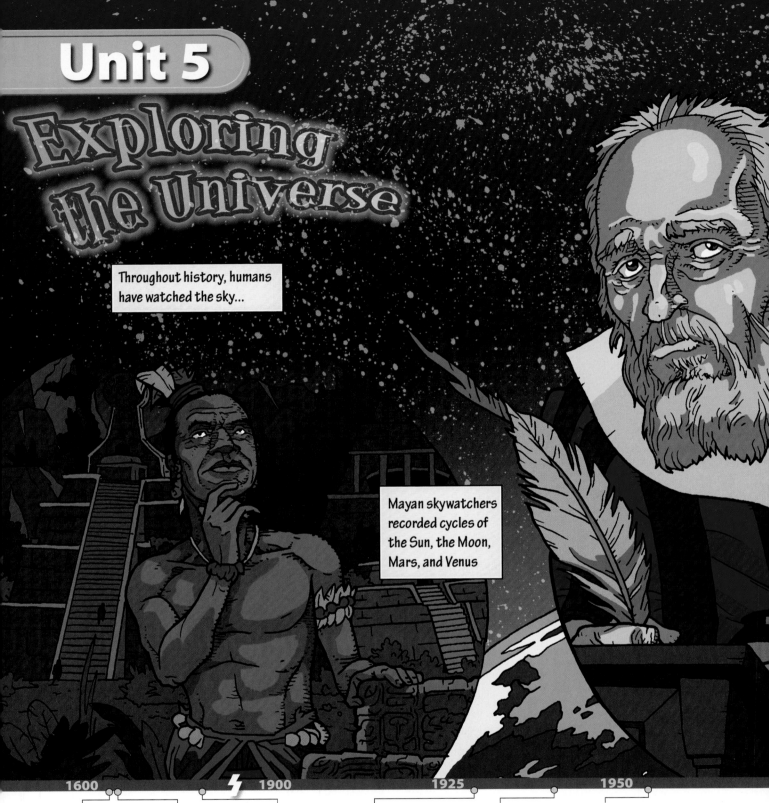

Unit 5

Exploring The Universe

Throughout history, humans have watched the sky...

Mayan skywatchers recorded cycles of the Sun, the Moon, Mars, and Venus

1600 **1900** **1925** **1950**

1608
A Dutch scientist named Hans Lippershey invents the first telescope. Galileo improves on the design soon after.

1610
Galileo observes the moons of Jupiter, Saturn's rings, individual stars in the Milky Way, and the phases of Venus via telescope.

1687
Sir Isaac Newton publishes a book outlining the law of universal gravitation and laws for the motions of the planets.

1926
Robert H. Goddard launches the first liquid-fueled rocket. Its flight lasts less than 3 seconds, but sets the stage for the U.S. rocket program.

1942
The German VS-2 rocket is the first human-made object to achieve suborbital spaceflight.

1957
The *Sputnik 1* spacecraft is the first artificial satellite to successfully orbit around Earth.

...Galileo built the first astronomical telescope to get a closer look at the heavens from Earth.

...Today, astronomers can send telescopes into space to see far beyond our world.

1961
Yuri Gagarin becomes the first human to enter space and the first to orbit Earth.

1969
Neil Armstrong is the first person to walk on the Moon, followed by Edwin "Buzz" Aldrin.

1977
Voyager 2 spacecraft is launched to explore the outer planets of the solar system, including Jupiter, Saturn, Uranus, and Neptune.

1990
The *Hubble Space Telescope* is carried into orbit by a space shuttle. This mission is the first for NASA's Great Observatories Program and is the largest, most versatile, and well-known of the space telescopes.

? Inquiry
Visit ConnectED for this unit's **STEM** activity.

Technology

It may sound strange, but some of the greatest benefits of the space program are benefits to life here on Earth. Devices ranging from hand-held computers to electric socks rely on technologies first developed for space exploration. **Technology** is the practical application of science to commerce or industry. Space technologies have increased our understanding of Earth and our ability to locate and conserve resources.

Problems, such as how best to explore the solar system and outer space, often send scientists on searches for new knowledge. Engineers use scientific knowledge to develop new technologies for space. Then, some of those technologies are modified to solve problems on Earth. For example, lightweight solar panels on the outside of a spacecraft convert the Sun's energy into electricity that powers the spacecraft for long space voyages. Similar but smaller, flexible solar panels, as shown in **Figure 1** are now available for consumers to purchase. They can be used to power small electronics when traveling. **Figure 2** shows how other technologies from space help conserve natural resources.

Figure 1 Lightweight, flexible solar cells developed for spacecraft help to conserve Earth's resources.

This image was taken by the *Terra* satellite and shows fires burning in California. The image helps firefighters see the size and the location of the fires. It also helps scientists study the effect of fires on Earth's atmosphere. ▼

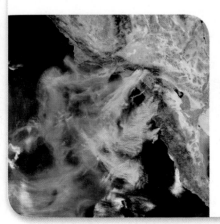

Some portable water purification kits use technologies developed to provide safe, clean drinking water for astronauts. This kit can provide clean, safe drinking water for an entire village in a remote area or supply drinking water after a natural disaster.

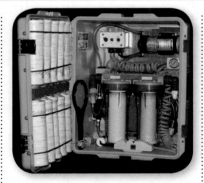

Engineers developed glass spheres about the size of a grain of flour to insulate super-cold spacecraft fuel lines. Similar microspheres act as insulators when mixed with paints. This technology can help reduce the energy needed to heat and cool buildings. ▼

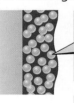

Wet paint often is mixed with tiny ceramic microspheres.

As the paint dries and the water evaporates, the microspheres pack together tightly, creating a layer of insulation.

Figure 2 Some technologies developed as part of the space program have greatly benefited life on Earth.

Figure 3 The satellite image on the left is similar to what you would see with your eyes from space. A satellite sensor that detects other wavelengths of light produced the colored satellite image on the right. It shows the locations of nearly a dozen different minerals.

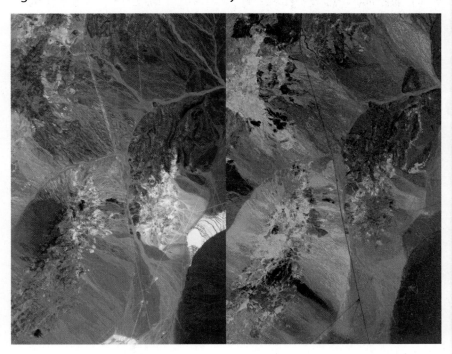

Solving Problems and Improving Abilities

Science and technology depend on each other. For example, images from space greatly improve our understanding of Earth. **Figure 3** above shows a satellite image of a Nevada mine. The satellite is equipped with sensors that detect visible light, much like your eyes do. The image on the right shows a satellite image of the same site taken with a sensor that detects wavelengths of light your eyes cannot see. This image provides information about the types of minerals in the mine. Each color in the image on the right shows the location of a different mineral, reducing the time it takes geologists to locate mineral deposits.

Scientists use other kinds of satellite sensors for different purposes. Engineers have modified space technology to produce satellite images of cloud cover over Earth's surface, as shown in **Figure 4.** Images like this one improve global weather forecasting and help scientists understand changes in Earth's atmosphere. Of course, science can answer only some of society's questions, and technology cannot solve all problems. But together, they can improve the quality of life for all.

Figure 4 This satellite image shows reflection of sunlight (yellow), deep clouds (white), low clouds (pale yellow), high clouds (blue), vegetation (green), and sea (dark).

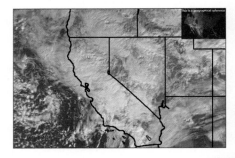

Chapter 19

Motion, Forces, and
Newton's Laws

THE BIG IDEA **In what ways do forces affect an object's motion?**

inquiry **How did they get up there?**

When you stop a video of moving acrobats, they sometimes look as if they are frozen in the air. A still photo of an acrobat in midair can help you analyze exactly what is happening.

- What are some ways you could describe the motion of the acrobats in the air?

- What caused the acrobats to fly high into the air?

- In what ways do forces affect the motion of the acrobats?

Get Ready to Read

What do you think?

Before you read, decide if you agree or disagree with each of these statements. As you read this chapter, see if you change your mind about any of the statements.

1 You must use a reference point to describe an object's motion.

2 An object that is accelerating must be speeding up.

3 Objects must be in contact with one another to exert a force.

4 Gravity is a force that depends on the masses of two objects and the distance between them.

5 All forces change the motion of objects.

6 The net force on an object is equal to the mass of the object times the acceleration of the object.

ConnectED Your one-stop online resource

connectED.mcgraw-hill.com

Video	**WebQuest**
Audio	**Assessment**
Review	**Concepts in Motion**
Inquiry	**Multilingual eGlossary**

Describing Motion

Reading Guide

Key Concepts 🔑
ESSENTIAL QUESTIONS

- What information do you need to describe the motion of an object?
- How are speed, velocity, and acceleration related?
- How can a graph help you understand the motion of an object?

Vocabulary

motion p. 691

reference point p. 691

distance p. 692

displacement p. 692

speed p. 692

velocity p. 693

acceleration p. 694

 Multilingual eGlossary

 Video BrainPOP®

WINNER EVERY GAME

SMALL
MEDIUM
LARGE
JUMBO
CHOICE

ONE
BALL $2.00

THREE
BALLS $5.00

VERY BA INNER!

Inquiry **Where is the white ball?**

In an arcade, many games involve something moving. Objects speed up, slow down, and change direction. How would you describe the position of the white ball in this game at any moment in time? How is its motion different from the motion of the other balls? What words could you use to describe the motion of the ball?

How can you describe motion?

You see things move in many ways each day. You might see a train moving along a track or raindrops falling to the ground. What information do you need to describe an object's motion?

1. Read and complete a lab safety form.
2. Choose a **small object,** such as a ball or a pencil. Move the object in some way.
3. Have a partner write a short description of the movement in the Science Journal.
4. Exchange objects and descriptions with several other pairs of students. Each time, use the description to try to duplicate the original motion.

Think About This

1. **Contrast** Why were some descriptions more useful than others when you tried to duplicate the motion?

2. 🔑 **Key Concept** What information do you think you need to accurately describe an object's motion?

Motion

Suppose you have been playing a shuffleboard game in an arcade. You decide to try something new, so you walk to a racing game. As you walk to the new game, your position in the room changes. **Motion** *is the process of changing position.* If the games are 5 m apart, you could say that your position changed by 5 m.

Motion and Reference Points

How would you describe your motion to a friend? You could say that you walked 5 m away from the shuffleboard game. Or you could say that you moved 5 m toward the racing game. *The starting point you use to describe the motion or the position of an object is called the* **reference point.** Motion is described differently depending on the reference point you choose. You can choose any point as a reference point. Both the racing game and the shuffleboard game can be reference points.

In addition to using a reference point to describe motion, you also need a direction. For example, the puck is moving away from the girl in **Figure 1.** Other descriptions of direction might include east or west, or up or down.

✔ **Reading Check** Describe your motion as you walk from your desk to the door. Use a reference point and a direction.

Figure 1 🔑 A description of the motion of the puck depends on the reference point you choose.

✔ **Visual Check** Name three different reference points you could choose in order to describe the motion of the puck.

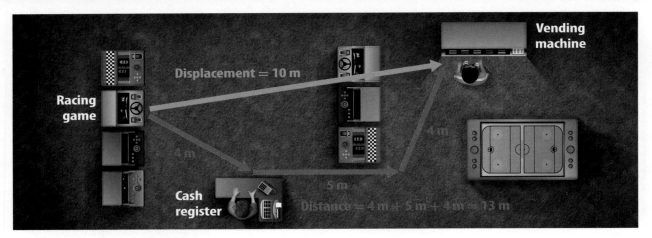

▲ **Figure 2** The distance traveled and the displacement from the game to the vending machine differ.

Distance and Displacement

Suppose you finish playing the racing game, and you go to the cash register to get more tokens. Then, you go to the vending machine for a snack. Your path is shown by the red arrows in **Figure 2.** How far did you travel? **Distance** *is the total length of your path.* The total distance you traveled is 4 m + 5 m + 4 m = 13 m.

Your **displacement** *is the distance between your initial, or starting, position and your final position.* Displacement is represented with a straight arrow extending from the starting point to the ending point. The displacement between the racing game where you started and the vending machine where you stopped is shown by the blue arrow in **Figure 2.** Your displacement is 10 m. To give a complete description of your motion, you must include a reference point, your displacement, and your direction from the reference point.

 Key Concept Check What information do you need to describe an object's motion?

Speed

Suppose you run out of tokens and leave the arcade. Walking slowly, it takes a long time to get to the end of the block. When you realize that you need to meet a friend at the library in 15 minutes, you start to run. When running, you travel the distance of the next block in a shorter time. How does your motion differ in the two blocks? Since you traveled the second block in less time than the first, your speed was different. **Speed** *is the distance an object moves divided by the time it took to move that distance.*

Constant and Changing Speed

Speed can be constant or changing. Look at **Figure 3.** The stopwatches above the girl show her motion every second for 6 seconds. In the first 4 seconds, the girl moves with constant, or unchanging, speed because she travels the same distance during each second. When the girl starts running, the distance she travels each second gets larger and larger. The girl's speed changes.

▼ **Figure 3** 🗝 The girl's speed begins to change between seconds 4 and 5.

Average Speed

Suppose you want to figure out how fast you ran from the arcade to the library. As you ran, your speed probably changed from second to second. Therefore, in order to describe the speed you traveled, you describe the average speed of the entire trip. Average speed is the ratio of the distance an object moves to the time it takes for the object to move that distance. If it takes you 15 minutes, or 0.25 h, to run the 1 km to the library, your average speed was 1 km/0.25 h, or 4 km/h.

Velocity

If you tell your friend that you traveled about 4 km/h, you are describing your speed. You could give your friend a better description of your motion if you also told him or her the direction in which you are moving. **Velocity** is *the speed and direction of an object's motion.*

Often, velocity is shown by using an arrow, as shown in **Figure 4.** The length of the arrow represents the speed of an object, while the direction in which the arrow points represents the direction in which the object is moving.

Constant Velocity

Velocity is constant, or does not change, when an object's speed and direction of movement do not change. If you use an arrow to describe velocity, you can divide the arrow into segments to show whether velocity is constant. Look at the skateboarding arrow in **Figure 4.** Each segment of the arrow shows the distance and the direction you move in a given unit of time. Because each segment is the same length, you are moving the same distance and in the same direction during each interval of time. Because both your speed and direction of movement are constant, you are moving at a constant velocity.

WORD ORIGIN · · · · · · · · · · · ·

velocity
from Latin *velocitatem*, means "swiftness or speed"

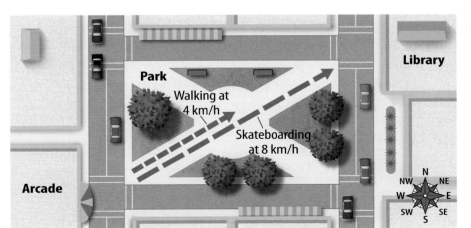

Figure 4 🔑 Your skateboarding velocity is greater than your walking velocity. Both velocities are constant because they represent a constant speed in a constant direction.

Park

Walking at 4 km/h

Skateboarding at 8 km/h

Library

Arcade

N NE NW W E SW SE S

| Speed changes, direction remains constant | Speed remains constant, direction changes | Speed changes, direction changes |

0 m/s — 0 s
10 m/s — 1 s
20 m/s — 2 s
30 m/s — 3 s
40 m/s — 4 s
50 m/s — 5 s

Figure 5 🗝 The velocity of an object changes if the speed changes, the direction changes, or both the speed and the direction change.

(Inquiry) MiniLab
15 minutes

How can velocity change?
The velocity of an object can change in two ways. Can you recognize the ways velocity changes?

1. Read and complete a lab safety form.

2. Toss a **one-hole stopper** to your partner. Observe and record the motion of the stopper.

3. Hold the stopper above a table. Release it. Record your observations in your Science Journal.

4. Tie one end of a **50-cm string** to the stopper. Gently swing the stopper at a constant speed in a horizontal circle near the floor.

Analyze and Conclude

1. **Analyze** the speed and the direction of the stopper each time you moved it. Which of these changed and which stayed the same each time?

2. 🗝 **Key Concept** How were changes in the motion of the stopper related to changes in velocity?

Changing Velocity

Velocity can change even if the speed of an object remains constant. Recall that velocity includes both an object's speed and its direction of travel. **Figure 5** shows several examples of changing velocity.

In the first panel, the ball drops toward the ground in a straight line, or constant direction. The increased length of each arrow shows that the speed of the ball increases as it falls. As speed changes, velocity changes.

In the second panel, each arrow is the same length. This tells you that the Ferris-wheel cars travel around a circle at a constant speed. However, each arrow points in a different direction. This tells you that the cars are changing direction. As direction changes, velocity changes.

The third panel of **Figure 5** shows the path of a ball thrown into the air. The arrows show that both the ball's speed and direction change, so its velocity changes.

When either an object's speed or velocity changes, the object is accelerating. **Acceleration** *is the measure of how quickly the velocity of an object changes.*

🗝 **Key Concept Check** Can an object traveling at a constant speed have a changing velocity? Why or why not?

Calculating Acceleration

When a ball is dropped, as in the first panel of **Figure 5,** its speed increases as it falls toward the ground. The velocity of the ball is changing. Therefore, the ball is accelerating. You can calculate average acceleration using the following equation:

$$\overline{a} = \frac{v_f - v_i}{t}$$

Notice that this equation refers only to a change in speed, not direction. The symbol for average acceleration is $\overline{a}$. The symbol v_f represents the final velocity, and the symbol v_i represents the initial, or starting, velocity. The symbol t stands for the time it takes to make that change in velocity.

 Key Concept Check How does acceleration differ from velocity?

Positive Acceleration

When an object, such as a falling ball, speeds up, its final velocity is greater than its initial velocity. If you calculate the ball's average acceleration, the numerator (final velocity minus initial velocity) is positive, so the average acceleration is positive. In other words, when an object speeds up, it has positive acceleration.

Negative Acceleration

If a ball is thrown straight up into the air, it slows down as it travels upward. The initial velocity of the ball is greater than its final velocity. The numerator in the equation is negative, so the average acceleration is negative. In other words, as an object slows down, it has negative acceleration. Some people refer to this as deceleration.

Math Skills ×÷+ **Solve One-Step Equation**

Solve for Average Acceleration A skateboarder moves at 2 m/s as he or she starts moving down a ramp. As the skateboarder heads down the ramp, he or she accelerates to a speed of 6 m/s in 4 seconds. What is the skateboarder's acceleration?

1 **This is what you know:**

final velocity:	$v_f = 6$ m/s
initial velocity:	$v_i = 2$ m/s
time:	$t = 4$ s

2 **This is what you need to find out:** average acceleration: $\overline{a}$

3 **Use this formula:**

$$\overline{a} = \frac{v_f - v_i}{t}$$

4 **Substitute:**
the values for v_f, v_i, and t

$$\frac{6 \text{ m/s} - 2 \text{ m/s}}{4 \text{ s}}$$

subtract

$$\frac{4 \text{ m/s}}{4 \text{ s}}$$

and divide

$$= 1 \text{ m/s}^2$$

Answer: The average acceleration is 1 m/s².

 Review

- **Math Practice**
- **Personal Tutor**

Practice

As the skateboarder starts moving up the other side of the ramp, his or her velocity drops from 6 m/s to 0 m/s in 3 seconds. What was his or her acceleration?

Using Graphs to Represent Motion

How can you track the motion of an animal that can move hundreds of miles without being seen by humans? In order to understand the movements of animals, such as the polar bear in **Figure 6,** biologists put tracking devices on them. These devices constantly send information about the position of the animal to **satellites.** Biologists download the data from the satellites and create graphs of motion such as those shown in **Figures 7** and **8.**

Displacement-Time Graphs

Figure 7 is a displacement-time graph of a polar bear's motion. The x-axis shows the time, and the y-axis shows the displacement of the polar bear from a reference point.

The line on a displacement-time graph represents the average speed the bear at that particular moment in time. It does not show the actual path of motion. As the average speed of the bear changes, the slope of the line on the graph changes. Because of this, you can use a displacement-time graph to describe the motion of an object.

▲ **Figure 6** Tracking devices help scientists record the movement of animals, such as polar bears.

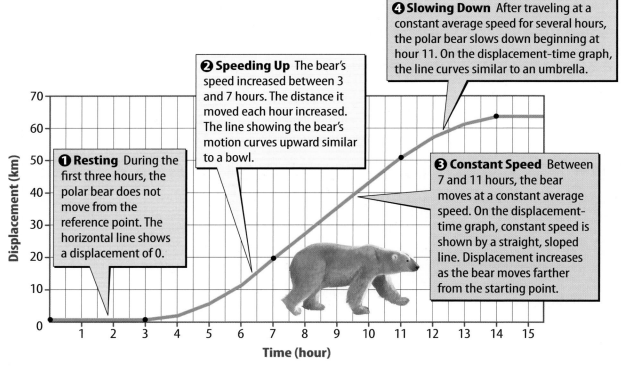

❹ **Slowing Down** After traveling at a constant average speed for several hours, the polar bear slows down beginning at hour 11. On the displacement-time graph, the line curves similar to an umbrella.

❷ **Speeding Up** The bear's speed increased between 3 and 7 hours. The distance it moved each hour increased. The line showing the bear's motion curves upward similar to a bowl.

❶ **Resting** During the first three hours, the polar bear does not move from the reference point. The horizontal line shows a displacement of 0.

❸ **Constant Speed** Between 7 and 11 hours, the bear moves at a constant average speed. On the displacement-time graph, constant speed is shown by a straight, sloped line. Displacement increases as the bear moves farther from the starting point.

▲ **Figure 7** 🗝 The displacement-time graph shows the bear's speed and distance from the reference point at any point in time.

🔍 **Visual Check** What was the average speed of the bear between hours 7 and 11?

[Review] **Personal Tutor**

Figure 8 The speed-time graph shows the speed of the bear at any given time during its journey. A horizontal line on a speed-time graph shows an object with a constant speed.

✓ **Visual Check** What happened to the bear's speed between hours 5 and 6?

Concepts in Motion Animation

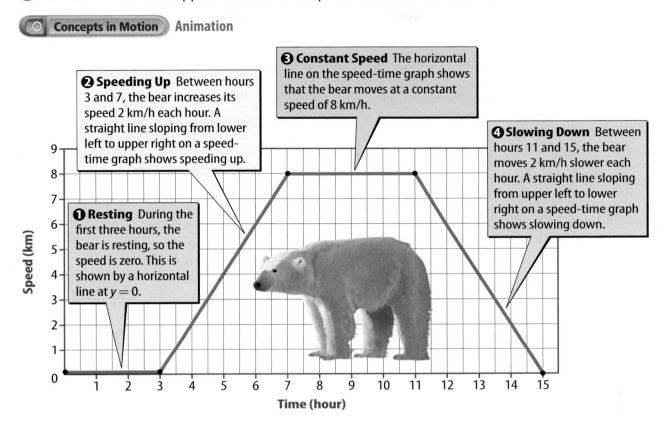

❷ **Speeding Up** Between hours 3 and 7, the bear increases its speed 2 km/h each hour. A straight line sloping from lower left to upper right on a speed-time graph shows speeding up.

❸ **Constant Speed** The horizontal line on the speed-time graph shows that the bear moves at a constant speed of 8 km/h.

❹ **Slowing Down** Between hours 11 and 15, the bear moves 2 km/h slower each hour. A straight line sloping from upper left to lower right on a speed-time graph shows slowing down.

❶ **Resting** During the first three hours, the bear is resting, so the speed is zero. This is shown by a horizontal line at $y = 0$.

Speed (km) — vertical axis: 0, 1, 2, 3, 4, 5, 6, 7, 8, 9
Time (hour) — horizontal axis: 1, 2, 3, 4, 5, 6, 7, 8, 9, 10, 11, 12, 13, 14, 15

Speed-Time Graphs

Figure 8 is a speed-time graph of the polar bear's motion. The *x*-axis shows the time, and the *y*-axis shows the speed of the bear. Notice that, in this case, the line shows how the speed, rather than the displacement, changes as the bear moves. A horizontal line at $y = 0$ means the bear is at rest because its speed is 0 km/hr. Notice that a horizontal line at $y = 0$ on either a displacement-time graph or a speed-time graph represents the bear at rest.

Keep in mind that *constant speed* is describing average speed. The bear might have sped up or slowed down slightly each second. But, during hours 7–11, you could describe that the bear's average speed remained constant since it covered the same distance each hour.

Interpreting the lines on graphs can provide you with a lot of information about the motion of an object.

🔑 **Key Concept Check** How can a graph help you understand an object's motion?

ACADEMIC VOCABULARY

satellite
(*noun*) an object in orbit around another object

Lesson 1 Review

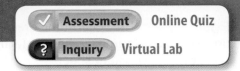
Visual Summary

A description of an object's motion includes a reference point, a direction from the reference point, and a distance.

Speed is the distance traveled by an object in a unit of time. Velocity includes both speed and direction of motion.

Acceleration is a change in velocity. Velocity changes when either the speed, the direction, or both the speed and the direction change.

FOLDABLES

Use your lesson Foldable to review the lesson. Save your Foldable for the project at the end of the chapter.

What do you think NOW?

You first read the statements below at the beginning of the chapter.

1. You must use a reference point to describe an object's motion.

2. An object that is accelerating must be speeding up.

Did you change your mind about whether you agree or disagree with the statements? Rewrite any false statements to make them true.

Use Vocabulary

1 **Describe** in your own words how you would choose a reference point.

2 **Distinguish** between the terms *distance* and *displacement*.

Understand Key Concepts

3 **Describe** the motion of a book as you lift it from the table and place it on a shelf.

4 Which of the following does NOT cause an object to accelerate?
   A. change in direction
   B. constant velocity
   C. slowing down
   D. speeding up

5 **Apply** Draw a speed-time graph of a parade float that accelerates from rest to 0.5 km/hr in 1 min and then moves at a constant speed for 10 min.

Interpret Graphics

6 **Draw** The table below includes information about the motion of an elevator. Draw a displacement-time graph using the data, and explain the elevator's motion.

Displacement	Time
0 m	0 s
1 m	1 s
4 m	2 s
10 m	3 s
10 m	4 s

Critical Thinking

7 **Analyze** whether you could have a vertical line on a displacement-time graph. Why or why not?

Math Skills

— Math Practice —

8 What is the average acceleration of a track star who goes from a velocity of 0 m/s to a velocity of 9 m/s to the east in 3 s?

It's Moving!

Fooling the Eye

You know that you describe an object's motion by explaining how its position changes. Did you know that you can use this concept to make a movie that shows nonmoving objects in motion! It's called stop-motion photography. How does it work?

1. First, set an object in a scene and take a picture of it. Keep changing the position of the object in the scene, taking a picture after each change.

2. Now use software to link all the pictures into a video. When you view the video, it will appear as if the object moved on its own. Of course, it's just an illusion. The illusion works because of the way your eye works. Motion is a change of position, and that's exactly what your eyes are seeing with stop-motion photography.

It's Your Turn

EXPERIMENT Set up your own stop-motion photography studio. If you don't have a camera, make sketches of each change. When you are finished, make a flip book of your sketches or photographs.

Reading Guide

Key Concepts 🔑
ESSENTIAL QUESTIONS

- How do different types of forces affect objects?
- What factors affect the way gravity acts on objects?
- How do balanced and unbalanced forces differ?

Vocabulary

force p. 701

contact force p. 702

noncontact force p. 702

friction p. 703

gravity p. 703

balanced forces p. 705

unbalanced forces p. 705

🅖 **Multilingual eGlossary**

🎞 **Video**

What's Science Got to do With It?

Forces

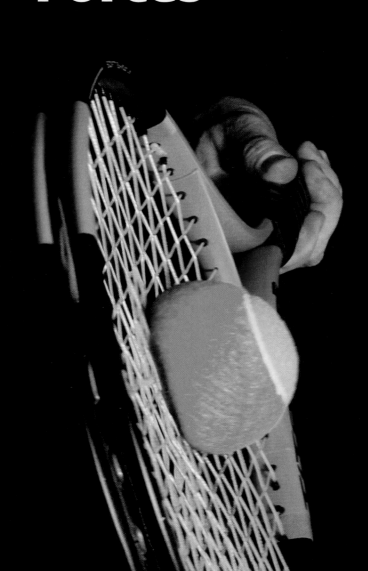

Inquiry **Why is one side of the ball flat?**

A ball, such as this tennis ball, is usually round. Its shape lets it roll farther and travel farther in the air. What could cause part of a ball to become flat like this one? Does the same thing happen when a baseball hits a bat? Or when a golf club hits a golf ball?

How can you change an object's shape and motion?

You probably can think of many ways that things change. For example, paper can change from a flat sheet to a crumpled ball. A sailboat changes its location as it moves across a lake. How can you change an object's shape and motion?

1 Read and complete a lab safety form.

2 Observe and record in your Science Journal how you make the following changes. Change the shape of a handful of **clay** several times.

3 Mold the clay into a log. Cause the log to roll, and then cause it to stop rolling.

4 Cause the log to roll so that its speed changes. Then change the log's direction of motion. Observe and record in your Science Journal how you make these changes.

Think About This

1. **Describe** what you did to change the shape of the clay.

2. **Explain** how you changed the motion of the clay.

3. 🔑 **Key Concept** How was your interaction with the clay similar when you changed its shape and when you changed its motion?

What are forces?

What do typing on a computer, lifting a bike, and putting on a sweater have in common? They all involve an interaction between you and another object. You push on the keys. You push or pull on the bike. You pull on the sweater. *A push or pull on an object is a* **force.**

A force has both size and direction. In **Figure 9,** the length of the arrow represents the size of the force. The direction in which the arrow points represents the direction of the force. The unit of force is the newton (N). It takes about 4 N of force to lift a can of soda.

There are two ways a force can affect an object. A force can change an object's speed. It also can change the direction in which the object is moving. In other words, a force can cause acceleration. Recall that acceleration is a change in an object's velocity—its speed and/or its direction in a given time. When you apply a force to a tennis ball, such as the one shown in the picture on the previous page, the force first stops the motion of the ball. The force then causes the ball to accelerate in the opposite direction, changing both its speed and direction.

Figure 9 The arrows show forces with very different sizes acting in opposite directions.

✓ **Reading Check** In what ways can forces affect objects?

Types of Forces

Some forces are easy to recognize. You can see a hammer applies a force as it hits a nail. Other forces seem to act on objects without touching them. For example, what force causes your ice cream to fall toward the ground if it slips out of the cone?

Contact Forces

The top left image of **Figure 10** shows a baker pushing his hand into dough, causing the top of the dough to accelerate downward. You can see the baker's hand and the dough come into contact with each other. *A **contact force** is a push or a pull applied by one object to another object that is touching it.* Contact forces also are called mechanical forces. The top half of **Figure 10** also shows other types of contact forces.

Noncontact Forces

The bottom left image of **Figure 10** shows a girl's hair being pulled toward the slide even though it isn't touching the slide. *A force that pushes or pulls an object without touching it is a **noncontact force.*** The force that pulls the girl's hair is an electric force. The bottom half of **Figure 10** shows other noncontact forces, such as magnetism and gravity.

Key Concept Check What is the difference between the way contact and noncontact forces affect objects?

Figure 10 The pictures in the top row show examples of various types of contact forces. The ones in the bottom row show examples of several types of noncontact forces.

A **contact**, or mechanical, force is a force exerted by a physical object that touches another object.

An **applied force** is a force in which one object directly pushes or pulls on another object.

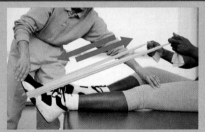

An **elastic** or spring force is the force exerted by a compressed or stretched object.

A **normal force** is the support force exerted on an object that touches another stable object.

A **noncontact**, or field, force is a force exerted when there is no visible object exerting the force.

Electric forces cause the girl's hair to stick out.

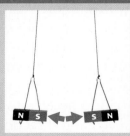

Magnetic forces hold these magnets apart.

Gravity is the force that pulls these divers toward the water.

How does friction affect an object's motion?

Air resistance is a force that opposes the motion of an object moving through air.

1. Read and complete a lab safety form.

2. Make a model parachute from **tissue paper, string, tape,** and a **metal washer.**

3. Use a **meterstick** to measure heights of 1, 2, 3, and 4 m on a nearby wall. Mark them with **tape.**

4. Drop the parachute from the 4-m mark. Your partner should start a **stopwatch** as soon as you drop the parachute and should stop the stopwatch when the washer passes the 3-m mark. Repeat this step three more times stopping the stopwatch at the 2-m mark, the 1-m mark, and the ground. Record the times in your Science Journal.

5. Remove the washer from the parachute. Measure and record the time for the washer to fall from the 4-m mark to the floor without the parachute.

Think About This

1. **Graph** the motion of the parachute on a distance-time graph.

2. **Calculate** the average speed of the washer with and without the parachute.

3. **Key Concept** How did friction affect the speed of the parachute and the washer?

Friction

Why does the baseball player in **Figure 11** slow down as he slides toward the base? **Friction** *is a contact force that resists the sliding motion between two objects that are touching.* The force of friction acts in the opposite direction of the motion, as shown by the blue arrow. Rougher surfaces produce greater friction than smooth surfaces. Other factors, such as the weight of an object, also affect the force of friction.

Gravity

Is there anywhere on Earth where you could drop a pencil and not have it fall? No! **Gravity** *is a noncontact attractive force that exists between all objects that have mass.*

Mass is the amount of matter in an object. Both your pencil and Earth have mass. They exert a gravitational pull on each other. In fact, they exert the same gravitational force on each other. Why doesn't your pencil pull Earth toward it? It actually does! The pencil has very little mass, so the force of gravity causes it to rapidly accelerate downward toward Earth's surface. Earth "falls" upward toward the pencil at the same time, but because of its mass, Earth's motion is too small to see.

Figure 11 The player must overcome friction or he won't reach the base.

WORD ORIGIN
gravity
from Latin *gravitare*, means to unite, join together

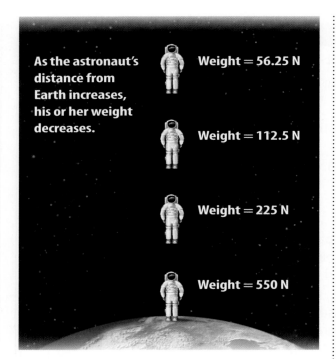

As the astronaut's distance from Earth increases, his or her weight decreases.

Weight = 56.25 N

Weight = 112.5 N

Weight = 225 N

Weight = 550 N

▲ **Figure 12** 🔑 Gravitational force (weight) decreases as the distance between the centers of the objects increases.

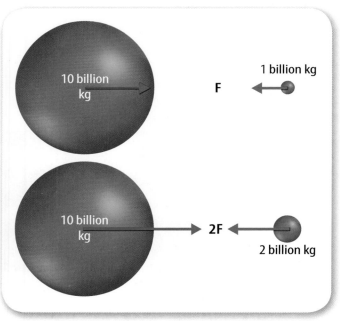

10 billion kg

1 billion kg

F

10 billion kg

2F

2 billion kg

▲ **Figure 13** The force of attraction between the bottom two objects is twice as much as between the top two objects.

✅ **Visual Check** Describe the acceleration of the bottom spheres due to the gravitational force between them.

 **Review** **Personal Tutor**

Distance and Gravity

You may have heard that astronauts become weightless in space. This is not true. Astronauts do have some weight in space, but it is much less in space than their weight on Earth. Weight is a measure of the force of gravity acting on an object. As two objects get farther apart, the gravitational force between the objects decreases. **Figure 12** shows how the weight of an astronaut changes as he or she moves farther from Earth.

You know that all objects exert a force of gravity on all other objects. If the astronaut drops a hammer on the Moon, will it fall toward Earth? No, the attraction between the Moon and the hammer is stronger than the attraction between Earth and the hammer because the hammer is very close to the Moon and very far from Earth. The hammer will fall down toward the Moon.

Mass and Gravity

Another factor that affects the force of gravity between two objects is the mass of the objects. As the mass of one or both objects increases, the gravitational force between them increases. For example, in **Figure 13**, *F* stands for the gravitational force. As the figure shows, doubling the mass of one of the objects doubles the force of attraction.

The effect of mass on the force of gravity is most noticeable when one object is very massive, such as a planet, and the other object has much less mass, such as a person. Even though the force of gravity acts equally on both objects, the less massive object accelerates more quickly due to its smaller mass. Because the planet accelerates so slowly, all you observe is the object with less mass "falling" toward the object with greater mass.

🔑 **Key Concept Check** What factors affect the way gravity acts on objects?

Combining Forces

Have you ever played tug-of-war? If you alone pull against a team, you will probably be pulled over the line. However, if you are on a team, your team might pull the rope hard enough to cause the other team to move in your direction. When several forces act on an object, the forces combine to act as a single force. The sum of the forces acting on an object is called the net force.

Forces in the Same Direction

When different forces act on an object in the same direction, you can find the net force by adding the forces together. In **Figure 14,** each team member pulls in the same direction. The net force on the rope is 110 N + 90 N + 100 N = 300 N.

Forces in Opposite Directions

When forces act in opposite directions, you must include the direction of the force when you add them. Like numbers on a number line, forces in the direction to the right are normally considered to be positive values. Forces to the left are negative values. In the first panel of **Figure 15,** the team on the right pulls with a force of 300 N. The team on the left pulls with a force of −300 N. The net force is 300 N + (−300 N) = 0.

Balanced and Unbalanced Forces

The net force on the rope in the top of **Figure 15** is 0. *When the net force on an object is 0 N, the forces acting on it are* **balanced forces.** If the forces acting on an object are balanced, the object's motion does not change. *When the net force acting on an object is not 0, the forces acting on the object are* **unbalanced forces.** The forces acting on the rope in the bottom of **Figure 15** are unbalanced. Unbalanced forces cause objects to change their motion, or accelerate.

 Key Concept Check How do balanced and unbalanced forces differ?

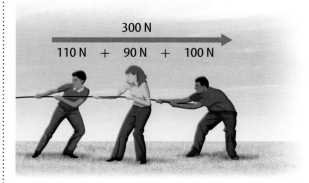
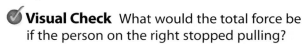

▲ **Figure 14** Forces in the same direction act as a single force.

 **Visual Check** What would the total force be if the person on the right stopped pulling?

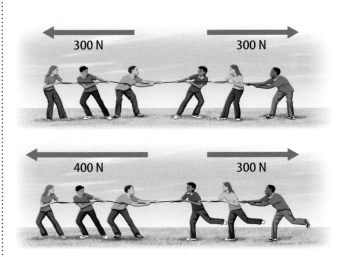

▲ **Figure 15** 🔑 No change in motion takes place when forces on an object are balanced. Unbalanced forces cause the team on the right to accelerate to the left.

Review Personal Tutor

Lesson 2 Review

Visual Summary

Forces are pushes and pulls exerted by objects on each other. Contact forces occur when objects are touching. Noncontact forces act from a distance.

Gravity is a force of attraction between two objects. The amount of gravitational force depends on the mass of the objects and the distance between them.

Balanced forces do not affect motion. Unbalanced forces change motion.

FOLDABLES

Use your lesson Foldable to review the lesson. Save your Foldable for the project at the end of the chapter.

What do you think **NOW?**

You first read the statements below at the beginning of the chapter.

3. Objects must be in contact with one another to exert a force.

4. Gravity is a force that depends on the masses of two objects and the distance between them.

Did you change your mind about whether you agree or disagree with the statements? Rewrite any false statements to make them true.

Use Vocabulary

1 **Describe** friction in your own words.

2 Two examples of _____ are gravity and magnetism.

Understand Key Concepts

3 As the distance between two objects increases, the gravitational force between the objects

   A. increases.   **C.** creates friction.

   B. decreases.   **D.** stays the same.

4 **Describe** the forces acting on a cyclist who is slowing down as he or she climbs a hill.

5 **Identify** any balanced and unbalanced forces acting on a book resting on a table.

Interpret Graphics

6 **Copy and complete** the graphic organizer to explain how distance and mass affect the force of gravity.

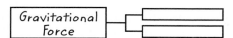

7 **Analyze** the four forces acting on the airplane flying at an altitude of 3,000 m, as shown below. How do the forces affect the plane's motion?

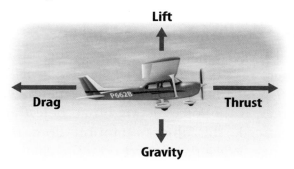

Critical Thinking

8 **Construct** a diagram that shows three forces acting on an object in the same direction and two forces acting in the opposite direction. Give the forces values that would cause no change in motion.

What factors affect friction?

When you push or pull an object across a surface, the force of friction resists the object's motion. If the friction is strong, you need a greater force to move the object. How does manipulating variables, such as mass and surface texture, affect friction?.

Learn It

In any experiment, it is important to **identify and manipulate variables.** The independent variable is the factor that you change during the experiment. The variable that might change as a result of the independent variable is called the dependent variable. Changing only one variable at a time helps you focus clearly on what is causing the dependent variable to change.

Try It

1. Read and complete a lab safety form.

2. You will test the effect that mass, surface area, and surface texture have on the force needed to pull a block across a surface. Discuss the investigation with your partner. Predict whether each of the three variables will affect friction between a block and the surface.

3. Think about how you can test your prediction. Consider the following questions:

- What tests will you perform? For each test, identify the independent variable and the dependent variable.

- What materials will you use?

- What type of data table will you construct to record your data?

4. Test several methods for moving your object that you think might work. Based on your results, write a plan for your teacher to approve.

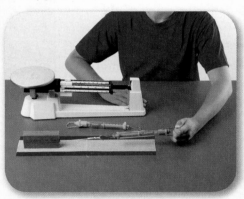

Apply It

5. Work with your partner to carry out your experiment. Record your results in your Science Journal.

6. Describe the independent variable and the dependent variables you used for each test you performed.

7. **Key Concept** Did your tests support your prediction about the effects of mass, surface area, and surface texture on friction? Explain.

Lesson 3

Reading Guide

Key Concepts 🔑
ESSENTIAL QUESTIONS

- How do unbalanced forces affect an object's motion?
- How are the acceleration, the net force, and the mass of an object related?
- What happens to an object when another object exerts a force on it?

Vocabulary

inertia p. 709

Newton's first law of motion p. 709

Newton's second law of motion p. 712

Newton's third law of motion p. 713

force pair p. 713

 Multilingual eGlossary

 Video BrainPOP®

Newton's Laws of Motion

Inquiry How does this feel?

Rides like this are called thrill rides because the riders feel as if they are going to crash, fall, or take off into space. How do forces cause these sensations? Why are the bars that hold the riders in place so important?

15 minutes

How are forces and motion related?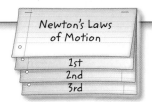

In the last lesson, you read about different forces acting on objects. Sometimes forces can produce unexpected results. In this lab, you will observe the effect of forces on an object's motion.

1. Read and complete a lab safety form.

2. Place an **index card** on a **plastic jar.** Center a **nickel** on top of the card.

3. Flick the card away horizontally. Observe the motion of the nickel. Record your observations about the motion in your Science Journal.

4. Spread a sheet of **newspaper** on the table with about 10 cm hanging over the edge.

5. Place a **book,** a **pen,** and a **paper clip** on top of the paper. Then quickly pull the edge of the paper straight down. Record your observations in your Science Journal.

Think About This

1. **Identify** the forces acting on the objects in steps 3 and 5.

2. 🔑 **Key Concept** How do you think forces are related to the motion of the objects?

Newton's Laws

Recall that forces are measured in a unit called a newton (N), named after English scientist Isaac Newton, who studied the motion of objects. Newton summarized his findings in three laws of motion. You demonstrate Newton's laws when you run to catch a baseball or ride your bike. How could you use Newton's laws to explain how the rides and the games at an amusement park work?

Newton's First Law

What causes the motion of amusement park rides to give riders a thrill? Without protective devices to hold you in your seat, you could fly off the ride! *The tendency of an object to resist a change in motion is called* **inertia.** Inertia acts to keep you at rest when the ride starts moving. It also keeps you moving in a straight line when the ride stops or changes direction. Your safety belt keeps you in the seat and moving with the ride.

Newton's first law of motion *states that if the net force acting on an object is zero, the motion of the object does not change.* In other words, an object remains at rest or in constant motion unless an outside, unbalanced force acts on it. Newton's first law of motion is sometimes called the law of inertia.

FOLDABLES

Use two sheets of paper to make a layered book. Label it as shown. Use it to organize your notes on Newton's laws.

Newton's Laws of Motion
1st
2nd
3rd

SCIENCE USE V. COMMON USE

inertia
Science Use the tendency to resist a change in motion

Common Use lack of action

Force of cables
Force of gravity

At Rest

Force of cables
Force of gravity

Constant Speed

Figure 16 The free-fall car's velocity is constant in both images because the forces are balanced.

How does inertia affect an object?

1. Read and complete a lab safety form.

2. Attach one end of a 20-cm long **string** to the **eye hook** that is attached to one end of a **wooden block.**

3. Half fill a **large test tube** with **colored water. Stopper** the tube tightly. Use **transparent tape** to attach the test tube to the block.

4. Use the string to pull the block. Observe the water when the block is at rest, as its velocity changes, and when its velocity is constant. Record your observations in your Science Journal.

Analyze and Conclude

1. **Describe** the motion of the water in the tube when the tube is at rest, accelerating, and moving at a constant velocity.

2. 🔑 **Key Concept** How does Newton's first law explain your observations?

Effects of Balanced Forces

Suppose you are at an amusement park and you want to ride a free-fall car, such as the one shown in **Figure 16.** How does the ride illustrate Newton's first law of motion? Recall that when the forces acting on an object are balanced, the object is either at rest or moving with a constant velocity.

Objects at Rest At the top of the ride, the force of the cable pulling upward on the car is equal to the force of gravity pulling downward on the car. Gravity and the cables pull on the car equally, but in opposite directions, so the forces are balanced. The car is at rest, as shown in the first panel of **Figure 16.** As long as the forces remain balanced, the car remains at rest.

Objects in Motion To lift the car to the top of the ride, the cable pulls upward. After a short acceleration, the car moves upward at a constant speed. The force of the cable pulling upward is the same size as the force of gravity pulling downward. With the forces once again balanced, the car rises to the top of the ride at a constant velocity. This is shown in the second panel of **Figure 16.** Newton's first law describes the car's motion when the forces applied to it are balanced.

Balanced forces act on the car only when it is at rest or moving with a constant velocity. When the car reaches the top of the ride, it doesn't remain at rest for long. When the operator releases the upward pull on the cable, the forces become unbalanced. Gravity causes the car to accelerate toward the ground. Because inertia tends to keep you at rest, the car feels as if it falls out from under you. Your safety belt acts as an outside force to keep you attached to the car.

✓ **Reading Check** What happens to the velocity of the car when the upward pull of the cable is greater than the downward pull of gravity as the car rises toward the top?

Key
→ Force
→ Acceleration

Speeding Up   **Slowing Down**

Effects of Unbalanced Forces

You continue your visit to the amusement park with a ride on the reverse bungee jump. According to Newton's first law of motion, the motion of an object changes only when a net force acts on it. This ride gives you two chances to experience what a net force can do.

Speeding Up After the ride attendant releases you, the upward force of the bungee cord is greater than the downward force of gravity. The forces are unbalanced as shown by the blue arrows in the left image in **Figure 17.** The net force acting on you is upward, and you **accelerate** upward as shown by the green arrow.

Slowing Down As you approach the top of your bungee jump, the cords become slack, as shown on the right in **Figure 17.** The blue arrows show that the upward force becomes less than the downward force of gravity. Even though you are still are moving upward because of inertia, the net force is now due to the downward force of gravity. You slow down, or decelerate.

 Key Concept Check If one force on an object is 5 N upward and the other is 10 N downward, what is the object's motion?

Changing Direction Your next stop is a swing ride such as the one shown in **Figure 18.** When the ride starts to turn, the force of the cables pulls your chair toward the center of the ride. The force of gravity acts downward. Because these forces don't act in opposite directions, the unbalanced force constantly changes your direction. You accelerate as you move in a circle.

The designers of amusement-park rides use inertia to create excitement. Much of what makes a swing ride fun is the feeling that you might fly off the ride with constant velocity if your safety belt didn't hold you in place.

WORD ORIGIN ···········

accelerate
from Latin *celer*, means "swift"

▲ **Figure 18** 🔑 The unbalanced force of the cable pulling toward the center causes acceleration in a circle.

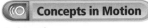 **Concepts in Motion**
Animation

Figure 19 Using a large force to throw the ball gives you the best chance of knocking over the bottles.

Newton's Second Law of Motion

Suppose you play a game in which you throw a baseball to knock over wooden milk bottles, as shown in **Figure 19.** You have seen that unbalanced forces cause objects to accelerate. To knock over the bottles, the ball has to accelerate fast enough to overcome the forces keeping the bottles upright. On your first try, you don't throw the ball with enough force. On your second try, you throw the ball as hard as you can. When you begin to throw the ball and the ball is still in your hand, it accelerates. The ball flies out of your hand rapidly and knocks over the bottles, winning you a prize. You won the prize by using another of Newton's laws.

Newton described the relationship between an object's acceleration (change in velocity) and the net force exerted on the object. **Newton's second law of motion** *states that the acceleration of an object is equal to the net force exerted on the object divided by the object's mass.* The direction of acceleration is the same as the direction of the net force.

$$\text{acceleration} = \frac{\text{force}}{\text{mass}} \qquad a = \frac{F}{m}$$

Calculating Acceleration

You can use the equation to calculate the acceleration of the ball. If you apply a force of **1.5 N** to a ball with a mass of **0.3 kg**, what is the ball's acceleration?

$$\text{acceleration} = \frac{\text{force}}{\text{mass}} \qquad \text{acceleration} = \frac{1.5 \text{ N}}{0.3 \text{ kg}} = \frac{5 \text{ m}}{\text{s}^2}$$

What do you think would happen to the acceleration if you double the force on the ball? The equation tells you!

$$\text{acceleration} = \frac{\text{force}}{\text{mass}} \qquad \text{acceleration} = \frac{3.0 \text{ N}}{0.3 \text{ kg}} = \frac{10 \text{ m}}{\text{s}^2}$$

When you double the force, the acceleration also doubles.

Changing the Mass

What would happen to the acceleration if the force you apply stays the same, but the mass of the ball changes? Instead of 0.3 kg, the ball has a mass of **0.6 kg**.

$$\text{acceleration} = \frac{\text{force}}{\text{mass}} \qquad \text{acceleration} = \frac{1.5 \text{ N}}{0.6 \text{ kg}} = \frac{2.5 \text{ m}}{\text{s}^2}$$

A ball with twice the mass has half the acceleration. Newton's second law lets you predict what combination of force and mass you need to get the acceleration you need.

Key Concept Check How are the acceleration, the net force, and the mass of an object related?

◀ **Figure 20** Each car exerts a force of the same size on the other car. The amount that each car accelerates depends on its mass.

Newton's Third Law

Suppose you are driving bumper cars with a friend, like in **Figure 20.** What happens when you crash into each other? **Newton's third law of motion** *says that when one object exerts a force on a second object, the second object exerts a force of the same size, but in the opposite direction, on the first object.* According to Newton's third law, the bumper cars apply forces to each other that are equal but are in opposite directions.

Action and Reaction Forces

When two objects apply forces on each other, one of the forces is called the action force, and the other is called the reaction force. For example, if the left car hits the right car in **Figure 20,** then the force exerted by the left car is the action force. The force exerted by the right car is the reaction force.

 Key Concept Check What happens when one object exerts a force on a second object?

Force Pairs

As you walk, your shoes push against the ground. If the ground did not push back with equal force, gravity would pull you down into the ground! *When two objects exert forces on each other, the two forces are a* **force pair.** The opposite forces of the bumper cars hitting each other in **Figure 20** are a force pair. Force pairs are not the same as balanced forces. Balanced forces combine or cancel each other out because they act on the same object. Each force in a force pair acts on a different object.

In **Figure 21,** the girl exerts a force on the ball. The ball exerts an equal but opposite force on the girl. Why does the ball's motion change more than the girl's motion? Newton's laws work together. Newton's first law explains that a force is needed to change an object's motion. His third law describes the action-reaction forces. Newton's second law explains why the effect of the force is greater on the ball. The mass of the ball is much less than the mass of the girl. A force of the same size produces a greater acceleration in an object with less mass—the ball.

Figure 21 The opposite forces of the girl's head and the ball are a force pair. ▼

Visual Check If the force of the girl's head on the ball is 1.5 N upward, what is the force of the ball on the girl's head?

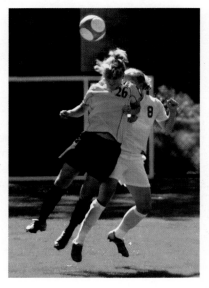

Newton's Laws in Action

Newton's laws do not apply to all motion in the universe. For example, they don't correctly predict the motion of very tiny objects, such as atoms or electrons. They do not work for objects that approach the speed of light.

However, because Newton's laws apply to the moving objects you observe each day, from amusement park rides to the movement of stars and planets, they are extremely useful. Using Newton's laws, humans have traveled to other planets and invented many useful tools and machines. You can often see the effects of all three laws at the same time. **Table 1** gives you some everyday examples of Newton's laws in action. Think about Newton's laws as you move through your day.

Table 1 🔑 Newton's laws explain the motions you experience every day.

✅ **Visual Check** How do you know that the table is exerting a force on the bowl of fruit?

Table 1 Newton's Laws in Action

Example	Newton's First Law	Newton's Second Law	Newton's Third Law
Resting Mass = 2 kg	The upward and downward forces on the bowl are balanced. The motion of the bowl is not changing. It is at rest.	Because the bowl is at rest, its acceleration is 0 m/s^2. You can use Newton's second law to calculate the net force on the bowl: $F = m \times a$ $F = 2 \text{ kg} \times 0 \text{ m/s}^2$ $F = 0 \text{ N}$	The force of gravity pulls the bowl down so it exerts a force on the table. The table pushes up on the bowl with a force that is the same size, but in the opposite direction.
Walking	The forces acting on the man and the woman are balanced. Their inertia keeps them moving at a constant speed in a straight line.	When an object moves at a constant velocity, there is no acceleration. A net force would have to act on the people before they would speed up or slow down.	The woman's feet push against the sand as she walks. The sand pushes on the woman's feet with equal force, moving her forward. The same is true of the man.
Skateboarding	Inertia keeps the dog and the skateboard at rest until the dog produces a net force by pushing its paw on the road.	When net forces act on the dog and the road, or Earth, the dog will accelerate at a much greater rate because its mass is much less than that of Earth.	The dog's paw exerts a backward force on the road. The road exerts an equal but opposite force on the dog's paw, pushing it forward.

Lesson 3 Review

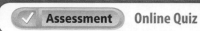

Visual Summary

Newton's first law of motion states that the motion of an object remains constant unless acted on by an outside force. This also is called the law of inertia.

Newton's second law of motion relates an object's acceleration to its mass and the net force applied to the object.

Newton's third law of motion states that for every action force, there is an equal but opposite reaction force. The two forces are called a force pair.

FOLDABLES®

Use your lesson Foldable to review the lesson. Save your Foldable for the project at the end of the chapter.

What do you think NOW?

You first read the statements below at the beginning of the chapter.

5. All forces change the motion of objects.

6. The net force on an object is equal to the mass of the object times the acceleration of the object.

Did you change your mind about whether you agree or disagree with the statements? Rewrite any false statements to make them true.

Use Vocabulary

1 **Describe** an example of Newton's third law of motion.

2 **Distinguish** between Newton's first and second laws of motion.

Understand Key Concepts

3 In order to accelerate, an object must be acted on by
   A. a force pair.   C. balanced forces.
   B. a large mass.   D. unbalanced forces.

4 **Interpret** A bicyclist rides with a constant velocity of 8 m/s. What would you need to know to calculate the net force on the rider?

Interpret Graphics

5 **Analyze** The diagram below shows the forces acting on a box. Describe the motion of the box.

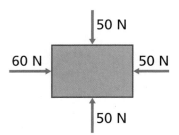

6 **Copy and complete** the graphic organizer by describing each of Newton's laws.

Critical Thinking

7 **Apply** Why does a box on the seat of a car slide around on the seat when the car speeds up, slows down, or turns a corner?

8 **Predict** what would happen if two people with equal mass standing on skateboards pushed against each other.

9 **Solve** A hockey player hits a 0.2-kg puck that accelerates at a rate of 20 m/s². What force did the player exert on the puck?

Materials

wood board
(1-m x 20-cm)

masking tape

string

tennis ball

large rubber
band

foam tubing

Also needed:
marble

Safety

Design an amusement park attraction using Newton's laws

What is your favorite ride or game at an amusement park? You may think that amusement parks are just for fun, but Newton's laws are important in the design of every ride and game. Work with a group to design and build a ride or game that applies Newton's laws.

Ask a Question

How do Newton's laws describe an amusement-park ride or game?

Make Observations

1 Read and complete a lab safety form.

2 Discuss different rides and games with your group. Think about how Newton's laws explain each attraction.

3 Your model ride or game must be a working model. You will not use motors, but your ride must use several different forces to make it work. If you design a game, it must demonstrate one or more of Newton's laws. Test several ideas with your group. If one idea does not work, adjust the design or try a different idea.

4 Based on your tests, choose one model ride or game to build.

5 Decide on the materials you will use. You may use some from the list or others approved by your teacher.

6 Write a design, along with a sketch, for your ride or game. List the materials, and describe how the ride or game will work. Ask your teacher to approve your design.

Form a Hypothesis

7 Based on your observations, formulate a hypothesis that explains why your ride or game will work according to one or more of Newton's laws.

Test Your Hypothesis

8 Build your ride or game according to your approved design.

9 Use your ride or game to test your design. In your Science Journal, identify which of Newton's laws are demonstrated by each part of your model. Also record details about your tests and your results.

Analyze and Conclude

10 **Evaluate** Did your ride or game clearly model Newton's laws? Explain.

11 **Analyze** Which of Newton's laws of motion most describes the way your ride or game works? Why?

12 **Compare** your ride or game to those built by other groups. Which do you think is the best example of each of Newton's laws? Explain your opinion.

13 **BIG IDEA** **The Big Idea** Describe the relationship between the forces and the motion for the ride or the game you built.

Communicate Your Results

Demonstrate your ride or game for the class. Explain how one or more of Newton's laws influence the way the ride or game works.

Inquiry Extension

Work with others in your group to write a brochure titled *Newton's Amusement Park*. The brochure should include descriptions and illustrations of the various rides and games, along with a brief explanation of how Newton's laws affect each ride or game.

Lab **TIPS**

☑ Test different parts of your design idea to be sure each part works before you settle on one design to build.

☑ Avoid making your ride or game too complicated. A simple ride or game might be better.

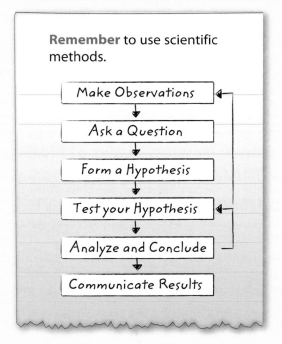

Remember to use scientific methods.

Make Observations

Ask a Question

Form a Hypothesis

Test your Hypothesis

Analyze and Conclude

Communicate Results

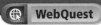
 THE BIG IDEA

Forces are pushes and pulls that may change the motion of an object. Balanced forces result in an object remaining at rest or moving at a constant speed. Unbalanced forces result in the acceleration of an object.

Key Concepts Summary 🔑

Vocabulary

Lesson 1: Describing Motion

- An object's **motion** depends on how it changes position. Motion can be described using **speed, velocity,** or **acceleration.**

- Speed is how fast an object moves. Velocity describes an object's speed and the direction it moves. Acceleration describes the rate at which an object's velocity changes.

- A graph can show you how either the displacement or the speed of an object changes over time.

motion p. 691
reference point p. 691
distance p. 692
displacement p. 692
speed p. 692
velocity p. 693
acceleration p. 694

Lesson 2: Forces

- A **force** is a push or pull on an object. **Contact forces** include **friction** and applied forces. **Noncontact forces** include **gravity,** electricity, and magnetism.

- Gravity is a force of attraction between any two objects. Gravitational force increases as the masses of the objects increase and decreases as the distance between the objects increases.

- **Balanced forces** acting on an object cause no change in the motion of the object. When **unbalanced forces** act on an object, the sum of the forces is not equal to zero. Unbalanced forces cause acceleration.

force p. 701
contact force p. 702
noncontact force p. 702
friction p. 703
gravity p. 703
balanced forces p. 705
unbalanced forces p. 705

Lesson 3: Newton's Laws of Motion

- **Inertia** is the tendency of an object to resist a change of motion. **Newton's first law of motion** states that an object will remain at rest or in constant straight-line motion unless unbalanced forces act on the object.

- **Newton's second law of motion** states that the acceleration of an object increases as the force acting on it increases and decreases as the mass of the object increases.

- **Newton's third law of motion** states that for every action force, there is an equal but opposite reaction force. The action-reaction forces are called a **force pair.**

inertia p. 709
Newton's first law of motion p. 709
Newton's second law of motion p. 712
Newton's third law of motion p. 713
force pair p. 713

FOLDABLES® Chapter Project

Assemble your lesson Foldables as shown to make a Chapter Project. Use the project to review what you have learned in this chapter.

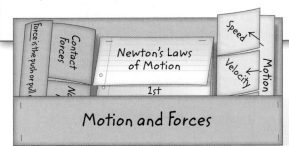

Use Vocabulary

1 An object's _____ is the difference between a object's final position and its starting position.

2 Give a specific example of motion.

3 Name two forces that may act on objects at a distance.

4 Explain what must happen to an object in order for it to accelerate.

5 What kinds of things can you predict using Newton's second law of motion?

6 The law of inertia is another name for _____.

7 You can explain the forces that act when you push against a wall using _____.

8 A _____ describes two forces that act on different objects.

Link Vocabulary and Key Concepts

Concepts in Motion Interactive Concept Map

Copy this concept map, and then use vocabulary terms and other terms from the chapter to complete the concept map.

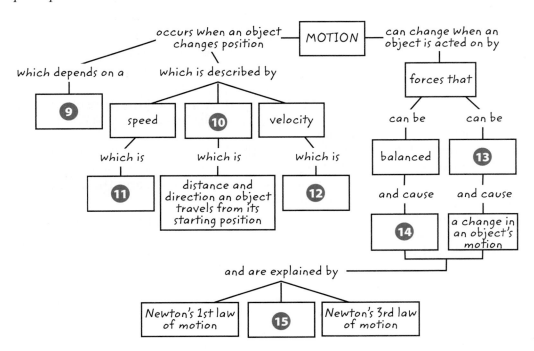

Understand Key Concepts

1 In which motions are the distance and the displacement the same?

A. A bird flies from its nest to the ground and back to its nest.

B. A dog chases its tail in a circle four times.

C. A fish swims all the way across a pond and then halfway back.

D. A worm moves 5 cm along a straight crack in a sidewalk.

2 The graph below represents the motion of a swimmer. Which statement best describes the swimmer's motion?

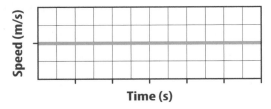

A. The swimmer is at rest.

B. The swimmer is in constant motion.

C. The swimmer's velocity is changing.

D. The swimmer is accelerating.

3 An airplane travels 290 km between Austin and Dallas in 1 h 15 min. What is its average speed?

A. 160 km/h

B. 200 km/h

C. 232 km/h

D. 250 km/h

4 Which represents a force pair?

A. A book pushes down on a table, and gravity pulls the book toward the floor.

B. A boy's foot pushes down on a bicycle pedal. The pedal pushes up on his foot.

C. A golf club hits a golf ball. Gravity pulls the ball back down to Earth.

D. A person's foot pushes on the floor, and the person's weight pushes on the floor.

Use the figure below to answer questions 5–7.

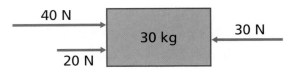

5 What is the net force on the object?

A. 30 N to the right

B. 30 N to the left

C. 60 N to the right

D. 90 N to the left

6 Which statement best describes the motion of the object?

A. It accelerates to the right.

B. It remains at rest.

C. It doesn't change speed but changes its direction of motion.

D. It moves at constant velocity to the right.

7 What is the acceleration of the object?

A. 0 m/s^2

B. 1.0 m/s^2 to the right

C. 1.6 m/s^2 to the right

D. 3 m/s^2 to the left

8 Which is a contact force?

A. A girl pulls the plug of an electric hair dryer from the socket.

B. A leaf falls to the ground because of Earth's gravitational force.

C. A magnet pulls on a nail 2 cm away.

D. A small bit of paper is pulled toward an electrically charged comb.

9 Which best describes the relationship between the force acting on an object, the object's mass, and the acceleration of the object?

A. Newton's first law of motion

B. Newton's law of inertia

C. Newton's second law of motion

D. Newton's third law of motion

Critical Thinking

10 **Contrast** the force of gravity between these pairs of objects: a 1-kg mass and a 2-kg mass that are 1 m apart; a 1-kg mass and a 2-kg mass that are 2 m apart; and two 2-kg masses that are 1 m apart.

11 **Construct** Ed rides an escalator moving at a constant speed to the second floor, which is 12 m above the first floor. The ride takes 15 s. Draw a displacement-time graph and a speed-time graph of his ride.

12 **Calculate** A marathon runner covers 42.0 km in 3 h 45 min. What was the runner's average speed?

13 **Justify** An astronomer measures the velocity of an object in space and decides that there is no net force acting on the object. Which of Newton's laws helped the astronomer make this decision?

14 **Analyze** The photo shows an astronaut tethered to a spacecraft. Use Newton's laws to describe what will happen when the astronaut pushes against the spacecraft.

Writing in Science

15 **Write** A driver followed a van with a surfboard strapped on top. The driver claims that the van stopped so quickly that the surfboard flew backward, hitting his car and causing damage. He wants the driver of the van to pay for damage to his vehicle and medical costs. You are the judge in the case. Use Newton's laws of motion to write a judgment in the case.

REVIEW THE BIG IDEA

16 While carrying a heavy box up the stairs, you set the box on a step and rest. Then you pick up the box and carry it to the top of the stairs. Describe these actions in terms of balanced and unbalanced forces acting on the box.

17 In what ways did balanced and unbalanced forces affect the motion of the acrobats in the air. What forces caused them to rise into the air? What forces are acting on them in the picture?

Math Skills

 Review

— Math Practice —

Solve One-Step Equations

18 A runner covers a distance of 1,500 m in 4 min. What is the runner's average speed?

19 Leaving the starting block, the runner accelerates from a velocity of 0 m/s to a velocity of 2 m/s in 3 s. What is the runner's acceleration?

20 What acceleration is produced when a 3,000-N force acts on a 1,200-kg car? Ignore any friction.

21 What force would a bowler have to exert on a 6-kg bowling ball to cause it to accelerate at the rate of 4 m/s^2?

Standardized Test Practice

Record your answers on the answer sheet provided by your teacher or on a sheet of paper.

Multiple Choice

1 Which is the result of an object's motion?

  A a change in mass

  B a change in position

  C a change in reference point

  D a change in volume

2 Which would be used to calculate an object's acceleration?

  A change in its speed divided by time

  B change in its velocity divided by time

  C change in its speed divided by velocity

  D change in its velocity divided by speed

Use the table below to answer questions 3 and 4.

Car	Initial Velocity (m/s)	Final Velocity (m/s)	Time (s)
A	0	25	10
B	25	15	10
C	15	25	20
D	10	10	25

3 Which car had a negative acceleration?

  A car A

  B car B

  C car C

  D car D

4 Which car or cars had an acceleration greater than 2 m/s^2?

  A car A only

  B car B only

  C cars A and C

  D cars A, C, and D

Use the graph to answer questions 5 and 6.

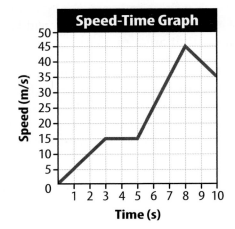

5 During which time period did the object slow down?

  A 0–3 seconds

  B 3–5 seconds

  C 5–8 seconds

  D 8–10 seconds

6 Which term describes the motion in the time period from 3 to 5 seconds?

  A at rest

  B constant speed

  C slowing down

  D speeding up

7 Which is a contact force?

  A gravity

  B friction

  C magnetic force

  D electrical force

8 Which can cause the force of gravity between two objects to increase?

   A if both objects start to spin

   B if one object increases in mass

   C if both objects decrease in mass

   D if the objects move farther apart

9 Which could be the net force acting on an object when the forces are balanced?

   A −10 N

   B 0 N

   C 2 N

   D 10 N

Use the diagram to answer question 10.

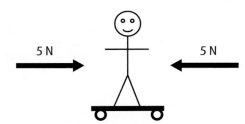

10 A skateboarder is traveling at a constant speed to the left. Suddenly the two forces shown act on him. Which describes the motion of the skateboarder when the two forces shown suddenly act on him?

   A His motion stops.

   B His speed increases.

   C His speed decreases.

   D His motion stays the same.

Constructed Response

Use the blank graph to answer questions 11 and 12.

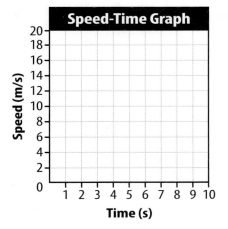

11 Describe how a period of constant acceleration would appear on a speed-time graph.

12 Describe how a period of nonconstant, positive acceleration would appear on a speed-time graph.

13 How does increasing the mass of an object affect the acceleration of an object if the forces acting on the object remain the same.? Explain.

14 According to Newton's third law of motion, what happens when you push on a sturdy wall with a force of 10 N?

NEED EXTRA HELP?													
If You Missed Question...	1	2	3	4	5	6	7	8	9	10	11	12	13	14
Go to Lesson...	1	1	1	1	1	1	2	2	2	3	1	1	3	3

Chapter 20

The Sun-Earth-Moon System

 What natural phenomena do the motions of Earth and the Moon produce?

Inquiry **Sun Bites?**

Look at this time-lapse photograph. The "bites" out of the Sun occurred during a solar eclipse. The Sun's appearance changed in a regular, predictable way as the Moon's shadow passed over a part of Earth.

- How does the Moon's movement change the Sun's appearance?
- What predictable changes does Earth's movement cause?
- What other natural phenomena do the motions of Earth and the Moon cause?

Get Ready to Read

What do you think?

Before you read, decide if you agree or disagree with each of these statements. As you read this chapter, see if you change your mind about any of the statements.

1 Earth's movement around the Sun causes sunrises and sunsets.

2 Earth has seasons because its distance from the Sun changes throughout the year.

3 The Moon was once a planet that orbited the Sun between Earth and Mars.

4 Earth's shadow causes the changing appearance of the Moon.

5 A solar eclipse happens when Earth moves between the Moon and the Sun.

6 The gravitational pull of the Moon and the Sun on Earth's oceans causes tides.

 **ConnectED** Your one-stop online resource

connectED.mcgraw-hill.com

 Video

 WebQuest

 Audio

 Assessment

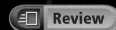 Review

 Concepts in Motion

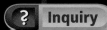 Inquiry

Multilingual eGlossary

Lesson 1

Reading Guide

Key Concepts
ESSENTIAL QUESTIONS

- How does Earth move?
- Why is Earth warmer at the equator and colder at the poles?
- Why do the seasons change as Earth moves around the Sun?

Vocabulary

orbit p. 728

revolution p. 728

rotation p. 729

rotation axis p. 729

solstice p. 733

equinox p. 733

g Multilingual eGlossary

Inquiry Floating in Space?

From the *International Space Station*, Earth might look like it is just floating, but it is actually traveling around the Sun at more than 100,000 km/h. What phenomena does Earth's motion cause?

Does Earth's shape affect temperatures on Earth's surface?

Temperatures near Earth's poles are colder than temperatures near the equator. What causes these temperature differences?

1. Read and complete a lab safety form.
2. Inflate a **spherical balloon** and tie the balloon closed.
3. Using a **marker,** draw a line around the balloon to represent Earth's equator.
4. Using a **ruler**, place a lit **flashlight** about 8 cm from the balloon so the flashlight beam strikes the equator straight on.
5. Using the marker, trace around the light projected onto the balloon.
6. Have someone raise the flashlight vertically 5–8 cm without changing the direction that the flashlight is pointing. Do not change the position of the balloon. Trace around the light projected onto the balloon again.

Think About This

1. Compare and contrast the shapes you drew on the balloon.

2. At which location on the balloon is the light more spread out? Explain your answer.

3. 🔑 **Key Concept** Use your model to explain why Earth is warmer near the equator and colder near the poles.

Earth and the Sun

If you look outside at the ground, trees, and buildings, it does not seem like Earth is moving. Yet Earth is always in motion, spinning in space and traveling around the Sun. As Earth spins, day changes to night and back to day again. The seasons change as Earth travels around the Sun. Summer changes to winter because Earth's motion changes how energy from the Sun spreads out over Earth's surface.

The Sun

The nearest star to Earth is the Sun, which is shown in **Figure 1.** The Sun is approximately 150 million km from Earth. Compared to Earth, the Sun is enormous. The Sun's diameter is more than 100 times greater than Earth's diameter. The Sun's mass is more than 300,000 times greater than Earth's mass.

Deep inside the Sun, nuclei of atoms combine, releasing huge amounts of energy. This process is called nuclear fusion. The Sun releases so much energy from nuclear fusion that the temperature at its core is more than 15,000,000°C. Even at the Sun's surface, the temperature is about 5,500°C. A small part of the Sun's energy reaches Earth as light and thermal energy.

Figure 1 The Sun is a giant ball of hot gases that emits light and energy.

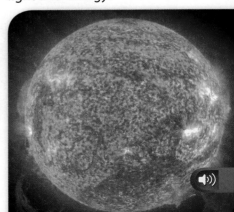

What keeps Earth in orbit?

Why does Earth move around the Sun and not fly off into space?

1. Read and complete a lab safety form.
2. Tie a piece of **strong thread** securely to a **plastic, slotted golf ball.**
3. Swing the ball in a horizontal circle above your head.

Analyze and Conclude

1. **Predict** what would happen if you let go of the thread.

2. **Key Concept** Which part of the experiment represents the force of gravity between Earth and the Sun?

Earth's Orbit

As shown in **Figure 2,** Earth moves around the Sun in a nearly circular path. *The path an object follows as it moves around another object is an* **orbit.** *The motion of one object around another object is called* **revolution.** Earth makes one complete revolution around the Sun every 365.24 days.

The Sun's Gravitational Pull

Why does Earth orbit the Sun? The answer is that the Sun's gravity pulls on Earth. The pull of gravity between two objects depends on the masses of the objects and the distance between them. The more mass either object has, or the closer together they are, the stronger the gravitational pull.

The Sun's effect on Earth's motion is illustrated in **Figure 2.** Earth's motion around the Sun is like the motion of an object twirled on a string. The string pulls on the object and makes it move in a circle. If the string breaks, the object flies off in a straight line. In the same way, the pull of the Sun's gravity keeps Earth revolving around the Sun in a nearly circular orbit. If the gravity between Earth and the Sun were to somehow stop, Earth would fly off into space in a straight line.

Key Concept Check What produces Earth's revolution around the Sun?

Figure 2 Earth moves in a nearly circular orbit. The pull of the Sun's gravity on Earth causes Earth to revolve around the Sun.

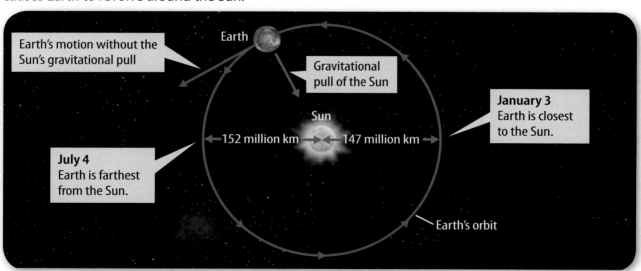

Earth's motion without the Sun's gravitational pull

Earth

Gravitational pull of the Sun

January 3
Earth is closest to the Sun.

Sun

152 million km — 147 million km

July 4
Earth is farthest from the Sun.

Earth's orbit

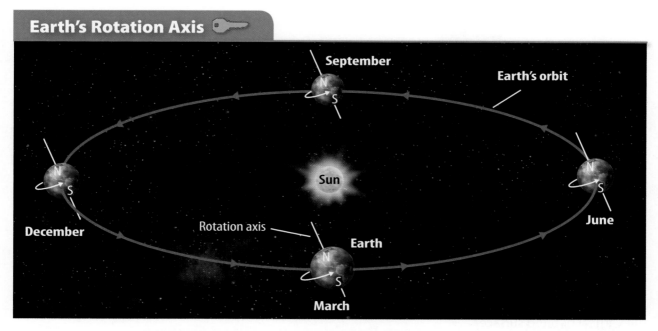

September

Earth's orbit

Sun

December

Rotation axis

Earth

June

March

Figure 3 This diagram shows Earth's orbit, which is nearly circular, from an angle. Earth spins on its rotation axis as it revolves around the Sun. Earth's rotation axis always points in the same direction.

Visual Check Between which months is the north end of Earth's rotation axis away from the Sun?

Earth's Rotation

As Earth revolves around the Sun, it spins. *A spinning motion is called* **rotation.** Some spinning objects rotate on a rod or axle. Earth rotates on an imaginary line through its center. *The line on which an object rotates is the* **rotation axis.**

Suppose you could look down on Earth's North Pole and watch Earth rotate. You would see that Earth rotates on its rotation axis in a counterclockwise direction, from west to east. One complete rotation of Earth takes about 24 hours. This rotation helps produce Earth's cycle of day and night. It is daytime on the half of Earth facing toward the Sun and nighttime on the half of Earth facing away from the Sun.

The Sun's Apparent Motion Each day the Sun appears to move from east to west across the sky. It seems as if the Sun is moving around Earth. However, it is Earth's rotation that causes the Sun's apparent motion.

Earth rotates from west to east. As a result, the Sun appears to move from east to west across the sky. The stars and the Moon also seem to move from east to west across the sky due to Earth's west to east rotation.

To better understand this, imagine riding on a merry-go-round. As you and the ride move, people on the ground appear to be moving in the opposite direction. In the same way, as Earth rotates from west to east, the Sun appears to move from east to west.

Reading Check What causes the Sun's apparent motion across the sky?

The Tilt of Earth's Rotation Axis As shown in **Figure 3,** Earth's rotation axis is tilted. The tilt of Earth's rotation axis is always in the same direction by the same amount. This means that during half of Earth's orbit, the north end of the rotation axis is toward the Sun. During the other half of Earth's orbit, the north end of the rotation axis is away from the Sun.

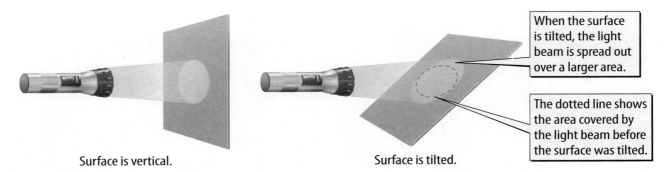

Surface is vertical.

Surface is tilted.

When the surface is tilted, the light beam is spread out over a larger area.

The dotted line shows the area covered by the light beam before the surface was tilted.

Figure 4 The light energy on a surface becomes more spread out as the surface becomes more tilted relative to the light beam.

 Visual Check Is the light energy more spread out on the vertical or tilted surface?

Temperature and Latitude

As Earth orbits the Sun, only one half of Earth faces the Sun at a time. A beam of sunlight carries energy. The more sunlight that reaches a part of Earth's surface, the warmer that part becomes. Because Earth's surface is curved, different parts of Earth's surface receive different amounts of the Sun's energy.

Energy Received by a Tilted Surface

Suppose you shine a beam of light on a flat card, as shown in **Figure 4.** As you tilt the card relative to the direction of the light beam, light becomes more spread out on the card's surface. As a result, the energy that the light beam carries also spreads out more over the card's surface. An area on the surface within the light beam receives less energy when the surface is more tilted relative to the light beam.

The Tilt of Earth's Curved Surface

Instead of being flat like a card, Earth's surface is curved. Relative to the direction of a beam of sunlight, Earth's surface becomes more tilted as you move away from the **equator.** As shown in **Figure 5,** the energy in a beam of sunlight tends to become more spread out the farther you travel from the equator. This means that regions near the poles receive less energy than regions near the equator. This makes Earth colder at the poles and warmer at the equator.

Key Concept Check Why is Earth warmer at the equator and colder at the poles?

Figure 5 Energy from the Sun becomes more spread out as you move away from the equator.

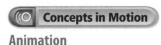 **Concepts in Motion**

Animation

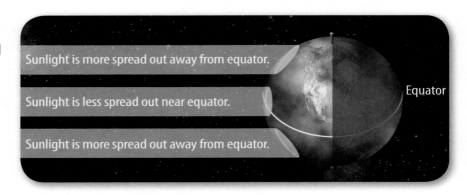

Sunlight is more spread out away from equator.

Sunlight is less spread out near equator.

Sunlight is more spread out away from equator.

Equator

North end of rotation axis is away from the Sun.

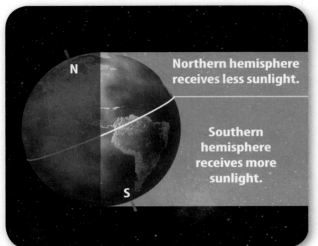

North end of rotation axis is toward the Sun.

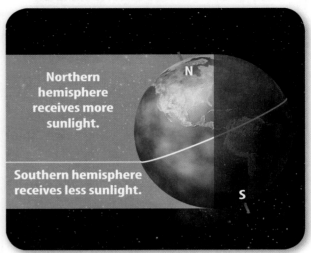

Figure 6 The northern hemisphere receives more sunlight in June, and the southern hemisphere receives more sunlight in December.

Seasons

You might think that summer happens when Earth is closest to the Sun, and winter happens when Earth is farthest from the Sun. However, seasonal changes do not depend on Earth's distance from the Sun. In fact, Earth is closest to the Sun in January! Instead, it is the tilt of Earth's rotation axis, combined with Earth's motion around the Sun, that causes the seasons to change.

Spring and Summer in the Northern Hemisphere

During one half of Earth's orbit, the north end of the rotation axis is toward the Sun. Then, the northern hemisphere receives more energy from the Sun than the southern hemisphere, as shown in **Figure 6.** Temperatures increase in the northern hemisphere and decrease in the southern hemisphere. Daylight hours last longer in the northern hemisphere, and nights last longer in the southern hemisphere. This is when spring and summer happen in the northern hemisphere, and fall and winter happen in the southern hemisphere.

Fall and Winter in the Northern Hemisphere

During the other half of Earth's orbit, the north end of the rotation axis is away from the Sun. Then, the northern hemisphere receives less solar energy than the southern hemisphere, as shown in **Figure 6.** Temperatures decrease in the northern hemisphere and increase in the southern hemisphere. This is when fall and winter happen in the northern hemisphere, and spring and summer happen in the southern hemisphere.

 Key Concept Check How does the tilt of Earth's rotation axis affect Earth's weather?

Math Skills

Convert Units

When Earth is 147,000,000 km from the Sun, how far is Earth from the Sun in miles? To calculate the distance in miles, multiply the distance in km by the conversion factor

$$147{,}000{,}000 \text{ km} \times \frac{0.62 \text{ miles}}{1 \text{ km}}$$
$$= 91{,}100{,}000 \text{ miles}$$

Practice

When Earth is 152,000,000 km from the Sun, how far is Earth from the Sun in miles?

 Review

- **Math Practice**
- **Personal Tutor**

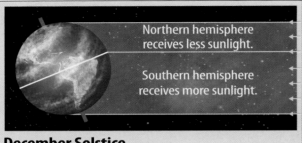

December Solstice

The December solstice is on December 21 or 22.
On this day

- the north end of Earth's rotation axis is away from the Sun;
- days in the northern hemisphere are shortest and nights are longest; winter begins;
- days in the southern hemisphere are longest and nights are shortest; summer begins.

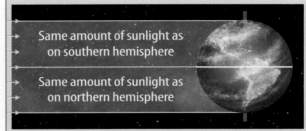

September Equinox

The September equinox is on September 22 or 23.
On this day

- the north end of Earth's rotation axis leans along Earth's orbit;
- there are about 12 hours of daylight and 12 hours of darkness everywhere on Earth;
- autumn begins in the northern hemisphere;
- spring begins in the southern hemisphere.

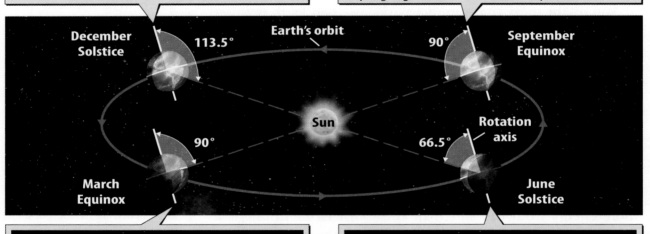

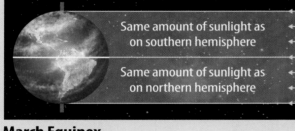

March Equinox

The March equinox is on March 20 or 21.
On this day

- the north end of Earth's rotation axis leans along Earth's orbit;
- there are about 12 hours of daylight and 12 hours of darkness everywhere on Earth;
- spring begins in the northern hemisphere;
- autumn begins in the southern hemisphere.

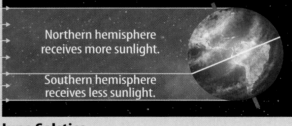

June Solstice

The June solstice is on June 20 or 21.
On this day

- the north end of Earth's rotation axis is toward the Sun;
- days in the northern hemisphere are longest and nights are shortest; summer begins;
- days in the southern hemisphere are shortest and nights are longest; winter begins.

Figure 7 The seasons change as Earth moves around the Sun. Earth's motion around the Sun causes Earth's tilted rotation axis to be leaning toward the Sun and away from the Sun.

Solstices, Equinoxes, and the Seasonal Cycle

Figure 7 shows that as Earth travels around the Sun, its rotation axis always points in the same direction in space. However, the amount that Earth's rotation axis is toward or away from the Sun changes. This causes the seasons to change in a yearly cycle.

There are four days each year when the direction of Earth's rotation axis is special relative to the Sun. *A* **solstice** *is a day when Earth's rotation axis is the most toward or away from the Sun. An* **equinox** *is a day when Earth's rotation axis is leaning along Earth's orbit, neither toward nor away from the Sun.*

March Equinox to June Solstice When the north end of the rotation axis gradually points more and more toward the Sun, the northern hemisphere gradually receives more solar energy. This is spring in the northern hemisphere.

June Solstice to September Equinox The north end of the rotation axis continues to point toward the Sun but does so less and less. The northern hemisphere starts to receive less solar energy. This is summer in the northern hemisphere.

September Equinox to December Solstice The north end of the rotation axis now points more and more away from the Sun. The northern hemisphere receives less and less solar energy. This is fall in the northern hemisphere.

December Solstice to March Equinox The north end of the rotation axis continues to point away from the Sun but does so less and less. The northern hemisphere starts to receive more solar energy. This is winter in the northern hemisphere.

Changes in the Sun's Apparent Path Across the Sky

Figure 8 shows how the Sun's apparent path through the sky changes from season to season in the northern hemisphere. The Sun's apparent path through the sky in the northern hemisphere is lowest on the December solstice and highest on the June solstice.

FOLDABLES

Make a bound book with four full pages. Label the pages with the names of the solstices and equinoxes. Use each page to organize information about each season.

WORD ORIGIN · · · · · · · · · · ·

equinox
from Latin *equinoxium*, means "equality of night and day"

Figure 8 As the seasons change, the path of the Sun across the sky changes. In the northern hemisphere, the Sun's path is lowest on the December solstice and highest on the June solstice.

✓ **Visual Check** When is the Sun highest in the sky in the northern hemisphere?

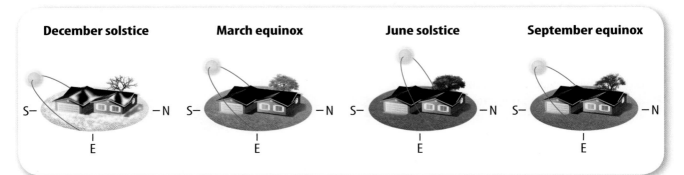

December solstice   March equinox   June solstice   September equinox

Visual Summary

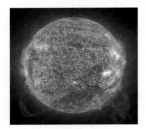

The gravitational pull of the Sun causes Earth to revolve around the Sun in a near-circular orbit.

Earth's rotation axis is tilted and always points in the same direction in space.

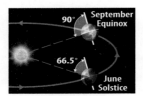

Equinoxes and solstices are days when the direction of Earth's rotation axis relative to the Sun is special.

FOLDABLES®

Use your lesson Foldable to review the lesson. Save your Foldable for the project at the end of the chapter.

What do you think NOW?

You first read the statements below at the beginning of the chapter.

1. Earth's movement around the Sun causes sunrises and sunsets.

2. Earth has seasons because its distance from the Sun changes throughout the year.

Did you change your mind about whether you agree or disagree with the statements? Rewrite any false statements to make them true.

Use Vocabulary

1 **Distinguish** between Earth's rotation and Earth's revolution.

2 The path Earth follows around the Sun is Earth's _____.

3 When a(n) _____ occurs, the northern hemisphere and the southern hemisphere receive the same amount of sunlight.

Understand Key Concepts

4 What is caused by the tilt of Earth's rotational axis?
   A. Earth's orbit   C. Earth's revolution
   B. Earth's seasons   D. Earth's rotation

5 **Contrast** the amount of sunlight received by an area near the equator and a same-sized area near the South Pole.

6 **Contrast** the Sun's gravitational pull on Earth when Earth is closest to the Sun and when Earth is farthest from the Sun.

Interpret Graphics

7 **Summarize** Copy and fill in the table below for the seasons in the northern hemisphere.

Season	Starts on Solstice or Equinox?	How Rotation Axis Leans
Summer		
Fall		
Winter		
Spring		

Critical Thinking

8 **Defend** The December solstice is often called the winter solstice. Do you think this is an appropriate label? Defend your answer.

Math Skills ×÷+

— Math Practice —

9 The Sun's diameter is about 1,390,000 km. What is the Sun's diameter in miles?

Materials

large foam ball

wooden skewer

foam cup

masking tape

flashlight

marker

Safety

How does Earth's tilted rotation axis affect the seasons?

The seasons change as Earth revolves around the Sun. How does Earth's tilted rotation axis change how sunlight spreads out over different parts of Earth's surface?

Learn It

Using a flashlight as the Sun and a foam ball as Earth, you can model how solar energy spreads out over Earth's surface at different times during the year. This will help you **draw conclusions** about Earth's seasons.

Try It

1. Read and complete a lab safety form.

2. Insert a wooden skewer through the center of a foam ball. Draw a line on the ball to represent Earth's equator. Insert one end of the skewer into an upside-down foam cup so the skewer tilts.

3. Prop a flashlight on a stack of books about 0.5 m from the ball. Turn on the flashlight and position the ball so the skewer points toward the flashlight, representing the June solstice.

4. In your Science Journal, draw how the ball's surface is tilted relative to the light beam.

5. Under your diagram, state whether the upper (northern) or lower (southern) hemisphere receives more light energy.

6. With the skewer always pointing in the same direction, move the ball around the flashlight. Turn the flashlight to keep the light on the ball. At the three positions corresponding to the equinoxes and other solstice, make drawings like those in step 4 and statements like those in step 5.

Apply It

7. How did the tilt of the surfaces change relative to the light beam as the ball circled the flashlight?

8. How did the amount of light energy on each hemisphere change as the ball moved around the flashlight?

9. **Key Concept** Draw conclusions about how Earth's tilt affects the seasons.

Reading Guide

Key Concepts 🔑
ESSENTIAL QUESTIONS

- How does the Moon move around Earth?
- Why does the Moon's appearance change?

Vocabulary
maria p. 738

phase p. 740

waxing phase p. 740

waning phase p. 740

g Multilingual eGlossary

Earth's Moon

Inquiry Two Planets?

The smaller body is Earth's Moon, not a planet. Just as Earth moves around the Sun, the Moon moves around Earth. The Moon's motion around Earth causes what kinds of changes to occur?

Why does the Moon appear to change shape?

The Sun is always shining on Earth and the Moon. However, the Moon's shape seems to change from night to night and day to day. What could cause the Moon's appearance to change?

1 Read and complete a lab safety form.

2 Place a **ball** on a level surface.

3 Position a **flashlight** so that the light beam shines fully on one side of the ball. Stand behind the flashlight.

4 Make a drawing of the ball's appearance in your Science Journal.

5 Stand behind the ball, facing the flashlight, and repeat step 4.

6 Stand to the left of the ball and repeat step 4.

Think About This

1. What caused the ball's appearance to change?

2. 🔑 **Key Concept** What do you think produces the Moon's changing appearance in the sky?

Seeing the Moon

Imagine what people thousands of years ago thought when they looked up at the Moon. They might have wondered why the Moon shines and why it seems to change shape. They probably would have been surprised to learn that the Moon does not emit light at all. Unlike the Sun, the Moon is a solid object that does not emit its own light. You only see the Moon because light from the Sun reflects off the Moon and into your eyes. Some facts about the Moon, such as its mass, size, and distance from Earth, are shown in **Table 1.**

FOLDABLES

Use two sheets of paper to make a bound book. Use it to organize information about the lunar cycle. Each page of your book should represent one week of the lunar cycle.

Table 1 Moon Data				
Mass	Diameter	Average distance from Earth	Time for one rotation	Time for one revolution
1.2% of Earth's mass	27% of Earth's diameter	384,000 km	27.3 days	27.3 days

Figure 9 The Moon probably formed when a large object collided with Earth 4.5 billion years ago. Material ejected from the collision eventually clumped together and became the Moon.

(((◉) **Concepts in Motion**   **Animation**

An object the size of Mars crashes into the semi-molten Earth about 4.5 billion years ago.

The impact ejects vaporized rock into space. As the rock cools, it forms a ring of particles around Earth.

The particles gradually clump together and form the Moon.

WORD ORIGIN ·

maria
from Latin *mare*, means "sea"
· ·

The Moon's Formation

The most widely accepted idea for the Moon's formation is the giant impact hypothesis, shown in **Figure 9.** According to this hypothesis, shortly after Earth formed about 4.6 billion years ago, an object about the size of the planet Mars collided with Earth. The impact ejected vaporized rock that formed a ring around Earth. Eventually, the material in the ring cooled, clumped together, and formed the Moon.

The Moon's Surface

The surface of the Moon was shaped early in its history. Examples of common features on the Moon's surface are shown in **Figure 10.**

Craters The Moon's craters were formed when objects from space crashed into the Moon. Light-colored streaks called rays extend outward from some craters.

Most of the impacts that formed the Moon's craters occurred more than 3.5 billion years ago, long before dinosaurs lived on Earth. Earth was also heavily bombarded by objects from space during this time. However, on Earth, wind, liquid water, and plate tectonics erased the craters. The Moon has no atmosphere, liquid water, or plate tectonics, so craters formed billions of years ago on the Moon have hardly changed.

Maria *The large, dark, flat areas on the Moon are called* **maria** (MAR ee uh). The maria formed after most impacts on the Moon's surface had stopped. Maria formed when lava flowed up through the Moon's crust and solidified. The lava covered many of the Moon's craters and other features. When this lava solidified, it was dark and flat.

✔ **Reading Check** How were maria produced?

Highlands The light-colored highlands are too high for the lava that formed the maria to reach. The highlands are older than the maria and are covered with craters.

The Moon's Surface Features

Highlands

The impacts of many objects helped shape the highlands. The highlands are the oldest and most highly-cratered regions on the Moon.

Rays

The bright streaks around this crater are rays. The impacts that formed craters also blasted out the material that formed rays.

Maria

This region is one of the Moon's maria. Its smooth surface is solid lava.

Craters

On the Moon's surface are millions of craters of many sizes. The diameter of the largest crater in this image is about 76 km.

▲ **Figure 10** The Moon's surface features include craters, rays, maria, and highlands.

The Moon's Motion

While Earth is revolving around the Sun, the Moon is revolving around Earth. The gravitational pull of Earth on the Moon causes the Moon to move in an orbit around Earth. The Moon makes one revolution around Earth every 27.3 days.

 Key Concept Check What produces the Moon's revolution around Earth?

The Moon also rotates as it revolves around Earth. One complete rotation of the Moon also takes 27.3 days. This means the Moon makes one rotation in the same amount of time that it makes one revolution around Earth. **Figure 11** shows that, because the Moon makes one rotation for each revolution of Earth, the same side of the Moon always faces Earth. This side of the Moon is called the near side. The side of the Moon that cannot be seen from Earth is called the far side of the Moon.

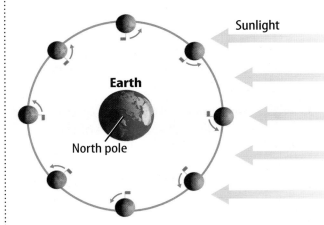

▲ **Figure 11** The Moon rotates once on its axis and revolves around Earth in the same amount of time. As a result, the same side of the Moon always faces Earth.

How can the Moon be rotating if the same side of the Moon is always facing Earth?

The Moon revolves around Earth. Does the Moon also rotate as it revolves around Earth?

1. Choose a partner. One person represents the Moon. The other represents Earth.

2. While Earth is still, the Moon moves slowly around Earth, always facing the same wall.

3. Next, the Moon moves around Earth always facing Earth.

Analyze and Conclude

1. For which motion was the Moon rotating?

2. For each type of motion, how many times did the Moon rotate during one revolution around Earth?

3. 🔑 **Key Concept** How is the Moon actually rotating if the same side of the Moon is always facing Earth?

SCIENCE USE V. COMMON USE······················

phase
Science Use how the Moon or a planet is lit as seen from Earth

Common Use a part of something or a stage of development

Phases of the Moon

The Sun is always shining on half of the Moon, just as the Sun is always shining on half of Earth. However, as the Moon moves around Earth, usually only part of the Moon's near side is lit. *The portion of the Moon or a planet reflecting light as seen from Earth is called a* **phase.** As shown in **Figure 12,** the motion of the Moon around Earth causes the phase of the Moon to change. The sequence of phases is the lunar cycle. One lunar cycle takes 29.5 days or slightly more than four weeks to complete.

🔑 **Key Concept Check** What produces the phases of the Moon?

Waxing Phases

During the **waxing phases,** *more of the Moon's near side is lit each night.*

Week 1—First Quarter As the lunar cycle begins, a sliver of light can be seen on the Moon's western edge. Gradually the lit part becomes larger. By the end of the first week, the Moon is at its first quarter phase. In this phase, the Moon's entire western half is lit.

Week 2—Full Moon During the second week, more and more of the near side becomes lit. When the Moon's near side is completely lit, it is at the full moon phase.

Waning Phases

During the **waning phases,** *less of the Moon's near side is lit each night.* As seen from Earth, the lit part is now on the Moon's eastern side.

Week 3—Third Quarter During this week, the lit part of the Moon becomes smaller until only the eastern half of the Moon is lit. This is the third quarter phase.

Week 4—New Moon During this week, less and less of the near side is lit. When the Moon's near side is completely dark, it is at the new moon phase.

Figure 12 As the Moon revolves around Earth, the part of the Moon's near side that is lit changes. The figure below shows how the Moon looks at different places in its orbit.

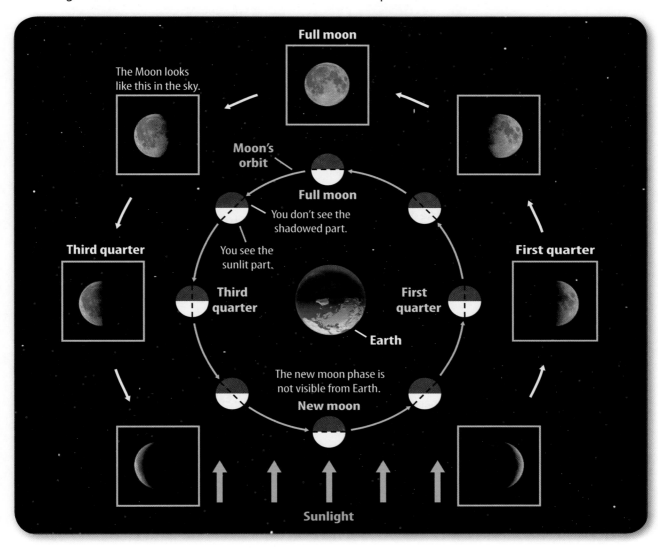

The Moon at Midnight

The Moon's motion around Earth causes the Moon to rise, on average, about 50 minutes later each day. The figure below shows how the Moon looks at midnight during three phases of the lunar cycle.

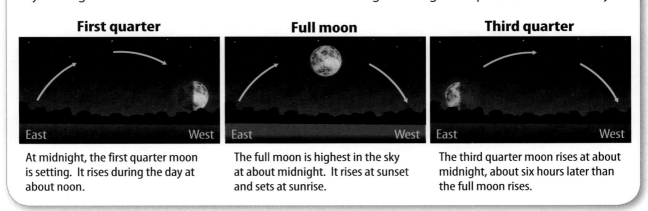

First quarter

At midnight, the first quarter moon is setting. It rises during the day at about noon.

Full moon

The full moon is highest in the sky at about midnight. It rises at sunset and sets at sunrise.

Third quarter

The third quarter moon rises at about midnight, about six hours later than the full moon rises.

Lesson 2 Review

Visual Summary

According to the giant impact hypothesis, a large object collided with Earth about 4.5 billion years ago to form the Moon.

Features like maria, craters, and highlands formed on the Moon's surface early in its history.

The Moon's phases change in a regular pattern during the Moon's lunar cycle.

FOLDABLES

Use your lesson Foldable to review the lesson. Save your Foldable for the project at the end of the chapter.

What do you think NOW?

You first read the statements below at the beginning of the chapter.

3. The Moon was once a planet that orbited the Sun between Earth and Mars.

4. Earth's shadow causes the changing appearance of the Moon.

Did you change your mind about whether you agree or disagree with the statements? Rewrite any false statements to make them true.

Use Vocabulary

1 The lit part of the Moon as viewed from Earth is a(n) _____.

2 For the first half of the lunar cycle, the lit part of the Moon's near side is _____.

3 For the second half of the lunar cycle, the lit part of the Moon's near side is _____.

Understand Key Concepts

4 Which phase occurs when the Moon is between the Sun and Earth?
A. first quarter   C. new moon
B. full moon   D. third quarter

5 **Reason** Why does the Moon have phases?

Interpret Graphics

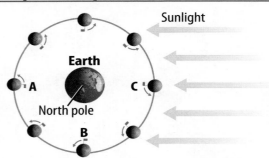

6 **Draw** how the Moon looks from Earth when it is at positions A, B, and C in the diagram above.

7 **Organize Information** Copy and fill in the table below with details about the lunar surface.

Crater	
Ray	
Maria	
Highland	

Critical Thinking

8 **Reflect** Imagine the Moon rotates twice in the same amount of time the Moon orbits Earth once. Would you be able to see the Moon's far side from Earth?

Return to the Moon

Exploring Earth's Moon is a step toward exploring other planets and building outposts in space.

The United States undertook a series of human spaceflight missions from 1961–1975 called the Apollo program. The goal of the program was to land humans on the Moon and bring them safely back to Earth. Six of the missions reached this goal. The Apollo program was a huge success, but it was just the beginning.

NASA began another space program that had a goal to return astronauts to the Moon to live and work. However, before that could happen, scientists needed to know more about conditions on the Moon and what materials are available there.

Collecting data was the first step. In 2009, NASA launched the *Lunar Reconnaissance Orbiter (LRO)* spacecraft. The *LRO* spent a year orbiting the Moon's two poles. It collected detailed data that scientists can use to make maps of the Moon's features and resources, such as deep craters that formed on the Moon when comets and asteroids slammed into it billions of years ago. Some scientists predicted that these deep craters contain frozen water.

One of the instruments launched with the *LRO* was the *Lunar Crater Observation and Sensing Satellite (LCROSS)*. LCROSS observations confirmed the scientists' predictions that water exists on the Moon. A rocket launched from *LCROSS* impacted the Cabeus crater near the Moon's south pole. The material that was ejected after the rocket's impact included water.

NASA's goal of returning astronauts to the Moon was delayed, and their missions now focus on exploring Mars instead. But the discoveries made on the Moon will help scientists develop future missions that could take humans farther into the solar system.

Apollo SPACE PROGRAM

The Apollo Space Program included 17 missions. Here are some milestones:

January 27 1967
Apollo 1 Fire killed all three astronauts on board during a launch simulation for the first piloted flight to the Moon.

December 21–27 1968
Apollo 8 First manned spacecraft orbits the Moon.

July 16–24 1969
Apollo 11 First humans, Neil Armstrong and Buzz Aldrin, walk on the Moon.

July 1971
Apollo 15 Astronauts drive the first rover on the Moon.

December 7–19 1972
Apollo 17 The first phase of human exploration of the Moon ended with this last lunar landing mission.

It's Your Turn

BRAINSTORM As a group, brainstorm the different occupations that would be needed to successfully operate a base on the Moon or another planet. Discuss the tasks that a person would perform in each occupation.

Eclipses and Tides

Reading Guide

Key Concepts 🔑
ESSENTIAL QUESTIONS

- What is a solar eclipse?
- What is a lunar eclipse?
- How do the Moon and the Sun affect Earth's oceans?

Vocabulary

umbra p. 745
penumbra p. 745
solar eclipse p. 746
lunar eclipse p. 748
tide p. 749

 Multilingual eGlossary

 Video

- BrainPOP®
- Science Video

Inquiry What is this dark spot?

Cosmonauts took this photo from aboard the *Mir* orbiting space station. An eclipse caused the shadow that you see. Do you know what kind of eclipse?

How do shadows change?

You can see a shadow when an object blocks a light source. What happens to an object's shadow when the object moves?

1 Read and complete a lab safety form.

2 Select an **object** provided by your teacher.

3 Shine a **flashlight** on the object, projecting its shadow on the wall.

4 While holding the flashlight in the same position, move the object closer to the wall— away from the light. Then, move the object toward the light. Record your observations in your Science Journal.

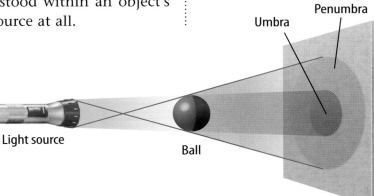

Think About This

1. Compare and contrast the shadows created in each situation. Did the shadows have dark parts and light parts? Did these parts change?

2.  **Key Concept** Imagine you look at the flashlight from behind your object, looking from the darkest and lightest parts of the object's shadow. How much of the flashlight could you see from each location?

Shadows—the Umbra and the Penumbra

A shadow results when one object blocks the light that another object emits or reflects. When a tree blocks light from the Sun, it casts a shadow. If you want to stand in the shadow of a tree, the tree must be in a line between you and the Sun.

If you go outside on a sunny day and look carefully at a shadow on the ground, you might notice that the edges of the shadow are not as dark as the rest of the shadow. Light from the Sun and other wide sources casts shadows with two distinct parts, as shown in **Figure 13**. *The* **umbra** *is the central, darker part of a shadow where light is totally blocked. The* **penumbra** *is the lighter part of a shadow where light is partially blocked.* If you stood within an object's penumbra, you would be able to see only part of the light source. If you stood within an object's umbra, you would not see the light source at all.

WORD ORIGIN

penumbra
from Latin *paene*, means "almost"; and *umbra*, means "shade, shadow"

Figure 13 The shadow that a wide light source produces has two parts— the umbra and the penumbra. The light source cannot be seen from within the umbra. The light source can be partially seen from within the penumbra.

Penumbra

Umbra

Light source

Ball

What does the Moon's shadow look like?

Like every shadow cast by a wide light source, the Moon's shadow has two parts.

1 Read and complete a lab safety form.

2 Working with a partner, use a **pencil** to connect two **foam balls**. One ball should be one-fourth the size of the other.

3 While one person holds the balls, the other should stand 1 m away and shine a **flashlight** or **desk lamp** on the balls. The balls and light should be in a direct line, with the smallest ball closest to the light.

4 Sketch and describe your observations in your Science Journal.

Analyze and Conclude

1. **Key Concept** Explain the relationship between the two types of shadows and solar eclipses.

Solar Eclipses

As the Sun shines on the Moon, the Moon casts a shadow that extends out into space. Sometimes the Moon passes between Earth and the Sun. This can only happen during the new moon phase. When Earth, the Moon, and the Sun are lined up, the Moon casts a shadow on Earth's surface, as shown in **Figure 14.** You can see the Moon's shadow in the photo at the beginning of this lesson. *When the Moon's shadow appears on Earth's surface, a* **solar eclipse** *is occurring.*

Key Concept Check Why does a solar eclipse occur only during a new moon?

As Earth rotates, the Moon's shadow moves along Earth's surface, as shown in **Figure 14.** The type of eclipse you see depends on whether you are in the path of the umbra or the penumbra. If you are outside the umbra and penumbra, you cannot see a solar eclipse at all.

Total Solar Eclipses

You can only see a total solar eclipse from within the Moon's umbra. During a total solar eclipse, the Moon appears to cover the Sun completely, as shown in **Figure 15** on the next page. Then, the sky becomes dark enough that you can see stars. A total solar eclipse lasts no longer than about 7 minutes.

Solar Eclipse

Figure 14 A solar eclipse occurs only when the Moon moves directly between Earth and the Sun. The Moon's shadow moves across Earth's surface.

Visual Check Why would a person in North America not see the solar eclipse shown here?

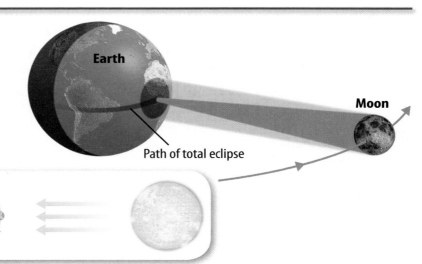

Earth

Moon

Path of total eclipse

Penumbra

Umbra

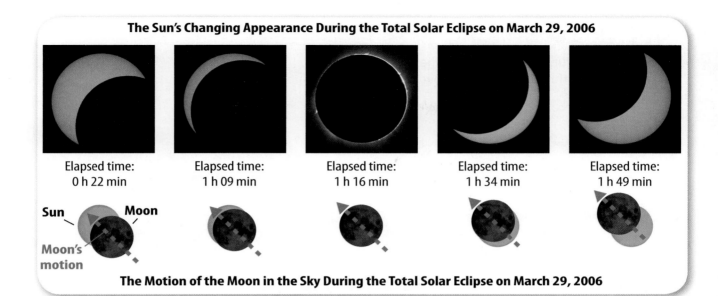

The Sun's Changing Appearance During the Total Solar Eclipse on March 29, 2006

| Elapsed time: 0 h 22 min | Elapsed time: 1 h 09 min | Elapsed time: 1 h 16 min | Elapsed time: 1 h 34 min | Elapsed time: 1 h 49 min |

Sun — **Moon**

Moon's motion

The Motion of the Moon in the Sky During the Total Solar Eclipse on March 29, 2006

Partial Solar Eclipses

You can only see a total solar eclipse from within the Moon's umbra, but you can see a partial solar eclipse from within the Moon's much larger penumbra. The stages of a partial solar eclipse are similar to the stages of a total solar eclipse, except that the Moon never completely covers the Sun.

Why don't solar eclipses occur every month?

Solar eclipses only can occur during a new moon, when Earth and the Sun are on opposite sides of the Moon. However, solar eclipses do not occur during every new moon phase. **Figure 16** shows why. The Moon's orbit is tilted slightly compared to Earth's orbit. As a result, during most new moons, Earth is either above or below the Moon's shadow. However, every so often the Moon is in a line between the Sun and Earth. Then the Moon's shadow passes over Earth and a solar eclipse occurs.

Figure 15 This sequence of photographs shows how the Sun's appearance changed during a total solar eclipse in 2006.

☑ **Visual Check** How much time elapsed from the start to the finish of this sequence?

Figure 16 A solar eclipse occurs only when the Moon crosses Earth's orbit and is in a direct line between Earth and the Sun.

((○ Concepts in Motion Animation

The Moon's Tilted Orbit

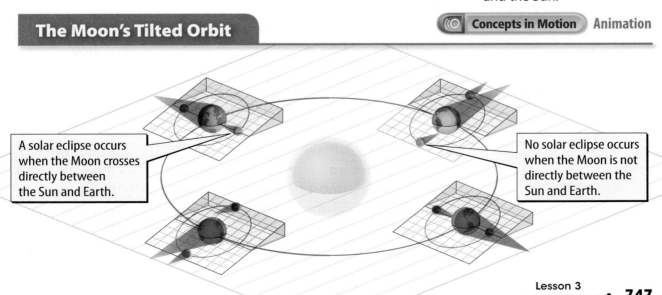

A solar eclipse occurs when the Moon crosses directly between the Sun and Earth.

No solar eclipse occurs when the Moon is not directly between the Sun and Earth.

Figure 17 A lunar eclipse occurs when the Moon moves through Earth's shadow.

Penumbra

Umbra

✏️ **Visual Check** Why would more people be able to see a lunar eclipse than a solar eclipse?

Lunar Eclipses

Just like the Moon, Earth casts a shadow into space. As the Moon revolves around Earth, it sometimes moves into Earth's shadow, as shown in **Figure 17.** A **lunar eclipse** *occurs when the Moon moves into Earth's shadow.* Then Earth is in a line between the Sun and the Moon. This means that a lunar eclipse can occur only during the full moon phase.

Like the Moon's shadow, Earth's shadow has an umbra and a penumbra. Different types of lunar eclipses occur depending on which part of Earth's shadow the Moon moves through. Unlike solar eclipses, you can see any lunar eclipse from any location on the side of Earth facing the Moon.

🔑 **Key Concept Check** When can a lunar eclipse occur?

Total Lunar Eclipses

When the entire Moon moves through Earth's umbra, a total lunar eclipse occurs. **Figure 18** on the next page shows how the Moon's appearance changes during a total lunar eclipse. The Moon's appearance changes as it gradually moves into Earth's penumbra, then into Earth's umbra, back into Earth's penumbra, and then out of Earth's shadow entirely.

You can still see the Moon even when it is completely within Earth's umbra. Although Earth blocks most of the Sun's rays, Earth's atmosphere deflects some sunlight into Earth's umbra. This is also why you can often see the unlit portion of the Moon on a clear night. Astronomers often call this Earthshine. This reflected light has a reddish color and gives the Moon a reddish tint during a total lunar eclipse.

FOLDABLES

Make a two-tab book from a sheet of notebook paper. Label the tabs *Solar Eclipse* and *Lunar Eclipse.* Use it to organize your notes on eclipses.

Solar Eclipse   Lunar Eclipse

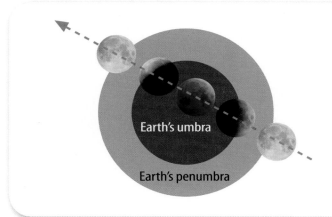

Figure 18 If the entire Moon passes through Earth's umbra, the Moon gradually darkens until a dark shadow covers it completely.

Earth's umbra

Earth's penumbra

✓ **Visual Check** How would a total lunar eclipse look different from a total solar eclipse?

Partial Lunar Eclipses

When only part of the Moon passes through Earth's umbra, a partial lunar eclipse occurs. The stages of a partial lunar eclipse are similar to those of a total lunar eclipse, shown in **Figure 18,** except the Moon is never completely covered by Earth's umbra. The part of the Moon in Earth's penumbra appears only slightly darker, while the part of the Moon in Earth's umbra appears much darker.

Why don't lunar eclipses occur every month?

Lunar eclipses can only occur during a full moon phase, when the Moon and the Sun are on opposite sides of Earth. However, lunar eclipses do not occur during every full moon because of the tilt of the Moon's orbit with respect to Earth's orbit. During most full moons, the Moon is slightly above or slightly below Earth's penumbra.

Tides

The positions of the Moon and the Sun also affect Earth's oceans. If you have spent time near an ocean, you might have seen how the ocean's height, or sea level, rises and falls twice each day. A **tide** is *the daily rise and fall of sea level.* Examples of tides are shown in **Figure 19.** It is primarily the Moon's gravity that causes Earth's oceans to rise and fall twice each day.

Figure 19 In the Bay of Fundy, high tides can be more than 10 m higher than low tides.

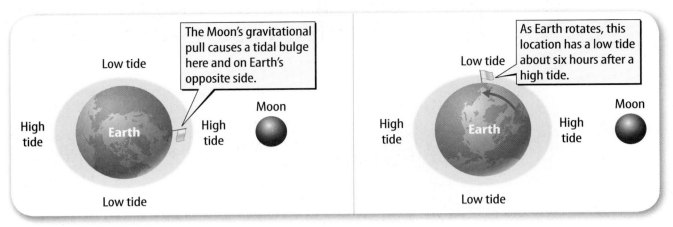

Figure 20 🔑 In this view down on Earth's North Pole, the flag moves into a tidal bulge as Earth rotates. A coastal area has a high tide about once every 12 hours.

The Moon's Effect on Earth's Tides

The difference in the strength of the Moon's gravity on opposite sides of Earth causes Earth's tides. The Moon's gravity is slightly stronger on the side of Earth closer to the Moon and slightly weaker on the side of Earth opposite the Moon. These differences cause tidal bulges in the oceans on opposite sides of Earth, shown in **Figure 20**. High tides occur at the tidal bulges, and low tides occur between them.

The Sun's Effect on Earth's Tides

Because the Sun is so far away from Earth, its effect on tides is about half that of the Moon. **Figure 21** shows how the positions of the Sun and the Moon affect Earth's tides.

Spring Tides During the full moon and new moon phases, spring tides occur. This is when the Sun's and the Moon's gravitational effects combine and produce higher high tides and lower low tides.

Neap Tides A week after a spring tide, a neap tide occurs. Then the Sun, Earth, and the Moon form a right angle. When this happens, the Sun's effect on tides reduces the Moon's effect. High tides are lower and low tides are higher at neap tides.

🔑 **Key Concept Check** Why is the Sun's effect on tides less than the Moon's effect?

Figure 21 A spring tide occurs when the Sun, Earth, and the Moon are in a line. A neap tide occurs when the Sun and the Moon form a right angle with Earth.

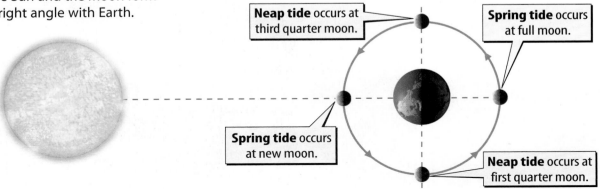

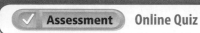
Visual Summary

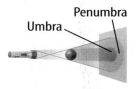

Penumbra
Umbra

Shadows from a wide light source have two distinct parts.

The Moon's shadow produces solar eclipses. Earth's shadow produces lunar eclipses.

Low tide
High tide Earth High tide
Moon
Low tide

The positions of the Moon and the Sun in relation to Earth cause gravitational differences that produce tides.

FOLDABLES

Use your lesson Foldable to review the lesson. Save your Foldable for the project at the end of the chapter.

What do you think NOW?

You first read the statements below at the beginning of the chapter.

5. A solar eclipse happens when Earth moves between the Moon and the Sun.

6. The gravitational pull of the Moon and the Sun on Earth's oceans causes tides.

Did you change your mind about whether you agree or disagree with the statements? Rewrite any false statements to make them true.

Use Vocabulary

1 **Distinguish** between an umbra and a penumbra.

2 **Use the term** *tide* in a sentence.

3 The Moon turns a reddish color during a total _____ eclipse.

Understand Key Concepts

4 **Summarize** the effect of the Sun on Earth's tides.

5 **Illustrate** the positions of the Sun, Earth, and the Moon during a solar eclipse and during a lunar eclipse.

6 **Contrast** a total lunar eclipse with a partial lunar eclipse.

7 Which could occur during a total solar eclipse?
   A. first quarter moon   **C.** neap tide
   B. full moon   **D.** spring tide

Interpret Graphics

8 **Conclude** What type of eclipse does the figure above illustrate?

9 **Categorize Information** Copy and fill in the graphic organizer below to identify two bodies that affect Earth's tides.

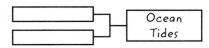

Ocean Tides

Critical Thinking

10 **Compose** a short story about a person long ago viewing a total solar eclipse.

11 **Research** ways to view a solar eclipse safely. Summarize your findings here.

Phases of the Moon

Materials

foam ball

pencil

lamp

stool

Safety

The Moon appears slightly different every night of its 29.5-day lunar cycle. The Moon's appearance changes as Earth and the Moon move. Depending on where the Moon is in relation to Earth and the Sun, observers on Earth see only part of the light the Moon reflects from the Sun.

Question

How do the positions of the Sun, the Moon, and Earth cause the phases of the Moon?

Procedure

1. Read and complete a lab safety form.

2. Hold a foam ball that represents the Moon. Make a handle for the ball by inserting a pencil about two inches into the ball. Your partner will represent an observer on Earth. Have your partner sit on a stool and record observations during the activity.

3. Place a lamp on a desk or other flat surface. Remove the shade from the lamp. The lamp represents the Sun.

4. Turn on the lamp and darken the lights in the room.
   ⚠ *Do not touch the bulb or look directly at it after the lamp is turned on.*

5. Position the Earth observer's stool about 1 m from the Sun. Position the Moon 0.5–1 m from the observer so that the Sun, Earth, and the Moon are in a line. The student holding the Moon holds the Moon so it is completely illuminated on one half. The observer records the phase and what the phase looks like in a data table.

6. Move the Moon clockwise about one-eighth of the way around its "orbit" of Earth. The observer swivels on the stool to face the Moon and records the phase.

7. Continue the Moon's orbit until the Earth observer has recorded all the Moon's phases.

8. Return to your positions as the Moon and Earth observer. Choose a part in the Moon's orbit that you did not model. Predict what the Moon would look like in that position, and check if your prediction is correct.

Analyze and Conclude

9. **Explain** Use your observations to explain how the positions of the Sun, the Moon, and Earth produce the different phases of the Moon.

10. **The Big Idea** Why is half of the Moon always lit? Why do you usually see only part of the Moon's lit half?

11. **Draw Conclusions** Based on your observations, why is the Moon not visible from Earth during the new moon phase?

12. **Summarize** Which parts of your model were waxing phases? Which parts were waning phases?

13. **Think Critically** During which phases of the Moon can eclipses occur? Explain.

Communicate Your Results

Create a poster of the results from your lab. Illustrate various positions of the Sun, the Moon, and Earth and draw the phase of the Moon for each. Include a statement of your hypothesis on the poster.

inquiry Extension

The Moon is not the only object in the sky that has phases when viewed from Earth. The planets Venus and Mercury also have phases. Research the phases of these planets and create a calendar that shows when the various phases of Venus and Mercury occur.

Lab Tips

☑ Make sure the observer's head does not cast a shadow on the Moon.

☑ The student holding the Moon should hold the pencil so that he or she always stands on the unlit side of the Moon.

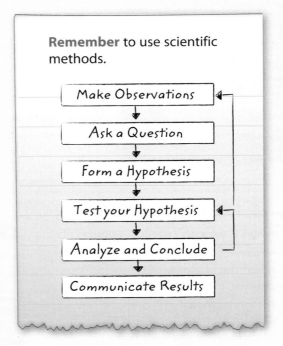

Remember to use scientific methods.

Make Observations
↓
Ask a Question
↓
Form a Hypothesis
↓
Test your Hypothesis
↓
Analyze and Conclude
↓
Communicate Results

Chapter 20 Study Guide

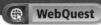

 THE BIG IDEA

Earth's motion around the Sun causes seasons. The Moon's motion around Earth causes phases of the Moon. Earth and the Moon's motions together cause eclipses and ocean tides.

Key Concepts Summary

Lesson 1: Earth's Motion

- The gravitational pull of the Sun on Earth causes Earth to revolve around the Sun in a nearly circular **orbit.**

- Areas on Earth's curved surface become more tilted with respect to the direction of sunlight the farther you travel from the equator. This causes sunlight to spread out closer to the poles, making Earth colder at the poles and warmer at the equator.

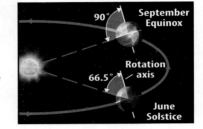

- As Earth revolves around the Sun, the tilt of Earth's **rotation axis** produces changes in how sunlight spreads out over Earth's surface. These changes in the concentration of sunlight cause the seasons.

Lesson 2: Earth's Moon

- The gravitational pull of Earth on the Moon makes the Moon revolve around Earth. The Moon rotates once as it makes one complete orbit around Earth.

- The lit part of the Moon that you can see from Earth—the Moon's **phase**—changes during the lunar cycle as the Moon revolves around Earth.

Lesson 3: Eclipses and Tides

- When the Moon's shadow appears on Earth's surface, a **solar eclipse** occurs.

- When the Moon moves into Earth's shadow, a **lunar eclipse** occurs.

- The gravitational pull of the Moon and the Sun on Earth produces **tides**, the rise and fall of sea level that occurs twice each day.

Vocabulary

orbit p. 728
revolution p. 728
rotation p. 729
rotation axis p. 729
solstice p. 733
equinox p. 733

maria p. 738
phase p. 740
waxing phase p. 740
waning phase p. 740

umbra p. 745
penumbra p. 745
solar eclipse p. 746
lunar eclipse p. 748
tide p. 749

FOLDABLES® Chapter Project

Assemble your Lesson Foldables as shown to make a Chapter Project. Use the project to review what you have learned in this chapter.

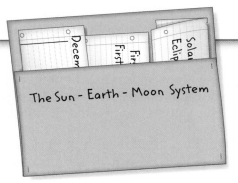

The Sun - Earth - Moon System

Use Vocabulary

Distinguish between the terms in the each of the following pairs.

1. revolution, orbit
2. rotation, rotation axis
3. solstice, equinox
4. waxing phases, waning phases
5. umbra, penumbra
6. solar eclipse, lunar eclipse
7. tide, phase

Link Vocabulary and Key Concepts

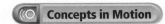
Concepts in Motion Interactive Concept Map

Copy this concept map, and then use vocabulary terms from the previous page to complete the concept map.

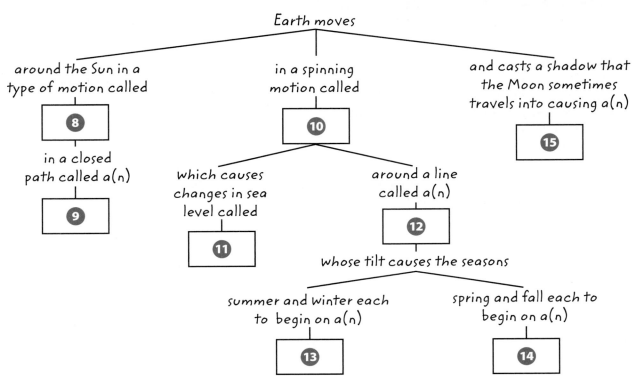

Earth moves

around the Sun in a type of motion called [8]

in a closed path called a(n) [9]

in a spinning motion called [10]

which causes changes in sea level called [11]

around a line called a(n) [12]

whose tilt causes the seasons

summer and winter each to begin on a(n) [13]

spring and fall each to begin on a(n) [14]

and casts a shadow that the Moon sometimes travels into causing a(n) [15]

Chapter 20 Review

Understand Key Concepts

1 Which property of the Sun most affects the strength of gravitational attraction between the Sun and Earth?

A. mass
B. radius
C. shape
D. temperature

2 Which would be different if Earth rotated from east to west but at the same rate?

A. the amount of energy striking Earth
B. the days on which solstices occur
C. the direction of the Sun's apparent motion across the sky
D. the number of hours in a day

3 In the image below, which season is the northern hemisphere experiencing?

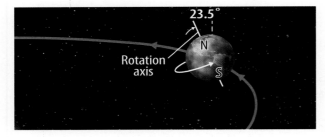

A. fall
B. spring
C. summer
D. winter

4 Which best explains why Earth is colder at the poles than at the equator?

A. Earth is farther from the Sun at the poles than at the equator.
B. Earth's orbit is not a perfect circle.
C. Earth's rotation axis is tilted.
D. Earth's surface is more tilted at the poles than at the equator.

5 How are the revolutions of the Moon and Earth alike?

A. Both are produced by gravity.
B. Both are revolutions around the Sun.
C. Both orbits are the same size.
D. Both take the same amount of time.

6 Which moon phase occurs about one week after a new moon?

A. another new moon
B. first quarter moon
C. full moon
D. third quarter moon

7 Why is the same side of the Moon always visible from Earth?

A. The Moon does not revolve around Earth.
B. The Moon does not rotate.
C. The Moon makes exactly one rotation for each revolution around Earth.
D. The Moon's rotation axis is not tilted.

8 About how often do spring tides occur?

A. once each month
B. once each year
C. twice each month
D. twice each year

9 If a coastal area has a high tide at 7:00 A.M., at about what time will the next low tide occur?

A. 11:00 A.M.
B. 1:00 P.M.
C. 3:00 P.M.
D. 7:00 P.M.

10 Which type of eclipse would a person standing at point X in the diagram below see?

A. partial lunar eclipse
B. partial solar eclipse
C. total lunar eclipse
D. total solar eclipse

Critical Thinking

11 **Outline** the ways Earth moves and how each affects Earth.

12 **Create** a poster that illustrates and describes the relationship between Earth's tilt and the seasons.

13 **Contrast** Why can you see phases of the Moon but not phases of the Sun?

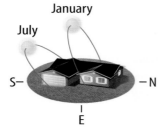
January
July
S—
—N
E

14 **Interpret Graphics** The figure above shows the Sun's position in the sky at noon in January and July. Is the house located in the northern hemisphere or the southern hemisphere? Explain.

15 **Illustrate** Make a diagram of the Moon's orbit and phases. Include labels and explanations with your drawing.

16 **Differentiate** between a total solar eclipse and a partial solar eclipse.

17 **Generalize** the reason that solar and lunar eclipses do not occur every month.

18 **Role Play** Write and present a play with several classmates that explains the causes and types of tides.

Writing in Science

19 **Survey** a group of at least ten people to determine how many know the cause of Earth's seasons. Write a summary of your results, including a main idea, supporting details, and a concluding sentence.

REVIEW **THE BIG IDEA**

20 At the South Pole, the Sun does not appear in the sky for six months out of the year. When does this happen? What is happening at the North Pole during these months? Explain why Earth's poles receive so little solar energy.

21 A solar eclipse, shown in the time-lapse photo below, is one phenomenon that the motions of Earth and the Moon produce. What other phenomena do the motions of Earth and the Moon produce?

Math Skills ✕ ÷ +

 Review
— Math Practice —

Convert Units

22 When the Moon is 384,000 km from Earth, how far is the Moon from Earth in miles?

23 If you travel 205 mi on a train from Washington D.C. to New York City, how many kilometers do you travel on the train?

24 The nearest star other than the Sun is about 40 trillion km away. About how many miles away is the nearest star other than the Sun?

Standardized Test Practice

Record your answers on the answer sheet provided by your teacher or on a sheet of paper.

Multiple Choice

1 Which is the movement of one object around another object in space?

   A axis

   B orbit

   C revolution

   D rotation

Use the diagram below to answer question 2.

Time 1

Time 2

2 What happens between times *1* and *2* in the diagram above?

   A Days grow shorter and shorter.

   B The season changes from fall to winter.

   C The region begins to point away from the Sun.

   D The region gradually receives more solar energy.

3 How many times larger is the Sun's diameter than Earth's diameter?

   A about 10 times larger

   B about 100 times larger

   C about 1,000 times larger

   D about 10,000 times larger

4 Which diagram illustrates the Moon's third quarter phase?

   A

   B

   C

   D

5 Which accurately describes Earth's position and orientation during summer in the northern hemisphere?

   A Earth is at its closest point to the Sun.

   B Earth's hemispheres receive equal amounts of solar energy.

   C The north end of Earth's rotational axis leans toward the Sun.

   D The Sun emits a greater amount of light and heat energy.

6 Which are large, dark lunar areas formed by cooled lava?

   A craters

   B highlands

   C maria

   D rays

7 During one lunar cycle, the Moon

   A completes its east-to-west path across the sky exactly once.

   B completes its entire sequence of phases.

   C progresses only from the new moon phase to the full moon phase.

   D revolves around Earth twice.

Standardized Test Practice

Use the diagram below to answer question 8.

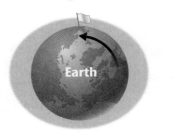

Moon

8 What does the flag in the diagram above represent?

A high tide

B low tide

C neap tide

D spring tide

9 During which lunar phase might a solar eclipse occur?

A first quarter moon

B full moon

C new moon

D third quarter moon

10 Which does the entire Moon pass through during a partial lunar eclipse?

A Earth's penumbra

B Earth's umbra

C the Moon's penumbra

D the Moon's umbra

Constructed Response

Use the diagram below to answer questions 11 and 12.

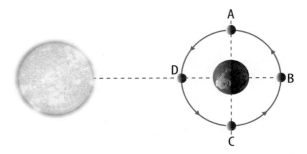

11 Where are neap tides indicated in the above diagram? What causes neap tides? What happens during a neap tide?

12 Where are spring tides indicated in the above diagram? What causes spring tides? What happens during a spring tide?

13 How would Earth's climate be different if its rotational axis were not tilted?

14 Why can we see only one side of the Moon from Earth? What is the name given to this side of the Moon?

15 What is a lunar phase? How do waxing and waning phases differ?

16 Why don't solar eclipses occur monthly?

NEED EXTRA HELP?																
If You Missed Question...	1	2	3	4	5	6	7	8	9	10	11	12	13	14	15	16
Go to Lesson...	1	1	1	2	1	2	2	3	3	3	3	3	1	2	2	3

Chapter 21

Exploring Our Solar System

THE BIG IDEA How and where do scientists look for life in the solar system?

Inquiry Can life exist here?

Mars is a harsh, desolate place. Yet scientists think it is possible that liquid water exists below the Martian surface. Does this mean that life might exist on Mars?

- What do scientists know about our solar system, and how do they study it?

- What conditions must be met before humans can travel into space?

- How and where do scientists look for life in the solar system?

Get Ready to Read

What do you think?

Before you read, decide if you agree or disagree with each of these statements. As you read this chapter, see if you change your mind about any of the statements.

1 Our solar system has eight planets.

2 Earth's atmosphere is mostly oxygen.

3 Earth's atmosphere protects life on Earth from dangerous solar radiation.

4 Scientists think conditions for life might exist on some moons in the solar system.

5 Astronauts float in space because there is no gravity above Earth's atmosphere.

6 The United States is the only country with a human space-flight program.

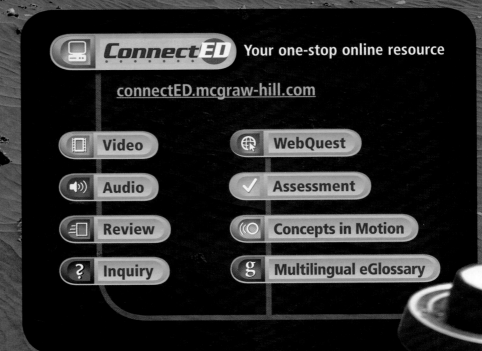

ConnectED Your one-stop online resource

connectED.mcgraw-hill.com

Video

WebQuest

Audio

Assessment

Review

Concepts in Motion

Inquiry

Multilingual eGlossary

Lesson 1

Our Solar System

Reading Guide

Key Concepts 🔑
ESSENTIAL QUESTIONS

- How do objects in the solar system move?
- How did distance from the Sun affect the makeup of objects in the solar system?
- What objects are in the solar system?

Vocabulary

astronomical unit p. 764
planet p. 766
dwarf planet p. 768
satellite p. 768
meteoroid p. 769
meteor p. 769
meteorite p. 769

g Multilingual eGlossary

▣ Video
- BrainPOP®
- Science Video

Inquiry From Beyond Earth?

Yes! This rock is the Hoba meteorite in Namibia, Africa. Scientists think it fell to Earth about 80,000 years ago. Where do you think this meteorite came from?

What makes planets layered?

Earth has a crust, a mantle, and a core. The crust is the least dense part of Earth, and the core is the most dense. Other planets also have layers of different density and makeup. Why do the planets form layers?

1. Read and complete a lab safety form.

2. Examine **balloons** in an **aquarium.** Carefully handle the balloons so they do not break. In your Science Journal, explain how you think the balloons differ. Predict what will happen when you add water to the aquarium.

3. Use a **pitcher** to fill the aquarium with water. Pour the water gently to avoid splashing.

4. Observe and record what happens to the balloons.

Think About This

1. How do you think the balloons are like the materials that make up the planets?

2. 🔑 **Key Concept** Why do you think the planets have layered structures? Do you think the materials that make up of the inner planets differ from the materials of the outer planets?

Origin and Structure of Our Solar System

A solar system is a group of objects that revolves around a star. You might be familiar with our solar system—the eight planets and other objects that revolve around the Sun. Our Sun is not the only star in the universe that has a solar system, but our solar system is the only one scientists can study in detail.

Formation of the Solar System

Did you know there are clouds in space? Unlike clouds in Earth's atmosphere, clouds in space are made mostly of hydrogen gas. Five billion years ago, our solar system formed from a spinning cloud of hydrogen gas and dust, probably similar to the cloud in **Figure 1.** When gravity caused the cloud to collapse, the cloud began to spin faster. The cloud also got hotter. When the center of the cloud became hot enough for nuclear reactions to occur, a star formed—the Sun. The stars you see in the night sky formed in much the same way.

As the cloud containing the Sun continued to spin, it flattened, with the Sun at its center. Small pieces of ice and rock orbiting the Sun clumped together and formed small, rocky or icy bodies called planetesimals (pla ne TE sih mulz). Gravity pulled some of the planetesimals together. They merged and grew in size, forming planets, asteroids, and other objects.

✓ **Reading Check** How did planetesimals become planets?

Figure 1 The dark patch is not a hole in space. It is a cloud of hydrogen gas and dust that blocks the light from stars behind it.

(◎) **Concepts in Motion**

Animation

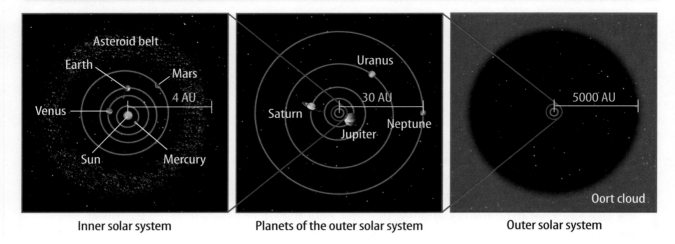

Inner solar system | Planets of the outer solar system | Outer solar system

Figure 2 The solar system often is divided into regions depending on distance from the Sun.

 Visual Check Which planet is farthest from the Sun?

Distances Within the Solar System

Our solar system is so large that scientists rarely use kilometers to measure distances within it. Instead, they use astronomical units. *An* **astronomical unit** (AU) *is Earth's average distance from the Sun, nearly 150 million km.* As shown in **Figure 2,** the inner solar system extends to about 4 AU from the Sun. It contains the planets closest to the Sun—Mercury, Venus, Earth, and Mars. It also contains the asteroid belt between Mars and Jupiter, 2–4 AU from the Sun.

The four planets farthest from the Sun—Jupiter, Saturn, Uranus, and Neptune—are part of the outer solar system. The outer solar system extends from Jupiter's orbit to the Oort (ORT) cloud. The Oort cloud is a large, spherical shell of icy planetesimals that scientists think orbits the Sun from about 5,000 AU to as far as 50,000 AU away.

✓ **Reading Check** How far does the solar system extend?

Makeup of the Solar System

You live on a planet with a solid, rocky surface. But the solar system is made mostly of hydrogen gas. Ices, rocks, and metals make up less than 2 percent of our solar system's mass.

Hydrogen gas is the least dense material in the solar system. Only the Sun, Jupiter, and Saturn are massive enough to have large amounts of hydrogen. Ices are denser than gases. They are made mostly of water, carbon dioxide, methane, or ammonia. Rocks are denser than ices and are made mostly of **silicates.** Metals, such as iron, are the solar system's densest materials.

REVIEW VOCABULARY ·····

silicates
compounds composed mostly of silicon and oxygen, with smaller amounts of other atoms

Movement in the Solar System

The Sun contains 99 percent of the solar system's mass. Most other objects in the solar system revolve around it, held in orbit by the Sun's enormous gravitational pull. Revolution is the movement of one object around another object. The closer objects are to the Sun, the faster they revolve. While they revolve, most objects in the solar system also rotate, or spin, on their axes. Both of these movements—rotation and revolution—are regular and predictable.

Direction of Motion

The motion of an object around the Sun is like the motion of an object whirled on a string. The string pulls on the object just as the Sun's gravity pulls on planets and other objects that revolve around it.

Recall that our solar system formed from a spinning cloud of gas and dust. Objects that formed from this material spun in the same direction. Planets and most other solar system objects still revolve around the Sun in the same direction. If you were above Earth's North Pole and looked down on the solar system, you would see objects revolving in a counterclockwise direction. The Sun and six of the eight planets, including Earth, also rotate in a counterclockwise direction. Venus and Uranus rotate clockwise.

 Key Concept Check Why do most objects in our solar system move in the same direction?

The View from Earth

Earth rotates on its axis once every 24 h. But on Earth, it seems as though you are standing still, and the Sun, the Moon, and stars move around you. While Earth rotates from west to east, objects in the sky appear to move from east to west. The same is true on a merry-go-round. As you move in one direction on a merry-go-round, objects around you seem to move in the opposite direction.

 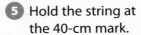

Objects in Our Solar System

Our solar system contains billions of objects. Scientists group these objects into categories based on makeup, size, distance from the Sun, and whether the object orbits the Sun or another object.

Recall that the solar system formed from a cloud of gas and dust that was extremely hot at its center, where the Sun formed. As regions beyond the Sun cooled, some of the gases solidified into ices, rocks, and metals. Ices formed far from the Sun, where temperatures were extremely cold. Closer to the Sun, temperatures were too high for ices to form. Most gases there solidified into rocks and metals. The densest matter, the metals, sank to the centers of the largest objects throughout the solar system.

Key Concept Check How did distance from the Sun affect the makeup of objects in the solar system?

The Sun

The Sun is made mostly of hydrogen gas. It also contains helium and tiny amounts of other elements. It is the only star in our solar system, and it the largest object in the solar system. The Sun's diameter is 10 times that of Jupiter and more than 100 times that of Earth.

Planets

A **planet** *orbits the Sun, is large enough to be nearly spherical in shape, and has no other large object in its orbital path.* **Figure 3** shows the eight planets in their order from the Sun.

The four inner planets formed from rocks and metals. They are smaller than the outer planets, have few or no moons, and rotate slowly. The four outer planets formed mostly from gas and ice. They are large, have many moons, rotate quickly, and have rings. The planets are described in **Table 1**.

Figure 3 The Sun is the largest object in the solar system.

Visual Check How do the inner and outer planets differ?

Sun • Mercury • Venus • Earth • Mars • Jupiter • Saturn • Uranus • Neptune

Distances not to scale

Planet	Interior	Makeup and Atmosphere
Mercury 0.39 AU from the Sun 4,900 km diameter		Mercury has a large metal core under its small rocky mantle. Its surface is covered with craters and looks much like the surface of Earth's moon. Mercury has no permanent atmosphere.
Venus 0.72 AU from the Sun 12,100 km diameter		Venus is similar to Earth in size and makeup. Its rocky mantle surrounds a molten or partially molten metal core. Its thick carbon dioxide atmosphere traps thermal energy, making Venus's surface the hottest of all the planets.
Earth 1 AU from the Sun 12,800 km diameter		Most of Earth is covered by a thin layer of liquid water. Earth has a rocky mantle, a molten outer metal core, and a solid inner metal core. Its atmosphere is 80 percent nitrogen and 20 percent oxygen.
Mars 1.5 AU from the Sun 6,800 km diameter		Mars has a rocky mantle and a partially molten metal core. Its thin atmosphere is mostly carbon dioxide. Iron in its surface rock gives the planet a reddish color. The surface of Mars has ice but no liquid water.
Jupiter 5.2 AU from the Sun 143,000 km diameter		Jupiter is more massive than all the other planets combined. Under its atmosphere of hydrogen gas is a layer of liquid hydrogen. Rock and metal have sunk to its core. A thin, barely visible ring system surrounds it.
Saturn 9.6 AU from the Sun 121,000 km diameter		Saturn's makeup is similar to that of Jupiter, but its atmosphere is hazier. Saturn's rings, the most distinctive rings of all the outer planets, are made mostly of small particles of ice.
Uranus 19 AU from the Sun 51,100 km diameter		Uranus has a hydrogen gas outer layer; a fluid inner layer made of water, methane, and ammonia; and a rocky core. Uranus has a blue-green color because of the small amount of methane in its cloud layers. It has thin rings.
Neptune 30 AU from the Sun 49,500 km diameter		Neptune is more massive than Uranus, but is slightly smaller. Neptune and Uranus have similar makeup. Neptune is a deeper blue because it has more methane in its atmosphere. Like the other outer planets, it has thin rings.

Earth's
Moon

Eris

Pluto

Ceres

▲ Figure 4 Dwarf planets are small. Even the largest known dwarf planet, Eris, is smaller than Earth's moon.

Figure 5 The asteroid Ida is about 50 km long. The small object at the right is Ida's satellite, Dactyl. ▼

Dwarf Planets

Dwarf planets *orbit the Sun and are nearly spherical in shape, but they share their orbital paths with other objects of similar size.* All known dwarf planets, including Pluto, are smaller than Earth's moon, as shown in **Figure 4.** There are at least five dwarf planets in our solar system. However, scientists think the solar system might contain hundreds of dwarf planets, most of them orbiting the Sun beyond Neptune. At least one dwarf planet, Ceres (SIHR eez), shown in **Figure 4 ,** orbits the Sun between the orbits of Mars and Jupiter, in the asteroid belt.

Asteroids

Asteroids are small, rocky or metallic objects that are remnants from the solar system's formation. There are hundreds of thousands of asteroids in the asteroid belt, but they are so small that their total mass is less than the mass of Earth's moon. Though most asteroids exist in the asteroid belt, some are found elsewhere in the solar system. Most asteroids, such as Ida, shown in **Figure 5,** have irregular shapes and craters.

✓ **Reading Check** Where do most asteroids exist?

Natural Satellites

A **satellite** *is an object that orbits a larger object other than a star.* Natural satellites are also known as moons. There are over 170 moons in the solar system. Most of them orbit planets or dwarf planets, but some moons orbit smaller objects, such as asteroids. As shown in **Figure 5,** Ida has a moon. It is called Dactyl (DAK tul). Ida is 20 times larger than Dactyl. Similar to planets, satellites have different makeups depending on their location. The satellites of the inner planets are mostly rock. Jupiter's moons are a mixture of rock and ice. The moons around the three outermost planets are mostly ice.

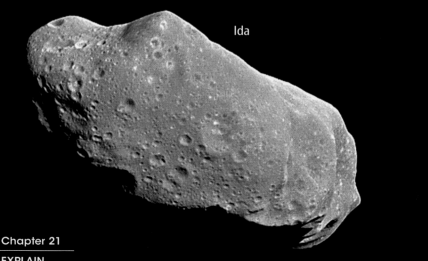

Ida

Dactyl

Kuiper Belt and Comets

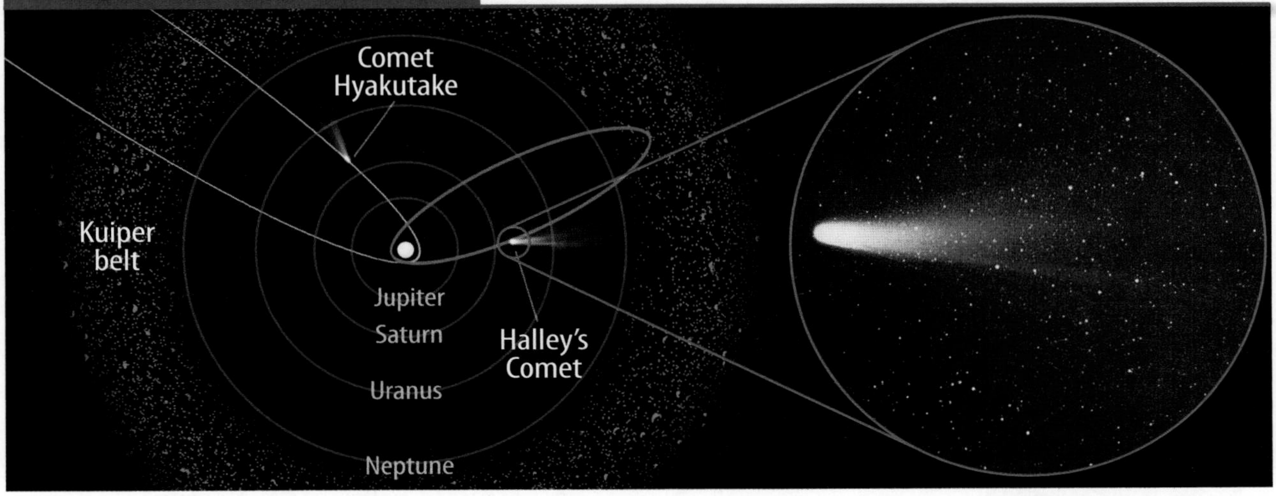

Figure 6 Kuiper belt objects orbit the Sun beyond Neptune. Some comets originate in the Kuiper belt. Others originate farther away.

☑ **Visual Check** Which direction does a comet's tail point?

Kuiper Belt Objects

You read that the asteroid belt lies between the orbits of Mars and Jupiter. The Kuiper (KI pur) belt, illustrated in **Figure 6,** is a similar but much larger belt of objects between 30 and 50 AU from the Sun. Like the asteroid belt, the Kuiper belt contains remnants from the solar system's formation. While asteroids are primarily rock and metal, Kuiper belt objects are mostly ice. Pluto is the best-known object in the Kuiper belt.

Comets

Comets are small objects made mostly of ice. Most originate in the Kuiper belt or the Oort cloud. Comets revolve around the Sun with long, stretched-out orbits, as shown in **Figure 6.** As a comet nears the Sun, some of its ice becomes gas and forms the comet's tail. The tails of some comets extend millions of kilometers into space. A comet loses mass with each orbit. After a certain number of orbits, the comet breaks up.

Meteoroids, Meteors, and Meteorites

A **meteoroid** *is a solar system object that is smaller than an asteroid or a comet.* Some meteoroids are the results of collisions between asteroids. Some are debris from comets. *A* **meteor** *is the streak of light created when a meteoroid enters Earth's atmosphere.* Most meteoroids are no bigger than a grain of rice and are only visible when they heat up and glow as they pass through Earth's atmosphere. A few large meteoroids become meteorites. *A* **meteorite** *is a meteoroid that strikes Earth.* The meteorite shown in the photo at the beginning of this lesson is the largest known meteorite on Earth.

WORD ORIGIN ············

meteor
from Greek *meteoros*, means "high up"

🔑 **Key Concept Check** What objects are in our solar system?

Lesson 1 Review

Visual Summary

The inner solar system extends from the Sun through the asteroid belt.

Even some asteroids have satellites.

Comet tails can be millions of kilometers long.

FOLDABLES

Use your lesson Foldable to review the lesson. Save your Foldable for the project at the end of the chapter.

What do you think NOW?

You first read the statements below at the beginning of the chapter.

1. Our solar system has eight planets.

2. Earth's atmosphere is mostly oxygen.

Did you change your mind about whether you agree or disagree with the statements? Rewrite any false statements to make them true.

Use Vocabulary

1 **Use the term** *astronomical unit* in a sentence.

2 **Distinguish** between a meteoroid and a meteorite.

3 Pluto is a(n) _____.

Understand Key Concepts 🔑

4 Which is farthest from the Sun?
- **A.** asteroid belt
- **C.** Neptune
- **B.** Kuiper belt
- **D.** Oort cloud

5 **Describe** the motions of objects in our solar system.

6 **Relate** how distance from the Sun affects the makeup of an object.

Interpret Graphics

7 **Compare and Contrast** How are these planets similar? How are they different?

8 **Summarize** Copy and fill in the graphic organizer below. List the materials that make up our solar system in order of their density, from low to high.

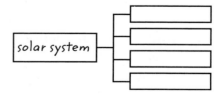

Critical Thinking

9 **Assess** how Earth might be different if it had formed five times farther from the Sun.

10 **Conclude** why Earth has a metal core.

Detecting Space Junk

Earth is surrounded by debris.

AMERICAN MUSEUM OF NATURAL HISTORY

Since *Sputnik I*, the first human-made satellite, orbited Earth in 1957, humans have launched thousands of satellites into orbit around Earth. Many of these satellites are no longer used, yet they remain in orbit. Humans also have sent many other spacecraft into space. Parts of the rockets that launched these crafts, as well as parts of the crafts themselves, also remain in orbit around Earth. As a result, the space around our planet is becoming littered with junk.

Space junk, or orbital debris, is any piece of human-made material that orbits Earth and serves no function. It could be almost anything: an abandoned satellite, a chunk of old rocket, or a chip of paint. Whizzing through space at 26,000 km/h, even a tiny object could cause serious damage to a spacecraft or an astronaut. To avoid dangerous collisions, scientists at NASA and elsewhere track space junk. The U.S. Space Surveillance Network uses a variety of optical and radar-imaging tools across the globe to track over 13,000 Earth-orbiting objects with diameters greater than 10 cm . To detect smaller pieces of orbital debris, scientists rely mostly on the Haystack radar. How does it work?

▲ Millions of pieces of space junk circle Earth. Most are smaller than a marble. In this image, the dots represent Earth-orbiting objects larger than 10 cm in diameter. Of these, only 5 percent are working satellites.

The Haystack Long Range Imaging Radar in Massachusetts generally monitors objects smaller than 30 cm in diameter in low-Earth orbit. This is a region of space between 160 km and 2,000 km above Earth's surface, where it is most crowded with space junk.

2 Radio waves sent out by the antenna bounce off objects and return to the antenna.

1 The radar antenna focuses on one area in the sky and detects objects that pass through its field of view.

3 The radio-wave data are fed into computers and used by scientists to estimate the number, size, speed, and location of the objects.

▲ To monitor small objects that orbit Earth, NASA relies mostly on the Haystack radar in Massachusetts.

It's Your Turn

PROBLEM SOLVING With a group, research ideas for cleaning up Earth's orbit. Decide which one you think would be most effective. Draw a diagram of how the process would work. Share your research and diagram with the class.

Reading Guide

Key Concepts
ESSENTIAL QUESTIONS

- What conditions on Earth enable life to exist?

- What conditions on other bodies in the solar system might enable life to exist?

- Where might life possibly exist beyond Earth?

Vocabulary
astrobiology p. 773

organic p. 775

geyser p. 777

 Multilingual eGlossary

Video

- **What's Science Got to do With It?**

Life in the Solar System

Inquiry Extraterrestrial Life?

No, scientists have not yet found life beyond Earth. This is a methane ice worm that burrows deep into methane ice in the Gulf of Mexico. How do you think studying organisms that live in extreme environments on Earth helps scientists search for life beyond Earth?

Inquiry Launch Lab

10 minutes

How are Earth's organisms protected from harmful solar energy?

The Sun's ultraviolet, or UV, radiation is useful to many organisms on Earth. However, too much UV radiation can be harmful.

1. Read and complete a lab safety form.
2. Use **scissors** to cut out a **cardboard circle** about 5 cm wide.
3. Use **tape** to attach the circle to the piece of **special paper** provided by your teacher.
4. Place your paper, with the cardboard circle on top, on a sunny windowsill for 3 min.
5. Take the paper back to your work area. With the room darkened, remove the cardboard circle from the paper. Observe the paper, and draw what you see in your Science Journal.

Think About This

1. The cardboard circle models the part of Earth that protects the planet's organisms from harmful UV energy. Which part of Earth do you think the circle models?

2. **Key Concept** What conditions on Earth do you think enable life to exist?

Conditions for Life on Earth

Life exists in nearly every environment on Earth. Some environments have conditions so extreme that humans cannot live in them. These places might have extreme temperatures, high salt levels, total darkness, or little water. Even though humans cannot live in these places, other organisms can.

Despite the extreme conditions in which some organisms live, all of Earth's life-forms need the same basic things to survive: a source of energy, liquid water, and nourishment. Scientists have not yet found life anywhere else in the solar system. But by studying the conditions that support life on Earth, they are learning about conditions that might support life elsewhere. **Astrobiology** *is the study of the origin, development, distribution, and future of life in the universe.*

 Key Concept Check What do organisms on Earth need to survive?

Energy from the Sun

The Sun is the source of almost all energy on Earth. Sunlight provides light and thermal energy. It also provides energy for plants, which are at the base of most food chains. However, a small percentage of organisms on Earth receive energy from chemicals or from Earth itself, such as the animals shown in **Figure 7.**

WORD ORIGIN

astrobiology
from Greek *astron*, means "star"; Greek *bios*, means "life"; and Greek *logia*, means "study of"

Figure 7 A variety of animals live in complete darkness near hot water jets in the ocean floor.

Protection by the Atmosphere

Earth's moon receives about the same amount of sunlight as Earth. Yet conditions on the Moon are more extreme than they are on Earth. The Moon's surface temperature can rise to 100°C during the day and drop to −150°C at night. Temperatures are extreme on the Moon because the Moon, unlike Earth, does not have an atmosphere.

Maintains Temperatures Earth's atmosphere is like a blanket around Earth. It absorbs sunlight during the day and keeps heat from escaping into space during the night. It maintains Earth's average surface temperature at a comfortable 14°C.

 Reading Check How is Earth's atmosphere like a blanket?

Absorbs Harmful Radiation Have you ever had a painful sunburn? Sunburns are caused by the Sun's ultraviolet light. Even though you cannot see ultraviolet light, you can feel its effects. Too much ultraviolet light can harm you. Fortunately, Earth's atmosphere absorbs most of the Sun's ultraviolet light, as well as X-rays and other potentially harmful light from the Sun. The atmosphere also helps protect Earth from highly charged particles that erupt from the Sun in powerful storms.

Burns Up Meteoroids Earth's atmosphere also protects Earth's surface from meteoroids. Millions of meteoroids strike Earth's atmosphere every day. But almost all of them burn up in the atmosphere before they can hit Earth's surface.

Inquiry MiniLab

15 minutes

What is one factor that makes Earth "just right" for life?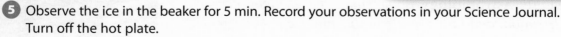

Do you know the story of Goldilocks, who finds the baby bear's food, chair, and bed "just right"? In this lab, you will explore one reason why Earth is "just right" for life.

1. Read and complete a lab safety form.

2. Your group will perform one part of this three-part activity. Determine which part your group will do.

3. Position a **beaker** over a **hot plate** at the position determined for your group.

4. Quickly transfer **ice** from the **cooler** to the beaker. Immediately turn the hot plate on high.
⚠ **CAUTION:** Be careful when using the hot plate.

5. Observe the ice in the beaker for 5 min. Record your observations in your Science Journal. Turn off the hot plate.

Analyze and Conclude

1. **Identify Variables** What were the independent and dependent variables in this experiment?

2. **Draw Conclusions** How was the ice in your group's beaker affected by the hot plate? How was the ice in the other groups' beakers affected? Why do you think they were different?

3. 🔑 **Key Concept** Based on your observations, what is one factor that makes Earth "just right" for life to exist on it?

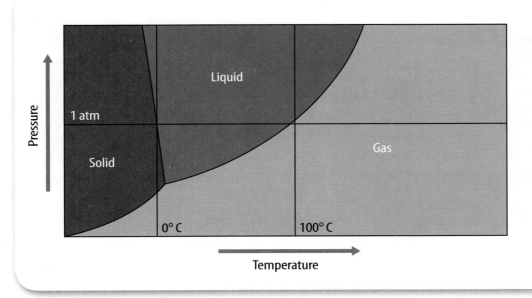

Figure 8 Water changes from liquid into gas or into solid as temperatures and pressures change.

✓ **Visual Check**
What happens to water when temperatures are high?

Liquid Water

Liquid water is necessary for all life on Earth. Water dissolves minerals and transports molecules in cells. Without liquid water, cells could not function and life would not exist. Earth's atmosphere keeps pressures and temperatures on Earth's surface within a range where water can exist as a liquid.

Depending on temperature and pressure on Earth, water is solid, liquid, or gas, as shown in **Figure 8**. At sea level on Earth (1 atm of pressure), water is liquid between 0°C and 100°C. Above 100°C, water boils and becomes water vapor. Below 0°C, it freezes into ice. However, at different altitudes on Earth, such as on the top of a mountain, the boiling and freezing temperatures of water change slightly because the pressure in the atmosphere changes. Without Earth's atmosphere, pressures on Earth's surface would be too low for water to be liquid. Water would exist only as water vapor or ice.

 Key Concept Check What would happen to water on Earth's surface if Earth had no atmosphere?

Nourishment

Living things are nourished by nutrients they take from the air, water, and land around them. They use the nutrients for energy, growth, and other processes, such as reproduction and cellular repair. All molecules that provide nourishment for life on Earth contain carbon. They are organic molecules. **Organic** *refers to a class of chemical compounds in living organisms that are based on carbon.* Though it is possible that inorganic life could exist elsewhere, astrobiologists are most interested in places beyond Earth where water is liquid and carbon is plentiful.

FOLDABLES

Make a three-tab Venn book. Label it as shown. Use it to compare and contrast the ability of Earth and the Moon to sustain life.

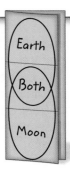

SCIENCE USE v. COMMON USE

organic
Science Use relating to carbon compounds in living organisms

Common Use relating to food grown without fertilizers, pesticides, or antibiotics

Looking for Life Elsewhere

In 1835, a New York newspaper published articles claiming that herds of bison and furry, winged bat-men had been observed on Earth's moon. Many people were fooled. Today, people know the Moon is airless, and scientists have yet to find life there. Because liquid water is essential for life on Earth, scientists look for places in our solar system where liquid water might exist or might have existed in the past. In 2009, scientists discovered water on the Moon. Although water might not exist on the surface of a planet or a moon, it might exist beneath the surface.

Mars

Other than Earth, Mars is the planet scientists think is most likely to have liquid water. On the surface of Mars, pressures probably are too low for water to be liquid; water would likely evaporate quickly in the thin, dry atmosphere. Temperatures are also low. They generally range from −87°C to −5°C, though they can reach a high of 20°C during the Martian summer.

Scientists have sent many uncrewed spacecraft to Mars, but none has detected liquid water. However, there is abundant evidence for water vapor and water ice on the Martian surface. And photographs show surface features on Mars that appear to have been carved by moving water. The channels shown in **Figure 9** look like streambeds. It is possible that water from an underground ocean seeped to the surface and flowed as rivers or floods before evaporating. How much water was in these channels and how long ago it flowed are still unknown.

 Key Concept Check Why do scientists think liquid water once might have existed on Mars?

Figure 9 Scientists hypothesize that these Martian channels could be ancient streambeds.

Other Planets

Mercury and Venus are too hot for water to be liquid on or near their surfaces. The four outer planets are too cold. The outer planets also are too gaseous. They have no solid surfaces on which liquid water could form. Though some liquid water might exist deep in the interiors of the outer planets, it is unlikely that the water could support life.

Natural Satellites

Scientists continue to look for further evidence of water on Earth's moon and on the moons of other planets. Even though temperatures in the outer solar system are extremely cold, scientists have found that as a satellite orbits a massive planet, the planet's gravity can cause the satellite's interior to heat. This might provide enough thermal energy to allow liquid water to exist near their icy surfaces.

Several moons in the outer solar system have surface features that indicate the presence of liquid water not far below. For example, scientists suggest that the ridges on Europa (yuh ROH puh), one of Jupiter's moons, shown in **Figure 10,** could be cracks in the ice where liquid water has seeped to the surface and frozen solid. Callisto and Ganymede, two other moons of Jupiter, and Titan, a moon of Saturn, show similar surface features.

Several other moons in the outer solar system, including Enceladus (en SEL uh dus), a moon of Saturn, and Triton, a moon of Neptune, show evidence of geysers (GI zurz). *A* **geyser** *is a warm spring that sometimes ejects a jet of liquid water or water vapor into the air.* The massive geysers on Enceladus shown in **Figure 11** are hundreds of kilometers high. Two other moons of Saturn, Tethys (TEE thus) and Dione (di OH nee), also have geyserlike plumes.

 Key Concept Check Where might life exist in the solar system beyond Earth?

▲ **Figure 10** 🔑 The lines on Europa might be cracks where liquid water has risen to the surface.

Figure 11 Geysers on Enceladus are evidence that liquid water might exist beneath the moon's icy surface. ▼

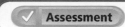

Visual Summary

Most—but not all— life on Earth receives energy from the Sun.

Features on Mars that look like stream-beds might be evidence that liquid water once existed on the surface.

Oceans of liquid water might be below the surface of some moons in the outer solar system.

FOLDABLES

Use your lesson Foldable to review the lesson. Save your Foldable for the project at the end of the chapter.

What do you think NOW?

You first read the statements below at the beginning of the chapter.

3. Earth's atmosphere protects life on Earth from dangerous solar radiation.

4. Scientists think conditions for life might exist on some moons in the solar system.

Did you change your mind about whether you agree or disagree with the statements? Rewrite any false statements to make them true.

Use Vocabulary

1 **Define** *astrobiology* in your own words.

2 Organisms on Earth use nutrients that contain _____ molecules.

Understand Key Concepts

3 Which do scientists think is most likely to support life?
- **A.** Jupiter
- **B.** Earth's moon
- **C.** Mars
- **D.** Venus

4 **Explain** why scientists follow the water when looking for life in the solar system.

5 **Construct** a diagram of the solar system. Identify objects where life might exist.

Interpret Graphics

6 **Analyze** The feature indicated by the arrow exists on some moons. How is it evidence of liquid water?

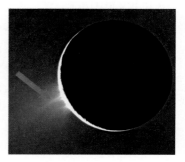

7 **Summarize** Copy and fill in the graphic organizer below to summarize the conditions necessary for life on Earth.

Life

Critical Thinking

8 **Evaluate** How does the study of life on Earth help scientists look for life elsewhere in the solar system?

Math Skills

Review
Math Practice

9 What is the mean of the following temperatures taken at a single location on Mars's surface over four days: $-25.4°C$, $-24.7°C$, $-28.1°C$, and $-28.7°C$?

What solar system objects beyond Earth might have conditions that support life?

Materials

colored pencils

data table

All life on Earth has the same basic needs: a source of energy, liquid water, and nourishment. These same requirements are helpful in determining whether other objects in space might be able to support life. Analyze data about different objects in the solar system. Then use the data to conclude whether you think life on any of the solar system objects is possible.

Learn It

When you analyze and interpret scientific data, you must think critically to determine what the data mean. Look for patterns and cause-and-effect relationships among groups of data. Compare and contrast the data with similar data. Once you analyze the data, summarize the data and **draw conclusions.** Conclusions are statements or decisions that are supported by data.

Try It

1 Obtain and review a copy of the data table *Habitability Factors of Various Objects in Our Solar System.* The word *habitability* means "capable or suitable to be lived on or inhabited."

2 Use a green pencil to mark an *X* next to each habitability factor you think would likely enable life to exist. Use a red pencil to mark an *X* next to each factor you think would likely not enable life to exist. Note that a life-form can be even a simple bacterium.

4 Based on your analysis of the data, choose three solar system objects you think are most likely able to support life-forms. Record the names of these objects on separate pages in your Science Journal.

5 Below the name of each object, write a paragraph explaining how you used the data from the table to conclude that life might be able to exist on the object.

6 Next to each paragraph, use the colored pencils to draw and label a detailed image of a life-form that you think could live on the solar system object. Include labels that explain the life-form's characteristics and how they enable it to live in the environment you analyzed.

Apply It

7 What types of data did you use to choose your three objects?

8 How did you conclude which objects might support life-forms?

9 🔑 **Key Concept** Where beyond Earth might life exist in the solar system?

Lesson 3

Reading Guide

Key Concepts
ESSENTIAL QUESTIONS

- What technology has allowed humans to explore and travel into space?
- What factors must humans consider when traveling into space?

Vocabulary
artificial satellite p. 781
rocket p. 781
space probe p. 782

 Multilingual eGlossary

 Video Science Video

Human Space Travel

Inquiry Astronaut Under Water?

When you are in water, you float. Astronauts in space float, too. Because floating in water is like floating in space, astronauts spend many hours under water preparing for space missions. Do you know why astronauts float in space?

Launch Lab

20 minutes

How well can you work under pressure?

Astronauts wear spacesuits when they work outside a spacecraft. The suits, which include gloves, are pressurized to protect the astronauts' bodies. What is it like to work in space wearing spacesuit gloves?

1. Read and complete a lab safety form.

2. Have a partner use a **stopwatch** to see how long it takes you to build a rectangular object using **20 plastic building blocks**. Record the time in your Science Journal. Then break down the object.

3. Put on a pair of **disposable gloves.** Blow a small amount of air into each glove using a **plastic straw.** Add just enough air to slightly inflate the fingers. Have your partner seal each inflated glove by wrapping **masking tape** around your wrist.

4. Repeat step 2.

5. Switch roles and repeat the activity.

Think About This

1. How did the time required to make an object from plastic blocks differ in steps 2 and 4?

2. Why do you think all parts of a spacesuit, including the gloves, are pressurized?

3. 🔑 **Key Concept** What factors do you think humans must consider when traveling into space?

Technology and Early Space Travel

You have lived your entire life in the space age. Most people consider the launch of *Sputnik I* in 1957 by the former Soviet Union to be the beginning of the space age. *Sputnik I* was the first artificial satellite sent into orbit around Earth. *An* **artificial satellite** *is any human-made object placed in orbit around a body in space.* Today, hundreds of artificial satellites operate in orbit around Earth. Some artificial satellites are communication satellites. Some observe Earth. A few observe stars and other objects in distant space.

Escaping Gravity

How do artificial satellites and other spacecraft reach space? You know that when you jump up into the air, you land back on the ground because of Earth's gravity. But if you could jump fast enough and high enough, you would launch into space! Only a rocket can travel fast enough and far enough to escape Earth's gravity. *A* **rocket** *is a vehicle propelled by the exhaust made from burning fuel.* As its exhaust is forced out, the rocket accelerates forward, as shown in **Figure 12.** Most rockets that travel long distances carry two or more tanks of fuel to be able to travel far enough to escape Earth's gravity.

Figure 12 🔑 The exhaust from a rocket launch propels a rocket skyward.

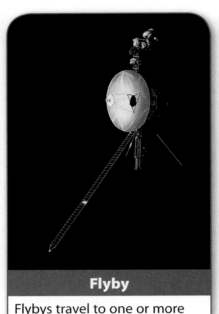

Flyby

Flybys travel to one or more distant space objects and fly by without orbiting or landing.

Orbiter

Orbiters travel to a distant space object and are placed into orbit around the object.

Lander

Landers travel to a distant space object and land on the surface.

Figure 13 Some space probes pass by an object. Others land on the surface.

Visual Check Which probe would transport a rover, a craft that moves on the surface of an object?

ACADEMIC VOCABULARY

transmit
(verb) to send something from one person, place, or thing to another

Robotic Space Probes

The Moon is the farthest object from Earth that humans have visited. However, scientists have sent robotic missions to every planet, as well as to some moons, asteroids, dwarf planets, and comets. *A* **space probe** *is an uncrewed vehicle that travels to and obtains information about objects in space.* Examples of the three main types of space probes are shown in **Figure 13.** Probes do not return to Earth. They are equipped with cameras and scientific instruments that **transmit** data back to Earth.

There are many reasons to send probes instead of people into space. It costs less and is often safer to send probes. Also, objects in space are very far away. A visit to Mars and back would take more than a year. A round trip to Saturn could take 15 years. Robotic missions are dangerous, too. Only half of the missions sent to Mars have been successful. Space probes that do arrive at their destinations experience harsh conditions and often do not survive long.

The National Aeronautics and Space Administration (NASA) is the U. S. government agency responsible for most space missions and space-flight technology. Other nations also have space programs. Astronauts from more than 30 countries have traveled to space, and several countries have sent robotic missions to the Moon and beyond.

Key Concept Check How do space probes help scientists explore space?

Challenges for Humans in Space

When **astronauts** travel into space, they must bring their environments and life-support systems with them. Otherwise, they could not withstand the temperatures, the pressures, and the other extreme conditions that exist in space.

Solar Radiation

One threat to astronauts is harmful radiation from the Sun. You read that Earth's atmosphere protects life on Earth from most of the Sun's dangerous radiation. However, as astronauts travel in space, they move far beyond Earth's atmosphere. They must rely on their spacecraft and spacesuits to shield them from dangerous solar radiation and solar particles.

Oxygen

Humans need oxygen. Outside Earth's atmosphere, there is not enough oxygen for humans to survive. Air circulation systems inside spacecraft supply oxygen and keep carbon dioxide, which people breathe out, from accumulating. The air humans breathe on Earth is a mixture of nitrogen and oxygen. For short trips into space, spacecraft carry tanks of oxygen and nitrogen, which are mixed into the proper proportions onboard. For long trips, oxygen is supplied by passing an electric current through water. This separates water's hydrogen and oxygen atoms.

 Key Concept Check What factors must humans consider when traveling into space?

WORD ORIGIN

astronaut
from Greek *astron*, means "star"; and Greek *nautes*, means "sailor"

FOLDABLES®

Make a vertical four-tab book. Label it as shown. Use it to organize your notes on the challenges for humans in space.

Solar Radiation

Oxygen

Temperatures and Pressures

Microgravity

Inquiry MiniLab   **20 minutes**

How hard is it to hit a target?

Escaping Earth's gravity is a challenge for scientists launching rockets. Landing a spacecraft on a solar system object also can be difficult.

1. Read and complete a lab safety form.

2. Use **tape** to make a circle about 2 m in diameter on the floor. At the center, make a target circle about 15 cm in diameter.

3. Test launch a **toy vehicle** toward the target starting from the larger circle. Make several attempts. You are successful only if your vehicle has all its wheels inside the target circle. In your Science Journal, record what percentage of your test launches were successful.

4. Launch your vehicle. As a class, determine the success rate of the launches.

Analyze and Conclude

1. **Model** What technology does the toy in this activity represent?

2. **Key Concept** What factors must be considered when launching a spacecraft?

▲ **Figure 14**
An EMU suit enables an astronaut to spend up to eight hours outside a spacecraft.

Temperature and Pressure Extremes

Most places in the solar system are either extremely cold or extremely hot. Pressures in space are also extreme. In most places, pressure is much lower than the pressure humans experience on Earth. Environmental control systems in spacecraft protect astronauts from temperature and pressure extremes. Outside their spacecraft, astronauts wear Extravehicular Mobility Unit (EMU) suits, as shown in **Figure 14.** EMU suits provide oxygen, protect astronauts from radiation and meteoroids, and enable astronauts to talk to each other.

✓ **Reading Check** What are the purposes of an EMU suit?

Microgravity

You might think astronauts are weightless in space. But astronauts in orbit around Earth are subjected to almost the same gravity as they are on Earth's surface. Then why is the astronaut shown in **Figure 15** floating? As their spacecraft orbits Earth, the astronauts inside are continually falling toward Earth. But because their spacecraft is moving, they do not actually fall. They float. If their spacecraft suddenly stopped moving, they would plunge downward.

The space environment that astronauts experience is often called microgravity. In microgravity, objects seem to be weightless. This can be an advantage. No matter how much something weighs on Earth, it can be moved with ease in space. Microgravity also makes some tasks, such as turning a screwdriver, more difficult. If an astronaut is not careful, instead of the screw turning, he or she might turn instead.

On Earth, working against gravity helps keep your muscles, bones, and heart strong and healthy. But in microgravity, astronauts' bones and muscles don't need to work as hard, and they begin to lose mass and strength. Astronauts in space must exercise each day to keep their bodies healthy.

Figure 15 Astronaut Eileen Collins floats because she is constantly falling toward Earth as her spacecraft orbits Earth. ▶

🖳 **Review**
Personal Tutor

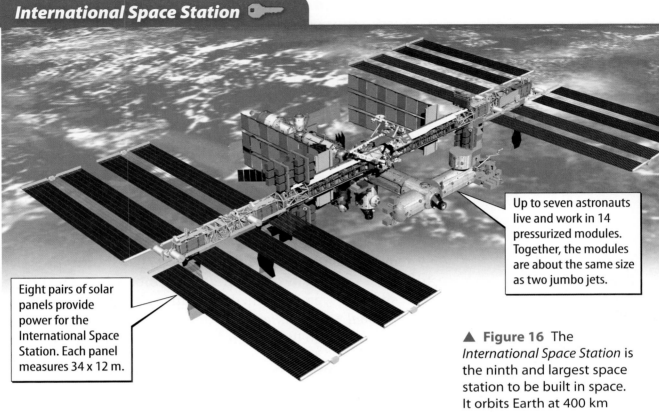

Eight pairs of solar panels provide power for the International Space Station. Each panel measures 34 x 12 m.

Up to seven astronauts live and work in 14 pressurized modules. Together, the modules are about the same size as two jumbo jets.

▲ **Figure 16** The *International Space Station* is the ninth and largest space station to be built in space. It orbits Earth at 400 km above Earth's surface.

✓**Visual Check** What is the combined area of the *ISS's* eight pairs of solar panels?

Living and Working in Space

Even when protected from the extremes of space, astronauts still face many challenges when living and working in space. Life in space is dramatically different from life on Earth.

International Space Station

The International Space Station (ISS), shown in **Figure 16,** is a large, artificial satellite that orbits Earth. People work and live on the *ISS* for up to six months at a time. Constructed by astronauts from over 15 nations, the *ISS* has been continuously occupied since the first crew arrived in 2000.

The *ISS* crew conducts scientific and medical experiments. These include experiments to learn how microgravity affects people's health and how it affects plants. People living in space for long periods might need to grow plants for food and oxygen. In the future, in addition to being an orbiting research laboratory, the *ISS* eventually might serve as a testing and repair station for missions to the Moon and beyond.

Living in space is not easy. For example, astronauts must place a clip on a book to hold it open to the right page. They eat packaged food, using magnetized trays and tableware. Toilets flush with air instead of water. And astronauts must be strapped down while they sleep, as shown in **Figure 17.** Otherwise, they would drift and bump into things.

Figure 17 European astronaut Paolo Nespoli sleeps strapped into a sleeping bag. ▼

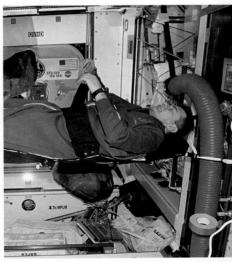

Transportation Systems

Space transportation systems are the rockets, the shuttles, and the other spacecraft that deliver cargo and humans to space. **Figure 18** shows the progression of NASA's space transportation systems. Early rockets and spacecraft, such as those used to transport astronauts to the Moon, were used only once. The early rockets and spacecraft included three programs. The Mercury program obtained the goal of putting a human into orbit. The Gemini program goals were to test the spacecraft and crews' thresholds before beginning the Apollo program. The Apollo program achieved the goal of placing the first human on the Moon.

The Space Shuttle was NASA's first reusable transportation system. First launched in 1981, there were five shuttles designed to hold a maximum of seven crew members in each shuttle. The shuttles were designed to transport astronauts to the *International Space Station* to conduct experiments and to service uncrewed satellites, such as the *Hubble Space Telescope*.

Reading Check How has the human space flight program at NASA changed over the years?

Future Space Exploration

Space engineers are continuously working on new technologies and ideas that could advance human space flight and exploration. The last image in **Figure 18** shows what future space transportation systems may look like.

Figure 18 NASA has launched many different types of spacecraft carrying humans into space. This timeline shows the variety of the spacecraft.

◄ **Mercury Missions** 1961–1963

▲ **Gemini Missions** 1965–1966

▼ **Skylab** 1973–1979

◄ **Apollo Missions** 1968–1972

▲ *International Space Station* 1998–present

Future Missions ▶

Lesson 3 Review

Visual Summary

When wearing an EMU suit, an astronaut can spend up to eight hours outside a spacecraft.

Sleeping can be a challenge for astronauts in space.

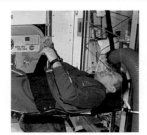

The space shuttle was NASA's first reusable transportation system.

FOLDABLES®

Use your lesson Foldable to review the lesson. Save your Foldable for the project at the end of the chapter.

What do you think NOW?

You first read the statements below at the beginning of the chapter.

5. Astronauts float in space because there is no gravity above Earth's atmosphere.

6. The United States is the only country with a human space-flight program.

Did you change your mind about whether you agree or disagree with the statements? Rewrite any false statements to make them true.

Use Vocabulary

1 Any object made by humans and placed in orbit around a body in space is a(n) _____.

2 **Distinguish** between a rocket and a space probe.

Understand Key Concepts 🔑

3 Which is NOT a type of space probe?
- **A.** flyby
- **B.** lander
- **C.** orbiter
- **D.** shuttle

4 **Compare** how a microgravity environment is both an advantage and a disadvantage for space travelers.

5 **Explain** how the *International Space Station* is used now and how it might be used in the future.

Interpret Graphics

6 **Identify** The photo at right shows the *International Space Station*. What purpose does the part labeled X serve?

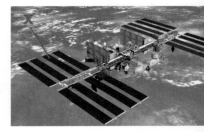

7 **Summarize** Copy and fill in the graphic organizer below to summarize challenges for humans traveling in space.

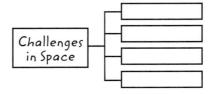

Critical Thinking

8 **Imagine** you are an astronaut on a year-long voyage. How would your life be different from your life on Earth? How would it be the same?

9 **Evaluate** the following statement: Astronauts in orbit are weightless because they are so far from Earth's surface.

Materials

creative
building
materials

raw egg in shell

plastic,
self-sealing
sandwich bag

office supplies

data table

Also needed
construction
materials
approved by
your teacher,
variety of cups,
video camera

Safety

Design Your Own Egg Lander

You have a summer job with a company in the space industry. You are to work with a team to design and test a model spacecraft that will land on one of the planets in our solar system. The spacecraft will contain a rover that has fragile scientific instruments. Here is your assignment:

- Analyze data to choose the best target planet.

- Design a lander space probe from available materials.

- Test the spacecraft to make sure it will
  (1) hit its target, which will be at least 3 m away;
  (2) withstand its landing;
  (3) protect its cargo—a rover, represented by a raw egg in its shell.

- Videotape the landing of the craft and its cargo. In your team's video, you will
  (1) discuss how your team chose your target planet;
  (2) describe the features of your egg lander;
  (3) show the launch and landing of your egg lander.

Ask a Question

How will you decide which planet is the best target? How will you design and construct your craft to meet the criteria of your assignment?

Make Observations

1. Read and complete a lab safety form.

2. Obtain a copy of the data table. Read the data and decide which planet you think would be the best target for exploration by a rover.
   Note: Data for Earth should be used for comparison only.

3. Discuss the possible targets with your team. As a team, decide which planet will be your target. Make sure you are able to explain why you chose the planet you did.

4. You and each member of your team will construct an individual trial lander, using the materials provided or approved by your teacher. Before you begin, make a detailed drawing of your trial lander in your Science Journal, including its intended cargo. Label the different types of materials you plan to use. Have your teacher approve your design.

5 Construct your trial lander. Perform at least five practice landings without the cargo. Make design changes as necessary.

6 Once you are sure your cargo will likely survive its fall, meet again with your team. Test all of the landers and decide which features of each worked the best.

Form a Hypothesis

7 As a team, consider the successes and failures of the individual trial landers. Form a hypothesis about the best way to design and build your final lander.

Test Your Hypothesis

8 As a team, design and construct a new lander using the best features of each trial lander. Place the egg in a plastic, self-sealing bag. Seal the bag completely, and place it in your team's lander.

9 Before you proceed, read the information under **Communicate Your Results.** Prepare your video script and setting. Record the first part of the video, where you describe your intended target and the features of your lander.

10 Take your team's spacecraft to the designated launch area and attempt to land it on the target. Record the launch and landing. Be sure to show how well your lander protected its cargo!

Analyze and Conclude

11 **Analyze Data** How did you decide on your target planet? What determined your team's final design?

12 **Discuss** How would your lander be different if it were designed to carry humans?

13 **Assess** What information would the lander's rover gather to search for life on your target planet?

14 **The Big Idea** How do spacecraft help scientists look for life in the solar system?

Communicate Your Results

Make a short video newscast in which you and your team explain how you chose your target and the lander design. Include footage with the actual landing of your craft.

Combine your video with those of your classmates and make a highlights video. What were the best characteristics of the lander designs? Which team had the most successful landing? Which team had the most spectacular crash?

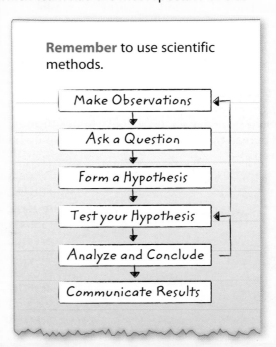

Remember to use scientific methods.

Make Observations →
Ask a Question →
Form a Hypothesis →
Test your Hypothesis →
Analyze and Conclude →
Communicate Results

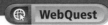
 THE BIG IDEA

Scientists use space probes, artificial satellites, and human transportation systems to explore our solar system and look for evidence of life on Mars and some moons.

Key Concepts Summary 🔑	Vocabulary
Lesson 1: Our Solar System • Our solar system formed from a spinning cloud of gas and dust. Most of the **planets** and other solar system objects orbit the Sun in the same direction the original cloud rotated. • The Sun is made mostly of hydrogen gas. Objects close to the Sun are made mostly of rock and metals. Objects farther away are made mostly of ices and gases. • The solar system includes the Sun, planets, **dwarf planets,** comets, natural **satellites,** asteroids, and **meteoroids.** 	**astronomical unit** p. 764 **planet** p. 766 **dwarf planet** p. 768 **satellite** p. 768 **meteoroid** p. 769 **meteor** p. 769 **meteorite** p. 769
Lesson 2: Life in the Solar System • Life on Earth requires a source of energy, liquid water, and nourishment. • Scientists search for places in the solar system where liquid water and **organic** carbon might exist. • Scientists think life could possibly exist on Mars and/or some satellites of the outer planets. 	**astrobiology** p. 773 **organic** p. 775 **geyser** p. 777
Lesson 3: Human Space Travel • Humans have developed **rockets, space probes, artificial satellites,** and human transportation systems to help them explore and travel into space. 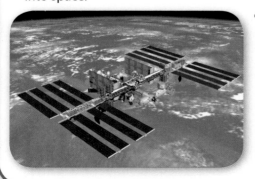 • When traveling, working, and living in space, humans must be protected from radiation, temperature and pressure extremes, and meteoroids. They must provide their own oxygen, and they must be prepared for a microgravity environment.	**artificial satellite** p. 781 **rocket** p. 781 **space probe** p. 782

Study Guide

Review
- **Personal Tutor**
- **Vocabulary eGames**
- **Vocabulary eFlashcards**

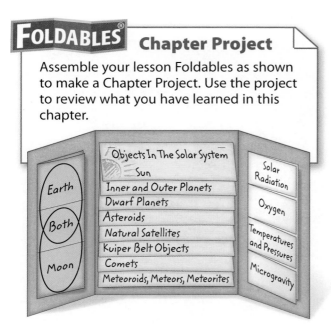

FOLDABLES® Chapter Project

Assemble your lesson Foldables as shown to make a Chapter Project. Use the project to review what you have learned in this chapter.

Objects In The Solar System
- *Sun*
- *Inner and Outer Planets*
- *Dwarf Planets*
- *Asteroids*
- *Natural Satellites*
- *Kuiper Belt Objects*
- *Comets*
- *Meteoroids, Meteors, Meteorites*

Earth / *Both* / *Moon*

Solar Radiation / *Oxygen* / *Temperatures and Pressures* / *Microgravity*

Use Vocabulary

Each of the following sentences is false. Make each sentence true by replacing the italicized term with the correct vocabulary term.

1 A *meteorite* is a streak of light in Earth's atmosphere.

2 A *rocket* is an uncrewed vehicle that gathers information about space objects.

3 Europa is a *dwarf planet* of Jupiter.

4 *Astronomy* is the study of life in the universe.

5 A *meteorite* is a small body in space, often no larger than a grain of rice.

6 All organisms on Earth need liquid water, *oxygen* molecules, and an energy source to survive.

Link Vocabulary and Key Concepts

Concepts in Motion Interactive Concept Map

Copy this concept map, and then use vocabulary terms from the previous page to complete the concept map.

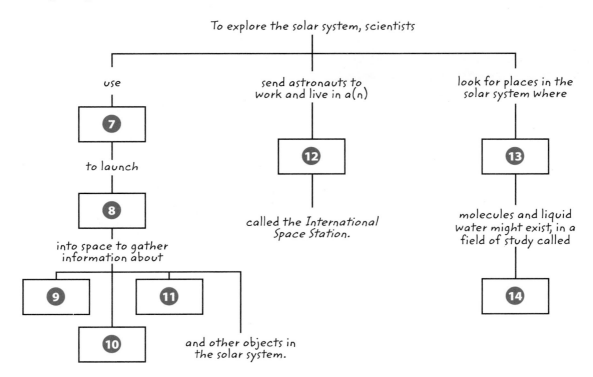

To explore the solar system, scientists

use → **7** → **to launch** → **8** → **into space to gather information about** → **9** / **11** / **10** → **and other objects in the solar system.**

send astronauts to work and live in a(n) → **12** → **called the *International Space Station*.**

look for places in the solar system where → **13** → **molecules and liquid water might exist, in a field of study called** → **14**

Understand Key Concepts

1 Where is the asteroid belt?
A. above Earth's atmosphere
B. between the orbits of Mars and Jupiter
C. between the orbits of Saturn and Jupiter
D. beyond the orbit of Neptune

2 Which is NOT a good place to look for life?
A. Enceladus
B. Europa
C. Mars
D. Venus

3 Which is critical for all life on Earth?
A. average temperatures above 0°C
B. energy from the Sun
C. liquid water
D. oxygen to breathe

4 Which type of solar system object is shown in the photo below?

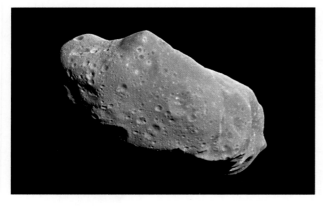

A. asteroid
B. comet
C. dwarf planet
D. meteoroid

5 Which best describes how astronauts are affected by microgravity?
A. They are constantly falling toward Earth.
B. They are lifted by oxygen in the air.
C. They are not subjected to an atmosphere.
D. They are not subjected to gravity.

6 What is the purpose of the object shown below?

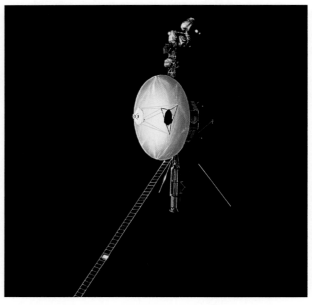

A. to be used as a lunar outpost
B. to be used as a research station
C. to explore planets up close
D. to transport astronauts

7 Which objects in the solar system contain the most hydrogen?
A. the moons of the inner planets
B. the moons of the outer planets
C. the Sun and the inner planets
D. the Sun and the outer planets

8 Which explains why most objects orbit the Sun in the same direction?
A. Earth rotates from west to east.
B. The objects are pulled by gravity.
C. The solar system formed from a spinning cloud.
D. The Sun contains most of the matter in the solar system.

9 How is oxygen provided on a long space voyage?
A. by collecting molecules from space
B. by harnessing nuclear reactions
C. by passing electricity through water
D. by recycling carbon dioxide

Critical Thinking

10 **Summarize** the role of gravity in the formation of the solar system.

11 **Explain** why solar radiation is a challenge for astronauts traveling in space.

12 **Deduce** why the *International Space Station* is being assembled in space and not on Earth.

13 **Interpret Graphics** Copy Mercury and Uranus, shown below, and label their interior parts. Explain why their makeups differ.

Writing in Science

14 You are an astronaut on the *International Space Station*. Write a letter of at least six lines to your best friend describing some of the challenges you face on the *ISS*.

REVIEW THE BIG IDEA

15 How do scientists use their understanding of life on Earth in their search for evidence that life has existed, or does exist, on Mars or on other objects in the solar system?

16 How and where do scientists look for life in the solar system?

Math Skills ✕÷+

Hour	Temp (°C)	Hour	Temp (°C)
2	−88.8	14	−23.4
4	−90.7	16	−27.1
6	−75.7	18	−54.9
8	−50.7	20	−63.9
10	−36.7	22	−71.7
12	−31.0	24	−76.2

 Review

— Math Practice —

Use Statistics

The table at left shows temperature data gathered by a Mars lander over a 24-h period. Use the table to answer the questions.

17 What was the mean temperature from hour 2 through hour 6?

18 What was the mean temperature for the first 12 hours?

19 What was the mean temperature over the 24-h period?

Record your answers on the answer sheet provided by your teacher or on a sheet of paper.

Multiple Choice

1 How are most planets and moons in the solar system alike?

  A Most have about the same mass.

  B Most revolve at about the same speed.

  C Most rotate at about the same speed.

  D Most rotate in a counterclockwise direction.

Use the graph below to answer question 2.

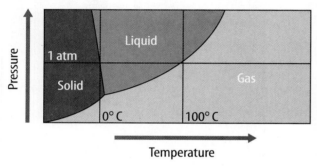

2 What information does the graph above illustrate?

  A conditions in which different forms of water exist

  B conditions in which life can exist

  C the pressures at different parts of Earth's atmosphere

  D the temperatures at different locations on Mars

3 What advantage does an uncrewed space probe have over a crewed mission?

  A Space probes can be safer and cost less.

  B Space probes can orbit Earth.

  C Space probes can travel faster.

  D Space probes can work in microgravity.

4 What feature is evidence that liquid water might exist on Enceladus?

  A clouds in its atmosphere

  B dry streambeds

  C geysers on the surface

  D the speed of its orbit

Use the diagrams below to answer question 5.

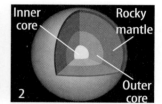
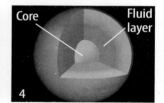
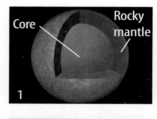
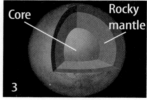

5 Which of these diagrams most likely represents an outer planet?

  A 1

  B 2

  C 3

  D 4

6 How are the asteroid belt and the Kuiper belt alike?

  A Both are about the same distance from the Sun.

  B Both are about the same size.

  C Both contain objects made mostly of rock and metal.

  D Both contain objects that are remnants of the solar system's formation.

7 Which element is a part of all organic compounds?

  A carbon

  B iron

  C nitrogen

  D sodium

8 Where might liquid water exist on Mars?

  A at its poles

  B in surface lakes and oceans

  C in rivers and channels

  D under its surface

Use the diagram below to answer question 9.

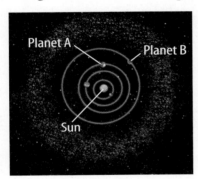

9 Based on their distances from the Sun, what prediction can you make about the movements of Planet A and Planet B?

  A Planet A and Planet B revolve at the same speed.

  B Planet A and Planet B revolve in a clockwise direction.

  C Planet A revolves faster than Planet B.

  D Planet B revolves faster than Planet A.

Constructed Response

10 Imagine you are a NASA scientist designing a crewed mission to search for life on another planet. What three properties should the astronauts look for? Explain why these properties are important.

Use the diagrams below to answer question 11.

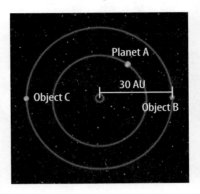 

11 Imagine that scientists have discovered a solar system around a nearby star. There are three objects orbiting the star, as shown above. Object A is a planet. How would you classify objects B and C? Explain your reasoning.

12 Some space probes have successfully reached Mars, but others have failed. What challenges do scientists face when sending a spacecraft to this planet?

13 The satellites of the outer planets are very far from the Sun, yet scientists think liquid water might exist on some of them. Explain what force might make temperatures and pressures on these objects suitable for liquid water.

NEED EXTRA HELP?													
If You Missed Question...	1	2	3	4	5	6	7	8	9	10	11	12	13
Go to Lesson...	1	2	3	2	1	1	2	2	1	2	1	3	2

Student Resources

For Students and Parents/Guardians

These resources are designed to help you achieve success in science. You will find useful information on laboratory safety, math skills, and science skills. In addition, science reference materials are found in the Reference Handbook. You'll find the information you need to learn and sharpen your skills in these resources.

Table of Contents

Figure 1 The Internet can be a valuable research tool.

Scientific Methods

Scientists use an orderly approach called the scientific method to solve problems. This includes organizing and recording data so others can understand them. Scientists use many variations in this method when they solve problems.

Identify a Question

The first step in a scientific investigation or experiment is to identify a question to be answered or a problem to be solved. For example, you might ask which gasoline is the most efficient.

Gather and Organize Information

After you have identified your question, begin gathering and organizing information. There are many ways to gather information, such as researching in a library, interviewing those knowledgeable about the subject, and testing and working in the laboratory and field. Fieldwork is investigations and observations done outside of a laboratory.

Researching Information Before moving in a new direction, it is important to gather the information that already is known about the subject. Start by asking yourself questions to determine exactly what you need to know. Then you will look for the information in various reference sources, like the student is doing in **Figure 1.** Some sources may include textbooks, encyclopedias, government documents, professional journals, science magazines, and the Internet. Always list the sources of your information.

Evaluate Sources of Information Not all sources of information are reliable. You should evaluate all of your sources of information, and use only those you know to be dependable. For example, if you are researching ways to make homes more energy efficient, a site written by the U.S. Department of Energy would be more reliable than a site written by a company that is trying to sell a new type of weatherproofing material. Also, remember that research always is changing. Consult the most current resources available to you. For example, a 1985 resource about saving energy would not reflect the most recent findings.

Sometimes scientists use data that they did not collect themselves, or conclusions drawn by other researchers. This data must be evaluated carefully. Ask questions about how the data were obtained, if the investigation was carried out properly, and if it has been duplicated exactly with the same results. Would you reach the same conclusion from the data? Only when you have confidence in the data can you believe it is true and feel comfortable using it.

Interpret Scientific Illustrations As you research a topic in science, you will see drawings, diagrams, and photographs to help you understand what you read. Some illustrations are included to help you understand an idea that you can't see easily by yourself, like the tiny particles in an atom in **Figure 2.** A drawing helps many people to remember details more easily and provides examples that clarify difficult concepts or give additional information about the topic you are studying. Most illustrations have labels or a caption to identify or to provide more information.

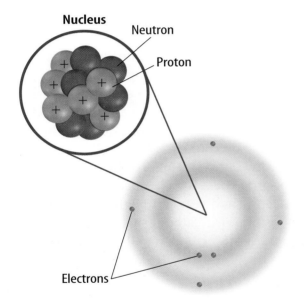

Figure 2 This drawing shows an atom of carbon with its six protons, six neutrons, and six electrons.

Concept Maps One way to organize data is to draw a diagram that shows relationships among ideas (or concepts). A concept map can help make the meanings of ideas and terms more clear, and help you understand and remember what you are studying. Concept maps are useful for breaking large concepts down into smaller parts, making learning easier.

Network Tree A type of concept map that not only shows a relationship, but how the concepts are related is a network tree, shown in **Figure 3.** In a network tree, the words are written in the ovals, while the description of the type of relationship is written across the connecting lines.

When constructing a network tree, write down the topic and all major topics on separate pieces of paper or notecards. Then arrange them in order from general to specific. Branch the related concepts from the major concept and describe the relationship on the connecting line. Continue to more specific concepts until finished.

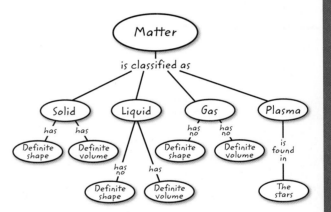

Figure 3 A network tree shows how concepts or objects are related.

Events Chain Another type of concept map is an events chain. Sometimes called a flow chart, it models the order or sequence of items. An events chain can be used to describe a sequence of events, the steps in a procedure, or the stages of a process.

When making an events chain, first find the one event that starts the chain. This event is called the initiating event. Then, find the next event and continue until the outcome is reached, as shown in **Figure 4** on the next page.

SCIENCE SKILL HANDBOOK

MATH SKILL HANDBOOK

FOLDABLES HANDBOOK

REFERENCE HANDBOOK

GLOSSARY/ GLOSARIO

INDEX

SCIENCE SKILL HANDBOOK

MATH SKILL HANDBOOK

FOLDABLES HANDBOOK

REFERENCE HANDBOOK

GLOSSARY/ GLOSARIO

INDEX

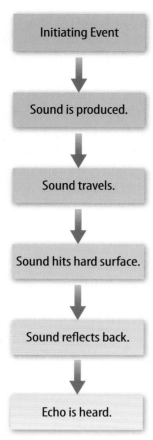

Figure 4 Events-chain concept maps show the order of steps in a process or event. This concept map shows how a sound makes an echo.

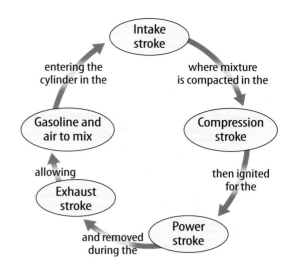

Figure 5 A cycle map shows events that occur in a cycle.

Spider Map A type of concept map that you can use for brainstorming is the spider map. When you have a central idea, you might find that you have a jumble of ideas that relate to it but are not necessarily clearly related to each other. The spider map on sound in **Figure 6** shows that if you write these ideas outside the main concept, then you can begin to separate and group unrelated terms so they become more useful.

Cycle Map A specific type of events chain is a cycle map. It is used when the series of events do not produce a final outcome, but instead relate back to the beginning event, such as in **Figure 5.** Therefore, the cycle repeats itself.

To make a cycle map, first decide what event is the beginning event. This is also called the initiating event. Then list the next events in the order that they occur, with the last event relating back to the initiating event. Words can be written between the events that describe what happens from one event to the next. The number of events in a cycle map can vary, but usually contain three or more events.

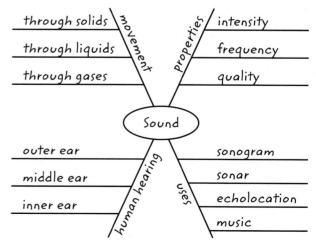

Figure 6 A spider map allows you to list ideas that relate to a central topic but not necessarily to one another.

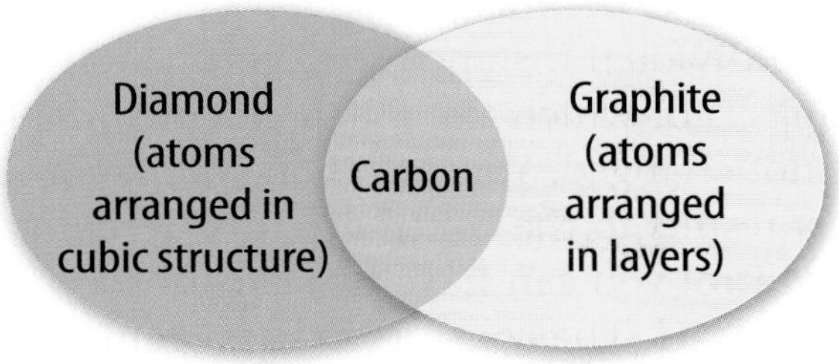

Figure 7 This Venn diagram compares and contrasts two substances made from carbon.

Venn Diagram To illustrate how two subjects compare and contrast you can use a Venn diagram. You can see the characteristics that the subjects have in common and those that they do not, shown in **Figure 7.**

To create a Venn diagram, draw two overlapping ovals that are big enough to write in. List the characteristics unique to one subject in one oval, and the characteristics of the other subject in the other oval. The characteristics in common are listed in the overlapping section.

Make and Use Tables One way to organize information so it is easier to understand is to use a table. Tables can contain numbers, words, or both.

To make a table, list the items to be compared in the first column and the characteristics to be compared in the first row. The title should clearly indicate the content of the table, and the column or row heads should be clear. Notice that in **Table 1** the units are included.

Table 1 Recyclables Collected During Week			
Day of Week	**Paper (kg)**	**Aluminum (kg)**	**Glass (kg)**
Monday	5.0	4.0	12.0
Wednesday	4.0	1.0	10.0
Friday	2.5	2.0	10.0

Make a Model One way to help you better understand the parts of a structure, the way a process works, or to show things too large or small for viewing is to make a model. For example, an atomic model made of a plastic-ball nucleus and chenille stem electron shells can help you visualize how the parts of an atom relate to each other. Other types of models can be devised on a computer or represented by equations.

Form a Hypothesis

A possible explanation based on previous knowledge and observations is called a hypothesis. After researching gasoline types and recalling previous experiences in your family's car you form a hypothesis—our car runs more efficiently because we use premium gasoline. To be valid, a hypothesis has to be something you can test by using an investigation.

Predict When you apply a hypothesis to a specific situation, you predict something about that situation. A prediction makes a statement in advance, based on prior observation, experience, or scientific reasoning. People use predictions to make everyday decisions. Scientists test predictions by performing investigations. Based on previous observations and experiences, you might form a prediction that cars are more efficient with premium gasoline. The prediction can be tested in an investigation.

Design an Experiment A scientist needs to make many decisions before beginning an investigation. Some of these include: how to carry out the investigation, what steps to follow, how to record the data, and how the investigation will answer the question. It also is important to address any safety concerns.

SCIENCE SKILL HANDBOOK

MATH SKILL HANDBOOK

FOLDABLES HANDBOOK

REFERENCE HANDBOOK

GLOSSARY/ GLOSARIO

INDEX

SCIENCE SKILL HANDBOOK

MATH SKILL HANDBOOK

FOLDABLES HANDBOOK

REFERENCE HANDBOOK

GLOSSARY/ GLOSARIO

INDEX

Test the Hypothesis

Now that you have formed your hypothesis, you need to test it. Using an investigation, you will make observations and collect data, or information. This data might either support or not support your hypothesis. Scientists collect and organize data as numbers and descriptions.

Follow a Procedure In order to know what materials to use, as well as how and in what order to use them, you must follow a procedure. **Figure 8** shows a procedure you might follow to test your hypothesis.

Procedure	
Step 1	Use regular gasoline for two weeks.
Step 2	Record the number of kilometers between fill-ups and the amount of gasoline used.
Step 3	Switch to premium gasoline for two weeks.
Step 4	Record the number of kilometers between fill-ups and the amount of gasoline used.

Figure 8 A procedure tells you what to do step-by-step.

Identify and Manipulate Variables and Controls In any experiment, it is important to keep everything the same except for the item you are testing. The one factor you change is called the independent variable. The change that results is the dependent variable. Make sure you have only one independent variable, to assure yourself of the cause of the changes you observe in the dependent variable. For example, in your gasoline experiment the type of fuel is the independent variable. The dependent variable is the efficiency.

Many experiments also have a control—an individual instance or experimental subject for which the independent variable is not changed. You can then compare the test results to the control results. To design a control you can have two cars of the same type. The control car uses regular gasoline for four weeks. After you are done with the test, you can compare the experimental results to the control results.

Collect Data

Whether you are carrying out an investigation or a short observational experiment, you will collect data, as shown in **Figure 9.** Scientists collect data as numbers and descriptions and organize them in specific ways.

Observe Scientists observe items and events, then record what they see. When they use only words to describe an observation, it is called qualitative data. Scientists' observations also can describe how much there is of something. These observations use numbers, as well as words, in the description and are called quantitative data. For example, if a sample of the element gold is described as being "shiny and very dense" the data are qualitative. Quantitative data on this sample of gold might include "a mass of 30 g and a density of 19.3 g/cm^3."

Figure 9 Collecting data is one way to gather information directly.

Figure 10 Record data neatly and clearly so it is easy to understand.

When you make observations you should examine the entire object or situation first, and then look carefully for details. It is important to record observations accurately and completely. Always record your notes immediately as you make them, so you do not miss details or make a mistake when recording results from memory. Never put unidentified observations on scraps of paper. Instead they should be recorded in a notebook, like the one in **Figure 10.** Write your data neatly so you can easily read it later. At each point in the experiment, record your observations and label them. That way, you will not have to determine what the figures mean when you look at your notes later. Set up any tables that you will need to use ahead of time, so you can record any observations right away. Remember to avoid bias when collecting data by not including personal thoughts when you record observations. Record only what you observe.

Estimate Scientific work also involves estimating. To estimate is to make a judgment about the size or the number of something without measuring or counting. This is important when the number or size of an object or population is too large or too difficult to accurately count or measure.

Sample Scientists may use a sample or a portion of the total number as a type of estimation. To sample is to take a small, representative portion of the objects or organisms of a population for research. By making careful observations or manipulating variables within that portion of the group, information is discovered and conclusions are drawn that might apply to the whole population. A poorly chosen sample can be unrepresentative of the whole. If you were trying to determine the rainfall in an area, it would not be best to take a rainfall sample from under a tree.

Measure You use measurements every day. Scientists also take measurements when collecting data. When taking measurements, it is important to know how to use measuring tools properly. Accuracy also is important.

Length To measure length, the distance between two points, scientists use meters. Smaller measurements might be measured in centimeters or millimeters.

Length is measured using a metric ruler or meterstick. When using a metric ruler, line up the 0-cm mark with the end of the object being measured and read the number of the unit where the object ends. Look at the metric ruler shown in **Figure 11.** The centimeter lines are the long, numbered lines, and the shorter lines are millimeter lines. In this instance, the length would be 4.50 cm.

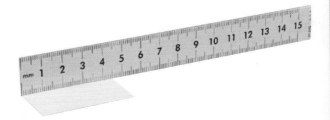

Figure 11 This metric ruler has centimeter and millimeter divisions.

SCIENCE SKILL HANDBOOK

MATH SKILL HANDBOOK

FOLDABLES HANDBOOK

REFERENCE HANDBOOK

GLOSSARY/ GLOSARIO

INDEX

SCIENCE SKILL HANDBOOK

MATH SKILL HANDBOOK

FOLDABLES HANDBOOK

REFERENCE HANDBOOK

GLOSSARY/ GLOSARIO

INDEX

Mass The SI unit for mass is the kilogram (kg). Scientists can measure mass using units formed by adding metric prefixes to the unit gram (g), such as milligram (mg). To measure mass, you might use a triple-beam balance similar to the one shown in **Figure 12.** The balance has a pan on one side and a set of beams on the other side. Each beam has a rider that slides on the beam.

When using a triple-beam balance, place an object on the pan. Slide the largest rider along its beam until the pointer drops below zero. Then move it back one notch. Repeat the process for each rider proceeding from the larger to smaller until the pointer swings an equal distance above and below the zero point. Sum the masses on each beam to find the mass of the object. Move all riders back to zero when finished.

Instead of putting materials directly on the balance, scientists often take a tare of a container. A tare is the mass of a container into which objects or substances are placed for measuring their masses. To find the mass of objects or substances, find the mass of a clean container. Remove the container from the pan, and place the object or substances in the container. Find the mass of the container with the materials in it. Subtract the mass of the empty container from the mass of the filled container to find the mass of the materials you are using.

Figure 12 A triple-beam balance is used to determine the mass of an object.

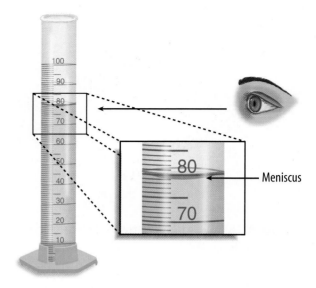

Meniscus

Figure 13 Graduated cylinders measure liquid volume.

Liquid Volume To measure liquids, the unit used is the liter. When a smaller unit is needed, scientists might use a milliliter. Because a milliliter takes up the volume of a cube measuring 1 cm on each side it also can be called a cubic centimeter ($cm^3 = cm \times cm \times cm$).

You can use beakers and graduated cylinders to measure liquid volume. A graduated cylinder, shown in **Figure 13,** is marked from bottom to top in milliliters. In lab, you might use a 10-mL graduated cylinder or a 100-mL graduated cylinder. When measuring liquids, notice that the liquid has a curved surface. Look at the surface at eye level, and measure the bottom of the curve. This is called the meniscus. The graduated cylinder in **Figure 13** contains 79.0 mL, or 79.0 cm^3, of a liquid.

Temperature Scientists often measure temperature using the Celsius scale. Pure water has a freezing point of 0°C and boiling point of 100°C. The unit of measurement is degrees Celsius. Two other scales often used are the Fahrenheit and Kelvin scales.

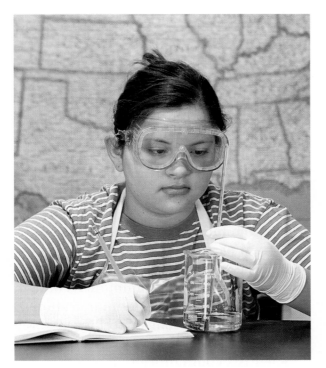

Figure 14 A thermometer measures the temperature of an object.

Scientists use a thermometer to measure temperature. Most thermometers in a laboratory are glass tubes with a bulb at the bottom end containing a liquid such as colored alcohol. The liquid rises or falls with a change in temperature. To read a glass thermometer like the thermometer in **Figure 14,** rotate it slowly until a red line appears. Read the temperature where the red line ends.

Form Operational Definitions An operational definition defines an object by how it functions, works, or behaves. For example, when you are playing hide and seek and a tree is home base, you have created an operational definition for a tree.

Objects can have more than one operational definition. For example, a ruler can be defined as a tool that measures the length of an object (how it is used). It can also be a tool with a series of marks used as a standard when measuring (how it works).

Analyze the Data

To determine the meaning of your observations and investigation results, you will need to look for patterns in the data. Then you must think critically to determine what the data mean. Scientists use several approaches when they analyze the data they have collected and recorded. Each approach is useful for identifying specific patterns.

Interpret Data The word *interpret* means "to explain the meaning of something." When analyzing data from an experiment, try to find out what the data show. Identify the control group and the test group to see whether changes in the independent variable have had an effect. Look for differences in the dependent variable between the control and test groups.

Classify Sorting objects or events into groups based on common features is called classifying. When classifying, first observe the objects or events to be classified. Then select one feature that is shared by some members in the group, but not by all. Place those members that share that feature in a subgroup. You can classify members into smaller and smaller subgroups based on characteristics. Remember that when you classify, you are grouping objects or events for a purpose. Keep your purpose in mind as you select the features to form groups and subgroups.

Compare and Contrast Observations can be analyzed by noting the similarities and differences between two or more objects or events that you observe. When you look at objects or events to see how they are similar, you are comparing them. Contrasting is looking for differences in objects or events.

SCIENCE SKILL HANDBOOK

MATH SKILL HANDBOOK

FOLDABLES HANDBOOK

REFERENCE HANDBOOK

GLOSSARY/ GLOSARIO

INDEX

SCIENCE SKILL HANDBOOK

MATH SKILL HANDBOOK

FOLDABLES HANDBOOK

REFERENCE HANDBOOK

GLOSSARY/ GLOSARIO

INDEX

Recognize Cause and Effect A cause is a reason for an action or condition. The effect is that action or condition. When two events happen together, it is not necessarily true that one event caused the other. Scientists must design a controlled investigation to recognize the exact cause and effect.

Draw Conclusions

When scientists have analyzed the data they collected, they proceed to draw conclusions about the data. These conclusions are sometimes stated in words similar to the hypothesis that you formed earlier. They may confirm a hypothesis, or lead you to a new hypothesis.

Infer Scientists often make inferences based on their observations. An inference is an attempt to explain observations or to indicate a cause. An inference is not a fact, but a logical conclusion that needs further investigation. For example, you may infer that a fire has caused smoke. Until you investigate, however, you do not know for sure.

Apply When you draw a conclusion, you must apply those conclusions to determine whether the data supports the hypothesis. If your data do not support your hypothesis, it does not mean that the hypothesis is wrong. It means only that the result of the investigation did not support the hypothesis. Maybe the experiment needs to be redesigned, or some of the initial observations on which the hypothesis was based were incomplete or biased. Perhaps more observation or research is needed to refine your hypothesis. A successful investigation does not always come out the way you originally predicted.

Avoid Bias Sometimes a scientific investigation involves making judgments. When you make a judgment, you form an opinion. It is important to be honest and not to allow any expectations of results to bias your judgments. This is important throughout the entire investigation, from researching to collecting data to drawing conclusions.

Communicate

The communication of ideas is an important part of the work of scientists. A discovery that is not reported will not advance the scientific community's understanding or knowledge. Communication among scientists also is important as a way of improving their investigations.

Scientists communicate in many ways, from writing articles in journals and magazines that explain their investigations and experiments, to announcing important discoveries on television and radio. Scientists also share ideas with colleagues on the Internet or present them as lectures, like the student is doing in **Figure 15.**

Figure 15 A student communicates to his peers about his investigation.

These safety symbols are used in laboratory and field investigations in this book to indicate possible hazards. Learn the meaning of each symbol and refer to this page often. *Remember to wash your hands thoroughly after completing lab procedures.*

PROTECTIVE EQUIPMENT   Do not begin any lab without the proper protection equipment.

GOGGLES Proper eye protection must be worn when performing or observing science activities that involve items or conditions as listed below.	**APRON** Wear an approved apron when using substances that could stain, wet, or destroy cloth.	**SOAP** Wash hands with soap and water before removing goggles and after all lab activities.	**GLOVES** Wear gloves when working with biological materials, chemicals, animals, or materials that can stain or irritate hands.

LABORATORY HAZARDS

Symbols	Potential Hazards	Precaution	Response
DISPOSAL	contamination of classroom or environment due to improper disposal of materials such as chemicals and live specimens	• DO NOT dispose of hazardous materials in the sink or trash can. • Dispose of wastes as directed by your teacher.	• If hazardous materials are disposed of improperly, notify your teacher immediately.
EXTREME TEMPERATURE	skin burns due to extremely hot or cold materials such as hot glass, liquids, or metals; liquid nitrogen; dry ice	• Use proper protective equipment, such as hot mitts and/or tongs, when handling objects with extreme temperatures.	• If injury occurs, notify your teacher immediately.
SHARP OBJECTS	punctures or cuts from sharp objects such as razor blades, pins, scalpels, and broken glass	• Handle glassware carefully to avoid breakage. • Walk with sharp objects pointed downward, away from you and others.	• If broken glass or injury occurs, notify your teacher immediately.
ELECTRICAL	electric shock or skin burn due to improper grounding, short circuits, liquid spills, or exposed wires	• Check condition of wires and apparatus for fraying or uninsulated wires, and broken or cracked equipment. • Use only GFCI-protected outlets	• DO NOT attempt to fix electrical problems. Notify your teacher immediately.
CHEMICAL	skin irritation or burns, breathing difficulty, and/or poisoning due to touching, swallowing, or inhalation of chemicals such as acids, bases, bleach, metal compounds, iodine, poinsettias, pollen, ammonia, acetone, nail polish remover, heated chemicals, mothballs, and any other chemicals labeled or known to be dangerous	• Wear proper protective equipment such as goggles, apron, and gloves when using chemicals. • Ensure proper room ventilation or use a fume hood when using materials that produce fumes. • NEVER smell fumes directly. • NEVER taste or eat any material in the laboratory.	• If contact occurs, immediately flush affected area with water and notify your teacher. • If a spill occurs, leave the area immediately and notify your teacher.
FLAMMABLE	unexpected fire due to liquids or gases that ignite easily such as rubbing alcohol	• Avoid open flames, sparks, or heat when flammable liquids are present.	• If a fire occurs, leave the area immediately and notify your teacher.
OPEN FLAME	burns or fire due to open flame from matches, Bunsen burners, or burning materials	• Tie back loose hair and clothing. • Keep flame away from all materials. • Follow teacher instructions when lighting and extinguishing flames. • Use proper protection, such as hot mitts or tongs, when handling hot objects.	• If a fire occurs, leave the area immediately and notify your teacher.
ANIMAL SAFETY	injury to or from laboratory animals	• Wear proper protective equipment such as gloves, apron, and goggles when working with animals. • Wash hands after handling animals.	• If injury occurs, notify your teacher immediately.
BIOLOGICAL	infection or adverse reaction due to contact with organisms such as bacteria, fungi, and biological materials such as blood, animal or plant materials	• Wear proper protective equipment such as gloves, goggles, and apron when working with biological materials. • Avoid skin contact with an organism or any part of the organism. • Wash hands after handling organisms.	• If contact occurs, wash the affected area and notify your teacher immediately.
FUME	breathing difficulties from inhalation of fumes from substances such as ammonia, acetone, nail polish remover, heated chemicals, and mothballs	• Wear goggles, apron, and gloves. • Ensure proper room ventilation or use a fume hood when using substances that produce fumes. • NEVER smell fumes directly.	• If a spill occurs, leave area and notify your teacher immediately.
IRRITANT	irritation of skin, mucous membranes, or respiratory tract due to materials such as acids, bases, bleach, pollen, mothballs, steel wool, and potassium permanganate	• Wear goggles, apron, and gloves. • Wear a dust mask to protect against fine particles.	• If skin contact occurs, immediately flush the affected area with water and notify your teacher.
RADIOACTIVE	excessive exposure from alpha, beta, and gamma particles	• Remove gloves and wash hands with soap and water before removing remainder of protective equipment.	• If cracks or holes are found in the container, notify your teacher immediately.

SCIENCE SKILL HANDBOOK

MATH SKILL HANDBOOK

FOLDABLES HANDBOOK

REFERENCE HANDBOOK

GLOSSARY/ GLOSARIO

INDEX

SCIENCE SKILL HANDBOOK

MATH SKILL HANDBOOK

FOLDABLES HANDBOOK

REFERENCE HANDBOOK

GLOSSARY/ GLOSARIO

INDEX

Safety in the Science Laboratory

Introduction to Science Safety

The science laboratory is a safe place to work if you follow standard safety procedures. Being responsible for your own safety helps to make the entire laboratory a safer place for everyone. When performing any lab, read and apply the caution statements and safety symbol listed at the beginning of the lab.

General Safety Rules

1. Complete the *Lab Safety Form* or other safety contract BEFORE starting any science lab.

2. Study the procedure. Ask your teacher any questions. Be sure you understand safety symbols shown on the page.

3. Notify your teacher about allergies or other health conditions that can affect your participation in a lab.

4. Learn and follow use and safety procedures for your equipment. If unsure, ask your teacher.

5. Never eat, drink, chew gum, apply cosmetics, or do any personal grooming in the lab. Never use lab glassware as food or drink containers. Keep your hands away from your face and mouth.

6. Know the location and proper use of the safety shower, eye wash, fire blanket, and fire alarm.

Prevent Accidents

1. Use the safety equipment provided to you. Goggles and a safety apron should be worn during investigations.

2. Do NOT use hair spray, mousse, or other flammable hair products. Tie back long hair and tie down loose clothing.

3. Do NOT wear sandals or other open-toed shoes in the lab.

4. Remove jewelry on hands and wrists. Loose jewelry, such as chains and long necklaces, should be removed to prevent them from getting caught in equipment.

5. Do not taste any substances or draw any material into a tube with your mouth.

6. Proper behavior is expected in the lab. Practical jokes and fooling around can lead to accidents and injury.

7. Keep your work area uncluttered.

Laboratory Work

1. Collect and carry all equipment and materials to your work area before beginning a lab.

2. Remain in your own work area unless given permission by your teacher to leave it.

3. Always slant test tubes away from your-self and others when heating them, adding substances to them, or rinsing them.

4. If instructed to smell a substance in a container, hold the container a short distance away and fan vapors toward your nose.

5. Do NOT substitute other chemicals/sub-stances for those in the materials list unless instructed to do so by your teacher.

6. Do NOT take any materials or chemi-cals outside of the laboratory.

7. Stay out of storage areas unless instructed to be there and supervised by your teacher.

Laboratory Cleanup

1. Turn off all burners, water, and gas, and disconnect all electrical devices.

2. Clean all pieces of equipment and return all materials to their proper places.

3. Dispose of chemicals and other materi-als as directed by your teacher. Place broken glass and solid substances in the proper containers. Never discard materi-als in the sink.

4. Clean your work area.

5. Wash your hands with soap and water thoroughly BEFORE removing your goggles.

Emergencies

1. Report any fire, electrical shock, glass-ware breakage, spill, or injury, no matter how small, to your teacher immediately. Follow his or her instructions.

2. If your clothing should catch fire, STOP, DROP, and ROLL. If possible, smother it with the fire blanket or get under a safety shower. NEVER RUN.

3. If a fire should occur, turn off all gas and leave the room according to estab-lished procedures.

4. In most instances, your teacher will clean up spills. Do NOT attempt to clean up spills unless you are given permission and instructions to do so.

5. If chemicals come into contact with your eyes or skin, notify your teacher immediately. Use the eyewash, or flush your skin or eyes with large quantities of water.

6. The fire extinguisher and first-aid kit should only be used by your teacher unless it is an extreme emergency and you have been given permission.

7. If someone is injured or becomes ill, only a professional medical provider or someone certified in first aid should perform first-aid procedures.

SCIENCE SKILL HANDBOOK

MATH SKILL HANDBOOK

FOLDABLES HANDBOOK

REFERENCE HANDBOOK

GLOSSARY/ GLOSARIO

INDEX

Math Review

SCIENCE SKILL HANDBOOK

MATH SKILL HANDBOOK

FOLDABLES HANDBOOK

REFERENCE HANDBOOK

GLOSSARY/ GLOSARIO

INDEX

Use Fractions

A fraction compares a part to a whole. In the fraction $\frac{2}{3}$, the 2 represents the part and is the numerator. The 3 represents the whole and is the denominator.

Reduce Fractions To reduce a fraction, you must find the largest factor that is common to both the numerator and the denominator, the greatest common factor (GCF). Divide both numbers by the GCF. The fraction has then been reduced, or it is in its simplest form.

Example

Twelve of the 20 chemicals in the science lab are in powder form. What fraction of the chemicals used in the lab are in powder form?

Step 1 Write the fraction.

$$\frac{\text{part}}{\text{whole}} = \frac{12}{20}$$

Step 2 To find the GCF of the numerator and denominator, list all of the factors of each number.

Factors of 12: 1, 2, 3, 4, 6, 12 (the numbers that divide evenly into 12)

Factors of 20: 1, 2, 4, 5, 10, 20 (the numbers that divide evenly into 20)

Step 3 List the common factors.

1, 2, 4

Step 4 Choose the greatest factor in the list. The GCF of 12 and 20 is 4.

Step 5 Divide the numerator and denominator by the GCF.

$$\frac{12 \div 4}{20 \div 4} = \frac{3}{5}$$

In the lab, $\frac{3}{5}$ of the chemicals are in powder form.

Practice Problem At an amusement park, 66 of 90 rides have a height restriction. What fraction of the rides, in its simplest form, has a height restriction?

Add and Subtract Fractions with Like Denominators To add or subtract fractions with the same denominator, add or subtract the numerators and write the sum or difference over the denominator. After finding the sum or difference, find the simplest form for your fraction.

Example 1

In the forest outside your house, $\frac{1}{8}$ of the animals are rabbits, $\frac{3}{8}$ are squirrels, and the remainder are birds and insects. How many are mammals?

Step 1 Add the numerators.

$$\frac{1}{8} + \frac{3}{8} = \frac{(1 + 3)}{8} = \frac{4}{8}$$

Step 2 Find the GCF.

$$\frac{4}{8} \text{ (GCF, 4)}$$

Step 3 Divide the numerator and denominator by the GCF.

$$\frac{4 \div 4}{8 \div 4} = \frac{1}{2}$$

$\frac{1}{2}$ of the animals are mammals.

Example 2

If $\frac{7}{16}$ of the Earth is covered by freshwater, and $\frac{1}{16}$ of that is in glaciers, how much freshwater is not frozen?

Step 1 Subtract the numerators.

$$\frac{7}{16} - \frac{1}{16} = \frac{(7 - 1)}{16} = \frac{6}{16}$$

Step 2 Find the GCF.

$$\frac{6}{16} \text{ (GCF, 2)}$$

Step 3 Divide the numerator and denominator by the GCF.

$$\frac{6 \div 2}{16 \div 2} = \frac{3}{8}$$

$\frac{3}{8}$ of the freshwater is not frozen.

Practice Problem A bicycle rider is riding at a rate of 15 km/h for $\frac{4}{9}$ of his ride, 10 km/h for $\frac{2}{9}$ of his ride, and 8 km/h for the remainder of the ride. How much of his ride is he riding at a rate greater than 8 km/h?

Add and Subtract Fractions with Unlike Denominators To add or subtract fractions with unlike denominators, first find the least common denominator (LCD). This is the smallest number that is a common multiple of both denominators. Rename each fraction with the LCD, and then add or subtract. Find the simplest form if necessary.

Example 1

A chemist makes a paste that is $\frac{1}{2}$ table salt (NaCl), $\frac{1}{3}$ sugar ($C_6H_{12}O_6$), and the remainder is water (H_2O). How much of the paste is a solid?

Step 1   Find the LCD of the fractions.

$$\frac{1}{2} + \frac{1}{3} \text{ (LCD, 6)}$$

Step 2   Rename each numerator and each denominator with the LCD.

Step 3   Add the numerators.

$$\frac{3}{6} + \frac{2}{6} = \frac{(3 + 2)}{6} = \frac{5}{6}$$

$\frac{5}{6}$ of the paste is a solid.

Example 2

The average precipitation in Grand Junction, CO, is $\frac{7}{10}$ inch in November, and $\frac{3}{5}$ inch in December. What is the total average precipitation?

Step 1   Find the LCD of the fractions.

$$\frac{7}{10} + \frac{3}{5} \text{ (LCD, 10)}$$

Step 2   Rename each numerator and each denominator with the LCD.

Step 3   Add the numerators.

$$\frac{7}{10} + \frac{6}{10} = \frac{(7 + 6)}{10} = \frac{13}{10}$$

$\frac{13}{10}$ inches total precipitation, or $1\frac{3}{10}$ inches.

Practice Problem On an electric bill, about $\frac{1}{8}$ of the energy is from solar energy and about $\frac{1}{10}$ is from wind power. How much of the total bill is from solar energy and wind power combined?

Example 3

In your body, $\frac{7}{10}$ of your muscle contractions are involuntary (cardiac and smooth muscle tissue). Smooth muscle makes $\frac{3}{15}$ of your muscle contractions. How many of your muscle contractions are made by cardiac muscle?

Step 1   Find the LCD of the fractions.

$$\frac{7}{10} - \frac{3}{15} \text{ (LCD, 30)}$$

Step 2   Rename each numerator and each denominator with the LCD.

$$\frac{7 \times 3}{10 \times 3} = \frac{21}{30}$$

$$\frac{3 \times 2}{15 \times 2} = \frac{6}{30}$$

Step 3   Subtract the numerators.

$$\frac{21}{30} - \frac{6}{30} = \frac{(21 - 6)}{30} = \frac{15}{30}$$

Step 4   Find the GCF.

$$\frac{15}{30} \text{ (GCF, 15)}$$

$$\frac{1}{2}$$

$\frac{1}{2}$ of all muscle contractions are cardiac muscle.

Example 4

Tony wants to make cookies that call for $\frac{3}{4}$ of a cup of flour, but he only has $\frac{1}{3}$ of a cup. How much more flour does he need?

Step 1   Find the LCD of the fractions.

$$\frac{3}{4} - \frac{1}{3} \text{ (LCD, 12)}$$

Step 2   Rename each numerator and each denominator with the LCD.

$$\frac{3 \times 3}{4 \times 3} = \frac{9}{12}$$

$$\frac{1 \times 4}{3 \times 4} = \frac{4}{12}$$

Step 3   Subtract the numerators.

$$\frac{9}{12} - \frac{4}{12} = \frac{(9 - 4)}{12} = \frac{5}{12}$$

$\frac{5}{12}$ of a cup of flour

Practice Problem Using the information provided to you in Example 3 above, determine how many muscle contractions are voluntary (skeletal muscle).

SCIENCE SKILL HANDBOOK

MATH SKILL HANDBOOK

FOLDABLES HANDBOOK

REFERENCE HANDBOOK

GLOSSARY/ GLOSARIO

INDEX

Multiply Fractions To multiply with fractions, multiply the numerators and multiply the denominators. Find the simplest form if necessary.

Example

Multiply $\frac{3}{5}$ by $\frac{1}{3}$.

Step 1 Multiply the numerators and denominators.

$$\frac{3}{5} \times \frac{1}{3} = \frac{(3 \times 1)}{(5 \times 3)}\ \frac{3}{15}$$

Step 2 Find the GCF.

$$\frac{3}{15}\ (\text{GCF, 3})$$

Step 3 Divide the numerator and denominator by the GCF.

$$\frac{3 \div 3}{15 \div 3} = \frac{1}{5}$$

$\frac{3}{5}$ multiplied by $\frac{1}{3}$ is $\frac{1}{5}$.

Practice Problem Multiply $\frac{3}{14}$ by $\frac{5}{16}$.

Find a Reciprocal Two numbers whose product is 1 are called multiplicative inverses, or reciprocals.

Example

Find the reciprocal of $\frac{3}{8}$.

Step 1 Inverse the fraction by putting the denominator on top and the numerator on the bottom.

$$\frac{8}{3}$$

The reciprocal of $\frac{3}{8}$ is $\frac{8}{3}$.

Practice Problem Find the reciprocal of $\frac{4}{9}$.

Divide Fractions To divide one fraction by another fraction, multiply the dividend by the reciprocal of the divisor. Find the simplest form if necessary.

Example 1

Divide $\frac{1}{9}$ by $\frac{1}{3}$.

Step 1 Find the reciprocal of the divisor. The reciprocal of $\frac{1}{3}$ is $\frac{3}{1}$.

Step 2 Multiply the dividend by the reciprocal of the divisor.

$$\frac{\frac{1}{9}}{\frac{1}{3}} = \frac{1}{9} \times \frac{3}{1} = \frac{(1 \times 3)}{(9 \times 1)} = \frac{3}{9}$$

Step 3 Find the GCF.

$$\frac{3}{9}\ (\text{GCF, 3})$$

Step 4 Divide the numerator and denominator by the GCF.

$$\frac{3 \div 3}{9 \div 3} = \frac{1}{3}$$

$\frac{1}{9}$ divided by $\frac{1}{3}$ is $\frac{1}{3}$.

Example 2

Divide $\frac{3}{5}$ by $\frac{1}{4}$.

Step 1 Find the reciprocal of the divisor. The reciprocal of $\frac{1}{4}$ is $\frac{4}{1}$.

Step 2 Multiply the dividend by the reciprocal of the divisor.

$$\frac{\frac{3}{5}}{\frac{1}{4}} = \frac{3}{5} \times \frac{4}{1} = \frac{(3 \times 4)}{(5 \times 1)} = \frac{12}{5}$$

$\frac{3}{5}$ divided by $\frac{1}{4}$ is $\frac{12}{5}$ or $2\frac{2}{5}$.

Practice Problem Divide $\frac{3}{11}$ by $\frac{7}{10}$.

Use Ratios

When you compare two numbers by division, you are using a ratio. Ratios can be written 3 to 5, 3:5, or $\frac{3}{5}$. Ratios, like fractions, also can be written in simplest form.

Ratios can represent one type of probability, called odds. This is a ratio that compares the number of ways a certain outcome occurs to the number of possible outcomes. For example, if you flip a coin 100 times, what are the odds that it will come up heads? There are two possible outcomes, heads or tails, so the odds of coming up heads are 50:100. Another way to say this is that 50 out of 100 times the coin will come up heads. In its simplest form, the ratio is 1:2.

Example 1

A chemical solution contains 40 g of salt and 64 g of baking soda. What is the ratio of salt to baking soda as a fraction in simplest form?

Step 1 Write the ratio as a fraction.

$$\frac{salt}{baking\ soda} = \frac{40}{64}$$

Step 2 Express the fraction in simplest form. The GCF of 40 and 64 is 8.

$$\frac{40}{64} = \frac{40 \div 8}{64 \div 8} = \frac{5}{8}$$

The ratio of salt to baking soda in the sample is 5:8.

Example 2

Sean rolls a 6-sided die 6 times. What are the odds that the side with a 3 will show?

Step 1 Write the ratio as a fraction.

$$\frac{number\ of\ sides\ with\ a\ 3}{number\ of\ possible\ sides} = \frac{1}{6}$$

Step 2 Multiply by the number of attempts.

$$\frac{1}{6} \times 6\ attempts = \frac{6}{6}\ attempts = 1\ attempt$$

1 attempt out of 6 will show a 3.

Practice Problem Two metal rods measure 100 cm and 144 cm in length. What is the ratio of their lengths in simplest form?

Use Decimals

A fraction with a denominator that is a power of ten can be written as a decimal. For example, 0.27 means $\frac{27}{100}$. The decimal point separates the ones place from the tenths place.

Any fraction can be written as a decimal using division. For example, the fraction $\frac{5}{8}$ can be written as a decimal by dividing 5 by 8. Written as a decimal, it is 0.625.

Add or Subtract Decimals When adding and subtracting decimals, line up the decimal points before carrying out the operation.

Example 1

Find the sum of 47.68 and 7.80.

Step 1 Line up the decimal places when you write the numbers.

$$\begin{array}{r} 47.68 \\ + 7.80 \end{array}$$

Step 2 Add the decimals.

$$\begin{array}{r} 47.68 \\ + 7.80 \\ \hline 55.48 \end{array}$$

The sum of 47.68 and 7.80 is 55.48.

Example 2

Find the difference of 42.17 and 15.85.

Step 1 Line up the decimal places when you write the number.

$$\begin{array}{r} 42.17 \\ -15.85 \end{array}$$

Step 2 Subtract the decimals.

$$\begin{array}{r} 42.17 \\ -15.85 \\ \hline 26.32 \end{array}$$

The difference of 42.17 and 15.85 is 26.32.

Practice Problem Find the sum of 1.245 and 3.842.

SCIENCE SKILL HANDBOOK

MATH SKILL HANDBOOK

FOLDABLES HANDBOOK

REFERENCE HANDBOOK

GLOSSARY/ GLOSARIO

INDEX

Multiply Decimals To multiply decimals, multiply the numbers like numbers without decimal points. Count the decimal places in each factor. The product will have the same number of decimal places as the sum of the decimal places in the factors.

Example

Multiply 2.4 by 5.9.

Step 1   Multiply the factors like two whole numbers.

$24 \times 59 = 1416$

Step 2   Find the sum of the number of decimal places in the factors. Each factor has one decimal place, for a sum of two decimal places.

Step 3   The product will have two decimal places.

14.16

The product of 2.4 and 5.9 is 14.16.

Practice Problem Multiply 4.6 by 2.2.

Divide Decimals When dividing decimals, change the divisor to a whole number. To do this, multiply both the divisor and the dividend by the same power of ten. Then place the decimal point in the quotient directly above the decimal point in the dividend. Then divide as you do with whole numbers.

Example

Divide 8.84 by 3.4.

Step 1   Multiply both factors by 10.

$3.4 \times 10 = 34, 8.84 \times 10 = 88.4$

Step 2   Divide 88.4 by 34.

$$
\begin{array}{r}
2.6 \\
34\overline{)88.4} \\
-68 \\
\hline
204 \\
-204 \\
\hline
0
\end{array}
$$

8.84 divided by 3.4 is 2.6.

Practice Problem Divide 75.6 by 3.6.

Use Proportions

An equation that shows that two ratios are equivalent is a proportion. The ratios $\frac{2}{4}$ and $\frac{5}{10}$ are equivalent, so they can be written as $\frac{2}{4} = \frac{5}{10}$. This equation is a proportion.

When two ratios form a proportion, the cross products are equal. To find the cross products in the proportion $\frac{2}{4} = \frac{5}{10}$, multiply the 2 and the 10, and the 4 and the 5. Therefore $2 \times 10 = 4 \times 5$, or $20 = 20$.

Because you know that both ratios are equal, you can use cross products to find a missing term in a proportion. This is known as solving the proportion.

Example

The heights of a tree and a pole are proportional to the lengths of their shadows. The tree casts a shadow of 24 m when a 6-m pole casts a shadow of 4 m. What is the height of the tree?

Step 1   Write a proportion.

$$\frac{\text{height of tree}}{\text{height of pole}} = \frac{\text{length of tree's shadow}}{\text{length of pole's shadow}}$$

Step 2   Substitute the known values into the proportion. Let h represent the unknown value, the height of the tree.

$$\frac{h}{6} \times \frac{24}{4}$$

Step 3   Find the cross products.

$h \times 4 = 6 \times 24$

Step 4   Simplify the equation.

$4h \times 144$

Step 5   Divide each side by 4.

$$\frac{4h}{4} \times \frac{144}{4}$$

$h = 36$

The height of the tree is 36 m.

Practice Problem The ratios of the weights of two objects on the Moon and on Earth are in proportion. A rock weighing 3 N on the Moon weighs 18 N on Earth. How much would a rock that weighs 5 N on the Moon weigh on Earth?

SCIENCE SKILL HANDBOOK

MATH SKILL HANDBOOK

FOLDABLES HANDBOOK

REFERENCE HANDBOOK

GLOSSARY/ GLOSARIO

INDEX

Use Percentages

The word *percent* means "out of one hundred." It is a ratio that compares a number to 100. Suppose you read that 77 percent of Earth's surface is covered by water. That is the same as reading that the fraction of Earth's surface covered by water is $\frac{77}{100}$. To express a fraction as a percent, first find the equivalent decimal for the fraction. Then, multiply the decimal by 100 and add the percent symbol.

Example 1

Express $\frac{13}{20}$ as a percent.

Step 1 Find the equivalent decimal for the fraction.

$$
\begin{array}{r}
0.65 \\
20\overline{)13.00} \\
\underline{12\,0} \\
1\,00 \\
\underline{1\,00} \\
0
\end{array}
$$

Step 2 Rewrite the fraction $\frac{13}{20}$ as 0.65.

Step 3 Multiply 0.65 by 100 and add the % symbol.

$$0.65 \times 100 = 65 = 65\%$$

So, $\frac{13}{20} = 65\%$.

This also can be solved as a proportion.

Example 2

Express $\frac{13}{20}$ as a percent.

Step 1 Write a proportion.

$$\frac{13}{20} = \frac{x}{100}$$

Step 2 Find the cross products.

$$1300 = 20x$$

Step 3 Divide each side by 20.

$$\frac{1300}{20} = \frac{20x}{20}$$

$$65\% = x$$

Practice Problem In one year, 73 of 365 days were rainy in one city. What percent of the days in that city were rainy?

Solve One-Step Equations

A statement that two expressions are equal is an equation. For example, $A = B$ is an equation that states that A is equal to B.

An equation is solved when a variable is replaced with a value that makes both sides of the equation equal. To make both sides equal the inverse operation is used. Addition and subtraction are inverses, and multiplication and division are inverses.

Example 1

Solve the equation $x - 10 = 35$.

Step 1 Find the solution by adding 10 to each side of the equation.

$$x - 10 = 35$$
$$x - 10 + 10 = 35 - 10$$
$$x = 45$$

Step 2 Check the solution.

$$x - 10 = 35$$
$$45 - 10 = 35$$
$$35 = 35$$

Both sides of the equation are equal, so $x = 45$.

Example 2

In the formula $a = bc$, find the value of c if $a = 20$ and $b = 2$.

Step 1 Rearrange the formula so the unknown value is by itself on one side of the equation by dividing both sides by b.

$$a = bc$$
$$\frac{a}{b} = \frac{bc}{b}$$
$$\frac{a}{b} = c$$

Step 2 Replace the variables a and b with the values that are given.

$$\frac{a}{b} = c$$
$$\frac{20}{2} = c$$
$$10 = c$$

Step 3 Check the solution.

$$a = bc$$
$$20 = 2 \times 10$$
$$20 = 20$$

Both sides of the equation are equal, so $c = 10$ is the solution when $a = 20$ and $b = 2$.

Practice Problem In the formula $h = gd$, find the value of d if $g = 12.3$ and $h = 17.4$.

SCIENCE SKILL HANDBOOK

MATH SKILL HANDBOOK

FOLDABLES HANDBOOK

REFERENCE HANDBOOK

GLOSSARY/ GLOSARIO

INDEX

Use Statistics

The branch of mathematics that deals with collecting, analyzing, and presenting data is statistics. In statistics, there are three common ways to summarize data with a single number—the mean, the median, and the mode.

The **mean** of a set of data is the arithmetic average. It is found by adding the numbers in the data set and dividing by the number of items in the set.

The **median** is the middle number in a set of data when the data are arranged in numerical order. If there were an even number of data points, the median would be the mean of the two middle numbers.

The **mode** of a set of data is the number or item that appears most often.

Another number that often is used to describe a set of data is the range. The **range** is the difference between the largest number and the smallest number in a set of data.

Example

The speeds (in m/s) for a race car during five different time trials are 39, 37, 44, 36, and 44.

To find the mean:

Step 1 Find the sum of the numbers.

$$39 + 37 + 44 + 36 + 44 = 200$$

Step 2 Divide the sum by the number of items, which is 5.

$$200 \div 5 = 40$$

The mean is 40 m/s.

To find the median:

Step 1 Arrange the measures from least to greatest.

36, 37, 39, 44, 44

Step 2 Determine the middle measure.

36, 37, 39, 44, 44

The median is 39 m/s.

To find the mode:

Step 1 Group the numbers that are the same together.

44, 44, 36, 37, 39

Step 2 Determine the number that occurs most in the set.

44, 44, 36, 37, 39

The mode is 44 m/s.

To find the range:

Step 1 Arrange the measures from greatest to least.

44, 44, 39, 37, 36

Step 2 Determine the greatest and least measures in the set.

44, 44, 39, 37, 36

Step 3 Find the difference between the greatest and least measures.

$$44 - 36 = 8$$

The range is 8 m/s.

Practice Problem Find the mean, median, mode, and range for the data set 8, 4, 12, 8, 11, 14, 16.

A **frequency table** shows how many times each piece of data occurs, usually in a survey. **Table 1** below shows the results of a student survey on favorite color.

Table 1	Student Color Choice	
Color	Tally	Frequency
red	IIII	4
blue	IIII	5
black	II	2
green	III	3
purple	IIII II	7
yellow	IIII I	6

Based on the frequency table data, which color is the favorite?

Use Geometry

The branch of mathematics that deals with the measurement, properties, and relationships of points, lines, angles, surfaces, and solids is called geometry.

Perimeter The **perimeter** (P) is the distance around a geometric figure. To find the perimeter of a rectangle, add the length and width and multiply that sum by two, or $2(l + w)$. To find perimeters of irregular figures, add the length of the sides.

Example 1

Find the perimeter of a rectangle that is 3 m long and 5 m wide.

Step 1 You know that the perimeter is 2 times the sum of the width and length.

$$P = 2(3\ m + 5\ m)$$

Step 2 Find the sum of the width and length.

$$P = 2(8\ m)$$

Step 3 Multiply by 2.

$$P = 16\ m$$

The perimeter is 16 m.

Example 2

Find the perimeter of a shape with sides measuring 2 cm, 5 cm, 6 cm, 3 cm.

Step 1 You know that the perimeter is the sum of all the sides.

$$P = 2 + 5 + 6 + 3$$

Step 2 Find the sum of the sides.

$$P = 2 + 5 + 6 + 3$$
$$P = 16$$

The perimeter is 16 cm.

Practice Problem Find the perimeter of a rectangle with a length of 18 m and a width of 7 m.

Practice Problem Find the perimeter of a triangle measuring 1.6 cm by 2.4 cm by 2.4 cm.

Area of a Rectangle The **area** (A) is the number of square units needed to cover a surface. To find the area of a rectangle, multiply the length times the width, or $l \times w$. When finding area, the units also are multiplied. Area is given in square units.

Example

Find the area of a rectangle with a length of 1 cm and a width of 10 cm.

Step 1 You know that the area is the length multiplied by the width.

$$A = (1\ cm \times 10\ cm)$$

Step 2 Multiply the length by the width. Also multiply the units.

$$A = 10\ cm^2$$

The area is 10 cm^2.

Practice Problem Find the area of a square whose sides measure 4 m.

Area of a Triangle To find the area of a triangle, use the formula:

$$A = \frac{1}{2}(base \times height)$$

The base of a triangle can be any of its sides. The height is the perpendicular distance from a base to the opposite endpoint, or vertex.

Example

Find the area of a triangle with a base of 18 m and a height of 7 m.

Step 1 You know that the area is $\frac{1}{2}$ the base times the height.

$$A = \frac{1}{2}(18\ m \times 7\ m)$$

Step 2 Multiply $\frac{1}{2}$ by the product of 18 × 7. Multiply the units.

$$A = \frac{1}{2}(126\ m^2)$$
$$A = 63\ m^2$$

The area is 63 m^2.

Practice Problem Find the area of a triangle with a base of 27 cm and a height of 17 cm.

SCIENCE SKILL HANDBOOK

MATH SKILL HANDBOOK

FOLDABLES HANDBOOK

REFERENCE HANDBOOK

GLOSSARY/ GLOSARIO

INDEX

SCIENCE SKILL HANDBOOK

MATH SKILL HANDBOOK

FOLDABLES HANDBOOK

REFERENCE HANDBOOK

GLOSSARY/ GLOSARIO

INDEX

Circumference of a Circle The **diameter** (*d*) of a circle is the distance across the circle through its center, and the **radius** (r) is the distance from the center to any point on the circle. The radius is half of the diameter. The distance around the circle is called the **circumference** (C). The formula for finding the circumference is:

$$C = 2\pi r \text{ or } C = \pi d$$

The circumference divided by the diameter is always equal to 3.1415926… This nonterminating and nonrepeating number is represented by the Greek letter π (pi). An approximation often used for π is 3.14.

Example 1

Find the circumference of a circle with a radius of 3 m.

Step 1 You know the formula for the circumference is 2 times the radius times π.

$$C = 2\pi(3)$$

Step 2 Multiply 2 times the radius.

$$C = 6\pi$$

Step 3 Multiply by π.

$$C \approx 19 \text{ m}$$

The circumference is about 19 m.

Example 2

Find the circumference of a circle with a diameter of 24.0 cm.

Step 1 You know the formula for the circumference is the diameter times π.

$$C = \pi(24.0)$$

Step 2 Multiply the diameter by π.

$$C \approx 75.4 \text{ cm}$$

The circumference is about 75.4 cm.

Practice Problem Find the circumference of a circle with a radius of 19 cm.

Area of a Circle The formula for the area of a circle is: $A = \pi r^2$

Example 1

Find the area of a circle with a radius of 4.0 cm.

Step 1 $A = \pi(4.0)^2$

Step 2 Find the square of the radius.

$$A = 16\pi$$

Step 3 Multiply the square of the radius by π.

$$A \approx 50 \text{ cm}^2$$

The area of the circle is about 50 cm^2.

Example 2

Find the area of a circle with a radius of 225 m.

Step 1 $A = \pi(225)^2$

Step 2 Find the square of the radius.

$$A = 50625\pi$$

Step 3 Multiply the square of the radius by π.

$$A \approx 159043.1$$

The area of the circle is about 159043.1 m^2.

Example 3

Find the area of a circle whose diameter is 20.0 mm.

Step 1 Remember that the radius is half of the diameter.

$$A = \pi\left(\frac{20.0}{2}\right)^2$$

Step 2 Find the radius.

$$A = \pi(10.0)^2$$

Step 3 Find the square of the radius.

$$A = 100\pi$$

Step 4 Multiply the square of the radius by π.

$$A \approx 314 \text{ mm}^2$$

The area of the circle is about 314 mm^2.

Practice Problem Find the area of a circle with a radius of 16 m.

Volume The measure of space occupied by a solid is the **volume** (V). To find the volume of a rectangular solid multiply the length times width times height, or $V = l \times w \times h$. It is measured in cubic units, such as cubic centimeters (cm^3).

Example

Find the volume of a rectangular solid with a length of 2.0 m, a width of 4.0 m, and a height of 3.0 m.

Step 1   You know the formula for volume is the length times the width times the height.

$$V = 2.0 \text{ m} \times 4.0 \text{ m} \times 3.0 \text{ m}$$

Step 2   Multiply the length times the width times the height.

$$V = 24 \text{ m}^3$$

The volume is 24 m^3.

Practice Problem   Find the volume of a rectangular solid that is 8 m long, 4 m wide, and 4 m high.

To find the volume of other solids, multiply the area of the base times the height.

Example 1

Find the volume of a solid that has a triangular base with a length of 8.0 m and a height of 7.0 m. The height of the entire solid is 15.0 m.

Step 1   You know that the base is a triangle, and the area of a triangle is $\frac{1}{2}$ the base times the height, and the volume is the area of the base times the height.

$$V = \left[\frac{1}{2}(b \times h)\right] \times 15$$

Step 2   Find the area of the base.

$$V = \left[\frac{1}{2}(8 \times 7)\right] \times 15$$

$$V = \left(\frac{1}{2} \times 56\right) \times 15$$

Step 3   Multiply the area of the base by the height of the solid.

$$V = 28 \times 15$$

$$V = 420 \text{ m}^3$$

The volume is 420 m^3.

Example 2

Find the volume of a cylinder that has a base with a radius of 12.0 cm, and a height of 21.0 cm.

Step 1   You know that the base is a circle, and the area of a circle is the square of the radius times π, and the volume is the area of the base times the height.

$$V = (\pi r^2) \times 21$$

$$V = (\pi 12^2) \times 21$$

Step 2   Find the area of the base.

$$V = 144\pi \times 21$$

$$V = 452 \times 21$$

Step 3   Multiply the area of the base by the height of the solid.

$$V \approx 9{,}500 \text{ cm}^3$$

The volume is about 9,500 cm^3.

Example 3

Find the volume of a cylinder that has a diameter of 15 mm and a height of 4.8 mm.

Step 1   You know that the base is a circle with an area equal to the square of the radius times π. The radius is one-half the diameter. The volume is the area of the base times the height.

$$V = (\pi r^2) \times 4.8$$

$$V = \left[\pi\left(\frac{1}{2} \times 15\right)^2\right] \times 4.8$$

$$V = (\pi 7.5^2) \times 4.8$$

Step 2   Find the area of the base.

$$V = 56.25\pi \times 4.8$$

$$V \approx 176.71 \times 4.8$$

Step 3   Multiply the area of the base by the height of the solid.

$$V \approx 848.2$$

The volume is about 848.2 mm^3.

Practice Problem   Find the volume of a cylinder with a diameter of 7 cm in the base and a height of 16 cm.

SCIENCE SKILL HANDBOOK

MATH SKILL HANDBOOK

FOLDABLES HANDBOOK

REFERENCE HANDBOOK

GLOSSARY/ GLOSARIO

INDEX

Science Applications

SCIENCE SKILL HANDBOOK

MATH SKILL HANDBOOK

FOLDABLES HANDBOOK

REFERENCE HANDBOOK

GLOSSARY/ GLOSARIO

INDEX

Measure in SI

The metric system of measurement was developed in 1795. A modern form of the metric system, called the International System (SI), was adopted in 1960 and provides the standard measurements that all scientists around the world can understand.

The SI system is convenient because unit sizes vary by powers of 10. Prefixes are used to name units. Look at **Table 2** for some common SI prefixes and their meanings.

Table 2	Common SI Prefixes		
Prefix	Symbol	Meaning	
kilo–	k	1,000	thousandth
hecto–	h	100	hundred
deka–	da	10	ten
deci–	d	0.1	tenth
centi–	c	0.01	hundreth
milli–	m	0.001	thousandth

Example

How many grams equal one kilogram?

Step 1 Find the prefix *kilo–* in **Table 2**.

Step 2 Using **Table 2**, determine the meaning of *kilo–*. According to the table, it means 1,000. When the prefix *kilo–* is added to a unit, it means that there are 1,000 of the units in a "kilounit."

Step 3 Apply the prefix to the units in the question. The units in the question are grams. There are 1,000 grams in a kilogram.

Practice Problem Is a milligram larger or smaller than a gram? How many of the smaller units equal one larger unit? What fraction of the larger unit does one smaller unit represent?

Dimensional Analysis

Convert SI Units In science, quantities such as length, mass, and time sometimes are measured using different units. A process called dimensional analysis can be used to change one unit of measure to another. This process involves multiplying your starting quantity and units by one or more conversion factors. A conversion factor is a ratio equal to one and can be made from any two equal quantities with different units. If 1,000 mL equal 1 L then two ratios can be made.

$$\frac{1{,}000 \text{ mL}}{1 \text{ L}} = \frac{1 \text{ L}}{1{,}000 \text{ mL}} = 1$$

One can convert between units in the SI system by using the equivalents in **Table 2** to make conversion factors.

Example

How many cm are in 4 m?

Step 1 Write conversion factors for the units given. From **Table 2,** you know that 100 cm = 1 m. The conversion factors are

$$\frac{100 \text{ cm}}{1 \text{ m}} \text{ and } \frac{1 \text{ m}}{100 \text{ cm}}$$

Step 2 Decide which conversion factor to use. Select the factor that has the units you are converting from (m) in the denominator and the units you are converting to (cm) in the numerator.

$$\frac{100 \text{ cm}}{1 \text{ m}}$$

Step 3 Multiply the starting quantity and units by the conversion factor. Cancel the starting units with the units in the denominator. There are 400 cm in 4 m.

$$4 \text{ m} = \frac{100 \text{ cm}}{1 \text{ m}} = 400 \text{ cm}$$

Practice Problem How many milligrams are in one kilogram? (Hint: You will need to use two conversion factors from **Table 2.**)

Table 3 Unit System Equivalents

Type of Measurement	Equivalent
Length	1 in = 2.54 cm 1 yd = 0.91 m 1 mi = 1.61 km
Mass and weight*	1 oz = 28.35 g 1 lb = 0.45 kg 1 ton (short) = 0.91 tonnes (metric tons) 1 lb = 4.45 N
Volume	$1 \text{ in}^3 = 16.39 \text{ cm}^3$ 1 qt = 0.95 L 1 gal = 3.78 L
Area	$1 \text{ in}^2 = 6.45 \text{ cm}^2$ $1 \text{ yd}^2 = 0.83 \text{ m}^2$ $1 \text{ mi}^2 = 2.59 \text{ km}^2$ 1 acre = 0.40 hectares
Temperature	$°C = \dfrac{(°F - 32)}{1.8}$ $K = °C + 273$

*Weight is measured in standard Earth gravity.

Convert Between Unit Systems **Table 3** gives a list of equivalents that can be used to convert between English and SI units.

Example

If a meterstick has a length of 100 cm, how long is the meterstick in inches?

Step 1 Write the conversion factors for the units given. From **Table 3,** 1 in = 2.54 cm.

$$\frac{1 \text{ in}}{2.54 \text{ cm}} \text{ and } \frac{2.54 \text{ cm}}{1 \text{ in}}$$

Step 2 Determine which conversion factor to use. You are converting from cm to in. Use the conversion factor with cm on the bottom.

$$\frac{1 \text{ in}}{2.54 \text{ cm}}$$

Step 3 Multiply the starting quantity and units by the conversion factor. Cancel the starting units with the units in the denominator. Round your answer to the nearest tenth.

$$100 \text{ cm} \times \frac{1 \text{ in}}{2.54 \text{ cm}} = 39.37 \text{ in}$$

The meterstick is about 39.4 in long.

Practice Problem 1 A book has a mass of 5 lb. What is the mass of the book in kg?

Practice Problem 2 Use the equivalent for in and cm (1 in = 2.54 cm) to show how $1 \text{ in}^3 \approx 16.39 \text{ cm}^3$.

Sidebar: SCIENCE SKILL HANDBOOK | MATH SKILL HANDBOOK | FOLDABLES HANDBOOK | REFERENCE HANDBOOK | GLOSSARY/GLOSARIO | INDEX

Precision and Significant Digits

When you make a measurement, the value you record depends on the precision of the measuring instrument. This precision is represented by the number of significant digits recorded in the measurement. When counting the number of significant digits, all digits are counted except zeros at the end of a number with no decimal point such as 2,050, and zeros at the beginning of a decimal such as 0.03020. When adding or subtracting numbers with different precision, round the answer to the smallest number of decimal places of any number in the sum or difference. When multiplying or dividing, the answer is rounded to the smallest number of significant digits of any number being multiplied or divided.

Example

The lengths 5.28 and 5.2 are measured in meters. Find the sum of these lengths and record your answer using the correct number of significant digits.

Step 1 Find the sum.

5.28 m   2 digits after the decimal
+ 5.2 m   1 digit after the decimal
10.48 m

Step 2 Round to one digit after the decimal because the least number of digits after the decimal of the numbers being added is 1.

The sum is 10.5 m.

Practice Problem 1 How many significant digits are in the measurement 7,071,301 m? How many significant digits are in the measurement 0.003010 g?

Practice Problem 2 Multiply 5.28 and 5.2 using the rule for multiplying and dividing. Record the answer using the correct number of significant digits.

Scientific Notation

Many times numbers used in science are very small or very large. Because these numbers are difficult to work with scientists use scientific notation. To write numbers in scientific notation, move the decimal point until only one non-zero digit remains on the left. Then count the number of places you moved the decimal point and use that number as a power of ten. For example, the average distance from the Sun to Mars is 227,800,000,000 m. In scientific notation, this distance is 2.278×10^{11} m. Because you moved the decimal point to the left, the number is a positive power of ten.

The mass of an electron is about 0.000 000 000 000 000 000 000 000 000 000 911 kg. Expressed in scientific notation, this mass is 9.11×10^{-31} kg. Because the decimal point was moved to the right, the number is a negative power of ten.

Example

Earth is 149,600,000 km from the Sun. Express this in scientific notation.

Step 1 Move the decimal point until one non-zero digit remains on the left.

1.496 000 00

Step 2 Count the number of decimal places you have moved. In this case, eight.

Step 2 Show that number as a power of ten, 10^8.

Earth is 1.496×10^8 km from the Sun.

Practice Problem 1 How many significant digits are in 149,600,000 km? How many significant digits are in 1.496×10^8 km?

Practice Problem 2 Parts used in a high performance car must be measured to 7×10^{-6} m. Express this number as a decimal.

Practice Problem 3 A CD is spinning at 539 revolutions per minute. Express this number in scientific notation.

Make and Use Graphs

Data in tables can be displayed in a graph—a visual representation of data. Common graph types include line graphs, bar graphs, and circle graphs.

Line Graph A line graph shows a relationship between two variables that change continuously. The independent variable is changed and is plotted on the x-axis. The dependent variable is observed, and is plotted on the y-axis.

Example

Draw a line graph of the data below from a cyclist in a long-distance race.

Table 4 Bicycle Race Data	
Time (h)	**Distance (km)**
0	0
1	8
2	16
3	24
4	32
5	40

Step 1 Determine the x-axis and y-axis variables. Time varies independently of distance and is plotted on the x-axis. Distance is dependent on time and is plotted on the y-axis.

Step 2 Determine the scale of each axis. The x-axis data ranges from 0 to 5. The y-axis data ranges from 0 to 50.

Step 3 Using graph paper, draw and label the axes. Include units in the labels.

Step 4 Draw a point at the intersection of the time value on the x-axis and corresponding distance value on the y-axis. Connect the points and label the graph with a title, as shown in **Figure 8.**

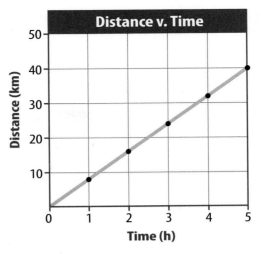

Figure 8 This line graph shows the relationship between distance and time during a bicycle ride.

Practice Problem A puppy's shoulder height is measured during the first year of her life. The following measurements were collected: (3 mo, 52 cm), (6 mo, 72 cm), (9 mo, 83 cm), (12 mo, 86 cm). Graph this data.

Find a Slope The slope of a straight line is the ratio of the vertical change, rise, to the horizontal change, run.

$$\text{Slope} = \frac{\text{vertical change (rise)}}{\text{horizontal change (run)}} = \frac{\text{change in } y}{\text{change in } x}$$

Example

Find the slope of the graph in **Figure 8**.

Step 1 You know that the slope is the change in y divided by the change in x.

$$\text{Slope} = \frac{\text{change in } y}{\text{change in } x}$$

Step 2 Determine the data points you will be using. For a straight line, choose the two sets of points that are the farthest apart.

$$\text{Slope} = \frac{(40 - 0) \text{ km}}{(5 - 0) \text{ h}}$$

Step 3 Find the change in y and x.

$$\text{Slope} = \frac{40 \text{ km}}{5 \text{ h}}$$

Step 4 Divide the change in y by the change in x.

$$\text{Slope} = \frac{8 \text{ km}}{\text{h}}$$

The slope of the graph is 8 km/h.

SCIENCE SKILL HANDBOOK

MATH SKILL HANDBOOK

FOLDABLES HANDBOOK

REFERENCE HANDBOOK

GLOSSARY/ GLOSARIO

INDEX

SCIENCE SKILL HANDBOOK

MATH SKILL HANDBOOK

FOLDABLES HANDBOOK

REFERENCE HANDBOOK

GLOSSARY/ GLOSARIO

INDEX

Bar Graph To compare data that does not change continuously you might choose a bar graph. A bar graph uses bars to show the relationships between variables. The *x*-axis variable is divided into parts. The parts can be numbers such as years, or a category such as a type of animal. The *y*-axis is a number and increases continuously along the axis.

Example

A recycling center collects 4.0 kg of aluminum on Monday, 1.0 kg on Wednesday, and 2.0 kg on Friday. Create a bar graph of this data.

Step 1 Select the *x*-axis and *y*-axis variables. The measured numbers (the masses of aluminum) should be placed on the *y*-axis. The variable divided into parts (collection days) is placed on the *x*-axis.

Step 2 Create a graph grid like you would for a line graph. Include labels and units.

Step 3 For each measured number, draw a vertical bar above the *x*-axis value up to the *y*-axis value. For the first data point, draw a vertical bar above Monday up to 4.0 kg.

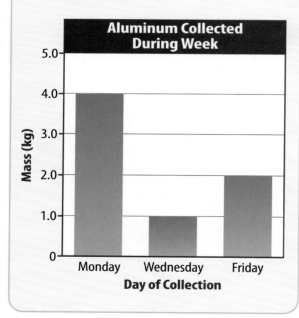

Practice Problem Draw a bar graph of the gases in air: 78% nitrogen, 21% oxygen, 1% other gases.

Circle Graph To display data as parts of a whole, you might use a circle graph. A circle graph is a circle divided into sections that represent the relative size of each piece of data. The entire circle represents 100%, half represents 50%, and so on.

Example

Air is made up of 78% nitrogen, 21% oxygen, and 1% other gases. Display the composition of air in a circle graph.

Step 1 Multiply each percent by 360° and divide by 100 to find the angle of each section in the circle.

$$78\% \times \frac{360°}{100} = 280.8°$$

$$21\% \times \frac{360°}{100} = 75.6°$$

$$1\% \times \frac{360°}{100} = 3.6°$$

Step 2 Use a compass to draw a circle and to mark the center of the circle. Draw a straight line from the center to the edge of the circle.

Step 3 Use a protractor and the angles you calculated to divide the circle into parts. Place the center of the protractor over the center of the circle and line the base of the protractor over the straight line.

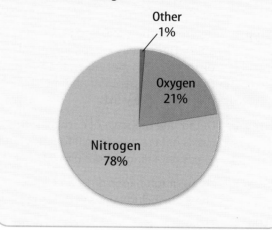

Practice Problem Draw a circle graph to represent the amount of aluminum collected during the week shown in the bar graph to the left.

Student Study Guides & Instructions
By Dinah Zike

1. You will find suggestions for Study Guides, also known as Foldables or books, in each chapter lesson and as a final project. Look at the end of the chapter to determine the project format and glue the Foldables in place as you progress through the chapter lessons.

2. Creating the Foldables or books is simple and easy to do by using copy paper, art paper, and internet printouts. Photocopies of maps, diagrams, or your own illustrations may also be used for some of the Foldables. Notebook paper is the most common source of material for study guides and 83% of all Foldables are created from it. When folded to make books, notebook paper Foldables easily fit into 11″ × 17″ or 12″ × 18″ chapter projects with space left over. Foldables made using photocopy paper are slightly larger and they fit into Projects, but snugly. Use the least amount of glue, tape, and staples needed to assemble the Foldables.

3. Seven of the Foldables can be made using either small or large paper. When 11″ × 17″ or 12″ × 18″ paper is used, these become projects for housing smaller Foldables. Project format boxes are located within the instructions to remind you of this option.

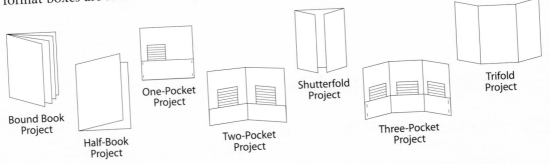

Bound Book Project

Half-Book Project

One-Pocket Project

Two-Pocket Project

Shutterfold Project

Three-Pocket Project

Trifold Project

4. Use one-gallon self-locking plastic bags to store your projects. Place strips of two-inch clear tape along the left, long side of the bag and punch holes through the taped edge. Cut the bottom corners off the bag so it will not hold air. Store this Project Portfolio inside a three-hole binder. To store a large collection of project bags, use a giant laundry-soap box. Holes can be punched in some of the Foldable Projects so they can be stored in a three-hole binder without using a plastic bag. Punch holes in the pocket books before gluing or stapling the pocket.

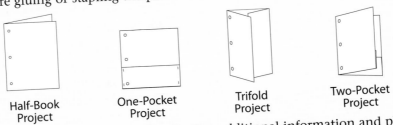

Half-Book Project

One-Pocket Project

Trifold Project

Two-Pocket Project

5. Maximize the use of the projects by collecting additional information and placing it on the back of the project and other unused spaces of the large Foldables.

SCIENCE SKILL HANDBOOK

MATH SKILL HANDBOOK

FOLDABLES HANDBOOK

REFERENCE HANDBOOK

GLOSSARY/ GLOSARIO

INDEX

Half-Book Foldable® By Dinah Zike

Step 1 Fold a sheet of notebook or copy paper in half.

Label the exterior tab and use the inside space to write information.

PROJECT FORMAT
Use 11″ × 17″ or 12″ × 18″ paper on the horizontal axis to make a large project book.

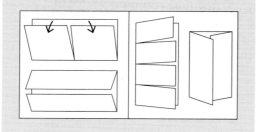

Variations

Paper can be folded horizontally, like a *hamburger* or vertically, like a *hot dog*.

A

B

C Half-books can be folded so that one side is ½ inch longer than the other side. A title or question can be written on the extended tab.

- -

Worksheet Foldable or Folded Book® By Dinah Zike

Step 1 Make a half-book (see above) using work sheets, internet print-outs, diagrams, or maps.

Step 2 Fold it in half again.

Variations

A This folded sheet as a small book with two pages can be used for comparing and contrasting, cause and effect, or other skills.

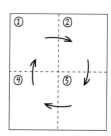

B When the sheet of paper is open, the four sections can be used separately or used collectively to show sequences or steps.

SCIENCE SKILL HANDBOOK

MATH SKILL HANDBOOK

FOLDABLES HANDBOOK

REFERENCE HANDBOOK

GLOSSARY/ GLOSARIO

INDEX

Two-Tab and Concept-Map Foldable® By Dinah Zike

Step 1 Fold a sheet of notebook or copy paper in half vertically or horizontally.

Step 2 Fold it in half again, as shown.

Step 3 Unfold once and cut along the fold line or valley of the top flap to make two flaps.

Variations

A Concept maps can be made by leaving a ½ inch tab at the top when folding the paper in half. Use arrows and labels to relate topics to the primary concept.

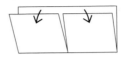

B Use two sheets of paper to make multiple page tab books. Glue or staple books together at the top fold.

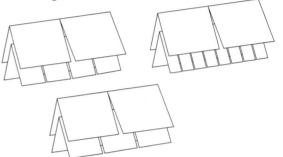

Three-Quarter Foldable® By Dinah Zike

Step 1 Make a two-tab book (see above) and cut the left tab off at the top of the fold line.

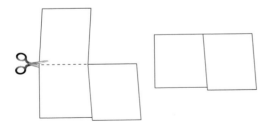

Variations

A Use this book to draw a diagram or a map on the exposed left tab. Write questions about the illustration on the top right tab and provide complete answers on the space under the tab.

B Compose a self-test using multiple choice answers for your questions. Include the correct answer with three wrong responses. The correct answers can be written on the back of the book or upside down on the bottom of the inside page.

SCIENCE SKILL HANDBOOK

MATH SKILL HANDBOOK

FOLDABLES HANDBOOK

REFERENCE HANDBOOK

GLOSSARY/ GLOSARIO

INDEX

Three-Tab Foldable® By Dinah Zike

Step 1 Fold a sheet of paper in half horizontally.

Step 2 Fold into thirds.

Step 3 Unfold and cut along the folds of the top flap to make three sections.

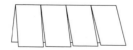

Variations

A Before cutting the three tabs draw a Venn diagram across the front of the book.

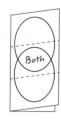

B Make a space to use for titles or concept maps by leaving a ½ inch tab at the top when folding the paper in half.

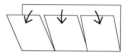

Four-Tab Foldable® By Dinah Zike

Step 1 Fold a sheet of paper in half horizontally.

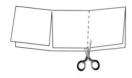

Step 2 Fold in half and then fold each half as shown below.

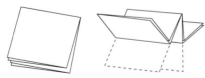

Step 3 Unfold and cut along the fold lines of the top flap to make four tabs.

Variations

A Make a space to use for titles or concept maps by leaving a ½ inch tab at the top when folding the paper in half.

B Use the book on the vertical axis, with or without an extended tab.

Folding Fifths for a Foldable® By Dinah Zike

Step 1 Fold a sheet of paper in half horizontally.

Step 2 Fold again so one-third of the paper is exposed and two-thirds are covered.

Step 3 Fold the two-thirds section in half.

Step 4 Fold the one-third section, a single thickness, backward to make a fold line.

Variations

A Unfold and cut along the fold lines to make five tabs.

B Make a five-tab book with a ½ inch tab at the top (see two-tab instructions).

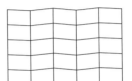

C Use 11″ × 17″ or 12″ × 18″ paper and fold into fifths for a five-column and/or row table or chart.

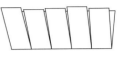

Folded Table or Chart, and Trifold Foldable® By Dinah Zike

Step 1 Fold a sheet of paper in the required number of vertical columns for the table or chart.

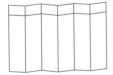

Step 2 Fold the horizontal rows needed for the table or chart.

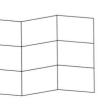

Variations

A Make a trifold by folding the paper into thirds vertically or horizontally.

PROJECT FORMAT
Use 11″ × 17″ or 12″ × 18″ paper and fold it to make a large trifold project book or larger tables and charts.

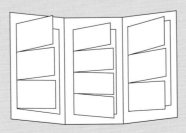

B Make a trifold book. Unfold it and draw a Venn diagram on the inside.

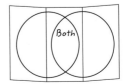

SCIENCE SKILL HANDBOOK

MATH SKILL HANDBOOK

FOLDABLES HANDBOOK

REFERENCE HANDBOOK

GLOSSARY/ GLOSARIO

INDEX

SCIENCE SKILL HANDBOOK

MATH SKILL HANDBOOK

FOLDABLES HANDBOOK

REFERENCE HANDBOOK

GLOSSARY/ GLOSARIO

INDEX

Two or Three-Pockets Foldable® By Dinah Zike

Step 1 Fold up the long side of a horizontal sheet of paper about 5 cm.

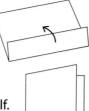

Step 2 Fold the paper in half.

Step 3 Open the paper and glue or staple the outer edges to make two compartments.

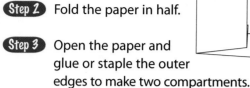

Variations

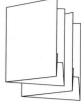

A Make a multi-page booklet by gluing several pocket books together.

B Make a three-pocket book by using a trifold (see previous instructions).

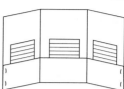

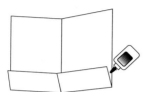

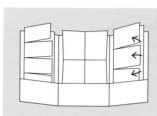

PROJECT FORMAT
Use 11″ × 17″ or 12″ × 18″ paper and fold it horizontally to make a large multi-pocket project.

- -

Matchbook Foldable® By Dinah Zike

Step 1 Fold a sheet of paper almost in half and make the back edge about 1–2 cm longer than the front edge.

Step 4 Close the book and fold the tab over the short side.

Variations

A Make a single-tab matchbook by skipping Steps 2 and 3.

Step 2 Find the midpoint of the shorter flap.

Step 3 Open the paper and cut the short side along the midpoint making two tabs.

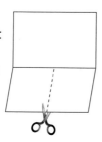

B Make two smaller matchbooks by cutting the single-tab matchbook in half.

Shutterfold Foldable® By Dinah Zike

Step 1 Begin as if you were folding a vertical sheet of paper in half, but instead of creasing the paper, pinch it to show the midpoint.

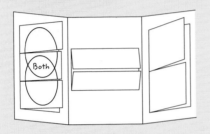

PROJECT FORMAT
Use 11″ × 17″ or 12″ × 18″ paper and fold it to make a large shutterfold project.

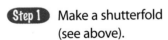

Step 2 Fold the top and bottom to the middle and crease the folds.

Variations

A Use the shutterfold on the horizontal axis.

B Create a center tab by leaving .5–2 cm between the flaps in Step 2.

Four-Door Foldable® By Dinah Zike

Step 1 Make a shutterfold (see above).

Step 2 Fold the sheet of paper in half.

Step 3 Open the last fold and cut along the inside fold lines to make four tabs.

Variations

A Use the four-door book on the opposite axis.

B Create a center tab by leaving .5–2 cm between the flaps in Step 1.

SCIENCE SKILL HANDBOOK

MATH SKILL HANDBOOK

FOLDABLES HANDBOOK

REFERENCE HANDBOOK

GLOSSARY/ GLOSARIO

INDEX

Bound Book Foldable® By Dinah Zike

Step 1 Fold three sheets of paper in half. Place the papers in a stack, leaving about .5 cm between each top fold. Mark all three sheets about 3 cm from the outer edges.

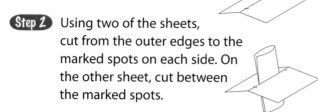

Step 2 Using two of the sheets, cut from the outer edges to the marked spots on each side. On the other sheet, cut between the marked spots.

Step 3 Take the two sheets from Step 1 and slide them through the cut in the third sheet to make a 12-page book.

Step 4 Fold the bound pages in half to form a book.

Variation

A Use two sheets of paper to make an eight-page book, or increase the number of pages by using more than three sheets.

PROJECT FORMAT
Use two or more sheets of 11" × 17" or 12" × 18" paper and fold it to make a large bound book project.

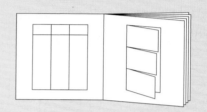

Accordian Foldable® By Dinah Zike

Step 1 Fold the selected paper in half vertically, like a *hamburger*.

Step 2 Cut each sheet of folded paper in half along the fold lines.

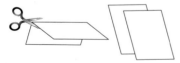

Step 3 Fold each half-sheet almost in half, leaving a 2 cm tab at the top.

Step 4 Fold the top tab over the short side, then fold it in the opposite direction.

Variations

A Glue the straight edge of one paper inside the tab of another sheet. Leave a tab at the end of the book to add more pages.

B Tape the straight edge of one paper to the tab of another sheet, or just tape the straight edges of nonfolded paper end to end to make an accordian.

C Use whole sheets of paper to make a large accordian.

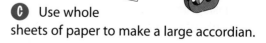

Science Skill Handbook

Math Skill Handbook

Foldables Handbook

Reference Handbook

Glossary/ Glosario

Index

Layered Foldable® By Dinah Zike

Step 1 Stack two sheets of paper about 1–2 cm apart. Keep the right and left edges even.

Step 2 Fold up the bottom edges to form four tabs. Crease the fold to hold the tabs in place.

Step 3 Staple along the folded edge, or open and glue the papers together at the fold line.

Variations

A Rotate the book so the fold is at the top or to the side.

B Extend the book by using more than two sheets of paper.

- -

Envelope Foldable® By Dinah Zike

Step 1 Fold a sheet of paper into a *taco*. Cut off the tab at the top.

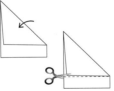

Step 2 Open the *taco* and fold it the opposite way making another *taco* and an X-fold pattern on the sheet of paper.

Step 3 Cut a map, illustration, or diagram to fit the inside of the envelope.

Step 4 Use the outside tabs for labels and inside tabs for writing information.

Variations

A Use 11″ × 17″ or 12″ × 18″ paper to make a large envelope.

B Cut off the points of the four tabs to make a window in the middle of the book.

SCIENCE SKILL HANDBOOK

MATH SKILL HANDBOOK

FOLDABLES HANDBOOK

REFERENCE HANDBOOK

GLOSSARY/ GLOSARIO

INDEX

Sentence Strip Foldable® By Dinah Zike

Step 1 Fold two sheets of paper in half vertically, like a *hamburger*.

Step 2 Unfold and cut along fold lines making four half sheets.

Step 3 Fold each half sheet in half horizontally, like a *hot dog*.

Step 4 Stack folded horizontal sheets evenly and staple together on the left side.

Step 5 Open the top flap of the first sentence strip and make a cut about 2 cm from the stapled edge to the fold line. This forms a flap that can be raised and lowered. Repeat this step for each sentence strip.

Variations

A Expand this book by using more than two sheets of paper.

B Use whole sheets of paper to make large books.

Pyramid Foldable® By Dinah Zike

Step 1 Fold a sheet of paper into a *taco*. Crease the fold line, but do not cut it off.

Step 2 Open the folded sheet and refold it like a *taco* in the opposite direction to create an X-fold pattern.

Step 3 Cut one fold line as shown, stopping at the center of the X-fold to make a flap.

Step 4 Outline the fold lines of the X-fold. Label the three front sections and use the inside spaces for notes. Use the tab for the title.

Step 5 Glue the tab into a project book or notebook. Use the space under the pyramid for other information.

Step 6 To display the pyramid, fold the flap under and secure with a paper clip, if needed.

Single-Pocket or One-Pocket Foldable® By Dinah Zike

Step 1 Using a large piece of paper on a vertical axis, fold the bottom edge of the paper upwards, about 5 cm.

Step 2 Glue or staple the outer edges to make a large pocket.

PROJECT FORMAT
Use 11″ × 17″ or 12″ × 18″ paper and fold it vertically or horizontally to make a large pocket project.

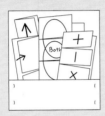

Variations

A Make the one-pocket project using the paper on the horizontal axis.

B To store materials securely inside, fold the top of the paper almost to the center, leaving about 2–4 cm between the paper edges. Slip the Foldables through the opening and under the top and bottom pockets.

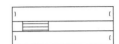

- -

Multi-Tab Foldable® By Dinah Zike

Step 1 Fold a sheet of notebook paper in half like a *hot dog*.

Step 2 Open the paper and on one side cut every third line. This makes ten tabs on wide ruled notebook paper and twelve tabs on college ruled.

Step 3 Label the tabs on the front side and use the inside space for definitions or other information.

Variation

A Make a tab for a title by folding the paper so the holes remain uncovered. This allows the notebook Foldable to be stored in a three-hole binder.

SCIENCE SKILL HANDBOOK

MATH SKILL HANDBOOK

FOLDABLES HANDBOOK

REFERENCE HANDBOOK

GLOSSARY/ GLOSARIO

INDEX

PERIODIC TABLE OF THE ELEMENTS

Element — Hydrogen
Atomic number — 1
Symbol — H
Atomic mass — 1.01

State of matter

Gas
Liquid
Solid
Synthetic

A column in the periodic table is called a **group.**

A row in the periodic table is called a **period.**

1

1 Hydrogen 1 H 1.01

2 Lithium 3 Li 6.94 | Beryllium 4 Be 9.01

2

3 Sodium 11 Na 22.99 | Magnesium 12 Mg 24.31

3 **4** **5** **6** **7** **8** **9**

4 Potassium 19 K 39.10 | Calcium 20 Ca 40.08 | Scandium 21 Sc 44.96 | Titanium 22 Ti 47.87 | Vanadium 23 V 50.94 | Chromium 24 Cr 52.00 | Manganese 25 Mn 54.94 | Iron 26 Fe 55.85 | Cobalt 27 Co 58.93

5 Rubidium 37 Rb 85.47 | Strontium 38 Sr 87.62 | Yttrium 39 Y 88.91 | Zirconium 40 Zr 91.22 | Niobium 41 Nb 92.91 | Molybdenum 42 Mo 95.96 | Technetium 43 Tc (98) | Ruthenium 44 Ru 101.07 | Rhodium 45 Rh 102.91

6 Cesium 55 Cs 132.91 | Barium 56 Ba 137.33 | Lanthanum 57 La 138.91 | Hafnium 72 Hf 178.49 | Tantalum 73 Ta 180.95 | Tungsten 74 W 183.84 | Rhenium 75 Re 186.21 | Osmium 76 Os 190.23 | Iridium 77 Ir 192.22

7 Francium 87 Fr (223) | Radium 88 Ra (226) | Actinium 89 Ac (227) | Rutherfordium 104 Rf (267) | Dubnium 105 Db (268) | Seaborgium 106 Sg (271) | Bohrium 107 Bh (272) | Hassium 108 Hs (270) | Meitnerium 109 Mt (276)

The number in parentheses is the mass number of the longest lived isotope for that element.

Lanthanide series — Cerium 58 Ce 140.12 | Praseodymium 59 Pr 140.91 | Neodymium 60 Nd 144.24 | Promethium 61 Pm (145) | Samarium 62 Sm 150.36 | Europium 63 Eu 151.96

Actinide series — Thorium 90 Th 232.04 | Protactinium 91 Pa 231.04 | Uranium 92 U 238.03 | Neptunium 93 Np (237) | Plutonium 94 Pu (244) | Americium 95 Am (243)

Metal

Metalloid

Nonmetal

Recently discovered

18

					13	**14**	**15**	**16**	**17**	Helium 2 **He** 4.00
					Boron 5 **B** 10.81	Carbon 6 **C** 12.01	Nitrogen 7 **N** 14.01	Oxygen 8 **O** 16.00	Fluorine 9 **F** 19.00	Neon 10 **Ne** 20.18
10	**11**	**12**			Aluminum 13 **Al** 26.98	Silicon 14 **Si** 28.09	Phosphorus 15 **P** 30.97	Sulfur 16 **S** 32.07	Chlorine 17 **Cl** 35.45	Argon 18 **Ar** 39.95
Nickel 28 **Ni** 58.69	Copper 29 **Cu** 63.55	Zinc 30 **Zn** 65.38	Gallium 31 **Ga** 69.72	Germanium 32 **Ge** 72.64	Arsenic 33 **As** 74.92	Selenium 34 **Se** 78.96	Bromine 35 **Br** 79.90	Krypton 36 **Kr** 83.80		
Palladium 46 **Pd** 106.42	Silver 47 **Ag** 107.87	Cadmium 48 **Cd** 112.41	Indium 49 **In** 114.82	Tin 50 **Sn** 118.71	Antimony 51 **Sb** 121.76	Tellurium 52 **Te** 127.60	Iodine 53 **I** 126.90	Xenon 54 **Xe** 131.29		
Platinum 78 **Pt** 195.08	Gold 79 **Au** 196.97	Mercury 80 **Hg** 200.59	Thallium 81 **Tl** 204.38	Lead 82 **Pb** 207.20	Bismuth 83 **Bi** 208.98	Polonium 84 **Po** (209)	Astatine 85 **At** (210)	Radon 86 **Rn** (222)		
Darmstadtium 110 **Ds** (281)	Roentgenium 111 **Rg** (280)	Copernicium 112 **Cn** (285)	Ununtrium * 113 **Uut** (284)	Ununquadium * 114 **Uuq** (289)	Ununpentium * 115 **Uup** (288)	Ununhexium * 116 **Uuh** (293)		Ununoctium * 118 **Uuo** (294)		

* The names and symbols for elements 113-116 and 118 are temporary. Final names will be selected when the elements' discoveries are verified.

Gadolinium 64 **Gd** 157.25	Terbium 65 **Tb** 158.93	Dysprosium 66 **Dy** 162.50	Holmium 67 **Ho** 164.93	Erbium 68 **Er** 167.26	Thulium 69 **Tm** 168.93	Ytterbium 70 **Yb** 173.05	Lutetium 71 **Lu** 174.97
Curium 96 **Cm** (247)	Berkelium 97 **Bk** (247)	Californium 98 **Cf** (251)	Einsteinium 99 **Es** (252)	Fermium 100 **Fm** (257)	Mendelevium 101 **Md** (258)	Nobelium 102 **No** (259)	Lawrencium 103 **Lr** (262)

SCIENCE SKILL HANDBOOK

MATH SKILL HANDBOOK

FOLDABLES HANDBOOK

REFERENCE HANDBOOK

GLOSSARY/ GLOSARIO

INDEX

Topographic Map Symbols

Topographic Map Symbols

▬▬▬▬	Primary highway, hard surface	∿	Index contour
▬▬▬▬	Secondary highway, hard surface	⋯⋯	Supplementary contour
═════	Light-duty road, hard or improved surface	⌒	Intermediate contour
=========	Unimproved road	⬭	Depression contours
++++++	Railroad: single track		
╫╫╫╫	Railroad: multiple track	▬ ▬ ▬	Boundaries: national
┿┿┿┿	Railroads in juxtaposition	▬ ▬ ▬	State
		▬ ▬ ▬	County, parish, municipal
▪▐█▓	Buildings	▬ ▬ ▬	Civil township, precinct, town, barrio
♪♪⊞ cem	Schools, church, and cemetery	▬·▬··▬	Incorporated city, village, town, hamlet
▫▭▨	Buildings (barn, warehouse, etc.)	·▬·▬·▬	Reservation, national or state
○ ○	Wells other than water (labeled as to type)	----------	Small park, cemetery, airport, etc.
●●●⊘	Tanks: oil, water, etc. (labeled only if water)	▬·▬···	Land grant
⊙ ⚥	Located or landmark object; windmill	▬▬▬▬	Township or range line, U.S. land survey
⤬ ⤬	Open pit, mine, or quarry; prospect	▬ ▬ ▬	Township or range line, approximate location

▦	Marsh (swamp)		
▦	Wooded marsh	⧆	Perennial streams
▩	Woods or brushwood	→ ←	Elevated aqueduct
⋮⋮⋮	Vineyard	○ ⟳	Water well and spring
▨	Land subject to controlled inundation	⤳	Small rapids
▦	Submerged marsh	⤳	Large rapids
▨	Mangrove	≈≈≈	Intermittent lake
⋮⋮⋮	Orchard	⤳	Intermittent stream
▦	Scrub	→⬚⬚⬚←	Aqueduct tunnel
▨	Urban area	⬲	Glacier
		∿	Small falls
x7369	Spot elevation	▤	Large falls
670	Water elevation	⬚	Dry lake bed

SCIENCE SKILL HANDBOOK

MATH SKILL HANDBOOK

FOLDABLES HANDBOOK

REFERENCE HANDBOOK

GLOSSARY/ GLOSARIO

INDEX

Rocks

Rocks		
Rock Type	**Rock Name**	**Characteristics**
Igneous (intrusive)	Granite	Large mineral grains of quartz, feldspar, hornblende, and mica. Usually light in color.
	Diorite	Large mineral grains of feldspar, hornblende, and mica. Less quartz than granite. Intermediate in color.
	Gabbro	Large mineral grains of feldspar, augite, and olivine. No quartz. Dark in color.
Igneous (extrusive)	Rhyolite	Small mineral grains of quartz, feldspar, hornblende, and mica, or no visible grains. Light in color.
	Andesite	Small mineral grains of feldspar, hornblende, and mica or no visible grains. Intermediate in color.
	Basalt	Small mineral grains of feldspar, augite, and possibly olivine or no visible grains. No quartz. Dark in color.
	Obsidian	Glassy texture. No visible grains. Volcanic glass. Fracture looks like broken glass.
	Pumice	Frothy texture. Floats in water. Usually light in color.
Sedimentary (detrital)	Conglomerate	Coarse grained. Gravel or pebble-size grains.
	Sandstone	Sand-sized grains 1/16 to 2 mm.
	Siltstone	Grains are smaller than sand but larger than clay.
	Shale	Smallest grains. Often dark in color. Usually platy.
Sedimentary (chemical or organic)	Limestone	Major mineral is calcite. Usually forms in oceans and lakes. Often contains fossils.
	Coal	Forms in swampy areas. Compacted layers of organic material, mainly plant remains.
Sedimentary (chemical)	Rock Salt	Commonly forms by the evaporation of seawater.
Metamorphic (foliated)	Gneiss	Banding due to alternate layers of different minerals, of different colors. Parent rock often is granite.
	Schist	Parallel arrangement of sheetlike minerals, mainly micas. Forms from different parent rocks.
	Phyllite	Shiny or silky appearance. May look wrinkled. Common parent rocks are shale and slate.
	Slate	Harder, denser, and shinier than shale. Common parent rock is shale.
Metamorphic (nonfoliated)	Marble	Calcite or dolomite. Common parent rock is limestone.
	Soapstone	Mainly of talc. Soft with greasy feel.
	Quartzite	Hard with interlocking quartz crystals. Common parent rock is sandstone.

SCIENCE SKILL HANDBOOK

MATH SKILL HANDBOOK

FOLDABLES HANDBOOK

REFERENCE HANDBOOK

GLOSSARY/ GLOSARIO

INDEX

Minerals

SCIENCE SKILL HANDBOOK

MATH SKILL HANDBOOK

FOLDABLES HANDBOOK

REFERENCE HANDBOOK

GLOSSARY/ GLOSARIO

INDEX

Minerals

Mineral (formula)	Color	Streak	Hardness Pattern	Breakage Properties	Uses and Other
Graphite (C)	black to gray	black to gray	1–1.5	basal cleavage (scales)	pencil lead, lubricants for locks, rods to control some small nuclear reactions, battery poles
Galena (PbS)	gray	gray to black	2.5	cubic cleavage perfect	source of lead, used for pipes, shields for X rays, fishing equipment sinkers
Hematite (Fe_2O_3)	black or reddish-brown	reddish-brown	5.5–6.5	irregular fracture	source of iron; converted to pig iron, made into steel
Magnetite (Fe_3O_4)	black	black	6	conchoidal fracture	source of iron, attracts a magnet
Pyrite (FeS_2)	light, brassy, yellow	greenish-black	6–6.5	uneven fracture	fool's gold
Talc ($Mg_3 Si_4O_{10}$ $(OH)_2$)	white, greenish	white	1	cleavage in one direction	used for talcum powder, sculptures, paper, and tabletops
Gypsum ($CaSO_4 \cdot 2H_2O$)	colorless, gray, white, brown	white	2	basal cleavage	used in plaster of paris and dry wall for building construction
Sphalerite (ZnS)	brown, reddish-brown, greenish	light to dark brown	3.5–4	cleavage in six directions	main ore of zinc; used in paints, dyes, and medicine
Muscovite (KAl_3Si_3 $O_{10}(OH)_2$)	white, light gray, yellow, rose, green	colorless	2–2.5	basal cleavage	occurs in large, flexible plates; used as an insulator in electrical equipment, lubricant
Biotite ($K(Mg,Fe)_3$ $(AlSi_3O_{10})$ $(OH)_2$)	black to dark brown	colorless	2.5–3	basal cleavage	occurs in large, flexible plates
Halite (NaCl)	colorless, red, white, blue	colorless	2.5	cubic cleavage	salt; soluble in water; a preservative

Minerals

Minerals

Mineral (formula)	Color	Streak	Hardness	Breakage Pattern	Uses and Other Properties
Calcite ($CaCO_3$)	colorless, white, pale blue	colorless, white	3	cleavage in three directions	fizzes when HCl is added; used in cements and other building materials
Dolomite ($CaMg(CO_3)_2$)	colorless, white, pink, green, gray, black	white	3.5–4	cleavage in three directions	concrete and cement; used as an ornamental building stone
Fluorite (CaF_2)	colorless, white, blue, green, red, yellow, purple	colorless	4	cleavage in four directions	used in the manufacture of optical equipment; glows under ultraviolet light
Hornblende $(CaNa)_{2-3}$ $(Mg,Al,Fe)_5-(Al,Si)_2$ Si_6O_{22} $(OH)_2)$	green to black	gray to white	5–6	cleavage in two directions	will transmit light on thin edges; 6-sided cross section
Feldspar ($KAlSi_3O_8$) ($NaAlSi_3O_8$), ($CaAl_2Si_2O_8$)	colorless, white to gray, green	colorless	6	two cleavage planes meet at 90° angle	used in the manufacture of ceramics
Augite $((Ca,Na)$ (Mg,Fe,Al) $(Al,Si)_2 O_6)$	black	colorless	6	cleavage in two directions	square or 8-sided cross section
Olivine $((Mg,Fe)_2 SiO_4)$	olive, green	none	6.5–7	conchoidal fracture	gemstones, refractory sand
Quartz (SiO_2)	colorless, various colors	none	7	conchoidal fracture	used in glass manufacture, electronic equipment, radios, computers, watches, gemstones

SCIENCE SKILL HANDBOOK

MATH SKILL HANDBOOK

FOLDABLES HANDBOOK

REFERENCE HANDBOOK

GLOSSARY/ GLOSARIO

INDEX

Weather Map Symbols

Sample Station Model

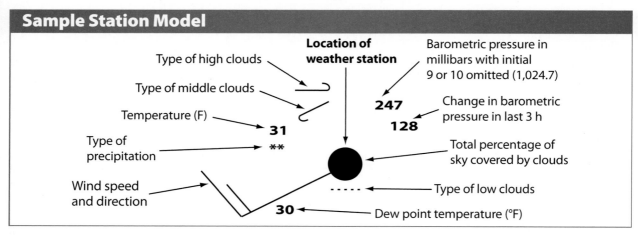

Type of high clouds	Location of weather station	Barometric pressure in millibars with initial 9 or 10 omitted (1,024.7)
Type of middle clouds	**247**	Change in barometric pressure in last 3 h
Temperature (F) **31**	**128**	Total percentage of sky covered by clouds
Type of precipitation **		Type of low clouds
Wind speed and direction	**30**	Dew point temperature (°F)

Sample Plotted Report at Each Station

Precipitation		Wind Speed and Direction		Sky Coverage		Some Types of High Clouds	
≡	Fog	○	0 calm	○	No cover	⌐⟍	Scattered cirrus
★	Snow	╱	1–2 knots	◐	1/10 or less	⟍⟍	Dense cirrus in patches
●	Rain	⤙	3–7 knots	◔	2/10 to 3/10	⟋⟍⟍	Veil of cirrus covering entire sky
⌐⊼	Thunderstorm	⤛	8–12 knots	◑	4/10	⟍⌐	Cirrus not covering entire sky
'	Drizzle	⤜	13–17 knots	◑	–		
▽	Showers	⤝	18–22 knots	◕	6/10		
		⤞	23–27 knots	◕	7/10		
		⤟	48–52 knots	◓	Overcast with openings		
		1 knot = 1.852 km/h		●	Completely overcast		

Some Types of Middle Clouds		Some Types of Low Clouds		Fronts and Pressure Systems	
╱	Thin altostratus layer	⌒	Cumulus of fair weather	Ⓗ or High Ⓛ or Low	Center of high- or low-pressure system
╱╱	Thick altostratus layer	⌣	Stratocumulus	▲▲▲▲	Cold front
⟋⌐	Thin altostratus in patches	-----	Fractocumulus of bad weather	●●●●	Warm front
⟋⌐	Thin altostratus in bands	—	Stratus of fair weather	▲●▲●	Occluded front
				●▲●▽	Stationary front

SCIENCE SKILL HANDBOOK

MATH SKILL HANDBOOK

FOLDABLES HANDBOOK

REFERENCE HANDBOOK

GLOSSARY/ GLOSARIO

INDEX

Use and Care of a Microscope

Eyepiece Contains magnifying lenses you look through.

Arm Supports the body tube.

Low-power objective Contains the lens with the lowest power magnification.

Stage clips Hold the microscope slide in place.

Coarse adjustment Focuses the image under low power.

Fine adjustment Sharpens the image under high magnification.

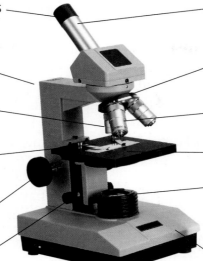

Body tube Connects the eyepiece to the revolving nosepiece.

Revolving nosepiece Holds and turns the objectives into viewing position.

High-power objective Contains the lens with the highest magnification.

Stage Supports the microscope slide.

Light source Provides light that passes upward through the diaphragm, the specimen, and the lenses.

Base Provides support for the microscope.

Caring for a Microscope

1. Always carry the microscope holding the arm with one hand and supporting the base with the other hand.
2. Don't touch the lenses with your fingers.
3. The coarse adjustment knob is used only when looking through the lowest-power objective lens. The fine adjustment knob is used when the high-power objective is in place.
4. Cover the microscope when you store it.

Using a Microscope

1. Place the microscope on a flat surface that is clear of objects. The arm should be toward you.
2. Look through the eyepiece. Adjust the diaphragm so light comes through the opening in the stage.
3. Place a slide on the stage so the specimen is in the field of view. Hold it firmly in place by using the stage clips.

4. Always focus with the coarse adjustment and the low-power objective lens first. After the object is in focus on low power, turn the nosepiece until the high-power objective is in place. Use ONLY the fine adjustment to focus with the high-power objective lens.

Making a Wet-Mount Slide

1. Carefully place the item you want to look at in the center of a clean, glass slide. Make sure the sample is thin enough for light to pass through.
2. Use a dropper to place one or two drops of water on the sample.
3. Hold a clean coverslip by the edges and place it at one edge of the water. Slowly lower the coverslip onto the water until it lies flat.
4. If you have too much water or a lot of air bubbles, touch the edge of a paper towel to the edge of the coverslip to draw off extra water and draw out unwanted air.

SCIENCE SKILL HANDBOOK

MATH SKILL HANDBOOK

FOLDABLES HANDBOOK

REFERENCE HANDBOOK

GLOSSARY/ GLOSARIO

INDEX

Diversity of Life: Classification of Living Organisms

A six-kingdom system of classification of organisms is used today. Two kingdoms—Kingdom Archaebacteria and Kingdom Eubacteria—contain organisms that do not have a nucleus and that lack membrane-bound structures in the cytoplasm of their cells. The members of the other four kingdoms have a cell or cells that contain a nucleus and structures in the cytoplasm, some of which are surrounded by membranes. These kingdoms are Kingdom Protista, Kingdom Fungi, Kingdom Plantae, and Kingdom Animalia.

Kingdom Archaebacteria

one-celled; some absorb food from their surroundings; some are photosynthetic; some are chemosynthetic; many are found in extremely harsh environments including salt ponds, hot springs, swamps, and deep-sea hydrothermal vents

Kingdom Eubacteria

one-celled; most absorb food from their surroundings; some are photosynthetic; some are chemosynthetic; many are parasites; many are round, spiral, or rod-shaped; some form colonies

Kingdom Protista

Phylum Euglenophyta one-celled; photosynthetic or take in food; most have one flagellum; euglenoids

Kingdom Eubacteria
Bacillus anthracis

Phylum Chlorophyta
Desmids

Phylum Bacillariophyta one-celled; photosynthetic; have unique double shells made of silica; diatoms

Phylum Dinoflagellata one-celled; photosynthetic; contain red pigments; have two flagella; dinoflagellates

Phylum Chlorophyta one-celled, many-celled, or colonies; photosynthetic; contain chlorophyll; live on land, in freshwater, or salt water; green algae

Phylum Rhodophyta most are many-celled; photosynthetic; contain red pigments; most live in deep, saltwater environments; red algae

Phylum Phaeophyta most are many-celled; photosynthetic; contain brown pigments; most live in saltwater environments; brown algae

Phylum Rhizopoda one-celled; take in food; are free-living or parasitic; move by means of pseudopods; amoebas

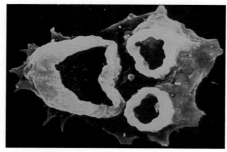

Amoeba

SCIENCE SKILL HANDBOOK

MATH SKILL HANDBOOK

FOLDABLES HANDBOOK

REFERENCE HANDBOOK

GLOSSARY/ GLOSARIO

INDEX

Phylum Zoomastigina one-celled; take in food; free-living or parasitic; have one or more flagella; zoomastigotes

Phylum Ciliophora one-celled; take in food; have large numbers of cilia; ciliates

Phylum Sporozoa one-celled; take in food; have no means of movement; are parasites in animals; sporozoans

Phylum Myxomycota
Slime mold

Phyla Myxomycota and Acrasiomycota one- or many-celled; absorb food; change form during life cycle; cellular and plasmodial slime molds

Phylum Oomycota many-celled; are either parasites or decomposers; live in freshwater or salt water; water molds, rusts and downy mildews

Kingdom Fungi

Phylum Zygomycota many-celled; absorb food; spores are produced in sporangia; zygote fungi; bread mold

Phylum Ascomycota one- and many-celled; absorb food; spores produced in asci; sac fungi; yeast

Phylum Basidiomycota many-celled; absorb food; spores produced in basidia; club fungi; mushrooms

Phylum Deuteromycota members with unknown reproductive structures; imperfect fungi; *Penicillium*

Phylum Mycophycota organisms formed by symbiotic relationship between an ascomycote or a basidiomycote and green alga or cyanobacterium; lichens

Phylum Oomycota
Phytophthora infestans

Lichens

SCIENCE SKILL HANDBOOK

MATH SKILL HANDBOOK

FOLDABLES HANDBOOK

REFERENCE HANDBOOK

GLOSSARY/ GLOSARIO

INDEX

Science Skill Handbook

Math Skill Handbook

Foldables Handbook

Reference Handbook

Glossary/ Glosario

Index

Kingdom Plantae

Divisions Bryophyta (mosses), **Anthocerophyta** (hornworts), **Hepaticophyta** (liverworts), **Psilophyta** (whisk ferns) many-celled non-vascular plants; reproduce by spores produced in capsules; green; grow in moist, land environments

Division Lycophyta many-celled vascular plants; spores are produced in conelike structures; live on land; are photosynthetic; club mosses

Division Arthrophyta vascular plants; ribbed and jointed stems; scalelike leaves; spores produced in conelike structures; horsetails

Division Pterophyta vascular plants; leaves called fronds; spores produced in clusters of sporangia called sori; live on land or in water; ferns

Division Ginkgophyta deciduous trees; only one living species; have fan-shaped leaves with branching veins and fleshy cones with seeds; ginkgoes

Division Cycadophyta palmlike plants; have large, featherlike leaves; produces seeds in cones; cycads

Division Coniferophyta deciduous or evergreen; trees or shrubs; have needlelike or scalelike leaves; seeds produced in cones; conifers

Division Anthophyta
Tomato plant

**Phylum
Platyhelminthes**
Flatworm

Division Gnetophyta shrubs or woody vines; seeds are produced in cones; division contains only three genera; gnetum

Division Anthophyta dominant group of plants; flowering plants; have fruits with seeds

Kingdom Animalia

Phylum Porifera aquatic organisms that lack true tissues and organs; are asymmetrical and sessile; sponges

Phylum Cnidaria radially symmetrical organisms; have a digestive cavity with one opening; most have tentacles armed with stinging cells; live in aquatic environments singly or in colonies; includes jellyfish, corals, hydra, and sea anemones

Phylum Platyhelminthes bilaterally symmetrical worms; have flattened bodies; digestive system has one opening; parasitic and free-living species; flatworms

Division Bryophyta
Liverwort

Phylum Chordata

Phylum Nematoda round, bilaterally symmetrical body; have digestive system with two openings; free-living forms and parasitic forms; roundworms

Phylum Mollusca soft-bodied animals, many with a hard shell and soft foot or footlike appendage; a mantle covers the soft body; aquatic and terrestrial species; includes clams, snails, squid, and octopuses

Phylum Annelida bilaterally symmetrical worms; have round, segmented bodies; terrestrial and aquatic species; includes earthworms, leeches, and marine polychaetes

Phylum Arthropoda largest animal group; have hard exoskeletons, segmented bodies, and pairs of jointed appendages; land and aquatic species; includes insects, crustaceans, and spiders

Phylum Echinodermata marine organisms; have spiny or leathery skin and a water-vascular system with tube feet; are radially symmetrical; includes sea stars, sand dollars, and sea urchins

Phylum Chordata organisms with internal skeletons and specialized body systems; most have paired appendages; all at some time have a notochord, nerve cord, gill slits, and a post-anal tail; include fish, amphibians, reptiles, birds, and mammals

SCIENCE SKILL HANDBOOK

MATH SKILL HANDBOOK

FOLDABLES HANDBOOK

REFERENCE HANDBOOK

GLOSSARY/ GLOSARIO

INDEX

Glossary/Glosario

Cómo usar el glosario en español:
1. Busca el término en inglés que desees encontrar.
2. El término en español, junto con la definición, se encuentran en la columna de la derecha.

Pronunciation Key

Use the following key to help you sound out words in the glossary:

a . . . . back (BAK)	ew. . . . . food (FEWD)
ay . . . . day (DAY)	yoo . . . . pure (PYOOR)
ah . . . . father (FAH thur)	yew. . . . . few (FYEW)
ow . . . . flower (FLOW ur)	uh. . . . . comma (CAH muh)
ar . . . . car (CAR)	u (+ con) . . . . rub (RUB)
e . . . . less (LES)	sh . . . . shelf (SHELF)
ee . . . . leaf (LEEF)	ch . . . . nature (NAY chur)
ih . . . . trip (TRIHP)	g . . . . gift (GIHFT)
i (i + con + e) . idea (i DEE uh)	j . . . . gem (JEM)
oh . . . . go (GOH)	ing . . . . sing (SING)
aw . . . . soft (SAWFT)	zh . . . . vision (VIH zhun)
or . . . . orbit (OR buht)	k . . . . cake (KAYK)
oy . . . . coin (COYN)	s . . . . seed, cent (SEED, SENT)
oo . . . . foot (FOOT)	z . . . . zone, raise (ZOHN, RAYZ)

English — A — Español

abiotic factor/adaptation

abiotic factor (ay bi AH tihk • FAK tuhr): a nonliving thing in an ecosystem. (p. 309)

acceleration: a measure of the change in velocity during a period of time. (p. 694)

accuracy: a description of how close a measurement is to an accepted or true value. (p. NOS 12)

acid precipitation: precipitation that has a lower pH than that of normal rainwater (5.6). (pp. 549, 599)

active transport: the movement of substances through a cell membrane using the cell's energy. (p. 64)

adaptation (a dap TAY shun): an inherited trait that increases an organism's chance of surviving and reproducing in a particular environment. (p. 207)

factor abiótico/adaptación

factor abiótico: componente no vivo de un ecosistema. (pág. 309)

aceleración: medida del cambio de velocidad durante un período de tiempo. (pág. 694)

exactitud: descripción de qué tan cerca está una medida a un valor aceptable. (pág. NOS 12)

precipitación ácida: precipitación que tiene un pH más bajo que el del agua de la lluvia normal (pH 5.6). (pág. 549, 599)

transporte activo: movimiento de sustancias a través de la membrana celular usando la energía de la célula. (pág. 64)

adaptación: rasgo heredado que aumenta la oportunidad de un organismo de sobrevivir y reproducirse en su medioambiente. (pág. 207)

air mass: a large area of air that has uniform temperature, humidity, and pressure. (p. 624)

air pollution: the contamination of air by harmful substances including gases and smoke. (p. 598)

air pressure: the pressure that a column of air exerts on the air or a surface below it. (p. 616)

alkali (AL kuh li) metal: an element in group 1 on the periodic table. (p. 403)

alkaline (AL kuh lun) earth metal: an element in group 2 on the periodic table. (p. 403)

allele (uh LEEL): a different form of a gene. (p. 164)

alternation of generations: process that occurs when the life cycle of an organism alternates between diploid and haploid generations. (p. 290)

analogous (uh NAH luh gus) structures: body parts that perform a similar function but differ in structure. (p. 215)

artificial satellite: any human-made object placed in orbit around a body in space. (p. 781)

asexual reproduction: a type of reproduction in which one parent organism produces offspring without meiosis and fertilization. (p. 129)

asthenosphere (as THE nuh sfir): the partially melted portion of the mantle below the lithosphere. (p. 487)

astrobiology: the study of the origin, development, distribution, and future of life on Earth and in the universe. (p. 773)

astronomical unit (AU): the average distance from Earth to the Sun—about 150 million km. (p. 764)

atmosphere (AT muh sfihr): a thin layer of gases surrounding Earth. (p. 573)

atom: a small particle that is the building block of matter. (p. 349)

masa de aire: gran área de aire que tiene temperatura, humedad y presión uniformes. (pág. 624)

polución del aire: contaminación del aire por sustancias dañinas, como gases y humo. (pág. 598)

presión del aire: presión que una columna de aire ejerce sobre el aire o sobre la superficie debajo de ella. (pág. 616)

metal alcalino: elemento del grupo 1 de la tabla periódica. (pág. 403)

metal alcalinotérreo: elemento del grupo 2 de la tabla periódica. (pág. 403)

alelo: forma diferente de un gen. (pág. 164)

alternancia de generaciones: proceso que ocurre cuando el ciclo de vida de un organismo se alterna entre generaciones diploides y haploides. (pág. 290)

estructuras análogas: partes del cuerpo que ejecutan una función similar pero tienen una estructura distinta. (pág. 215)

satélite artificial: cualquier objeto hecho por el ser humano y puesto en órbita alrededor de un cuerpo en el espacio. (pág. 781)

reproducción asexual: tipo de reproducción en la cual un organismo parental produce crías sin mitosis ni fertilización. (pág. 129)

astenosfera: parte parcialmente fundida del manto debajo de la litosfera. (pág. 487)

astrobiología: estudio del origen, desarrollo, distribución y futuro de la vida en la Tierra y el universo. (pág. 773)

unidad astronómica: distancia promedio desde la Tierra al Sol, aproximadamente 150 millones de km. (pág. 764)

atmósfera: capa delgada de gases que rodean la Tierra. (pág. 573)

átomo: partícula pequeña que es el componente básico de la materia. (pág. 349)

B

balanced forces: forces acting on an object that combine and form a net force of zero. (p. 705)

fuerzas en equilibrio: fuerzas que actúan sobre un objeto, se combinan y forman una fuerza neta de cero. (pág. 705)

SCIENCE SKILL HANDBOOK

MATH SKILL HANDBOOK

FOLDABLES HANDBOOK

REFERENCE HANDBOOK

GLOSSARY/GLOSARIO

INDEX

SCIENCE SKILL HANDBOOK

MATH SKILL HANDBOOK

FOLDABLES HANDBOOK

REFERENCE HANDBOOK

GLOSSARY/ GLOSARIO

INDEX

binomial nomenclature: a naming system that gives each organism a two-word scientific name. (p. 21)

biological evolution: the change over time in populations of related organisms. (p. 199)

biome: a geographic area on Earth that contains ecosystems with similar biotic and abiotic features. (p. 312)

biotic factor (bi AH tihk • FAK tuhr): a living or once-living thing in an ecosystem. (p. 311)

biotic potential: the potential growth of a population if it could grow in perfect conditions with no limiting factors. (p. 319)

blizzard: a violent winter storm characterized by freezing temperatures, strong winds, and blowing snow. (p. 631)

budding: the process during which a new organism grows by mitosis and cell division on the body of its parent. (p. 131)

nomenclatura binomial: sistema de nombrar que le da a cada organismo un nombre científico de dos palabras. (pág. 21)

evolución biológica: cambio a través del tiempo en las poblaciones de organismos relacionados. (pág. 199)

bioma: área geográfica en la Tierra que contiene ecosistemas con características bióticas y abióticas similares. (pág. 312)

factor biótico: ser vivo o que una vez estuvo vivo en un ecosistema. (pág. 311)

potencial biótico: crecimiento potencial de una población si puede crecer en condiciones perfectas sin factores limitantes. (pág. 319)

ventisca: tormenta violenta de invierno caracterizada por temperaturas heladas, vientos fuertes, y nieve que sopla. (pág. 631)

germinación: proceso durante el cual un organismo nuevo crece por medio de mitosis y división celular en el cuerpo de su progenitor. (pág. 131)

C

caldera (kal DER uh): large depression formed when the top of a volcano collapses. (p. 515)

Calorie: the amount of energy it takes to raise the temperature of 1 kg of water by 1°C. (p. 233)

camouflage (KAM uh flahj): an adaptation that enables a species to blend in with its environment. (p. 208)

carbohydrate (kar boh HI drayt): a macromolecule made up of one or more sugar molecules, which are composed of carbon, hydrogen, and oxygen; usually the body's major source of energy. (p. 47)

carbon cycle: the series of processes that continuously move carbon among Earth systems. (p. 472)

carrying capacity: the largest number of individuals of one species that an ecosystem can support over time. (p. 319)

cast: a fossil copy of an organism made when a mold of the organism is filled with sediment or mineral deposits. (p. 195)

caldera: depresión grande formada cuando la parte de arriba de un volcán colapsa. (pág. 515)

caloría: cantidad de energía necesaria para aumentar la temperatura de 1 kg de agua en 1°C. (pág. 233)

camuflaje: adaptación que permite a las especies mezclarse con su medioambiente. (pág. 208)

carbohidrato: macromolécula constituida de una o más moléculas de azúcar, las cuales están compuestas de carbono, hidrógeno y oxígeno; usualmente es la mayor fuente de energía del cuerpo. (pág. 47)

ciclo del carbono: serie de procesos que mueven continuamente el carbono entre los sistemas terrestres. (pág. 472)

capacidad de carga: número mayor de individuos de una especie que un medioambiente puede mantener con el tiempo. (pág. 319)

contramolde: copia fósil de un organismo compuesto en un molde de el organismo está lleno de sedimentos o los depósitos de minerales. (pág. 195)

cell: the smallest unit of life. (p. 10)

cell cycle: a cycle of growth, development, and division that most cells in an organism go through. (p. 85)

cell differentiation (dihf uh ren shee AY shun): the process by which cells become different types of cells. (p. 99)

cell membrane: a flexible covering that protects the inside of a cell from the environment outside the cell. (p. 52)

cell theory: the theory that states that all living things are made of one or more cells, the cell is the smallest unit of life, and all new cells come from preexisting cells. (p. 44)

cell wall: a stiff structure outside the cell membrane that protects a cell from attack by viruses and other harmful organisms. (p. 52)

cellular respiration: a series of chemical reactions that convert the energy in food molecules into a usable form of energy called ATP. (pp. 69, 274)

centromere: a structure that holds sister chromatids together. (p. 88)

chemical change: a change in matter in which the substances that make up the matter change into other substances with different chemical and physical properties. (p. 375)

chemical energy: energy that is stored in and released from the bonds between atoms. (p. 428)

chemical property: the ability or inability of a substance to combine with or change into one or more new substances. (p. 374)

chemical weathering: the process that changes the composition of rocks and minerals due to exposure to the environment. (p. 520)

chloroplast (KLOR uh plast): a membrane-bound organelle that uses light energy and makes food—a sugar called glucose—from water and carbon dioxide in a process known as photosynthesis. (p. 57)

célula: unidad más pequeña de vida. (pág. 10)

ciclo celular: ciclo de crecimiento, desarrollo y división por el que pasan la mayoría de células de un organismo. (pág. 85)

diferenciación celular: proceso por el cual las células se convierten en diferentes tipos de células. (pág. 99)

membrana celular: cubierta flexible que protege el interior de una célula del ambiente externo de la célula. (pág. 52)

teoría celular: teoría que establece que todos los seres vivos están constituidos de una o más células (la célula es la unidad más pequeña de vida) y que las células nuevas provienen de células preexistentes. (pág. 44)

pared celular: estructura rígida en el exterior de la membrana celular que protege la célula del ataque de virus y otros organismos dañinos. (pág. 52)

respiración celular: serie de reacciones químicas que convierten la energía de las moléculas de alimento en una forma de energía utilizable llamada ATP. (pág. 69, 274)

centrómero: estructura que mantiene unidas las cromátidas hermanas. (pág. 88)

cambio químico: cambio de la materia en el cual las sustancias que componen la materia se transforman en otras sustancias con propiedades químicas y físicas diferentes. (pág. 375)

energía química: energía almacenada en y liberada por los enlaces entre los átomos. (pág. 428)

propiedad química: capacidad o incapacidad de una sustancia para combinarse con o transformarse en una o más sustancias. (pág. 374)

meteorización química: proceso que cambia la composición de las rocas y los minerales debido a la exposición al medioambiente. (pág. 520)

cloroplasto: organelo limitado por una membrana que usa la energía lumínica para producir alimento –un azúcar llamado glucosa– del agua y del dióxido de carbono en un proceso llamado fotosíntesis. (pág. 57)

SCIENCE SKILL HANDBOOK

MATH SKILL HANDBOOK

FOLDABLES HANDBOOK

REFERENCE HANDBOOK

GLOSSARY/ GLOSARIO

INDEX

SCIENCE SKILL HANDBOOK

MATH SKILL HANDBOOK

FOLDABLES HANDBOOK

REFERENCE HANDBOOK

GLOSSARY/ GLOSARIO

INDEX

cladogram: a branched diagram that shows the relationships among organisms, including common ancestors. (p. 23)

cleavage: the breaking of a mineral along a smooth, flat surface. (p. 482)

climate: the long-term average weather conditions that occur in a particular region. (p. 651)

cloning: a type of asexual reproduction performed in a laboratory that produces identical individuals from a cell or a cluster of cells taken from a multicellular organism. (p. 134)

closed system: a system that does not exchange matter or energy with the environment. (p. 439)

codominance: an inheritance pattern in which both alleles can be observed in a phenotype. (p. 168)

community: all the populations living in an ecosystem at the same time. (p. 311)

compact bone: the hard outer layer of bone. (p. 246)

comparative anatomy: the study of similarities and differences among structures of living species. (p. 214)

compound: a substance containing atoms of two or more different elements chemically bonded together. (p. 352)

compound microscope: a light microscope that uses more than one lens to magnify an object. (p. 28)

compression: the squeezing force at a convergent boundary. (p. 505)

computer model: detailed computer programs that solve a set of complex mathematical formulas. (p. 638)

concentration: the amount of a particular solute in a given amount of solution. (p. 378)

conduction (kuhn DUK shun): the transfer of thermal energy due to collisions between particles. (pp. 449, 585)

constants: the factors in an experiment that remain the same. (p. NOS 18)

consumer: an organism that cannot make its own food and gets energy by eating other organisms. (p. 327)

cladograma: diagrama de brazos que muestra las relaciones entre los organismos, incluidos los ancestros comunes. (pág. 23)

exfoliación: rompimiento de un mineral en láminas o superficies planas. (pág. 482)

clima: promedio a largo plazo de las condiciones del tiempo atmosférico de una región en particular. (pág. 651)

clonación: tipo de reproducción asexual realizada en un laboratorio que produce individuos idénticos a partir de una célula o grupo de células tomadas de un organismo pluricelular. (pág. 134)

sistema cerrado: sistema que no intercambia materia o energía con el ambiente. (pág. 439)

condominante: patrón heredado en el cual los dos alelos se observan en un fenotipo. (pág. 168)

comunidad: todas las poblaciones que viven en un ecosistema al mismo tiempo. (pág. 311)

hueso compacto: capa externa y dura del hueso. (pág. 246)

anatomía comparativa: estudio de las similitudes y diferencias entre las estructuras de las especies vivas. (pág. 214)

compuesto: sustancia que contiene átomos de dos o más elementos diferentes unidos químicamente. (pág. 352)

microscopio compuesto: microscopio de luz que usa más de un lente para aumentar la imagen de un objeto. (pág. 28)

compresión: tensión en un límite convergente. (pág. 505)

modelo de computadora: programas de computadora que resuelven un conjunto de fórmulas matemáticas complejas. (pág. 638)

concentración: cantidad de cierto soluto en una cantidad dada de solución. (pág. 378)

conducción: transferencia de energía térmica debido a colisiones entre partículas. (pág. 449, 585)

constantes: factores en un experimento que permanecen iguales. (pág. NOS 18)

consumidor: organismo que no elabora su propio alimento y obtiene energía comiendo otros organismos. (pág. 327)

contact force: a push or a pull on one object by another object that is touching it. (p. 702)

continental drift: Wegener's hypothesis that suggests the continents are in constant motion on Earth's surface. (p. 502)

convection: the transfer of thermal energy by the movement of particles from one part of a material to another. (p. 449); the circulation of particles within a material caused by differences in thermal energy and density. (p. 585)

convergent boundary: the boundary between two plates that move toward each other. (p. 504)

core: the dense, metallic center of Earth. (p. 488)

critical thinking: comparing what you already know with information you are given in order to decide whether you agree with it. (p. NOS 8)

crust: the brittle, rocky outer layer of Earth. (p. 486)

cytokinesis (si toh kuh NEE sus): a process during which the cytoplasm and its contents divide. (p. 89)

cytoplasm: the liquid part of a cell inside the cell membrane; contains salts and other molecules. (p. 53)

cytoskeleton: a network of threadlike proteins joined together that gives a cell its shape and helps it move. (p. 53)

fuerza de contacto: empuje o arrastre ejercido sobre un objeto por otro que lo está tocando. (pág. 702)

deriva continental: hipótesis de Wegener que sugiere que los continentes están en constante movimiento en la superficie de la Tierra. (pág. 502)

convección: transferencia de energía térmica por el movimiento de partículas de una parte de la materia a otra. (pág. 449); circulación de partículas en el interior de un material causada por diferencias en la energía térmica y la densidad. (pág. 585)

límite convergente: límite entre dos placas que se acercan una hacia la otra. (pág. 504)

núcleo: centro de la Tierra denso y metálico. (pág. 488)

pensamiento crítico: comparación que se hace cuando se sabe algo acerca de información nueva, y se decide si se está o no de acuerdo con ella. (pág. NOS 8)

corteza: capa delgada de roca de la parte más externa de la Tierra. (pág. 486)

citocinesis: proceso durante el cual el citoplasma y sus contenidos se dividen. (pág. 89)

citoplasma: fluido en el interior de una célula que contiene sales y otras moléculas. (pág. 53)

citoesqueleto: red de proteínas en forma de filamentos unidos que le da forma a la célula y le ayuda a moverse. (pág. 53)

D

daughter cells: the two new cells that result from mitosis and cytokinesis. (p. 89)

deforestation: the removal of large areas of forests for human purposes. (p. 671)

density: the mass per unit volume of a substance. (p. 361)

dependent variable: the factor a scientist observes or measures during an experiment. (p. NOS 18)

deposition: the laying down or settling of eroded material. (p. 523)

células hija: las dos células nuevas que resultan de la mitosis y la citocinesis. (pág. 89)

deforestación: eliminación de grandes áreas de bosques con propósitos humanos. (pág. 671)

densidad: cantidad de masa por unidad de volumen de una sustancia. (pág. 361)

variable dependiente: factor que el científico observa o mide durante un experimento. (pág. NOS 18)

deposición: establecimiento o asentamiento de material erosionado. (pág. 523)

SCIENCE SKILL HANDBOOK

MATH SKILL HANDBOOK

FOLDABLES HANDBOOK

REFERENCE HANDBOOK

GLOSSARY/ GLOSARIO

INDEX

description: a spoken or written summary of an observation. (p. NOS 10)

dew point: temperature at which air is saturated and condensation can occur. (p. 617)

dichotomous key: a series of descriptions arranged in pairs that leads the user to the identification of an unknown organism. (p. 22)

diffusion: the movement of substances from an area of higher concentration to an area of lower concentration. (p. 62)

diploid: a cell that has pairs of chromosomes. (p. 118)

displacement: the difference between the initial, or starting, position and the final position of an object that has moved. (p. 692)

dissolve: to form a solution by mixing evenly. (p. 353)

distance: the total length of your path. (p. 692)

divergent boundary: the boundary between two plates that move away from each other. (p. 504)

DNA: the abbreviation for deoxyribonucleic (dee AHK sih ri boh noo klee ihk) acid, an organism's genetic material. (p. 174)

dominant (DAH muh nunt) trait: a genetic factor that blocks another genetic factor. (p. 159)

Doppler radar: a specialized type of radar that can detect precipitation as well as the movement of small particles, which can be used to approximate wind speed. (p. 636)

drought: a period of below-average precipitation. (p. 665)

ductility (duk TIH luh tee): the ability to be pulled into thin wires. (p. 402)

dwarf planet: an object that orbits the Sun and is nearly spherical in shape, but shares its orbital path with other objects of similar size. (p. 768)

descripción: resumen oral o escrito de una observación de. (pág. NOS 10)

punto de rocío: temperatura en la cual el aire está saturado y occure la condensación. (pág. 617)

clave dicotómica: serie de descripciones organizadas en pares que dan al usuario la identificación de un organismo desconocido. (pág. 22)

difusión: movimiento de sustancias de un área de mayor concentración a un área de menor concentración. (pág. 62)

diploide: célula que tiene pares de cromosomas. (pág. 118)

desplazamiento: diferencia entre la posición inicial, o salida, y la final de un objeto que se ha movido. (pág. 692)

disolver: preparar una solución mezclando de manera homogénea. (pág. 353)

distancia: longitud total de un trayecto. (pág. 692)

límite divergente: límite entre dos placas que se alejan la una de la otra. (pág. 504)

ADN: abreviatura para ácido desoxirribo-nucleico, material genético de un organismo. (pág. 174)

rasgo dominante: factor genético que bloquea otro factor genético. (pág. 159)

radar Dopler: tipo de radar especializado que detecta tanto la precipitación como el movimiento de partículas pequeñas, que se pueden usar para determinar la velocidad aproximada del viento. (pág. 636)

sequía: período de bajo promedio de precipitación. (pág. 665)

ductilidad: capacidad para formar alambres delgados (pág. 402)

planeta enano: objeto de forma casi esférica que orbita el Sol y que comparte el recorrido de la órbita con otros objetos de tamaño similar. (pág. 768)

E

earthquake: vibrations caused by the rupture and sudden movement of rocks along a break or a crack in Earth's crust. (p. 509)

ecosystem: all the living things and nonliving things in a given area. (p. 309)

egg: the female reproductive, or sex, cell; forms in an ovary. (p. 117)

electric energy: energy carried by an electric current. (p. 429)

electron microscope: a microscope that uses a magnetic field to focus a beam of electrons through an object or onto an object's surface. (p. 29)

element: a substance that consists of only one type of atom. (p. 351)

El Niño/Southern Oscillation: the combined ocean and atmospheric cycle that results in weakened trade winds across the Pacific Ocean. (p. 664)

embryo: an immature diploid plant that develops from the zygote. (p. 292)

embryology (em bree AH luh jee): the science of the development of embryos from fertilization to birth. (p. 216)

endocytosis (en duh si TOH sus): the process during which a cell takes in a substance by surrounding it with the cell membrane. (p. 64)

energy: the ability to cause change. (p. 427)

energy pyramid: a model that shows the amount of energy available in each link of a food chain. (p. 329)

energy transfer: the process of moving energy from one object to another without changing form. (p. 437)

energy transformation: the conversion of one form of energy to another. (p. 437)

equinox: when Earth's rotation axis is tilted neither toward nor away from the Sun. (p. 733)

erosion: the moving of weathered material, or sediment, from one location to another. (p. 519)

terremoto: vibraciones causadas por la ruptura y el movimiento repentino de las rocas a lo largo de una fractura o grieta en la corteza de la Tierra. (pág. 509)

ecosistema: todos los seres vivos y los componentes no vivos de un área dada. (pág. 309)

óvulo: célula reproductiva femenina o sexual; forma en un ovario. (pág. 117)

energía eléctrica: energía transportada por una corriente eléctrica. (pág. 429)

microscopio electrónico: microscopio que usa un campo magnético para enfocar un haz de electrones a través de un objeto o sobre la superficie de un objeto. (pág. 29)

elemento: sustancia que consiste de un sólo tipo de átomo. (pág. 351)

El Niño/Oscilación meridional: ciclo atmosférico y oceánico combinado que produce el debilitamiento de los vientos alisios en el Océano Pacífico. (pág. 664)

embrión: planta diploide inmadura que se desarrolla de un zigoto. (pág. 292)

embriología: ciencia que trata el desarrollo de embriones desde la fertilización hasta el nacimiento. (pág. 216)

endocitosis: proceso durante el cual una célula absorbe una sustancia rodeándola con la membrana celular. (pág. 64)

energía: capacidad de ocasionar cambio. (pág. 427)

pirámide energética: modelo que explica la cantidad de energía disponible en cada vínculo de una cadena alimentaria. (pág. 329)

transferencia de energía: proceso por el cual se mueve energía de un objeto a otro sin cambiar de forma. (pág. 437)

transformación de energía: conversión de una forma de energía a otra. (pág. 437)

equinoccio: cuando el eje de rotación de la Tierra se inclina sin acercarse ni alejarse del Sol. (pág. 733)

erosión: transporte de material meteorizado, o de sedimento, de un lugar a otro. (pág. 519)

SCIENCE SKILL HANDBOOK

MATH SKILL HANDBOOK

FOLDABLES HANDBOOK

REFERENCE HANDBOOK

GLOSSARY/ GLOSARIO

INDEX

exocytosis (ek soh si TOH sus): the process during which a cell's vesicles release their contents outside the cell. (p. 64)

explanation: an interpretation of observations. (p. NOS 10)

extinction (ihk STINGK shun): event that occurs when the last individual organism of a species dies. (p. 198)

F

facilitated diffusion: the process by which molecules pass through a cell membrane using special proteins called transport proteins. (p. 63)

fault: a crack or a fracture in Earth's lithosphere along which movement occurs. (p. 509)

fermentation: a reaction that eukaryotic and prokaryotic cells can use to obtain energy from food when oxygen levels are low. (p. 70)

fertilization (fur tuh luh ZAY shun): a reproductive process in which a sperm joins with an egg. (pp. 117, 255)

fission: cell division that forms two genetically identical cells. (p. 130)

food chain: a model that shows how energy flows in an ecosystem through feeding relationships. (p. 328)

food web: a model of energy transfer that can show how the food chains in a community are interconnected. (p. 328)

force: a push or a pull on an object. (p. 701)

force pair: the forces two objects apply to each other. (p. 713)

fossil record: record of all the fossils ever discovered on Earth. (p. 193)

fracture: the breaking of a mineral along a rough or irregular surface. (p. 482)

friction: a contact force that resists the sliding motion of two surfaces that are touching. (p. 703)

front: a boundary between two air masses. (p. 626)

exocitosis: proceso durante el cual las vesículas de una célula liberan sus contenidos fuera de la célula. (pág. 64)

explicación: interpretación que se hace de las observaciones. (pág. NOS 10)

extinción: evento que ocurre cuando el último organismo individual de una especie muere. (pág. 198)

difusión facilitada: proceso por el cual las moléculas pasan a través de la membrana celular usando proteínas especiales, llamadas proteínas de transporte. (pág. 63)

falla: grieta o fractura en la litosfera de la Tierra a lo largo de la cual ocurre el movimiento. (pág. 509)

fermentación: reacción que las células eucarióticas y procarióticas usan para obtener energía del alimento cuando los niveles de oxígeno son bajos. (pág. 70)

fertilización: proceso reproductivo en el cual un espermatozoide se une con un óvulo. (pág. 117, 255)

fisión: división celular que forma dos células genéticamente idénticas. (pág. 130)

cadena alimentaria: modelo que explica cómo la energía fluye en un ecosistema a través de relaciones alimentarias. (pág. 328)

red alimentaria: modelo de transferencia de energía que explica cómo las cadenas alimentarias están interconectadas en una comunidad. (pág. 328)

fuerza: empuje o arrastre ejercido sobre un objeto. (pág. 701)

par de fuerzas: fuerzas que dos objetos se aplican entre sí. (pág. 713)

registro fósil: registro de todos los fósiles descubiertos en la Tierra. (pág. 193)

fractura: rompimiento de un mineral en una superficie desigual o irregular. (pág. 482)

fricción: fuerza de contacto que resiste el movimiento de dos superficies que están en contacto. (pág. 703)

frente: límite entre dos masas de aire. (pág. 626)

fruit: plant structure that contains one or more seeds; develops from the ovary and sometimes other parts of the flower. (p. 295)

fruta: estructura de la planta que contiene una o más semillas; se desarrolla del ovario y algunas veces de otras partes de la flor. (pág. 295)

G

gamete (GA meet): human reproductive cell. (p. 255)

gene (JEEN): a section of DNA on a chromosome that has genetic information for one trait. (p. 164)

genetics: the study of how traits are passed from parents to offspring. (p. 153)

genotype (JEE nuh tipe): the alleles of all the genes on an organism's chromosomes; controls an organism's phenotype. (p. 164)

genus (JEE nus): a group of similar species. (p. 21)

geologic time scale: a chart that divides Earth's history into different time units based on changes in the rocks and fossils. (p. 197)

geothermal energy: thermal energy from Earth's interior. (p. 543)

geyser (GI zur): a warm spring that sometimes ejects a jet of liquid water or water vapor into the air. (p. 777)

global climate model: a set of complex equations used to predict future climates. (p. 673)

global warming: an increase in the average temperature of Earth's surface. (pp. 549, 670)

glycolysis: a process by which glucose, a sugar, is broken down into smaller molecules. (p. 69)

gravity: an attractive force that exists between all objects that have mass. (p. 703)

greenhouse gas: a gas in the atmosphere that absorbs Earth's outgoing infrared radiation. (pp. 474, 670)

group: a column on the periodic table. (p. 396)

gameto: célula reproductora humana. (pág. 255)

gen: parte del ADN en un cromosoma que contiene información genética para un rasgo. (pág. 164)

genética: estudio de cómo los rasgos pasan de los padres a los hijos. (pág. 153)

genotipo: de los alelos de todos los genes en los cromosomas de un organismo, los controles de fenotipo de un organismo. (pág. 164)

género: grupo de especies similares. (pág. 21)

escala de tiempo geológico: tabla que divide la historia de la Tierra en diferentes unidades de tiempo, basado en los cambios en las rocas y fósiles. (pág. 197)

energía geotérmica: energía térmica del interior de la Tierra. (pág. 543)

géiser: manantial caliente que algunas veces expulsa un chorro de agua líquida o vapor de agua al aire. (pág. 777)

modelo de clima global: conjunto de ecuaciones complejas para predecir climas futuros. (pág. 673)

calentamiento global: incremento en la temperatura promedio de la superficie de la Tierra. (pág. 549, 670)

glucólisis: proceso por el cual la glucosa, un azúcar, se divide en moléculas más pequeñas. (pág. 69)

gravedad: fuerza de atracción que existe entre todos los objetos que tienen masa. (pág. 703)

gas de invernadero: gas en la atmósfera que absorbe la salida de radiación infrarroja de la Tierra. (pág. 474, 670)

grupo: columna en la tabla periódica. (pág. 396)

H

habitat: the place within an ecosystem where an organism lives; provides the biotic and abiotic factors an organism needs to survive and reproduce. (p. 320)

hábitat: lugar en un ecosistema donde vive un organismo; proporciona los factores bióticos y abióticos que un organismo necesita para vivir y reproducirse. (pág. 320)

SCIENCE SKILL HANDBOOK

MATH SKILL HANDBOOK

FOLDABLES HANDBOOK

REFERENCE HANDBOOK

GLOSSARY/ GLOSARIO

INDEX

halogen (HA luh jun): an element in group 17 on the periodic table. (p. 411)

haploid: a cell that has only one chromosome from each pair. (p. 119)

heat: the movement of thermal energy from a region of higher temperature to a region of lower temperature. (p. 448)

heredity (huh REH duh tee): the passing of traits from parents to offspring. (p. 153)

heterogeneous mixture: a mixture in which substances are not evenly mixed. (p. 353)

heterozygous (he tuh roh ZI gus): a genotype in which the two alleles of a gene are different. (p. 165)

high-pressure system: a large body of circulating air with high pressure at its center and lower pressure outside of the system. (p. 623)

homeostasis (hoh mee oh STAY sus): an organism's ability to maintain steady internal conditions when outside conditions change. (pp. 13, 231)

homogeneous mixture: a mixture in which two or more substances are evenly mixed but not bonded together. (p. 353)

homologous (huh MAH luh gus) chromosomes: pairs of chromosomes that have genes for the same traits arranged in the same order. (p. 118)

homologous (huh MAH luh gus) structures: body parts of organisms that are similar in structure and position but different in function. (p. 214)

homozygous (hoh muh ZI gus): a genotype in which the two alleles of a gene are the same. (p. 165)

hormone: a chemical signal that is produced by an endocrine gland in one part of an organism and carried in the bloodstream to another part of the organism. (p. 251)

hot spot: a location where volcanoes form far from plate boundaries. (p. 512)

humidity (hyew MIH duh tee): the amount of water vapor in the air. (p. 616)

hurricane: an intense tropical storm with winds exceeding 119 km/h. (p. 630)

halógeno: elemento del grupo 17 de la tabla periódica. (pág. 411)

haploide: célula que tiene solamente un cromosoma de cada par. (pág. 119)

calor: movimiento de energía térmica desde una región de alta temperatura a una región de baja temperatura. (pág. 448)

herencia: paso de rasgos de los padres a los hijos. (pág. 153)

mezcla heterogénea: mezcla en la cual las sustancias no están mezcladas de manera uniforme. (pág. 353)

heterocigoto: genotipo en el cual los dos alelos de un gen son diferentes. (pág. 165)

sistema de alta presión: gran cuerpo de aire circulante con presión alta en el centro y presión más baja fuera del sistema. (pág. 623)

homeostasis: capacidad de un organismo de mantener las condiciones internas estables cuando las condiciones externas cambian. (pág. 13, 231)

mezcla homogénea: mezcla en la cual dos o más sustancias están mezcladas de manera uniforme, pero no están unidas químicamente. (pág. 353)

cromosomas homólogos: pares de cromosomas que tienen genes de iguales rasgos dispuestos en el mismo orden. (pág. 118)

estructuras homólogas: partes del cuerpo de los organismos que son similares en estructura y posición pero diferentes en función. (pág. 214)

homocigoto: genotipo en el cual los dos alelos de un gen son iguales. (pág. 165)

hormona: señal química producido por una glándula endocrina en una parte de un organismo y llevado en la corriente sanguínea a otra parte del organismo. (pág. 251)

punto caliente: lugar lejos de los límites de las placas donde se forman volcanes. (pág. 512)

humedad: cantidad de vapor de agua en el aire. (pág. 616)

huracán: tormenta tropical intensa con vientos que exceden los 119 km/h. (pág. 630)

hypothesis: a possible explanation for an observation that can be tested by scientific investigations. (p. NOS 4)

hipótesis: explicación posible de una observación que se puede probar por medio de investigaciones científicas. (pág. NOS 4)

I

ice age: a period of time when a large portion of Earth's surface is covered by glaciers. (p. 660)

era del hielo: período de tiempo cuando los glaciares cubren una gran porción de la superficie de la Tierra. (pág. 660)

immunity: the resistance to specific pathogens, or disease-causing agents. (p. 240)

inmunidad: resistencia a patógenos específicos o a agentes causantes de enfermedades. (pág. 240)

incomplete dominance: an inheritance pattern in which an offspring's phenotype is a combination of the parents' phenotypes. (p. 168)

dominancia incompleta: patrón heredado en el cual el fenotipo de un hijo es una combinación de los fenotipos de los padres. (pág. 168)

independent variable: the factor that is changed by the investigator to observe how it affects a dependent variable. (p. NOS 18)

variable independiente: factor que el investigador cambia para observar cómo afecta la variable dependiente. (pág. NOS 18)

inertia (ihn UR shuh): the tendency of an object to resist a change in motion. (p. 709)

inercia: tendencia de un objeto a resistir un cambio en el movimiento. (pág. 709)

inexhaustible resource: a natural resource that will not run out, no matter how much of it people use. (p. 542)

recurso inagotable: recurso natural que no se acabará, sin importar cuánto lo usen las personas. (pág. 542)

inference: a logical explanation of an observation that is drawn from prior knowledge or experience. (p. NOS 4)

inferencia: explicación lógica de una observación que se extrae de un conocimiento previo o experiencia. (pág. NOS 4)

interglacial: a warm period that occurs during an ice age or between ice ages. (p. 660)

interglacial: período tibio que ocurre durante una era del hielo o entre las eras del hielo. (pág. 660)

International System of Units (SI): the internationally accepted system of measurement. (p. NOS 10)

Sistema Internacional de Unidades (SI): sistema de medidas aceptado internacionalmente. (pág. NOS 10)

interphase: the period during the cell cycle of a cell's growth and development. (p. 86)

interfase: período durante el ciclo celular del crecimiento y desarrollo de una célula. (pág. 86)

ionosphere: a region within the mesosphere and thermosphere containing ions. (p. 577)

ionosfera: región entre la mesosfera y la termosfera que contiene iones. (pág. 577)

isobar: lines that connect all places on a map where pressure has the same value. (p. 637)

isobara: línea que conectan todos los lugares en un mapa donde la presión tiene el mismo valor. (pág. 637)

J

jet stream: a narrow band of high winds located near the top of the troposphere. (p. 593)

corriente de chorro: banda angosta de vientos fuertes cerca de la parte superior de la troposfera. (pág. 593)

SCIENCE SKILL HANDBOOK

MATH SKILL HANDBOOK

FOLDABLES HANDBOOK

REFERENCE HANDBOOK

GLOSSARY/ GLOSARIO

INDEX

SCIENCE SKILL HANDBOOK
MATH SKILL HANDBOOK
FOLDABLES HANDBOOK
REFERENCE HANDBOOK
GLOSSARY/GLOSARIO
INDEX

K

kinetic energy: energy due to motion. (p. 429)

energía cinética: energía debida al movimiento. (pág. 429)

L

land breeze: a wind that blows from the land to the sea due to local temperature and pressure differences. (p. 594)

lava flow: long stream of molten rock. (p. 514)

law of conservation of energy: law that states that energy can be transformed from one form to another, but it cannot be created or destroyed. (p. 436)

light microscope: a microscope that uses light and lenses to enlarge an image of an object. (p. 28)

limiting factor: a factor that can limit the growth of a population. (p. 319)

lipid: a large macromolecule that does not dissolve in water. (p. 47)

lithosphere (LIH thuh sfihr): the rigid outermost layer of Earth that includes the uppermost mantle and crust. (p. 487)

low-pressure system: a large body of circulating air with low pressure at its center and higher pressure outside of the system. (p. 623)

lunar eclipse: an occurrence during which the Moon moves into Earth's shadow. (p. 748)

luster: the way a mineral reflects or absorbs light at its surface. (pp. 401, 482)

lymphocyte (LIHM fuh site): a type of white blood cell that is made in the thymus, the spleen, and bone marrow. (p. 239)

brisa terrestre: viento que sopla desde la tierra hacia el mar debido a diferencias en la temperatura local y la presión. (pág. 594)

flujo de lava: corriente larga de roca derretida. (pág. 514)

ley de la conservación de la energía: ley que plantea que la energía puede transformarse de una forma a otra, pero no puede crearse ni destruirse. (pág. 436)

microscopio de luz: microscopio que usa luz y lentes para aumentar la imagen de un objeto. (pág. 28)

factor limitante: factor que limita el crecimiento de una población. (pág. 319)

lípido: macromolécula extensa que no se disuelve en agua. (pág. 47)

litosfera: capa rígida más externa de la Tierra formada por el manto superior y la corteza. (pág. 487)

sistema baja presión: gran cuerpo de aire circulante con presión baja en el centro y presión más alta fuera del sistema. (pág. 623)

eclipse lunar: ocurrencia durante la cual la Luna se mueve hacia la zona de sombra de la Tierra. (pág. 748)

brillo: forma en que un mineral refleja o absorbe la luz en su superficie. (pág. 401, 482)

linfocito: tipo de glóbulos blancos que se producen en el timo, el bazo y la médula del hueso. (pág. 239)

M

macromolecule: substance that forms from joining many small molecules together. (p. 45)

malleability (ma lee uh BIH luh tee): the ability of a substance to be hammered or rolled into sheets. (p. 402)

mantle: the thick middle layer in the solid part of Earth. (p. 487)

macromolécula: sustancia que se forma al unir muchas moléculas pequeñas. (pág. 45)

maleabilidad: capacidad de una sustancia de martillarse o laminarse para formar hojas. (pág. 402)

manto: capa delgada central de la parte sólida de la Tierra. (pág. 487)

maria (MAR ee uh): the large, dark, flat areas on the Moon. (p. 738)

mass: the amount of matter in an object. (p. 360)

matter: anything that has mass and takes up space. (p. 349)

mechanical energy: sum of the potential energy and the kinetic energy in a system. (p. 430)

meiosis: a process in which one diploid cell divides to make four haploid sex cells. (p. 119)

metal: an element that is generally shiny, is easily pulled into wires or hammered into thin sheets, and is a good conductor of electricity and thermal energy. (p. 401)

metalloid (MEH tul oyd): an element that has physical and chemical properties of both metals and nonmetals. (p. 413)

meteor: a meteoroid that has entered Earth's atmosphere and produces a streak of light. (p. 769)

meteorite: a meteoroid that strikes a planet or a moon. (p. 769)

meteoroid: a small, rocky particle that moves through space. (p. 769)

microclimate: a localized climate that is different from the climate of the larger area surrounding it. (p. 655)

mid-ocean ridge: a long, narrow mountain range on the ocean floor; formed by magma at divergent plate boundaries. (p. 512)

mimicry (MIH mih kree): an adaptation in which one species looks like another species. (p. 208)

mitosis (mi TOH sus): a process during which the nucleus and its contents divide. (p. 89)

mixture: matter that can vary in composition. (p. 353)

mold: the impression of an organism in a rock. (p. 195)

monsoon: a wind circulation pattern that changes direction with the seasons. (p. 665)

motion: the process of changing position. (p. 691)

multicellular: a living thing that is made up of two or more cells. (p. 10)

mares: áreas extensas, oscuras y planas en la Luna. (pág. 738)

masa: cantidad de materia en un objeto. (pág. 360)

materia: cualquier cosa que tiene masa y ocupa espacio. (pág. 349)

energía mecánica: suma de la energía potencial y de la energía cinética en un sistema. (pág. 430)

meiosis: proceso en el cual una célula diploide se divide para constituir cuatro células sexuales haploides. (pág. 119)

metal: elemento que generalmente es brillante, fácilmente puede estirarse para formar alambres o martillarse para formar hojas delgadas y es buen conductor de electricidad y energía térmica. (pág. 401)

metaloide: elemento que tiene las propiedades físicas y químicas de metales y no metales. (pág. 413)

meteoro: meteoroide que ha entrado a la atmósfera de la Tierra y produce un haz de luz. (pág. 769)

meteorito: meteoroide que impacta un planeta o una luna. (pág. 769)

meteoroide: partícula rocosa pequeña que se mueve por el espacio. (pág. 769)

microclima: clima localizado que es diferente del clima de área más extensa que lo rodea. (pág. 655)

dorsal oceánica: cordillera larga y angosta en el lecho del océano, formada por magma en los límites de las placas divergentes. (pág. 512)

mimetismo: una adaptación en el cual una especie se parece a otra especie. (pág. 208)

mitosis: proceso durante el cual el núcleo y sus contenidos se divide. (pág. 89)

mezcla: materia cuya composición puede variar. (pág. 353)

molde: impresión de un organismo en una roca. (pág. 195)

monsón: patrón de viento circulante que cambia de dirección con las estaciones. (pág. 665)

movimiento: proceso de cambiar de posición. (pág. 691)

pluricelular: ser vivo formado por dos o más células. (pág. 10)

SCIENCE SKILL HANDBOOK

MATH SKILL HANDBOOK

FOLDABLES HANDBOOK

REFERENCE HANDBOOK

GLOSSARY/GLOSARIO

INDEX

SCIENCE SKILL HANDBOOK

MATH SKILL HANDBOOK

FOLDABLES HANDBOOK

REFERENCE HANDBOOK

GLOSSARY/ GLOSARIO

INDEX

mutation (myew TAY shun): a permanent change in the sequence of DNA, or the nucleotides, in a gene or a chromosome. (p. 179)

mutación: cambio permanente en la secuencia de ADN, de los nucleótidos, en un gen o en un cromosoma. (pág. 179)

naturalist: a person who studies plants and animals by observing them. (p. 203)

naturalista: persona que estudia las plantas y los animales por medio de la observación. (pág. 203)

natural resource: part of the environment that supplies material useful or necessary for the survival of living things. (p. 537)

recurso natural: parte del medioambiente que suministra material útil o necesario para que los seres vivos sobrevivan. (pág. 537)

natural selection: the process by which organisms with variations that help them survive in their environment live longer, compete better, and reproduce more than those that do not have the variations. (p. 206)

selección natural: proceso por el cual los organismos con variaciones que las ayudan a sobrevivir en sus medioambientes viven más, compiten mejor y se reproducen más que aquellas que no tienen esas variaciones. (pág. 206)

neuron (NOO rahn): the basic functioning unit of the nervous system; a nerve cell. (p. 248)

neurona: unidad básica de funcionamiento del sistema nervioso; célula nerviosa. (pág. 248)

Newton's first law of motion: law that states that if the net force acting on an object is zero, the motion of the object does not change. (p. 709)

primera ley del movimiento de Newton: ley que establece que si la fuerza neta ejercida sobre un objeto es cero, el movimiento de dicho objeto no cambia. (pág. 709)

Newton's second law of motion: law that states that the acceleration of an object is equal to the net force exerted on the object divided by the object's mass. (p. 712)

segunda ley del movimiento de Newton: ley que establece que la aceleración de un objeto es igual a la fuerza neta que actúa sobre él divida por su masa. (pág. 712)

Newton's third law of motion: law that states that for every action there is an equal and opposite reaction. (p. 713)

tercera ley del movimiento de Newton: ley que establece que para cada acción hay una reacción igual en dirección opuesta. (pág. 713)

niche (NICH): the way a species interacts with abiotic and biotic factors to obtain food, find shelter, and fulfill other needs. (p. 320)

nicho: forma como una especie interactúa con los factores abióticos y bióticos para obtener alimento, encontrar refugio y satisfacer otras necesidades. (pág. 320)

noble gas: an element in group 18 on the periodic table. (p. 412)

gas noble: elemento del grupo 18 de la tabla periódica. (pág. 412)

noncontact force: a force that one object applies to another object without touching it. (p. 702)

fuerza de no contacto: fuerza que un objeto puede aplicar sobre otro sin tocarlo. (pág. 702)

nonmetal: an element that has no metallic properties. (p. 409)

no metal: elemento que tiene propiedades no metálicas. (pág. 409)

nonrenewable energy resource: an energy resource that is available in limited amounts or that is used faster than it can be replaced in nature. (p. 442)

recurso energético no renovable: recurso energético disponible en cantidades limitadas o que se usa más rápido de lo que se repone en la naturaleza. (pág. 442)

nonrenewable resource: a natural resource that is used up faster than it can be replaced by natural processes. (p. 538)

nuclear energy: energy stored in and released from the nucleus of an atom. (p. 428)

nucleic acid: a macromolecule that forms when long chains of molecules called nucleotides join together. (p. 46)

nucleotide (NEW klee uh tide): a molecule made of a nitrogen base, a sugar, and a phosphate group. (p. 175)

nucleus: part of a eukaryotic cell that directs cell activity and contains genetic information stored in DNA. (p. 55)

nutrient: a part of food used by the body to grow and survive. (p. 233)

recurso no renovable: recurso natural que se usa más rápidamente de lo que se puede reemplazar por procesos naturales. (pág. 538)

energía nuclear: energía almacenada en y liberada por el núcleo de un átomo. (pág. 428)

ácido nucléico: macromolécula que se forma cuando cadenas largas de moléculas llamadas nucleótidos se unen. (pág. 46)

nucelótido: molécula constituida de una base de nitrógeno, azúcar y un grupo de fosfato. (pág. 175)

núcleo: parte de la célula eucariótica que gobierna la actividad celular y contiene la información genética almacenada en el ADN. (pág. 55)

nutriente: parte del alimento que el cuerpo usa para crecer y vivir. (pág. 233)

O

observation: the act of using one or more of your senses to gather information and take note of what occurs. (p. NOS 4)

open system: a system that exchanges matter or energy with the environment. (p. 439)

orbit: the path an object follows as it moves around another object. (p. 728)

organ: a group of different tissues working together to perform a particular job. (p. 102)

organelle: membrane-surrounded component of a eukaryotic cell with a specialized function. (p. 54)

organic: a class of chemical compounds in living organisms that are based on carbon. (p. 775)

organism: something that has all the characteristics of life. (p. 9)

organ system: a group of organs that work together and perform a specific task. (pp. 103, 231)

osmosis: the diffusion of water molecules only through a membrane. (p. 62)

ovary: structure located at the base of the style of a flower that contains one or more ovules. (p. 294)

observación: acción de usar uno o más sentidos para reunir información y tomar notar de lo que ocurre. (pág. NOS 4)

sistema abierto: sistema que intercambia materia o energía con el ambiente. (pág. 439)

órbita: trayectoria que un objeto sigue a medida que se mueve alrededor de otro objeto. (pág. 728)

órgano: grupo de diferentes tejidos que trabajan juntos para realizar una función específica. (pág. 102)

organelo: componente de una célula eucariótica rodeado de una membrana con una función especializada. (pág. 54)

orgánico: clase de compuestos químicos en los organismos vivos con base en carbono. (pág. 775)

organismo: algo que tiene todas las características de la vida. (pág. 9)

sistema de órganos: grupo de órganos que trabajan juntos para realizar una tarea específica. (pág. 103, 231)

ósmosis: difusión de las moléculas de agua únicamente a través de una membrana. (pág. 62)

ovario: estructura situado en la base del estilo de una flor que contiene uno o más óvulos. (pág. 294)

SCIENCE SKILL HANDBOOK

MATH SKILL HANDBOOK

FOLDABLES HANDBOOK

REFERENCE HANDBOOK

GLOSSARY/GLOSARIO

INDEX

Science Skill Handbook

Math Skill Handbook

Foldables Handbook

Reference Handbook

Glossary/Glosario

Index

ovule: female reproductive structure of a seed plant where the haploid egg develops. (p. 292)

ovum (OH vum): female reproductive cell, or gamete. (p. 255)

ozone layer: the layer of atmosphere that prevents most harmful ultraviolet (UV) radiation from reaching Earth. (p. 548); the area of the stratosphere with a high concentration of ozone. (p. 576)

óvulo: estructura reproductiva femenina de la semilla de una planta donde el huevo haploide se desarrolla. (pág. 292)

óvulo: célula reproductora femenina, o gameto. (pág. 255)

capa de ozono: capa de la atmósfera que evita que la mayor parte de la radiación ultravioleta dañina llegue a la Tierra. (pág. 548); área de la estratosfera con gran concentración de ozono. (pág. 576)

P

particulate (par TIH kyuh lut) matter: the mix of both solid and liquid particles in the air. (p. 600)

passive transport: the movement of substances through a cell membrane without using the cell's energy. (p. 61)

penumbra: the lighter part of a shadow where light is partially blocked. (p. 745)

period: a row on the periodic table. (p. 396)

periodic table: a chart of the elements arranged into rows and columns according to their physical and chemical properties. (p. 391)

phase: the portion of the Moon or a planet reflecting light as seen from Earth. (p. 740)

phenotype (FEE nuh tipe): how a trait appears or is expressed. (p. 164)

phosphorus (FAHS fuh rus) cycle: the series of processes that move phosphorus among Earth systems. (p. 475)

photochemical smog: air pollution that forms from the interaction between chemicals in the air and sunlight. (pp. 548, 599)

photoperiodism: a plant's response to the number of hours of darkness in its environment. (p. 282)

photosynthesis (foh toh SIHN thuh sus): a series of chemical reactions that convert light energy, water, and carbon dioxide into the food-energy molecule glucose and give off oxygen. (pp. 71, 272)

physical change: a change in the size, shape, form, or state of matter that does not change the matter's identity. (p. 367)

partículas en suspensión: mezcla de partículas tanto sólidas como líquidas en el aire. (pág. 600)

transporte pasivo: movimiento de sustancias a través de una membrana celular sin usar la energía de la célula. (pág. 61)

penumbra: parte más clara de una sombra donde la luz se bloquea parcialmente. (pág. 745)

periodo: hilera en la tabla periódica. (pág. 396)

tabla periódica: cuadro en que los elementos están organizados en hileras y columnas según sus propiedades físicas y químicas. (pág. 391)

fase: parte de la Luna o de un planeta que refleja la luz que se ve desde la Tierra. (pág. 740)

fenotipo: forma como aparece o se expresa un rasgo. (pág. 164)

ciclo del fósforo: serie de procesos que mueven el fósforo entre los sistemas terrestres. (pág. 475)

smog fotoquímico: polución del aire que se forma de la interacción entre los químicos en el aire y la luz solar. (pág. 548, 599)

fotoperiodismo: respuesta de una planta al número de horas de oscuridad en su medioambiente. (pág. 282)

fotosíntesis: serie de reacciones químicas que convierte la energía lumínica, el agua y el dióxido de carbono en glucosa, una molécula de energía alimentaria, y libera oxígeno. (pág. 71, 272)

cambio físico: cambio en el tamaño, la forma o el estado de la materia en el que no cambia la identidad de la materia. (pág. 367)

physical property: a characteristic of matter that you can observe or measure without changing the identity of the matter. (p. 358)

physical weathering: the process of breaking down rocks and minerals without changing their compositions. (p. 519)

pistil: female reproductive organ of a flower. (p. 294)

planet: an object that orbits the Sun, is large enough to be nearly spherical in shape, and has no other large object in its orbital path. (p. 766)

plant hormone: a substance that acts as a chemical messenger within a plant. (p. 283)

plate tectonics: theory that Earth's surface is broken into large, rigid pieces that move with respect to each other. (p. 501)

polar easterlies: cold winds that blow from the east to the west near the North Pole and South Pole. (p. 593)

pollen (PAH lun) grain: spore that forms from tissue in a male reproductive structure of a seed plant. (p. 292)

pollination (pah luh NAY shun): the process that occurs when pollen grains land on a female reproductive structure of a plant that is the same species as the pollen grains. (p. 292)

pollution: the contamination of the environment with substances that are harmful to life. (p. 547)

polygenic inheritance: an inheritance pattern in which multiple genes determine the phenotype of a trait. (p. 169)

population: all the organisms of the same species that live in the same area at the same time. (p. 311)

potential (puh TEN chul) energy: stored energy due to the interactions between objects or particles. (p. 427)

precipitation: water, in liquid or solid form, that falls from the atmosphere. (p. 619)

precision: a description of how similar or close measurements are to each other. (p. NOS 12)

propiedad física: característica de la materia que puede observarse o medirse sin cambiar la identidad de la materia. (pág. 358)

meteorización física: proceso por el cual se rompen las rocas y los minerales, sin cambiar su composición. (pág. 519)

pistilo: órgano reproductor femenino de una flor. (pág. 294)

planeta: objeto que orbita el Sol, lo suficientemente grande para tener forma casi esférica, y que no tiene otro objeto grande en el recorrido de su órbita. (pág. 766)

fitohormona: sustancia que actúa como mensajero químico dentro de una planta. (pág. 283)

tectónica de placas: teoría que afirma que la superficie de la Tierra está divida en piezas enormes y rígidas que se mueven una con respecto a la otra. (pág. 501)

brisas polares: vientos fríos que soplan del este al oeste cerca del Polo Norte y del Polo Sur. (pág. 593)

grano de polen: espora que se forma de tejido en una estructura reproductiva masculina de una planta de semilla. (pág. 292)

polinización: proceso que ocurre cuando los granos de polen posan sobre una estructura reproductiva femenina de una planta que es de la misma especie que los granos de polen. (pág. 292)

polución: contaminación del medioambiente con sustancias dañinas para la vida. (pág. 547)

herencia poligénica: patrón de herencia en el cual genes múltiples determinan el fenotipo de un rasgo. (pág. 169)

población: todos los organismos de la misma especie que viven en la misma área al mismo tiempo. (pág. 311)

energía potencial: energía almacenada debido a las interacciones entre objetos o partículas. (pág. 427)

precipitación: agua, de forma líquida o sólida, que cae de la atmósfera. (pág. 619)

precisión: sescripción de qué tan similar o cercana están las mediciones una de otra. (pág. NOS 12)

prediction: a statement of what will happen next in a sequence of events. (p. NOS 5)

producer: an organism that uses an outside energy source, such as the Sun, and produces its own food. (p. 326)

protein: a long chain of amino acid molecules; contains carbon, hydrogen, oxygen, nitrogen, and sometimes sulfur. (p. 47)

Punnett square: a model that is used to show the probability of all possible genotypes and phenotypes of offspring. (p. 166)

predicción: afirmación de lo que ocurrirá a continuación en una secuencia de eventos. (pág. NOS 5)

productor: organismo que usa una fuente de energía externa, como el Sol, para elaborar su propio alimento. (pág. 326)

proteína: larga cadena de aminoácidos; contiene carbono, hidrógeno, oxígeno, nitrógeno y, algunas veces, sulfuro. (pág. 47)

cuadro de Punnett: modelo que se utiliza para demostrar la probabilidad de que todos los genotipos y fenotipos posibles de cría. (pág. 166)

R

radiant energy: energy carried by an electromagnetic wave. (p. 432)

radiation: the transfer of thermal energy by electromagnetic waves. (pp. 449, 582)

rain shadow: an area of low rainfall on the downwind slope of a mountain. (p. 653)

recessive (rih SE sihv) trait: a genetic factor that is blocked by the presence of a dominant factor. (p. 159)

recycling: manufacturing new products out of used products. (p. 560)

reference point: the starting point you use to describe the motion or the position of an object. (p. 691)

reflex: an automatic movement in response to a stimulus. (p. 249)

regeneration: a type of asexual reproduction that occurs when an offspring grows from a piece of its parent. (p. 132)

relative humidity: the amount of water vapor present in the air compared to the maximum amount of water vapor the air could contain at that temperature. (p. 617)

renewable energy resource: an energy resource that is replaced as fast as, or faster than, it is used. (p. 440)

renewable resource: a natural resource that can be replenished by natural processes at least as quickly as it is used. (p. 540)

energía radiante: energía que transporta una onda electromagnética. (pág. 432)

radiación: transferencia de energía térmica por ondas electromagnéticas. (pág. 449, 582)

sombra de lluvia: área de baja precipitación en la ladera de sotavento de una montaña. (pág. 653)

rasgo recesivo: factor genético bloqueado por la presencia de un factor dominante. (pág. 159)

reciclaje: fabricación de productos nuevos hechos de productos usados. (pág. 560)

punto de referencia: punto que se escoge para describir el movimiento o posición de un objeto. (pág. 691)

reflejo: movimiento automático en respuesta a un estímulo. (pág. 249)

regeneración: tipo de reproducción asexual que ocurre cuando un organismo se origina de una parte de su progenitor. (pág. 132)

humedad relativa: cantidad de vapor de agua presente en el aire comparada con la cantidad máxima de vapor de agua que el aire podría contener en esa temperatura. (pág. 617)

recurso energético renovable: recurso energético que se repone tan rápido, o más rápido, de lo que se consume. (pág. 440)

recurso renovable: recurso natural que se reabastece mediante procesos naturales tan rápidamente como se usa. (pág. 540)

SCIENCE SKILL HANDBOOK

MATH SKILL HANDBOOK

FOLDABLES HANDBOOK

REFERENCE HANDBOOK

GLOSSARY/ GLOSARIO

INDEX

replication: the process of copying a DNA molecule to make another DNA molecule. (p. 176)

reproduction: the process by which new organisms are produced. (p. 255)

revolution: the orbit of one object around another object. (p. 728)

RNA: ribonucleic acid, a type of nucleic acid that carries the code for making proteins from the nucleus to the cytoplasm. (p. 177)

rocket: a vehicle propelled by the exhaust made from burning fuel. (p. 781)

rotation: the spin of an object around its axis. (p. 729)

rotation axis: the line on which an object rotates. (p. 729)

replicación: proceso por el cual se copia una molécula de ADN para hacer otra molécula de ADN. (pág. 176)

reproducción: proceso por el cual se producen nuevos organismos. (pág. 255)

revolución: movimiento de un objeto alrededor de otro objeto. (pág. 728)

ARN: ácido ribonucleico, un tipo de ácido nucléico que contiene el código para hacer proteínas del núcleo para el citoplasma. (pág. 177)

cohete: vehículo propulsado por gases de escape producidos por la ignición de combustible. (pág. 781)

rotación: movimiento giratorio de un objeto sobre su eje. (pág. 729)

eje de rotación: línea sobre la cual un objeto rota. (pág. 729)

S

satellite: any small object that orbits a larger object other than a star. (p. 768)

science: the investigation and exploration of natural events and of the new information that results from those investigations. (p. NOS 2)

scientific law: a rule that describes a pattern in nature. (p. NOS 7)

scientific theory: an explanation of observations or events that is based on knowledge gained from many observations and investigations. (p. NOS 7)

sea breeze: a wind that blows from the sea to the land due to local temperature and pressure differences. (p. 594)

sediment: rock material that forms when rocks are broken down into smaller pieces or dissolved in water as rocks erode. (p. 522)

seed: a plant embryo, its food supply, and a protective covering. (p. 292)

selective breeding: the selection and breeding of organisms for desired traits. (p. 209)

semiconductor: a substance that conducts electricity at high temperatures but not at low temperatures. (p. 413)

satélite: cualquier objeto pequeño que orbita un objeto más grande diferente de una estrella. (pág. 768)

ciencia: la investigación y exploración de los eventos naturales y de la información nueva que es el resultado de estas investigaciones. (pág. NOS 2)

ley científica: regla que describe un patrón dado en la naturaleza. (pág. NOS 7)

teoría científica: explicación de observaciones o eventos con base en conocimiento obtenido de muchas observaciones e investigaciones. (pág. NOS 7)

brisa marina: viento que sopla del mar hacia la tierra debido a diferencias en la temperatura local y la presión. (pág. 594)

sedimento: material rocoso formado cuando las rocas se rompen en piezas pequeñas o se disuelven en agua al erosionarse. (pág. 522)

semilla: embrión de una planta, su suministro de alimento y cubierta protectora. (pág. 292)

cría selectiva: selección y la cría de organismos para las características deseadas. (pág. 209)

semiconductor: sustancia que conduce electricidad a altas temperaturas, pero no a bajas temperaturas. (pág. 413)

SCIENCE SKILL HANDBOOK

MATH SKILL HANDBOOK

FOLDABLES HANDBOOK

REFERENCE HANDBOOK

GLOSSARY/ GLOSARIO

INDEX

sexual reproduction: type of reproduction in which the genetic material from two different cells—a sperm and an egg—combine, producing an offspring. (p. 117)

shear: parallel forces acting in opposite directions at a transform boundary. (p. 505)

significant digits: the number of digits in a measurement that are known with a certain degree of reliability. (p. NOS 13)

sister chromatids: two identical chromosomes that make up a duplicated chromosome. (p. 88)

soil: a mixture of weathered rock, rock fragments, decayed organic matter, water, and air. (p. 521)

solar eclipse: an occurrence during which the Moon's shadow appears on Earth's surface. (p. 746)

solstice: when Earth's rotation axis is tilted directly toward or away from the Sun. (p. 733)

solubility: the maximum amount of solute that can dissolve in a given amount of solvent at a given temperature and pressure. (p. 362)

sound energy: energy carried by sound waves. (p. 431)

space probe: an uncrewed spacecraft sent from Earth to explore objects in space. (p. 782)

species (SPEE sheez): a group of organisms that have similar traits and are able to produce fertile offspring. (p. 21)

specific heat: the amount of thermal energy (joules) needed to raise the temperature of 1 kg of material 1°C. (p. 653)

speed: the distance an object moves divided by the time it takes to move that distance. (p. 692)

sperm: a male reproductive, or sex, cell; forms in a testis. (pp. 117, 255)

spongy bone: the interior region of bone that contains many tiny holes. (p. 246)

spore: a daughter cell produced from a haploid structure. (p. 290)

reproducción sexual: tipo de reproducción en la cual el material genético de dos células diferentes de un espermatozoide y un óvulo se combinan, produciendo una cría. (pág. 117)

cizalla: fuerzas paralelas que actúan en direcciones opuestas en un límite transformante. (pág. 505)

cifras significativas: número de dígitos que se conoce con cierto grado de fiabilidad en una medida. (pág. NOS 13)

cromátidas hermanas: dos cromosomas idénticos que constituyen un cromosoma duplicado. (pág. 88)

suelo: mezcla de roca meteorizada, fragmentos de rocas, materia orgánica descompuesta, agua y aire. (pág. 521)

eclipse solar: acontecimiento durante el cual la sombra de la Luna aparece sobre la superficie de la Tierra. (pág. 746)

solsticio: cuando el eje de rotación de la Tierra se inclina acercándose o alejándose del Sol. (pág. 733)

solubilidad: cantidad máxima de soluto que puede disolverse en una cantidad dada de solvente a temperatura y presión dadas. (pág. 362)

energía sonora: energía que transportan las ondas sonoras. (pág. 431)

sonda espacial: nave espacial sin tripulación enviada desde la Tierra para explorar objetos en el espacio. (pág. 782)

especie: grupo de organismos que tienen rasgos similares y que están en capacidad de producir crías fértiles. (pág. 21)

calor específico: cantidad de energía térmica (julios) requerida para subir la temperatura de 1 kg de materia a 1°C. (pág. 653)

rapidez: distancia que un objeto recorre dividida por el tiempo que éste tarda en recorrer dicha distancia. (pág. 692)

espermatozoide: célula reproductora masculina o sexual; forma en un testículo. (pág. 117, 255)

hueso esponjoso: región interior de un hueso que contiene muchos huecos diminutos. (pág. 246)

espora: célula hija producida de una estructura haploide. (pág. 290)

stability: whether circulating air motions will be strong or weak. (p. 586)

stamen: the male reproductive organ of a flower. (p. 294)

stem cell: an unspecialized cell that is able to develop into many different cell types. (p. 100)

stimulus (STIHM yuh lus): a change in an organism's environment that causes a response. (p. 279)

stratosphere (STRA tuh sfihr): the atmospheric layer directly above the troposphere. (p. 576)

streak: the color of a mineral's powder. (p. 482)

subduction zone: the area where one plate slides under another plate. (p. 504)

substance: matter with a composition that is always the same. (p. 351)

succession: the gradual change from one community to another community in an area. (p. 313)

surface report: a description of a set of weather measurements made on Earth's surface. (p. 635)

sustainability: meeting human needs in ways that ensure future generations also will be able to meet their needs. (p. 559)

symbiosis: a close, long-term relationship between two species that usually involves an exchange of food or energy. (p. 320)

estabilidad: condición en la que los movimientos del aire circulante pueden ser fuertes o débiles. (pág. 586)

estambre: órgano reproductor masculino de una flor. (pág. 294)

célula madre: célula no especializada que tiene la capacidad de desarrollarse en diferentes tipos de células. (pág. 100)

estímulo: un cambio en el medioambiente de un organismo que causa una respuesta. (pág. 279)

estratosfera: capa atmosférica justo arriba de la troposfera. (pág. 576)

raya: color del polvo de un mineral. (pág. 482)

zona de subducción: área donde una placa se desliza debajo de otra placa. (pág. 504)

sustancia: materia cuya composición es siempre la misma. (pág. 351)

sucesión: cambio gradual de una comunidad a otra comunidad en un área. (pág. 313)

informe de superficie: descripción de un conjunto de mediciones del tiempo realizadas en la superficie de la Tierra. (pág. 635)

sostenibilidad: satisfacción de las necesidades humanas de forma que se asegure que las generaciones futuras también podrán satisfacer sus necesidades. (pág. 559)

simbiosis: relación estrecha a largo plazo entre dos especies que generalmente involucra intercambio de alimento o energía. (pág. 320)

T

technology: the practical use of scientific knowledge, especially for industrial or commercial use. (p. NOS 6)

temperature: the measure of the average kinetic energy of the particles in a material. (p. 447)

temperature inversion: a temperature increase as altitude increases in the troposphere. (p. 587)

tension: the pulling force at a divergent boundary. (p. 505)

thermal conductor: a material through which thermal energy flows quickly. (p. 452)

tecnología: uso práctico del conocimiento científico, especialmente para uso industrial o comercial. (pág. NOS 6)

temperatura: medida de la energía cinética promedio de las partículas de un material. (pág. 447)

inversión de temperatura: aumento de la temperatura en la troposfera a medida que aumenta la altitud. (pág. 587)

tensión: fuerza de tracción en un límite divergente. (pág. 505)

conductor térmico: material en el cual la energía térmica se mueve con rapidez. (pág. 452)

SCIENCE SKILL HANDBOOK

MATH SKILL HANDBOOK

FOLDABLES HANDBOOK

REFERENCE HANDBOOK

GLOSSARY/GLOSARIO

INDEX

thermal energy: the sum of the kinetic energy and the potential energy of the particles that make up an object. (p. 430)

thermal insulator: a material through which thermal energy flows slowly. (p. 452)

tide: the periodic rise and fall of the ocean's surface caused by the gravitational force between Earth and the Moon, and Earth and the Sun. (p. 749)

tissue: a group of similar types of cells that work together to carry out specific tasks. (p. 101)

tornado: a violent, whirling column of air in contact with the ground. (p. 629)

trace fossil: the preserved evidence of the activity of an organism. (p. 195)

trade winds: steady winds that flow from east to west between 30°N latitude and 30°S latitude. (p. 593)

transcription: the process of making mRNA from DNA. (p. 177)

transform boundary: the boundary between two plates that slide past each other. (p. 504)

transition element: an element in groups 3–12 on the periodic table. (p. 404)

translation: the process of making a protein from RNA. (p. 178)

tropism (TROH pih zum): plant growth toward or away from an external stimulus. (p. 280)

troposphere (TRO puh sfihr): the atmospheric layer closest to Earth's surface. (p. 576)

energía térmica: suma de la energía cinética y potencial de las partículas que componen un objeto. (pág. 430)

aislante térmico: material a través del cual la energía térmica fluye con lentitud. (pág. 452)

marea: ascenso y descenso periódico de la superficie del océano causados por la fuerza gravitacional entre la Tierra y la Luna, y entre la Tierra y el Sol. (pág. 749)

tejido: grupo de tipos similares de células que trabajan juntas para llevar a cabo diferentes funciones. (pág. 101)

tornado: columna de aire violenta y rotativa en contacto con el suelo. (pág. 629)

traza fósil: evidencia conservada de la actividad de un organismo. (pág. 195)

vientos alisios: vientos constantes que soplan del este al oeste entre 30°N de latitud y 30°S de latitud. (pág. 593)

transcripción: proceso por el cual se hace mARN de ADN. (pág. 177)

límite transformante: límite entre dos placas que se deslizan una con respecto a la otra. (pág. 504)

elemento de transición: elemento de los grupos 3–12 de la tabla periódica. (pág. 404)

traslación: proceso por el cual se hacen proteínas a partir de ARN. (pág. 178)

tropismo: crecimiento de las plantas hacia o lejos de un estímulo externo. (pág. 280)

troposfera: capa atmosférica más cercana a la Tierra. (pág. 576)

U

umbra: the central, darker part of a shadow where light is totally blocked. (p. 745)

unbalanced forces: forces acting on an object that combine and form a net force that is not zero. (p. 705)

unicellular: a living thing that is made up of only one cell. (p. 10)

upper-air report: a description of wind, temperature, and humidity conditions above Earth's surface. (p. 635)

umbra: parte central más oscura de una sombra donde la luz está completamente bloqueada. (pág. 745)

fuerzas no balanceadas: fuerzas que actúan sobre un objeto, se combinan y forman una fuerza neta diferente de cero. (pág. 705)

unicelular: ser vivo formado por una sola célula. (pág. 10)

informe del aire superior: descripción de las condiciones del viento, de la temperatura y de la humedad por encima de la superficie de la Tierra. (pág. 635)

V

vaporization: the change of state from a liquid to a gas. (p. 451)

variable: any factor that can have more than one value. (p. NOS 18)

variation (ver ee AY shun): a slight difference in an inherited trait among individual members of a species. (p. 205)

vegetative reproduction: a form of asexual reproduction in which offspring grow from a part of a parent plant. (p. 133)

velocity: the speed and the direction of a moving object. (p. 693)

vestigial (veh STIH jee ul) structure: body part that has lost its original function through evolution. (p. 215)

volcanic ash: tiny particles of pulverized volcanic rock and glass. (p. 514)

vaporización: cambio de estado líquido a gaseoso. (pág. 451)

variable: cualquier factor que tenga más de un valor. (pág. NOS 18)

variación: ligera diferencia en un rasgo hereditario entre los miembros individuales de una especie. (pág. 205)

reproducción vegetativa: forma de reproducción asexual en la cual el organismo se origina a partir de una planta parental. (pág. 133)

velocidad: rapidez y dirección de un objeto en movimiento. (pág. 693)

estructura vestigial: parte del cuerpo que a través de la evolución perdió la función original. (pág. 215)

ceniza volcánica: partículas diminutas de roca y vidrio volcánicos pulverizados. (pág. 514)

W

waning phases: phases of the Moon during which less of the Moon's near side is lit each night. (p. 740)

water cycle: the series of natural processes by which water continually moves throughout the hydrosphere. (p. 619)

water vapor: water in its gaseous form. (p. 574)

wave: a disturbance that transfers energy from one place to another without transferring matter. (p. 431)

waxing phases: phases of the Moon during which more of the Moon's near side is lit each night. (p. 740)

weather: the atmospheric conditions, along with short-term changes, of a certain place at a certain time. (p. 615)

weathering: the mechanical and chemical processes that change Earth's surface over time. (p. 519)

westerlies: steady winds that flow from west to east between latitudes 30°N and 60°N, and 30°S and 60°S. (p. 593)

wind: the movement of air from areas of high pressure to areas of low pressure. (p. 591)

fases menguantes: fases de la Luna durante las cuales el lado cercano de la Luna está menos iluminado cada noche. (pág. 740)

ciclo del agua: serie de procesos naturales por los que el cual el agua se mueve continuamente en toda la hidrosfera. (pág. 619)

vapor de agua: agua en forma gaseosa. (pág. 574)

onda: perturbación que transfiere energía de un lugar a otro sin transferir materia. (pág. 431)

fases crecientes: fases de la Luna durante las cuales el lado cercano de la Luna está más iluminado cada noche. (pág. 740)

tiempo atmosférico: condiciones atmosféricas, junto con cambios a corto plazo, de un lugar determinado a una hora determinada. (pág. 615)

meteorización: procesos mecánicos y químicos que con el paso del tiempo cambian la superficie de la Tierra. (pág. 519)

vientos del oeste: vientos constantes que soplan de oeste a este entre latitudes 30°N y 60°N, y 30°S y 60°S. (pág. 593)

viento: movimiento del aire desde áreas de alta presión hasta áreas de baja presión. (pág. 591)

SCIENCE SKILL HANDBOOK　MATH SKILL HANDBOOK　FOLDABLES HANDBOOK　REFERENCE HANDBOOK　GLOSSARY/GLOSARIO　INDEX

work: the amount of energy used as a force moves an object over a distance. (p. 437)

trabajo: cantidad de energía usada como fuerza que mueve un objeto a cierta distancia. (pág. 437)

Z

zygote (ZI goht): the cell that forms when a sperm cell fertilizes an egg cell. (pp. 117, 255)

zigoto: célula nueva que se forma cuando un espermatozoide fertiliza un óvulo. (pág. 117, 255)

SCIENCE SKILL HANDBOOK

MATH SKILL HANDBOOK

FOLDABLES HANDBOOK

REFERENCE HANDBOOK

GLOSSARY/ GLOSARIO

INDEX

Index

Italic numbers = illustration/photo   **Bold numbers** = vocabulary term
lab = indicates entry is used in a lab on this page

SCIENCE SKILL HANDBOOK

MATH SKILL HANDBOOK

FOLDABLES HANDBOOK

REFERENCE HANDBOOK

GLOSSARY/ GLOSARIO

INDEX

SCIENCE SKILL HANDBOOK

MATH SKILL HANDBOOK

FOLDABLES HANDBOOK

REFERENCE HANDBOOK

GLOSSARY/ GLOSARIO

INDEX

SCIENCE SKILL HANDBOOK

MATH SKILL HANDBOOK

FOLDABLES HANDBOOK

REFERENCE HANDBOOK

GLOSSARY/ GLOSARIO

INDEX

SCIENCE SKILL HANDBOOK

MATH SKILL HANDBOOK

FOLDABLES HANDBOOK

REFERENCE HANDBOOK

GLOSSARY/ GLOSARIO

INDEX

SCIENCE SKILL HANDBOOK

MATH SKILL HANDBOOK

FOLDABLES HANDBOOK

REFERENCE HANDBOOK

GLOSSARY/ GLOSARIO

INDEX

SCIENCE SKILL HANDBOOK

MATH SKILL HANDBOOK

FOLDABLES HANDBOOK

REFERENCE HANDBOOK

GLOSSARY/ GLOSARIO

INDEX

SCIENCE SKILL HANDBOOK

MATH SKILL HANDBOOK

FOLDABLES HANDBOOK

REFERENCE HANDBOOK

GLOSSARY/ GLOSARIO

INDEX

Credits

Art Acknowledgments: The McGraw-Hill Companies, Argosy, Epigraphics, MCA+, Articulate Graphics, John E Kaufmann Graphic Arts, Mapping Specialists

Photo Credits

137 (t)Dr. Brad Mogen/Visuals Unlimited, (c)Roslin Institute, (b)Mark Steinmetz; 138–139 Hutchings Photography/Digital Light Source; 140 (t)Science Pictures Ltd./Photo Researchers, (b)CNRI/Photo Researchers; 143 Bill Coster/Getty Images; 150–151 Tom and Pat Leesson; 152 Nigel Cattlin/Visuals Unlimited; 153 (tl)Geoff du Feu/Alamy, (tr)Ken Karp/McGraw-Hill Companies, (cl)Glow Images/Getty Images, (cr)Getty Images, (bl)(br)The McGraw-Hill Companies; 154 Wally Eberhart/Visuals Unlimited; 156–157 (purple flower)DEA PICTURE LIBRARY/Photolibrary, (white flower) WILDLIFE/Peter Arnold, Inc.; 160 (t)Tom and Pat Leesson, (bl)DEA PICTURE LIBRARY/Photolibrary, (br)WILDLIFE/Peter Arnold, Inc.; 161 (t)Pixtal/age Fotostock, (c)World History Archive/age Fotostock, (b)Tek Image/Photo Researchers, (bkgd)Jason Reed/Getty Images; 162 (l to r, t to b)(2)Getty Images, (3)Punchstock, (4)CORBIS, (5)(6)(7)(8)(9) Getty Images, (10)CMCD/Getty Images, (11)(12)CORBIS; 163 (t)Richard Hutchings/Digital Light Source, (b)Biophoto Associates/Photo Researchers; 165 Martin Shields/Photo Researchers; 167 (l)Maya Barnes/The Image Works, (r)Bill Aron/PhotoEdit; 168 (l)Peter Smithers/CORBIS, (tr)Geoff Bryant/Photo Researchers, (c)Bill Ross/CORBIS, (bl)J. Schwanke/Alamy, (bc)(br)Yann Arthus-Bertrand/CORBIS; 169 Chris Clinton/Getty Images; 170 (tl)Picture Net/Corbis, (tr)June Green/Alamy Images, (c)Carolyn A. McKeone/Photo Researchers, (bl)(br)Hania Arensten Berdys/www.gardensafari.net; 171 Chris Clinton/Getty Images; 172 Kristina Yu/Exploratorium Store; 173 PHOTOTAKE Inc./Alamy; 174 182 183 Hutchings Photography/Digital Light Source, (tr)The McGraw-Hill Companies; 184 (white flower)WILDLIFE/Peter Arnold, Inc., (purple flower)DEA PICTURE LIBRARY/Photolibrary; 187 Tom and Pat Leesson; 190–191 William Osborn/Minden Pictures; 192 Florida Museum of Natural History photo by Eric Zamora ©2008; 193 Hutchings Photography/Digital Light Source; 194 (l)B.A.E. Inc./Alamy, (r)The Natural History Museum/Alamy; 195 (l)Mark Steinmetz, (c)Tom Bean/CORBIS, (r)Dorling Kindersely/Getty Images; 199 Hutchings Photography/Digital Light Source; 200 Dorling Kindersely/Getty Images; 201 (t)Ryan McVay/Getty Images, (c)C Squared Studios/Getty Images, (b)Richard Broadwell/Alamy, (2)Tim O'Hara/Corbis; 202 DLILLC/CORBIS; 203 Hutchings Photography/Digital Light Source; 204 (l)Jeffrey Greenberg/Photo Researchers, (c)David Hosking/Alamy, (r)Mark Jones/Photolibrary; 205 Chip Clark; 207 (l)Carey Alan & Sandy/Photolibrary, (c)Robert Shantz/Alamy, (r)Stan Osolinski/Photolibrary; 208 (l)Paul Sutherland/National Geographic/Getty Images, (t)NHPA/Photoshot, (r)Kay Nietfeld/dpa/CORBIS, (b)Mitsuhiko Imamori/Minden Pictures; 209 (l)ARCO/D. Usher/age Fotostock, (t)Photodisc/Getty Images, (r)Joe Blossom/NHPA/Photoshot, (b)Hutchings Photography/Digital Light Source; 210 (t)Mark Jones/Photolibrary, (tc)Chip Clark, (b)Don Mammoser/Bruce Coleman, Inc./Photoshot, (bc)Paul Sutherland/National Geographic/Getty Images; 211 (t)D. Parer & E. Parer-Cook, (b)B. Rosemary Grant/AP Images, (bkgd)Skan9/Getty Images; 212 Joseph Van Os/Getty Images; 213 Hutchings Photography/Digital Light Source; 218 T. Daeschler/The Academy of Natural Sciences/VIREO; 219 Joseph Van Os/Getty Images; 220 (tr)USGS, (br)Mary Plage/Photolibrary, (t to b)Hutchings Photography/Digital Light Source, (2)(3)(4)Macmillan/McGraw-Hill, (5)Hutchings Photography/Digital Light Source; 223 Mark Steinmetz; 225 William Osborn/Minden Pictures; 228–229 The Visible Human Project, U.S. National Library of Medicine; 230 Ralph Hutchings/Visuals Unlimited/Getty Images; 231 (t)Hutchings Photography/Digital Light Source, (b)John Terence Turner/Alamy; 232 Hutchings Photography/Digital Light Source; 233 Jill Braaten/The McGraw-Hill Companies; 234 Thinkstock/Getty Images; 235 Holger Winkler/zefa/CORBIS; 237 Image Source/Jupiterimages; 239 C Squared Studios/Getty Images; 240 Omikron/Photo Researchers; 242 (l)Omikron/Photo Researchers, (r)Mark Steinmetz; 243 (l to r, t to b)Stockdisc/Getty Images, (1)The McGraw-Hill Companies/Ken Karp photographer, (2)(3)(4)Hutchings Photography/Digital Light Source, (5)(6)Macmillan/McGraw-Hill; 244 YOSHIKAZU TSUNO/AFP/Getty Images; 245 Hutchings Photography/Digital

Light Source; 246 CMCD/Getty Images; 247 (t)(c)Dr, Gladden Willis/Visuals Unlimited/Getty Images, (b)Innerspace Imaging/Photo Researchers; 248 Doug Pensinger/Getty Images; 250 Hutchings Photography/Digital Light Source; 251 Mark Andersen/Getty Images; 252 (t)CMCD/Getty Images, (b)Innerspace Imaging/Photo Researchers; 253 (t)Andrew Paul Leonard/Photo Researchers, (b)National Marrow Donor Program, (bkgd)Steve Gschmeissner/Photo Researchers; 254 David M. Phillips/Photo Researchers; 255 Hutchings Photography/Digital Light Source; 257 (l)Dr G. Moscoso/Photo Researchers, (t)(r)Anatomical Travelogue/Photo Researchers, (c)Neil Bromhall/Photolibrary, (b)Biophoto Associates/Photo Researchers; 258 Digital Vision; 259 (l)Biophoto Associates/Photo Researchers, (r)PhotoLink/Getty Images; 260 (t)The McGraw-Hill Companies, (b)Macmillan/McGraw-Hill; 261 Hutchings Photography/Digital Light Source; 262 (t)C Squared Studios/Getty Images, (c)CMCD/Getty Images, (b)Anatomical Travelogue/Photo Researchers; 265 The Visible Human Project, U.S. National Library of Medicine; 268–269 Mark Moffett/Minden Pictures; 270 Dr. Dennis Kunkel/Visuals Unlimited/Getty Images; 271 (t)Hutchings Photography/Digital Light Source, (b)The McGraw-Hill Companies; 272 Siede Preis/Getty Images; 273 The McGraw-Hill Companies; 274 Hutchings Photography/Digital Light Source; 276 The McGraw-Hill Companies; 277 (t)Jacques Jangoux/Photolibrary, (b)MARK EDWARDS/Peter Arnold, Inc., (bkgd)Digital Vision/Getty Images; 278 NHPA/Photoshot; 279 (t)Hutchings Photography/Digital Light Source, (b)Cathlyn Melloan/Getty Images; 281 (l)(r)Martin Shields/Photo Researchers, (c)Stephen Dalton/Minden Pictures, (r)Ed Reschke/Peter Arnold Inc.; 282 Eye of Science/Photo Researchers; 283 (t)Andrew Brown; Ecoscene/CORBIS, (b)Dr. Daniel Nickrent/Southern Illinois University; 284 Sylvan Wittwer/Visuals Unlimited; 285 (tl)Monty Rakusen/Getty Images, (tr)Nigel Cattlin/Photo Researchers, (c)Nigel Cattlin/Alamy, (bl)B. ANTHONY STEWART/National Geographic Stock, (br)GardenPhotos.com/Alamy; 286 Martin Shields/Photo Researchers; 287 (l to r, t to b)Hutchings Photography/Digital Light Source, (1)(2)Macmillan/McGraw-Hill, (3)The McGraw-Hill Companies, (4)(5)Hutchings Photography/Digital Light Source, (6)Macmillan/McGraw-Hill, (7)Hutchings Photography/Digital Light Source; 288 Bjorn Rorslett/Photo Researchers; 289 (t)Ingram Publishing/SuperStock, (b)Wally Eberhart/Visuals Unlimited/Getty Images; 291 CARR CLIFTON/ MINDEN PICTURES/National Geographic Stock; 294 Hutchings Photography/Digital Light Source; 296 Stockdisc/PunchStock, (t)CARR CLIFTON/ MINDEN PICTURES/National Geographic Stock, (t)Don Klumpp/Getty Images, (b)Eye of Science/Photo Researchers, (1)Alan & Linda Detrick/Photo Researchers, (2)Comstock/PunchStock, (3)Brad Mogen/Visuals Unlimited, (4)Walter H. Hodge/Peter Arnold, Inc., (5)Siede Preis/Getty Images, (6)Renee Morris/Alamy, (7)Elizabeth Whiting & Associates/Alamy, (8)USDA/Photo Researchers, (9)Adam Jones/Photo Researchers, (10)CORBIS, (11)Ingram Publishing/SuperStock, (12)Gary Meszaros/Visuals Unlimited/Getty Images, (r)Hutchings Photography/Digital Light Source; 300 The McGraw-Hill Companies; 302 (l)Scott Camazine/Photo Researchers, (r)Mark Steinmetz; 303 (l)Cathlyn Melloan/Getty Images, (r)Mark Moffett/Minden Pictures; 306–307 Dave Welling; 308 John Warburton-Lee/DanitaDelimont.com; 309 (t)Hutchings Photography/Digital Light Source, (b)Brand X Pictures; 310 (l)Edmond Van Hoorick/Getty Images, (r)Tom Bean; 312 (c)Image by Reto Stockli, NASA/Goddard Space Flight Center. Enhancements by Robert Simmon, (cw from top)CORBIS, (2)Ron Sanford/CORBIS, (3)Tim Fitzharris/Minden Pictures/Getty Images, (4)Digital Vision/Getty Images, (5)Jim Brandenburg/Minden Pictures, (6)Philippe Bourseiller/Getty Images; 313 (t)(c)U.S. Geological Survey, (b)Jamie & Judy Wild/DanitaDelimont.com, (bkgd)InterNetwork Media/Getty Images; 314 (tl)Brand X Pictures, (tcl)Image by Reto Stockli, NASA/Goddard Space Flight Center. Enhancements by Robert Simmon, (bl)Edmond Van Hoorick/Getty Images, (bcl)Jamie & Judy Wild/DanitaDelimont.com, (br)Tom Bean; 315 (t)Deborah Gordon, (c)Dale Ward, (b)Robert J. Erwin/Photo Researchers; 316 Bryan Mullennix/Getty Images; 317–318 Hutchings Photography/Digital Light Source; 321

Credits

(t)Mark Moffett/Minden Pictures, (c)Thomas Hallstein/Alamy, (b)Mark Steinmetz; **322** Thomas Hallstein/Alamy; **323** Steve Austin/Papilio/CORBIS; **324** Tom & Pat Leeson; **327** (l)B. Drake/PhotoLink/Getty Images, (cl)David Young-Wolff/PhotoEdit, (cr)The McGraw-Hill Companies, (r)Paul Debois/Getty Images; **329** Hutchings Photography/Digital Light Source; **333** David Young-Wolff/PhotoEdit; **334** (br)(t to b)(2)Hutchings Photography/Digital Light Source, (3)Macmillan/McGraw-Hill, (4)(5) Hutchings Photography/Digital Light Source, (6)Macmillan/McGraw-Hill, (7) Hutchings Photography/Digital Light Source; **335** Hutchings Photography/Digital Light Source; **336** (c)Image by Reto Stockli, NASA/Goddard Space Flight Center. Enhancements by Robert Simmon, (cw from top)CORBIS, (2)Ron Sanford/CORBIS, (3)Tim Fitzharris/Minden Pictures/Getty Images, (4)Digital Vision/Getty Images, (5)Jim Brandenburg/Minden Pictures, (6)Philippe Bourseiller/Getty Images; **338** Tim Fitzharris/Minden Pictures/Getty Images; **339** (l)Mark Steinmetz; **339** (r)Dave Welling; **344** (t)Roger Ressmeyer/CORBIS, (c)Getty Images, (b)American Museum of Natural History; **346–347** Gregory G.Dimijian/Photo Reasearchers, Inc; **348** VEER Alison Shaw/Getty Images; **349** (t)Hutchings Photography/Digital Light Source, (b)The McGraw-Hill Companies, Inc./Doug Sherman, photographer; **350** Hutchings Photography/Digital Light Source; **352** Charles D. Winters/Photo Researchers, Inc.; **353** (l)Mark Steinmetz, (cl)Ken Karp for MMH, (cr)Foodfolio/age fotostock, (r)Photo Spin/Getty Images; **355** Mark Steinmetz; **356** Don Farrall/Getty Images; **357** (inset)Mike Perry/Alamy, (bkgd)Alaska Stock Images/age fotostock; **358** Hutchings Photography/Digital Light Source; **359** Rob Rae/age fotostock; **360** (t)Royalty-Free/CORBIS, (b)Hutchings Photography/Digital Light Source; **362** (l)Royalty-Free/CORBIS, (c)Getty Images, (r)ULTRA.F/Getty Images; **363** (l to r)Brand X/Corbis, (2)Studio/age fotostock, (3)Steve Shott/Steve Shott/Getty Images, (4)Dorling Kindersley/Getty Images, (5)Crawford/Dorling Kindersley/Getty Images; **364** (t)Hutchings Photography/Digital Light Source, (b)Crawford/Dorling Kindersley/Getty Images; **365** (t to b)Macmillan/McGraw-Hill, (2)(3)(4)Hutchings Photography/Digital Light Source; **366** John Terence Turner/Taxi/Getty Images; **367** Hutchings Photography/Digital Light Source; **368** Perennou Nuridsany/Photo Researchers, Inc.; **369** (l)Digital Vision/Getty Images, (c)BananaStock/AGEfotostock, (r)IT Stock/ age fotostock; **370** GK Hart/Vikki Hart/Getty Images; **371** IT Stock/age fotostock; **372** Hutchings Photography/Digital Light Source; **373** Luis Calabor/Getty Images; **374** Hutchings Photography/Digital Light Source; **375** (l)Brand X Pictures/Punchstock, (cl)IMAGEMORE Co.,Ltd./Getty Images, (cr)Martin Hospach/Getty Images, (r)Siede Preis/Getty Images; **376** Hutchings Photography/Digital Light Source; **378** (tl)(tr)The McGraw-Hill Companies, (cl)The Print Collector/age fotostock, (cr)Ambient Images Inc./Alamy, (bl)(br)The McGraw-Hill Companies; **379** (t)Siede Preis/Getty Images, (b)Brand X Pictures/Punchstock; **380** (t to b)(2)(3)Hutchings Photography/Digital Light Source, (4)Macmillan/McGraw-Hill; **381** Hutchings Photography/Digital Light Source; **382** (t)Digital Vision/Getty Images, (b)Martin Hospach/Getty Images; **385** Gregory G.Dimijian/Photo Reasearchers, Inc; **388–389** Nick Caloyianis/National Geographic/Getty Images; **390** P.J. Stewart/Photo Researchers, Inc.; (t)(c)(b)Hutchings Photography/Digital Light Source; **393** (tl)DEA/A.RIZZI/De Agostini Picture Library/Getty Images, (tr)Astrid & Hanns-Frieder Michler/Photo Researchers, Inc., (cl)Visuals Unlimited/Ken Lucas/Getty Images, (cr)Richard Treptow/Photo Researchers, Inc., (bl)CORBIS, (br)ImageState/Alamy; **396** (l)David J. Green/Alamy, (c)WILDLIFE/Peter Arnold Inc., (r)Mark Schneider/Visuals Unlimited/Getty Images; **397** (l)LBNL/Photo Researchers, Inc., (c)Boyer/Roger Viollet/Getty Images, (r)ullstein bild/Peter Arnold, Inc.; **399** Hutchings Photography/Digital Light Source; **400** Paul Katz/photolibrary.com; **401** Hutchings Photography/Digital Light Source; **402** (tl)The McGraw-Hill Companies, (tc)Paul Katz/Getty Images, (tr)Egyptian National Museum, Cairo, Egypt, Photo © Boltin Picture Library/The Bridgeman Art Library International, (bl)NASA, (bc)Hutchings Photography/Digital Light Source, (br)Charles Stirling/Alamy; **403** (l)The McGraw-Hill Companies, Inc./Stephen Frisch,

photographer, (c)sciencephotos/Alamy, (r)Martyn Chillmaid/Oxford Scientific (OSF)/photolibrary.com; **404** (l)Royalty-Free/CORBIS, (cl)Dr. Parvinder Sethi, (cr)Joel Arem/Photo Researchers, Inc., (r)Ingram Publishing/SuperStock; **405** Hutchings Photography/Digital Light Source; **406** (t)Egyptian National Museum, Cairo, Egypt, Photo © Boltin Picture Library/The Bridgeman Art Library International, (c)The McGraw-Hill Companies, Inc./Stephen Frisch, photographer, (b)Paul Katz/Getty Images; **407** Jeff Hunter/Getty Images; **408** E.O. lawrence Berkely National Laboratory, University of California, U.S. Department of Energy; **409** Hutchings Photography/Digital Light Source; **410** (tl)Ted Foxx/Alamy, (tr)Richard Treptow/Photo Researchers, Inc., (c)Hutchings Photography/Digital Light Source, (bl)Photodisc/Getty Images, (br)Charles D. Winters/Photo Researchers, Inc.; **411** sciencephotos/Alamy; **412** NASA-JPL; **413** (l)Ingemar Aourell/Getty Images, (cl)Don Farrall/Getty Images, (cr)Gabe Palmer/Alamy, (r)Henrik Sorensen/Getty Images; **414** (t)PhotoLink/Getty Images, (b)Hutchings Photography/Digital Light Source; **415** (t)Richard Treptow/Photo Researchers, Inc., (c)sciencephotos/Alamy, (b)PhotoLink/Getty Images; **416–417** Hutchings Photography/Digital Light Source; **418** (t)David J. Green/Alamy, (b)Mark Schneider/Visuals Unlimited/Getty Images; **421** Nick Caloyianis/National Geographic/Getty Images; **424–425** Frank Lukasseck/Corbis; **425** Juniors Bildarchiv/photolibrary.com; **427** (t)Hutchings Photography/Digital Light Source, (b)Dimitri Vervitsiotis/Getty Images; **428** (t)David Young-Wolff/Getty Images, (b)Digital Vision/PunchStock; **429** (tl)(tc)i love images/Alamy (tr)Asia Images Group Pte Ltd/Alamy, (b)Hutchings Photography/Digital Light Source; **430** David Stoecklein/CORBIS; **431** (l)Royalty-Free/CORBIS, (r)Hutchings Photography/Digital Light Source; **433** (tl)Dimitri Vervitsiotis/Getty Images, (r)Tommaso di Girolamo/age fotostock, (bl)David Stoecklein/CORBIS; **434** (t)(b) Hutchings Photography/Digital Light Source, (c)The McGraw-Hill Companies; **435** David Tipling/Minden Pictures; **436** Hutchings Photography/Digital Light Source; **437** MOODBOARD/age fotostock; **438** Hutchings Photography/Digital Light Source; **439** (l)Transtock, (inset)Frederic Charpentier/Alamy; **440** (t)JOSE MANUEL RIBEIRO/Reuters/Landov, (b)Thinkstock/Masterfile; **441** (t)Harald Sund/Getty Images, (c)Roger Ressmeyer/CORBIS, (b)AP Photo/Heribert Proepper; **443** (tl)MOODBOARD/age fotostock, (cl)Transtock, (r)The McGraw-Hill Companies, (bl)Thinkstock/Masterfile; **444** (tl)Adrian Sherratt/Alamy, (bl)MIXA/Alamy, (br)Brand X Pictures/PunchStock, (bkgd)Richard Clark/photolibrary.com; **445** Adam Jones/Visuals Unlimited, Inc.; **446** Foodfolio/age fotostock; **447** (l)Sue Flood/Getty Images, (r)David Taylor/Photo Researchers, Inc.; **448** foodfolio/Alamy; **450** The McGraw-Hill Companies.; **451** (l)Charles D. Winters/Photo Researchers, Inc., (r)Thomas Sbampato/Alaskastock/photolibrary.com; **452** (l)Scientifica/Visuals Unlimited, Inc., (r)Hutchings Photography/Digital Light Source; **453** (tl)Adam Jones/Visuals Unlimited, Inc., (r)Thomas Sbampato/photolibrary.com, (bl)Scientifica/Visuals Unlimited, Inc.; **454–455** Hutchings Photography/Digital Light Source; **456** (t)Juniors Bildarchiv/photolibrary.com, (c)JOSE MANUEL RIBEIRO/Reuters/Landov, (b)Thomas Sbampato/photolibrary.com; **458** (l)Gavin Hellier/Getty Images, (r)Jupiterimages/Getty Images; **459** (l)David Madison/Getty Images, (r)Frank Lukasseck/Corbis; **466–467** Mark Steinmetz; **468** Caren Brinkema/Alamy, (t to b)Hutchings Photography/Digital Light Source, (2)Brand X Pictures/PunchStock, (3)Gary Vestal/Getty Images, (4)Corbis/SuperStock, (5)age fotostock/SuperStock; **473** Hutchings Photography/Digital Light Source; **474** (l)Henry Georgi/Getty Images, (tr)Designpics.com/PunchStock, (br)Joel W. Rogers/CORBIS; **476** (l)José Manuel Sanchis Calvete/CORBIS, (r)Smithsonian Institution/Corbis; **477** Doug Menuez/Getty Images; **478** (t)Henry Georgi/Getty Images, (b)Designpics.com/PunchStock; **479** (inset)Luther Linkhart/SuperStock, (bkgd)CORBIS; **480** TIM FITZHARRIS/MINDEN PICTURES/National Geographic Stock; **481** (t)Hutchings Photography/Digital Light Source, (b)Doug Sherman/Geofile; **483** (l)Jason Edwards/Getty Images, (r)AfriPics.com/Alamy; **484** Matthew Ward/Getty Images; **485** (bkgd)StockTrek/Getty

Images; **489** Doug Sherman/Geofile; **490** (t to b)Hutchings Photography/Digital Light Source, (2)(3)The McGraw-Hill Companies, (4)(5)Hutchings Photography/Digital Light Source; **492** TIM FITZHARRIS/MINDEN PICTURES/National Geographic Stock; **494** (r)StockTrek/Getty Images; **495** (l)Mike Johnson/Alamy, (r)Mark Steinmetz; **496** (r)StockTrek/Getty Images; **498–499** Altrendo Travel/Getty Images; **500** COLIN MONTEATH/ HEDGEHOG HOUSE/ MINDEN PICTURES/National Geographic Stock; **501** Hutchings Photography/Digital Light Source; **502** age fotostock/SuperStock; **503** Hutchings Photography/Digital Light Source; **505** Roger Ressmeyer/CORBIS; **506** age fotostock/SuperStock; **507** (tl)Hutchings Photography/Digital Light Source; **507** (cl)Macmillan/McGraw-Hill, (r)Hutchings Photography/Digital Light Source, (bl)Macmillan/McGraw-Hill; **508** Cyrus Read/AVO/USGS; **509** (t)Hutchings Photography/Digital Light Source, (b)Lloyd Cluff/CORBIS; **510** Lloyd Cluff/CORBIS; **511** Hutchings Photography/Digital Light Source; **512** National Oceanic and Atmospheric Administration (NOAA); **514** (t)Dinodia Images/Alamy, (b)Carsten Peter/Getty Images; **515** (l)Roger Ressmeyer/CORBIS, (r)U.S. Geological Survey; **516** (t)Lloyd Cluff/CORBIS, (c)U.S. Geological Survey, (b)Lloyd Cluff/CORBIS; **517** (inset)American Museum of Natural History, (bkgd)Doug Sherman/Geofile/The McGraw-Hill Companies; **518** Massimo Ripani/Grand Tour/Corbis; **519** (t)Hutchings Photography/Digital Light Source, (b)Gunnar Kullenberg/PhotoLibrary; **520** (t)GEORGE AND MONSERRATE SCHWARTZ/Alamy, (b)Jerry Dodrill/Getty Images; **522** (t)Steve Terrill/CORBIS, (c)Michael Blann/Getty Images, (b)George Steinmetz/Corbis; **523** (t)James P Blair/PhotoLibrary, (c)Lloyd Cluff/CORBIS, (b)Jochen Schlenker/Getty Images; **524** (t)Hutchings Photography/Digital Light Source, (b)Image courtesy Liam Gumley, Space Science and Engineering Center, University of Wisconsin-Madison and the MODIS science team/N; **525** (t)Jerry Dodrill/Getty Images, (c)Michael Blann/Getty Images, (bl)Lloyd Cluff/CORBIS, (br)Siegfried Layda/Getty Images; **526** (t to b) Hutchings Photography/Digital Light Source, (2)(3)The McGraw-Hill Companies, (4)Arctic-Images/Corbis, (5)Pacific Stock/SuperStock, (6)Bob Stefko/Getty Images, (7)CORBIS; **527** (t)Brand X Pictures/Punchstock, (b)Adam Jones/Getty Images; **528** (t)Carsten Peter/Getty Images, (b)George Steinmetz/Corbis; **530** Digital Vision/Getty Images; **531** (l)Roger Ressmeyer/CORBIS, (r)Altrendo Travel/Getty Images; **534–535** Sarah Leen/National Geographic/Getty Images; **536** Ron Chapple Stock/Alamy; **537** Creatas/PunchStock; **538** (l)Mark Steinmetz, (r)ImageState/age fotostock, CHRIS JAMES/Peter Arnold, Inc.; **539** Gary Braasch/Getty Images; **543** inga spence/Alamy; **544** (t)Creatas/PunchStock, (b)Ron Chapple Stock/Alamy; **545** PhotoLink/Getty Images; **546** Thomas R. Fletcher/Alamy; **547** (t)Hutchings Photography/Digital Light Source, (b)NATALIE B. FOBES/National Geographic Stock; **550** Hutchings Photography/Digital Light Source; **551** (l)Scott Bauer/USDA, (r)Harrison Shull/Getty Images; **552** (t)NATALIE B. FOBES/National Geographic Stock, (c)Robert Jureit/Getty Images, (b)Harrison Shull/Getty Images; **553** (t)The McGraw-Hill Companies, (c)Rob Melnychuk/Brand X/CORBIS, (b)Hutchings Photography/Digital Light Source; **554** NASA Jet Propulsion Laboratory (NASA-JPL); **555** Butch Martin/Getty Images; **556** (t)JAMES L. STANFIELD/National Geographic Stock, (b)Mark Steinmetz; **559** (l)Jonathan Nourok/PhotoEdit, (c)Steve Skjold/Alamy Images, (r)David Young-Wolff/PhotoEdit; **560** (l)Peter Starman/Getty Images, (r)Tony Craddock/Photo Researchers; **561** (t)NASA Jet Propulsion Laboratory (NASA-JPL), (b)Tony Craddock/Photo Researchers; **563** Mark Steinmetz; **564** (t)Mark Steinmetz, (c)Thomas R. Fletcher/Alamy, (b)JAMES L. STANFIELD/National Geographic Stock; **567** Sarah Leen/National Geographic/Getty Images; **570–571** Daniel H. Bailey/Alamy; **572** CORBIS; **573–574** Hutchings Photography/Digital Light Source; **575** (t)PhotoLink/Getty Images, (b)C. Sherburne/PhotoLink/Getty Images; **577** Per Breiehagen/Getty Images; **578** CORBIS; **579** Robert Jureit/Getty Images, (t)PhotoLink/Getty Images, (c)Per Breiehagen/Getty Images, (r)Hutchings Photography/Digital Light Source; **580** (t)American Museum of Natural History, (b)Pedro Guzman, (bkgd)PhotoLink/Getty Images; **581** John King/

Alamy; **582** Hutchings Photography/Digital Light Source; **583** Eric James/Alamy; **586** (t)C. Sherburne/PhotoLink/Getty Images, (b)James Brunker/Alamy; **588** Eric James/Alamy; **589** (r)(t to b)Hutchings Photography/Digital Light Source, (2)Macmillan/McGraw-Hill, (3)(4)(5)(6)(7)Hutchings Photography/Digital Light Source, (8)Macmillan/McGraw-Hill, (9)(10)(11) Hutchings Photography/Digital Light Source; **590** Lester Lefkowitz/Getty Images; **591** Hutchings Photography/Digital Light Source; **593** (t)CORBIS, (b)Hutchings Photography/Digital Light Source; **595** Lester Lefkowitz/Getty Images; **596** Hutchings Photography/Digital Light Source; **597** Reuters/CORBIS; **598** C. Sherburne/PhotoLink/Getty Images; **599** MICHAEL S. YAMASHITA/National Geographic Image Collection; **602** (t to b)Digital Vision Ltd./SuperStock, (2)Masterfile, (3)Creatas/PictureQuest, (4)C. Sherburne/PhotoLink/Getty Images; **603** (t)C. Sherburne/PhotoLink/Getty Images, (b)MICHAEL S. YAMASHITA/National Geographic Image Collection; **604** (r)(t to b)Hutchings Photography/Digital Light Source, (2)Macmillan/McGraw-Hill, (3)Hutchings Photography/Digital Light Source, (4)Macmillan/McGraw-Hill, (5)(6)(7)(8)(9)Hutchings Photography/Digital Light Source; **605** Hutchings Photography/Digital Light Source; **606** (t)PhotoLink/Getty Images, (b)C. Sherburne/PhotoLink/Getty Images; **608** supershoot/Alamy; **609** Daniel H. Bailey/Alamy; **612–613** George Frey/Getty Images; **614** Peter de Clercq/Alamy; **615** Hutchings Photography/Digital Light Source; **616** (l)Jan Tadeusz/Alamy, (r)matthias engelien/Alamy; **618** (l)WIN-Initiative/Getty Images, (c)MIMOTITO/Getty Images, (r)age fotostock/SuperStock; **620** (t)WIN-Initiative/Getty Images, (b)Jan Tadeusz/Alamy; **621** (t)NASA/Jeff Schmaltz, MODIS Land Rapid Response Team, (b)Jocelyn Augustino/FEMA; **622** Kyle Niemi/U.S. Coast Guard via Getty Images; **623** **625** Hutchings Photography/Digital Light Source; **628** (l)Amazon-Images/Alamy, (c)Roger Coulam/Alamy, (r)mediacolor's/Alamy; **629** Eric Nguyen/CORBIS; **630** StockTrek/Getty Images; **631** AP Photo/Dick Blume, Syracuse Newspapers; **632** Eric Nguyen/CORBIS; **634** Gene Rhoden/Visuals Unlimited; **635** Hutchings Photography/Digital Light Source; **636** National Oceanic and Atmospheric Administration (NOAA); **638** Dennis MacDonald/Alamy; **639** (t)National Oceanic and Atmospheric Administration (NOAA), (b)Dennis MacDonald/Alamy; **640** (t to b)Aaron Haupt, (2)(3)(4)(5)Hutchings Photography/Digital Light Source; **642** (t)Peter de Clercq/Alamy, (c)AP Photo/Dick Blume, Syracuse Newspapers, (b)matthias engelien/Alamy; **645** George Frey/Getty Images; **648–649** Ashley Cooper/CORBIS; **650** Hemis.fr/SuperStock; **654** (tl)Rolf Hicker/age fotostock, (tc)Brand X Pictures/PunchStock, (tr)Jeremy Woodhouse/Getty Images, (bl)Andoni Canela/age fotostock, (br)Steve Cole/Getty Images; **656** age fotostock/SuperStock; **657** (t)Rolf Hicker/age fotostock, (b)Digital Vision/Getty Images; **658** (r)(t to b) (2)(3)(4)Hutchings Photography/Digital Light Source, (5)Macmillan/McGraw-Hill, (6)Hutchings Photography/Digital Light Source; **659** J. A. Kraulis/Masterfile; **660** (t)Hutchings Photography/Digital Light Source, (b)Nick Cobbing/Alamy; **666** Nick Cobbing/Alamy; **667** American Museum of Natural History; **668** Ashley Cooper/Alamy; **669** Hutchings Photography/Digital Light Source; **671** Chris Cheadle/Getty Images; **672** Steve McCutcheon/Visuals Unlimited, Inc.; **674** Bruce Harber/age fotostock; **675** (t)Chris Cheadle/Getty Images, (b)Bruce Harber/age fotostock; **676** (r)Hutchings Photography/Digital Light Source, (t to b)(2)(3)Macmillan/McGraw-Hill, (4)(5)(6)Hutchings Photography/Digital Light Source, (7)Macmillan/McGraw-Hill; **678** (t)Andoni Canela/age fotostock, (c)J. A. Kraulis/Masterfile, (b)Steve McCutcheon/Visuals Unlimited, Inc.; **681** Ashley Cooper/CORBIS; **686** (t)courtesy of PowerFilm, Inc., (c)courtesy of Seldon Technologies, (b)Jacques Descloitres, MODIS Rapid Response Team, NASA/GSFC; **687** (t)USGS, (b)NOAA; **688–689** AFP/Getty Images; **690** Hill Street Studios/Getty Images; **691** (t)Hutchings Photography/Digital Light Source, (t)Nancy Ney/Getty Images; **694** Hutchings Photography/Digital Light Source; **696** PAUL NICKLEN/National Geographic Stock; **698** Nancy Ney/Getty Images; **700** Visuals Unlimited/Corbis; **701** (t)Hutchings Photography/Digital Light Source, (c)Anthony-Masterson/Getty Images,

Credits

(b)Per Breiehagen/Getty Images; **702** (tl)Visual Cuisines/Getty Images, (tc)Photodisc/Alamy (tr)Floresco Productions/Corbis, (bl)Mark Spowart/Alamy, (bc)sciencephotos/Alamy, (br)Steve Casimiro/Getty Images; **703** (t)Hutchings Photography/Digital Light Source, (b)David Madison/Getty Images; **706** sciencephotos/Alamy; **707** Hutchings Photography/Digital Light Source; **708** imac/Alamy; **709** Hutchings Photography/Digital Light Source; **710** (tl)(tr)Bernd Mellmann/Alamy, (b)Hutchings Photography/Digital Light Source; **711** (tl)(tr)camera lucida lifestyle/Alamy, (b)Richard Green/Alamy; **712** skip caplan/Alamy; **713** (t)Uwe Krejci/Getty Images, (b)The McGraw-Hill Companies, Inc./Jill Braaten, photographer; **714** (t)DEX IMAGE/Getty Images, (c)Masterfile, (b)Nikki O'Keefe Images/Getty Images; **715** (t)Per Breiehagen/Getty Images, (c)skip caplan/Alamy, (b)The McGraw-Hill Companies, Inc./Jill Braaten, photographer; **716** (r)(t to b)(2)(3)(4) Hutchings Photography/Digital Light Source, (5)(6)The McGraw-Hill Companies; **717** Hutchings Photography/Digital Light Source; **718** (t)sciencephotos/Alamy, (b)skip caplan/Alamy; **721** (l)NASA, (r)AFP/Getty Images; **724–725** O. Alamany & E. Vicens/CORBIS; **726** NASA Human Spaceflight Collection; **727** (t)Hutchings Photography/Digital Light Source, (b)SOHO (ESA & NASA); **728** Hutchings Photography/Digital Light Source; **734** SOHO (ESA & NASA); **735** (r)(t to b)(2)(3)(4)(6)Hutchings Photography/Digital Light Source, (7)Macmillan/McGraw-Hill, (8)Hutchings Photography/Digital Light Source; **736** NASA; **737** Hutchings Photography/Digital Light Source; **739** (tl)Lunar and Planetary Institute, (tr)Lunar and Planetary Institute, (c)NASA/JPL/USGS, (bl)ClassicStock/Alamy, (br)Lunar and Planetary Institute; **740** Hutchings Photography/Digital Light Source; **741** Eckhard Slawik/Photo Researchers, Inc.; **742** (t)Lunar and Planetary Institute, (b)Eckhard Slawik/Photo Researchers, Inc.; **743** NASA; **744** UPI Photo/NASA/Landov; **745–746** Hutchings Photography/Digital Light Source; **747** (l to r)Copyright by Fred Espenak, www.MrEclipse.com; **749** Robert Estall photo agency/Alamy; **751** Copyright by Fred Espenak, www.MrEclipse.com; **752** (r)(t to b)Hutchings Photography/Digital Light Source, (2)The McGraw-Hill Companies, (3)(4)Hutchings Photography/Digital Light Source; **753** (l to r)Copyright by Fred Espenak, www.MrEclipse.com; **754** (l)Eckhard Slawik/Photo Researchers, Inc., (r)Copyright by Fred Espenak, www.MrEclipse.com; **757** O. Alamany & E. Vicens/CORBIS; **760–761** Cornell University/JPL/NASA; **762** Scholz, F./Peter Arnold Inc.; **763** (t)Hutchings Photography/Digital Light Source, (b)European Southern Observatory/Photo Researchers, Inc.; **765** Hutchings Photography/Digital Light Source; **768** NASA, Jet Propulsion Laboratory; **769** NASA; **770** (t)NASA, Jet Propulsion Laboratory, (b)NASA; **771** JSC/NASA; **772** NOAA/NASA; **773** (t)Hutchings Photography/Digital Light Source, (b)DR KEN MACDONALD/SCIENCE PHOTO LIBRARY; **774** Hutchings Photography/Digital Light Source; **776** NASA/JPL/University of Arizona; **777** (t)Galileo Project/JPL/NASA, (b)NASA/JPL/Space Science Institute, (inset)NASA/JPL/University of Arizona/University of Colorado; **778** (t)DR KEN MACDONALD/SCIENCE PHOTO LIBRARY, (cl)NASA/JPL/University of Arizona, (r)NASA/JPL/Space Science Institute, (bl)Galileo Project/JPL/NASA; **779** (l)(cl)The McGraw-Hill Companies, (r)Hutchings Photography/Digital Light Source; **780** NASA; **781** (t)Hutchings Photography/Digital Light Source, (b)StockTrek/Getty Images; **782** (l)Atlas Photo Bank/Photo Researchers, Inc, (c)AP Images, (r)NASA/JPL; **783** Hutchings Photography/Digital Light Source; **784** (t)Adastra/Getty Images, (b)NASA; **785** NASA; **786** (b)(bl)(bcl)NASA Marshall Space Flight Center Collection, (bcl)Stocktrek/age fotostock, (bc)NASA Marshall Space Flight Center Collection, (bcr)NASA Great Images in Nasa Collection, (br)NASA Johnson Space Center Collection; **787** (t)Adastra/Getty Images, (other)NASA; **788** (t to b)The McGraw-Hill Companies, (2)Hutchings Photography/Digital Light Source, (3)Macmillan/McGraw-Hill, (4)(5)The McGraw-Hill Companies, (6)Hutchings Photography/Digital Light Source; **789** Hutchings Photography/Digital Light Source; **790** (t)NASA/JPL/University of Arizona, (b)NASA; **792** (l)NASA, Jet Propulsion Laboratory, (r)Atlas Photo Bank/Photo Researchers, Inc; **793** Cornell University/JPL/NASA; **SR-00–SR-01** Gallo Images - Neil Overy/Getty Images; **SR-10** Aaron Haupt; **SR-2** Hutchings Photography/Digital Light Source; **SR-6** Michell D. Bridwell/PhotoEdit; **SR-7** (t)The McGraw-Hill Companies, (b)Dominic Oldershaw; **SR-8** StudiOhio; **SR-9** Timothy Fuller; **SR-47** Matt Meadows; **SR-48** (c)NIBSC/Photo Researchers, Inc., (r)Science VU/Drs. D.T. John & T.B. Cole/Visuals Unlimited, Inc. Stephen Durr; **SR-49** (t)Mark Steinmetz, (r)Andrew Syred/Science Photo Library/Photo Researchers, (br)Rich Brommer; **SR-50** David Fleetham/Visuals Unlimited/Getty Images, (l)Lynn Keddie/Photolibrary, (tr)G.R. Roberts; **SR-51** Gallo Images/CORBIS.

PERIODIC TABLE OF THE ELEMENTS

Element — Hydrogen
Atomic number — 1
Symbol — **H**
Atomic mass — 1.01
State of matter

Gas
Liquid
Solid
⊙ Synthetic

A column in the periodic table is called a **group**.

A row in the periodic table is called a **period**.

	1	**2**	**3**	**4**	**5**	**6**	**7**	**8**	**9**
1	Hydrogen 1 **H** 1.01								
2	Lithium 3 **Li** 6.94	Beryllium 4 **Be** 9.01							
3	Sodium 11 **Na** 22.99	Magnesium 12 **Mg** 24.31							
4	Potassium 19 **K** 39.10	Calcium 20 **Ca** 40.08	Scandium 21 **Sc** 44.96	Titanium 22 **Ti** 47.87	Vanadium 23 **V** 50.94	Chromium 24 **Cr** 52.00	Manganese 25 **Mn** 54.94	Iron 26 **Fe** 55.85	Cobalt 27 **Co** 58.93
5	Rubidium 37 **Rb** 85.47	Strontium 38 **Sr** 87.62	Yttrium 39 **Y** 88.91	Zirconium 40 **Zr** 91.22	Niobium 41 **Nb** 92.91	Molybdenum 42 **Mo** 95.96	Technetium 43 ⊙ **Tc** (98)	Ruthenium 44 **Ru** 101.07	Rhodium 45 **Rh** 102.91
6	Cesium 55 **Cs** 132.91	Barium 56 **Ba** 137.33	Lanthanum 57 **La** 138.91	Hafnium 72 **Hf** 178.49	Tantalum 73 **Ta** 180.95	Tungsten 74 **W** 183.84	Rhenium 75 **Re** 186.21	Osmium 76 **Os** 190.23	Iridium 77 **Ir** 192.22
7	Francium 87 **Fr** (223)	Radium 88 **Ra** (226)	Actinium 89 **Ac** (227)	Rutherfordium 104 ⊙ **Rf** (267)	Dubnium 105 ⊙ **Db** (268)	Seaborgium 106 ⊙ **Sg** (271)	Bohrium 107 ⊙ **Bh** (272)	Hassium 108 ⊙ **Hs** (270)	Meitnerium 109 ⊙ **Mt** (276)

The number in parentheses is the mass number of the longest lived isotope for that element.

Lanthanide series	Cerium 58 **Ce** 140.12	Praseodymium 59 **Pr** 140.91	Neodymium 60 **Nd** 144.24	Promethium 61 ⊙ **Pm** (145)	Samarium 62 **Sm** 150.36	Europium 63 **Eu** 151.96
Actinide series	Thorium 90 **Th** 232.04	Protactinium 91 **Pa** 231.04	Uranium 92 **U** 238.03	Neptunium 93 ⊙ **Np** (237)	Plutonium 94 ⊙ **Pu** (244)	Americium 95 ⊙ **Am** (243)